THE ASQ METROLOGY HANDBOOK

THE ASQ METROLOGY HANDBOOK

Third Edition

Heather A. Wade, Editor

ASQExcellence
Milwaukee, Wisconsin

Published by ASQExcellence, Milwaukee, WI
Produced and distributed by Quality Press, ASQ, Milwaukee, WI

© 2013, 2022 by ASQExcellence

Library of Congress Cataloging-in-Publication Data

Names: Wade, Heather A., editor.
Title: The ASQ metrology handbook , third edition / Heather A. Wade, editor.
Description: Includes bibliographical references and index. | Milwaukee, WI: ASQExcellence, 2022.
Identifiers: LCCN: 2022946261 | ISBN: 978-1-63694-019-9 (hardcover) | 978-1-63694-020-5 (pdf) | 978-1-63694-021-2 (epub)
Subjects: LCSH Measurement—Handbooks, manuals, etc. | Calibration—Handbooks, manuals, etc. | Quality assurance—Handbooks, manuals, etc. | BISAC MATHEMATICS / Measurement | TECHNOLOGY & ENGINEERING / Measurement
Classification: LCC T50 .M423 2022 | DDC 620/.0044—dc23

LC record available at https://lccn.loc.gov/2021036478. A87 2021 | DDC 363.19/2—dc23

No part of this book may be reproduced in any form or by any means, electronic, mechanical, photocopying, recording, or otherwise, without the prior written permission of the publisher.

ASQ and ASQExcellence advance individual, organizational, and community excellence worldwide through learning, quality improvement, and knowledge exchange.

Attention bookstores, wholesalers, schools, and corporations: Quality Press and ASQExcellence books are available at quantity discounts with bulk purchases for business, trade, or educational uses. For information, please contact Quality Press at 800-248-1946 or books@asq.org.

To place orders or browse the selection of ASQExcellence and Quality Press titles, visit our website at: http://www.asq.org/quality-press.

Printed in the United States.

26 25 24 23 22 GP 5 4 3 2 1

Quality Press
600 N. Plankinton Ave.
Milwaukee, WI 53203-2914
Email: books@asq.org

ASQ Excellence Through Quality™

ASQE™ Inspire Excellence

Disclaimers

Certain commercial entities, equipment, or materials may be identified in this book to *adequately* describe a procedure or concept. Views, thoughts, and opinions expressed in this book belong solely to the author, and not necessarily to the author's employer, organization, committee, or other group or individual. Such identification is not intended to imply recommendation or endorsement by ASQ, ASQE, or by the authors of this book, nor is it intended to imply that the entities, materials, or equipment are necessarily the best available for the purpose.

The references cited in this book were cited in their currently published version (unless noted otherwise). Due to the ever-evolving nature of technology, standards, and publications, the readers should always refer to the original source for the most up-to-date versions.

Table of Contents

List of Figures and Tables... xv
Foreword xxv
Preface xxvii
Acknowledgments xxix
Introduction xxxi

Part I Background

Chapter 1 History and Philosophy of Metrology/Calibration 2
   Ancient Measurement.. 2
   Measurement Progress in the Last 2000 Years 5
   Standards, Commerce, and Metrology................................. 6
   History of Quality Standards....................................... 8
   Measurement and the Industrial Revolution.......................... 9
   Milestones in U.S. Food and Drug Law History...................... 11

Part II Quality Systems

Chapter 2 The Basics of a Quality System.............................. 18
   Quality Tools 19
   Process Improvement Techniques.................................... 21

Chapter 3 Quality Standards and Their Evolution 23
   What Are Quality Standards and Why Are They Important? 23
   The Role of ISO... 24
   Evolution of Quality Standards.................................... 26
   History of Quality Standards: 1900–1940........................... 27
   History of Quality Standards—1940 to the Present 27

Chapter 4 Quality Documentation 32

Chapter 5 Calibration Procedures...................................... 35
   Introduction 35
   Calibration Procedure Contents 37
   Other SOPs and Competencies 38
   Safety and Handling Considerations 38
   Method Validation and Verification 41
   Summary... 42

viii   Table of Contents

Chapter 6   Calibration Records 43
   Record Integrity and Retention 46

Chapter 7   Calibration Certificates 48

Chapter 8   Management Systems and Quality Manuals 56

Chapter 9   Metrological Traceability 63
   Traceability Definitions 63
   National Metrology Institutes as a Source of Traceability 64
   International Agreements and Arrangements: Recommended
      Calibration Providers 65
   Evaluating Metrological Traceability 67
   Documenting Calibration Hierarchies 69
   Circular Calibrations 71
   Reverse Traceability 72
   Tools for Assessing Traceability 73
   The Importance of Traceability 73

Chapter 10   Calibration Programs 75
   Delay Dating 82

Chapter 11   The International System of Units (SI) and Measurement Standards 85
   The International System of Units: Defining, Realizing,
      and Disseminating 85
   Presentation of Measurement Results: Units and Symbols 91
   Measurement Standard Types 91
   Accuracy Matters—Understanding and Communicating Measurement
      Uncertainties 94
   Write It Right: Understanding Nuances of Metrics in Technical Writing 98

Chapter 12   Audit Requirements 101

Chapter 13   Scheduling and Recall Systems 106

Chapter 14   Labels and Equipment Status 108

Chapter 15   Training and Competency 112
   Background 112
   Competence Requirements 113
   Job Descriptions 115
   Training Plan 118
   Training Sources 119
   Formal Training 120
   Informal Training 121
   Summary 122

Chapter 16   Environmental Controls 123

Chapter 17   Industry-Specific Requirements 131
   ISO/IEC 17025 131
   FDA Quality System Regulations 139
   Other Industry Requirements 147

Chapter 18   Computers, Software, and Software Validation **151**
    Background. 151
    Calibration Management Software Systems 155
    Security 162
    Instrumentation Software Classifications 164
    Automated Calibration Procedure Software 165
    Spreadsheets. 166

Part III   Metrology Concepts

Chapter 19   A General Understanding of Metrology. **172**

Chapter 20   Measurement Methods, Systems, Capabilities, and Data. **179**
    Measurement Methods. 179
    Measurement Systems. 181
    Measurement Capabilities 185
    Measurement Data. 195
    Calibration Methods and Techniques. 197

Chapter 21   Specifications **199**
    Types of Specification Limits. 201

Chapter 22   Substituting Calibration Standards. **213**
    Measurement Standard Substitution 213
    Equipment Substitution and Calibration Procedures 221

Chapter 23   Proficiency Testing, Interlaboratory Comparisons,
                and Measurement Assurance Programs **222**
    Proficiency Testing 222
    Measurement Assurance Programs 234

Part IV   Mathematics and Statistics: Their Use in Measurement

Chapter 24   Number Formatting. **244**
    Significant Digits. 244
    Notation Methods 246
    Standard Notation 246
    SI Prefix System 253
    ISO Preferred Numbers 254
    Number Rounding Methods 255
    Other Number Formatting Issues. 257
    Date and Time Formats. 258

Chapter 25   Unit Conversions **261**
    SI Units 261
    Coherence of SI Units 262
    SI-Derived Units. 263
    Units Not to Be Used Within the SI System of Units 266
    SI Units Arranged by Unit Category. 268
    Conversion Factors and Their Uses. 273

Uncertainties for Some Fundamental Units (Standard and Relative) 273
　　　Number Bases. 274
　　　Conversion Factors for Converting Between Equivalent
　　　　　Measurement Units. 276

Chapter 26　Ratios **277**
　　　Decibel Measures. 277
　　　Logarithms in Microsoft Excel. 281
　　　Types of Linearizing Transformations 282
　　　Graphs 282

Chapter 27　Statistics. **284**
　　　Fundamental Measurement Assumptions 284
　　　Degrees of Freedom. 284
　　　Residuals. 285
　　　Central Tendency. 285
　　　Bimodal Distributions. 286
　　　Central Limit Theorem 286
　　　Confidence Interval. 288
　　　Root Mean Square 289
　　　Sum of Squares. 289
　　　Root Sum of Squares 289
　　　Population Variance 290
　　　Population Standard Deviation. 290
　　　Sample Variance. 290
　　　Sample Standard Deviation 291
　　　Standard Error of the Mean 291
　　　Measures of Skewness and Kurtosis. 292
　　　Skewness. 292
　　　Kurtosis. 292
　　　Correlation 293
　　　Linear Relationships 295
　　　Zero and Span Relationships 297
　　　Interpolation. 297
　　　Formats of Tabular Data 297
　　　Linear Interpolation Methods 298
　　　Interpolation Methods for Nonlinear Data 299
　　　Types of Distributions and Their Properties 301
　　　Standard Normal (Gaussian) Z-Distribution. 301
　　　Autocorrelation 306

Chapter 28　Mensuration, Volume, and Surface Areas **308**
　　　Mensuration, Lengths, and Angles. 308
　　　Length 308
　　　Circle 311
　　　Ellipse 311
　　　Angle 312
　　　Plane Area. 314
　　　Perimeter. 317
　　　Ellipse Perimeter (Circumference) 318

Volume... 319
Surface Area... 323

Part V Uncertainty in Measurement

Chapter 29 Measurement Uncertainty......................... 326
Measurement Uncertainty............................... 328
Evaluating Measurement Uncertainty.................... 329
Measurement Uncertainty Evaluation Process............ 329
Other Measurement Uncertainty Considerations.......... 345
Managing Measurement Uncertainty...................... 346

Chapter 30 Measurement Risk and Decision Rules.............. 349
Measurement Risk...................................... 349
Test Uncertainty Ratio (TUR).......................... 360
A 4:1 TUR... 367
Conclusion.. 367
Annex A... 368

Part VI Measurement Parameters

Chapter 31 Introduction to Measurement Parameters........... 370

Chapter 32 DC and Low Frequency............................ 372
DC and AC... 372
Low Frequency... 372
Measurement Parameters................................ 373

Chapter 33 Radio Frequency and Microwave.................... 393
RF and Microwave Frequencies.......................... 394
Measurement Parameters................................ 396
Measurement Methods................................... 410

Chapter 34 Mass and Weighing................................ 416
SI Unit of Mass....................................... 417
Mass and Weighing..................................... 422
Good Weighing Procedures.............................. 428
Weighing Instruments—Types, Documentary Standards, Calibration
 Procedures... 434
Mass Calibration Standards (Weights).................. 438
Summary... 443

Chapter 35 Force.. 444
Basic Introduction.................................... 444
The SI Units for Force................................ 445
Why Force Measurement Is Important.................... 445
Types of Force Instruments............................ 446
S-Beam (S-Type)....................................... 448
Shear Web... 449
Button Load Cell...................................... 450
Single-Column or High-Stress Load Cells............... 451

Multi-Column Load Cells. 452
Hysteresis 457
Nonrepeatability 458
Static Error Band 459
Methods/Standards. 459
Common Error Sources in Force Calibration 460
Calibration Using Different Setups for Compression and Tension 461
Adapters Affect Mechanical Interactions 464
Tension Clevis Adapters for Tension Links, Crane Scales, and
   Dynamometers. 468
Four-Wire and Six-Wire Systems. 469
Equipment Design. 471
Uncertainty Contributors. 471
Type A Uncertainty Contributions 472
Type B Uncertainty Contributions 472
Other Error Sources. 472
Guidance for Type A Uncertainty Contributions When ASTM E74
   Is Used as the Standard 473
Guidance for Type B Uncertainty Contributors When ASTM E74
   Is Used as the Standard 474
Conclusion 475

Chapter 36 Dimensional and Mechanical Parameters 476
Length Measures. 477
Angle Measurement—Definitions 478
Levels. 482

Chapter 37 Other Parameters: Electro-Optical and Radiation 485
Optics. 485
Colorimetry 489
Ionizing Radiation. 490
Optical Radiation. 492
pH. 495

Chapter 38 Chemical and Biological Measurements and Uncertainties. 498
Introduction 498
Organizations. 499
Traceability Fundamentals. 500
Establishing Traceability for Chemical Tests. 501
Evaluating Measurement Uncertainty for Chemical Tests 507
Contributors to Uncertainty in Chemical Testing 510
Conclusion 511

PART VII Managing a Metrology Department or Calibration Laboratory

Chapter 39 Getting Started. 514

Chapter 40 Best Practices. 517
Customer Service (Lab Liaisons). 518
Using Metrics for Department/Laboratory Management. 521

  Preventive Maintenance Programs................................... 524
  Surveys and Customer Satisfaction.................................. 526

Chapter 41 Process Workflow 528

Chapter 42 Budgeting and Resource Management 534
  Budget 534

Chapter 43 Vendors and Suppliers 536

Chapter 44 Housekeeping and Safety 541
  All Industries 541
  5S Methodology 541
  Biotechnology and Pharmaceutical Industry 542
  Electronics or High-Voltage/Current Industries 542
  Airline Industry 543
  Physical-Dimensional Calibration 543
  Computer Industry 544

Appendix A Further Resources and Publications 545

Appendix B Acronyms and Abbreviations 550

Appendix C Glossary of Terms 566
  Introduction 566
  Glossary 568

Appendix D Common Conversions 591

Endnotes *619*
Bibliography/References *661*
Index *679*
About the Contributors *695*

List of Figures and Tables

Figure 1.1	Early dimensional standard at the Royal Observatory, Greenwich, England.	7
Figure 7.1	Nonaccredited calibration certificate.	52
Figure 7.2	Robust accredited calibration certificate.	53
Figure 8.1	An example format for an ISO/IEC 17025:2017-based quality manual.	60
Table 9.1	Recommended sources of calibration	66
Figure 9.1	Traceability pathways to the SI.	66
Table 9.2	Essential elements of metrological traceability	67
Figure 9.2	Lab pyramid.	69
Figure 9.3	Standards pyramid.	70
Figure 9.4	Lab hierarchy.	70
Figure 9.5	Standards hierarchy.	70
Figure 9.6	Circular calibration.	72
Table 10.1	Calibration program components	77
Table 11.1	Seven base units in the International System of Units (SI).	86
Table 11.2	The seven defining constants of the SI and the seven corresponding units they define	86
Table 11.3	The 2019 definitions of the seven base units using the defining constants.	87
Table 11.4	The 22 SI units with special names and symbols.	89
Table 11.5	The SI prefixes for multiples and submultiples of units	90
Figure 11.1	The SI unit diagram.	90
Table 11.6	Definitions and examples of standards.	92
Figure 11.2	Standards and uncertainty growth.	94
Table 11.7	Unit name translations.	99
Table 12.1	The three different types of audits.	104
Figure 15.1	"KASI": Knowledge, Attitude, Skills, Interpersonal Skills.	114

Figure 15.2	Competence (KASI criteria).	115
Table 15.1	Training and competency matrix example	118
Table 15.2	Formal vs. informal training.	119
Table 16.1	SA-TR52.00.01-2006 temperature and humidity limits	126
Table 16.2	General-purpose calibration laboratories	127
Table 16.3	Standards calibration laboratories or higher-accuracy requirements	127
Figure 18.1	NIST GLP 15, Figure 1—Software Life Cycle.	153
Figure 18.2	Software selection process.	156
Figure 18.3	Software implementation process.	157
Table 18.1	Advantages and disadvantages for converting data	157
Table 18.2	Advantages and disadvantages for using existing workflows.	158
Figure 18.4	Maintenance lifecycle of a software system.	161
Figure 18.5	Configuration management change control form example.	163
Table 19.1	Frequently used constants.	173
Table 19.2	Common measurement parameters.	174
Table 19.3	Common measurands and equipment used to source them	176
Table 19.4	Common measurands and some of their associated formulas	177
Table 20.1	Frequently used VIM terms	179
Table 20.2	Seven categories of measurement methods	180
Table 20.3	Measurement method category examples.	180
Table 20.4	Measurement systems VIM definitions.	182
Table 20.5	NASA 10-stage sequence in defining measurement requirements	183
Figure 20.1	Normal frequency distribution curve—1 standard deviation.	184
Table 20.6	Normal frequency distribution curves and associated standard deviation	184
Table 20.7	Control chart zones.	189
Table 20.8	X-bar and R-bar control chart calculations	190
Table 20.9	Table for calculating the control limits	191
Figure 20.2	X-bar chart for example of time interval measurements.	191
Figure 20.3	R-bar chart for the example of time interval measurement.	192

Table 20.10	Range R&R study	193
Figure 20.4	Example of appraiser variations.	195
Table 22.1	Specifications of meters A and B	215
Table 22.2	Measurement unit values at 28 V AC.	217
Table 22.3	Example test set manual's measurement requirements	220
Figure 23.1	Measurement comparison scheme.	223
Table 23.1	Assigning a reference value	224
Table 23.2	Measurement comparison scheme: raw data and calculations	225
Figure 23.2	Measurement comparison scheme: mean ± 3 standard deviations.	227
Figure 23.3	Measurement comparison scheme: mean ± U ($k = 2$).	227
Figure 23.4	Uncertainty overlap plot.	228
Table 23.3	Interlaboratory testing comparison data.	229
Figure 23.5	Interlaboratory testing comparison data.	229
Table 23.4	Derivation of consensus or reference value	230
Table 23.5	Split-sample data analysis	231
Figure 23.6	Analysis of variance.	231
Figure 23.7	Interpretation of one-way ANOVA data.	232
Table 23.6	Control chart constants	235
Table 23.7	Data for individual measurement	236
Figure 23.8	Individual measurements chart.	238
Figure 23.9	Moving range chart.	238
Table 23.8	Multiple measurement data.	239
Figure 23.10	X-bar chart.	241
Figure 23.11	Range chart.	241
Table 24.1	Numbers reported and significant digits	244
Table 24.2	Numbers and number of significant digits.	244
Table 24.3	ISO (Renard's) series of numbers	254
Table 24.4	Recommended spacing and use of decimal marker	258
Table 25.1	Seven base units in the International System of Units (SI)	262
Table 25.2	Some non-SI units accepted for use with SI units	263
Table 25.3	SI units derived from base units.	263
Table 25.4	The 22 SI units with special names and symbols.	264
Table 25.5	Other derived units	264

Table 25.6	The SI prefixes for multiples and submultiples of units	266
Table 25.7	SI-named units that are multiples of SI base or derived units	267
Table 25.8	Units not to be used within the SI system of units	267
Table 25.9	SI units arranged by unit category	268
Table 25.10	Conversion matrix: plane angle units	274
Table 25.11	Frequently used constants	275
Figure 27.1	Bimodal distribution	287
Figure 27.2	Normal distributions	287
Figure 27.3	Confidence intervals	288
Figure 27.4	Example of positive and negative skewness	292
Figure 27.5	Example of positive and negative kurtosis	293
Figure 27.6	Examples of correlation graphs	294
Figure 27.7	Pearson's correlation coefficient formula	294
Figure 27.8	Slope and intercept graph	296
Figure 27.9	Normal probability density function	301
Figure 27.10	Calculating z-score	301
Figure 27.11	F-distribution formula	302
Figure 27.12	Chi-square distribution formula	302
Figure 27.13	Hypergeometric distribution formula	303
Figure 27.14	Binomial distribution formula	303
Figure 27.15	Binomial distribution plots	304
Figure 27.16	Poisson distribution formula	304
Figure 27.17	Autocorrelation function	307
Figure 28.1	Circle general equation	311
Figure 28.2	Right triangle	313
Figure 28.3	Rectangle area	314
Figure 28.4	Parallelogram area	315
Figure 28.5	Oblique triangle area	316
Figure 28.6	Rectangular prism volume	319
Figure 28.7	Sphere volume	319
Figure 28.8	Ellipsoid volume	320
Figure 28.9	Pyramid volume	320
Figure 28.10	Truncated pyramid volume	321
Figure 28.11	Cone volume	321
Figure 28.12	Truncated cone volume	322
Figure 28.13	Rectangular prism surface area	323

Figure 28.14	Sphere surface area.	323
Figure 28.15	Prolate ellipsoid of revolution surface area.	324
Figure 28.16	Oblate ellipsoid of revolution surface area..	324
Table 29.1	Type A uncertainty example.	331
Figure 29.1	Population standard deviation.	332
Figure 29.2	Sample standard deviation.	332
Figure 29.3	Population standard deviation of the mean.	332
Figure 29.4	Sample standard deviation of the mean..	332
Table 29.2	Examples of Type A uncertainty evaluation calculation	333
Figure 29.5	Example measurement uncertainty budget.	337
Table 29.3	Correction factors for non-normal distributions	338
Figure 29.6	Root sum square for Type A contributors.	338
Figure 29.7	Root sum square for Type B contributors..	338
Figure 29.8	Root sum square for combined Type A and B contributors.	338
Figure 29.9	Combined Type A uncertainty example.	339
Figure 29.10	Combined Type B uncertainty example..	339
Figure 29.11	Combined Type A and B uncertainty example.	339
Table 29.4	Confidence level with associated coverage factor (k)	341
Figure 29.12	Welch-Satterthwaite formula.	346
Figure 29.13	*JCGM 100:2008* Table G.2, Student's *t*-Distribution Table.	347
Figure 30.1	Three pillars of measurement.	350
Figure 30.2	Guard band and acceptance interval illustration found in *ILAC G8:09/2019*.	351
Figure 30.3	Error types.	353
Figure 30.4	Scenario 1 data.	355
Figure 30.5	Scenario 1 graph.	355
Figure 30.6	Scenario 2 data.	356
Figure 30.7	Scenario 2 graph.	356
Figure 30.8	Scenario 3 data where PFA is 0.000 %.	358
Figure 30.9	Scenario 3 graph where PFA is 0.000 %.	358
Figure 30.10	Scenario 3 data where PFA is larger than 2.5 %.	359
Figure 30.11	Scenario 3 graph where PFA is larger than 2.5 %.	359
Figure 30.12	TUR formula (*UUT*, unit under test).	361
Figure 30.13	Example of a TUR formula (adapted from the ANSI/NCSL Z540.3 *Handbook*)..	362
Figure 30.14	TUR formula numerator.	363

Figure 30.15	TUR formula with the numerator value included.	363
Figure 30.16	CMC portion of the denominator.	363
Figure 30.17	TUR formula with CMC value included.	364
Figure 30.18	Resolution portion of the denominator.	364
Figure 30.19	TUR formula with the resolution value included.	364
Figure 30.20	Repeatability portion of the denominator.	364
Figure 30.21	TUR formula with repeatability value added.	365
Figure 30.22	Other uncertainty contribution sources in the denominator.	365
Figure 30.23	TUR formula with all uncertainty contribution sources included.	365
Figure 30.24	TUR calculated-formula.	365
Figure 30.25	TUR calculated value.	366
Figure 30.26	TUR data.	366
Figure 30.27	TUR graph.	366
Figure 30.28	*ANSI/NCSL Z540.3 Handbook* Method 5 and 6 guard bands.	368
Table 31.1	The six major measurement parameters	371
Table 32.1	Direct current parameters	375
Table 32.2	Direct voltage parameters	375
Table 32.3	Seebeck coefficients: Thermoelectric effects from connector materials.	377
Table 32.4	Resistance parameters	378
Table 32.5	Alternating voltage parameters	380
Table 32.6	Alternating current parameters	382
Table 32.7	Capacitance parameters.	383
Table 32.8	Inductance parameters.	384
Table 32.9	Time interval and frequency parameters	385
Table 32.10	Phase angle parameters.	390
Table 32.11	Electrical power parameters	392
Table 33.1	Common frequency bands and names	395
Table 33.2	Common waveguide bands.	396
Table 33.3	RF power parameters.	397
Figure 33.1	Power sensor calibration system—block diagram.	402
Table 33.4	Attenuation or insertion loss parameters	403
Table 33.5	Reflection coefficient, standing wave ratio parameters	405
Figure 33.2	Smith chart.	406

List of Figures and Tables

Table 33.6	RF voltage parameters	408
Table 33.7	Modulation parameters	409
Table 33.8	Noise figure, excess noise ratio parameters	410
Figure 33.3	A scalar network analyzer measures magnitude only. The reflection coefficient phase varies with frequency and causes an area of uncertainty.	412
Figure 33.4	A vector network analyzer measures magnitude and phase at each frequency, so magnitude uncertainty is reduced.	412
Figure 33.5	A general two-port network. This can be used to represent a device such as a coaxial attenuator.	414
Figure 33.6	S-parameter flowgraph of a two-port network.	414
Figure 34.1	The International Prototype Kilogram (IPK).	418
Figure 34.2	NIST Kibble balance.	420
Figure 34.3	Silicon sphere.	421
Figure 34.4	Kilogram metrological traceability pyramid.	421
Figure 34.5	Density block and air buoyancy illustration.	424
Table 34.1	Example air buoyancy on 1 cm^3 density blocks	425
Table 34.2	Example air buoyancy on different materials	425
Figure 34.6	Density difference by block illustration.	426
Figure 34.7	Stainless steel vs. aluminum kilograms.	427
Table 34.3	Reference conditions for conventional mass.	428
Figure 34.8	Cause-and-effect diagram.	430
Table 34.4	Weighing instrument classifications common in legal metrology	437
Table 34.5	Comparison of mass standards classifications.	439
Table 34.6	Mass calibration procedure recommendations	441
Figure 35.1	Proving rings.	447
Figure 35.2	Load cells.	447
Figure 35.3	Force gauge.	448
Figure 35.4	S-beam load cell.	449
Figure 35.5	Shear web load cells.	450
Figure 35.6	Button load cell.	451
Figure 35.7	Single-column load cell.	452
Figure 35.8	Lightweight 600k (26 lbf) multi-column load cell.	453
Figure 35.9	Compression calibration examples.	454
Figure 35.10	Tension calibration examples.	454

Figure 35.11	Deadweight calibrating machine.	455
Table 35.1	Basic force conversions.	456
Figure 35.12	Non-linearity expressed graphically.	457
Figure 35.13	Hysteresis example.	458
Figure 35.14	Universal calibrating machine with a unit under test in compression (left) and in tension (right).	462
Figure 35.15	Tension adapters designed using recommendations from ISO 376.	463
Figure 35.16	Compression adapters designed using recommendations from ISO 376.	463
Figure 35.17	Universal testing machine.	464
Figure 35.18	Ultra-precision shear web load cell showing eccentric forces.	465
Figure 35.19	S-beam load cell with slight misalignment producing a 0.752 % error.	466
Figure 35.20	Proper way to thread an alignment plug: Thread is past flush and into cell.	467
Figure 35.21	Button and washer load cell adapters.	467
Figure 35.22	Tension members with two ball nuts and two ball cups.	468
Figure 35.23	Tension link difference in output with pin size.	469
Figure 35.24	Clevis kits (U.S. Patent 11,078,052).	469
Figure 35.25	4-wire cable and diagram.	470
Figure 35.26	6-wire cable and diagram.	471
Figure 35.27	Example uncertainty budget contributions.	473
Table 36.1	Conversion factors for plane angle units.	479
Figure 37.1	Converging (positive) lens. Convex lenses are those that are wider in the center than they are on the top and bottom. They are used to converge light.	485
Figure 37.2	Diverging (negative) lens. Concave lenses are those that are wider at the top and bottom and narrower in the center. They are used to diverge light.	486
Figure 37.3	Concave mirrors reflect light inward, and convex mirrors reflect light outward.	486
Figure 37.4	Refracting telescopes use lenses to focus light.	486
Figure 37.5	Reflecting telescopes use mirrors to focus light.	486
Figure 37.6	Penetrating powers of alpha and beta particles, and gamma rays.	491
Figure 38.1	IUPAC Periodic Table of the Elements, 1 Dec 2018.	502
Figure 38.2	Basic process for establishing traceability in a chemical test.	503

Figure 38.3	Two results obtained in two different tests using different reference materials...	506
Figure 38.4	Two results obtained in two different tests using the same reference materials...	506
Figure 40.1	Forecast of scheduled calibrations due...	522
Figure 41.1	A sample business process interaction diagram...	532

Foreword

My introduction to real metrology was quite rude. I was applying for a job as an analytical chemist, and the laboratory director interviewing me asked how to run an infrared spectrum. After rather proudly reciting what I had been taught in college, I was immediately deflated when he asked me how accurate the measurement result would be. As I stammered my non-answer, I realized that job was not going to become mine. Oh, if I had only had *The ASQ Metrology Handbook* at that time!

Over the course of the past few decades, metrology has changed significantly. Concepts have been defined more rigorously. Electronics, computers, microtechnology, lasers, and many other technological developments have pushed measurement capability to almost unbelievable accuracy. The industrial world has realized that measurement is fundamental to product and process quality. Consumers demand product characteristics and functionality realizable only with today's measurement accuracy. It is the job of the metrologist to meet the need for greater measurement accuracy by using all tools now available.

In this third edition of *The ASQ Metrology Handbook*, the American Society for Quality (ASQ) has assembled the basic components of the practice of metrology as it is known today. For those who want to be part of the ever-growing metrology community, it introduces the fundamental concepts in a clear and precise way. For those who are already metrology professionals, it is an ideal companion to supplement and expand your knowledge. And for those who work in or manage metrology and calibration laboratories—either today or tomorrow—*The ASQ Metrology Handbook* covers the essentials of operating a respected laboratory.

The practice of metrology is both constant and changing. Metrology is *constant* in that it relies on a strong sense of the fundamental. Good measurement is based on understanding what must be controlled, what must be reported, and what must be done to ensure that repeated measurements continue to maintain accuracy. Metrology is always *changing* as scientists, engineers, and technicians learn more about how to make good measurements. In the latter part of the last century, concepts such as uncertainty, quality systems, statistics, and good metrology laboratory management underwent significant changes, resulting in better metrological practice.

This third edition of *The ASQ Metrology Handbook* provides an accessible and comprehensive summary of these new developments. It is essential that every metrology professional be familiar with these new approaches and use them daily in their work. A metrology professional must continue to expand their knowledge through courses, conferences, and study. Through the work of organizations such as ASQ and *The ASQ Metrology Handbook*, the task of keeping up with advances has been made much simpler.

Dr. John Rumble, Jr.
R&R Data Services, Inc.
Gaithersburg, Maryland
November 2021

Preface

I am and have been incredibly lucky in my professional career. The worldwide metrology community is a welcoming and supportive environment, even among competitors. I am among the metrologists who did not receive military training in metrology and calibration; I learned on the job and through conferences and external trainings. I found my way into the field of metrology when I stumbled across an open position at my company for someone to calibrate and manage all calibration and test equipment for the company's microbiology, chemical, and physical engineering test labs. I had strong and diverse educational training from my biology degree coursework at the University of Michigan and had already worked as a microbiologist, an engineering test technician, an extraction and analytical chemist, and in quality functions. This position was written for me! I was able to apply scientific principles from mathematics, physics, and chemistry. I practiced the scientific method as I learned the different measurement areas and developed new and improved existing calibration procedures. Upon recommendation from accreditation assessors, I started attending metrology conferences and trainings. It was there that I learned about ASQ, the Certified Calibration Technician (CCT) certification, and *The [ASQ] Metrology Handbook*. Eager to learn from the best, I sought out Jay Bucher and Dilip Shah, both co-authors of *The [ASQ] Metrology Handbook* (1st ed.). They welcomed me and offered guidance, feedback, and opportunities. I found that other conference and training attendees also had a similar eagerness and enthusiasm for metrology. We hungered to learn more.

The [ASQ] Metrology Handbook (1st ed.) was such a great all-around reference that it was useful beyond just studying for the ASQ-CCT exam. It was applicable beyond just bench-level knowledge and application. It was the reference that I, and so many others, needed and used. Once I passed the CCT exam, I was in the pool of CCTs to be invited to CCT exam workshops. From my time participating in these workshops and serving on the ASQ Certification Board, I learned about the rigor of ASQ certification exams. The professionals at ASQ who support the exam development and update processes use sound statistical, quality, and psychometric practices to uphold the rigor and integrity of the ASQ Certification exams. There is good reason that ASQ certifications are so highly regarded around the world!

I was fortunate to be asked by Jay Bucher to contribute to the second edition of *The [ASQ] Metrology Handbook* for its publication in 2012. I was able to share some of what I had learned at conferences and trainings in that *Handbook*. It felt good to share useful knowledge and practices in a format and medium that would be accessible by so many. These activities continued and I made contributions to other publications and organizations as well. Shortly after Jay Bucher passed away, the CCT exam went through another update cycle. Just after that five-year exam update cycle was completed, one of the most frequently referenced publications, ISO/IEC 17025:2017, was updated and published. This caused headaches for the ASQ-CCT Certification subcommittee, as the CCT exam still referenced the 2005 version of ISO/IEC 17025 and folks were now using the 2017 version. The 2017 version would be addressed in the next CCT exam update cycle. As the next CCT exam update cycle approached, Jay was no longer around to pull together the other contributing authors. In addition to ISO/IEC 17025, there had been remarkable changes in other international standards, measurement technologies, and quality practices, as well as in the redefinition of the entire International System of Units (SI). There needed to be an update to the most popular reference for the ASQ-CCT certification exam as well as to make it more attractive to even more professionals in more disciplines than just physical-dimensional.

As chair of ASQ-Measurement Quality Division and with so many professional connections in the global metrology community, I felt the responsibility to lead the effort to update *The ASQ Metrology Handbook* to its third and most comprehensive edition to date. The large and diverse group of contributing authors and supporting reviewers are top-notch leaders in their fields and have contributed the best of their knowledge and applications. The expertise of each of the contributing authors adds depth, clarification, and readily useful knowledge for readers.

Our purpose in writing this handbook was to develop a practical metrology reference for calibration professionals as well as those in manufacturing, testing, and quality groups. It is our intent that this handbook will provide answers, instruction, and guidance, and spark thought-provoking discussions and a desire for more knowledge.

Acknowledgments

Jay Bucher's acknowledging words from the second edition are so appropriate that I am using them again for the third edition with only very minor changes. This handbook is a synthesis of the education, training, experience, knowledge, and effort of several individuals. The authors acknowledge and are grateful for the support of our friends, families, communities, and employers during this project. This book has touched the lives of friends, coworkers, supervisors, clients, and families. We would be remiss if we did not acknowledge their part in allowing us to do our part. Their patience and understanding made this task a great deal easier. To all of you, we collectively say, "Thank you, from the bottom of our hearts. We could not have done this without your understanding and support."

To the metrology community, this book was written with your needs and requirements in mind. On behalf of ASQ's Measurement Quality Division, we solicit your observations, comments, and suggestions for making this a better book in the future. Calibration and metrology are living entities that continue to grow and evolve. What we do would not be much fun if they did not.

And finally, to this unique team of coauthors: thank you. Two simple words that say it all. This book is the result of cooperation, understanding, and devotion to the calibration/metrology community and a burning desire to share what we have learned, experienced, and accomplished in our individual careers. I thank you, the rest of the team thanks you, and the metrology community surely will thank you.

Heather A. Wade, ASQ-CCT, ASQ-CQA
Editor, *The ASQ Metrology Handbook, Third Edition*

ADDITIONAL ACKNOWLEDGMENTS

The following individuals contributed their time and efforts to *The [ASQ] Metrology Handbook, First Edition*: Keith Bennett, Hershal Brewer, David Brown, Jay L. Bucher, Christopher Grachanen, Emil Hazarian, Graeme C. Payne, and Dilip Shah.

The following individuals contributed their time and efforts to *The [ASQ] Metrology Handbook, Second Edition*: Jay L. Bucher, Christopher Grachanen, Emil Hazarian, Dilip Shah, and Heather Wade.

The following individuals volunteered and contributed their time and efforts to review, update, and contribute to the third edition of *The ASQ Metrology Handbook*: Dilip Shah, Georgia Harris, Paul Keep, Henry Zumbrun, Heather Sandoe, Tony Hamilton, Pamela Wright, Walter Nowocin, Jane Weitzel, Elizabeth Benham, Michael A. Lombardi, Val Miller, Mark Ruefenacht, and Sharry Masarek. On the Measurement Quality Division's challenge coin it states, "Volunteers aren't paid . . . not because they're worthless, but because they're priceless." That is a most appropriate sentiment when applied to this team of volunteers. Without their assistance to accomplish the time-consuming task of checking for updates, reviewing what the current standards state, gathering permissions, and adding relevant material since the second-edition update 10 years ago, we could not have put together this new edition.

I would like to give special thanks to Dilip Shah, Georgia Harris, and the late Jay Bucher for your leadership, input, guidance, knowledge, wisdom, and mentorship. I would not be the metrologist I am without you. I also believe that the entire metrology community is better because of you.

Introduction

When one pauses to consider how much has changed in the world and in the world of metrology in the 10 years that have passed since the publication of the second edition of *The [ASQ] Metrology Handbook* in 2012, one might gasp in wonder and awe. The International System of Units was redefined; now all base SI units are defined based on natural physical constants. There is currently research and development at National Metrology Institute levels to provide the fundamental traceability of the SI units through the exact values of quantum constants. In the not-too-distant future, industrial-level laboratories may have their own quantum standards and realize SI units in their own labs, thus re-leveling the metrological traceability pyramid. Published international standards have been updated, with many of these shifting from being prescriptive (i.e., what and how to do) to focusing on outputs and risks. The freedom to determine one's own risks and even one's own competency criteria brings with it the need for ideas and guidance on *how* to meet these evolving requirements. Our purpose in writing this handbook was to develop a practical metrology reference for calibration professionals as well as those in manufacturing, testing, and quality groups. It is our intent that this book will provide answers, instruction, and guidance, and spark thought-provoking discussions and a desire for more knowledge.

In addition to changes in reference standards and publications, there have been advancements in the use of computers and software. Whereas once a computer workstation might have been shared by multiple people in a laboratory, technicians now use handheld tablets and automatic data-capturing software and transmit data over the internet to improve efficiency in performing calibration activities. Software and computers have been tools in laboratories for a long time. Software has always been considered "equipment," but there has been little guidance on how to track, validate, and verify software in metrology labs. New in-depth chapters have been added, addressing decision rules and measurement risk and the measurement parameter of force. Several chapters have been expanded, including those on metrological traceability; calibration programs; the International System of Units (SI); software selection, validation, and verification; measurement uncertainty; and mass and weight.

Previous editions of *The ASQ Metrology Handbook* have been heavily focused on physical-dimensional measurements. There is a need for information about chemical and biological measurements (such as in life science) and their measurement uncertainties. These areas are discussed in their own chapters. The previous editions of *The ASQ Metrology Handbook* have also been heavily U.S.-centric. We have tried to shift beyond a U.S. point of view and include a more inclusive, global perspective. We would like your input for how we can make the next editions even better.

Metrology, the science of measurement, affects everyone daily, whether we realize it or not. There is so much in our world that would not be possible without metrology. Just look at the pages in this book. Software was used to compile and format the pages and images using linear dimensions. The computers, tablets, and software used to view the electronic content have all had to meet prescribed requirements to function correctly. If the book you are holding is a hard-copy version, the tools used to cut the pages had to be measured so that each page was the same size. We are surrounded by measurements. Look at any object around you; somehow measurements played a part in its creation.

Whether you use this handbook to further your education, change disciplines in your career, accept more responsibilities as a supervisor or manager, train your fellow calibration practitioners, or use it to prepare for ASQ's Certified Calibration Technician exam, we hope this handbook provides the information, guidance, and knowledge to help you achieve your goals.

Although each chapter could be its own book, we believe we have given enough information to provide solid foundations in each area, as well as providing an extensive list of supplemental resources. The expertise of each of the contributing authors brings depth, clarification, and readily useful knowledge for the readers. We hope you take what you learn, build with it, build upon it, and make your own impacts in the world of metrology. With all our work and impacts, it will be interesting to see what the world and metrology are like in another 10 years.

Part I
Background

Chapter 1 History and Philosophy of Metrology/Calibration

Chapter 1
History and Philosophy of Metrology/Calibration

> *Weights and measures may be ranked among the necessaries of life to every individual of human society. They enter into the economical arrangements and daily concerns of every family. They are necessary to every occupation of human industry; to the distribution and security of every species of property; to every transaction of trade and commerce; to the labors of the husbandman; to the ingenuity of the artificer; to the studies of the philosopher; to the researches of the antiquarian; to the navigation of the mariner; and the marches of the soldier; to all the exchanges of peace, and all the operations of war. The knowledge of them, as in established use, is among the first elements of education, and is often learned by those who learn nothing else, not even to read and write. This knowledge is riveted in the memory by the habitual application of it to the employments of men throughout life.*
>
> John Quincy Adams, Report to the Congress, 1821

ANCIENT MEASUREMENT

Weights and measures were some of the earliest tools invented. From the beginning of humankind there was a need to standardize measurements used in everyday life, such as construction of weapons used for hunting and protection, gathering and trading of food and clothing, and territorial divisions. The units of specific measurements, such as length, were defined as the length of an individual's arm, for example (other parts of the human anatomy were also used). Weight probably would have been defined as the amount a human could lift or the weight of a stone the size of a hand. Time was defined by the length of a day and days between the cycles of the moon. Cycles of the moon were used to determine seasons. Although definitions of these measurements were rudimentary, they were sufficient to meet local requirements. Little is known about the details of any of these measurements, but artifacts more than 20,000 years old indicate some form of timekeeping.

The earliest Egyptian calendar was based on the moon's cycles, but later the Egyptians realized that the Dog Star in Canis Major, which we call Sirius, rose next to the sun every 365 days, about when the annual inundation of the Nile began. Based on this knowledge, they devised a 365-day calendar that seems to have begun in 4236 BCE, which appears to be one of the earliest years recorded in history.

As civilizations grew, they became more sophisticated, and better definitions for measurement were required. The history of Egypt begins in approximately 3000 BCE. This is when Upper and Lower Egypt became unified under one ruler, Menes, and when the first pyramids were being built. At this time, trade of goods was common, as were levies or taxes on them. Definitions for liquid and dry measures were important. Length was also an important measure, as was time. Some key Babylonian units were the *Kush* (cubit) for length, *Sar* (garden-plot) for area and volume, *Sila* for capacity, and *Mana* for weight. At the base of the system is the barleycorn (*She*), used for the smallest unit in length, area, volume, and weight.

Length

The smallest unit of length is the *She* (barleycorn), which equals about 1/360 meter.

6 *She*	=	1 *Shu-Si* (finger)
30 *Shu-Si*	=	1 *Kush* (cubit—about ½ m)
6 *Kush*	=	1 *Gi / Ganu* (reed)
12 *Kush*	=	1 *Nindan / GAR* (rod—6 m)
10 *Nindan*	=	1 *Eshe* (rope)
60 *Nindan*	=	1 *Ush* (360 m)
30 *USH*	=	1 *Beru* (10.8 km)

Area and Volume

The basic area unit is the *Sar*, an area of 1 square *Nindan*, or about 36 square meters. The area units *She* and *Gin* are used as generalized fractions of this basic unit.

180 *She*	=	1 *Gin*
60 *Gin*	=	1 *Sar* (garden plot 1 sq. *Nindan*—36 sq. m)
50 *Sar*	=	1 *Ubu*
100 *Sar*	=	1 *Iku* (1 sq. *Eshe*—0.9 acre, 0.36 Ha)
6 *Iku*	=	1 *Eshe*
18 *Iku*	=	1 *Bur*

Capacity

These units were used for measuring volumes of grain, oil, beer, and so on. The basic unit is the *Sila*, about 1 liter. The semistandard Old Babylonian system used in mathematical texts is derived from the ferociously complex mensuration systems used in the Sumerian period.

180 *She*	=	1 *Gin*
60 *Gin*	=	1 *Sila* (1 liter)
10 *Sila*	=	1 *Ban*
6 *Ban*	=	1 *Bariga*
5 *Bariga*	=	1 *Gur*

Weight

The basic unit of weight is the *Mana*, about ½ kilogram.

180 *She*	=	1 *Gin/Shiqlu* (shekel)
60 *Gin*	=	1 *Mana* (*Mina*—500 g)
60 *Mana*	=	1 *Gu/Biltu* (talent, load—30 kg)

Royal Cubit Stick

The *royal cubit* (524 millimeters or 20.62 inches) was subdivided in an extraordinarily complicated way. The basic sub-unit was the *digit*, doubtlessly a finger's breadth, of which there were 28 in the royal cubit.

- Four digits equaled a *palm*, five a *hand*.
- Twelve digits, or three palms, equaled a *small span*.
- Fourteen digits, or one-half a cubit, equaled a *large span*.
- Sixteen digits, or four palms, made *one t'ser*.
- Twenty-four digits, or six palms, were a *small cubit*.

The Egyptians studied the science of geometry to assist them in the construction of the pyramids. The royal Egyptian cubit was decreed to be equal to the length of the forearm from the bent elbow to the tip of the extended middle finger plus the width of the palm of the hand of the pharaoh or king ruling at that time.

The royal cubit master was carved out of a block of granite to endure for all times. Slaves and workers engaged in building tombs, temples, pyramids, and so on, were supplied with cubits made of wood or granite. The royal architect or foreperson of the construction site was responsible for maintaining and transferring the unit of length to workers' instruments. They were required to bring back their cubit sticks at each full moon to be compared to the royal cubit master. Failure to do so was punishable by death. Though the punishment prescribed was severe, the Egyptians had anticipated the spirit of the present-day system of legal metrology, standards, traceability, and calibration recall.

With this standardization and uniformity of length, the Egyptians achieved surprising accuracy. Thousands of workers were engaged in building the Great Pyramid of Giza. Using cubit sticks, they achieved an accuracy of 0.05%. In roughly 756 feet or 9069.4 inches, they were within 4½ inches.[1]

Digit

The digit was in turn subdivided. Reading from right to left in the upper register, the 14th digit on a cubit stick was marked off into 16 equal parts. The next digit was divided into 15 parts, and so on, to the 28th digit, which was divided into two equal parts. Thus, measurement could be made to digit fractions with any denominator from 2 through 16. The smallest division, ¹⁄₁₆ of a digit, was equal to $\frac{1}{448}$ part of a royal cubit. Although the Egyptians achieved very good standardization of length, this standardization was regional. There were multiple standards for the cubit, which varied greatly due to the standard they were based on, the length from the tip of the middle finger to the elbow of the current ruler. Variations of the cubit are as follows:

- Arabian (black) cubit of 21.3 inches
- Arabian (hashimi) cubit of 25.6 inches
- Assyrian cubit of 21.6 inches
- Ancient Egyptian cubit of 20.6 inches
- Ancient Israeli cubit of 17.6 inches
- Ancient Grecian cubit of 18.3 inches
- Ancient Roman cubit of 17.5 inches

Of these seven cubits, the variation from the longest to the shortest was 8.1 inches, with an average value of 20.36 inches. These variations made trade between different regions difficult. As time elapsed, the need for trade on a regional basis became greater. The need for more sophistication and accuracy thus also became greater.

MEASUREMENT PROGRESS IN THE LAST 2000 YEARS

Efforts at standardizing measurement evolved around the world, not just in Egypt. English, French, and American leaders strived to bring order to their marketplaces and governments:

732–King of Kent—The measurement of an acre is in common use.

960–Edgar the Peaceful decrees, "All measures must agree with standards kept in London and Winchester."

1215–King John agrees to have national standards of weights and measures incorporated into the Magna Carta.

1266–Henry III declares in an act that:

- One penny should weigh the same as 32 grains of wheat.
- There should be 20 pennies to the ounce.
- There should be 12 ounces to the pound.
- There should be 8 pounds to the weight of 1 gallon of wine.

1304–Edward I declares in a statute that:

- For medicines, 1 pound equals 12 ounces (apothecaries' measure, still used in the United States).
- For all other liquid and dry measures, 1 pound equals 15 ounces.
- One ounce still equals 20 pennies.

1585–In his book *The Tenth*, Simon Stevin suggests that a decimal system should be used for weights and measures, coinage, and divisions of the degree of arc.

1670–Authorities give credit for originating the metric system to Gabriel Mouton, a French vicar.

1790–Thomas Jefferson proposes a decimal-based measurement system for the United States. France's Louis XVI authorizes scientific investigations aimed at a reform of French weights and measures. These investigations lead to the development of the first metric system.

1792–The U.S. Mint is formed to produce the world's first decimal currency (the U.S. dollar, consisting of 100 cents).

1795–France officially adopts the metric system.

1812–Napoleon temporarily suspends the compulsory provisions of the 1795 metric system adoption.

1824–George IV, in a Weights and Measures Act (5 Geo IV c. 74), establishes the "Imperial System of Weights and Measures," which is still used.

1840–The metric system is reinstated as the compulsory system in France.

1866–The use of the metric system is made legal (but not mandatory) in the United States by the (Kasson) Metric Act of 1866. This law also makes it unlawful to refuse to trade or deal in metric quantities.

STANDARDS, COMMERCE, AND METROLOGY

Standards of measurement before the 1700s were local and often arbitrary, making trade between countries—and even cities—difficult. The need for standardization as an aid to commerce became apparent during the Industrial Revolution. Early standardization and metrology needs were based on military requirements, especially those of large maritime powers such as Great Britain and the United States. A major task of navies in the eighteenth and nineteenth centuries was protection of their country's international trade, much of which was carried by merchant ships. Warships would sail with groups of merchant ships to give protection from pirates, privateers, and ships of enemy nations; or they would sail independently to "show the flag" and enforce the right of free passage on the seas.

A typical ship is the frigate USS *Constitution*. She was launched in 1797 and armed with thirty-four 24-pound cannons and twenty 32-pound cannons. (The size of cannon in that era was determined by the mass, and therefore the diameter, of the spherical cast-iron shot that would fit in the bore.) For reasons related to accuracy, efficiency, and economy, the bores of any particular size of cannon all had to be the same diameter. Likewise, the iron shot had to be the same size. If a cannonball were too large, it would not fit into the muzzle; too small, and it would follow an unpredictable trajectory when fired. The requirements of ensuring that dimensions were the same led to the early stages of a modern metrology system, with master gages, transfer standards, and regular comparisons. Figure 1.1 is an example of a length standard that dates from this era. This one, mounted on the

Figure 1.1 Early dimensional standard at the Royal Observatory, Greenwich, England. (Photo by Graeme C. Payne.)

wall of the Royal Observatory at Greenwich, England, is one of several that were placed in public places by the British government for the purpose of standardizing dimensional measurements. It has standards for three and six inches, one and two feet, and one yard. In use, a transfer standard would be placed on the supports, and it should fit snugly between the flats of the large posts. The actual age of this standard is unknown.

Over time, as measurements became standardized within countries, the need arose to standardize measurements between and among countries. A significant milestone in this effort was the adoption of the Convention of the Metre Treaty in 1875. This treaty set the framework for, and still governs, the international system of weights and measures. It can be viewed as one of the first voluntary standards with international acceptance, and possibly the most important to science, industry, and commerce. The United States was one of the first seventeen nations to adopt the Metre Convention:

1875–The Convention of the Metre is signed in Paris by seventeen nations, including the United States. The Metre Convention, often called the Treaty of the Meter in the United States, provides for improved metric weights and measures and the establishment of the General Conference on Weights and Measures (CGPM) devoted to international agreement on matters of weights and measures.[2]

1878–Queen Victoria declares the Troy pound illegal. Commercial weights can only be of the quantity of 56 lbs, 28 lbs, 14 lbs, 7 lbs, 4 lbs, 2 lbs, 1 lbs, 8 oz, 4 oz, 2 oz, and so on.

1889–As a result of the Metre Convention, the United States receives a prototype meter and kilogram to be used as measurement standards.

1916–The Metric Association is formed as a nonprofit organization advocating adoption of the metric system in U.S. commerce and education.

The organizational name, started as the American Metric Association, is changed to the U.S. Metric Association (USMA) in 1974.

1954–The International System of Units (SI) begins its development at the tenth CGPM. Six of the new metric base units are adopted.

1958–A conference of English-speaking nations agrees to unify their standards of length and mass, and to define them in terms of metric measures. The American yard is shortened, and the imperial yard is lengthened as a result. The new conversion factors are announced in 1959 in the *Federal Register*.[3]

HISTORY OF QUALITY STANDARDS

In some respects, the concept we call *quality* has been with humankind through the ages. Originally, aspects of quality were passed on by word of mouth, from parent to child, and later from craftsperson to apprentice. As the growth of agriculture progressed, people started settling in villages, and resources were available to support people who were skilled at crafts other than farming and hunting. These craftspersons and artisans would improve their skill by doing the same work repeatedly and make improvements based on feedback from customers. The main pressure for quality was social because the communities were small and trade was local. A person's welfare was rooted in their reputation as an honest person who delivered a quality product—and the customers were all neighbors, friends, or family.

The importance and growth of quality and measurement increased with the development of the first cities and towns, about 6000 to 7000 years ago. Astronomy, mathematics, and surveying were important trades. Standard weights and measures were all developed as needed. All of these were important for at least three activities: commerce, construction, and taxation. The systems of measuring also represented an important advancement in human thought because it is a shift from simple counting to representing units that can be subdivided at least to the resolution available to the unaided eye. In Egypt, systems of measurement were good enough 5000 years ago to survey and construct the pyramids at Giza with dimensional inaccuracies of only about 0.05 %. This was made possible using regular calibration and traceability. Remember, the cubit rules used by the builders were required to be compared periodically to the pharaoh's granite master cubit. Juran mentions evidence of written specifications for products around 3500 years ago,[4] and Bernard Grun notes regulations about the sale of products (beer, in this case) about that same time. Standardization of measurements increased gradually as well, reaching a peak in the later years of the Roman Empire. (It is often said, humorously, that the standard-gauge spacing of railway tracks in Europe and North America is directly traceable to the wheel spacing of Roman war chariots.)

Measurement science and quality experienced a resurgence as Europe emerged from the Dark Ages. Again, some of the driving forces were commerce, construction, and taxation. Added to these were military requirements and the needs of the emerging fields of science. Many of the quality aspects were assumed by trade and craft guilds. The guilds set specifications and standards for their trades and developed a training system that persists to this day: the system of apprentices, journeymen, and master craftsmen. Guilds exercised quality control through inspection, but often stifled quality improvement and product innovation.

The Industrial Revolution accelerated the growth of both quality and measurement. Quantities of manufactured items increased, but each part was still essentially custom made. Even while referring to a master template, it was difficult to construct parts that could be randomly selected and assembled into a functioning device. By the mid-1700s the capability of a craftsperson to produce substantially identical parts was demonstrated in France. In 1789, Eli Whitney used this capability in the United States when he won a government contract to provide 10,000 muskets with interchangeable parts, but it took him 10 years to fulfill the contract. By 1850 it was possible for a skilled machinist to make hundreds of repeated parts with dimensional uncertainties of no more than ± 0.002 inch. Nevertheless, those parts were still largely handmade, one at a time.

MEASUREMENT AND THE INDUSTRIAL REVOLUTION

The Industrial Revolution began around 1750. Technology evolved quickly regarding materials, energy, time, architecture, and in humanity's relationship with the Earth. As a result, industries also quickly evolved. With a growing population, the need for clothing, transportation, medicines, and food drove industry to find better, more efficient methods to support this need. Technology had evolved sufficiently to support this growth.

During this time there was tremendous growth of discoveries in quantum mechanics and molecular, atomic, nuclear, and particle physics. These discoveries laid the groundwork for much of the seven base units of the current International System of Units (SI). These seven units are well-defined and dimensionally independent: the meter (m), kilogram (kg), second (s), ampere (A), kelvin (K), mole (mol), and candela (cd).

During the mid-1800s there was rapid progress in temperature and thermodynamics developments. Note the speed of discovery:

1714–Daniel Gabriel Fahrenheit invents mercury and alcohol thermometers.

1821–Thermocouples are invented by Thomas Johann Seebeck. The discovery that metals have a positive temperature coefficient of resistance, which leads to the use of platinum as a temperature indicator (PRT), is made by Humphrey Davy.

1834–Lord Kelvin formulates the second law of thermodynamics.

1843–The mechanical equivalent of heat is discovered.

1848–Lord Kelvin discovers the absolute zero point of temperature (0 K).

1889–Platinum thermometers are defined by many different freezing points and boiling points of ultrapure substances, such as the freezing point of H_2O at 0 °C, the boiling point of H_2O at 100 °C, and the boiling point of sulfur at 444.5 °C.

1900–Max Planck formulates the Planck Radiation Law, which governs the intensity of radiation emitted by unit surface area into a fixed direction from the blackbody as a function of wavelength for a fixed temperature.

1927–The Seventh CGPM adopts the International Temperature Scale of 1927.

There were many more discoveries during this period that covered electromagnetic emissions, radioactivity, nuclear chain reactions, superconductivity, and others. The following is a list of some of the properties and quantities of electricity that were defined:

1752–Benjamin Franklin proves that lightning and the spark from amber are the same thing.

1792–Alessandro Volta shows that when moisture comes between two different metals, electricity is created. This leads him to invent the first electric battery, the voltaic pile, which he makes from thin sheets of copper and zinc separated by moist pasteboard. The unit of electrical potential, the *volt*, is named after Volta.

1826–George Simon Ohm states Ohm's law of electrical resistance.

1831–Michael Faraday discovers the first method of generating electricity by means of motion in a magnetic field.

1832–Faraday discovers the laws of electrolysis.

1882–A New York street is lit by electric lamps.

1887–Heinrich R. Hertz discovers the photoelectric effect.

1897–J. J. Thomson discovers the electron.

1930–Paul A. M. Dirac introduces the electron hole theory.

The discoveries and descriptions of these physical phenomena led to technological advancements and the development of new products. In turn, better measurements were needed. To build a single machine, an inventor could get by without much standardization of measurements, but to build multiple machines, using parts from multiple suppliers, measurements had to be standardized to a common entity.

Metrology is keeping up with physics and industry. Recent successes in industry that would not be possible without metrology include the following: Fission and fusion are being refined both as weapons and as a source of energy. Semiconductors are being developed, refined, and applied on a larger scale with the invention of the integrated circuit, solar cells, light-emitting diodes, and liquid crystal displays. Communication technology has developed through application of satellites and fiber optics. Lasers have been invented and applied in useful technologies including communication, medicine, and industrial applications. The ability to place a human-made object outside the Earth's atmosphere started the space race and has led to placing a human on the Moon, satellites used for multiple purposes, probes to other planets, the space shuttle, and space stations. A few spacecraft have passed beyond the farthest known planets and are headed into interstellar space. One thing is certain: As technology continues to grow, measurement challenges will grow proportionally.

MILESTONES IN U.S. FOOD AND DRUG LAW HISTORY

From the beginnings of civilization people have been concerned about the quality and safety of foods and medicines. In 1202, King John of England proclaimed the first English food law, the Assize of Bread, which prohibited adulteration of bread with such ingredients as ground peas or beans. Regulation of food in the United States dates from early colonial times. Federal controls over the drug supply began with inspection of imported drugs in 1848. The following chronology describes some of the milestones in the history of food and drug regulation in the United States:[5]

1820–Eleven physicians meet in Washington, D.C., to establish the U.S. Pharmacopeia, the first compendium of uniform set of guidelines for medicine quality for the United States.

1848–The Drug Importation Act passed by Congress requires U.S. Customs Service inspection to stop entry of adulterated drugs from overseas.

1862–President Lincoln appoints a chemist, Charles M. Wetherill, to serve in the new Department of Agriculture. This is the beginning of the Bureau of Chemistry, the predecessor of the Food and Drug Administration.

1880–Peter Collier, chief chemist, U.S. Department of Agriculture, recommends passage of a national food and drug law, following his own food adulteration investigations. The bill is defeated, but during the next 25 years more than 100 food and drug bills are introduced in Congress.

1883–Dr. Harvey W. Wiley becomes chief chemist, expanding the Bureau of Chemistry's food adulteration studies. Campaigning for a federal law, Wiley is called the Crusading Chemist and Father of the Pure Food and Drugs Act. He retired from government service in 1912 and died in 1930.

1902–The Biologics Control Act is passed to ensure purity and safety of serums, vaccines, and similar products used to prevent or treat diseases in humans.

Congress appropriates $5000 to the Bureau of Chemistry to study chemical preservatives and colors and their effects on digestion and health. Wiley's studies draw widespread attention to the problem of food adulteration. Public support for passage of a federal food and drug law grows.

1906–The original Food and Drugs Act is passed by Congress on June 30 and signed by President Theodore Roosevelt. It prohibits interstate commerce in misbranded and adulterated foods, drinks, and drugs.

The Meat Inspection Act is passed the same day. Shocking disclosures of unsanitary conditions in meat-packing plants, the use of poisonous preservatives and dyes in foods, and cure-all claims for worthless and dangerous patent medicines are the major problems leading to the enactment of these laws.

1927–The Bureau of Chemistry is reorganized into two separate entities. Regulatory functions are allotted to the Food, Drug, and Insecticide Administration, and nonregulatory research is overseen by the Bureau of Chemistry and Soils.

1930–The name of the Food, Drug, and Insecticide Administration is shortened to Food and Drug Administration (FDA) under an agricultural appropriations act.

1933–The FDA recommends a complete revision of the obsolete 1906 Food and Drugs Act. The first bill is introduced into the Senate, launching a five-year legislative battle.

1937–Elixir of Sulfanilamide, containing the poisonous solvent diethylene glycol, kills 107 persons, many of whom are children, dramatizing the need to establish drug safety before marketing and to enact the pending food and drug law.

1938–The Federal Food, Drug, and Cosmetic (FDC) Act of 1938 is passed by Congress, containing new provisions:

- Extending control to cosmetics and therapeutic devices
- Requiring new drugs to be shown safe before marketing—starting a new system of drug regulation
- Eliminating the Sherley Amendment requirement to prove intent to defraud in drug misbranding cases
- Providing that safe tolerances be set for unavoidable poisonous substances
- Authorizing standards of identity, quality, and fill-of-container for foods
- Authorizing factory inspection
- Adding the remedy of court injunctions to the previous penalties of seizures and prosecutions

Under the Wheeler-Lea Act, the Federal Trade Commission is charged with overseeing advertising associated with products otherwise regulated by The FDA, except for prescription drugs.

1943–In *United States v. Dotterweich*, the Supreme Court rules that the responsible officials of a corporation, as well as the corporation itself, may be prosecuted for violations. It need not be proven that the officials intended, or even knew of, the violations.

1949–The FDA publishes guidance to industry for the first time. This guidance, "Procedures for the Appraisal of the Toxicity of Chemicals in Food," comes to be known as the *black book*.

1951–The Durham-Humphrey Amendment defines the kinds of drugs that cannot be safely used without medical supervision and restricts their sale to prescription by a licensed practitioner.

1958–Food Additives Amendment is enacted, requiring manufacturers of new food additives to establish safety. The Delaney proviso prohibits the approval of any food additive shown to induce cancer in humans or animals.

The FDA publishes in the *Federal Register* the first list of *Substances Generally Recognized as Safe* (GRAS). The list contains nearly 200 substances.

1959–The U.S. cranberry crop is recalled three weeks before Thanksgiving for FDA tests to check for aminotriazole, a weed killer found to cause cancer in laboratory animals. Cleared berries are allowed a label stating that they have been tested and have passed FDA inspection, the only such endorsement ever allowed by the FDA on a food product.

1962–Thalidomide, a new sleeping pill, is found to have caused birth defects in thousands of babies born in western Europe. News reports on the role of Dr. Frances Kelsey, FDA medical officer, in keeping the drug off the U.S. market arouse public support for stronger drug regulation.

Kefauver-Harris Drug Amendments are passed to ensure drug efficacy and greater drug safety. For the first time, drug manufacturers are required to prove to the FDA the effectiveness of their products before marketing them. The new law also exempts from the Delaney proviso animal drugs and animal feed additives shown to induce cancer, but which leave no detectable levels of residue in the human food supply.

President John F. Kennedy proclaims the Consumer Bill of Rights in a message to Congress. Included are the right to safety, the right to be informed, the right to choose, and the right to be heard.

1972–Over-the-counter drug review begins to enhance the safety, effectiveness, and appropriate labeling of drugs sold without prescription.

Regulation of biologics—including serums, vaccines, and blood products—is transferred from National Institutes of Health (NIH) to the FDA.

1976–Medical Device Amendments are passed to ensure safety and effectiveness of medical devices, including diagnostic products. The amendments require manufacturers to register with the FDA and follow quality control procedures. Some products must have premarket approval by the FDA; others must meet performance standards before marketing.

Vitamins and Minerals Amendments (Proxmire Amendments) stop the FDA from establishing standards limiting potency of vitamins and minerals in food supplements or regulating them as drugs based solely on potency.

1978–Good manufacturing practices become effective.

1979–Good laboratory practices become effective.

1983–The Orphan Drug Act is passed, enabling the FDA to promote research and marketing of drugs needed for treating rare diseases.

1984–Fines Enhancement Laws of 1984 and 1987 amend the U.S. Code to greatly increase penalties for all federal offenses. The maximum fine for individuals is now $100,000 for each offense and $250,000 if the violation is a felony or causes death. For corporations, the amounts are doubled.

1988–The Food and Drug Administration Act of 1988 officially establishes the FDA as an agency of the Department of Health and Human Services, with a Commissioner of Food and Drugs appointed by the president with the advice and consent of the Senate, and broadly spells out the responsibilities of the secretary and the commissioner for research, enforcement, education, and information.

The Prescription Drug Marketing Act bans the diversion of prescription drugs from legitimate commercial channels. Congress finds that the resale of such drugs leads to the distribution of mislabeled, adulterated, subpotent, and counterfeit drugs to the public. The new law requires drug wholesalers to be licensed by the states; restricts reimportation from other countries; and bans the sale, trade, or purchase of drug samples, and traffic or counterfeiting of redeemable drug coupons.

1990–The Safe Medical Devices Act is passed, requiring nursing homes, hospitals, and other facilities that use medical devices to report to the FDA incidents suggesting that a medical device probably caused or contributed to the death, serious illness, or serious injury of a patient. Manufacturers are required to conduct postmarket surveillance on permanently implanted devices whose failure might cause serious harm or death, and to establish methods for tracing and locating patients depending on such devices. The act authorizes the FDA to order device product recalls and other actions.

1995–The FDA declares cigarettes to be drug delivery devices. Restrictions are proposed on marketing and sales to reduce smoking by young people.

1996–Federal Tea Tasters Repeal Act repeals the Tea Importation Act of 1897 to eliminate the Board of Tea Experts and user fees for the FDA's testing of all imported tea. Tea itself is still regulated by the FDA.

1997–The Food and Drug Administration Modernization Act reauthorizes the Prescription Drug User Fee Act of 1992 and mandates the most wide-ranging reforms in agency practices since 1938. Provisions include measures to accelerate review of devices, regulate advertising of unapproved uses of approved drugs and devices, and regulate health claims for foods.

1998–The First phase to consolidate FDA laboratories nationwide from nineteen facilities to nine by 2014 includes dedication of the first of five new regional laboratories.

1999–The website www.ClinicalTrials.gov is founded to provide the public with updated information on enrollment in federally and privately supported clinical research, thereby expanding patient access to studies of promising therapies. A final rule mandates that all over-the-counter drug labels must contain data in a standardized format. These drug facts are designed to provide the patient with easy-to-find information, analogous to the nutrition facts label for foods.

2000–Under the Data Quality Act, federal agencies are required to issue guidelines to maximize the quality, objectivity, utility, and integrity of the information they generate, and to provide a mechanism whereby those affected can secure correction of information that does not meet these guidelines.

2005–Formation of the Drug Safety Board is announced, consisting of FDA staff and representatives from the National Institutes of Health and the Veterans Administration. The Drug Safety Board will advise the director, Center for Drug Evaluation and Research, FDA, on drug safety issues and work with the agency in communicating safety information to health professionals and patients.

2011–The FDA Food Safety and Modernization Act (FSMA) provides FDA with new enforcement authorities related to food safety standards, gives FDA tools to hold imported foods to the same standards as domestic foods, and directs FDA to build an integrated national food safety system in partnership with state and local authorities.

2013–The Pandemic and All-Hazards Preparedness Reauthorization Act (PAHPRA) establishes and reauthorizes certain programs under the Public Health Service Act and the Food, Drug, and Cosmetic Act with respect to public health security and all-hazards preparedness and response.

The Drug Quality Safety and Security Act (DQSA) is enacted by Congress following an outbreak in 2012 of an epidemic of fungal meningitis linked to a compounded steroid. Among other provisions, DQSA outlines steps for an electronic and interoperable system to identify and trace certain prescription drugs throughout the United States.

Part II
Quality Systems

Chapter 2	The Basics of a Quality System
Chapter 3	Quality Standards and Their Evolution
Chapter 4	Quality Documentation
Chapter 5	Calibration Procedures
Chapter 6	Calibration Records
Chapter 7	Calibration Certificates
Chapter 8	Management Systems and Quality Manuals
Chapter 9	Metrological Traceability
Chapter 10	Calibration Programs
Chapter 11	The International System of Units (SI) and Measurement Standards
Chapter 12	Audit Requirements
Chapter 13	Scheduling and Recall Systems
Chapter 14	Labels and Equipment Status
Chapter 15	Training and Competency
Chapter 16	Environmental Controls
Chapter 17	Industry-Specific Requirements
Chapter 18	Computers, Software, and Software Validation

Chapter 2
The Basics of a Quality System

What is a quality system? What decides that a laboratory or department must have one and why? Let us begin by answering the second question. According to ISO/IEC 17025:2017, Clause 8.2.1, "Laboratory management shall establish, document, and maintain policies and objectives for the fulfilment of the purposes of this document (ISO/IEC 17025:2017) and shall ensure that the policies and objectives are acknowledged and implemented at all levels of the laboratory organization."[1] ISO 9001-2015, Clause 4.4.1 says, "The organization shall establish, implement, maintain and continually improve a quality management system, including processes needed and their interactions, in accordance with the requirements of this Internal Standard."[2] ANSI/NCSL Z540.1-1994, Part 1, Clause 5.2 states: "The quality manual and related documentation shall state the laboratory's policies and operational procedures established in order to meet the requirements of this Standard."[3] ANSI/NCSL Z540.3-2006 says, in Clause 5.1, "The organization shall include all measuring and test equipment in the calibration system that have an influence on the quality of the organization's product."[4] ISO 10012:2003, Clause 4 states that "[t]he measurement management system shall ensure that specified metrological requirements are established."[5] NCSL International Recommended Practice–6: *Recommended Practice for Calibration Quality Systems for the Healthcare Industries*, RP-6, 2015 states, in Clause 5.1: "The calibration quality system should include the necessary elements for the control of M&TE."[6] All of these quality organizations, which have major roles in metrology systems, require that a quality system *shall*, not *should*, be maintained.

Now, to answer the first question: The basic premise and foundation of a good quality system is to *say what you do, do what you say, record what you did, check the results, and act on the difference*. In simple terms, *say what you do* means write, in detail, how to do your job. This includes calibration procedures, standard operating procedures (SOPs), protocols, work instructions, work forms, and so on. *Do what you say* means follow the documented procedures or instructions every time you calibrate, validate, or perform a function that follows specific written instructions. *Record what you did* means accurately and completely record the results of your measurements and adjustments, including what your standard(s) indicated both before and after adjustment. *Check the results* means make certain the inspection, measurement, and test equipment (IM&TE) meets the tolerances, accuracies, or upper/lower limits specified in your procedures or instructions. *Act on the difference* means if the IM&TE is out of tolerance, does not meet the specified accuracies, or exceeds the upper/lower test limits written in your procedures, then you are required to inform the IM&TE user because they may have to reevaluate manufactured goods, change a process, or recall a product and/or previously calibrated equipment that used the particular standard.

To help ensure that all operations throughout a metrology department, calibration laboratory, or work area where calibrations are accomplished occur in a stable manner, one needs to establish a quality management system. The effective operation of such a system should result in stable processes and, therefore, in a consistent output from those processes. Once stability and consistency are achieved, then it is possible to initiate improvements. Each calibration technician must follow the calibration procedure as it is written, collect the data as they are found, and document the data each time. Once data are collected, trends can then be evaluated, intervals changed, and improvements to the processes and procedures implemented. This is the basic foundation of a quality system. Although these steps check your process for correctness, they never validate your process (which you define) for correctness relative to accepted practices in that specific field. For example, suppose a quality system states: "Reference standards calibration will be performed every five years." If these reference standards are calibrated every five years, then doing so will meet the requirement of the quality system. However, it will not meet accepted metrology practices of defining calibration intervals based on performance of the standard in question unless you use the historical data, check the results, and act on the difference when compiling your information. Trends will either support the current calibration interval or give sufficient data to increase or decrease those intervals.[7,8]

Some of the quality tools that help determine the status of information, data, intervals, and so on are check sheets, Pareto charts, flowcharts, cause-and-effect diagrams (also called fishbone diagrams), histograms, scatter diagrams, and control charts.

QUALITY TOOLS

Various quality tools[9] are available to assist a company, department, or laboratory to continually improve, update, and adapt its programs, policies, and procedures to maintain a business advantage, identify problems before they affect the bottom line, and help keep the quality system alive and healthy.

Check Sheet

The function of a check sheet is to present information in an efficient, graphical format. This may be accomplished with a simple listing of items; however, the utility of the check sheet may be significantly enhanced, in some instances, by incorporating a depiction of the system under analysis into the form.

Pareto Chart

Pareto charts are extremely useful because they can be used to identify those factors that have the greatest cumulative effect on the system and thus screen out the less significant factors in an analysis. Ideally, this allows the user to focus attention on a few crucial factors in a process. Pareto charts are created by plotting the cumulative frequencies of the relative frequency data (event count data) in descending order. When this is done, the most essential factors for the analysis are graphically apparent in an orderly format.

Flowchart

Flowcharts are pictorial representations of a process. By breaking the process down into its constituent steps, flowcharts can be useful in identifying where errors are likely to be found in the system.

Cause-and-Effect Diagram

This diagram, also called an *Ishikawa diagram* (or *fishbone diagram*), is used to associate multiple potential causes with a single effect. Thus, given a particular effect, the diagram is constructed to identify and organize potential causes for it.

The primary branch represents the effect (the quality characteristic that is intended to be improved and controlled) and is typically labeled on the right side of the diagram. Each major branch of the diagram corresponds to a major cause (or class of causes) that relates to the effect. Minor branches correspond to more detailed causal factors. This type of diagram is useful in any analysis as it illustrates the relationship between cause and effect in a rational manner.

Histogram

Histograms provide a simple, graphical view of accumulated data, including their dispersion and central tendency. Histograms are simple to construct and provide the easiest way to evaluate the distribution of data.

Scatter Diagram

Scatter diagrams are graphical tools that attempt to depict the influence that one variable has on another. A common scatter diagram usually displays points representing the observed value of one variable corresponding to the value of another variable.

Control Chart

The control chart is the fundamental tool of statistical process control, as it indicates the range of variability that is built into a system (known as *common cause variation* or *random variation*). Thus, it helps determine whether a process is operating consistently or if a special cause or nonrandom event has occurred to change the process mean or variance. The process control chart may also be called a *process description chart*.

The bounds of the control chart are marked by upper and lower control limits that are calculated by applying statistical formulas to data from the process. Data points that fall outside these bounds represent variations due to special causes, which can typically be found and eliminated. In contrast, improvements in common cause variation require fundamental changes in the process, which leads to process improvement techniques.

Various tools are available for accomplishing process improvements. Any quality system should be a living entity: constantly changing, improving, and adapting to the business environment where it is used. Without process improvement, the system becomes stagnant and will fall behind the times.

PROCESS IMPROVEMENT TECHNIQUES[10]

PDCA

The plan–do–check–act (PDCA) or plan–do–study–act (PDSA) cycle was originally conceived by Walter Shewhart in the 1930s and later adopted by W. Edwards Deming. The model provides a framework for the improvement of a process or system. It can be used to guide the entire improvement project or to develop specific projects once target improvement areas have been identified.

The PDCA cycle is designed to be used as a dynamic model. The completion of one turn of the cycle flows into the beginning of the next. Following in the spirit of *continuous* quality improvement, the process can always be reanalyzed, and a new cycle of change can begin. This continual cycle of change is represented in the ramp of improvement. Using what we learn in one PDCA trial, we can begin another, more complex trial.

The first step is to *plan*. In this phase, analyze what you intend to improve, looking for areas that hold opportunities for change. Choose areas that offer the most return for the effort—the biggest bang for your buck. To identify these areas for change, consider using a flowchart or Pareto chart.

Next, *do* what is planned. Carry out the change or test, preferably on a small scale. The *check* or *study* phase is a crucial step in the PDCA cycle. After you have implemented the change for a time, it is critical to determine how well it is working. Is it really leading to improvement in the way you had hoped? Decide on several measures with which you can monitor the level of improvement. Run charts can be helpful with this measurement.

After planning a change, implementing it, and then monitoring it, you must decide whether it is worth continuing that change. If it consumed too much time, was difficult to adhere to, or even led to no improvement, consider aborting the change and planning a new one. If the change led to a desirable improvement or outcome, however, consider expanding the trial to a different area or slightly increasing the complexity. *Act* on your discovery. This sends you back into the *plan* phase of the cycle.

Brainstorming

Most problems are not solved by the first idea that comes to mind. To get to the best solution, it is important to consider many possible solutions. One of the best ways to do this is called *brainstorming*. Brainstorming is the act of defining a problem or idea and suggesting anything related to the topic, no matter how odd or remotely connected a suggestion may sound. All these ideas are recorded, but they are evaluated only after the brainstorming is completed.

To begin brainstorming, gather a group. Select a leader and a recorder (these may be the same person). Define the problem or idea to be brainstormed. Make sure everyone is clear on the topic being explored. Set up the rules for the session. They should include: the leader is in control; everyone can contribute; no one will insult, demean, or evaluate another participant or a response; no answer is wrong; each answer will be recorded unless it is a repeat; and a time limit will be set and adhered to.

Start the brainstorming! Have the leader select members of the group to share their answers. The recorder should write down all responses, if possible, so everyone can see them. Make sure not to evaluate or criticize any answers until the group is done with its brainstorming.

Once the brainstorming is finished, go through the results and begin evaluating the responses. Some initial qualities to look for when examining the responses include looking for any answers that are repeated or similar, grouping like concepts together, and eliminating responses that do not fit. Now that the list has been pruned, discuss the remaining responses as a group.

The mere use of quality control tools does not necessarily constitute a quality program. Thus, to achieve lasting improvements in quality, it is essential to establish a system that will continuously promote quality in all aspects of its operation.

The following chapters explain in detail what is required in calibration procedures, records, certificates, and a quality manual. It is one thing to say that a quality system is required; it is another to explain what it is, what needs to be in it, what the requirements are in different standards and regulations, and how to apply those requirements in a systematic approach that can be tailored for your individual situation. No two calibration laboratories or metrology departments/groups are the same; however, they may have similar guidelines for metrology. Each has specific requirements that must be met, while providing the same basic function for the company: traceable calibration measurements.

Chapter 3
Quality Standards and Their Evolution

WHAT ARE QUALITY STANDARDS AND WHY ARE THEY IMPORTANT?

Businesses exist for their outputs, whether those are products or services.[1] Products or services will be purchased if they meet the needs and requirements of the customer. Over time, certain business management practices have been observed that enable a business to consistently provide products and services at required quality levels. *Quality management system standards* provide standard vocabulary, guidelines, and best practices for fundamental quality assurance and management practices. In theory, a business that meets the requirements of a quality management standard should be capable of producing its products or providing its service at a consistent level of quality.

Be aware that even though the phrases *quality standard* or *quality system* are often used, they are shorthand forms of referring to the *quality management system standards* or the *quality management system* of an organization. These terms should *not* be taken to refer to the technical requirements of a product that constitute its product quality attributes.

There are two different classes of quality standards: those that are required by law or regulation and those that are voluntary. A government law, or a regulation of a government agency, may include or specify quality standard requirements. In these cases, a business *must* comply if it is either in a regulated industry or wishes to sell products or services to the government. In all other cases, a quality standard is technically voluntary. The Quality Management System (QMS) standards prescribe accepted good practices for an organization to follow, but compliance is voluntary. This means that a business *may* choose to follow or ignore it; however, the voice of the customer and the forces of competition have an effect otherwise. In the modern business environment, agreements with customers are typically governed by the terms of purchase orders or other contracts. Competition between companies is sensitive to conditions in the overall market. As a practical matter, some quality standards are voluntary only to the extent that the organization can afford to lose business by not following them. All of this results in several forces that drive the importance of voluntary quality standards.

- **Customer requirements.** On a retail level, customers will not buy if the product or service does not meet their requirements. Large customers, such as other businesses, can state specific quality requirements in a request for quote and then in a purchase order. In many cases these quality requirements include language stating that the supplier must provide proof that it meets the requirements of a specific quality standard. For example, the largest automobile manufacturers in North America and Europe currently require that their primary suppliers be registered to ISO/TS 16949.

- **Competitive advantage.** A second force is a desire to stand out in a competitive environment. In many industries, there is little difference to distinguish one supplier from another. A company may decide to adopt a quality standard to gain a marketing advantage.

- **Response to competition.** Another force is response to the pressures of competition. When a critical mass of businesses in an industry sector formally adopts a quality standard, this creates pressure on the others to do likewise. The alternative is often loss of business and the decline of the companies that do not respond. It is now a generally accepted axiom that if a business expects to be competitive on a national or international level, then registration to an appropriate quality standard is a minimum business necessity.

- **Government requirements.** Around the world, governments are both customers and regulators in certain industries. As customers, governmental agencies often require suppliers to adhere to quality standards. Other government agencies have their own quality requirements that must be followed by the industries they regulate. In the United States, the FDA regulates food and medical products, both domestic and those imported from more than 150 countries into the United States. In Canada, Health Canada regulates product safety, drug and health products, environmental and workplace health, and food and nutrition, among other health-related programs.[2] In Europe, most of this regulation occurs at the national level. The European Medicines Agency (EMA) monitors and supervises the safety of medicines that have been authorized for use by the European Union (EU).[3] These are a just a few of the national and international governmental organizations engaged in setting industry regulations and requirements.

THE ROLE OF ISO

Many standards documents are referred to as ISO standards. This indicates that the standard was developed by or in cooperation with the International Organization for Standardization (ISO), headquartered in Geneva, Switzerland.[4] ISO is a nongovernmental organization. At the time of the updating of this handbook to its third edition, ISO has a membership of 166 national standards bodies—and there is only one member per country.[5] ISO was created in 1946 as a replacement for an earlier organization (the International Federation of Standardizing Associations [ISA], founded in 1926), which ceased operations in 1942.[6]

The purpose of ISO is to remove technical barriers to trade and to otherwise aid and improve international commerce by harmonizing existing standards and developing new voluntary standards as needed. Many of the more than 24,000 ISO standards affect people every day because they cover areas as diverse as the size of credit cards, the properties of camera film, the threads on bolts in the engine of a car, the shipping containers used for international freight, and business management systems for quality, environmental safety, and social responsibility. One of the most important international standards for metrology is ISO/IEC 17025:2017, *General requirements for the competence of testing and calibration laboratories*.

Conformance to standards may be verified by conducting a compliance audit. There are three types of audits that may be conducted:

- First-party audit: An internal audit by the organization to ensure that it meets the compliance requirements of the standard that it is registered/accredited to meet.

- Second-party audit: An audit conducted by the customer of the organization to ensure that the organization meets the customer's requirements.

- Third-party audit: An audit conducted by an independent registration/Accreditation Body to ensure that the organization meets the requirements of the standard(s) with which it is allegedly complying.

It is important to note some differences in terminology with respect to third-party audits of quality systems. ISO defines "certification" and "accreditation" differently because they are different in how they are applied and what they mean. It is important to use the correct terms.

"Certification—the provision by an independent body of written assurance (a certificate) that the product, service or system in question meets specific requirements.

Accreditation—the formal recognition by an independent body, generally known as an Accreditation Body, that a certification body operates according to international standards."[7]

Regarding ISO quality standards, such as ISO 9001, an organization may say that it has "ISO 9001:2015 certification" or that it is "ISO 9001:2015 certified." Simply stating "ISO certified" or "ISO certification" is misleading because this wording does not clarify to which ISO standard one is "certified."[8] In some parts of the world, the terms *certified* and *registered* are used interchangeably. This is because the "Certification Body" (the one that issues the certification, sometimes called a *registrar*) *registers* the *certification*. These terms, *certified* and *registered*, both mean the same thing: a qualified third-party auditor has formally issued a certificate which states that the organization meets the requirements of the specified standard.

International organizations cooperate with ISO for accreditation purposes. "The International Accreditation Forum (IAF) is the world association of accreditation bodies and other bodies associated in conformity assessment in the fields of management systems, products, processes, services, personnel, validation and verification and other similar programmes of conformity assessment."[8]

"ILAC [the International Laboratory Accreditation Cooperation] is the international organisation for accreditation bodies operating in accordance with ISO/IEC 17011 and involved in the accreditation of conformity assessment bodies including calibration laboratories (using ISO/IEC 17025), testing laboratories (using ISO/IEC 17025), medical testing laboratories (using ISO 15189), inspection bodies (using ISO/IEC 17020) and proficiency testing providers (using ISO/IEC 17043)."[9]

When the conformance standard is in the ISO/IEC 17000-series (for conformity assessment), the organizations would be considered *accredited*. These conformity standards are developed by ISO's committee on conformity assessment (CASCO) in collaboration with the International Electrotechnical Commission (IEC).[10] This means that for any ISO/IEC 17000-series (conformity) standard, including ISO/IEC 17025, the organization is also always referred to as *accredited*. Accreditation of a testing or calibration organization always includes evaluation of technical competence to perform the work listed in the scope.

Voluntary product and service technical standards developed by ISO and other organizations—notably the International Electrotechnical Commission (IEC) and the International Telecommunications Union (ITU)—aid international commerce by specifying technical requirements.[11] These requirements are often expressed in terms of a physical quantity such as dimension, mass, voltage, temperature, or chemical composition. The role of metrology is to ensure that measurements of the physical quantities are the same no matter where the measurements are made or where the resulting products are produced or sold. The most important aid for this is that all measurements are now made with reference to a single set of measurement standards, the International System of Units (SI), that is defined and managed by the BIPM.[12] Because the measurements mean the same thing in different countries, international commerce is much easier, as these and other technical barriers to trade are removed. Under the ILAC Mutual Recognition Arrangement (ILAC-MRA), ILAC-MRA members agree to accept accredited technical test and calibration data for exported goods. The result is that a product manufactured in one country can be sold and used in any other country if it is otherwise suitable for the intended use.[13]

EVOLUTION OF QUALITY STANDARDS

Concepts of quality, and the business practices needed to achieve consistent quality, have evolved over time. The most dramatic changes have occurred since the start of the twentieth century. Then, companies set their own practices and standards, and production was all-important. Now, business has moved to the present state where quality is equally important, and international standards exist to define the minimum acceptable practices.

HISTORY OF QUALITY STANDARDS: 1900–1940

The quality control, quality assurance, and quality standards as we now know them have their immediate roots in the early twentieth century. This coincided with very rapid growth of mass production, early use of automation, and scientific research and development. Inspection as a means of quality control was already in use, but was inefficient and was often under control of the production departments. During the 1920s several people started systematic studies, applying statistical analysis for the first time to improve production quality. Most of this work was pioneered at Western Electric's Hawthorne Works in Chicago, with important work done by people including Walter Shewhart, Harold Dodge, and Joseph Juran. Statistical studies were used by Shewhart to provide a means of saving money by controlling a process, by Dodge for improving the inspection sampling process, and by Juran for evaluating quality improvements and educating top management. Inspection sampling systems, control charts, and other quality tools invented at the Hawthorne Works form the foundations of the modern statistical quality assurance system.[14]

HISTORY OF QUALITY STANDARDS—1940 TO THE PRESENT

When the United States entered World War II at the end of 1941, the production needs of the War Department marked the imposition of these new statistical quality control ideas on industry. This also brought the terms *military specification* and *military standard* into common use. The specifications defined what was needed, and many of the standards defined how to ensure that the delivered product was acceptable. They were product- and process-based and depended on inspection to ensure the quality of the final output. A significant number of the people recruited by the War Department came from the Western Electric system—notably Juran, Dodge, and Shewhart, as well as a colleague of Shewhart's from the Department of Agriculture, W. Edwards Deming. They brought with them inspection sampling methods, control charts, and other statistical tools. All of these were developed further during the war, and some became military standards, especially the inspection sampling tables MIL-STD-105 and MIL-STD-414. (These are now known as ANSI/ASQ Z1.4 and Z1.9, respectively.)

Effective use of this new statistical quality control (SQC) required that large numbers of engineers and other practitioners be trained in the new subject. Some of these people formed local groups to study and share experiences outside the work environment. In 1946 most of these local groups merged to create the American Society for Quality Control (ASQC) as a formal professional society.[15]

After World War II the United States started aiding Germany and Japan in rebuilding their devastated economies. As part of the rebuilding effort in Japan, a few people were invited to teach the new SQC and management methods to Japanese engineers and managers. The best known of these teachers were Deming and Juran. The Japanese took the teaching to heart and incorporated the new lessons into their industrial culture from the top down. Deming's teaching was so highly regarded that the Union of Japanese Scientists and Engineers (JUSE) established an annual quality award named in honor of him: the Deming Prize.[16]

The next significant event was the appearance of another U.S. military standard, MIL-Q-9858, *Quality Program Requirements*, in 1959. This standard was the first to include most of the elements of a modern quality management system. The current ISO quality management system standards can be traced to their roots in quality practices developed by the U.S. War Department during World War II. At some point those practices were incorporated into a military specification, MIL-Q-5923, *General Quality Control Requirements*. What appears to be the major step, though, occurred when MIL-Q-9858 was released in April 1959 as a replacement for the earlier specification. Other important documents followed over the next few years:

April 1959 MIL-Q-9858, *Quality Program Requirements*. This was replaced in December 1963 by a revised version (MIL-Q-9858A) that remained in force until being canceled in 1996.

October 1960 MIL-H-110 (Interim), *Quality Control and Reliability Handbook*. This was replaced by MIL-HDBK-50, *Evaluation of a Contractor's Quality Program*, in April 1965; this is still in force.

February 1962 MIL-C-45662, *Calibration System Requirements*. This was replaced by MIL-STD-45662 in 1980. The final revision (MIL-STD-45662A) was canceled in 1995 and replaced by ANSI/NCSL Z540-1-1994 and ISO 10012-Part 1-1992 and Part 2-1997.

Milestones in the evolution of metrology/calibration-related standards include:

- ANSI/ASQC M1-1987 by ASQC Metrology Technical Committee (predecessor to ASQ Measurement Quality Division)—reaffirmed in 1996

- ISO GUIDE 25:1990—replaced by ISO/IEC 17025:1999

- ISO/IEC 17025:1999

- ISO 10012:2003—combined ISO10012-1:1992 and ISO10012-2:1997 into one standard

- ISO/IEC 17025:2005

- ANSI/NCSL Z540.3-2006—replaced ANSI/NCSL Z540-1:1994 and withdrawn in 2020

- ISO/IEC 17025:2017

Even though the ANSI/NCSL Z540-1 and ANSI/NCSL Z540.3 standards were withdrawn, many long-term U.S. Department of Defense contracts are written to include compliance to the calibration requirements of those standards. Given that we live in a global economy and subcontractors to these contracts exist globally, it would be a good practice to be familiar with these standards. Key items to interpret from these standards from a metrology perspective are the use of the terms "test accuracy ratio (TAR)" and "test uncertainty ratio (TUR)." It should be noted that the concept of TUR existed in the 1962 version of MIL-STD-45662. Although there are better methods to evaluate conformance to these requirements, neither the TAR nor the TUR terms should be ignored.

Further consideration must be given to quantitative considerations required by calibration-related standards. Readers should familiarize themselves with topics such as measurement uncertainty, risk management, and statements of conformity covered in other chapters of this handbook and other guidance documents.

June 1963 MIL-Q-21549B, *Product Quality Program Requirements for Fleet Ballistic Missile Weapon System Contractors*. The original version was probably introduced in 1961.[17] This standard has since been canceled.

These standards constitute the roots of what are now considered to be the fundamentals of an effective quality management system. During the late 1960s the North Atlantic Treaty Organization (NATO) incorporated these standards into various Allied Quality Assurance Procedures (AQAP) documents. In the early 1970s there was a movement in British industrial circles to create a set of generic quality management system standards equivalent to the AQAPs but for commercial use. This led to the submission of a first draft standard by the Society of Motor Manufacturers and Traders to the British Standards Institution. This was circulated for public comment in 1973 and led to the publication of the three-part BS 5179 series of guidance standards in 1974:

BS 5179: *Guide to the operation and evaluation of quality assurance systems.*

> Part 1: *Final inspection system*
>
> Part 2: *Comprehensive inspection system*
>
> Part 3: *Comprehensive quality control system*[18]

BS 5179 was renumbered to BS 5750 in 1979. When ISO/TC176 started work on what was to become the first edition of ISO 9000, they drew on that and the related standards for background.[19]

Other elements were included in additional military standards that appeared over the next few years. As explained by Stanley Marash:

> In addition to documentation and auditing, other integral features of military and aerospace standards included management responsibility . . . corrective action processes, control of purchasing, flow-down of requirements to suppliers and subcontractors, control of measuring and test equipment, identification and segregation of nonconforming product, application of statistical methods, and other requirements we now take for granted.[20]

Use of these and other government quality management standards, however, was mostly confined to companies fulfilling defense and other government contracts, or companies in regulated industries where the standards had force.

Over this period many U.S. industries had forgotten much of the quality regime that had become mandatory during the war. They deemed production numbers and schedules more important than quality. After all, they simply had to respond to the public's increasing demand for more and better automobiles, appliances, stereos, and so on. This led to large amounts of rework and waste (again) and quality by inspection (again). It also led directly to such tragedies as the Apollo 1 fire that killed three astronauts in January 1967, which was traced directly to insufficient quality in the electrical wiring.[21]

Awareness of the need to improve quality management performance was growing across all industries. In 1968 ASQC introduced the Certified Quality Engineer program. The 1970s and early 1980s made American industry very aware of the competitive need for improved product quality. Since 1945, the quality lessons from wartime production had been largely forgotten or ignored, and suddenly products made in Japan were a major competitive threat. Compared to similar domestic products, the imports usually had better product quality and lower cost. In the late 1970s one of the authors of this chapter was working for a large consumer electronics retailer. At one point, another company was acquired and its products were integrated into the existing system, complete with going through receiving inspection at the warehouses. The other company had carried a much higher proportion of U.S.-made products. During the product-line integration period, it was not uncommon for the warehouses to reject 10 % or more of product from U.S. manufacturers, whereas the reject rate from Japanese plants was less than 1 % and steadily improving.

The economic pressure was not driven solely by competition from Japan and Taiwan. Inflation and resource scarcity (such as periodic shortages of petroleum) were also significant at the time. But there was evidence that things could be better and actually were in some places. In June 1980, U.S. industry got a major wake-up call when NBC News aired a 90-minute white paper program called *If Japan Can . . . Why Can't We?* (now hosted on YouTube by The Deming Institute).[22] Many authorities credit this television show with starting the resurgence in quality management in American business. About one-third of the program explored the management theory of Deming and the influence of his early 1950s teaching of SQC and his management theory (continual improvement of the whole process, which is a system) on Japanese industry. In addition to rediscovering Deming, American industry started exploring the work of Juran (another rediscovery), and new authorities such as Armand Feigenbaum, Philip Crosby, Kaoru Ishikawa, and others.

Another aid in quality management improvement in the United States arose in 1987 with the establishment of the Malcolm Baldrige National Quality Award (MBNQA) program in the U.S. Department of Commerce. Although not a conformance standard, this annual award for U.S. organizations is important because it has become a widely accepted benchmark for performance excellence of quality management systems. Many organizations use the MBNQA criteria as a guide for self-assessment, and they are the basis of many state and local quality award programs. The awards themselves, and related publications by the winners, may be viewed as a store of best practices used by a wide range of businesses and healthcare and educational institutions. The MBNQA[23] focuses on performance in five key areas:

1. Product and process outcomes
2. Customer outcomes
3. Workforce outcomes
4. Leadership and governance outcomes
5. Financial and market outcomes

Organizations do not receive the award for specific products or services. To receive the award, an organization must have a system that (1) ensures continuous improvement in overall performance in delivering products and/or services, and (2) provides an approach for satisfying and responding to customers and stakeholders.

Quality standards will continue to evolve. Business operations and strategies are changing at rapid paces. Nevertheless, even as these practices and standards evolve, quality fundamentals and standards will always be critical.

Chapter 4

Quality Documentation

We have all heard the phrase: "The job isn't done until the paperwork is complete." Within an effective quality system, more is required than simply "completing the paperwork." Another primary consideration is what happens when changes are made to the policies, procedures, certificates, records, database, software, or other quality management documents. How are these changes managed and communicated? Through document control.

Both ISO/IEC 17025:2017 and ISO 9001:2015 specify requirements that must be met to ensure document control compliance within the quality system. Effective document control ensures that the most current, authorized versions of calibration procedures, certificates, labels, forms, external documents, and quality management documents are available for use by those authorized to use them and accessible where needed. This reinforces the organization's ability to deliver to its customers the quality they deserve and reduce the risk of nonconformities. Furthermore, as documentation control systems evolve, documentation and records often become electronic-based instead of or in addition to paper-based.

Quality documentation encompasses at least an organization's policies, procedures, records, certificates, and quality manual. The following chapters go into specific details on each of these topics; however, each is related to the others in that one cannot occur or function without the others. You must have calibration procedures to follow to generate the data to record on the calibration certificates, calibration records, and labels. The management system (such as that set out in a quality manual) lays the groundwork for everything in the system and is a living document that continually changes as customer and company needs dictate.

According to ISO/IEC 17025:2017, Clause 8.3, Control of Management System Documents, "The laboratory shall control the documents (internal and external) that relate to the fulfilment of this document." The laboratory *shall* ensure that documents are approved, uniquely identified, periodically reviewed, have changes and current revision status identified, and are available at points of use where distribution is controlled.[1]

For laboratories using ISO 9001:2015 for their quality management system, ISO 9001:2015 Clause 7.5 states: "The organization's quality management system shall include": a) "documented information required by [ISO 9001:2015]" and b) "documented information determined by the organization as . . . necessary for the [QMS] effectiveness."[2] ISO 9001:2015 continues to stress the importance of the control of documented information in 7.5.3.1, stating that documents shall be controlled to ensure that they are available and suitable for use and that they are adequately protected.[3]

Neither ISO/IEC 17025:2017 nor ISO 9001:2015 requires a defined document control procedure anymore. However, the following outcomes for documents are specified across these two international standards and must be met:

- Periodic review and updating as necessary
- Approval for (appropriate) suitability and adequacy by authorized personnel
- Identification of changes and revision status
- Appropriate (ISO 9001:2015) or unique (ISO/IEC 17025:2017) identification of documents
- Relevant versions are available where and when they are needed, and are adequately protected, and
- The unintended use of obsolete documents is prevented.

Establishing an effective document control system is time-consuming, but once established and practiced it helps to ensure that the most up-to-date document versions are available and are the most appropriate for use.

Since the last update of this handbook, there has been a proliferation of commercial off-the-shelf software packages that are well designed to manage and control an organization's document system (see Chapter 18, "Computers, Software, and Software Validation"). A small business just starting may not be able to afford those packages, but may want to consider the cost/benefit of maintaining its own system versus a commercial or custom-designed system.

Here is a brief overview of a basic document control system using the resources an organization may already have. Each of these details should be followed, whether using a paper-based or electronic-based system. A "master list" with all of the pertinent information about each document is helpful to have to maintain official-level control and to account for every and all documents and each of the following details.

Each of the controlled documents (procedures, records, certificate templates, and so on) should have a unique title and a unique identification number (many systems use a number often known as a *control number*). There should also be a revision number. Whether on the document or on the master list, it is helpful to include the name of the approver or approval authority, the revision date, and a statement that only the revisions listed should be used. It is necessary to identify the changes to documents for each revision level. When identifying changes in documents, some systems may use a different font color for added (new) items, add black or red lines in the document borders to indicate where changes have been made, or annotate the changes in a change- or revision-control section of the document. Still others refer to comparisons of archived documents as the only reference to changes. Archiving copies of previous revisions can have benefits, but they must be easily identifiable as "obsolete" and stored in a location where they cannot be readily or inadvertently accessed.

Part of an organization's training program should include when and how to inform and train its staff of changes to the documents. Some organizations ensure that training and/or notification of changes has occurred before they allow the latest revisions to be posted. In some systems the new revision must be posted in order for the user to have access to the documents. Another approach would be to maintain all quality documentation electronically via an intranet. In this process any printed documents would be invalid. This process ensures that only the latest and authorized documents are available to all users to whom they pertain. Whichever way an organization's system works, it is vital that everyone involved be informed and trained when changes are made, and that only the latest revisions are available for their use. All training must be documented in the respective personnel's training records.

No matter which system or combination of systems is used, the documentation only has to meet the quality system requirements that have been set for the organization.

Chapter 5
Calibration Procedures

INTRODUCTION

All quality systems that address calibration require documented instructions for the calibration of inspection, measurement, and test equipment (IM&TE). Different standards have various requirements, and those are addressed in Chapter 17, "Industry-Specific Requirements," wherever they differ from the information in this chapter. Under a quality system, this is the *say what you do* portion. This refers to the need to document, in detail, how to perform laboratory activities, including calibration; this includes calibration procedures, standard operating procedures (SOPs), protocols, job aids, work instructions, work cards, and so on. Why follow formal and documented instructions or procedures? The answer is simple: To get consistent results from a calibration, one must be able to follow the same step-by-step instructions every time a calibration activity is performed.

The importance of documented procedures is included in multiple published standards. To begin with, ISO 10012-2003, Clause 6.2.1, states, "Measurement management system procedures shall be documented to the extent necessary and validated to ensure the proper implementation, their consistency of application, and the validity of measurement results. New procedures or changes to documented procedures shall be authorized and controlled. Procedures shall be current, available and provided when required."[1]

ANSI/ASQC M1-1996, Clause 4.9, states, "Documented procedures, of sufficient detail to ensure that calibrations are performed with repeatable accuracy, shall be utilized for the calibration of all ensembles."[2]

ISO 9001:2015 is focused on outputs. It is silent on the requirements for details *within* a calibration procedure other than to require that "[t]he organization shall determine and provide the resources needed to ensure valid and reliable results when monitoring or measuring is used to verify the conformity of products and services to requirements."[3]

ISO/IEC 17025:2017 is also silent on the requirements *within* a calibration procedure and is also focused on outputs. It does require a laboratory to "document its procedures to the extent necessary to ensure the consistent application of its laboratory activities and the validity of the results."[4]

The common theme across these standards is to document calibration method procedures with enough detail to ensure consistent application and valid results. So, what should be in a calibration method procedure to ensure these outputs? Some companies use original equipment manufacturer (OEM) procedures that are in service manuals as a starting point. Keep in mind that some service manuals have procedures for adjusting the IM&TE as well as (or instead of) the calibration (performance verification) process. Also, some OEM procedures are vague and lack the specific requirements needed to ensure a good calibration, such as equipment requirements, environmental conditions, consideration of uncertainty components, and so on. Finally, in many cases, the equipment manufacturer simply does not provide any calibration or service information as a matter of policy. By drafting your own procedures or using prewritten procedures for your own activities, you might save time by eliminating the adjustment process if it is not required to improve the outcome of the calibration.

There are multiple perspectives regarding the adjustment of IM&TE. On one end of the spectrum, some (particularly government regulatory agencies) require that IM&TE be adjusted at every calibration, whether or not it is necessary. At the other end of the spectrum, some argue that any adjustment is tampering with the natural system (per Deming), and what should be done is simply to record the values and make corrections to measurements. In between these two extremes exist varying decision criteria for when adjustment is necessary. An important note to remember is that no adjustment should be made until a complete *as-found* calibration has been performed and recorded. As the words imply, *as found* is the condition the IM&TE is received or found in before beginning any step of the calibration. Some activities, such as cleaning contact points, may be necessary to prevent contamination of reference standards and ensure safe workspaces. Only after all as-found results from the reference standards and the IM&TE have been recorded can any adjustment, alignment, or repair be performed. However, sometimes IM&TE is in a condition (i.e., broken) that prevents collection of *any* data until it is repaired.

This still leaves us with two questions: *what to include in a calibration procedure* and *how should the procedure be formatted*?

NCSL International, Recommended Practice–3: Calibration Procedure Requirements (RP-3) is an entire recommended practice concerning calibration procedure requirements, and it lists factors that could influence the technical adequacy of calibration results. "These factors include the following:

1. Customer or client requirements

2. Applicability of measuring and test equipment

3. Measurement requirements, including measurement quantities and tolerances

4. As-found performance and condition

5. Measurement uncertainty and decision risk

6. Measurement traceability

7. Calibration standards and equipment performance requirements

8. Measurement methods

9. Measurement assurance and sampling requirements

10. Human factors and suitability
11. Environmental and stabilization requirements
12. Calibration reporting requirements
13. Data recording, analysis, and presentation requirements
14. Handling and storage of calibration items
15. Labeling and sealing requirements
16. Configuration and change control, and
17. Safety of personnel and equipment."[5]

What could a calibration procedure look like? Formatting of how a calibration procedure appears can vary. With the exponential growth of information on the internet, examples can be found with a simple search in a search engine. Reputable sources, such as National Metrology Institutes and government agencies, may make their calibration procedures free to access. As a start, one can follow the formats of these procedures while also ensuring that the resulting procedure contains enough information and detail that one will be able to successfully and repeatedly perform a calibration to produce valid and sound results. Your own company may already have a designated format for procedures. Calibration procedures should be among the controlled documentation as specified in the quality management system.

As with any procedure or work instruction, there should be an order to the procedure where sections can build upon one another. It is helpful to reference the list of factors from NCSL RP-3 (provided earlier in this chapter) when adding detail.

CALIBRATION PROCEDURE CONTENTS

Title and Revision Control

For example, begin with the procedure title (and, if used, the associated numbering or abbreviation identifier) and information regarding the document control revision level and status. Whether in the procedure itself or elsewhere in the document control system, the revision history of the procedure should be included. If the revision history is to go in the calibration procedure, place it where it makes the most sense for your laboratory and use of the procedures.

Purpose

A section that describes the purpose of the procedure can be helpful. The purpose can include what this procedure is used for: What are the desired results or outcomes of the procedure? What is the type of IM&TE to be calibrated? What measurement results and equipment does it apply to? Are there published standards or other defined requirements that should apply to the measurement results? It is helpful to include these points in sufficient detail to ensure consistent application and valid results.

Environmental and Stabilization Requirements

What are the applicable environmental conditions and limits in which the calibration should be performed? What, if any, are the requirements for environmental stabilization of the reference standards, IM&TE, and calibration environment?

OTHER SOPS AND COMPETENCIES

What other SOPs and/or competencies are necessary for the performance of this calibration? There should be enough information about the referenced SOPs that the reader knows and understands what they are.

SAFETY AND HANDLING CONSIDERATIONS

What are any health and safety considerations for the personnel performing the calibration? How should the equipment involved in the calibration (reference standards, supporting equipment, apparatuses, reagents, IM&TE, etc.) be handled to ensure safety and integrity of the equipment? Items to consider include, but are not limited to, safety glasses, steel-toed shoes, hardhats, electrostatic discharge (ESD) straps, gloves, lab coats, respirators, and so on. This section could list the specific details or could reference the controlled document(s) that provide these details.

Handling instructions for equipment may be found in a number of sources, such as an all-encompassing procedure, within individual procedures, in equipment user or operational manuals, in supplemental materials, or elsewhere. When handling instructions are referenced in their own document, that document should be controlled in the document control system.

Reference Standards and Apparatus

List and describe the reference standards and any supporting equipment needed (including fixtures, tools, apparatuses, safety equipment, reagents, and other items). For this equipment, what are their requirements for parameters, ranges, accuracy, uncertainty, precision, resolution, etc.? Decide what and how much of this information to include in the procedure. Some labs will include the unique identification of the reference standards to be used. Although this may be helpful to easily identify which standard to use, if the list of reference standards is limited to *specific* reference standards and that specific standard is not available, then the lab would not be conforming to its procedure unless there is a caveat, such as that "or equivalent" standard can be used. When using "or equivalent" as a caveat, it would be helpful to define what "or equivalent" means regarding performance, accuracy, or other defined characteristics of the reference standard.

Symbols and Formulas

Are there unique symbols and/or formulas for this procedure? They should be clearly defined. Formulas for calculating data should be included. For complex formulas, it is good practice to define or describe the order of operations so that each person using the procedure applies the formula in the same manner.

Workspace Setup

Describe how the calibration station or area should be assembled and prepared. Does equipment have to be warmed up for a period of time? Does the equipment have to be exercised before use? Some equipment, such as those with load cells, springs, etc., may require steps to exercise the equipment before use. Including pictures and/or videos can be especially helpful where descriptive written words could be interpreted differently among people and different languages. Use of pictures and/or videos also meets the intent of "documenting procedures to the extent necessary to ensure consistent application and valid results."

Procedural Steps

What are the performance requirements of the IM&TE to be calibrated? Include applicable functions, parameters, and ranges as well as the assigned tolerances or specifications. What activities and measurements are to be performed and in what order? Use of numbered steps is a helpful guiding practice. Where necessary, use diagrams to indicate placement of measurements on the respective equipment. Again, including pictures and/or videos is helpful. What information about the IM&TE has to be recorded? What data have to be recorded? How and where should data be recorded? The details required to be recorded, as well as where to record them, should be clearly defined.

Measurement limits specifying adjustment (where possible) should be defined. If the measurements are used to make statements of conformity, what is the decision rule to be applied? (See Chapter 30, "Measurement Risk and Decision Rules," for more information on decision rules.)

Measurement results before any adjustment or repair (where the adjustment or repair could affect the validity of measurement results) are to be recorded. Information describing what was adjusted and how it was adjusted is a useful record. If the adjustment procedures are not already included in the procedure, describe them, or provide a reference to where those instructions can be found. Supplemental information can be helpful, such as descriptions of the effects of types of adjustments.

Often, the word *calibration* is unintentionally misused to mean *adjustment*, when in fact the two words have different definitions and intents. Not all IM&TE can be adjusted. It is important to remember the *VIM* definition of "calibration": nowhere in that definition is the aspect of "adjustment" included. For example, a steel ruler or a liquid-in-glass thermometer cannot be adjusted, but it certainly can be calibrated.

> **Calibration** (*VIM* 2.39): operation that, under specified conditions, in a first step, establishes a relation between the **quantity values** with **measurement uncertainties** provided by measurement standards and corresponding **indications** with associated measurement uncertainties and, in a second step, uses this information to establish a relation for obtaining a **measurement result** from an indication.
>
> Note 2—Calibration should not be confused with **adjustment of a measuring system**, often mistakenly called "self-calibration," nor with **verification** of a calibration.[6]

Measurement results after any adjustment or repair are also to be recorded. These often replicate the same steps as for measurements made before adjustment.

Measurement Uncertainty and Evaluation of Results

Whether they appear in individual calibration procedures, in its own procedure, or as documented in the measurement uncertainty budget, what are the components of measurement uncertainty of this calibration? If a decision rule is to be applied (for making adjustments and/or when making statements of conformity), follow the decision rule to apply the measurement uncertainty results of the calibration.

Who, When, and Where?

Where should it be recorded who performed the calibration and when and where the calibration occurred? It should be documented that anyone performing the calibration is competent and authorized to perform that activity. If secondary reviews are necessary, what portion(s) of the calibration activity record should be reviewed and how? Is it important to define this in the individual procedure, or is it defined elsewhere in the quality system?

Recording of Data and Results

Any of the procedure steps that specify a measurement should have a record of those results. These records may be recorded in hard copy or electronically. They also may be combined and summarized on a calibration report or calibration certificate. See Chapter 6, "Calibration Records," and Chapter 7, "Calibration Certificates."

End of Calibration Activities

Instructions for end-of-calibration activities can include aspects such as how equipment is to be labeled, safeguarded, packaged, stored, transported, and/or prepared for returning to use.

For labeling, define what is to be included on the calibration status label, if that is not already defined elsewhere in the quality system. If possible, previous calibration labels should be removed before placing a new label on equipment. Additionally, are there other labels, such as tamper-evident seals, which are to be applied to the IM&TE? If so, where should they be applied?

Are there steps that should be followed to safeguard the equipment? How should the equipment be prepared for returning to use? These considerations might also be found in the overall equipment handling procedure.

What is the process for submitting calibration results? These may also be defined in a separate procedure.

References (Bibliography)

What are all the referenced and related documents for this procedure? Having another place, whether in the individual procedure or elsewhere, where these are summarized can be useful. In addition, there may be linkages to other documents, equipment, and published standards (such as test methods, international standards) that were not specifically used in the performance of the calibration and that are also helpful to document.

METHOD VALIDATION AND VERIFICATION

It is not enough to simply draft a calibration procedure or download one from the internet and then begin using it. Many international standards require that, at a minimum, calibration methods be appropriate for the intended use and for the laboratory to verify that by performing the method they can produce repeatable and valid results.

ISO/IEC 17025:2017 7.2 is an entire clause about *Selection, verification, and validation of methods*. In this standard, "method" is considered synonymous with the term "measurement procedure" as defined in the *VIM* (JCGM 200-2012).[7]

How does one know when to "validate" versus "verify" a method? Let us begin by defining the words *validation* and *verification*.

Validation (*VIM* 2.45): *verification, where the specified requirements are adequate for intended use.*[8]

Verification (*VIM* 2.44): *provision of objective evidence that a given item fulfils specified requirements.*

Note 5 for Verification: *Not every verification is a **validation**.*[9]

At first glance, the definitions seem so similar. When does a method have to be validated rather than verified? In the case of methods that have been "published either in international, regional, or national standards, or by reputable technical organizations, or in relevant scientific texts or journals, or as specified by the manufacturer of the equipment,"[10] these methods are often considered to have been developed and validated for their intended use. In these instances, if a laboratory uses the method as intended without any modifications, then the laboratory would only need to *verify* that it can correctly and repeatedly perform the method as written and generate valid measurement results. Some ways that this verification can be substantiated is by performing the method in repeatable conditions, such as in an intra-laboratory comparison. Another way is by comparison with external labs, such as in inter-laboratory comparisons or proficiency tests. Other options are listed in ISO/IEC 17025:2017, Clause 7.7.[11]

If a laboratory develops its own method, uses a nonstandard method, or modifies or uses a standard method outside the intended scope, then the laboratory must first *validate* that the method is appropriate and adequate for use (see definition of *validation*). The validation must be "a planned activity and performed by competent personnel equipped with adequate resources."[12] In a planned activity, how the method is to be performed and the measures of validation are predetermined.

Validation is an example of applying the plan–do–check–act quality tool. The *plan* includes developing and documenting the protocol (i.e., the steps and expected results). The *do* is the execution of the plan (i.e., the validation protocol.) The *check* is comparing the results of the doing to the expected results in the plan. Based on the differences, changes may be made to the plan and the process repeated.

ISO/IEC 17025:2017, Clause 7.2.2.1, lists ways that a method can be validated, including statistical analysis, testing method robustness through variation of controlled parameters, comparison of results with other validated methods, and interlaboratory comparisons (and proficiency tests).[13]

Once a method has been validated, assessed, and proven that it is appropriate for its intended use, and that the intended use is relevant to the specified needs and requirements, then the laboratory must *verify* that it can repeatedly perform the method and generate valid results. A laboratory must verify that it can successfully perform the method before implementing the method in actual use. Whether validating or verifying, the results of these activities must be recorded and also include "a statement on the validity of the method, detailing its fitness for the intended use."[14]

SUMMARY

Remember, the purpose of the calibration procedure is to provide the sufficiently detailed information and directions necessary to ensure proper and consistent application so that the measurement results are repeatable and valid by those competent and authorized to perform the procedure. The calibration procedure is the document of *what is to be done*. The calibration results are a record of *what was done*. Laboratories shall verify and/or validate methods to prove that they have competent people and the appropriate equipment and can consistently generate required outcomes of the calibration and consistently produce valid data.

Chapter 6

Calibration Records

Auditors and assessors have a saying: "If you did not record it, it did not happen." There is truth to this statement. How do you prove what was done in a test or calibration? How do you communicate to another person what the test or calibration results were? Without records, you cannot. We will be focusing on calibration records in this chapter. Some readers might ignore this chapter altogether. For them, there are no requirements to keep or maintain records of their calibrations or of the maintenance performed on the inspection, measurement, and test equipment (IM&TE) that they support. For many of us, this would seem almost unbelievable, but it is true. If their IM&TE passes calibration—that is, meets the tolerances (staying within the defined lower and upper tolerance limits)—they may only be required to place a calibration label on the unit and update their database with the next date due for calibration.

For the rest of us, this is not an option. Documentation of the "as found" and, when needed, "as left" measurement results, along with the corresponding reference standard measurement, is as natural as using calibration procedures and traceable standards. There is also a saying, "A short pencil is better than a long memory." By wearing down the pencil, emptying the pen of ink, or filling the computer storage space, more will be recorded and preserved than what we think we can solely remember. Besides, how do we *prove* what happened from our own memory?

What, then, are the benefits of records? There is an opinion in metrology that "old is good." This concept is not readily understood by most, but when one understands the relationship of historical data and the stability of a standard, the need for good historical records becomes apparent. For example, a standard with 10 years of quality documentation related to its performance is much more valuable than a standard with no history. With historical data you can predict the stability of a particular instrument and build on it; without the history there is nothing to build upon. This is a common concept in metrology and also a reason why quality documentation and calibration records are invaluable.

Why do doctors, lawyers, dentists, and mechanics keep records? Some do so because it is required by law (record archiving for liability purposes has been around for a long time); others do so because they need a history to know what has happened previously and, perhaps, to compare the differences between then and now. When working with food, pharmaceuticals, cosmetics, aircraft, and other sectors of industry where IM&TE are used, previous circumstances have proven that the procedures, and the recorded data and history of the calibration and/or repair of the test equipment, are critical in finding out why accidents happened, defaults were accepted, or mistakes were made. Records are also important in forecasting how IM&TE will function over time. Without historical information we are bound to repeat mistakes and not make the process improvements we should all be striving for. Records play a crucial part in the calibration process, and their retention and availability should be an integral part of your quality system.

Here are some references that include specific requirements for documenting your (testing, sampling, or) calibration activities using records. ISO/IEC 17025:2017, Clause 7.5.1, states: "The laboratory shall ensure that technical records for each laboratory activity contain the results, report and sufficient information to facilitate, if possible, identification of factors affecting the measurement result and its associated measurement uncertainty and enable the repetition of the laboratory activity under conditions as close as possible to the original." This clause continues by stressing that the records are to include the date and the identity of the personnel responsible for each laboratory activity; it also specifies that "original observations, data, and calculations shall be recorded at the time they are made and shall be identifiable with the specific task."[1]

ANSI/ISO/ASQ Q10012-2003, Clause 6.2.3, states: "Records containing information required for the operation of the measurement management system shall be maintained. Documented procedures shall ensure the identification, storage, protection, retrieval, retention time and disposition of records."[2]

NCSL International Recommended Practice–6: *Recommended Practice for Calibration Quality Systems for the Healthcare Industries*, RP-6, 2015, section 5.13, "Records," states: "Records should be maintained for all M&TE that are included in the calibration quality system."[3]

ANSI/ASQC M1-1996, *Calibration Systems*, Clause 4.7, "Records," states that "records shall include, but not be limited to:

a. Description of equipment and unique identification,

b. Date most recent calibration was performed,

c. Indication of procedure used,

d. Calibration interval,

e. Calibration results obtained (i.e., in or out of tolerance),

f. By whom the ensemble was calibrated, and

g. Standards used."[4]

Specific record requirements for other standards or regulations can be found in Chapter 17, "Industry-Specific Requirements."

So, what do these standards ask you to do? They say to collect the IM&TE readings as they occur and record them, either on a hard-copy record or electronically, if that is how your system works. Data can also be recorded using computer process equipment or software designed and validated to perform that function.

For argument's sake, let us assume you must keep a record of your calibrations. What has to be in the record? What happens to the record when you are through with the calibration and data collection? Is the record saved, destroyed, archived, or used as scratch paper during the recycling process? Before we go any further, let us remember that, generally speaking, there are two types of users. The first is the group that calibrates IM&TE for their own company and does not have any external or commercial customers, nor any guiding published standards that already specify what they must record. The second group deals with (internal and/or) external customers who are paying for their calibration services. Both groups may require information contained in a record, and some of the details needed may be unique to each group. Some information is required because of the standard that covers your calibration activity, some is needed to meet customer requirements, and some is needed because it is prudent to have in the record. So, having said that, let us answer each of these questions one at a time.

For a calibration record to be valid, it should at least:

1. Identify which IM&TE is being calibrated. The record must have a unique identification number assigned to it; this could be the IM&TE serial number and/or asset number. Include the manufacturer and model information. Identify who owns it (when appropriate). If the record is for internal use only, possibly show the department, group, or cost center that owns it.

2. Include details as to the IM&TE's condition if that may have an influence on the results or be necessary for any other requirement.

3. Include details about the relevant environmental conditions during calibration (when applicable or required).

4. List the IM&TE's parameters, ranges, and tolerances (when applicable or required).

5. Ensure metrological traceability (back to the SI) for the reference standards you are using to perform the calibration by identifying your reference standards (and ensuring *their* metrological traceability) and including the calculated measurement uncertainty for the calibrated IM&TE.

6. Identify the method or procedure used for the calibration, and its revision level or number. This is important for being able to replicate a calibration or to determine possible contributors to a nonconformance, such as if the incorrect method or revision were used.

7. Ensure that the reference standard's measurement uncertainty or accuracy meets the tolerance requirements for the calibration. Remember, the reference standard must have a smaller measurement uncertainty and a smaller accuracy than the IM&TE being calibrated.

8. Include corresponding and complete measurement results for the reference standard and the IM&TE. The complete IM&TE measurement results should be recorded *before* any adjustments, alignments, or repairs (if possible). Sometimes IM&TE are brought for calibration in nonoperational or otherwise broken status and any *before-calibration* data are impossible to collect. For IM&TE that has been adjusted, aligned, or repaired after the collection of *before-calibration* (also called *as-found*) *data*, the measurement results *after*

these changes should also be recorded (also called *as-left data*). For IM&TE where no adjustments are possible, *before* and *after* data will be identical. When required, any out-of-tolerance readings should be presented in magnitude and direction so that an impact assessment can be made. Also see Chapter 30, "Measurement Risk and Decision Rules," for how out-of-tolerance decisions may be made.

9. If the calibration interval is known and a next calibration due date is requested, record it. More information on date formats can be found in Chapter 14, "Labels and Equipment Status."

10. Include comments or remarks, such as clarification of limits, repairs, or adjustments that occurred during the process of calibration.

11. The person who performed the calibration should record their name and the date of the activity(ies). Typically, for calibration activities that take more than one day to complete, the date when all calibration functions have been completed is recorded as the calibration date. This date is also the date used for calculating the next calibration due date.

12. In some organizations, records are also required to be reviewed by a different person than the person who performed the calibration. If this is the case, at least the name and review date should be recorded.

RECORD INTEGRITY AND RETENTION

Records should be maintained so that they are safe, secure from tampering or alteration, their confidentiality is ensured, and they are available when needed. What good is a record if you do not know that you have it or where it is? When changes are made to a written record, the error should be crossed out (not erased, made illegible, or removed/deleted) and the correct value or reading entered next to it. The person who made the change must also record their initials and the date of change either next to the entry or noted where this information can be referenced. There may not be sufficient space to legibly write this, so some type of identifier, such as an asterisk, must be entered so that it can be uniquely linked to this identifier elsewhere on the page. Some industries also include the reason the change had to be recorded.

If you record or store your records electronically, equivalent measures must be taken to avoid any loss or change in the original data. If changes are made to an electronic record, a new record must be added. The new record will contain the original data with the changes, as well as identification of who made the change and when the change occurred, and a link or reference to the old record. The original record should stay unchanged, but may also have a type of clear linkage or identification showing that it is archived and a link to the replacement record. Many robust software programs include a feature that automatically does this. This feature is referred to as an "audit trail," "audit tracking," "audit log," "change tracking," or similar wording.

Most organizations require records be maintained for a predetermined time period and list these requirements in their records retention policy. For labs conforming to or accredited to ISO/IEC 17025:2017, Clause 8.4.2 requires that controls be implemented for the "identification, storage, protection, back-up, archive, retrieval, retention time, and disposal of its records."[5] This clause also mandates that records shall be retained per time periods as stated in contractual obligations. Confidentiality of information in the records still needs to be maintained and balanced with the ability to readily access the records.

There are different approaches on how to store records for easy accessibility. You might consider storing records organized by the IM&TE's unique identification number or by the location of the unit, with all items in a specific room, lab, or department having their records stored in that location or file. When records are being stored electronically, there are even more options. You must be able to easily find the record during audit or review. One system that is useful with electronic storage is using a naming format that includes a unique identification number followed by the (ISO 8601 date format) date of calibration. In this case, an IM&TE's unique identification number, JJ8675309, and the respective device's calibration on May 20, 2022, would appear as JJ8675309-20220520. Not only does this identify the unit, it readily shows the date the unit was calibrated, distinguishing it from records of previous calibrations. It also makes each record easily searchable and sortable in an electronic system.

The recordkeeping is part of the "do" portion of plan–do–check–act, where the results are recorded at the time of the doing. The important point is to be systematic about and diligent in maintaining the practice.

Chapter 7
Calibration Certificates

If your calibration responsibilities include the creation and/or completion of calibration certificates, then this chapter is for you. There are instances, however, where data in calibration records are a direct substitute for certificates, such as when you only perform calibrations internally for your own customers and there are no other requirements for creating calibration certificates. In such circumstances, the combination of the calibration record and calibration label attached to the inspection, measurement, and test equipment (IM&TE) serves the same purpose. They provide all the information needed to prove that the item has undergone a traceable calibration. The information included in a calibration record/label can provide the same vital information as included on a calibration certificate. Refer to Chapter 6, "Calibration Records," for more information.

The updated ISO/IEC 17025:2017 added two general clauses to Clause 7.8, *Reporting of Results*. Clause 7.8.1.1 states that "[r]esults shall be reviewed and authorized prior to release."[1] Clause 7.8.1.2 states, "[w]hen agreed upon with the customer, the results may be reported in a simplified way."[2] It then adds that "[a]ny information listed in 7.8.2 to 7.8.7 that is not reported to the customer shall be readily available."[3] Common requirements for all reports shall include at least the following information unless valid reasons are presented by the performing laboratory (Clause 7.8.2)[4]:

- A title
- Name and address of the laboratory
- Location of performance of the laboratory activities, including when performed at a customer facility or at sites away from the laboratory's permanent facilities, or in associated temporary or mobile facilities
- Unique identification that all components of the report are recognized as a portion of a complete report and a clear identification of the end
- Name and contact information of the customer
- Identification of the method used
- A description, unambiguous identification, and, when necessary, the condition of the item
- Date of receipt of the test or calibration item(s), or the date of sampling, where this is critical to the validity and application of the results

- Date(s) of performance of the laboratory activity
- Date of issue of the report
- Statement to the effect that the results relate only to the items calibrated
- Results with (where appropriate) the units of measurement
- Additions to, deviations, or exclusions from the method
- Identification of the person(s) authorizing the report
- Clear identification when results are from external providers

The standard continues, in Clause 7.8.2.2, to state that "[t]he laboratory shall be responsible for all the information provided in the report, except when information is provided by the customer. Data provided by a customer shall be clearly identified. In addition, a disclaimer shall be put on the report when the information is supplied by the customer and can affect the validity of results...."[5] An example of this would be the density of a weight sent in for a mass calibration or even the weight's classification for calibration.

Information specifically required by the standard for calibration certificates is as follows (Clause 7.8.4):[6]

- Measurement uncertainty of the measurement result presented in the same unit as that of the measurand or in a term relative to the measurand (e.g., percent).
- Conditions (e.g., environmental) under which the calibrations were made that have an influence on the measurement results. (This could include such information as the laboratory's local gravity for the calibration of a deadweight tester's force plates or the discharge air density for the correction of a volumetric flow rate.)
- Statement identifying how the measurements are metrologically traceable.
- Results before and after any adjustment or repair, if available.
- Where relevant, a statement of conformity with requirements or specifications.
- Where appropriate, opinions and interpretations.
- Where the laboratory is responsible for the sampling activity, calibration certificates shall meet the requirements listed in Clause 7.8.5—*Reporting sampling—specific requirements* where necessary for the interpretation of test results.
- A calibration certificate or calibration label shall not contain any recommendation on the calibration interval except where this has been agreed with the customer.

It is important to keep in mind that when making statements of conformity (e.g., pass/fail, in tolerance/out of tolerance) in a calibration certificate, the laboratory must first clearly define the decision rule to the customer or agree upon a specification or standard at the time of the service request. If a statement of conformity is agreed upon, this must also be communicated in the calibration certificate (Clause 7.8.6 of ISO/IEC 17025:2017).[7]

Opinions and interpretations can also be part of the report, but can only be made by laboratory personnel authorized to make such statements. These statements must be based on and referenced back to the calibration data of the report, and all such communications must be documented.[8]

Finally, remember that these requirements are the minimum ISO/IEC 17025:2017 requirements. Additional information, such as the reference standard(s) used to perform the calibration, can always be included as well, but all information is required to be accurate and reported consistently. If the reference standards identifications are not listed on the calibration certificate, there must be a record of their use to prove and ensure metrological traceability.

Another new facet to the standard's approach to reporting results is the clause requiring a "date of issue" (Clause 7.8.2.1 j).[9] This may seem like just another excuse to revise your report template, but this makes sense once you read the clauses in Clause 7.8.8, *Amendments to Reports*.[10] The three clauses in Clause 7.8.8 make specific references to and requirements for "issued" reports. If there is another date on the report that is acting as an issue date, then that date must be expressly and clearly stated somewhere in the organization's quality system.

According to ILAC P14:09/2020, *ILAC Policy for Measurement Uncertainty in Calibration*, labs accredited to ISO/IEC 17025:2017 must at least meet the following requirements when stating measurement uncertainty on a calibration certificate (see ILAC P14:09/2020 for the actual wording of each requirement as well as supporting notes and interpretations):[11]

1. Measurement uncertainty must be reported in compliance with the *Guide to the Expression of Uncertainty* (*GUM*).

2. Measurement results shall include the measurement uncertainty value for each measurement result. The coverage factor and coverage probability shall also be stated.

3. The numerical value of the expanded uncertainty shall be reported to, at the most, two significant digits. Any rounding of results shall meet the rounding guidance in Clause 7.2.6 of the *GUM*.[12]

4. The measurement uncertainty value(s) reported shall include the relevant short-term contributions of their device. Typically, the reported measurement uncertainty will be larger than the corresponding Calibration and Measurement Capabilities (CMC) on the scope of accreditation. This also means that a reported measurement uncertainty cannot be smaller than the CMC for which the laboratory is accredited.

5. Lastly, the measurement uncertainty shall be presented in the same unit as that of the measurand or in a term relative to the measurand (e.g., percent).

It is noteworthy that nowhere in *ILAC-P14:09/2020, ILAC Policy for Measurement Uncertainty in Calibration*, does it state that the coverage factor, k, must be expressed as "k = 2". It just states that "[t]he reported expanded measurement uncertainty is stated as the standard measurement uncertainty multiplied by the coverage factor k such that the coverage probability corresponds to approximately 95 %."[13]

Use of a blanket statement similar to "Does not exceed a 4:1 TUR [test uncertainty ratio] unless noted otherwise" is discouraged and in some places, no longer allowed, unless each measurement value still includes its associated measurement uncertainty value (and the rest of the ILAC P14 requirements).

Calibration certificates can be in either hard copy or electronic format. Whether handwritten or computer-created, they must be legible, readable, and provide the necessary information requested or required by the user.

Individual ISO/IEC 17025-accreditation bodies may require details and formats for calibration certificates in addition to ISO/IEC 17025 and ILAC requirements. Some calibration providers may include information on their certificates that is not required but serves to better enable and educate the customer. Examples include accompanying graphical representations of data, other graphs or images, formulas, or even elaborate formulas and/or uncertainty analysis. Though no longer required by ISO/IEC 17025:2017, it is still noted to include a statement that the report shall not be reproduced except in full without approval of the laboratory. This helps to provide assurance that parts of a report are not taken out of context.

Figure 7.1 illustrates an example of a nonaccredited calibration certificate. Figure 7.2 illustrates an example of a robust ISO/IEC 17025:2017 accredited calibration certificate. Use the ISO/IEC 17025:2017, Clause 7.8 requirements to compare the differences between the certificates. For the nonaccredited calibration certificate, this is just one example of how the calibration certificate may appear and with which details. The level of detail in a nonaccredited calibration certificate can vary. It is the customer's responsibility to make sure they are receiving the calibration details that they need.

Note: Not all ISO/IEC 17025:2017 accredited calibration certificates may include the robustness of detail shown in this sample certificate. Refer to the ISO/IEC 17025:2017, Clause 7.8 requirements.

Some points about ISO/IEC 17025:2017 certificate requirements are worth stating again.

First, although it is a widespread practice for reference standards to be listed on an accredited calibration certificate, it is not a requirement to list them on the calibration certificate. What must be met is the requirement of Clause 6.5—Metrological Traceability of the ISO/IEC 17025:2017 that a laboratory must establish and ensure the traceability of its measurements; note, though, that in Clause 7.8—Reporting of Results of the same standard, it is not a requirement that the standards used in a calibration appear in the certificate or report. If not reported on the calibration certificate, there must be a record somewhere that includes these details.

There is a common misstatement or misuse of the phrases "NIST-traceable," "NIST-traceability," or "traceable to NIST." None of these are valid, nor are they scientifically possible. The traceability requirement is for *measurement results* to be "traceable to the SI (International System of Units)" and the traceability chain may go *through* measurement results by a competent laboratory, certified values or certified reference values, or direct realization of the SI units by comparison, directly or indirectly, with national or international standards.[14]

Certificate of Calibration: 12345678

Customer: ACME Chemicals, Inc.
123 Anystreet Blvd.
Abig Town, MD 11111 USA
1.555.123.4567

Corporate Primary Calibration Laboratory (CPCL)
805 N Calibration St
Somewhere, IN 99999
Phone: 1.463.555.5000
Fax: 1.463.555.5100

Asset Number: K654321	**Procedure:** CSL-WI-PPS001
Manufacturer: Really Good Gage Co.	**Calibration Technician:** Josephine Doe
Device Type: digital pressure gage	**Date Calibrated:** 2021-09-08
Model: DPG-1M-G	**Next Due Date:** 2022-09-07
Serial Number: 987654	**Results:** As Found – Fail / As Left – Pass

The information contained in this report only applies to the device identified above and used to verify its operability as defined by the specifications presented in the associated service request form at the time of the calibration. This certificate shall not be reproduced other than in full, without the specific written approval by the CPCL.

These calibration results are traceable to the International System of Units (SI) through recognized National Metrology Institutes measurement standards.

As Found

Environmental Conditions

Temperature: 20.7 °C	Humidity: 48 %RH

Reference [kPa]	UUT Reading [kPa]	Lower Limit [kPa]	Upper Limit ± [kPa]	TUR	Status Pass/Fail
0.0	0.0	-1.0	1.0	33:1	Pass
250.0	250.3	249.0	251.0	33:1	Pass
500.0	500.7	499.0	501.0	20:1	Pass
750.0	751.1	749.0	751.0	13:1	Fail
1 000.0	1 001.7	999.0	1 001.0	10:1	Fail
500.0	501.1	499.0	501.0	20:1	Fail

As Left

Environmental Conditions

Temperature: 20.2 °C	Humidity: 55 %RH

Reference [kPa]	UUT Reading [kPa]	Lower Limit [kPa]	Upper Limit ± [kPa]	TUR	Status Pass/Fail
0.0	0.0	-1.0	1.0	33:1	Pass
245.0	249.9	249.0	251.0	33:1	Pass
500.0	500.0	499.0	501.0	20:1	Pass
750.0	750.2	749.0	751.0	13:1	Pass
1 000.0	1 000.4	999.0	1 001.0	10:1	Pass
500.0	500.3	499.0	501.0	20:1	Pass

Figure 7.1 Nonaccredited calibration certificate.

The *ILAC Policy for Measurement Uncertainty in Calibration*, ILAC-P14:09/2020, Procedure part 5—ILAC Policy on Statements of Measurement Uncertainty on Calibration Certificates, clearly states in Clause 5.1 that "Accreditation Body shall ensure that an accredited calibration laboratory reports the measurement uncertainty in compliance with the *GUM*."[15] It continues in 5.2 to require that "[t]he coverage factor and the coverage probability shall be stated on the calibration certificate." Clause 5.2 then provides an example explanatory note for

CHAPTER 7: CALIBRATION CERTIFICATES   53

Corporate Primary Calibration Laboratory 805 N Calibration St Somewhere, IN 99999 Phone: 1.463.555.5000 Fax: 1.463.555.5100	CALIBRATION REPORT 12345678-001	NAP Accreditation Symbol Calibration NAP Lab Code 1234.01

Section 1: General Information

Customer:	ACME Chemicals, Inc. 123 Anystreet Blvd. Abig Town, MD 11111 USA 1.555.123.4567
Asset Number: K654321
Manufacturer: Really Good Gage Co.
Model: DPG-1M-G
Serial Number: 987654

Purchase Order Number: JJWCITT
Work Order Number: 8675309
Procedure: CSL-WI-PPS001
Calibration Technician: Josephine Doe
Date of Calibration: 2021-09-08
Date Issued: 2021-09-09

The information contained in this report only applies to the device identified above and is used to verify its operability as defined by the specifications presented in the associated service request form at the time of the calibration. Some data contained in this report may not be covered by the Scope of Accreditation. Where applicable, unaccredited test points are indicated by an asterisk (*) or stipulated in section 2.3 – Remarks. This report has been found to conform to ISO/IEC 17025:2017 by the National Accreditation Program (NAP), Calibration NAP Lab Code 1234.01. This certificate shall not be reproduced other than in full, without the specific written approval by the Corporate Primary Calibration Laboratory (CPCL). This report shall not be used to claim certification, approval, or product endorsement by NAP, any NMI, or other governmental organization or agency.

Pressure is derived from the base quantity of mass and the derived quantities of acceleration and area. Measurements are traceable to the International System of Units (SI) through recognized national metrology institute standards, ratio-metric techniques, and/or natural physical constants.
All calculations or corrections using local gravity use the value of 9.80077 ms^{-2} ± 0.00002 ms^{-2} as calculated by NOAA Surface Gravity Prediction – https://geodesy.noaa.gov/cgi-bin/grav_pdx.prl.

The Measurement Uncertainty (U_{meas}) is calculated in accordance with the *Guide to the Expression of Uncertainty in Measurement* (GUM) and its associated supplements and stated as the standard uncertainty of measurement multiplied by the coverage factor (k) such that the coverage probability (p) corresponds to approximately 95 %.

The tolerance is the maximum permissible error defined by the equipment owner and whose status is determined by the lab. Errors that are less in magnitude than the tolerance minus the associated measurement uncertainty are determined to have passed. Errors greater in magnitude than the tolerance plus the associated measurement uncertainty are determined to have failed.
Errors that fall within the associated U_{meas} of these limits are considered *undetermined*.

Approving Signature:

Digitally signed by
ANTHONY HAMILTON
Date: 2021.09.09 13:55:01 -04:00 UTC
Adobe Acrobat version: 2017.011.30204

Figure 7.2 Robust accredited calibration certificate.

54 Part II: Quality Systems

Report Number: 12345678-001

Section 2: Calibration Data

2.1 As Found Data

Date Performed: 2021-09-07 **Initial Offset:** 0.0 kPa
Pressure Mode: Gauge **Head Height:** 10.2 cm

Ambient Conditions during Testing

Temperature: (20.7 to 20.9) °C **Humidity:** (48 to 49) %rh

Applied Pressure [kPa]	Test Reading [kPa]	Error [kPa]	Tolerance ± [kPa]	U_{meas} ± [kPa]	Status Pass/Fail
0.000	0.0	0.0	1.0	0.073	Pass
249.999	250.3	0.3	1.0	0.073	Pass
500.000	500.7	0.7	1.0	0.087	Pass
750.000	751.1	1.1	1.0	0.11	*undetermined*
1 000.00	1 001.7	1.7	1.0	0.13	*Fail*
500.000	501.1	1.1	1.0	0.087	*undetermined*
0.000	0.1	0.1	1.0	0.073	Pass

2.2 As Left Data

Date Performed: 2021-09-08 **Initial Offset:** 0.0 kPa
Pressure Mode: Gauge **Head Height:** 10.2 cm

Ambient Conditions during Testing

Temperature: (20.2 to 20.3) °C **Humidity:** (55 to 55) %rh

Applied Pressure [kPa]	Test Reading [kPa]	Error [kPa]	Tolerance ± [kPa]	U_{meas} ± [kPa]	Status Pass/Fail
0.000	0.0	0.0	1.0	0.073	Pass
249.999	249.9	-0.1	1.0	0.073	Pass
500.000	500.0	0.0	1.0	0.087	Pass
750.000	750.0	0.0	1.0	0.11	Pass
1 000.00	1 000.2	0.2	1.0	0.13	Pass
500.000	500.4	0.4	1.0	0.87	Pass
0.000	0.3	0.3	1.0	0.73	Pass

2.3 Remarks: Tamper-evident seals intact. Display resolves full units. No damage.
Device setup in accordance with operations manual, Really Good Gage Co. DPG-1M (dated 2016-07)

Calibrated By: **Calibration Date:**
Josephine Doe 2021-09-08

Figure 7.2 *Continued*

---

calibration certificates, which says: "The reported expanded measurement uncertainty is stated as the standard measurement uncertainty multiplied by the coverage factor k such that the coverage probability corresponds to approximately 95 %."[16]

In the *Guide to the Expression of Uncertainty in Measurement* (*GUM*), JCGM 100:2008 elaborates on how to determine the coverage factor in Annex G—Degrees of Freedom and Levels of Confidence.[17] In section G.6—Summary and Conclusions, and even more specifically in G.6.6, the *GUM* admits that for most "practical measurements," the coverage factor (k) can be simply estimated as 2. This practice should be followed only in cases where the Type A uncertainty is a major contributor to the budget and/or is from a small sample size (i.e., a low degree of freedom). Today, with the computing power that is readily available, it is possible to perform the uncertainty calculations more easily than by hand.

In summary, the details included in a calibration certificate must meet specified requirements, either by internal policies and/or external requirements (customer, regulations, standards). One must also review one's calibration certificates to make sure they include the information and data required for the intended applications and requirements.

Chapter 8
Management Systems and Quality Manuals

Quality manuals are the summary of documentation for the management system of an organization, not just for the quality portion of the management system. They should describe how the organization is managed, how documentation and information flow and are controlled, how the customer is serviced and supported, and how the management system is monitored and adjusted. The most common international quality standards used by organizations performing laboratory activities are ISO/IEC 17025 and ISO 9001. ISO/IEC 17025:2017, Clause 1, Scope, "specifies the general requirements for the competence, impartiality, and *consistent operation* of the laboratory."[1] ISO 9001:2015, Clause 1, Scope "specifies the quality management systems requirements for when an organization:

   a) needs to demonstrate its ability to consistently provide products and services that meet customer and applicable statutory and regulatory requirements, and

   b) aims to enhance customer satisfaction through the effective application of the system, including processes for improvement of the system and the assurance of conformity to customer and applicable statutory and regulatory requirements."[2]

Although quality manuals are no longer required by ISO 9001:2015 or ISO/IEC 17025:2017, many organizations and laboratories still choose to use a quality manual to provide an easily referenced document that encompasses the requirements for an ISO 9001 and/or ISO/IEC 17025 management system. Laboratories that claim conformance to ANSI/NCSL Z540.1–1994 *are* required to have a quality manual. Even if an organization or laboratory chooses not to use a quality manual, they still need to ensure consistent implementation and operation to meet the intent and requirements of ISO 9001 and/or ISO/IEC 17025.

Quality manuals written for ISO/IEC 17025 often follow the format of the standard. Additionally, for Clause 8.1, Management System Requirements, ISO/IEC 17025:2017 allows the option to use an established and maintained management system in accordance with ISO 9001 as long as that ISO 9001 management system also supports and demonstrates consistent fulfillment of the requirements of ISO/IEC 17025:2017, Clauses 4 to 7 and also fulfills at least the intent of the management system requirements specified in ISO/IEC 17025:2017, Clauses 8.2 to 8.9.[3] For laboratories that do not have or use an established ISO 9001 management system, the minimum management system requirements of the laboratory shall address the following:

- Management system documentation (ISO/IEC 17025:2017, Clause 8.2)
- Control of management system documents (ISO/IEC 17025:2017, Clause 8.3)
- Control of records (ISO/IEC 17025:2017, Clause 8.4)
- Actions to address risks and opportunities (ISO/IEC 17025:2017, Clause 8.5)
- Improvement (ISO/IEC 17025:2017, Clause 8.6)
- Corrective actions (ISO/IEC 17025:2017, Clause 8.7)
- Internal audits (ISO/IEC 17025:2017, Clause 8.8)
- Management reviews (ISO/IEC 17025:2017, Clause 8.9)[4]

Regardless of the management system option chosen by an ISO/IEC 17025 laboratory, ISO/IEC 17025:2017, Clause 8.1.1 states that "the laboratory shall establish, document, implement and maintain a management system that is capable of supporting and demonstrating the consistent achievement of the requirements of this document and assuring the quality of the laboratory results."[5] Note that there is still the requirement to "document" the management system in ISO 9001:2015 and in ISO/IEC 17025:2017; however, this documentation no longer has to be in a defined "quality manual."

The laboratory should use the format that is most appropriate for its own needs. Regardless of the format that is used, the quality management system documentation must be easy to read and used by all personnel to help ensure consistent operations. The importance of ensuring consistent operation and achievement of the requirements of ISO/IEC 17025:2017 is so important that it is stated multiple times in the standard. ISO/IEC 17025:2017 specifies this in the beginning of Clause 1, Scope and reiterates it in Clause 8.1.1. For any management system documents, some critical issues to avoid include (1) a lack of page numbers, (2) a lack of section headings, and (3) a lack of references to other related documents.

While the listed issues are not requirements in ISO/IEC 17025:2017, they are nonetheless helpful. A quality manual and quality management documents lacking these items may be unclear and difficult to use. Additionally, quality manuals that are exceptionally verbose are difficult to use. Quality manuals that are difficult to use will likely not get used by the personnel who perform the day-to-day work of the laboratory.

Many alternative methods exist for quality manuals, including the use of flowcharts. Flowcharts are visual, typically simpler to understand, and therefore are more likely to be used by all personnel.

The management system documents are meant to be living documents. Organizations undergo change, and the management documents should reflect the changes and should reflect the philosophy of how the organization is managed on a day-to-day basis. Quality manuals are meant to reflect the specific organization or portion of an organization, so each manual is—or at least should be—unique.

A quality manual may contain different amounts and types of information depending on the needs of the organization. For example, a small laboratory may have one manual containing more descriptive information, rather than the several layers of procedural documentation that a large laboratory may have. Formats and layouts will likely be different for different manuals. The key aspects of the quality manual are that the quality manual meets the needs of the laboratory, accurately documents the management system used in the laboratory, has provisions to service the customer and safeguard customer information, and addresses any external requirements such as ISO/IEC 17025.

Every organization has information flowing through it all the time. For a testing and/or calibration laboratory, information is flowing in from the customer and back out to the customer. The flow of information takes many forms, from simple conversations to formal contracts or complaints. The quality manual should describe the protocols and controls that are used to direct, store, use, safeguard, and communicate the information in the manner best suited for the laboratory. The quality manual should include protocols for both physical and electronic documentation, and especially describe the protocols to safeguard confidentiality of customer information, whether it is transmitted electronically or stored in a file cabinet. Some examples of controls on information include a description of how records are maintained and how procedures, forms, and other documents (internal and external) are controlled.

Service and support of the customer have to be documented clearly in the quality management documentation. After all, the customer pays the bills! Some ways to service the customer are really communication methods. Documentation of the methods used to service the customer can be accomplished in different ways. Detailed descriptions can be provided in a quality manual or referred to and detailed elsewhere. The needs and size of the laboratory will dictate exactly how service to the customer is documented.

A management system has to be monitored, even in a small laboratory. Formal reviews of the management system must be held on some periodic basis. Audits can monitor the system; complaints are another tool. Various business metrics can also be monitored. Business metrics can range from the simple, from volume of a particular type of equipment over some time period, to the complex with detailed statistical analysis. Whether the laboratory is large or tiny, the quality manual should describe how the management system is monitored or should point to more specific documents such as procedures that describe the monitoring.

Quality manuals have been in existence in one form or another, and by various names, for a long time. The current term *quality manual* and the more precise definition of its layout and contents are much more recent. ISO 9001:1987 was the first truly international document to give the quality manual real definition and a more consistent purpose. Through evolution of the standards, ISO 9001:2015 and ISO/IEC 17025:2017 develop the purpose and requirements for the *outcomes* of a quality management system more completely than previous versions of those standards. For a testing and/or calibration laboratory that services customers in most fields of business, ISO/IEC 17025 may be the preferred primary document used to define the management system and the quality manual. ISO/IEC 17025:2017 has several prescriptive requirements for management review, corrective action, training, and other aspects of the management system. Many customers also require compliance to ISO 9001:2015, so the laboratory must include any additional requirements from that standard. Some industries have prescriptive requirements that may be unique to that particular industry. A testing and/or calibration laboratory that services customers in those industries should also include such prescriptive requirements in their quality management documentation.

ISO 9001 and ISO/IEC 17025 both require the establishment of a quality policy as part of a quality management system. The quality policy is simply a summary statement of the laboratory's interpretation of its goals for providing a quality service or product, servicing its customers, and ensuring compliance to standards.

In ISO 9001:2015, Clause 5.2.1 states: "Establishing the quality policy—Top management shall establish, implement and maintain a quality policy that:

a) is appropriate to the purpose and context of the organization and supports its strategic direction.

b) provides a framework for setting quality objectives.

c) includes a commitment to satisfy applicable requirements.

d) includes a commitment to continual improvement of the quality management system."[6]

In ISO/IEC 17025:2017, Clause 8.2.1 states: "Laboratory management shall establish, document, and maintain policies and objectives for the fulfilment of the purposes of this document and shall ensure that the policies and objectives are acknowledged and implemented at all levels of the laboratory organization";[7] Clause 8.2.2 specifies that "[t]he policies and objectives shall address the competence, impartiality, and consistent operation of the laboratory."[8]

Employees should be familiar with the management system documents and their contents. Refer to the management system documents whenever necessary instead of relying on memory. Referring to documents during regular operation should be considered an open-book test. Always referring to the current management system documents will reduce the chance someone will have forgotten or misinterpreted a critical instruction, as well as reduce the chance of using an out-of-date version. During assessments and audits, these practices will also demonstrate to any assessors or auditors that the management system documents ensure consistent operations and could even prevent nonconformities written against document control requirements. It is important to the organization for the principles to be consistently interpreted and applied across the personnel levels. Management should provide open communication and refresher trainings on the management system documents on a periodic basis. Management system documents should be updated as necessary to ensure that the system stays current and effective. If these steps are followed, the success of the quality manual and of the management system can be assured.

Figure 8.1 is an example format for an ISO/IEC 17025:2017–based quality manual, with explanation of the headings and sections.

Introduction

Quality Policy Statement

1 Scope. This is the technical scope of the laboratory's operations, that is, what services are provided to clients. If the laboratory is accredited, then the listing will specifically state what services are or are not under the scope of accreditation.

2 References.

3 Terms and Definitions.

4 General Requirements. This would explain how the organization maintains the impartiality of its activities and the confidentiality of its customer's information.

5 Structural Requirements. This is where the organization would define its legal standing and the management structure and responsibilities for its operation.

6 Resource Requirements. This section spells out the people, places, and things that the organization needs to provide adequate services.

6.1 General.

6.2 Personnel. Training plans must exist, and records of training qualifications must be maintained, as well as requirements for personnel supervision, authorization, and the monitoring of their competency.

6.3 Facilities and Environmental Conditions. Environmental conditions must be specified for both in-laboratory and on-site environments. Acceptable requirements may vary widely for some applications but generally will be consistent with requirements outlined by the related test or calibration methods.

6.4 Equipment. Equipment must be fit for use and purpose. Equipment must be calibrated if its measurements affect the validity of reported results.

6.5 Metrological Traceability. Traceability of measurement results is to be to the International System of Units (SI) units through national or international standards, whenever possible.

6.6 Externally Provided Products and Services. Laboratories must use suitable vendors for products and services. Communications between the laboratory and its vendors must be clear and complete, and records of those communications must be retained.

7 Process Requirements.

7.1 Review of Requests, Tenders, and Contracts. All contracts/orders from customers should be considered legal and binding contracts. They also should be clear, concise, and easily understood, providing the customer with satisfactory service. This system should include a formal means for record reviews on a regular basis, and as is done with other parts of a quality system, this review should be documented and saved for future reference. Whenever there is any deviation from a contract or service with a customer, it must be documented, and its record retained.

7.2 Selection, Verification, and Validation of Methods. Most test and calibration methods will come from sources such as standard methods (such as ISO, ASTM, AOAC, USP, etc.), the equipment manufacturer, and available military procedures (for example, USAF 33K, USN 17-20 series). Test or calibration procedures developed by the laboratory must be validated. This typically requires predefined criteria to measure the expected performance of the procedure and a report that describes the observed results of the use of the procedure and specific acceptance by authorized personnel.

Figure 8.1   An example format for an ISO/IEC 17025:2017-based quality manual.

7.3 Sampling. Test laboratories need to have a defined plan and method to sample from a group of test materials or products to ensure validity of results. For calibration labs, sampling documentation may be needed to demonstrate the reasoning behind the selection of calibration measurement points in a calibration method procedure. Sampling of calibrated equipment or products is typically not applicable for calibration laboratories, as most calibrations relate only to the item calibrated.

7.4 Handling of Test or Calibration Items. The laboratory must have procedures and facilities to ensure the integrity of customer test or calibration items.

7.5 Technical Records. This section details what data and records are required to be retained so that any necessary corrections or calculations can be made, and test or calibration results can be verified and traceable to the activity.

7.6 Evaluation of Measurement Uncertainty. No measurement is ever 100 % perfect; there is always measurement uncertainty in every measurement. The contributors to measurement uncertainty must be identified and evaluated.

7.7 Ensuring the Validity of Results. Ensuring validity of results is one way of demonstrating competency and reducing risk. Validity testing must be done using both internal monitoring and comparisons to other laboratories.

7.8 Reporting of Results. This section includes the prescriptive requirements for calibration certificates and test reports.

7.9 Complaints. A system for receiving and acting on complaints is required and also can contribute to improvements in the management system. The company must have a formal complaint system that covers both how complaints can be received and how they are managed, resolved, and documented.

7.10 Nonconforming Work. The company must establish and maintain a policy and procedures that are used whenever any aspect of its laboratory activities or results of that work do not conform to its own procedures or the agreed requirements of the customer. This may include, as examples, instances where equipment or environmental conditions are out of specified limits or when procedures are not consistently followed. Nonconforming work procedures should state that nonconforming product, services, calibrations, or any type of work performed for the customer are identified, segregated, and managed to prevent unintended use. All of this must be performed in writing, according to documented policies, with the affected customer being informed of all actions taken and problem resolution solutions. Records showing how the quality system was improved to prevent further occurrences must also be documented.

7.11 Control of Data and Information Management. This section specifies the necessary data integrity requirements of the organization.

8 Management System Requirements.

8.1 Options. These requirements allow for either a defined ISO/IEC 17025 management system meeting the requirements of Clauses 8.2 to 8.9 or the use of an established and maintained management system that is in accordance with ISO 9001 and meets the requirements of ISO/IEC 17025:2017, Clauses 4 to 7 and 8.2 to 8.9.

8.2 Management System Documentation. This section details how management is to document its policies to support the overall quality and output of its laboratory activities.

8.3 Control of Management System Documentation. This section describes how documentation, both internal and external, is organized and controlled within the organization.

Figure 8.1 *Continued*

8.4 Control of Records. Records that must be controlled include internal audit and management review records. Calibration information, including original handwritten observations, are records and must also be controlled.

8.5 Actions to Address Risks and Opportunities. One of the critical areas specifically defined in ISO/IEC 17025:2017, this section describes how the organization identifies, monitors, and evaluates risks and opportunities to its business activities.

8.6 Improvement. This section could be seen as mandating that a laboratory be proactive in finding solutions to problems and practice continual improvement.

8.7 Corrective Actions. This section details how a laboratory reacts to identified nonconformities. Any nonconformities must also be thoroughly documented, and a system must be in place for controlling corrective actions.

8.8 Internal Audits. Internal audits must be done on a preplanned schedule with defined criteria. Internal audits may also result in corrective actions.

8.9 Management Reviews. Periodic and planned management reviews must be performed. This section defines the minimum criteria to be evaluated and discussed in management reviews.

Figure 8.1 *Continued*

Chapter 9
Metrological Traceability

Traceability refers to the ability to identify, track, or trace something back to its source. We can compare measurement traceability to genealogy, where there is a process for being able to track from children to parents, to grandparents, to great-grandparents, and so on. Documenting a family tree in the study of genealogy is similar to documenting the hierarchy of measurement traceability. However, in the case of measurement traceability, or *metrological traceability* as it is officially defined, we are hoping to trace measurement results to the International System of Units (SI).

In the world of metrology, metrological traceability is often required, essential, or simply requested. It is required as a part of laboratory accreditation when implementing ISO/IEC 17025:2017, *General requirements for the competence of testing and calibration laboratories* (e.g., Clause 6.4 requires calibration of equipment and standards that can affect the measurement results). It is also often required in a regulated environment, and may simply be requested by customers, whether or not they know what it means.

TRACEABILITY DEFINITIONS

First, how is *metrological traceability* defined? Later we will also discuss how it can be implemented and assessed. The term *traceability* does not stand alone in the JCGM 200:2012 (also available as ISO/IEC Guide 99), *International vocabulary of metrology—Basic and general concepts and associated terms* (2012) *(VIM 3).*[1] *VIM 3* Clauses 2.41 (metrological traceability), 2.42 (metrological traceability chain), and 2.43 (metrological traceability to a measurement unit) are the three definitions we will consider here. In addition to the three definitions themselves, all the definitions have interpretive notes. The fact that there are so many interpretive notes associated with these definitions alerts you that there is a complexity to this concept that might not be readily apparent.

Metrological traceability is defined in Clause 2.41 of the *VIM 3* as the "property of a measurement result whereby the result can be related to a reference through a documented unbroken chain of calibrations, each contributing to the measurement uncertainty."[2] Eight interpretive notes are included in this definition! The notes include specific reference to the SI and to having calibration hierarchies in place to document the chain of calibrations. Note 7 of *VIM 3*, Clause 2.41 explicitly lists six essential elements according to the International Laboratory Accreditation

Cooperation (ILAC) policy (ILAC P-10:2002), which was updated in 2020 (ILAC P-10:07/2020).

The next definition of interest in the *VIM 3* is Clause 2.42, *Metrological traceability chain*, which is defined as "traceability chain sequence of measurement standards and calibrations that is used to relate a measurement result to a reference."[3] This definition contains three interpretive notes, one of which talks about hierarchies, and another of which reinforces the reference of traceability to measurement results.

The final definition in the *VIM 3* regarding metrological traceability is Clause 2.43, *Metrological traceability to a measurement unit*, which is defined as "metrological traceability to a unit metrological traceability where the reference is the definition of a measurement unit through its practical realization."[4] There is one note that specifies that the expression "traceability to the SI" means "metrological traceability to a measurement unit of the International System of Units."

So, these definitions and corresponding notes show that metrological traceability is specifically related to measurement results and corresponding measurement uncertainties, documented evidence supporting reference to the SI, and evidence of an unbroken chain of calibrations. These definitions and notes also demonstrate that a documented hierarchy should be in place. Examples of traceability pyramids and hierarchies (Figures 9.2 through 9.5) are provided as examples in this chapter and in the noted references.

In the context of these definitions and interpretive notes, it should also be clear that a laboratory is not traceable (accredited or not). A person is not traceable. A calibration or test report and assigned test number is not traceable. What *may* be traceable is the measurement result and uncertainty, provided that all the essential elements are in place without any breaks in the chain and without any breaks in the evidence supporting the calibration or test. Essential elements that might be considered or are recommended are listed in Table 9.2.

NATIONAL METROLOGY INSTITUTES AS A SOURCE OF TRACEABILITY

Once upon a time, nearly all high-level calibrations within a country were obtained by or referenced to the National Metrology Institute (NMI) designated within that country (e.g., the National Institute of Standards and Technology [NIST] in the United States, the Centro Nacional de Metrologia [CENAM] in Mexico, the National Research Council [NRC] in Canada, the National Physical Laboratory [NPL] in the United Kingdom, the Physikalisch-Technische Bundesanstalt [PTB] in Germany, National Metrology Institute of Japan [NMIJ], and so on). In the past, many laboratories and auditors alike wanted a "test number" from the NMI as evidence of traceability. Keep in mind that "test numbers," even those issued from an NMI, may contain measurement results and uncertainties, but such test numbers are not complete evidence to support the definitions and assessment requirements we have already noted. In fact, the NIST *Policy on Traceability*[5] discourages the use of "NIST test numbers" as evidence of traceability even though test numbers were something that auditors often (incorrectly) looked for.

NMIs are still the highest level of traceability for most countries. In some cases, other Designated Institutes (DI) might maintain the highest level of standards in a country. In most cases, NMIs realize the definitions of the SI as set forth by the Bureau International des Poids et Mesures (BIPM) and adopted by the International Committee of Weights and Measures (CIPM). The BIPM is responsible for coordinating the worldwide measurement system to ensure comparable and internationally accepted measurement results, as well as for evaluating equivalency among NMIs. Upon realization of the defined standards set forth by the BIPM, NMIs disseminate measurements to reference calibration laboratories.

As an alternative to obtaining calibrations from NMIs, calibrations can also be obtained from laboratories that people think are generally reliable and suitable, such as manufacturers and laboratories with well-known technical experts. However, the transitions to a global marketplace and implementation of laboratory accreditation changed everything regarding what was formerly confidently accepted from calibration providers.

INTERNATIONAL AGREEMENTS AND ARRANGEMENTS: RECOMMENDED CALIBRATION PROVIDERS

So, how is the concept of metrological traceability implemented and evaluated around the world now? Where can a laboratory obtain suitable calibrations and have confidence in obtaining adequate evidence of metrological traceability? Two international agreements/arrangements are generally referenced when discussing calibration options in the world of metrology. These agreements ("arrangements") are the International Committee of Weights and Measures (CIPM) Mutual Recognition Arrangement (MRA) and the International Laboratory Accreditation Cooperation (ILAC) Mutual Recognition Arrangement (MRA), often simply designated as the CIPM MRA and ILAC MRA.

The CIPM MRA is an arrangement at the level of the National Metrology Institutes. The ILAC MRA is an arrangement in the international accreditation system for testing and calibration laboratories and reference material providers.

The CIPM MRA is a "Mutual Recognition Arrangement of national measurement standards and of calibration and measurement certificates issued by National Metrology Institutes" and "is the framework through which NMIs demonstrate the international equivalence of their measurement standards and the calibration and measurement certificates they issue. Furthermore, it addresses the dissemination of the SI units through calibration and the provision of reference materials."[6] NMIs can sign the CIPM MRA and submit their Calibration and Measurement Capabilities (CMC) to the BIPM, which are then reviewed and if approved posted online in a database known as the Key Comparison Database (BIPM KCDB). As of 2021, more than 250 organizations from around the world were signatories to the CIPM MRA.[7]

The ILAC MRA "provides significant technical underpinning to the calibration, testing, medical testing and inspection results, provision of proficiency testing programs and production of the reference materials of the accredited conformity assessment bodies that in turn delivers confidence in the acceptance of services and results."[8] Accreditation Body (AB) signatories to the ILAC MRA agree to implement the policies of ILAC and then accredit calibration and testing laboratories and reference material providers. As of 2021, more than 100 accreditation bodies were signatories to the ILAC MRA.[9]

Table 9.1 Recommended sources of calibration.

Source	Status	
National Metrology Institute (NMI)	1. Signatory to the CIPM MRA with scope published in the BIPM key comparison database (KCDB).[10]	3a. Not a signatory or scope not published for the measurements, but suitable for use.
Calibration or testing laboratory	2. Accredited by an Accreditation Body (AB) that is a signatory to the ILAC Mutual Recognition Arrangement (ILAC MRA).[11]	3b. Laboratory not accredited or measurement results outside the scope of accreditation; measurements are suitable for use.

Table 9.1 summarizes the recommendations given in both the *Joint BIPM, OIML, ILAC, and ISO Declaration on Metrological Traceability*[12] and the *ILAC Policy on Metrological Traceability of Measurement Results* (P10:07/2020)[13] as implemented by Accreditation Body (AB) signatories to the ILAC MRA regarding calibration sources.

These changes in the world of measurement now mean that a laboratory can obtain "traceable" calibrations from any NMI as identified in Table 9.1, category 1, not just from those in their own country. Laboratories can also obtain a "traceable" calibration from any accredited laboratory in category 2 that has a suitable Calibration and Measurement Capability. In both cases (categories 1 and 2 in Table 9.1), the peer review in the case of NMIs or assessment process by accreditation bodies has already ensured that metrological traceability evidence is available to support reported measurement results and associated uncertainties. ISO/IEC 17025:2017, Appendix A, A.3 also references category 1 and 2 laboratories as suitable for establishing metrological traceability. Keep in mind that the laboratory providing a calibration service is responsible for working with the customer to provide relevant evidence supporting each calibration (that is, ILAC or the Accreditation Body are not responsible for providing evidence of traceability). Figure 9.1 illustrates traceability pathways to the SI and NMIs and on to accredited laboratories (and among each other) in the context of the CIPM MRA and the ILAC MRA. This figure also illustrates that "traceable to the SI" is

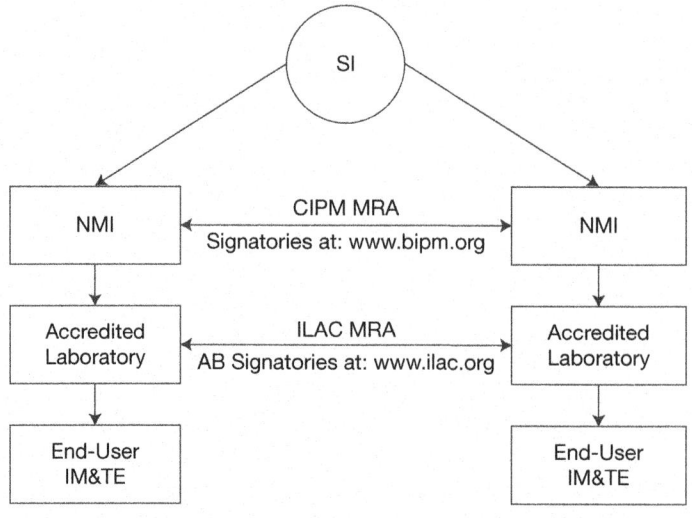

Figure 9.1 Traceability pathways to the SI.

the correct reference (i.e., not "NIST-traceable" or other "NMI-traceable" phrases or identifications, which are still often observed in marketing materials or even on calibration certificates).

EVALUATING METROLOGICAL TRACEABILITY

What about laboratories noted in categories 3a and 3b in Table 9.1? Can these laboratories provide valid metrological traceability? The answer is "it depends; evidence must be evaluated further." Mechanisms are in place to use calibration providers in categories 3a and 3b when services are not otherwise available. How can a laboratory simplify the recommendations or essential elements to consider 1) what must be implemented to ensure suitable evidence of metrological traceability (e.g., they want to ensure traceability but do not maintain accreditation), 2) identify what assessors should look for in laboratories that are not accredited (whether an Accreditation Body for a new applicant or a supplier evaluation process), and 3) consider what a laboratory should assess in NMIs that are not signatories to the CIPM MRA? Items on the list in Table 9.2 are considered as guidelines or as essential elements of metrological traceability by various parties such as BIPM, OIML, ILAC, and ISO. National laboratories also might have guidelines, like those of the NIST Office of Weights and Measures, which publishes Good Measurement Practices to support legal metrology in the United States where U.S. state laws require traceability to the SI through NIST. References to each item in the essential elements list in Table 9.2 note the source organization and applicable reference documents.

Table 9.2 Essential elements of metrological traceability.

Essential Elements	Description	Reference Documents
Realization of and reference to SI units	The measurand(s) must be defined. The primary national, international, or intrinsic standards must be primary standards for the realization of the SI. See the BIPM International System of Units brochure, NIST *Special Publication 330* (2019), and Chapter 11 in this handbook. Direct realization of the SI units may also be used by laboratories, provided they ensure direct or indirect comparison with national or international standards.	• Defined in *VIM 3*, Clause 2.43 • *Joint BIPM, OIML, ILAC, and ISO Declaration on Metrological Traceability*, Recommendations • *ISO/IEC 17025:2017*, Clause 6.5.2 c) and Annex A.2.1 a) • *NISTIR 6969 Good Measurement Practice (GMP) GMP 13*, 1.5.1
Unbroken chain of comparisons (documented hierarchy or pyramid)	The laboratory must maintain a documented system of comparisons with each step having the essential elements of metrological traceability going back to a standard acceptable to the parties (usually a national or international standard).	• Defined in *VIM 3*, Clause 2.41 • *ILAC P10:07/2020*, Appendix A, bullets 3 and 6 • *ISO/IEC 17025:2017*, Clause 6.5 and Annex A.2.1 b) • *NISTIR 6969 GMP 13*, Clause 1.5.2; example hierarchies illustrate unbroken chains to the SI

Continued

Table 9.2 *Continued*

Essential Elements	Description	Reference Documents
Documented calibration program	Calibrations of equipment and standards (where appropriate) must be repeated at established and appropriate intervals to preserve metrological traceability of the standard over time, related to uncertainty required, stability, frequency of use, stability of equipment, etc. Complete and up-to-date calibration certificates must be available for all equipment and standards (even for internal calibrations performed by the laboratory). See additional guidance for calibration programs in Chapter 10.	• *ILAC P10:07/2020*, Appendix A, bullets 3 and 6 • *ISO/IEC 17025:2017*, 6.4.7 and Annex A.2.1, b) • *NISTIR 6969 GMP 13*, 1.5.3 • *NISTIR 6969 GMP 11, Calibration Intervals*
Documented measurement uncertainty	The measurement uncertainty for each measurement in the traceability hierarchy must be calculated according to defined methods and must be stated so that an overall uncertainty for the whole chain may be calculated or estimated. Measurement uncertainty must/should (depending on evaluation) follow the principles established in the *Evaluation of measurement data—Guide to the expression of uncertainty in measurement (GUM)*.	• Defined in *VIM 3*, Clause 2.41. • *Joint BIPM, OIML, ILAC, and ISO Declaration on Metrological Traceability*, Recommendations • *ILAC P10:07/2020*, Appendix A, bullet 2 • *ISO/IEC 17025:2017*, 7.6 and Annex A.2.1, c) • *NISTIR 6969 GMP 13*, 1.5.4
Documented and validated measurement procedures	Each step in the chain must be performed according to documented and validated procedures, with measurement results recorded, and the reported measurement results with expanded uncertainties must be documented (e.g., in a calibration certificate).	• *ILAC P10:07/2020*, Appendix A, bullet 1 • *ISO/IEC 17025:2017*, 7.2 and Annex A, A.2.1, d) • *NISTIR 6969 GMP 13*, 1.5.5
Technical competence	The laboratories or bodies performing one or more steps in the chain must supply evidence of technical competence (e.g., by having personnel with technical expertise, maintaining appropriate training records, participating in interlaboratory comparisons). Laboratory accreditation is often accepted as demonstration of competence.	• *ILAC P10:07/2020*, Appendix A, bullet 5 • *ISO/IEC 17025:2017*, 6.2 and Annex A, A.2.1, e) • *NISTIR 6969 GMP 13*, 1.5.6

Continued

Table 9.2 *Continued*

Essential Elements	Description	Reference Documents
Measurement validity (measurement assurance)	A proper measurement assurance program must be established to ensure the validity of the measurement process and the accuracy of equipment and standards used in calibration or testing at the time of the measurement.	• *ILAC P10:07/2020*, Appendix A, bullet 4 • *ISO/IEC 17025:2017*, 7.7 • *NISTIR 6969 GMP 13*, 1.5.7
Facility and environmental conditions	There must be documentation and records of the facilities and environmental conditions that can influence measurement results.	• *ILAC P10:07/2020*, Appendix A, bullet 7 • *ISO/IEC 17025:2017*, 6.3
Audits	Audits of the calibration laboratory (internal and third party).	• *ILAC P10:07/2020*, Appendix A, bullet 8 • *ISO/IEC 17025:2017*, 6.6, 8.8

DOCUMENTING CALIBRATION HIERARCHIES

What might an unbroken chain, or series of calibrations related all the way to the SI, look like? Figures 9.2 through 9.5 show examples of pyramids and hierarchies from the SI to ending measuring equipment (IM&TE) or measurement standard. The figures illustrate the unbroken chain through a series of laboratories (Figure 9.2 and Figure 9.4) and through a series of measurement standards (Figure 9.3 and Figure 9.5). These examples reflect the fact that pyramids or hierarchy charts are both ways commonly used to represent an unbroken chain of calibrations.

As we have discussed, a laboratory does not have to send its standards to an NMI for calibrations, but it must be able to provide objective evidence that an unbroken chain of calibrations to the SI exists (e.g., components listed in Table 9.2 must be in place). In documenting the hierarchies in this or similar

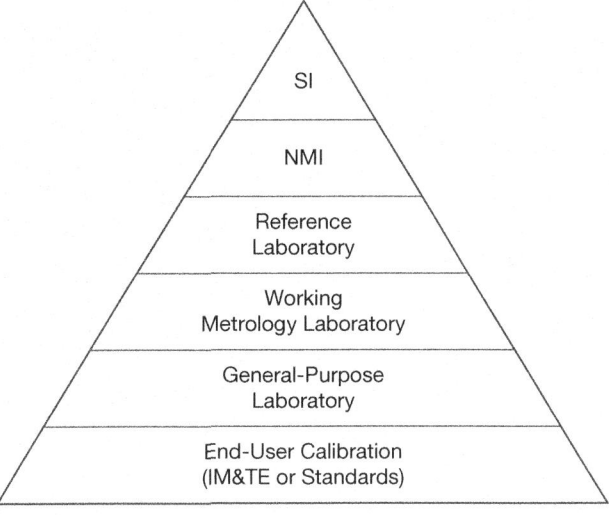

Figure 9.2 Lab pyramid.

70 Part II: Quality Systems

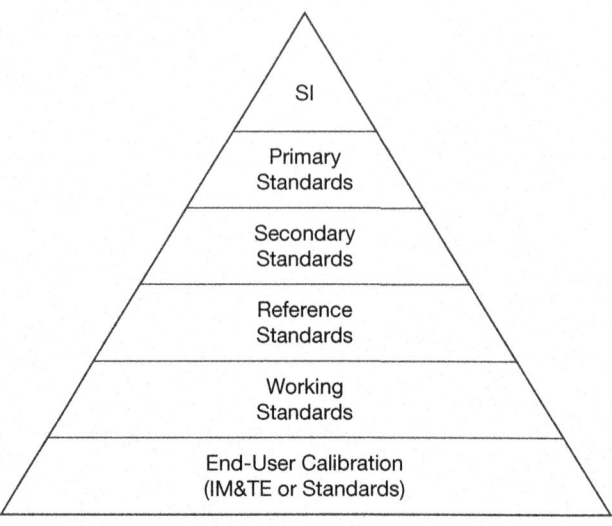

Figure 9.3 Standards pyramid.

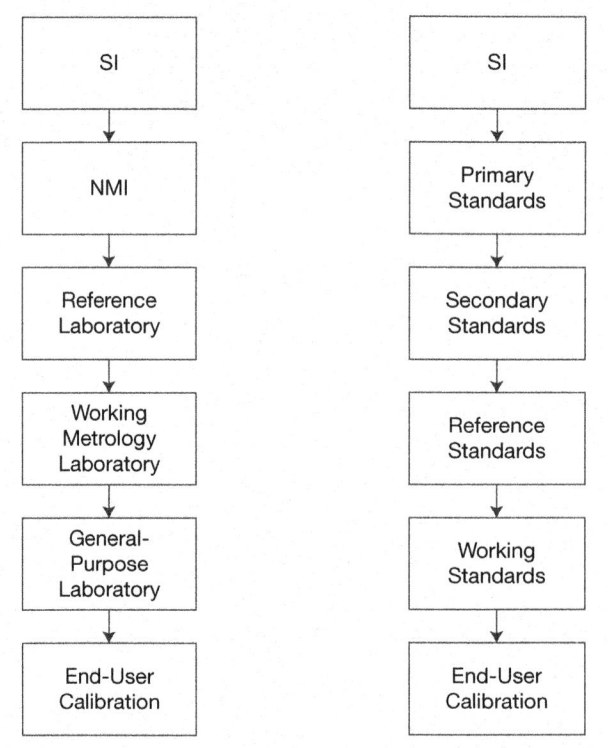

Figure 9.4 Lab hierarchy.　　**Figure 9.5** Standards hierarchy.

ways, a laboratory's customers can also show they have traceability of their measurement results and uncertainties to a national or international standard through their calibrations and those of any outside vendor whose services they may be using.

It is usually not feasible or desirable to have all equipment and standards calibrated by an NMI. Doing so would significantly increase costs while not giving the return on investment that would keep a business profitable. Also, it would be a burden on the NMI to service all its customers, as there would be such unrealistically small uncertainties that this would become an unnecessary and futile effort. Most IM&TE can be calibrated against a company's working standards and still maintain the desired traceability, and accuracy required, with sufficiently small uncertainties. In these cases, working standards may also be calibrated against internal reference standards or sent to an outside vendor that can provide traceable calibration to the SI.

Within an organization, one may have as many as two or three levels of standards in a documented hierarchy, such as those shown in Figures 9.3 and 9.5. Using the correct reference standard or working standard to perform a downstream calibration is critical. If one is using the highest-level standard with extremely small uncertainties to calibrate IM&TE that has a large resolution, one could be wasting valuable resources to achieve only a small cost benefit. For example, using a $5000 temperature standard that has a calibration uncertainty of ± 0.010 °C to calibrate a $25 temperature device that has a resolution of 1 °C could be a waste of time and money. In contrast, using an $80 thermometer that has been calibrated with traceable measurement results and uncertainties at the level of ± 0.1 °C might maintain adequate accuracy and traceability while maximizing resources and keeping cost to a minimum. In addition to considering measurement adequacy and costs, the proper care and handling of standards at each level in a hierarchy may be (and often is) different. In the case of mass standards at the reference standard levels of a calibration hierarchy, the use of gloves and/or forceps to prevent contamination (adding mass) of the mass standards is essential to ensure that mass values and uncertainties remain stable. However, at the lowest level in a mass hierarchy, such as in the case of cast-iron weights used to calibrate vehicle or rubbish scales, the use of gloves may not be needed other than to keep one's hands clean!

CIRCULAR CALIBRATIONS

Here is a cautionary note about a concept called *circular calibration*. This situation could arise when a particular standard is used to calibrate another unit, then that other unit is in turn used to calibrate the first standard. This can also occur if laboratories obtain calibrations from other (often accredited) laboratories, and they all get calibrations from each other with a weak link or even no link to the SI. Figure 9.6 illustrates this concept.

Circular calibrations do not provide an unbroken chain of calibrations to the SI because there is no link to a higher standard to show traceability. In this case, evidence of traceability does not exist for this standard.

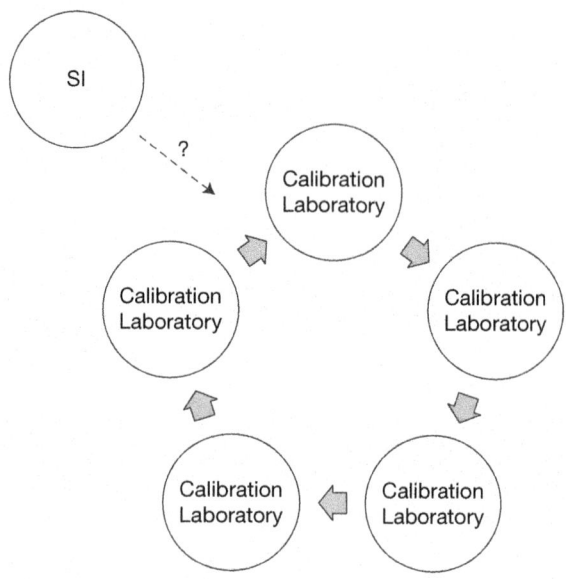

Figure 9.6 Circular calibration.

REVERSE TRACEABILITY

Another term that is often used is *reverse traceability*. Reverse traceability refers to working back through the traceability chain to identify all standards or equipment that either caused or were affected by nonconforming work.

If the standard being used to calibrate IM&TE is found to be nonconforming with documentary standards (e.g., outside of maximum permissible errors or tolerances), a determination must be made to account for all the IM&TE which used that particular standard between the time it was calibrated and the time it was found to be nonconforming. Some calibration software programs have reverse traceability reports built into them. Others do not. It can be very labor- and time-intensive to research historical records to ascertain what items were calibrated using a particular standard, but this is exactly what must occur when standards are found to be nonconforming. An assessment of how errors might affect product, processes, or manufacturing must be conducted, and a determination made as to recall of product or stoppage of production. If a reverse traceability program is available for use, it can save valuable time and resources.

Once the items that were calibrated by the nonconforming standard are identified, the impact is assessed: If the impact is significant, standards must be isolated to preclude any other or further usage. An impact assessment of nonconformance must be made on both the IM&TE and the product or process that may have been used during any manufacturing processes or procedures. Was the customer notified in writing? Was a product recall required or initiated? Was this accomplished because of written procedures or because it made sense? Was a root cause analysis done to find the reason for all the extra work and expense? Many questions must be answered when one of the company's standards is found to be nonconforming. There should be written procedures to address each of these situations, with documentation and records proving what was found, what was done, and who the approving authority was in each case. This is all part of a good quality system and cannot be overlooked. This evaluation of nonconforming work and potential recalls is also a requirement for laboratories accredited to or compliant with ISO/IEC 17025:2017 (e.g., Clauses 6.4.9 and 7.10).

TOOLS FOR ASSESSING TRACEABILITY

The list of essential elements in Table 9.2 provides a handy reference that can be used when assessing calibrations within your own laboratory or when performing supplier evaluations. In addition to obtaining evidence of accreditation and evaluating the CMC for available measurements and associated uncertainties, this list may give you some ideas of questions you might want to ask a calibration laboratory supplying calibration or test results. Additional sample auditing forms, which also may be used, are available in the appendices to NIST Good Measurement Practice 13.[14]

THE IMPORTANCE OF TRACEABILITY

Why is this unbroken chain of calibrations to the SI so important? Does it really benefit anyone or anything? Consider these examples and keep Figure 9.1 in mind while thinking about international measurement equivalency among NMIs and/or accredited laboratories. An aircraft loads its passengers and takes off from one country. It flies to its destination in another country. Service personnel find that the aircraft requires repairs and has to have some parts replaced before it can return home. If there is no traceability of the equipment and standards to equivalent measurements, the parts may not fit if manufactured by a different vendor. By having everything calibrated to a traceable standard, it makes no difference where the parts are manufactured, shipped, or replaced. The parts and measuring equipment used to verify installation will work if they meet the design specifications and tolerances. For example, torque wrenches used to secure the replacement parts require calibration, too. If the parts are not torqued to the correct specification, components could come loose in flight, either after initial installation or after replacement.

Here is another example that happens every day throughout the world. Scientists are looking for the cure for cancer. They experiment with new molecules and compounds to find the one magic bullet that will save millions of lives and make cancer a thing of the past. They use certain types of equipment and standards every day in their experiments. Pipettes, balances, centrifuges, autoclaves, spectrophotometers, and water baths are but a few of the IM&TE used by biotechnology and pharmaceutical companies to aid in the discovery of cures and manufacture their products. Without traceable calibrations, they might not be able to reproduce their results, either in the research and development environment or on the production line. One milliliter of product must meet the same exacting tolerances no matter where in the world it is made or used. Using traceably calibrated IM&TE is one means of getting repeatable, reliable measurement results.

It is important to review two of the important characteristics in the definition of *traceability*: 1) it is a property of the measurement result and 2) all associated calibrations must have stated uncertainties. Traceability applies to the measured value and its uncertainty, as a single entity. One without the other does not meet the definition of metrological traceability. This also means that only the *results* of a calibration may have the property of traceability. In the metrological sense, traceability never applies to a report document, a calibration procedure, a piece of equipment, or a laboratory. It also means that traceability is not assured by having a supplier's test report number on your calibration certificate. That may have some meaning for a purchase order, but it is meaningless in supporting metrological traceability.

Chapter 10
Calibration Programs

Just because your equipment (IM&TE) or standards were calibrated this morning does not mean they will be accurate or traceable this afternoon, or even next year! Many things can go wrong that affect IM&TE and/or standards. It could be something as simple as equipment being damaged in transport from one laboratory to another or something more catastrophic, such as IM&TE being knocked off a bench, or even something more complex, such as a calculation error in the procedure or software that was used to perform the calibration. Timelines[1] must also be considered when evaluating suitable calibration intervals for IM&TE and standards. One item might remain stable for three to five or even ten years, whereas another might only provide reliable data for six months. Another item might be stable sitting on a shelf for six months but drift the first time it is used. Knowledge of the operating characteristics of IM&TE and standards is essential for ensuring ongoing stability and validity of a laboratory's test and measurement results. Evaluating before and after calibration values helps ensure that damage has not occurred during handling and shipping of the equipment or standards and also helps ensure that evidence supporting ongoing metrological traceability is available. All the components that are part of ensuring ongoing stability and validity of the IM&TE and standards could be considered part of a rigorous calibration program.

ISO/IEC 17025:2017 states, in 6.4.7, that "the laboratory shall establish a calibration programme, which shall be reviewed and adjusted as necessary to maintain confidence in the status of calibration."[2] But a "calibration program" is not actually defined in the context of this current standard. In the *International Vocabulary of Metrology* (2012), the definition of metrological traceability (2.41), Note 7, specifies a reference to the International Laboratory Accreditation Cooperation (ILAC) policy on traceability[3] (ILAC P-10:2002). ILAC P-10 has since been updated, but ILAC P-10:2002 specified that one of the essential elements of traceability was a requirement for suitable calibration intervals.[4] However, documented and suitable calibration intervals are only one part of having a defensible calibration program. In the ISO/IEC 17025:2005 version (adopted without change in 2010), Clause 5.6.1 specified that "the laboratory shall have an established program and procedure for the calibration of its equipment"[5] and included a note to the effect that "such a program should include a system for selecting, using, calibrating, checking, controlling and maintaining measurement standards, reference materials used as measurement standards, and measuring and test equipment used to perform tests and calibrations."[6]

So, what is a program and what are the components of a program? In a standard definition, *program* as a noun might be defined as "a set of related measures or activities with a particular long-term aim"[7] and as a verb might be defined as "arranged according to a plan or schedule."[8] Both definitions are important for application to a calibration program. In the context of a calibration or testing laboratory, a long-term aim might be to ensure accurate, traceable, valid measurement or test results and have uncertainties reported to the end user. In the context of ISO/IEC 70125:2017, the purpose of the calibration program is to "maintain confidence in the status of calibration."[9] Given the flexibility in documentation in the latest ISO/IEC 17025 standard, one might also ask: Does the calibration program really have to be documented? However, in the context of a high-quality calibration laboratory, one might find that the existence of documentation helps monitor and ensure consistent applications in all aspects of the calibrations and the reported calibration or test results.

So then, what are the key components of a calibration program? A group of experienced metrologists[10] were asked this question in 2021: "What do you think are the top two or three requirements of a calibration program?" The responses included significantly more than setting appropriate calibration intervals, ensuring that calibrations are performed as scheduled, and seeing that intervals are adjusted as appropriate. A second question was asked: "What are the top risks related to your laboratory calibration program?" In the context of a risk-based quality system, identifying what can go wrong in a calibration program might be just as important as a list of components. Table 10.1 provides a consolidated summary of the responses submitted to the question of calibration program requirements with descriptive instructions. Considering the absence of these components might be another way to review risks to ensuring the validity of measurement results.

Given this insight as to what is important in a calibration program, most laboratories will continue focusing on setting and adjusting calibration intervals as the primary component of implementing a suitable calibration program. The remainder of this chapter will focus on approaches to managing calibration intervals.

IM&TE and standards that make a quantitative measurement require calibration. Most people will not argue with this statement, but how often do equipment and standards have to be calibrated? Is the interval between calibrations adjustable, or, once it is set, is it written in stone? Who determines the various calibration intervals for diverse types of IM&TE and for different types of companies? Let us begin the discussion by exploring why calibration intervals are required, which should help to answer how often calibration should be done; that is, what is the calibration interval?

The dependability of any IM&TE has little to do with its accuracy, measurement uncertainty, or the number of external knobs. The type of equipment, how often an item is used, its calibration history, the type of environment where it is used, and to what extent the user expects it to repeat measurements all play a major role in determining how often that piece of IM&TE requires calibration. For example, in the world of aircraft, the more often a plane is flown, the more reliable it becomes. This is because some systems stay reliable if they are regularly used. Other parts are rebuilt or replaced at defined intervals and the whole airframe

Table 10.1 Calibration program components.

Component	Component Description
1	Identify laboratory Calibration and Measurement Capability (CMC), with a scope that includes measurement parameters, ranges, and uncertainties; maintain a suitable inventory of IM&TE, standards, and associated calibration certificates to provide applicable calibrations or tests; and update CMC when appropriate. Also consider the internal scope that is needed for calibrations and tests within the laboratory as well.
2	Ensure that staff are competent through training, proficiency testing, and ongoing monitoring of competency in providing calibrations; document all training and monitoring of staff competency.
3	Ensure that suitable calibration intervals are planned and documented; determine an established baseline and actually do the monitoring; update if/as needed.
4	Schedule calibrations with supplier(s)—even if that is within your own laboratory; this will likely require evaluating your own workload and availability of standards.
5	Document and follow shipping, handling, use, storage, and maintenance procedure(s) to prevent damage. Identify and determine impact on previous measurements if damage occurs.
6	Schedule internal audits (in this case, specifically to assess "traceability" and the "calibration program"). Conduct assessments. Document observations from all steps in the calibration program.
7	Implement a procedure for calibration supplier selection and perform complete supplier evaluation (including maintaining and monitoring history and evaluating and saving supplier CMC prior to use).
8	Request budget approvals and process financial requests from management and laboratory administration.
9	Conduct contract review discussions with suppliers (including discussion and agreement of procedures to be used, uncertainties to be provided, decision rules and specification evaluations that will be conducted). Expect and plan for this step.
10	Evaluate returned calibrations and certificates; evaluate "calibration stickers" or "due dates" if present (request action from providers if due dates were not requested and the certificate(s) and/or labels include due dates).
11	Update supplier evaluation history; provide customer feedback to the supplier.
12	Document any corrective or preventive action taken based on the evaluation of returning IM&TE or standards; document observations and guidance for future use.
13	Update laboratory reference documents, software, and records (hierarchies, inventories, uncertainties, observations, corrective actions). File and retain new calibration certificates.
14	Conduct statistical evaluation and adjust intervals (if needed) using data and following documented procedures. This means adjustment of intervals must be a defendable and documented technical assessment, not a financial decision. Financial-based decisions about calibration adjustment intervals are a risk to impartiality.

is regularly inspected and maintained. (An equivalent operation for your family automobile would be to have it stripped of all removable items, have all those items inspected and repaired or replaced as needed, have the body and frame inspected and repaired, and everything put back together—every year.) With proper maintenance a modern aircraft will last indefinitely. If it sits on a ramp or in a hangar for extended periods of time, its components will have a higher rate of failure than those used on a regular basis. In some instances, this could be true for IM&TE; many perform more reliably with regular use. In other areas, state-of-the-art units seem to work better with less usage. There are also differences due to the usage environment. Two companies may have equal numbers of the same model of IM&TE but have different calibration intervals because one company's use is all indoors, and the other company uses theirs for field service in all kinds of weather. Therefore, most companies set their own calibration intervals using input from the IM&TE's manufacturer, historical data collected over several calibration cycles, their own risk appetite, and/or data retrieved from outside sources that are using the same or similar types of IM&TE. There are software solutions available that can reduce the number-crunching required to analyze data. Other companies find simple methods more conducive to their applications. Whether purchased software or user-developed methods are used, there can be no doubt that calibration intervals are an important part of any calibration program. Let us examine a simple method used by a newly implemented, fictitious metrology department.

The initial calibration interval for most IM&TE at the Acme Widget Company was set at 12 months. Most manufacturers recommend this for a couple of reasons. If their equipment will not hold calibration for at least a year under normal operating conditions, they likely will not stay in business for very long. Second, IM&TE may be sent back to the manufacturer to be calibrated, generating income and after-sale service. It is to the manufacturer's benefit to have the shortest interval while ensuring that its equipment continues to function properly between calibrations. The Acme Widget Company was monitoring how often each calibration performed met or did not meet specifications. No matter how far out of tolerance the IM&TE might be, the out-of-tolerance condition was recorded. Over a set time, a year or 18 months, the total number of calibrations for that particular type of IM&TE was tabulated, including the number of times the same type of IM&TE did not pass specification. The pass rate for that type of equipment was tallied and recorded.

Pass rate = (number of times passed calibration − number of times failed calibration)/total number of calibrations

This same exercise was conducted on each general type of IM&TE. When completed, the pass rates varied between 92.9 % and 100 %. For items with a pass rate greater than 98 %, their intervals were doubled. For items with a pass rate between 95 % and 98 %, their intervals were increased by 50 %. Items with pass rates less than 95 % were examined to see if their intervals should be reduced or monitored for another round of calibrations. There were instances where items exceeded 95 % or 98 % pass rates, but their calibration intervals were not lengthened per the formula. Those items were being used in critical areas of production, on a more frequent basis, or a combination of both. In these instances, it was determined to be more prudent and exercise caution rather than run the risk of having to recall product or remanufacture goods. This type of risk is known as *risk appetite* or *risk tolerance*, depending on applicable definition. Each company must make hard choices with a critical eye to the cost of doing unnecessary calibrations (too frequently for how the items are used) versus calibrating at extended intervals and running the risk of producing bad measurement results or bad products and spoiling its reputation to save a few dollars. By performing a thorough analysis of calibration intervals on a regular basis, an organization can get the most value from its calibration program while reducing the risk associated with lengthening intervals only for the sake of saving time and money.

Several publications are available on the topic of setting and adjusting calibration intervals. A joint guide from ILAC (ILAC-G24) and the International Organization for Legal Metrology (OIML), *D 10, Guidelines for the determination of calibration intervals of measuring instruments* (2007 edition) provides a list of recommendations for how initial calibration intervals are often set.[11] (Note that at the time of the writing of this handbook edition, ILAC G24 and OIML D 10 were being reviewed for revision. Visit www.ilac.org or www.oiml.org for the current versions of these and other guides and standards.) This joint guide also overlaps with recommendations in the *NISTIR 6969-2019, GMP 11: Good Measurement Practice 11* for initial calibration intervals.[12] A consolidated list for setting initial calibration intervals and approaches to adjusting calibration intervals might look like the following, which an experienced technical staff member should consider:

- Calibration history;
- Adjustment of (or change in) the individual instruments;
- Expected extent and severity of use;
- Influence of the environment;
- The required uncertainty in measurement;
- Maximum permissible errors or tolerances;
- Measurement assurance data;
- Interlaboratory comparisons;
- Data for the population of similar standards, equipment, or technologies;
- NMI recommendations;
- Statistical analysis methods with pooled or published data about the same or similar devices; and
- Manufacturer's recommendations.

Because there is no single best practice for adjusting intervals, and there are several interval analysis methodologies, the first Recommended Practice (RP) developed by NCSL International was *Recommended Practice-1* (RP-1), *Establishment and Adjustment of Calibration Intervals,* currently in its fourth edition (published 2010, copyright 2017).[13] Methods are categorized by their effectiveness, cost to implement, their suitability for large or small inventories, and other factors. One of the factors is the renewal, or adjustment, policy implemented by the calibration laboratory:

- **Renew always.** An equipment management policy or practice in which IM&TE parameters are adjusted or otherwise optimized (where possible) at every calibration.

- **Renew-if-failed.** An equipment management policy or practice in which IM&TE parameters are adjusted or otherwise optimized (if possible) only if found to be out of tolerance during calibration.

- **Renew-as-needed.** An equipment management policy or practice in which IM&TE parameters are adjusted or otherwise optimized (if necessary) if found to be outside safe adjustment limits.

The methods listed in RP-1 fall into two broad groups: those that use statistical tests and those that do not. Each has its benefits and drawbacks. The nonstatistical approaches are low-cost and can be easily implemented. However, the nonstatistical approaches are also slow to settle to a stable calibration interval. It is possible for an instrument to be on a different interval after each calibration, which can wreak havoc on the laboratory work scheduling. The statistical approaches can achieve a good calibration interval very quickly. However, they require a lot of historical data and range from moderate to expensive to implement. If an organization has a large inventory with large numbers of equivalent items (such as 300 of a single model of meter), then a statistical approach may be useful. A small inventory of many different models with only a few units of each may require time periods longer than the average life of the equipment for sufficient data to be collected for a statistically valid test.

A side topic of calibration intervals is how the next due date is determined within the calibration interval program. Some calculate the next due date by day, month, and year from the day the equipment was last calibrated. Others simply use only the month and year, with the due date falling on the last (business) day of the month. The following are some pros and cons for both options.

Using only the month and year allows organizations more flexibility as to when they can schedule the next calibrations. This option also helps reduce the number of overdue calibrations that show up during audits and inspections. Any item due for calibration during the month will not show up as overdue until the following month. Some laboratories have incorporated a policy that the lab has up to x days after the end-of-month due date to calibrate before the device is considered past due. If historical data support the reliability of the IM&TE using this system, it can also make a calibration function look better to upper management during audits. The downside to this system is encountered when working with IM&TE that has a short calibration interval of only one or two months. As an example, assume an item has a two-month calibration interval. It was calibrated on February 2, making it due for calibration during the month of April. If it is next calibrated on April 30, it will be almost three months between calibrations. If it was calibrated on February 27 and next calibrated on April 1, it was barely a month between calibrations. This is one of the problems encountered using the month/year method of stating the next calibration date.

When using the day/month/year system, the exact interval is applied to each IM&TE, which allows users to know exactly when they can no longer use it; at midnight of the calibration due date, it is considered out of calibration. This eliminates guessing or wondering on the part of the user or calibration staff. Most software packages used in calibration systems calculate the next due date according to these conventions, and consistency throughout the industry reduces the need for training on this topic as technicians move from company to company. If a company is using only month and year on their calibration labels, however, it might better suit its needs to calibrate all pieces of IM&TE while a production line is down for maintenance or repairs. That way, nothing becomes overdue during that month, no matter on which day of the month it was calibrated.

Another factor to consider is that time between calibrations is data used by interval analysis methods. When grouping items for statistical analysis, one of the key criteria is that they all must be on the same calibration interval, and all should have been on that interval long enough to have undergone at least two calibrations each. Historically, many organizations have recorded intervals in a variety of units—usually weeks, months, or years—that vary by individual item. It will be easier to manage interval analysis and adjustment if the same units (days, weeks, months, or years) are used across all items in the calibration program. It does not matter which unit is used if it results in whole numbers and is applied consistently for all items. Many organizations use months as the units; others use weeks. Some use years for everything, but that is practical only if all items are exceptionally durable and highly reliable and the laboratory's risk appetite is low.

Finally, the calibration procedure used is also a key factor in interval analysis. A requirement of statistical analysis methods is that all items in the group must have been calibrated using the same procedure. If a calibration procedure is changed (where the change affects the measurements), only the data collected before or after the change may be used in the analysis.

While on the topic of calibration intervals, it should be mentioned that most organizations use a software program, as opposed to a paper system, to trigger an effective reminder/recall for recalibration of their IM&TE and maintain the calibration history using that program. If you are still using a paper-based recall system, it is possible that many time-saving functions are being overlooked, and the chances of items falling through the cracks are significantly higher.

A laboratory—particularly an in-house laboratory—will often get pressure from customers to extend a calibration interval, usually for the customers' convenience. Extending calibration intervals for financial reasons or for convenience alone is often not technically valid, is a risk to impartiality, and should be vigorously resisted. A calibration interval should only be increased based on data showing that the reliability of the instruments supports its increase. It is much easier, and more acceptable, to shorten an interval based on limited data. For example, suppose that a company has a goal of 95 % reliability; that is, 95 % of all items should be in tolerance when calibrated. If a particular model of meters has failed five of the last ten calibrations, then it is clearly indicated to reduce the interval (or replace the equipment). An increase in the interval, however, could be justified only after the next 100 calibrations have been performed with no failures. (In this example, the interval was reduced, *and* it was suggested to the customer that it seek out a model with higher reliability for their needs.)

DELAY DATING

Another subset of interval analysis is *delay dating*. According to NCSL International's *Laboratory Management-19, Implementation of a Delayed Dating Approach to Calibration* (LM-19) (2018), "delay dating" is the practice of postponing "the calibration usage period by delaying the start of the calibration interval based on historical observations and/or rigorous analysis of an instrument or class of instruments."[14] By delaying the start of the calibration interval, eligible and approved IM&TE would not require calibration again until the end of its new calibration due date based on the interval after it is assigned into service. This can be beneficial as a time- and cost-saver for companies by not calibrating a piece of equipment solely for the exercise of calibrating it. This can yield extraordinary savings in money, time, and resources for some laboratories. However, while it can yield financial savings, there is an incremental increase in risk. Some "factors to consider are:

1. The incremental increase in risk of an Out-Of-Tolerance (OOT) condition if delayed interval start date is applied,

2. The decreased cost associated with a long-term reduction in the number of required calibrations, including the handling cost associated with pulling and transporting instruments from the field to the calibration facility,

3. The increase and timeliness of availability of instruments that can be pre-staged using delayed dating techniques, or

4. Long term stability analysis for M&TE for applicability."[15]

Initial Steps for a Delayed Dating Program

Once there is approval to explore a delayed dating program of a laboratory's own M&TE, the first steps consist of determining if the effort will be worth it.

1. Asset Evaluation[16]

    a. Categorize M&TE based on similarities: Manufacturer, model, and if necessary, measurement ranges within models.

    b. Apply the Pareto 80/20 principle on the categorized M&TE types in inventory. Begin by working on the first 20 % of the M&TE that make up 80 % of the total inventory.

2. Equipment Applicability[17]

    a. Classify M&TE based on *how* it makes measurements and displays measured values. Is it by mechanical or electrical means?

    b. Are there (moving) parts that might harden, degrade, or corrode if not used for extended periods of time?

    c. How is the M&TE powered? Is it powered by batteries, electrical connection, or not powered at all?

3. Usage Analysis[18]

    a. What are the current nominal calibration intervals?

    b. Which of these M&TE are currently stored unused before calibrating before being put into use? How long have they been stored?

    c. NCSL International LM-19, Section 5.1 provides an example benefits analysis based on Usage Analysis data.

4. Cost/Benefit Analysis[19]

    a. Perform using analysis information from NCSL International LM-19, section 4.

    b. Example cost/benefit analysis is also provided in NCSL International LM-19.

Other considerations for inclusion of IM&TE in a delayed dating program are original equipment manufacturer (OEM) recommendations and limitations. Some OEMs may have information on long-term storage stability and storage recommendations. What are the test methods, requirements, and/or regulations for which the IM&TE are used that might already dictate calibration intervals or exclude them from a delay-dating program?[20]

What are the environmental conditions in which the IM&TE are stored as well as used? The equipment evaluation should be based on the worst-case or most adverse conditions.[21]

How have these groups of IM&TE performed over time? What is their reliability history? If the IM&TE are not confidently reliable to make it to the end of their current calibration intervals, delay-dated intervals will not improve their reliability. How much are the IM&TE used during their calibration intervals? Is their usage heavy or sporadic? Are there interim accuracy or performance checks, and are those checks recorded? What is the laboratory's target reliability or pass rate? Why did the IM&TE fail its calibration? Was it an event not related to the IM&TE's regular stability, such as a catastrophic drop on the ground, poor handling, or unusual usage or environments?[22]

Which IM&TE are calibrated internally, and which are calibrated by an external vendor? If bulk external calibrations have been negotiated for discounts, then would staggering the calibrations affect those discounts?

For IM&TE approved for the delayed dating program, a system must be established for identification, tracking, and control.[23] A part of that system should be requirements and resources for storage. IM&TE should be stored in a controlled environment that does not adversely affect it. IM&TE should also be stored in a secure and protected space. Access should be restricted or limited to authorized personnel. The IM&TE also has to be readily identified as to its delay-dated status awaiting activation. Additionally, specific storage materials and resources might be needed, such as desiccants, padding, static-resistant bags/containers, vacuum sealing, spray protectant, or protective paper, wax or dip seal, and environmental controls. In some cases, use of tamper-evident seals can be applied to storage containers to readily reveal if the IM&TE was accessed. Replaceable batteries should be removed to prevent corrosion of contacts.[24]

Other details in the delayed dating program are policies and procedures that state the maximum time an instrument (type, group, class, etc.) may remain in delay-dated storage without activation until calibration, verification, or some other activity is needed. Once an instrument is activated, there should be a procedure that details how the respective records (including if and how calibration certificates should be updated) are updated and how the device is labeled. The laboratory will need to decide which date(s) are to be included on the calibration label (e.g., if the initial calibration date, the activation date, and new calibration due date should all be on the label). The basis for removal of any item from the delayed dating program should be clearly documented.[25] *NCSL International LM-19* includes more details and examples.

What one company can delay-date, another company may not be able to justify for delayed dating. Each company must evaluate (and reevaluate) its own equipment. Areas of evaluation that may be valuable include breakdowns of equipment use frequency, manufacturer model, serial numbers, accuracy needs, use conditions, storage conditions, calibration pass-rate history, history of adjustments, etc. Although delay dating may be a way of lowering costs and saving time, it should not be used as an excuse to avoid regular service, preventive maintenance, or calibrations that would prolong equipment life and performance.

How often to calibrate IM&TE is a balancing act between the costs associated with time and resources (buying standards, providing adequate facilities, hiring and training competent staff, etc.) and the risk of failure. This is true for organizations that perform in-house calibration functions as well as those that outsource all or part of their calibrations. By extending calibration intervals without enough historical data, an organization runs the risk of being on the ragged edge compared with making its interval adjustments in a systematic and calculated fashion, putting its program on the forefront in both reliability and dependability.

Chapter 11

The International System of Units (SI) and Measurement Standards

THE INTERNATIONAL SYSTEM OF UNITS: DEFINING, REALIZING, AND DISSEMINATING

The definitive international reference on the International System of Units (SI) is a booklet published by the International Bureau of Weights and Measures (BIPM, Bureau International des Poids et Mesures) and often referred to as the BIPM SI Brochure. Entitled *Le Système International d' Unités (SI)*, the booklet is in French followed by a text in English. The latest version of the official SI Brochure (9th edition as of this writing) and appendices are readily available on the BIPM website.[1] The SI Brochure lists the most significant Resolutions of the Conférence Générale des Poids et Mesures (the CGPM, known in English as the General Conference on Weights and Measures) and decisions of the Comité International des Poids et Mesures (the CIPM, known in English as the International Committee on Weights and Measures) that concern the metric system going back to the first meeting of the CGPM in 1889.

The BIPM SI Brochure has been translated into many languages around the world and is a guide to National Metrology Institutes (NMIs), scientists, engineers, educators—everyone! Many countries have adopted the BIPM SI Brochure and incorporated the adoption of the SI into their laws and regulations. As stated in the BIPM SI Brochure: "The International System of Units, the SI, has been used around the world as the preferred system of units, the basic language for science, technology, industry and trade since it was established in 1960 by a resolution at the 11th meeting of the CGPM."[2]

As an example, the U.S. secretary of commerce, acting through the director of the National Institute of Standards and Technology (NIST), is authorized by statute (15 U.S.C. § 272) under subsection (2) "to develop, maintain, and retain custody of the national standards of measurement, and provide the means and methods for making measurements consistent with those standards" and under subsection (9) "to assure the compatibility of United States national measurement standards with those of other nations."[3] Under this authority, the SI is interpreted or modified by the director of NIST for use in the United States. The secretary of commerce, acting through the NIST director, is designated to direct and coordinate efforts by federal departments and agencies to implement government metric usage in accordance with the Metric Conversion Act (15 U.S.C. § 205b)[4], as amended by the Omnibus Trade and Competitiveness Act of 1988.

The National Institute of Standards and Technology Special Publication 330, 2019 edition (NIST SP 330), is the United States version of the English text of the ninth edition of the BIPM SI Brochure (the most current), published in 2019. The 2019 edition of NIST SP 330 replaced its immediate predecessor, the 2008 edition, which was based on the eighth edition of the BIPM SI Brochure (published in 2006 and updated in 2014). NIST SP 330 is also readily available to the public.[5]

The International System of Units is defined in the *International Vocabulary of Metrology—Basic and General Concepts and Associated Terms* (*VIM 3*)[6] as a "system of units, based on the International System of Quantities, their names and symbols, including a series of prefixes and their names and symbols, together with rules for their use, adopted by the General Conference on Weights and Measures (CGPM)"; a *base unit* is defined as a "measurement unit that is adopted by convention for a base quantity." The seven familiar base units are shown in Table 11.1. These are the same base units shown in the *VIM 3* and restated in the SI Brochure, and should be familiar to most metrologists.

The year 2019 was an exciting year for the International System of Units! The entire system of units was redefined based on a new approach to clarifying the definitions of the seven base units by basing them on "defining constants" where the numerical values have been fixed as shown in Table 11.2. The new definitions of the seven base units, based on the defining constants, is shown in

Table 11.1 Seven base units in the International System of Units (SI).[7]

Base quantity name	Typical symbol	Base unit	Symbol
length	l, x, r, etc.	meter (metre)	m
mass	m	kilogram	kg
time	t	second	s
electric current	I, i	ampere	A
thermodynamic temperature	T	kelvin	K
amount of substance	n	mole	mol
luminous intensity	I_v	candela	cd

Table 11.2 The seven defining constants of the SI and the seven corresponding units they define.[8]

Defining constant	Symbol	Numerical value	Unit
hyperfine transition frequency of Cs	$\Delta\nu_{Cs}$	9 192 631 770	Hz
speed of light in vacuum	c	299 792 458	m s^{-1}
Planck constant	h	6.626 070 15 × 10^{-34}	J s
elementary charge	e	1.602 176 634 × 10^{-19}	C
Boltzmann constant	k	1.380 649 × 10^{-23}	J K^{-1}
Avogadro constant	N_A	6.022 140 76 × 10^{23}	mol^{-1}
luminous efficacy	K_{cd}	683	lm W^{-1}

Table 11.3 The 2019 definitions of the seven base units using the defining constants.[9]

Base quantity	Definitions
length	The meter, symbol m, is the SI unit of length. It is defined by taking the fixed numerical value of the speed of light in vacuum c to be 299 792 458 when expressed in the unit m s^{-1}, where the second is defined in terms of the cesium frequency $\Delta\nu_{Cs}$.
mass	The kilogram, symbol kg, is the SI unit of mass. It is defined by taking the fixed numerical value of the Planck constant h to be 6.626 070 15 × 10^{-34} when expressed in the unit J s, which is equal to kg m^2 s^{-1}, where the meter and the second are defined in terms of c and $\Delta\nu_{Cs}$.
time	The second, symbol s, is the SI unit of time. It is defined by taking the fixed numerical value of the cesium frequency $\Delta\nu_{Cs}$, the unperturbed ground-state hyperfine transition frequency of the cesium 133 atom, to be 9 192 631 770 when expressed in the unit Hz, which is equal to s^{-1}.
electric current	The ampere, symbol A, is the SI unit of electric current. It is defined by taking the fixed numerical value of the elementary charge e to be 1.602 176 634 × 10^{-19} when expressed in the unit C, which is equal to A s, where the second is defined in terms of $\Delta\nu_{Cs}$.
thermodynamic temperature	The kelvin, symbol K, is the SI unit of thermodynamic temperature. It is defined by taking the fixed numerical value of the Boltzmann constant k to be 1.380 649 × 10^{-23} when expressed in the unit J K^{-1}, which is equal to kg m^2 s^{-2} K^{-1}, where the kilogram, meter, and second are defined in terms of h, c, and $\Delta\nu_{Cs}$.
amount of substance	The mole, symbol mol, is the SI unit of amount of substance. One mole contains exactly 6.022 140 76 × 10^{23} elementary entities. This number is the fixed numerical value of the Avogadro constant, N_A, when expressed in the unit mol^{-1} and is called the Avogadro number. The amount of substance, symbol n, of a system is a measure of the number of specified elementary entities. An elementary entity may be an atom, a molecule, an ion, an electron, any other particle or specified group of particles.
luminous intensity	The candela, symbol cd, is the SI unit of luminous intensity in a given direction. It is defined by taking the fixed numerical value of the luminous efficacy of monochromatic radiation of frequency 540 × 10^{12} Hz, K_{cd}, to be 683 when expressed in the unit lm W^{-1}, which is equal to cd sr W^{-1}, or cd sr kg^{-1} m^{-2} s^3, where the kilogram, meter and second are defined in terms of h, c, and $\Delta\nu_{Cs}$.

Table 11.3. The definitions of the second (s), metre (m), and candela (cd) were not changed in technical content, but the wording of the definitions were revised to make the form and style consistent with the new definitions of the kilogram (kg), ampere (A), kelvin (K), and mole (mol).

The SI Brochure includes appendices for each of the units that contain the *mise en pratiques*, or the methods currently used to realize the definitions. There are three steps in the process from defining measurement units to standards that are practically used in the measurement system: defining, realizing, and disseminating. Let us look at what these terms mean.

Defining. The *VIM 3* definition of a measurement unit notes that it is a "real scalar quantity, defined and adopted by convention, with which any other quantity of the same kind can be compared to express the ratio of the two quantities as a number."[10] In this sense, the definitions of the base and derived units have been adopted by the CGPM/CIPM. The accuracy of a definition is important, as it is reflected in the accuracy of the measurements that can be achieved.

Realizing. The definition of the unit must be realized, or put into practice (*mise en pratique*), so that it can be used as a reference for measurement. This task is normally, but not solely, limited to NMIs. The units are realized in the form of experimental setups referred to as *mises en pratique* that may be revised whenever new experiments are developed. The current practices are published as appendices to the SI Brochure and may be found on the BIPM website.

Disseminating. The end users of measurements are trade, industry, and calibration laboratories. End users usually do not have access to the representations of the SI units held by NMIs, although with the new definitions in a practical sense, future developments may enable realization at levels other than NMIs. End users also need the values of the SI units for reference. This is accomplished through the process of dissemination, whereby the units are made available to the end users of measurement results.

The concept of national and international measurement systems has been accepted globally through international arrangements (as was discussed in Chapters 9 and 10 on metrological traceability and calibration programs). As noted earlier in the case of the United States, nations have given their NMIs the responsibility of realizing and maintaining national standards of measurement, which are the representations of the SI units for each parameter. These NMIs also take part in the process of dissemination of these units to the actual users. This task is carried out by the National Institute of Standards and Technology in the United States, the National Physical Laboratory in the United Kingdom, Physikalisch-Technische Bundesanstalt in Germany, the National Physical Laboratory in India, and others. The traceability of measurements is achieved using measurement standards in the calibration process. A wide variety of measurement standards may be considered; these are discussed further in the last section of this chapter.

There are derived units in addition to the seven base units. The SI Brochure introduces derived units as follows (the tables of this chapter are referenced in brackets): "Derived units are defined as products of powers of the base units. When the numerical factor of this product is one, the derived units are called coherent derived units. The base and coherent derived units of the SI form a coherent set, designated the set of coherent SI units. The word 'coherent' here means that equations between the numerical values of quantities take exactly the same form as the equations between the quantities themselves. Some of the coherent derived units in the SI are given special names. Table 4 [Table 11.4 in this handbook] lists 22 SI units with special names. Together with the seven base units in Table 2 [Table 11.1 in this handbook] they form the core of the set of SI units. All other SI units are combinations of some of these 29 units. It is important to note that any of the seven base units and 22 SI units with special names can be constructed directly from the seven defining constants. In fact, the units of the seven defining constants include both base and derived units."

Chapter 11: The International System of Units (SI) and Measurement Standards

Table 11.4 The 22 SI units with special names and symbols.[11]

Derived quantity	Unit (symbol)	Unit in terms of base units	Unit expressed in other SI units
plane angle	radian (rad)	m/m	
solid angle	steradian (sr)	m^2/m^2	
frequency	hertz (Hz)	s^{-1}	
force	newton (N)	$kg\ m\ s^{-2}$	
pressure, stress	pascal (Pa)	$kg\ m\ s^{-2}$	
energy, work, amount of heat	joule (J)	$kg\ m^2\ s^{-2}$	N m
power, radiant flux	watt (W)	$kg\ m^2\ s^{-3}$	$J\ s^{-1}$
electric charge	coulomb (C)	A s	
electric potential difference	volt (V)	$kg\ m^2\ s^{-3}\ A^{-1}$	W/A
capacitance	farad (F)	$kg^{-1}\ m^{-2}\ s^4\ A^2$	C/V
electric resistance	ohm (Ω)	$kg\ m^2\ s^{-3}\ A^{-2}$	V/A
electric conductance	siemens (S)	$kg^{-1}\ m^{-2}\ s^3\ A^2$	A/V
magnetic flux	weber (Wb)	$kg\ m^2\ s^{-2}\ A^{-1}$	V S
magnetic flux density	tesla (T)	$kg\ s^{-2}\ A^{-1}$	Wb/m^2
inductance	henry (H)	$kg\ m^2\ s^{-2}\ A^{-2}$	Wb/A
Celsius temperature	degree (°C)	°C = K	
luminous flux	lumen (lm)	cd sr	cd sr
illuminance	lux (lx)	$cd\ sr\ m^{-2}$	lm/m^2
activity (of a radionuclide)	becquerel (Bq)	s^{-1}	
absorbed dose, kerma	gray (Gy)	$m^2\ s^{-2}$	J/kg
dose equivalent	sievert (Sv)	$m^2\ s^{-2}$	J/kg
catalytic activity	katal (kat)	$mol\ s^{-1}$	

In addition to the base and derived units, additional derived quantities and units may be used. There are additional coherent derived units in the SI expressed in terms of base units as well. For example, area (represented by A) can be presented as m^2, volume (represented by V) can be presented as m^3, and speed or velocity (represented by v) is $m\ s^{-1}$, and so on. There are a few non-SI units accepted for use with the SI units and include the common use of units. The CIPM has accepted some of these units because they are widely used and are expected to be used into the future. A few examples include liter, minute, hour, and day. See the SI Brochure or NIST SP 330 for additional units of interest.

Each defined unit may be used with multiples and submultiples of the common prefixes shown in Table 11.5. To illustrate, here are a few examples commonly observed in daily life: kilometer (e.g., driving or running and

Table 11.5 The SI prefixes for multiples and submultiples of units.[12]

Factor	Prefix	Symbol	Factor	Prefix	Symbol
$10^{24} = (10^3)^8$	yotta	Y	10^{-1}	deci	d
$10^{21} = (10^3)^7$	zetta	Z	10^{-2}	centi	c
$10^{18} = (10^3)^6$	exa	E	$10^{-3} = (10^3)^{-1}$	milli	m
$10^{15} = (10^3)^5$	peta	P	$10^{-6} = (10^3)^{-2}$	micro	μ
$10^{12} = (10^3)^4$	tera	T	$10^{-9} = (10^3)^{-3}$	nano	n
$10^9 = (10^3)^3$	giga	G	$10^{-12} = (10^3)^{-4}$	pico	p
$10^6 = (10^3)^2$	mega	M	$10^{-15} = (10^3)^{-5}$	femto	f
$10^3 = (10^3)^1$	kilo	k	$10^{-18} = (10^3)^{-6}$	atto	a
10^2	hecto	h	$10^{-21} = (10^3)^{-7}$	zepto	z
10^1	deka	da	$10^{-24} = (10^3)^{-8}$	yocto	y

swimming competitions), and kilogram (e.g., purchased commodities, weight at the doctor's office). The SI prefixes can be used with several of the non-SI units, such as milliliter (mL) but not, for example, with the non-SI units of time; for example, mega-minute or micro-minute are inappropriate and sound rather silly!

See Figure 11.1 to get an idea of the system in terms of the definitions, base units, derived units, special units, and prefixes.

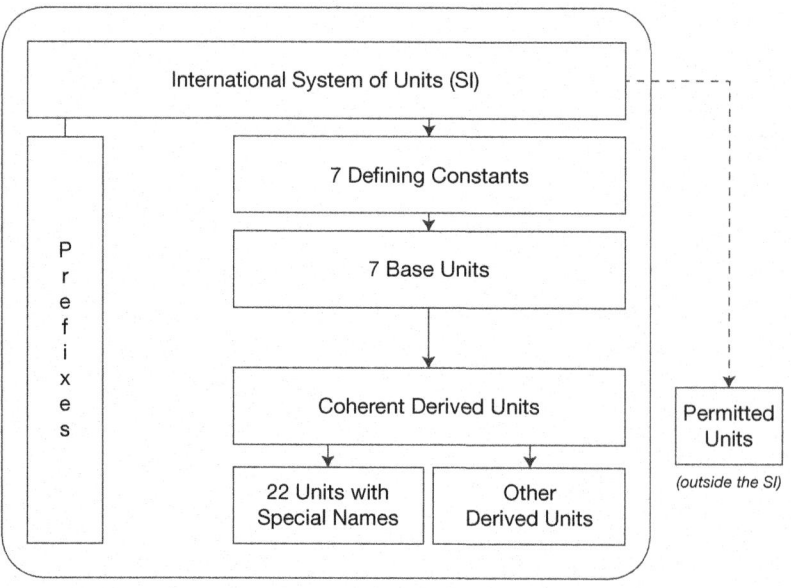

Figure 11.1 The SI unit diagram.

PRESENTATION OF MEASUREMENT RESULTS: UNITS AND SYMBOLS

Presentation of measurements and units follows rules and style conventions for presentation and publication to ensure accuracy, comprehension, and consistency of use. The SI Brochure and the NIST SP 330 both state the following: "General principles for the writing of unit symbols and numbers were first given by the 9th CGPM (1948, Resolution 7). These were subsequently elaborated by ISO, IEC, and other international bodies. Consequently, there now exists a general consensus on how unit symbols and names, including prefix symbols and names as well as quantity symbols should be written and used, and how the values of quantities should be expressed. Compliance with these rules and style conventions, the most important of which are presented in this chapter, supports the readability of scientific and technical papers."[13]

Common recommendations were included in a set of two ASQ's *Quality Progress* magazine "Measure for Measure" articles titled "Accuracy Matters: Understanding and Communicating Measurement Uncertainties"[14] and "Write It Right."[15] These articles are included in their entirety (with updated references) with permission from Elizabeth (Gentry) Benham and Georgia Harris, at the end of this chapter. See the SI Brochure or NIST SP 330, or use the "Check List for Reviewing Manuscripts" provided in NIST Special Publication 811 to ensure that you follow standard practice rules in presenting measurement results, especially whenever results are communicated to others, whether in papers, in educational and marketing materials, and especially on calibration and test certificates.

A key feature of the guidance provided in these international and national publications is reinforced in the ISO/IEC 17025:2017 documentary standard for calibration and testing laboratories that are reporting measurement results and uncertainties. When a measurement result of a quantity is reported, the estimated value of the measurand (the quantity to be measured), and the uncertainty associated with that value, are to be expressed in the same unit. It is a good practice for metrologists and scientists to study and follow the rules for presentation of measurement results, units, and symbols, because as Gentry (now Benham) and Harris indicated in their ASQ *Quality Progress* articles, "if we don't get it right, who will?"

MEASUREMENT STANDARD TYPES

The *VIM 3* lists several types of measurement standards with various descriptors, and the term *measurement standard* is used in the definition of metrological traceability and calibration, as we will discuss shortly. The general definition for measurement standard is: "realization of the definition of a given quantity, with stated quantity value and associated measurement uncertainty, used as a reference."[16]

A sampling of other types of measurement standards is shown in Table 11.6. Definitions are taken from the *VIM 3* and examples are included.

What we mean by "measurement standard" depends on the type of measurement to be made as well as the dissemination process down to the end user. It can be difficult to define the "end" in *end user*. It is useful to revisit the

Table 11.6 Definitions and examples of standards.[17]

Type of standard	Definition: A measurement standard	Example
International	Recognized by signatories to an international agreement and intended to serve worldwide.	New definition of the kilogram based on the Planck constant.
National	Recognized by national authority to serve in a state or economy as the basis for assigning quantity values to other measurement standards for the kind of quantity concerned.	Countries with designated kilograms may still refer to them as national standards, even if they can realize mass from the definition (e.g., the United States has several kilograms, notably K20 and K4, as the original national standards).
Primary	Established using a primary reference measurement procedure, or created as an artifact, chosen by convention.	For example, triple-point-of-water cell as a primary measurement standard of thermodynamic temperature.
Secondary	Established through calibration with respect to a primary measurement standard for a quantity of the same kind.	For example, a standard platinum resistance thermometer that is calibrated at referenced triple-points defined by the International Temperature Scale.
Reference	Designated for the calibration of other measurement standards for quantities of a given kind in each organization or at a given location.	Within the United States, legal metrology laboratories maintain kilogram standards calibrated by NIST or other accredited laboratories; these are the highest level of reference standards for them.
Working	That is used routinely to calibrate or verify measuring instruments or measuring systems; usually calibrated with respect to a reference standard.	Any standard that is used on a regular basis to disseminate to other standards or verify/calibrate measuring instruments. Field standards used in legal metrology are a type of working standard (e.g., volumetric working standards used to verify delivery of petroleum products through a meter).
Traveling	Sometimes of special construction, intended for transport between different locations.	Portable battery-operated cesium-133 frequency measurement standard.
Intrinsic	Based on an inherent and reproducible property of a phenomenon or substance.	Several examples are: • Triple-point-of-water cell as an intrinsic measurement standard of thermodynamic temperature • Intrinsic measurement standard of electric potential difference based on the Josephson effect • Intrinsic measurement standard of electric resistance based on the quantum Hall effect • Sample of copper as an intrinsic measurement standard of electric conductivity

Continued

Table 11.6 *Continued*

Type of standard	Definition: A measurement standard	Example
Reference material	Material, sufficiently homogeneous and stable with reference to specified properties, which has been established to be fit for its intended use in measurement or in examination of nominal properties. A "certified" reference material is accompanied by documentation issued by an authoritative body attesting to measurement values and uncertainties.	For example: • Water of stated density used for comparison in density measurements for gravimetric volume calibrations • Water of stated purity with dynamic viscosity used to calibrate viscometers • Human serum without an assigned quantity value for the amount-of-substance concentration of the inherent cholesterol, used only as a measurement precision control material
Reference data	Data related to a property of a phenomenon, body, or substance, or to a system of components of known composition or structure, obtained from an identified source, critically evaluated, and verified for accuracy. Standard reference data: reference data issued by a recognized authority (e.g., CODATA).	For example: • Reference data for solubility of chemical compounds as published by the IUPAC • Published and verified data sets used with analytical instruments such as a gas spectrophotometer or gas chromatograph
Transfer device	Device used as an intermediary to compare measurement standards.	A weighing instrument is used as a transfer device to compare reference mass standards and working mass standards.

definitions of *metrological traceability* and *calibration* as we discuss measurement standards (*VIM 3* reference numbers are given in parentheses).

Metrological traceability (2.41): "Property of a measurement result whereby the result can be related to a reference through a documented unbroken chain of *calibrations*, each contributing to the measurement uncertainty."[18]

Calibration (2.39): "Operation that, under specified conditions, in a first step, establishes a relation between the quantity values with measurement uncertainties provided by *measurement standards* and corresponding indications with associated measurement uncertainties and, in a second step, uses this information to establish a relation for obtaining a measurement result from an indication."[19]

By reading these two definitions along with the definition of measurement standard, you see that "realization" is a key part of the definition that we discussed earlier in this chapter. You can also see that a stated quantity value is part of that definition, and that associated measurement uncertainty is a part of *each* of these definitions. Further, as already noted, the measurement quantity value and its associated uncertainty must be in the same units. We will not discuss uncertainties associated with the measurement process in this chapter, as that subject is discussed thoroughly elsewhere in this handbook. However, in any discussion on selecting appropriate standards, measurement uncertainties and

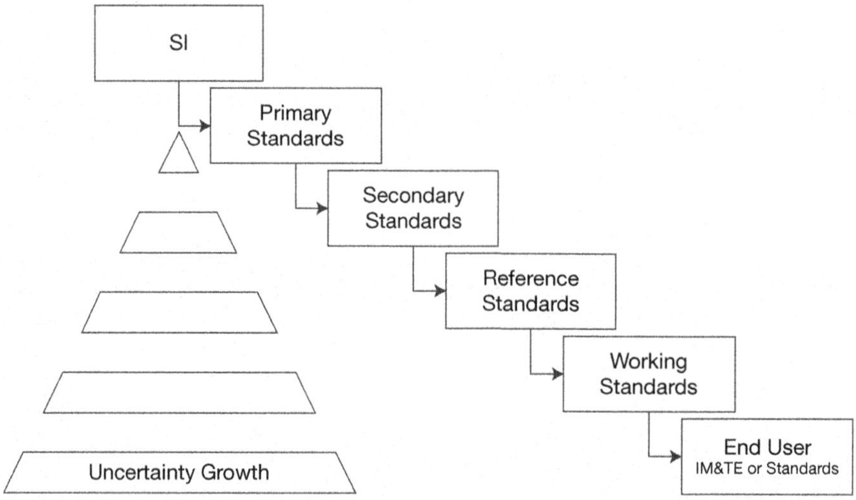

Figure 11.2 Standards and uncertainty growth.

decision risks must be considered to ensure that the user of the dissemination or calibration results obtains the level of measurement results and uncertainties needed for the next step in the calibration and traceability hierarchy.

In Chapter 9, metrological traceability was represented as a pyramid or flow chart, and Figures 9.3 and 9.5 showed the standards hierarchy from one level to the next. A similar approach is used in Chapter 34 to show the definition, realization, and dissemination process (Figure 34.4) for mass. You might reflect on the graphical representation of some of these definitions as measurement units are transferred from one to another and realize also that the measurement uncertainty grows with each subsequent step of transferring measurement results in a calibration (see Figure 11.2). Awareness of the International System of Units, dissemination steps, and types of measurement standards available in each measurement parameter of interest is critical knowledge for scientists, measurement experts, educators, and even the media. Correct presentation of measurement results, measurement units, and measurement symbols is an obligatory aspect of clearly and accurately communicating measurement results.

ACCURACY MATTERS—UNDERSTANDING AND COMMUNICATING MEASUREMENT UNCERTAINTIES

ASQ *Quality Progress*, vol. 49, Issue 5 (May 2016), "Measure for Measure" article by Elizabeth J. (Benham) Gentry and Georgia L. Harris

(Reprinted in its entirety with permission of the authors and with updated references and updated endnote numbering)

Many decisions are based on accurate measurement results, such as: "Should medicine be prescribed for high cholesterol or high glucose?" or "Should a measuring instrument or standard be adjusted to meet tolerances?"

CHAPTER 11: THE INTERNATIONAL SYSTEM OF UNITS (SI) AND MEASUREMENT STANDARDS   95

The answers are based on measurement results. And as a patient, scientist, citizen or policymaker, we make assumptions about the accuracy of measurement results in reports and calibration certificates. We assume they're good, right, or to say it more correctly, "They're accurate." But note that accuracy is often defined as hitting the center of a target or true value.

One of our colleagues regularly says, "The only true value is on a sign above a hardware store." But people who use measurements often trust the accuracy of their measurement results usually without question, believing the results are "good and right."

A measurement result alone is incomplete without some assessment and measure of reported uncertainty. People can estimate the temperature outside on a warm spring day within a few degrees based on their experience. But if we use a thermometer, our first hope is that it's accurate and gives us the correct or right temperature. After this, we must consider the resolution of the standard: "Is the readability of the thermometer 1 degree Celsius, 0.1 degree or 0.01 degree?"

Our confidence that the results are right will depend on the readability or resolution of the standard or measuring instrument. Our confidence shouldn't be based on a calculator or spreadsheet giving us a calculated value to 15 decimal places when the resolution or uncertainty is a fraction of that.

Repeatability

Repeatability of an instrument or standard also is a variable of concern. Many people naturally repeat measurements to get a sense of whether multiple values agree. We use simple measurements in daily life, such as stepping on a scale to monitor your weight or checking a vehicle's mileage to calculate fuel efficiency.

Assigning uncertainty to a measurement result is a rigorous, documented and validated process that is assessed nearly as often as the measurement results themselves. Measurement scientists often use internationally accepted procedures to obtain standardized measurement results. They also use the JCGM 100:2008 *Guide to the Expression of Uncertainty in Measurement* for evaluating and reporting associated uncertainties.[20]

The readability (or resolution and repeatability) of measurement results gives a sense of confidence (or lack thereof), and these also have associated measures of uncertainty. It's a wise practice to ask for the measurement uncertainty and use it to assess the quality and precision of a measurement result. Uncertainty values provide confidence in the measurement result: It quantifies the boundaries or limits within which a measurement result should agree with a true quantity value.

Terms and Communication

Accurate measurement results and associated uncertainties must be communicated. This could be in a newspaper, a scientific paper or on a calibration certificate. This also means it's critical to have accuracy in our words and measurement results.

Guiding documents help standardize communication: *The International Vocabulary of Metrology* (VIM) provides guidance on terms used with measurement and calibration results.[21] When measurement professionals use terms such as "accuracy," "traceability," "uncertainty" and "reference standards," they have specific meanings that should be used by every scientist.

For example, the VIM defines "accuracy"—as it's related to a measurement result—as the "closeness of agreement between a measured quantity value and a true quantity value of a measurand." According to the VIM, "measurand" is "the quantity intended to be measured." This definition of accuracy also includes three explanatory notes:

1. "The concept 'measurement accuracy' is not a quantity and is not given a numerical quantity value. A measurement is said to be more accurate when it offers a smaller measurement error."

2. "The term 'measurement accuracy' should not be used for measurement trueness, and the term 'measurement precision' should not be used for 'measurement accuracy', which, however, is related to both of these concepts."

3. "'Measurement accuracy' is sometimes understood as closeness of agreement between measured quantity values that are being attributed to the measurand."[22]

If measurement results between or among laboratories are compared, scientists must be able to talk about the same things. This is why standardized definitions are essential: They can prevent confusion in communicating measurement results.

Units, Symbols and Results

Measurement results must communicate proper quantities, units and symbols. Many countries adopt the International System of Units (SI, also known as the metric system) as the reference basis for measurement results. There also is a reference document for presenting measurement units, symbols and results.[23]

The U.S. Metric Program of the National Institute of Standards and Technology (NIST), Office of Weights and Measures, helps implement the national policy to establish the SI as the preferred system of weights and measures for U.S. trade and commerce. It provides leadership and assistance on SI use and conversion to federal agencies, state and local governments, businesses, trade associations, standards development organizations, educators and the general public.

NIST Special Publication (SP) 330[24] and NIST SP 811[25] provide the legal interpretation of and guidelines for SI use in the United States. These publications provide standardized guidance on how measurement units and results should be presented in writing.

Black Dots

We like to ask, "If the measurement scientists don't get the communication of measurement results right, who will?" Regularly reviewing measurement results and uncertainties on calibration certificates and in laboratory documents yield numerous errors that can negatively affect interpretations of results by users.

Errors are often observed in the following situations:

- Measurement uncertainties are not included, are incomplete, are inaccurate, or are not properly rounded.
- Incorrect terminology is used.
- Typos are left uncorrected.
- Unit conversions are wrong.
- Incorrect units and symbols are presented, or correct units are inconsistently used.

We refer to these errors as "black dots." To customers, a black dot on a clear page is what they notice. This blemish is what they will remember, regardless of the other accurate information presented. Errors in reporting results can lead to confusion or bad decisions by users—often with critical effects. Black dots can destroy laboratory credibility.

There are examples of black dots in daily life and news headlines, such as:

- In 1998, NASA's Mars climate orbiter was lost after a failure to communicate requirements and convert measurement units from two measurement systems.[26]
- In 2003, Disneyland Tokyo's Space Mountain roller coaster accident highlighted a scenario in which axle-and-bearing design specifications were converted to metric units and implemented in the ride. After time passed, routine maintenance called for bearing replacements. Instead of being replaced with metric-designed bearings, they were replaced with the incorrect size based on the original, nonmetric design. This created a gap between the axle and bearing. Eventually, the extra vibration and stress caused the axle to fail, derailing the roller coaster. Luckily, no passengers were injured.[27]

Document control, version control and archiving records are essential tools for ensuring changes made over time are effectively communicated to all personnel affected by a change. Failure to adequately control laboratory documents, such as calibration certificate templates, can be the root cause of black dots that are released to customers.

Avoiding black dots is fundamental to ensuring communication of accurate measurement results. Reviewing for typos, grammatical errors, accurate terminology, completeness, and use of appropriate measurement units and symbols is essential. The second part of this article will offer suggestions to improve the quality, accuracy and communication of measurement results.

WRITE IT RIGHT: UNDERSTANDING NUANCES OF METRICS IN TECHNICAL WRITING

ASQ *Quality Progress*, vol. 49, Issue 7 (July 2016), "Measure for Measure" article by Elizabeth J. (Benham) Gentry and Georgia L. Harris

(Reprinted in its entirety with permission of the authors and with updated references and updated endnote numbering)

Many questions arise while you're writing laboratory documents, clarifying measurement results or implementing measurement system best practices. The proper use of measurement units and symbols in laboratory documents—such as calibration reports, control charts, uncertainty tables or standard operating procedures—is critical to effectively communicate technical information.

The National Institute of Standards and Technology (NIST) was delegated the responsibility to interpret or modify the International System of Units (SI, also known as the metric system) for use in the United States. To accomplish this, NIST provides several SI resources to support sectors of science, technology, trade and commerce. It also serves as the U.S. technical representative to the International Bureau of Weights and Measures (BIPM) that defines the SI. These publications are used to guide the measurement unit style in technical and documentary standards.

NIST Special Publication (SP) 330[28] and *NIST SP 811*[29] provide the legal interpretation of and guidelines for SI use in the United States. *NIST SP 811* also provides detailed rules for SI writing style, including a useful editorial checklist.

Striving for Zero Errors

NIST SP 811 is written for technical audiences, such as engineers, scientists and academics. Appendix B provides rounding guidance and unit-conversion factors for a broad set of measurement units. NIST published several similar technical guides, including the *Metric Style Guide for the News Media*, which provides condensed SI content to highlight commonly used measurement information.[30] A convenient hub of SI style guidance also is available on the NIST metric program's website.[31]

Use a leading zero: For numbers less than one, a zero is written before the decimal point.[32] This ensures a quantity is appropriately interpreted and helps avoid consequences of a misplaced decimal point. Without a leading zero, a value like .25, for example, could be misinterpreted as 25, an error that makes it 100 times greater in magnitude. Such an error could seriously harm a patient if the quantity represented a medication dose.

Avoid unit-conversion errors: Using the SI reduces the number of errors associated with measurement conversions between U.S. customary units and the SI. Eliminating conversions altogether negates the need to document which conversion factors are being used and their sources. Conversion calculations require rigorous software validation, which is a time-consuming process. At best, conversion-calculation errors can cause expensive mistakes. At worst, their consequences can be a matter of life and death.

Ground Control to Accuracy

The 1999 crash of NASA's $125 million NASA Mars climate orbital spacecraft served as a wake-up call for potential errors related to working with multiple measurement systems. The mishap occurred because the spacecraft entered the Mars atmosphere on a trajectory that was too low.[33]

NASA later identified the root cause of the erroneous trajectory and velocity calculations: A contractor failed to use SI units of force (Newton, or N) as specified by NASA in the coding of a ground software file used in trajectory models. One corrective action that NASA recommended was to perform software audits to evaluate specification compliance on all data transferred between NASA and the contractor.[34]

Language Arts

Several helpful conversion-factor resources have been made available on the NIST metric program's website.[35] Caution is recommended to organizations developing unit-conversion software or using online calculators for technical purposes. It's important to conduct a rigorous validation and verification analysis before using unit-conversion software.

Spelling and pronunciation of measurement units—This can be challenging. Advantages of the SI over the many other historic and customary measurement unit systems is that the SI provides a coherent set of internationally accepted unit symbols that can be used to communicate across all languages. Table 1 [Table 11.7 in this handbook] provides examples of how unit names are translated in several languages:

Table 11.7 Unit name translations.[36]

| | Unit symbol | | |
Language	m	s	kg
English (U.S.)	meter	second	kilogram
Spanish	metro	segundo	kilogramo
Italian	metro	sécondo	chilogrammo

In *NIST SP 811*, words are spelled in accordance with the *U.S. Government Printing Office Style Manual* (U.S. Government Publishing Office, 2008), which follows *Webster's Third New International Dictionary of the English Language* (Merriam-Webster, 1993). The spellings "meter," "liter" and "deka" are used rather than "metre," "litre," and "deca" as in the original BIPM English text of the SI brochure.

The BIPM SI brochure is the definitive international reference on the SI.[37] The text is published in French and English and has been translated into many other languages.

Capitalization of units, symbols and prefixes—Unit names start with a lowercase letter except at the beginning of the sentence or title, such as "pascal," "becquerel," "newton" or "tesla." For degrees Celsius (symbol °C), the unit "degree" is lowercase. But the modifier "Celsius" is capitalized because it's a person's name. A space is left between the numerical value and the unit symbol, and values are not hyphenated. For example: 20 °C and 10 kg are correct; 20°c, 20° C, 10-kg or 10kg are incorrect. If a unit name is spelled out during use, normal grammar rules apply.

Unit symbols are written in lowercase letters (such as "m" for meter, "s" for second or "kg" for kilogram). But symbols for units derived from the name of a person are capitalized—such as W for watt, V for volt, Pa for pascal or K for kelvin. The recommended symbol for "liter" in the United States also is capitalized as L to avoid misinterpreting "l" with the number one. A period should not be used following a unit symbol or abbreviation. For example, gram is represented as "g" not "g." Symbols of prefixes that mean a million or more are capitalized, and those that are less than a million are lowercased. For example, M for mega (millions) and "m" for milli (thousandths).

U.S. customary units—After the SI was developed, many style requirements were applied to non-SI measurement systems, including U.S. customary units—such as inch, foot, yard, mile, ounce, pound, gill or gallon. Although NIST does not publish a style resource for U.S. customary units, appendix C of NIST's Handbook 44, "General Tables of Units of Measurement," is a good resource for U.S. customary units used in trade and commerce, their relationships, and unit-conversion factors.[38]

Because the SI is critical as an international standard, its use in product design, manufacturing, marketing and labeling is essential for the U.S. industry's success in the global marketplace. NIST's metric program encourages using the SI in all facets of education, including honing workers' skills.

The successful voluntary transition of the United States to the SI is a critical factor in the competitive economic success of industry.[39] Accuracy in terminology use, measurement results, and measurement units is necessary to avoid the embarrassment of having others find your "black dots" (errors that can negatively affect interpretations of your results in scientific communications). There are many resources that can help you avoid being responsible for inaccuracies in measurement reporting.

Chapter 12
Audit Requirements

Some consider audits to be among the most important activities performed for a quality management system. This may become evident when we consider the following explanation of a quality system. "The basic premise and foundation of a good quality system is to say what you do, do what you say, record what you did, check the results, and act on the difference. Also, for the entire system to work, the organization needs to establish a quality management system to ensure that all operations throughout the metrology department, calibration laboratory, or work area where calibrations are accomplished occur in a stable manner. The effective operation of such a system will result in stable processes and, therefore, in a consistent output from those processes. Once stability and consistency are achieved, then it's possible to initiate improvements."[1]

How does the organization know that everything is working correctly or according to documented policies and procedures? How does it know where improvements can/should be made? How does it know it is providing to customers the quality product or service for which they are paying? The answer is simple: perform an audit.

ISO/IEC 17025:2017 references terms and definitions from ISO/IEC Guide 99 (*VIM*) and ISO/IEC 17000. According to ISO/IEC 17000:2020, Conformity assessment—Vocabulary and general principles, section 6 Terms relating to selection and determination, 6.4, *audit* is defined as a "process for obtaining relevant information about an object of conformity assessment and evaluating it objectively to determine the extent to which specified requirements are fulfilled."[2]

In ISO 19011:2018, Guidelines for auditing management systems, *audit* is defined as "systematic, independent, and documented process for obtaining objective evidence and evaluating it objectively to determine the extent to which the audit criteria are fulfilled."[3]

Unfortunately, when the word "audit" is mentioned to a technician, manager, supervisor, or person in a quality position, their reaction is often one of disdain, fear, and stress. The fact is, these reactions are inappropriate and unhelpful. Any organization that is responsive to its customers, both internal and external, and desires to find problems before they affect its products, services, or customers will conduct audits based on its quality system on a regular basis.

There are a variety of psychological/sociological aspects to an external audit/assessment. It is natural for human beings to get nervous when an outside "expert" is critiquing their work. This apprehension can be minimized through mimicking the external audits, as closely as possible, during your organization's internal audits. Frequent internal audits can help identify issues in early stages as well as help make personnel more comfortable with following the procedures, encourage their use of those procedures, and get all personnel more comfortable with outsider oversight. Audits and assessments should always be considered "open-book tests" where the auditee is expected and encouraged to reference the currently published documents rather than strictly relying on memory. This experience and preparation of auditees will be noticed by an experienced assessor and be of benefit to the laboratory by increasing the probability of a smoother accreditation assessment.

An audit process asks questions, looks at how an organization is supposed to be conducting business, and checks if that organization is following its defined procedures. All this is done with a mindset of helping itself, its customer base, and its bottom line! An audit should not be viewed as a dreadful thing or a necessary evil. It costs little in both time and money compared to fixing problems after the problems have occurred. Audits afford an organization the opportunity to correct errors, make improvements, and find areas where it can change for the better before customers, quality, or certification are adversely affected. It is a self-policing effort.

Are audits a requirement within the various international standards? Absolutely. Here are some examples of those requirements. ISO/IEC 17025:2017, Clause 8.8.1, states, "The laboratory shall conduct internal audits at planned intervals to provide information on whether the management system:

a) conforms to:

- the laboratory's own requirements for its management system, including the laboratory activities.
- the requirements of this International Standard.

b) is effectively implemented and maintained."[4]

ISO/IEC 17025:2017 continues, in Clause 8.8.2, "The laboratory shall:

a) plan, establish, implement, and maintain an audit program including the frequency, methods, responsibilities, planning requirements and reporting, which shall take into consideration the importance of the laboratory activities concerned, changes affecting the laboratory, and the results of previous audits.

b) define the audit criteria and scope for each audit.

c) ensure that the results of the audits are reported to relevant management.

d) implement appropriate correction and corrective actions without undue delay.

e) retain records as evidence of the implementation of the audit program and the audit results."[5]

ISO 9001:2015, Clause 9.2.1, states: "The organization shall conduct internal audits at planned intervals to provide information on whether the quality management system: a) conforms to: 1) the organization's own requirements for its quality management system; 2) the requirements of this International Standard; b) is effectively implemented and maintained."[6]

ISO 10012:2003, Clause 8.2.3, states: "The metrological function shall plan and conduct audits of the measurement management system to ensure its continuing effective implementation and compliance with the specified requirements. Audit results shall be reported to affected parties within the organization's management. The results of all audits . . . shall be recorded."[7]

NCSL International Recommend Practice-6: *Recommended Practice for Calibration Quality Systems for the Healthcare Industries*, RP-6, 2015, section 5.16, states, "The calibration quality system should be subject to periodic audits conducted at a frequency and to a degree that will ensure compliance with all documented requirements. It is recommended that a procedure describing the audits and reporting thereof be available and includes:

- Function or group responsible for conducting audits
- Frequency or group responsible for conducting audits
- Frequency and extent of audits
- Description of the auditing methods used to ensure that measurements and calibration have been performed competently
- Reporting of (a) occurrence of deviation and (b) any corrective actions and/or preventive actions taken."[8]

As a minimum, if there is no internal audit function requirement, a self-inspection program could go a long way in preparing the organization for audits and inspections. By setting up a self-inspection program, the organization is:

- Making an effort to find problems
- Seeing where it is not meeting the quality system requirements
- Demonstrating a desire to continuously improve its program through self-initiative
- Finding opportunities before they are found by others
- Making itself proactive instead of reactive to problems and solutions

The repeated theme among these various international standards and recommended practices is that audits should be pre-planned, well-defined, documented, and performed by competent auditors, and that the results should be acted upon and reported to management.

ISO 19011:2018 *Guidelines for auditing management systems* is a useful standard for understanding the principles of auditing, how to manage an audit program, and conduct an audit, and for judging competence and evaluation of auditors.[9] In addition to this international standard, there are well-written books and other references that provide more in-depth information and guidance on how to audit effectively. It is not the intent of this handbook or chapter to teach all the skills and nuances to effectively perform audits.

How do organizations perform internal audits? One way to perform internal audits is to follow the practice of "say what you do, do what you say, record what you did, check the results, and act on the difference." Review objective evidence to confirm if the organization is following its quality policies and procedures. Do all the records contain the required information, and do they create a paper trail or audit trail for traceability purposes? If an item was found to be out of tolerance during calibration, was action taken? Was the customer informed, and does the organization have the records to reflect that the notification indeed occurred? All required records must be documented if they are to be used in an audit. The more specific the question, the easier it is to answer. Internal audits are an important part of continuous process improvement, but they take time, effort, and forthrightness by all involved.

There are three different types of audits: first-party, second-party, and third-party audits. ISO 19011:2018 concentrates on first-party audits (internal audits) and second-party audits (audits conducted by organizations on their external providers and other external interested parties). ISO/IEC 17021-1:2015 *Conformity assessment—Requirements for bodies providing audit and certification of management systems* is about third-party conformity assessment activities; bodies performing this activity are therefore third-party conformity assessment bodies.[10] These three different types of audits are illustrated in Table 12.1.

An internal audit (first-party audit) is conducted by competent and authorized personnel from within an organization, department, or function. These auditors examine the entity's system and record the results for internal use only. Internal audits are usually performed by a person who is independent of the department or process that is being audited, so as to avoid potential conflicts of interest. For some organizations and laboratories, hiring an outside consultant to perform internal audits fulfills the requirements and ensures impartiality and objectivity.

An external audit is one conducted by a customer (second-party audit) for the purpose of evaluating its supplier, such as for granting or continuation of business. An external audit could also be conducted by an auditing or assessing body (third-party audit) with the results provided to the management of the company, department, or organization. Most external audits are performed to confirm if an organization is complying or conforming with specific standard(s), guideline(s), or regulation(s). They can be either subjective or directive in nature. For example, if an organization were audited for compliance to current Good Manufacturing Practice (cGMP) requirements (Food and Drug Administration), it would be informed of any findings by use of the FDA's Form 483, which is part of a public record. Accreditation assessments typically require a demonstration of technical proficiency and competency where the assessor visually witnesses

Table 12.1 The three different types of audits.[11]

First-party audit	Second-party audit	Third-party audit
Internal audit	External provider audit	Certification and/or accreditation audit
——	Other external interested party audit	Statutory, regulatory, and similar audit

the performance of accredited activities. Depending on the audit or assessment criteria, it could also include review of the inspection, measurement, and test equipment (IM&TE), the technician performing the activity, the process (calibration procedure, records, documentation, and other outputs), or some or all these areas.

How often audits are conducted might depend on who is performing the audit and the purpose of the audit. An organization may receive an initial audit for ISO 9001 compliance and then be subject to surveillance audits on a regular basis as defined by the organization's registrar. Neither ISO/IEC 17025:2017 nor ISO 9001:2015 defines the frequency with which internal audits must be performed; instead, internal audit programs "shall" include the planned frequency as determined by the organization itself. Regardless of the internal audit frequency, the outcomes of internal audits should fulfill the standards requirements for internal audits.

Once an audit is conducted, the results must be documented and retained according to the organization's record retention policy. The key point here is that the records are retained for future reference. The records of follow-up audits to ensure that observations, findings, and/or write-ups have been corrected also must be reported and retained. These audit records shall include the identification of corrective and/or preventive actions taken based on the results of the audit. The audit records must meet the defined audit criteria and scope. Audit results are to be reported to the proper authority.

Reflective questions to consider include: Are the people using the quality system aware of the findings and updated on any changes to the system? Is there documentation that supports all of this? If procedures are changed, are the technicians, supervisors, and manager trained and authorized in the new or updated procedures and their training records updated accordingly? Is there an area in the audit for checking that training and/or competency records are properly maintained? (See Chapter 15 for more information on training and competency records.) It is important to assign custodial responsibility for audit deficiencies (i.e., specify the person assigned to follow through with the corrective action plan and to define the timetable for correcting discrepancies). How these details are to be determined, including at least timelines, the responsible party to whom the findings are assigned, and how long the records are maintained should be defined in the quality system.

Chapter 13

Scheduling and Recall Systems

The past, the present, and the future—what do they all have in common? In the world of calibration, you can access all three through your calibration management software or schedule. The *past* shows not only what you *have* calibrated, but also what you *have not* calibrated: the inspection, measurement, and test equipment (IM&TE) overdue for calibration.

The IM&TE waiting to be calibrated could be referred to as the *present*. It sits on the incoming shelf waiting for time, standards, a free technician, funds from its owner, or technical data, owner's manuals, parts, etc.

The *future* has another name in the metrology community: It is called the *schedule*. Depending on your system, requirements, resources, or directives from upper management, your schedule could forecast the future workload in intervals of days, weeks, months, or longer. The deciding factors may include whether the customer is internal or external and the availability of staff, standards, bench space, and the cables/accessories required to perform the calibration.

The importance of proper scheduling of test equipment cannot be overemphasized. By predicting what will be required of a department's staff, standards, space, and time, the needed work assignments, use of standards, and combining or scheduling similar items to be calibrated together, you can turn what might be chaos into an orderly schedule of events. There is still the unexpected to consider, though. Staff may be affected by illness or injury or a family emergency. Customers buy new equipment and they do not always eliminate old items. Some new equipment in your workload may require you to suddenly invest in a new measurement standard, validate and/or verify a measurement procedure, and train people to competently perform the calibration. The laboratory may gain or lose an important customer. In truth, none of these items is unexpected, but their probability is low enough that they are not considered as much, and they are admittedly hard to quantify for planning purposes. So, an orderly schedule can be devised and is a desirable goal, but the prudent manager should be prepared for variation and change in it. One measure that can be quantified is the percentage of expected or scheduled calibrations and the percentage of unexpected or unscheduled calibrations. From these data (collected over time), trends can be detected, and schedules and resources can be prepared ahead of time.

Some departments evaluate their workload both on a weekly basis and by their latest 30-day schedule. Any updates would have already been accomplished as work was completed and should be current in their system. Any changes that might affect work should also show up on the new schedule and can be easily included. See Chapter 40, "Best Practices," for more discussion on scheduling. It has also been found that calibrating like items can reduce the normal time it takes to accomplish one item done multiple times (see Chapter 41, "Process Workflow," for more on calibrating like items).

Is there a system in place in case of the dreaded R word—recall? What are the ramifications of equipment found out of tolerance (your standards, customer equipment, or notification from your outside vendors)? What are the ins and outs of recalling equipment and determining the necessity for recall through stated requirements? Are any of these situations included your quality system in case they occur? Who is responsible, and for what are those people responsible? The following suggestions might help.

Some software management systems have a reverse traceability function that allows the user to quickly identify the standard or system used for a particular function. Then, when queried, the system can identify all items that have been calibrated using a particular standard or system. Test labs also need a way to reverse-trace which IM&TE were used for which tests. Most standards have requirements for having a system in place, in writing, that identifies how to proceed if a recall is necessary. This could include recall of standards, products, calibrated test equipment, and/or test items.

Here is how a reverse traceability problem might occur and be solved. A company sends its measurement standard out for calibration. The vendor informs the company that the standard was out of tolerance when received and provides the as-found and as-left data. The customer company must then decide if the out-of-tolerance condition of its standard had an impact on the IM&TE it was used to calibrate and if the calibrated equipment also had an impact on the production or process it was used on. Without having the proper documentation available to trace when and where the items were used, it would be impossible to know this information. This is another reason to maintain thorough and accurate documentation records. Also, having the ability to generate a reverse traceability list using software reduces the time it takes to find the equipment involved (versus searching through paper records that could be in multiple and unknown locations) and remove it from service before it affects other product or processes.

Once the affected IM&TE is identified and found, it should be segregated from other IM&TE and labeled as such so that anyone would readily know the equipment was out of service. Some companies have designated areas for this type of equipment or products. Sometimes, however, the equipment cannot be removed from where it is used because it is either too big or too complex to move. In these cases, lockout boxes on electrical plugs and lockout notices posted on the equipment can serve the same result: segregation. Simply identifying the problem without identifying the problematic equipment and removing it from service is not enough. What is needed is a "paper" trail showing what was accomplished, when it happened, what was accomplished to preclude it from happening again, and what the ramifications of the entire process involved were. Whether on paper or electronically, all these records should be part of the permanent record for your company. Not only will an auditor want to see these records, but they can be used for future training within the company of what not to do and how not to do it. If we do not learn from our mistakes, we are bound to repeat them. In the world of metrology, this has proven to be absolutely true.

Chapter 14
Labels and Equipment Status

What is the status of your inspection, measurement, and test equipment (IM&TE)? Does the user know? Can an auditor tell? Do you have to go to your computer or printout to know? When effectively used, calibration labels and their status in your calibration management system could answer all these questions. Quick, simple, and easy to use is the intent, if not the requirement, in most systems.

Multiple national and international quality standards address the requirements for equipment labeling. ISO/IEC 17025:2017, Clause 6.4.8, states, "All equipment requiring calibration or which has a defined period of validity shall be labelled, coded or otherwise identified to allow the user of the equipment to readily identify the status of calibration or period of validity."[1] ISO 10012-2003 reflects the need for labels, as stated in paragraph 7.1.1, "General," which reads, "Information relevant to the metrological confirmation status of measuring equipment shall be readily available to the operator, including any limitations or special requirements."[2] ISO 9001:2015 has a general requirement in 7.1.5.2, "Measurement traceability": "When measurement traceability is a requirement, or is considered by the organization to be an essential part of providing confidence in the validity of measurement results, measuring equipment shall be: . . . b) identified in order to determine their status."[3] NCSL International Recommended Practice-6: *Recommended Practice for Calibration Quality Systems for the Healthcare Industries* (RP-6; 2015), states in section 5.13.5, "Labels," "All M&TE included in the calibration quality system should be clearly identified to alert the user of its calibration status. The calibration status should be indicated by a label or tag prominently affixed on the equipment or by some other means clearly evident to the user. Requirements for calibration labels may include unique identification of M&TE, date of calibration, due date of next calibration, conformance to specifications or limitations of use, the individual performing calibration or affixing the label."[4]

Sometimes equipment is of a particular size or shape that does not allow labels to be attached easily. In some cases, color codes may be used to indicate the equipment status when referenced to a related schedule that includes the color code references. In those cases, the system followed is well documented and there is a procedure or process that indicates which color references the time period or other status details.

What kinds of labels are referred to in this chapter? Typical references include *calibration stickers, no calibration required* (NCR) *stickers,* or *limited calibration stickers.* They all have the same things in common: For the IM&TE that the labels reference or are affixed to, the labels identify their respective unique identification number, the date the IM&TE was calibrated, and the date on which IM&TE will have to be recalibrated (usually called the *calibration due date*), and they include the name, stamp, or signature of the person who performed the calibration. For equipment that requires frequent calibration and is therefore calibrated by the user, labels that say "Calibrated by User"[5] fulfill the calibration label requirement.

In some practices, if there is not a calibration label attached to the IM&TE, then it is assumed that the IM&TE does not require calibration. In other systems, when there is no label, the IM&TE is assumed to be uncalibrated. In either of these situations, though, best practices require the attachment of a readily apparent *no calibration required* (NCR) label to the IM&TE. These IM&TE should have been justified and documented as having no effect upon the respective products, production processes, production equipment, or safety conditions.[6] It is helpful to also include the unique identification number of the test instrument somewhere on the NCR label. This prevents a second party from removing the label and placing it on another unit. Sometimes during audits and inspections, the equipment user has been known to take drastic measures to keep from being cited with a deficiency for its IM&TE. Good labeling practice keeps honest people honest.

Some companies also make use of *do not use, out of calibration* labels for items that are out of calibration, broken, or waiting for service of some type, and when they cannot be removed from their place of use. These brightly colored labels easily get the attention of any potential user and allow for quick identification of IM&TE that cannot be used.

Another useful label is the tamper-evident seal that says something like *"out of calibration if seal is broken"* or *"Calibration void if broken."* It is placed over screws or access areas securing covers or panels used to enclose test equipment. Such seals are also used to cover holes or access panels that have adjustment areas, screws, or knobs that require access to be limited to authorized personnel. These types of seals are one way to meet the requirements of ISO/IEC 17025:2017, Clause 6.4.12, "The laboratory shall take practicable measures to prevent unintended adjustments of equipment from invalidating results."[7] ISO 10012:2003, Clause 7.1.3, as well as many regulatory agencies and company policies, requires: "Equipment adjustment control . . . by . . . Access to adjusting means and devices on confirmed measuring equipment, whose setting affects the performance, shall be sealed or otherwise safeguarded to prevent unauthorized changes. Seals or safeguards shall be designed and implemented such that tampering will be detected."[8] Even if not required, use of such seals is still a good practice. A tamper-evident seal serves as a deterrent to inappropriate adjustment, whether accidental or purposeful, and is a visual indicator of the integrity of the calibration.

Just because there are calibration labels on your IM&TE does not mean that everything is fine with the system. The labels must match the information in the calibration system database (as well as the information on the calibration record or certificate), and be legible, easy to find, and up to date. Here are hints on how to manage your labels, whether completed by hand or printed by a printing device:

- Use black or dark ink (no pencil, Sharpies™, Magic Markers™, or crayons).

- Cover the label with tape to help preserve the data (such as chemical-resistant or UV-resistant tape, when applicable).

- Make a new label if an error is made (lineouts, white-out, or other gross correction practices are not acceptable).

- Never use another technician's stamp, name, or identifying mark on your work.

- Keep all labels in a secure, locked area, with access limited only to those authorized to use them.

- If the IM&TE is small or otherwise makes attaching a label difficult, use an alternate system (metal tag, color coding, manila tag, referencing in accessible database) and document how this alternate system will be managed and used.

Most laboratory software database systems have provisions for printing calibration labels from the results of a calibration event, and there are a number of printer and label systems available. This type of system reduces or eliminates many of the potential problems such as transcription errors or typographical errors. Dates, technician identification, and equipment identity number should never have errors because the data is pulled directly from the actual calibration and equipment record.

Other types of labels that may have to be used in different systems include *limited calibration* stickers, any type of chart or graph that displays data for the user's benefit, *calibrate before use* labels, radiation labels, and/or preventive maintenance labels.

Limited calibration labels have the same information as a regular calibration label, but also have an additional area that identifies either the range or tolerances of the IM&TE that are limited, or those that cannot be used. Charts or graphs that are attached or referred to on a limited (or regular) calibration label must also have the same information on them as are on a regular calibration label: the unique identification number for that particular unit, the date calibrated, the next calibration date, and the name of the person accomplishing the calibration.

Calibrate before use (CBU) stickers are attached to units that require calibration before they can be placed back in service. Examples of these would be items that are rarely used over an extended period of time or units that received a calibration and revert to CBU status after a defined period of time, with a calibration sticker and CBU sticker both attached to the unit. Most items selected for CBU status have limited or specialized use, and the time and money spent to continually keep them in service is not worth the cost.

Standardize before use stickers can be used on items such as pH meters and conductivity meters that require a standardization against buffers or solutions that have traceable qualities to certified reference materials (CRM). Such a label can also be used on any other type of equipment that requires standardization or normalization before it is used.

When a preventive maintenance inspection has been performed on IM&TE, it is sometimes advisable to place a label on the unit reflecting its status. The information helps the customer know that the preventive maintenance has been performed if the unit did not receive a calibration. When used for informational purposes, labels can be employed to assure customers of equipment status, dates that must be followed, and who to contact when more information is needed. This is not to say that the IM&TE should look like a well-used automobile bumper with a variety of stickers, but sometimes information that the customer uses on a regular/daily basis can be displayed on a label or sticker for immediate use.

Any information displayed on a label should also be recorded in the calibration record, management software database, and/or customer information folder. In some cases, it is better to be redundant with information than to lose it when a label is removed or lost.

Chapter 15
Training and Competency

Everybody recognizes the importance of training, yet training is often one of the first tasks delayed or canceled when resources are constrained. It takes a concerted effort by leaders to define, plan, and execute a successful training program. Training is part of our lives, both in expanding our knowledge and experience and in learning to adapt to new ideas, concepts, and problems. No one is born knowing how to calibrate equipment. People are taught through formal education and on-the-job training, or self-taught through the internet, home study, or correspondence courses. No matter how one receives new information, training is a lifelong endeavor.

Training is of particular importance to the calibration field. There are just not enough trained calibration technicians available to fill all the job openings in the United States. This is due to three main factors:

1. Delayed formal recognition by the U.S. Department of Labor Standard Occupational Classification for calibration job codes and descriptions

2. Reduced number of trained calibration technicians coming from the Department of Defense military services

3. The small number of existing college-level calibration technical degree programs

Therefore, leaders need to be more efficient and effective with establishing internal training programs that can provide both formal and informal training paths for new technicians and continuing development of existing technicians.

BACKGROUND

There are many source documents that describe compliance requirements and benefits for performing and documenting training. Here we discuss three of the most relevant training documents regarding calibration positions: (1) ISO/IEC 17025:2017, (2) FDA 21 CFR Part 820, and (3) NCSL International Laboratory Management and Recommended Practices.

ISO/IEC 17025, which is universally recognized as describing technical competencies for calibration laboratories, describes training requirements in Clause 6.2, Personnel: "6.2.2 The laboratory shall document the competence requirements for each function influencing the results of laboratory activities, including requirements for education, qualification, training, technical knowledge, skills and experience."[1] Additionally, ISO/IEC 17025 describes the training documentation requirements for calibration laboratories: "6.2.5 The laboratory shall have procedure(s) and retain records for:

- determining the competence requirements.
- selection of personnel.
- training of personnel.
- supervision of personnel.
- authorization of personnel.
- monitoring competence of personnel."[2]

The FDA imposes high expectations on regulated companies regarding performance and document training. In 21 CFR Part 820, training is highlighted in the Quality System Requirements section: "820.25 Personnel: (a) General. Each manufacturer shall have sufficient personnel with the necessary education, background, training, and experience to assure that all activities required by this part are correctly performed. (b) Training. Each manufacturer shall establish procedures for identifying training needs and ensure that all personnel are trained to adequately perform their assigned responsibilities. Training shall be documented.[3]"

NCSL International offers multiple training-related guidance documents in two categories: Laboratory Management and Recommended Practices. The documents listed here are worth reading, especially if you are developing your own training program or wanting to perform a gap analysis for improvement. Let us begin with NCSL International's *Laboratory Management – 14 (LM-14): Metrology Human Resources Handbook*.[4] LM-14 has the following content: Job Background, Job Descriptions, Benchmarking Data, and Education Sources. NCSL International's *Laboratory Management (LM-13): Calibration Laboratory Personnel Qualifications* advances to the next level and identifies personnel qualifications.[5] LM-13 has the following content: Job Descriptions, Traits and Attitudes, Management, Guidelines for Calibration Technicians, and Training Resources. Finally, NCSL International *Recommended Practice (RP-17), Documenting Metrology Education, Training and On-the-Job Training* provides guidance on how to document training.[6] RP-17 has the following content: Basic Training Requirements, Training Needs, Assessments, Personnel Training, and Training Resources; there are also several document checklists in the Appendix.

COMPETENCE REQUIREMENTS

Before designing a training program, we should first identify the competence requirements for calibration laboratory personnel. These are the expected knowledge, attitudes, skills, and interpersonal skills (KASI)[7] of personnel for a given role (see Figure 15.1). These competencies are used:

- As the basis for expectation of performance on the job
- When selecting new hires or newly promoted personnel, when training personnel, and when supervising training

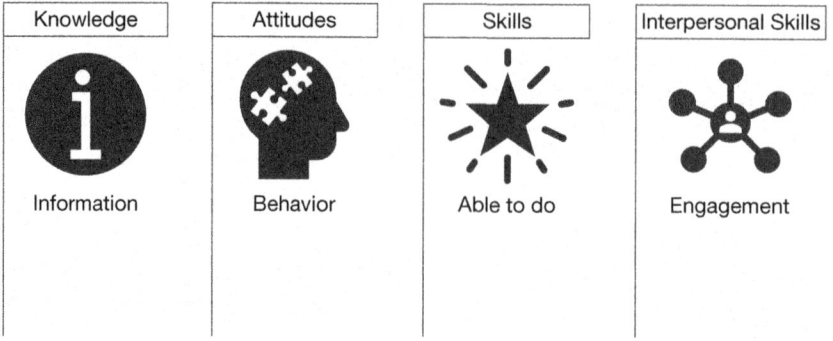

Figure 15.1 "KASI": Knowledge, Attitude, Skills, Interpersonal Skills.

- As the basis for authorizing personnel to work independently
- For continuous monitoring of competence

Knowledge includes all the information personnel must know in their assigned role. This may include prerequisite information required before hire and information required after hire to be successful in the role.[8] For example, you may require that a potential candidate technician have knowledge of quality fundamentals as a prerequisite for hiring, and once hired, knowledge of your quality steps when performing specific tasks.

Skills encompass what personnel must be able to do in their role.[9] Skills are divided into two areas: hard skills and soft skills. Hard skills are skills where proficiency can be measured. Examples include the ability to operate machinery, program computers, or speak a foreign language. These skills are taught using training materials and instruction, typically in a formal classroom, on the job, or both.[10]

Soft-skill proficiency is more difficult to measure and instruct. These skills involve the way you conduct yourself (attitude) and interact with others (interpersonal skill). Attitudes are what personnel must be willing to believe to achieve the desired culture of the laboratory, and they are reflected in behavior.[11] Examples are your willingness to be flexible during a period of change, to demonstrate positivity, and to manage your time. Interpersonal skills involve the way personnel interact with others and are also reflected in behavior. Examples are the way you communicate, solve problems, work on a team, and tolerate personality differences.[12] All these skills can be visualized as depicted in Figure 15.2.

When you think about identifying KASI for a calibration laboratory role, consider what you might identify if you were to advertise the position for hire. What prior areas of knowledge are needed for a laboratory technician? What skills are needed for a senior laboratory technician? What attitudes and interpersonal skills are needed for a laboratory manager? Once you have identified the KASI for each role in the laboratory, the next step is to document the information (usually in a job description), and then use it in your talent acquisition and maintenance process (hire, promote, train, supervise, monitor).

Figure 15.2 Competence (KASI criteria).

JOB DESCRIPTIONS

In the United States, the Standard Occupational Classification System (SOC) is currently used to classify all occupations in the economy, including private, public, and military occupations.[13] Information about occupations—such as employment levels and projections, pay and benefits, skills required, and demographic characteristics of job holders—is widely used by individuals, businesses, researchers, educators, and public-policy makers.[14] Without inclusion of "calibration" job codes in the SOC, U.S. educators and guidance counselors would not be aware of this field as a potential (and rewarding) career.

Last updated in 1991, and since replaced by the SOC, was the U.S. Department of Labor Office of Administrative Law Judges (OALJ) Law Library's *Dictionary of Occupational Titles (DOT)* (4th ed., Rev. 1991). The OALJ *DOT* had an occupational title listing for "Metrologist (profess. & kin.) 012.067-010. Develops and evaluates calibration systems that measure characteristics of objects, substances, or phenomena, such as length, mass, time, temperature, electric current, luminous intensity, and derived units of physical or chemical measure. Identifies magnitude of error sources contributing to uncertainty of results to determine reliability of measurement process in quantitative terms. Redesigns or adjusts measurement capability to minimize errors. Develops calibration methods and techniques based on principles of measurement science, technical analysis of measurement problems, and accuracy and precision requirements. Directs engineering, quality,

and laboratory personnel in design, manufacture, evaluation, and calibration of measurement standards, instruments, and test systems to ensure selection of approved instrumentation. Advises others on methods of resolving measurement problems and exchanges information with other metrology personnel through participation in government and industrial standardization committees and professional societies.
GOE: 05.01.04 STRENGTH: S GED: R6 M6 L6 SVP: 8 DLU: 77"[15]

There was also a 1991 OALJ Law Library occupational title of "019.281-010 CALIBRATION LABORATORY TECHNICIAN (aircraft mfg.; electron. comp.) alternate titles: engineering laboratory technician; quality assurance calibrator; standards laboratory technician; test equipment certification technician: Tests, calibrates, and repairs electrical, mechanical, electromechanical, and electronic measuring, recording, and indicating instruments and equipment for conformance to established standards, and assists in formulating calibration standards. Plans sequence of testing and calibration procedures for instruments and equipment, according to blueprints, schematics, technical manuals, and other specifications. Sets up standard and special purpose laboratory equipment to test, evaluate, and calibrate other instruments and test equipment. Disassembles instruments and equipment, using hand tools, and inspects components for defects. Measures parts for conformity with specifications, using micrometers, calipers, and other precision instruments. Aligns, repairs, replaces, and balances component parts and circuitry. Reassembles and calibrates instruments and equipment. Devises formulas to solve problems in measurements and calibrations. Assists engineers in formulating test, calibration, repair, and evaluation plans and procedures to maintain precision accuracy of measuring, recording, and indicating instruments and equipment.
GOE: 02.04.01 STRENGTH: L GED: R5 M5 L4 SVP: 7 DLU: 88"[16]

After 1991, the *OALJ Dictionary of Occupational Titles* was replaced by the United States Department of Labor's Standard Occupational Classification System. Surprisingly, no formal separate job codes for the calibration field were identified by the U.S. Department of Labor in the 2000 Standard Occupational Classification System. It took the collective efforts of the ASQ Measurement Quality Division and NCSL International members to perform the survey work necessary to formally submit a request to the Department of Labor to add calibration job codes into the Standard Occupational Classification document. After many years of dedicated and focused effort by the ASQ Measurement Quality Division and NCSL International, the Department of Labor added "calibration" as its own new job code, 17-3028 Calibration Technologist and Technicians, in the 2018 Standard Occupational Classification document:

> SOC Job Code 17-3028 Calibration Technologist and Technicians: Execute or adapt procedures and techniques for calibrating measurement devices, by applying knowledge of measurement science, mathematics, physics, chemistry, and electronics, sometimes under the direction of engineering staff. Determine measurement standard suitability for calibrating measurement devices. May perform preventive maintenance

on equipment. May perform corrective actions to address identified calibration problems.[17]

Two other job descriptions were submitted to the Department of Labor for inclusion in the 2010 Standard Occupational Classification document—Calibration Engineer and Metrologist—but were rejected as being too similar to existing engineer job codes:[18]

- **Calibration Engineer.** Apply measurement science, mathematics, physics, and engineering to design and develop systems, equipment, and methods to calibrate electrical, dimensional, optical, physical, mechanical, environmental, and/or chemical inspection, measurement, and test equipment (IM&TE). Analyze and solve calibration problems using advanced mathematical and engineering knowledge. Use statistics to analyze measurement standards and processes. May develop software to assist in calibration laboratory and/or departmental processes. Recommend calibration standards and IM&TE. Maintain calibration laboratory and/or quality systems. Perform laboratory and/or departmental administration and management.

- **Metrologist.** Apply measurement science, mathematics, and physics to develop, document, and maintain calibration systems, procedures, and methods for electrical, dimensional, optical, physical, mechanical, environmental, and/or chemical inspection, measurement, and test equipment (IM&TE) based on analysis of measurement problems and accuracy and precision requirements. Evaluate new calibration methods and procedures. Use statistics to analyze measurement standards and processes. May develop software to assist in calibration laboratory and/or departmental processes. Recommend calibration standards and IM&TE. Maintain calibration laboratory and/or departmental accreditation, and quality systems. Perform laboratory and/or departmental administration and management.

The U.S. Department of Defense (DoD) is well known for providing some of the best-trained calibration technicians and leaders. The DoD breaks down job descriptions into Military Occupational Specialty (MOS) codes. One example is from the U.S. Air Force and the Precision Measurement Equipment Laboratory (PMEL) 2P031 MOS[19]: "Performs and manages repair, calibration, and modification of test, measurement, and diagnostic equipment (TMDE), including precision measurement equipment laboratory (PMEL) standards and automatic test equipment. Supervises the process and use of TMDE to perform voltage, current, power, impedance, frequency, microwave, temperature, physical-dimensional, and optical measurements. Inspects, aligns, troubleshoots, and repairs PMEL standards, common and weapon system peculiar TMDE. Inspects TMDE for preventive maintenance, cleanliness, and safety requirements. Performs equipment maintenance using theories of operation, block diagrams, schematics, logic trees, and software diagnostics. Isolates malfunctions to component level. Calibrates and certifies TMDE to technical data specifications ensuring traceability to Air Force Reference Standards. Records and reports maintenance data; prepares technical order improvement reports, special training requests, training quality reports, and modification proposals. Tracks equipment warranties. Provides training and manages technical order distributions. Handles, labels, and disposes of hazardous materials and waste according to environmental standards. Plans, organizes, and coordinates mission support requirements. Collects and analyzes maintenance data and performs trend analysis. Identifies mission essential TMDE and its impact on workload. Coordinates lateral support, command certification, or contract services. Evaluates procedures for storage, inventory, and inspection of property. Provides training and assistance to TMDE users. Maintains PMEL automated management systems (PAMS). Develops and evaluates workload plans, budget, and support agreements. Manages PMEL quality program (QP). Submits reports to higher headquarters, maintains a safe working environment, and ensures laboratory certification."

TRAINING PLAN

While effectively written job descriptions are the foundation for successful training, a well-defined training plan is the key to having a successful and long-lasting training program. A training plan has three objectives:

1. To provide a method for documenting training.

2. To provide an individual road map to training that is targeted for the year or for a longer term.

3. To provide a visual display of the training that has been accomplished and what training is still needed within the job description or for development into a new role.

A training matrix is a tool that can meet all three training plan objectives. A typical training matrix is comprised of these areas (see Table 15.1):

- **Column.** Training task, Training reference/source, Training levels with dates and Trainer/Evaluator initials

- **Row.** Training tasks aligned to job description

Table 15.1 Training and competency matrix example.

Training Task	Reference	Introduction	Skilled	Trainer / SME
A.1 Calibration Standards	SOP-01	05-31-2001 WEN	11-01-2010 WEN	
B.1 Electronic	SOP-20	05-31-2001 WEN	07-15-2008 WEN	10-10-2010 WEN
C.1 Physical Measurement	SOP-15	FY22		
D.1 Dimensional	SOP-11			
E.1 Environmental	SOP-22	05-15-2008 WEN	FY22	
F.1 Writing Cal Procedure	SOP-02	06-10-2005 WEN	11-01-2007 WEN	10-10-2010 WEN
G.1 Audits	SOP-03	FY22		

Training and competency levels can be broken down into three primary areas:

- Level 1—Introduction: Identifies that the person has been introduced to the task, which signifies that they are in the process of learning the task.

- Level 2—Skilled: Identifies that the person has attained sufficient knowledge of and/or skill in the task and is authorized to perform the task unsupervised.

- Level 3—Trainer: Identifies that the person has sufficient knowledge and ability to be identified as a subject matter expert (SME) and is authorized to train others on this task.

Using such a training and competency matrix, the manager/supervisor can identify what training tasks are targeted for completion for the year by placing a "FYXX" notation within any task at any level to signify that this is the targeted training for the year identified, with the current year being the focus of the "training plan." To ensure that the training plan is maintained and updated, the training and competency matrix should be reviewed with the individual each time a performance review is scheduled or at time(s) otherwise determined. The expectation is that the individual is responsible for updating the training matrix to the targeted training tasks as part of the overall performance objectives for the year.

TRAINING SOURCES

Training sources can be broken into two main categories (see Table 15.2):

1. **Formal training.** This type of training is typically associated with having a facilitator to perform or facilitate training, an assessment to demonstrate competency, and a certificate to show that the training was completed successfully.

2. **Informal training.** This type of training is typically associated with being self-taught, is performed through individual research, and does not involve a facilitator.

Table 15.2 Formal vs. informal training.

FORMAL	INFORMAL
Accredited college/university	Online learning (videos, self-paced e-learning, podcasts, etc.)
Professional certification	Reading (magazines, books, articles, blogs)
Apprenticeship	Webinars
On-the-job training	
Seminar	
Accredited professional training	

FORMAL TRAINING

There are several categories of formal training:

- **Formal education.** This type of training is typically performed at a college (and in the United States, one that is accredited by the U.S. Department of Education or recognized by the state of residence). Technical colleges are included in this category. The NCSL International's *LM-14* includes a list of colleges and universities in the United States that provide calibration-related degrees. The U.S. Department of Defense also provides exceptional training within the military services through which equivalent college credits can be obtained and recognized by the U.S. Department of Education. Globally, National Metrology Institutes (known as NMIs, such as NIST, NPL, NRC, PTB, AIST, CENAM, etc.) may offer in-depth in-person and/or remote training (virtual instructor-led or instructor-facilitated e-learning).

- **Professional certification.** This type of training is typically performed by an organization that develops the criteria for evaluating a person's capability and knowledge to signify attainment of a certification level by passing an examination or completion of a course of study. This is usually combined with the individual having a minimum amount of work experience before being qualified to take the examination. One certification example is ASQ's Certified Calibration Technician (CCT) program.[20] The basic requirement for the CCT is to have five years of full-time, paid work experience in one or more areas included in the CCT "Body of Knowledge" and then to pass the certification exam. The ASQ CCT requires a recertification every three years.

- **Apprenticeship.** This type of training is for practitioners of a trade or profession with on-the-job training and some accompanying study. Most of this training is done while working for an employer who helps the apprentices learn their trade or profession, in exchange for their continued labor for an agreed period after they have achieved measurable competencies. In some cases, people who successfully complete an apprenticeship can reach the "journeyman" or professional certification level of competence and obtain a "license" or other signifying designation. In other cases, they can be offered a permanent job at the company that provided the placement.

- **On-the-job training.** This type of training can be considered a company's internal "apprenticeship" program for which the company develops its own training program. A training matrix is used as a road map to identify the training requirements in order to attain sufficient skill levels to meet the job description. It meets the formal training criteria in that there must be a trainer and there must be documentation to provide objective evidence of successful attainment of the training skill levels.

- **Seminars.** This type of training is typically performed in a classroom setting, although modern technology allows for a virtual classroom setting through a web-based connection. Typically, a certificate of training is provided after successful attendance of the seminar.

Measurement Science Conference (MSC), ASQ, and NCSL International provide frequent conferences, seminars, and tutorials on many topics related to quality, calibration, metrology, and testing. In particular, the ASQ Measurement Quality Division's (MQD) members and member leaders are a dependable source of education and training. They regularly present papers and tutorials at conferences and training events, including the ASQ World Conference on Quality and Innovation (WCQI), NCSL International, MSC, with other ASQ sections and technical communities, and several other presentation opportunities. ASQ-MQD member leaders lead the update to the *Metrology Handbook* editions. *Cal Lab: The International Journal of Metrology* (https://www.callabmag.com) is a quarterly publication that provides a directory of scheduled formal training on calibration-related topics.

- **Professional training events.** This type of training is typically performed in a classroom setting, or through a web-based connection for a virtual classroom setting, by an accredited training organization. With an accredited organization there is assurance that training content and assessments are instructionally sound, and successful completion is usually recognized with the formal conveyance of a training certificate. One example of accreditation for continuing education and training is International Accreditors for Continuing Education and Training (IACET, https://www.iacet.org). One can search for IACET-accredited training organizations through a variety of search options.

INFORMAL TRAINING

Here are some informal training categories:

- **Online training courses or information.** This type of training is self-taught, accessed over the internet, and may be associated with a final exam to provide objective evidence of knowledge attainment, but is not facilitated by an instructor or offered as an accredited course. Examples include manufacturer videos or podcasts on processes. In some cases, a certificate can be produced, which a manager/supervisor can sign and thereby attest to the accomplishment of the training task, and added to the individual's training matrix/record. A2LA WorkPlace Training e-Learning is one example that provides online calibration-related training courses with a final exam and certificate of accomplishment.[21]

- **Magazines/books/articles.** This type of training is self-taught, accessed through individual research, and seldom documented. Nevertheless, the wealth of information available is still vital to learners' continued journey in attaining knowledge that will help them perform better in their job role and prepare them for advancement. Here is a list of U.S.-based magazines that focus on calibration-related topics:

    o ASQ *Quality Progress*

    o *Cal Lab: The International Journal of Metrology* (Cal Lab Magazine)

- o NCSL International *Measure*
- o NCSL International *Metrologist*
- o *Quality Digest*

- **Webinars.** This type of training is related to seminars, but webinars are more informal in that they typically do not provide a certificate of accomplishment. Additionally, the training may be both shorter in duration and cover more narrowly focused topics. This type of training is seldom documented but nonetheless provides valuable information. As one example, equipment manufacturers offer webinars specific to metrology on their company websites or through various social media platforms.

SUMMARY

The topic of training presents a dichotomy in that on one hand everyone recognizes the importance of training and then on the other hand it is the first activity delayed or canceled when resources are constrained or hard to obtain. Thus, it is important that managers and supervisors focus efforts on identifying proper competencies and implementing an effective and efficient training program. This starts by having well-defined job descriptions and then building on that foundation by developing a comprehensive training plan using a training and competency matrix as a tool.

Managers and supervisors need to set the tone by stressing that both training and competency are important and that they will be continuously tracked and performed. They must do so not only because it is required by regulated companies and accreditation standards, but also because it is needed to develop individuals to be more successful in their job roles and to help with retention of hard-to-come-by skilled calibration professionals.

Chapter 16

Environmental Controls

The importance of the environment in which inspection, measurement, and test equipment (IM&TE) is calibrated usually does not come to mind immediately for most laboratory personnel. A major reason might be that most laboratory personnel have not been trained in the importance of environment and how different environmental factors may affect measurements. Some of the environmental factors to be considered include temperature, humidity, radio frequency, vibration, electrostatic discharge, dust, sunlight, and drafts. All IM&TE measurements have uncertainty, and the environment where an item is calibrated and/or used can be a significant contributor to the item's measurement uncertainty. A review of what the standards require concerning environmental controls should remove any doubt as to how critical a part such controls play in a company's measurement (test or calibration) processes.

Environmental controls are so important to processes and activities that multiple national and international standards include environmental controls in their requirements. According to ISO/IEC 17025:2017, Clause 5.3, Accommodation and environmental conditions, "The laboratory shall ensure that the environmental conditions do not invalidate the results or adversely affect the required quality of any measurement . . . . The laboratory shall monitor, control, and record environmental conditions . . . where they influence the quality of the results."[1] ANSI/NCSL Z540.3-2006, *Requirements for the Calibration of Measuring and Test Equipment*, Clause 5.3.6, states, in part, "All factors and conditions of the calibration area that adversely influence the calibration results shall be defined, monitored, recorded, and mitigated to meet calibration process requirements. Note: Influencing factors and conditions may include temperature, humidity, vibration, etc."[2]

ISO 10012-2003, Clause 6.3.1, states that "[m]easuring equipment shall be used in an environment that is controlled or known to the extent necessary to ensure valid measurement results. Measuring equipment used to monitor and record the influencing quantities shall be included in the measurement management system."[3] ANSI/ASQC M1-1996, *Calibration Systems*, Clause 4.4, Environmental Controls, states: "Environmental controls shall be established and monitored as necessary to assure that calibrations are performed in an environment suitable for the accuracy required."[4] ISO 9001:2015, Clause 6.4, Work environment, states, "The organization shall determine and manage the work environment needed to achieve conformity to product requirements."[5] Finally, NCSL International Recommended Practice-6: *Recommended Practice for Calibration Quality Systems for the Healthcare Industries, RP-6,* 2015, section 5.11, Environmental controls, reads, "The calibration environment need be controlled only to the extent required by the most environmentally sensitive measurement performed in the area. To show compliance with environmental requirements, environmental conditions should be monitored, and a record maintained of these conditions."[6]

It is obvious from the inclusion of environmental requirements in each of these standards that the conditions where IM&TE is calibrated and/or used are critical. How important those conditions are can be determined by the IM&TE's tolerances as specified by the manufacturer. Most manufacturers' operating manuals list the operational temperature and humidity limits, as well as any other limiting factors for that type or model of measurement equipment. When considering these manufacturers' prescribed limits, equipment users need to keep in mind that the Type B uncertainties provided by the manufacturers are only applicable to those environmental conditions. How do the device's characteristics change outside of those limits? Without conducting your own testing to find out, you do not know. Moreover, you are unlikely to find that out even when you send the standard out for calibration unless that provider can simulate your actual environment during the calibration.

One of the more demanding areas where the environment is critical is a 20 °C (68 °F) dimensional calibration room. With temperature fluctuations typically not allowed to exceed 0.5 °C, the maintenance, monitoring, and use of these areas is critical. Temperature (and humidity) recording devices are employed to monitor the area, with their data stored for immediate use and future reference. There is a saying that by making a measurement, one changes the measurement. Consider how the mere presence of a human being can change the ambient temperature. A typical adult at rest radiates the same temperature as a 100-Watt light bulb.[7] In some areas, an alternative source of heat (a light bulb) is turned on or off whenever a person leaves or enters the room so as to balance the heat loss and maintain a constant temperature. Additionally, the use of small tabletop fans creating air movement between a technician and the test or calibration setup can help dissipate the impact of body heat from the technician. (The use of a simple fan is credited to the now-retired physicist and dimensional measurement guru at NIST, Ted Doiron.) This practice is particularly helpful for dimensional measurements and other measurements that are not affected by air flow or vibrations. The measurement standards, units to be tested or calibrated, and work area should all be allowed to acclimate together. Clean gloves should be worn and/or rubber-tipped tongs should be used when handling reference standards (such as gage blocks or length standards) to minimize the transfer of temperature from the technician's hands and body.

In areas that require other types of controls, foresight in laboratory or facility design plans is necessary. It is far less expensive to initially design grounding systems, entry control points to minimize dust or contamination problems, and temperature/humidity controls than it is to upgrade an existing area or lab. Sticky mats, EMF grounding straps, shoe covers, lab coats, smocks, sterile gloves, and so on, are only a start to the list of items needed to meet certain regulatory requirements. Some companies can conform to these requirements by monitoring their heating/air conditioning systems or outlets; others are required to have monitoring devices in every room of their facility. A useful guide to designing laboratory environments is *NCSL International RP-7 Laboratory Design 2000* (2017).

In many laboratory environments, applicable standards or regulations require that the temperature (and relative humidity) in the laboratory be continually monitored and that the current values of those conditions be entered as part of a calibration record. The traditional method of doing this has been to use a circular seven-day chart recorder. Though useful, this method does have some problems and limitations. It is not possible to analyze the data, the accuracy and resolution are limited, there is more paper to file, and each technician must look at it to estimate the readings during each calibration. If the laboratory has a computer system, using a high-accuracy temperature/humidity data logger can avoid all this. The data logger is a small device that can be mounted in a convenient location in the laboratory and connected to a computer. The manufacturers have software that can either monitor the data constantly or download the accumulated data at intervals. Some software systems can also automatically notify assigned personnel if monitored environmental conditions have exceeded (or are close to) limits. If the data are stored in a database, the random and systematic variations of temperature and humidity can later be statistically evaluated to produce a Type A value for uncertainty analysis. If the laboratory has a network and a calibration management system, another level of automation can be added. The calibration management system software can be programmed to automatically read the data logger at the start of a calibration procedure and put the temperature and humidity into the record. This eliminates a potential error source and can be used to describe the environmental conditions in more detail than a discrete point: for example, "the environmental conditions during this calibration were between 20.28 °C and 20.63 °C and 42.5 % RH and 44.1 % RH."

The laboratory should consult standard methods (such as for applicable calibration or testing) and recommended practices such as *NCSL International Recommended Practice: Laboratory Design (RP-7)*, *NCSL International Recommended Practice: Guide to Selecting Standards Laboratory Environments (RP-14)*, *NCSL International Recommended Practice: Verification of Laboratory Environments (RP-16)*. Notably, NCSL International Recommended Practices (RPs) are internationally considered the benchmark for lab practices. The OIML's *G 13 (Ex P 7): Planning of Metrology and Testing Laboratories* (1989 edition) references *NCSL RP-7 Laboratory Design* and the U.S. national standard for dimensional measuring, (ANSI) ASME B89.6.2-1973 (R2017), Temperature and Humidity Environment for Dimensional Measurement.[8]

The laboratory must also evaluate the guidance in terms of the measurement standards it uses. For example, a common recommendation for the temperature of an electronics calibration laboratory is 23 °C $\pm$ 5 °C. If the laboratory has equipment such as a long-scale digital multimeter (one with a resolution of 1 mV or better on the 10 V range), the real temperature requirement of the equipment is more likely within 23 °C $\pm$ 1 °C.

On April 15, 1931, the CIPM meeting unanimously adopted the temperature of 20 °C "[a]s the normal temperature for adjustment of industrial standards...."[9] This same reference temperature was incorporated into the first ISO standard, published in 1951, "ISO 1—Standard reference temperature for industrial length measurements." This defined the standard temperature for industrial length measurements at 20 °C.[10] ISO 1:2016—*Geometrical product specifications (GPS)— Standard reference temperature for the specification of geometrical and dimensional properties* is also applicable to the definition of the measurand used in verification or calibration.[11] At the time this edition of the *Handbook* was being written, ISO 1:2016 was in the process of being updated.

International Society of Automation (ISA) has published a standard, ISA-TR52.00.01-2006, *Recommended Environments for Standards Laboratories*, where it defines distinct levels of measurement reference standards:

3.1 Echelon I—at the National Metrology Institute level.

3.2 Echelon II—further divided into two types:

Type I—upper-level standards calibrated by comparison to Echelon I standards.

Type II—lower-level standards used to calibrate standards in Echelon III.

3.3 Echelon III—production-level measuring instruments.[12]

ISA-TR52.00.01-2006 provides guidance on environmental conditions such as acoustic noise, dust particle count, electrical and magnetic fields, laboratory air pressure, lighting, relative humidity, temperature, vibration, and voltage regulation. ISA-TR52.00.01-2006 refers to NCSL International RP-14 for several environmental condition limits.[13]

The recommended temperature and humidity ranges and limits from ISA-TR52.00.01-2006 are listed in Table 16.1.

Some IM&TE have a standardization routine (often called self-calibration or auto-calibration) that requires the temperature to be within 1 °C of the temperature the last time that routine was performed, and it must be performed at least once every day if the instrument is being used at its tightest accuracy. Close temperature control, with the goal of minimizing variation, is also necessary to reduce

Table 16.1 SA-TR52.00.01-2006 temperature and humidity limits.[14]

Parameter	Echelon II type	Target temperature	Temperature tolerance	Temperature tolerance at gaging point	Relative humidity (RH) tolerance
Dimensional and optical	Type I	20 °C	± 0.3 °C	20 °C ± 0.1 °C	<45 % RH
	Type II	20 °C	± 1 °C	20 °C ± 0.3 °C	<45 % RH
Temperature, acceleration, DC (direct current), low-frequency, pressure-vacuum	Type I	23 °C	± 1 °C	N/A	35 to 55 % RH
	Type II	23 °C	± 1.5 °C	N/A	20 to 55 % RH
Flow, force, high-frequency, microwave	Type I	23 °C	± 1.5 °C	N/A	35 to 55 % RH
	Type II	23 °C	± 1.5 °C	N/A	20 to 55 % RH

Table 16.2 General-purpose calibration laboratories.

Measurement area	Temperature	Stability	Relative humidity (RH)
General dimensional calibration	20 °C ± 1.0 °C	± 0.5 °C per hour	20 to 45 % RH
Mass calibration	20 °C ± 0.5 °C	Depends on accuracy	40 to 50 % RH
Force calibration	23 °C ± 1.5 °C	Depends on accuracy	40 to 50 % RH
Optics calibration	23 °C ± 0.5 °C	Depends on accuracy	20 to 45 % RH
General DC electrical calibration General AC electrical calibration	23 °C ± 1 °C 23 °C ± 2 °C	± 0.5 °C per hour	> 20 % RH < 70 % RH
Most other physical disciplines: Temperature, volume, frequency	23 °C ± 2 °C	± 1 °C per hour	20 to 55 % RH

Table 16.3 Standards calibration laboratories or higher-accuracy requirements.

Measurement area	Temperature	Stability and uniformity	Relative humidity
Dimensional and optical	20 °C ± 0.5 °C	± 0.1 °C per hour	< 45 % RH
Precision mass calibration	20 °C ± 0.1 °C	± 0.1 °C per hour	35 to 55 % RH
Precision force	23 °C ± 0.25 °C	± 0.1 °C per hour	35 to 55 % RH
High-precision DC electrical	23 °C ± 0.25 °C	Depends on accuracy	35 to 55 % RH
Physical, mechanical	23 °C ± 0.5 °C	Depends on accuracy	35 to 55 % RH

thermo-electric effects at connections. Other standards, such as some vector network analyzers, must be standardized again if the temperature changes by 1 °C during the measurements. With these instruments, the dimensions of connectors are critical to quality measurements, and at high microwave frequencies the thermal expansion or contraction from a 1 °C change can be a significant effect. Reference Table 16.2 for suggested temperature and humidities for various general-purpose calibration laboratory activities. For laboratories performing higher-accuracy calibrations, reference Table 16.3 for suggested temperature and humidities.

There are some things to remember when creating or maintaining a laboratory environment. This information is based on NCSL International Recommended Practices, ISA, and UKAS LAB-36:

- 45 % relative humidity is an absolute maximum for dimensional areas to prevent rust and other corrosion.

- 20 % relative humidity is an absolute minimum for all areas to prevent equipment damage from electrostatic discharge.

- Temperature stability is the maximum variation over time. This is typically measured at the work surface height.

- Temperature uniformity is the maximum variation through the working volume of the laboratory. This is typically measured at several points over the floor area between the average work surface height and one meter higher.

- The air-handling system should be set up so that the air pressure inside the laboratory area is higher than in the surrounding area. This will reduce dust because air will flow out through doors and other openings.

- Ideally, a calibration lab should not be contiguous with exterior walls of a building and should have no windows. This will make temperature control much easier.

- Some measurement areas may have additional limits for vibration, dust particles, or specific ventilation requirements.

- It is important that the working volume of the laboratory be free from excessive drafts. The temperature should be stable and uniform, and any temperature gradients, measured vertically or horizontally, should be small. To achieve these conditions, at the standard temperature of 20 °C, good thermal insulation and air-conditioning with automatic temperature control is necessary.

- The temperature control necessary depends, to some extent, on the items to be calibrated and the uncertainties required. For general gage work, the temperature of the working volume should be maintained within 20 °C ± 2 °C.[15] Variations in temperature at any position should not exceed 2 °C per day and 1 °C per hour.[16] These are the minimum expectations for United Kingdom Accreditation Service (UKAS) accreditation.

- For higher-grade calibrations demanding smaller uncertainties, such as the calibration of gage blocks by comparison with standards, the temperature of the working volume should be maintained within 20 °C ± 1 °C. Variations in temperature at any position should not exceed 1 °C per day and 0.5 °C per hour.

- For the calibration of gage blocks by interferometry, the temperature within the interferometer should be maintained within 20 °C ± 0.5 °C. Variations in temperature shall not exceed 0.1 °C per hour.

- Within the laboratory, storage space should be provided in which items to be calibrated may be allowed to soak to reach the controlled temperature. It is most important that, immediately before calibration, time is allowed for further soaking adjacent to, or preferably on, the measuring equipment. Standards, gage blocks, and similar items should be laid flat and side by side on a metal plate for a minimum of 30 minutes before being compared. Large items should be set up and left overnight. This is to ensure that temperature differences between equipment, standards, and the item being measured are as small as possible.

One condition that seems to consistently get overlooked is local acceleration of gravity corrections. This can affect force, mass, and pressure measurements. Metal plates used in deadweight testers in force and pressure disciplines are often marked and calibrated to a weight or force and not mass. However, the lab that does the weight calibrations may be in a different location than where these plates are used. If the local acceleration of gravity of the laboratory that calibrated the plates is known or provided, and the organization using the plates to perform a force or pressure measurement knows its own local gravity, then a local acceleration of gravity correction can be accurately calculated using these measurements. Without this information, the pressure or force measurement will be off by an amount and in a direction that is not known. These uncorrected deviation values can make the difference between being in tolerance or out of tolerance! Obviously, the local gravity of your own facility does not have to be monitored because it is based on latitude, longitude, and altitude, but there might be situations in which it needs to be known. This can be determined in different ways. One is actually having an accredited organization perform a measurement of your facility's gravity. Another method, at least in the United States, is to acquire your longitude and latitude and elevation and input those into NGS SURFACE GRAVITY PREDICTION (noaa.gov) at https://geodesy.noaa.gov/cgi-bin/grav_pdx.prl.[17] In either case, take note of the uncertainty of the measurement.

Consider what conditions are appropriate for the measurement. Clause 7.5.1 under 7.5—Technical records in ISO/IEC 17025:2017 states: "The laboratory shall ensure that technical records for each laboratory activity contain the results, report and sufficient information to facilitate, if possible, identification of factors affecting the measurement result and its associated measurement uncertainty and enable the repetition of the laboratory activity under conditions as close as possible to the original."[18] Clause 7.8.1.2 states that "results shall be provided accurately, clearly, unambiguously and objectively, usually in a report (e.g. test report or a calibration certificate or report of sampling) and shall include all the information agreed with the customer and necessary for the interpretation of the results and all information required by the method used."[19]

Regardless of what conditions are implemented in a laboratory, one must note that limits are generally set to legitimize the Type B uncertainties associated with a testing or calibration system. When this is the case, the measurement uncertainty of the instruments used to monitor these conditions should be used to guard-band those same limits. For example, with a temperature sensor that has an expanded uncertainty of ± 0.25 °C and an upper limit of 25.0 °C, accredited activities should stop if the temperature goes up to 24.75 °C. This means that tighter limits will require tighter-accuracy instrumentation.

Standard test and calibration methods, equipment, customers, and funding budgets may dictate more specific or different environmental criteria. Manufacturers also set temperature and humidity limits for their equipment. Whatever criteria you follow, understand the impacts and be able to measure, monitor, and control them and compensate if necessary. When determining the appropriate environmental conditions and limits, it is critical to understand the process, methods, equipment, and desired result(s). Basic practices include "soaking" reference standards, units under test or calibration, and lab workspace in the same temperature and humidity so as to minimize temperature differences among the items. For example, placing an object or reference standard on the pan of an electronic balance or scale will create a thermal drift in the instrument that will, in turn, affect the accuracy of the balance or scale. For dimensional measurements and calibrations, there are enlightening videos on YouTube in which the growth in the length of a dimensional object caused by increases in temperature is readily apparent. Fluid volume calibrations are also affected by changes in temperature, humidity, and barometric pressure. Force and torque measurements are also affected by temperature, humidity, barometric pressure, and local acceleration of gravity. When the effects of these environmental conditions are not accounted for during measurement, it is possible that the increase in the measurement uncertainty could be larger than the accuracy tolerance. This increases the risk to the calibration or test provider as well as to their customers and to the users of products manufactured. Some of these differences can have an impact on health and safety. To say that calibration is the foundation of all measurements would not be an exaggeration. Proper calibration and application of metrological knowledge and principles can help reduce these risks and help improve the outcomes from our measurements.

Chapter 17
Industry-Specific Requirements

The phrase "different strokes for different folks" is very applicable to this chapter's subject matter. One might believe that calibration is calibration is calibration. In most cases, this might be true. But various industries, both in the United States and abroad, have their own particular requirements that must be met to conform to their standards. This chapter covers many of these industry requirements, reviews their specific demands in terms of calibration and recordkeeping, and identifies where unique emphasis is placed throughout their processes. Each area may have unique verbiage, acronyms, and guidelines that the calibration or metrology practitioner must come to know. The bottom line in all the standards is the same philosophy for any quality system: "say what you do, do what you say, record what you did, check the results, and act on the difference." This is also the quality tool known as plan–do–check–act (PDCA).[1] Meeting each of the directive's compliance requirements could determine whether or not a company passes an audit or assessment and then can advertise to the world that it is certified or accredited by a particular governing body or helping to bring new products or services to market faster. This chapter is by no means an attempt to reference all industries and their quality management standards. An entire book could be written on its own about quality standards across industries.

ISO/IEC 17025

ISO/IEC 17025:2017, "General requirements for the competence of testing and calibration laboratories," is *the* international standard for accreditation of both testing and calibration laboratories. Some industries require, by regulation, that laboratories performing testing and/or calibration be accredited to ISO/IEC 17025:2017. In other industries without this requirement, market pressure has pushed laboratories to become accredited in order to compete and stay in business. The 2017 version of ISO/IEC 17025 was realigned to match other ISO 17000-series standards and away from the previously ISO 9001-aligned format. The ISO 17000-series of standards and documents—i.e., any ISO document starting with 17xxx—are meant to be used for conformity assessment. The other main changes of the 2017 version compared to previous editions are as follows:

- Shift to risk-based thinking, which has enabled some reduction in prescriptive requirements and their replacement by performance-based requirements

- Greater flexibility in requirements for processes, procedures, documented information, and organizational responsibilities
- Definition of *laboratory* has been added[2]

The main differences between ISO/IEC 17025 and ISO 9001 are the technical requirements and more detailed requirements regarding defining, ensuring, and monitoring competence. After all, "competence" is in the title of ISO/IEC 17025:2017. Each clause has other differences compared to ISO 9001.

ISO/IEC 17025:2017 begins with Clause 1, Scope, which covers its purpose, applicability, and use. Clause 2, Normative References, includes documents that are referred to in ISO/IEC 17025:2017: *ISO/IEC Guide 99* (VIM)—also known as JCGM 200, and ISO/IEC 17000 Conformity Assessment—Vocabulary and general principles. Clause 3, Terms and Definitions, draws upon the VIM and ISO/IEC 17000 vocabulary with the following terms defined: *impartiality, complaint, interlaboratory comparison, intra-laboratory comparison, proficiency testing, laboratory, decision rule, verification,* and *validation*.

Clause 4 covers the general requirements of impartiality and confidentiality.[3] Impartiality is such an important consideration that it is both defined and a general requirement. Why is the commitment to and assurance of impartiality important? Because we, as customers and users of products and services affected by ISO/IEC 17025 laboratories' activities, must be able to have confidence in the integrity of the data. We must have confidence that the data generated are real and defendable. If a laboratory does not identify, review, and manage risks to its impartiality, its business and results are at risk of personnel being influenced to the point that they might be led into unethical business practices: for instance, giving preferential treatment to a customer or supplier or "manufacturing" results to meet customer demands instead of reporting the actual results—in other words, committing fraud. The other general requirement is *confidentiality*. This requirement includes protecting the laboratory's own information as well as protecting proprietary customer data and information. During an accreditation assessment, the assessor will typically examine the firewalls and other protections employed by the laboratory to safeguard impartiality and confidentiality.

Clause 5 covers structural requirements such as the legal responsibility for laboratory activities as well as the defined management structure and responsibilities.

Clause 6, Resource Requirements, describes the people (6.2, Personnel), places (6.3, Facilities and Environmental Conditions), things (6.4, Equipment; 6.5, Metrological Traceability), and (6.6) externally provided products and services necessary for being in conformance with this standard.

In ISO/IEC 17025:2017, the main focus of Clause 6.2, Personnel, is *competence*. Training is just one of the ways of becoming competent, but training and competence are not interchangeable. One can be improperly trained and not be able to generate valid measurement results. It takes competence to be able to generate valid measurement results. See Chapter 15 for more about training and competency.

Clause 6.3, Facilities and Environmental Conditions, is further covered in Chapter 16.

Clause 6.4, Equipment. As with the other Clause 6 parts, without the proper and properly calibrated equipment, the output (measurement results) of testing and calibration labs would not be possible.

Clause 6.5, Metrological Traceability. The International Vocabulary of Metrology's definition of *metrological traceability* is "property of a **measurement result** whereby the result can be **related to a reference** through a **documented unbroken chain of calibrations**, each contributing to the **measurement uncertainty**" (*VIM* 2.41).[4] A key point about metrological traceability is that it is *measurement results* that are traceable (i.e., linked) to a standard (national, international, consensus, natural physical constant, certified reference material). Metrological traceability is not to a lab or to an entity, such as "traceable to NIST"; it is to reference standards. Regardless of the industry, ensuring metrological traceability is critical to the basis of good measurements. See Chapter 9 of this book for more information on metrological traceability.

Clause 6.6. The sources and suitability of externally provided products and services must be evaluated. This pertains to *any* product or service that affects laboratory activities where the product or service is used.

Clause 7 covers process requirements, such as for contract negotiation with customers and for selection, verification, and validation of methods. Clauses 7.3–7.8 cover laboratory activities, including the performance of and outputs of sampling, testing, and calibration. Clause 7.9 covers complaints (whether from inside or outside the laboratory). Clause 7.10 covers nonconforming work. Clause 7.11 covers control of data and information management.

Clause 8 is the section regarding management system requirements. Whether a laboratory chooses the option to use its own established and maintained ISO 9001 management system or its own ISO/IEC 17025-specific management system, as a minimum the laboratory management system shall address all of the subclause requirements in Clause 8. Subclauses include 8.2, Management System Documentation; 8.3, Control of Management System Documents; 8.4, Control of Records; 8.5, Actions to Address Risk and Opportunities; 8.6, Improvement; 8.7, Corrective Actions; 8.8, Internal Audits; and 8.9, Management Reviews.

Management review is the main key to the successful operation of any quality management system. ISO/IEC 17025 has very prescriptive requirements for management review. The requirements include a review of audits and proficiency tests, and other factors, such as training and assessments by both internal and external auditors and assessors. The prescriptive nature of the management review section is meant to ensure that necessary laboratory operations are reviewed in a systematic manner. The projected volume and type of work is also included in the prescriptive requirements, in order to require the laboratory to examine operations from a business perspective, not just a quality perspective. Provisions regarding corrective and preventive actions and control of nonconforming calibration each specifically require procedures to be developed and implemented. Overall, risk must be evaluated in multiple ISO/IEC 17025:2017 clauses. What these risks are and how they are mitigated are up to laboratory.

ISO 9001

Before the last half of the twentieth century, the majority of products we encountered were fairly local: They were made in the same country or in a close neighbor. International trade did not have a large economic impact on the average person. Now, the majority of products or their components that we encounter are sourced across multiple economies. To ensure the quality of products and services all over the world, businesses need a way to be assured that those products and services are produced in a quality manner. Standards that are specific to a company, industry, or country are not sufficient in a global economy. There is a need for an internationally recognized standard for the minimum requirements of an effective quality management system. The ISO 9001 series of standards was developed to fill that role.

Global Trade

Although trade has been in evidence well into prehistory, for most businesses it had been a peripheral part of their operations. This has changed over the past 50 years or so, to the point where most products are multinational and even the concept of a country of origin is questionable. Raw materials, design, parts manufacturing, hardware, assembly, software, machine tools, agricultural products, and more are all part of the global economy. Consider these examples:

- A computer dealer in the United States sells computers under its private brand name. The dealer buys components from various places—China, Taiwan, Korea, Singapore, Japan, and Israel—and the operating system and software from a company in the United States. These items are assembled into a functioning computer system, and the last step in the process is to place a "Made in the USA" label on the back.

- An automobile manufacturer, a joint venture of an American and a Japanese manufacturer, is in the United States. The design was developed jointly in the United States and Japan. The parts come from other places in the United States, Canada, and Mexico. Some parts come from a U.S.-based factory of a German company. The engine is made in Hungary. The transmission is made in another country. A considerable number of buyers choose this brand because they are convinced the vehicles are made in America.

- A modern airliner, such as the Boeing 777, includes parts and assemblies from hundreds of suppliers from countries around the world. The plane is sold internationally to airline operators.

- A software company in Atlanta, Georgia, develops code during the day and every evening electronically sends it to a subsidiary in Bangladesh. That organization (during its workday) evaluates the code for software quality and conformance to technical requirements. The bug reports are sent back to Atlanta in time for the start of the next day's work there. When we reference business hours, we have to keep in mind the locations where the business operations are occurring.

- A continuous-cast steel mill receives ore and scrap metal from multiple locations, including from other countries. The finished products are shipped to locations all over the United States and to other countries.

- When you visit your favorite grocery store (or when groceries are shipped to you from your favorite online grocer), you may be buying grapes from Chile, corn from Nebraska, crabs from Japan, mussels from Canada, dates from Palestine, apricots from Turkey, and lamb from

New Zealand, all without realizing it. Yet without global commerce, these seasonal products would not be available in the customer's local area for most of the year, if at all.

It should be evident that quality standards that are specific to a single company, industry, or country are not sufficient. The ISO 9000 series of quality management system standards, as well as the earlier versions, exist to serve as an aid to international commerce. In line with the mission of ISO, they are a means of reducing technical barriers to trade. The standards provide an internationally recognized set of minimum practices for an effective quality management system. The standards describe what practices should exist and what they should achieve; they do *not* prescribe how to do it.

National Versions of International Standards

Many countries have their own national versions of the ISO 9001 series and other international standards. In most cases the national version is a translated edition, sometimes with additional introductory material, which is published by the relevant national standards body. In the United States, for example, the national version of ISO 9001:2015 is ANSI/ISO/ASQ Q9001:2015. While the specified text (usually French and/or English) of the international (ISO) version of a standard is the official authoritative version, most countries designate their national version as the legal equivalent of the international standard.

Important Features of ISO 9001

It is important to understand that a quality management system (QMS) standard only applies to the management system of an organization. It does not have anything to do (directly) with a product. Service-only organizations also follow ISO 9001:2015. Where ISO 9001:2008 used the term "product" to include all output categories, the 2015 version now uses "products and services" to include all output categories.[5] The QMS standard is general because it can apply to any organization in any line of business in any country. The product is the subject of separate specifications and technical requirements. The product specifications are specific because they apply to a particular product or service, and sometimes to a specific supplier. The quality standard and the product technical requirements are separate but complement each other. Both are needed in an agreement between a supplier and a customer. Another important feature of the ISO 9001 system is the concept of third-party evaluation of an organization's QMS. In a third-party audit, a company is evaluated by a qualified organization that is independent of the customer and supplier, but that is trusted by both. The evidence of conformity to the ISO 9001 requirements is the registration or certification by the auditor. A company may accept that certificate and thereby eliminate the cost of sending their own or contracted people to each supplier to perform quality audits. It is beneficial to suppliers because they are audited to a single set of requirements, and they do not have to host as many audits.

The list of ISO 9000 family standards is available on the ISO website (www.iso.org) and should be checked at intervals, but it is not necessarily fully up to date. For most industries, ISO 9001:2015 is the only conformance standard and is the only one an organization's quality management system can be audited against. In the automotive industry, ISO/TS 16949:2009 applies. All the other documents in the ISO 9000 family are guidance to aid in implementation.

ISO 9000 and Your Business

An organization will be audited against the requirements of one of the conformance standards if it wants its quality management system to be registered (or certified, in many countries). A full discussion of this is outside the scope of this book, but here are some pointers:

- The ISO 9000 system does not say how to do anything. It describes a set of results to be achieved for an effective QMS, describes processes that must be in place, and provides general guidance. An organization decides on the best way to accomplish these results in its own structure. There is no such thing as the ISO way of managing anything.

- A well-run business will only have to make minor adjustments, if any.

- An organization does not have to reshape its business management system to achieve the standard. All it must do is describe how its system—whatever that is—meets the requirements of the standard.

- The largest problem areas are documentation and corrective action and preventive action.

What the Standard Says About Calibration

Whereas ISO 9001:2008 was much more detailed and prescriptive for the topic of calibration in Clause 7.6, Control of monitoring and measuring equipment, ISO 9001:2015 is far less prescriptive. ISO 9001:2015 now covers calibration under Clause 7.1.5, Monitoring and measuring resources. It only mentions "measurement traceability" (7.1.5.2) and does not use the words "metrological traceability," though, it seems, metrological traceability is the intent. "Measuring equipment shall be calibrated or verified, or both, at specific intervals, or prior to use, against measurement standards traceable to international or national measurement standards."[6] A glaring difference between ISO/IEC 17025:2017 and ISO 9001:2015 requirements is that calibration prior to use is optional in ISO 9001:2015.

Manufacturers and test and calibration laboratories and consultants are very frequently asked one question: "Do I *really* need to have this (whatever it is) calibrated?" The standard says that the answer is "yes" if it (whatever it is) is making a measurement that provides evidence of conformity to requirements. But there are still questions that fall into indeterminate areas as far as the standard (or an auditor) is concerned. In 2000, Philip Stein suggested another test:

> Ask the question: Does it matter whether the answer from this measurement is correct?
>
> - If it does matter, then calibration is needed.
>
> - If it does not matter, then why is the measurement being made in the first place?[7]

Stein's test also leads to a risk assessment that can apply to any measurement situation: What can happen if the measurement is wrong, and what can happen if the measurement is not made at all? For example, some organizations say that electricians' voltmeters need not be calibrated because they are not used for any product realization processes. They are "only used for troubleshooting and repairing plant wiring." Applying a risk assessment reveals the flaw in that line of thinking. If a voltage measurement is wrong, an electrician, believing it is safe to handle the wire, could be injured or killed by a line that is still live. As another example, many organizations say that certain meters (voltmeters again, for this example) need not be calibrated because they are "only used for troubleshooting and repair of equipment," and calibrated tools are used for final test. Applying a risk assessment here shows that if the voltmeter is reading incorrectly, there may be a risk of rework because of units that fail inspection.

So, there are cases where a measuring instrument may not have to undergo calibration for a process per the QMS, but calibration may still be required for other reasons. In addition to safety and reduction of rework, other reasons include health and regulatory compliance.

A metrology or calibration organization (a stand-alone company or an in-house department) must be sure to understand the implications of ISO 9001:2015, Clause 7.1.5. It means something different to a calibration laboratory than it does to a factory turning out 10,000 widgets per hour.

A calibration laboratory (or metrology department or any other variation of these terms) is a service organization. It is providing a service to its customers; the product is the service of calibrating the customers' calibration and test instruments. This means that all the workload items are customer-owned property that is passing through the laboratory's process (Clause 8.5 of ISO 9001:2015). The monitoring and measuring devices of Clause 7.1.5 are often the calibration standards—the instruments the laboratory uses when calibrating customer items. This includes their reference and transfer standards. Out on the production floor it may be possible to argue about the "where necessary to ensure valid and reliable results"[8] phrase, but in the calibration laboratory no argument is possible. It is clear that the measurement standards are necessary and that their results must be valid. There are two potential problem areas here.

If a calibration shows that the *as received* condition was out of tolerance, the organization must assess the condition and its impact on any product, take appropriate action, and keep appropriate records including the results of the calibration. If a measurement standard is found to be out of tolerance when calibrated, the calibration laboratory must assess the condition and its effect on the output. It has to determine all of the items that were calibrated by that standard since it was last known to be in calibration. (The ability to do this is called *reverse traceability*.) The laboratory must compare the out-of-tolerance condition to the performance specifications of the units under test and determine if the error in the standard makes the calibration results invalid. (If there is more than one model number, this must be done for each model.) In each case where the result would be invalid, the laboratory must notify the customer about the problem and request return of the item for recalibration. The customer then must evaluate the impact on their own production. As for keeping the results of the calibration, a laboratory should be doing that anyway for all their measurement standards.

Another potential problem would be if computer software is used to monitor and measure resources. In this case the organization must prove that the software operates as intended and produces valid results before placing it in regular use. See Chapter 18, "Computers, Software, and Software Validation," for more information on computers, software, and software validation.

ISO 10012

A note in Clause 7.6 of ISO 9001:2000 referred to ISO 10012, *Measurement management systems—Requirements for measurement processes and measuring equipment*. This standard (actually, its predecessors ISO 10012-1:1992 and ISO 10012-2:1997) was suggested as a guidance reference. When the first edition of this book was published, the United States was in the process of adopting a national version of this standard. It has been approved and adopted as the United States National Standard and is known as ANSI/ISO/ASQ Q10012-2003, *Measurement management systems—Requirements for measurement processes and measuring equipment*.

This is not a stand-alone standard. It is intended to be used in conjunction with other quality management system standards. It gives generic guidance for two areas: management of measurement processes as part of a quality or environmental management system, and management of the calibration (metrological confirmation) system for the measuring instruments required by those processes.

Remember that metrology is the science of measurement. With that in mind, there are two recurring phrases in the standard that merit discussion beyond the definitions in that standard:

- Clause 3.6 defines the *metrological function* as the "function with administrative and technical responsibility for defining and implementing the measurement management system." Elsewhere, the standard states that measurement requirements are based on the requirements for the product or, ultimately, the customer requirements. The measurement management system model is similar to the process management model presented in Clause 8.5 of ISO 9001:2015. It begins with the measurement requirements of the customer.

    Instead of product realization, the main processes are the measurement process and metrological confirmation. It also includes management system analysis and improvement, management responsibility, and resource management. Effectively, this means that everything that touches product realization and requires or makes measurements is part of the measurement management system. It is not, as people may believe, limited to the calibration activity.

- Clause 3.5 defines *metrological confirmation* as the "set of operations required to ensure that measuring equipment conforms to the requirements for its intended use." This includes calibration of measuring equipment, which is what people in the calibration lab are focused on, but it also includes verification: processes to ensure that the measuring equipment is capable of making the measurements required by the product realization processes and that the measurements being made meet the requirements of the product and the customer.

Therefore, it includes even production-level measurement system analysis.

A measurement process may include one or more measuring instruments and may occur anywhere in the product realization processes. Measurement processes may exist in design (such as transforming customer requirements to product specifications and tolerances), testing (such as measuring the vibration forces applied to a prototype), production, inspection, and service.

From this, it is clear that the metrological function or measurement management system encompasses all of the product- or process-oriented measurement requirements within an organization. It is not limited to the calibration activity and the calibration recall system. Factors to be considered by the system include, but are not limited to, risk assessment, customer requirements, process requirements, process capability, measuring instrument capability compared to the requirements, resource allocation, training, calibration, measurement process design, environmental requirements of each measurement process and instrument, control of equipment, records, customer satisfaction, and much more.

The metrological function is also responsible for estimating and recording the measurement uncertainty of each measurement process in the measurement management system and ensuring that all measurements are traceable to SI units. This includes traditional gage repeatability and reproducibility studies. It may even require expanding their use, or it may require preparing uncertainty budgets for each measuring system in the product realization processes.

ANSI/ISO/ASQ Q10012:2003 specifically states in its scope statement that it is not a substitute for or merely an addition to ISO/IEC 17025.[9] This does not mean an accredited laboratory can ignore it, though. An accredited laboratory may be part of an organization that uses this standard's guidance to manage its overall measurement system, or an accredited laboratory may choose to apply parts of the standard to its own system.

FDA QUALITY SYSTEM REGULATIONS

The FDA is responsible for protecting public health by ensuring the safety, efficacy, and security of human drugs, biological products, and medical devices. It does this by establishing regulations for companies wanting to manufacture and market life sciences products in the United States. The *Code of Federal Regulations* (CFR) establishes the U.S. Food and Drug Administration's guidelines on nonclinical studies, finished pharmaceuticals, and medical devices. Each title of the CFR addresses a different regulated area; 21 CFR relates to "Pharmaceuticals and Medical Devices" in general. At a high level, regulations in the CFR are laws which ensure that companies establish good laboratory and manufacturing practices under which their nonclinical studies and manufacturing operations are controlled to a high-quality and reliable level with sufficient document and testing traceability for the intended use. The four most applicable CFRs are:

1. 21 CFR Part 11—Electronic Records; Electronic Signatures

2. 21 CFR Part 58—Good Laboratory Practice for Nonclinical Laboratory Studies

3. 21 CFR Part 211—Current Good Manufacturing Practice for Finished Pharmaceuticals

4. 21 CFR Part 820—Quality System (QS) Regulation for Medical Devices

An important note to understand about the FDA CFRs is that they only provide the *minimum* compliance requirements for companies to manufacture quality products. The CFRs provide the required end result instead of prescribing specifics on how a manufacturer is to comply with the regulations. The FDA expects companies to use good judgment when developing a quality system that adheres to the FDA CFRs in order to produce high-quality products. For example, the CFRs provide only very basic compliance requirements for calibration. Elements, such as schedules, labels, procedures, etc., are stated in very few sentences and fewer than a few paragraphs. This is insufficient information for companies to build a fully compliant calibration quality system that can support high-quality manufacturing. The FDA is aware of this limitation and has published guidance documents to assist manufacturers in building better quality systems. One of the more detailed FDA guidance documents that does contain sufficient information on what a quality system looks like for calibration is the *Medical Device Quality Systems Manual: A Small Entity Compliance Guide*. Though published in 1996, it is still cited as a useful resource for FDA inspectors and is therefore a useful resource for regulated companies.

The compliance guide has these elements in the "Equipment and Calibration" chapter:[10]

- Equipment GMP controls
- Manufacturing materials
- Automated production and QA systems
- Measuring equipment calibration
- Audit of calibration system
- Integrating measurements into the QA system

Equipment GMP Controls

A key area often lightly touched on in FDA regulations is the expectation that equipment has maintenance performed in addition to periodic calibration. FDA expectations for periodic maintenance are similar to those for calibration: schedules, procedures, records, etc. One particular compliance requirement just for maintenance equipment is: "where adjustment is necessary to maintain proper operation, post the inherent limitations and allowable tolerances of the equipment or make these readily available to personnel responsible for making the adjustments."

Manufacturing Materials

The FDA CFRs are typically focused on medical product manufacturing. This section of the compliance guide discusses areas typically not found within the CFRs themselves, such as the expectation that manufacturing materials should be carefully analyzed, and their use controlled to ensure that any manufacturing materials do not affect production equipment or are left as residue on the equipment to affect operations.

Automated Production and QA Systems

This section of the compliance guide goes into great detail regarding the FDA expectations for equipment and software used in automated production operations. Topics such as software validation guidelines, employee responsibility and training, formal development of software, commercial software and equipment, validation of automated equipment and processes, automated data collection and processing, and equipment controls and audits are explained in minute detail.

Measuring Equipment Calibration

According to the compliance guide, under calibration requirements the Quality System (QS) regulation requires (in Section 820.72(b)) that equipment be calibrated according to written procedures that include specific directions and limits for accuracy and precision. Good Manufacturing Practice (GMP) calibration requirements are:

- Routine calibration according to written procedures.
- Documentation of the calibration of each piece of equipment requiring calibration.
- Specification of accuracy and precision limits.
- Training of calibration personnel.
- Use of standards traceable to the National Institute of Standards and Technology (NIST), other recognizable standards, or when necessary, in-house standards.
    Note: Current traceability statements are to SI units.
- Provisions for remedial action to evaluate whether there was any adverse effect on the device's quality.[11]

The FDA seems to place more emphasis on remedial action for product and documentation requirements than does any other agency, standard, or regulation. The safety of the American public is critical during the manufacture of food, drugs, and cosmetics. The actions taken by the manufacturer when equipment is found to be out of tolerance during routine calibration is critically looked at by the inspectors and auditors. How a company manages this particular process can make the difference between having FDA approval and not being allowed to produce a particular product. Is recall of product performed? What process is used to make that determination? What effect did the equipment have on the process or production? Is there documentation at all levels of decision making? Is that documentation available for inspection?

Some companies use a two-pronged approach to this process and call it "alert and action." Whenever equipment is found to be out of tolerance, alert procedures go into place. Notification is given to the equipment owner; repairs and adjustments are recorded, and their historical records are archived for future reference. Depending on the criticality of the equipment, how it was used in the process, and how much impact it may have had on product quality, a second process might be implemented, called "action." In the case of action procedures, critical evaluation of product or process is performed to evaluate the quality of product or process, and all levels of supervision up the chain must sign off that quality was not affected by the equipment being out of tolerance. This is far more time-consuming and critical to the manufacturing process than an alert procedure. What system a company puts in place would depend on the criticality of the product produced, the effect on the product users, and the possible impact of poor quality for the customer. Of course, both processes would involve careful documentation and archiving of the records involved with the affected equipment.

An interesting point made in the compliance guide related to "Equipment Selection" is the recognition that equipment with accuracies much larger than the processes can be used longer without recalibration than equipment that only marginally meets the accuracy requirements of the process.

Pertaining to the compliance guide "Management of Metrology" section, here is a partial quotation from 21 CFR § 820.72, Inspection, measuring and test equipment: ". . . Each manufacturer shall establish and maintain procedures to ensure that equipment is routinely calibrated, inspected, checked, and maintained. . . . These activities shall be documented. (b) *Calibration*. Calibration procedures shall include specific directions and limits for accuracy and precision. . . . These activities shall be documented. (1) *Calibration standards*. Calibration standards used for inspection, measuring, and test equipment shall be traceable to national or international standards [the SI]. . . . (2) *Calibration records*. The equipment identification, calibration dates, the individual performing each calibration, and the next calibration date shall be documented."[12]

Proper documentation is critical, not only to the process, but also for future statistical analysis, review of calibration intervals, and production trends and root cause analysis when problems are identified. By having the proper documentation in place, you do not have to spend extra time for no reason whenever a person in a critical or supervisory position changes. Continuity is an important ingredient of any company's production or manufacturing processes and procedures.

According to the compliance guide, a typical equipment calibration procedure includes:

- Purpose and scope
- Frequency of calibration
- Equipment and standards required
- Limits for accuracy and precision
- Preliminary examinations and operations
- Calibration process description
- Remedial action for product
- Documentation requirements[13]

Most of these can be found in other standards or regulations (see Chapter 5, "Calibration Procedures"). Under frequency of calibration, the calibration interval must be stated and followed according to the guide. A list of the equipment and standards must also be in the procedure. Within this list, the range and accuracy of both the standards and the equipment should be listed. Many types of equipment must be set up before any type of calibration can even be started, and that is what the standards and regulations refer to as preliminary examinations and operations. The description of the calibration procedure must be in enough detail to allow all levels of calibration professionals to be able to follow the instructions as written without additional supervision once they are trained in the use of that procedure and type of equipment.

The compliance guide's "Management of Metrology" section goes on to state that "[m]anagers and administrators should understand the scope, significance, and complexity of a metrology program in order to effectively administer it. The selection and training of competent calibration personnel is an important consideration in establishing an effective metrology program. Personnel involved in calibration should ideally possess the following qualities:

- Technical education and experience in the area of job assignment
- Basic knowledge of metrology and calibration concepts
- An understanding of basic principles of measurement disciplines, data processing steps, and acceptance requirements
- Knowledge of the overall calibration program
- Ability to follow instructions regarding the maintenance and use of measurement equipment and standards
- Mental attitude which results in safe, careful, and exacting execution of their duties."[14]

The compliance guide's "Calibration Records" section states that "[c]alibration of each piece of equipment shall be documented to include equipment identification, the calibration date, the calibrator, and the date the next calibration is due." Measuring instruments should be calibrated at periodic intervals established on the basis of stability, purpose, and degree of usage of the equipment. A manufacturer should use a suitable method to remind employees that recalibration is due.[15]

The FDA has placed great emphasis on data integrity of quality system records to include calibration. They have defined ALCOA[16] as a data integrity acronym to ensure that data is complete, consistent, and accurate; it stands for:

- **A**ttributable—The raw data are traceable to an authorized person recording the data.
- **L**egible—The raw data are easy to read.
- **C**ontemporaneously Recorded—The raw data are recorded at the time of observation.
- **O**riginal or True Copy—The data are a firsthand observation.

- **A**ccurate—The data provide an unaltered and correct recording of the observation.

Here are some commonly recognized Good Documentation Practices (GDPs) associated with GMP documents:

- Use black or blue indelible ink—no pencil, marker, or similar.
- All fields are completed or filled in. Use "N/A" to represent Not Applicable and provide an explanation.
- Writing over errors is not allowed.
- Put a single line-out through errors so the original content can still be viewed.
- Sign and date all corrections and include an explanation of errors.
- Never document or sign another person's work or data.
- Enter data as it is recorded or immediately upon completion.
- Avoid abbreviations, catchphrases, or unprofessional remarks.
- Enter data as received or viewed—never guess or anticipate readings or data.
- Round numbers using accepted procedures.
- Never backdate a record or document.
- Any information added to the record must be signed and dated.

The compliance guide provides expectations for calibration standards. The Quality System regulation requires that standards used to calibrate equipment be traceable to national or international standards. Traceability can also be achieved through a contract calibration laboratory, which in turn uses NIST services for measurements that can show traceability to the SI.

The calibration environment is also included in the compliance guide. As appropriate, environmental controls should be established and monitored to assure that measuring instruments are calibrated and used in an environment that will not adversely affect the accuracy required. Consideration should be given to the effects of temperature, humidity, vibration, and cleanliness when purchasing, using, calibrating, and storing instruments.

Audit of Calibration System

The calibration program shall be included in the quality system audits required by the Quality System regulation. Usually, one of the first questions asked during an audit or inspection is, "May I see your overdue list?" The answer to this question can either generate more questions or give the auditor an overall good impression of the system that is in place. You should have inventory control (all items individually tagged for easy identification, and those items entered into an automated or computerized system); a scheduling system, which includes the calibration intervals set up for your system; and proof that the system has been validated. If there are items on your overdue list, have they been segregated from calibrated equipment to prevent use in production and/or manufacturing? If not, what is to keep them from being used? How long have they been overdue for calibration? What is the reason for their overdue status? Is there a lack of standards, staff, time, facilities, or a combination of these reasons? Have you documented that you know items are overdue and what you are doing about it, or is this a complete surprise? Being aware of what is going on within your system and having a plan in place to remedy the situation can go a long way in assuring the auditor that this is a one-time occurrence (if that is the case).

If your standards are overdue or out of tolerance, do you have a recall system for knowing which items those standards were used to calibrate since the last time the standard was certified? Can you readily produce a list of equipment to recall? Do you have a system in place for labeling or identifying items that have been recalled or are out of service? Is all of this written down in a procedure that anyone in your department or laboratory can follow?

Have all of your technicians been trained to perform the tasks they are assigned? Do you have documentation to prove who has been trained on what items and when that training was completed?

When it comes to records, can you provide a paper (even if you are paperless) trail showing traceability to a national/international standard? Is it easy to follow, and can your technicians as well as their supervisors accomplish it?

An audit of a calibration quality system has multiple benefits. First, the audit will identify areas for continuous improvement and provide confirmation of areas that are performing well. Second, the audit can provide a gap assessment of the program with regard to outside compliance regulations and to internal company policies and procedures. This will help minimize inspection findings from the FDA. Besides an audit, there is another way to assess how well your calibration quality system compares to those of other organizations. Because the FDA is a federal agency, it is required to share with the public any inspection results. The FDA performs this by providing two websites where you can search for released FDA Warning Letters and FDA Form 483 Inspection Findings:

- FDA Warning Letters:[17] https://www.fda.gov/inspections-compliance-enforcement-and-criminal-investigations/compliance-actions-and-activities/warning-letters

- FDA Form 483 Inspection Findings:[18] https://www.fda.gov/about-fda/office-regulatory-affairs/ora-foia-electronic-reading-room

Here is an example of a paper trail for a fictitious piece of test equipment. An auditor observes a scientist using a water bath to produce a product. The scientist copies down the identification number for the water bath, the date it was calibrated, when the calibration label shows it is next due for calibration, and who performed the calibration. The auditor who is inspecting the calibration system asks to see the traceability for the calibrated water bath. The calibration record for the water bath, showing the as-found data, as-left data (when applicable), dates calibrated and next due, and who performed the calibration, is produced for the auditor. There is a statement showing traceability back to a national or international standard (in most cases in the United States, via NIST). The record also shows which calibration procedure was used, along with its revision number, and gives an uncertainty statement (this could be a 4:1 ratio or uncertainty budget showing the actual error). The calibration standards are identified, and their calibration due dates are also shown on the record. The record for the standards used during the calibration of the water bath are then checked for this same information, and the record for the standards used for that calibration is also inspected. This continues for each standard used in the traceability chain. Certificates of calibration from outside vendors are also be inspected for the same information, until traceability to a national/international standard can be observed. As each record is produced, the accuracy of the item being calibrated against the standard used must show the test uncertainty ratio (TUR) or uncertainty budget for that particular calibration. When TURs of less than 4:1 occur, there should be proof that the user is aware of this and accepts this lower TUR.

The maintenance of this paper trail is valuable for more than just an audit. It provides information for use during recalls of equipment and evaluation of calibration intervals and allows calibration professionals to assess the uncertainty of their standards and equipment for future calibrations.

Integrating Measurement into the Quality Assurance (QA) System

The compliance guide touches on the concept that a proper and controlled calibration process can ensure acceptable levels between false accepts and false rejects of calibrated equipment. Additionally, it is understood that if the appropriate product-quality parameters are not checked, calibrated equipment will have little quality value. The FDA expects that a decent quality system includes proper and controlled calibration activities in a highly integrated manner.

Summary

These are the requirements according to the FDA regulations. For some, they are easy to understand and follow. For most of the general public, however, they are complicated and vague. By breaking each section into small pieces, it will be easier to understand what is asked of the common calibration technician or metrology manager. The overriding FDA calibration expectation is to answer the question: What does "manufacturers are responsible for ensuring the establishment of routine calibration" mean in plain English? Simply, an organization is responsible for having a calibration system in place that meets the requirements as outlined in the guide. The easiest way to comply with these requirements is to follow what was set out in Chapter 2: "say what you do, do what you say, record what you did, check the results, and act on the difference." GMP requirements break these down into specific demands. The biggest difference between what is required by the FDA for a compliant calibration system compared to other standards is the greater detail in documentation. This includes recordkeeping, change control for procedures, security and archiving of records, evaluation of out-of-tolerance equipment and how it affected product or processes, and so on. If a company is paperless, it also has to meet the requirements of 21 CFR Part 11, Electronic Records and Electronic Signatures.

OTHER INDUSTRY REQUIREMENTS

ISO 9001 is supposed to be a *generic* model for a quality management system, applicable to any industry. Because of specific requirements, many industries have developed QMS requirements that are based on ISO 9001 but have added industry-specific requirements. Other industries, particularly those subject to government regulation, may have QMS requirements that are completely separate from the ISO 9000 series. This has created a system where there are two types of industry-specific requirements: those based on some version of the ISO 9000 series and those that are independently developed.

Standards Based on ISO 9000

There are four major industry areas, or sectors, which use tailored quality management standards based on a version of the ISO 9000 series. The sector-specific standards contain the requirements of ISO 9001, interpret those requirements in terms of that industry's practice, and include additional sector-specific requirements. Three of the industry areas are aerospace manufacturing (AS9100A), automotive manufacturing (QS9000 and ISO/TS 16949), and telecommunications (TL9000). The fourth area, medical device manufacturing (ISO 13485:2016), was discussed in the previous section in conjunction with 21 CFR Part 820.72. Another industry area, computer software, has a standard for applying ISO 9001 to the software development process and a registration program to audit that.

Aerospace Manufacturing—SAE AS9100D

The aerospace manufacturing industry largely conforms to the AS9100D series of standards. This series of standards is based on the ISO 9001:2015 quality management system standard, with additional requirements specific to aerospace manufacturing. The AS/EN/JIS 9100 series is now widely accepted in the aerospace industry.[19]

Aerospace quality management standards are coordinated globally by the International Aerospace Quality Group (IAQG), established in 1998. This organization is a cooperative group sponsored by Stanford Applied Engineering (SAE) International, representing North, Central, and South America; the European Association of Aerospace Companies (AECMA), representing Europe and Africa; and the Society of Japanese Aerospace Companies (SJAC), representing the Asia-Pacific region. An important reason for the formation of IAQG was the realization of the members that, where safety and quality are concerned, cooperation has more importance than otherwise normal competition.[20] According to the IAQG charter, the organization was founded to establish and promote cooperation among international aerospace companies with respect to quality improvement and cost reduction. This is achieved by voluntary establishment of common quality standards, specifications, and techniques, continuous improvement processes, sharing results, and other actions.[21] IAQG members are representatives of aircraft and engine manufacturers, and major parts and component suppliers that agree to and sign the charter.

Automotive Manufacturing

From the early 1980s through 2003, quality management practices in the automotive industry evolved from supplier requirements set by individual manufacturers to requirements defined by the major manufacturers in a country working together by having an international technical specification based on ISO 9001:2008. For the benefit of readers who are not familiar with what is meant by the automotive industry, it can be defined as companies eligible to adopt ISO/TS 16949:2009. Graham Hills explained eligibility with regard to the manufacturers of finished automotive products, and

> suppliers that make or fabricate production materials, production or service parts or production part assemblies. It also applies to specific service-oriented suppliers—heat treating, welding, painting, plating or other finishing services. The customers for these types of suppliers must be manufacturers of automobiles, trucks (heavy, medium, and light duty), buses or motorcycles. [It] does not apply to manufacturing suppliers for off-highway, agricultural or mining OEMs. It also does not apply to service-oriented suppliers offering distribution, warehousing, sorting or non-value-added services. Nor does it apply to aftermarket parts manufacturers.[22]

ISO/TS 16949. After the ISO 9000 series of quality management system standards was first published in 1987, some companies started incorporating it into requirements for their suppliers. This is, of course, an intended application of the standard. By the mid-1990s, automotive manufacturers in particular had extensive requirements for their suppliers based on ISO 9001:1994. There were four major systems: AVSQ 94 in Italy, EAQF 94 in France, QS-9000 in the United States, and VDA 6.1 in Germany. Companies and trade associations started seeing a problem, though. Suppliers often had to become registered to two or more sets of requirements, and sometimes the requirements conflicted. For example, a parts manufacturer in South Carolina sells parts to the three U.S. auto manufacturers, a couple of German manufacturers with plants in the United States and Germany, and a U.S. plant of a Japanese manufacturer. The manufacturer would have to be registered to the automotive quality system of each country. (And that is better than even a few years earlier, when it had to be approved by each company.) By about 1997, the International Automotive Task Force (IATF) started a liaison with ISO technical committee 176 (ISO/TC 176) to establish a single quality management standard for the automotive sector. The first edition of ISO/TS 16949 was published in 1999. It is also a significant item because it marked the first time a sector-specific document was produced using this method.[23]

The current edition is ISO/TS 16949:2009, *Quality management systems— Particular requirements for the application of ISO 9001:2008 for automotive production and relevant service part organizations*. This version is aligned with ISO 9001:2008 and was developed by IATF and ISO/TC 176 with input from the Japan Automobile

Manufacturers Association (JAMA). Because non-ISO groups developed this document, it is a technical specification, not an international standard. It includes all of the text of ISO 9001, plus additional sector-specific requirements.[24] The intent is for ISO/TS 16949 to be a single industry quality management system that is accepted by all manufacturers regardless of location. Each major automotive manufacturer has its own timetable for mandatory conformance by its suppliers, which range from immediate to December 2006.[25] This can be viewed as a two-step process for companies: registration to ISO 9001:2008 since all the requirements of that standard are included, and then certification to the additional ISO/TS 16949:2009 requirements added by the IATF.[26]

ISO/TS 16949:2009 has an impact on calibration laboratories, as shown by the requirements of Clause 7.6.3, Laboratory requirements.[27] These requirements are:

- The scope of an organization's internal laboratory is a required part of the quality management system documentation. The scope must include the capability to perform all of the in-house calibrations with adequate procedures and qualified people. Also, the capability to perform the tests correctly is required, which implies participation in proficiency testing and interlaboratory comparisons. Laboratory accreditation to ISO/IEC 17025 is mentioned but is not a requirement for an in-house laboratory.

- If an external or independent calibration laboratory is doing work for an automotive manufacturer or supplier, it must be accredited to ISO/IEC 17025. There are only two exceptions. One is that the automotive customer can have evidence that the laboratory is acceptable, principally by fully auditing the supplying laboratory to the requirements of ISO/IEC 17025. The other exception is if the inspection, measurement, and test equipment (IM&TE) calibration activity is being calibrated to the original equipment manufacturer's (OEM's) requirements and procedure. In that case the automotive customer must ensure that the requirements that would apply to its internal laboratory apply.

Standards *Not* Based on ISO 9000

There are other quality management system requirements that are not based on the ISO 9000 standards; these are often sector-specific. They may have been developed within the industry or by a government regulatory agency. A couple are discussed here because of their broad impact on society: civil aviation and civil engineering.

Civil Aviation

Civil aviation has been subject to government regulation for most of the first 100 years of powered flight. Even the much-publicized deregulation of U.S. airlines in 1978 applied only to competition in domestic routes and fares; every other aspect of the industry is still highly regulated. Other nations also provide a high degree of regulation of air transportation, primarily in the interest of safety, and in many cases, because of significant government ownership of the companies.

In the United States, government involvement in civil (nonmilitary) aviation started in 1918, when the Post Office Department established an experimental air mail service. Starting in 1925, Congress passed laws that, by 1967, governed every aspect of operating an air transport company.[28] The regulations covered routes, fares, cargo, maintenance, aircraft design, airport operation, navigation, and so on. In many ways the airline industry was similar to a public utility in that it provided services to the public under heavy government regulation.[29] That was changed by the Airline Deregulation Act of 1978. This eliminated restrictions on domestic routes, virtually eliminated barriers to entering the business, allowed competition based on price, and removed other restrictions on both competition and cooperative marketing arrangements. Significantly, though, all other areas of regulation were unaffected. Of particular interest to people in the calibration industry, aircraft manufacture and maintenance remain heavily regulated, as do flight control, communication, and navigation systems.

The Federal Aviation Regulations (FARs, 14 CFR) are the U.S. civil aviation regulations.[30] In Europe, the Joint Aviation Authorities (JAA) regulations (JAR) are similar to the FARs in the United States. The FARs as a whole can be viewed as a type of quality management system, even though that phrase (or *quality manual*) does not appear anywhere in them. It also goes well beyond quality management by including detailed requirements for processes, methods, and products. It is a system that is very rigorously defined and demands full compliance. The FARs have rules on just about every aspect of the aerospace industry. The regulations that apply to areas that affect the largest number of people are those dealing with airlines and their maintenance operations. The necessity to comply with the myriad regulations tends to drive up costs. The effectiveness of the system, however, is shown in the amazingly low fatal accident rate compared to the national highway fatal accident rate.

Two sections of the FARs most directly apply to airline companies and maintenance operations:

- **14 CFR.** Part 121 has regulations about every person or operation who touches a plane operated by an airline, except those affected by deregulation. Examples include training of pilots, flight attendants, mechanics, and flight dispatchers, and requirements if the company maintains or repairs its own aircraft.[31]

- **14 CFR.** Part 145 has regulations for organizations that perform aircraft maintenance or repair for other parties. This includes independent repair stations, and also includes airlines that operate under Part 121 if they provide service for other airlines (which several do).[32]

Calibration is covered in both places. Under Part 121, an airline must have "[p]rocedures, standards, and limits necessary for . . . periodic inspection and calibration of precision tools, measuring devices, and test equipment" (14 CFR. 121, Subpart G, "Manual Requirements").[33] Under Part 145, a repair station must have a documented system that requires use of calibrated IM&TE (14 CFR. 145, Subpart C, "Equipment, Materials, and Data Requirements")[34] and maintain a documented calibration system (14 CFR. § 145.211, "Quality Control System").[35]

Chapter 18

Computers, Software, and Software Validation

Computers and software are prevalent in all of today's commercial, government, and organizational calibration laboratories. This is good news, as organizations can take advantage of the computer's capabilities to automate, calculate, collect, manipulate, and report out vast amounts of data. Automated data-capture and data-collection software helps to improve efficiencies and reduce transcription errors. Often, an automated test or calibration process can be started and then run processes on its own. This allows the test or calibration technician to perform other tasks at the same time. Laboratory information management systems (LIMS) can eliminate the use of paper records and automatically generate electronic reports and certificates, print labels, and update the calibration schedules. The downside, especially in governmentally regulated industries, is that the selection, implementation, and operation of calibration software systems is a complicated process taking many resources to properly operate them and much documentation to prove and support that operation. This chapter is structured to cover the four main aspects of calibration software systems: calibration management software systems, instrumentation software, automated calibration procedure software, and spreadsheets. But first, we should cover some background on the regulatory requirements for computer software and the topic of software validation.

BACKGROUND

There are many guides and standards that describe compliance requirements for and benefits of computer systems and software. We will discuss five of the most relevant documents: (1) ISO/IEC 17025:2017; (2) ISO 10012:2003; (3) NCSL International Recommended Practice-6, *Calibration Quality Systems for the Healthcare Industries (RP-6)*; (4) NCSL International Recommended Practice-13, *Computer Systems in Metrology (RP-13)*; (5) NISTIR 8250, *GLP 15: Good Laboratory Practice for Software Quality Assurance*; and (6) ISPE *GAMP 5 Guide: A Risk-Based Approach to Compliant GxP Computerized Systems* (2008).

ISO/IEC 17025:2017, *General requirements for the competence of testing and calibration laboratories*, considers software to be included as part of the "equipment" definition in Clause 6.4.1, "The laboratory shall have access to equipment (including, but not limited to, measuring instruments, software, measurement standards, reference materials, reference data, reagents, consumables or auxiliary apparatus) that is required for the correct performance of laboratory activities and that can influence the results."[1] ISO/IEC 17025:2017 states, in Clause 7.11.2, "The laboratory information management system(s) used for the collection, processing, recording, reporting, storage or retrieval of data shall be validated for functionality, including the proper functioning of interfaces within the laboratory information management system(s) by the laboratory before introduction. Whenever there are any changes, including laboratory software configuration or modifications to commercial off-the-shelf software, they shall be authorized, documented and validated before implementation."[2]

ISO 10012:2003, *Measurement management systems—Requirements for measurement processes and measuring equipment*, Clause 6.2.2, Software, states: "Software used in the measurement processes and calculation of results shall be documented, identified and controlled to ensure suitability for continued use. Software, and any revisions to it, shall be tested and/or validated prior to initial use, approved for use, and archived. Testing shall be to the extent necessary to ensure valid measurement results."[3]

NCSL International Recommend Practice-6, *Recommended Practice for Calibration Quality Systems for the Healthcare Industries, RP-6*, 2015, states in section 5.10, Computer Software Validation, "Computer software should have its development, validation, maintenance, and utilization managed and controlled. Some examples of calibration-related software:

- Calibration-management software

- Instrumentation, measuring, and test equipment control software and data-collection software

- Statistical software

- Test-procedure software."[4]

Validation, to an established protocol, is necessary to demonstrate compliant software. Software validation is as important as hardware calibration in ensuring the quality of measurements that are controlled or assisted by computer(s). Software changes should be controlled in the same manner as documented processes.

A validated software package adhering to FDA 21 CFR (Code of Federal Regulations) Part 11, *Electronic Records; Electronic Signatures* can "dramatically streamline efficiency and enhance the compliance of a calibration quality system."[5]

NCSL International Recommended Practice-13 (RP-13), *Computer Systems in Metrology—1996*, "presents an outline of the software program elements and guidelines that may be used to improve the quality, reliability and usability of Metrology Computer Systems. The guidelines presented herein, are applicable to all commercially acquired and in-house developed computer systems and automated equipment used for the capture, processing, manipulation, recording, reporting, storage or retrieval of calibration or test data, and that are depende[d] upon to implement requirements of the Metrology Quality Program."[6]

NIST GLP 15, "Good Laboratory Practice for Software Quality Assurance," is part of NISTIR 8250, *Calibration Procedures for Weights and Measures Laboratories* (2019). It is a procedure for "protecting, validating, and approving the accuracy of

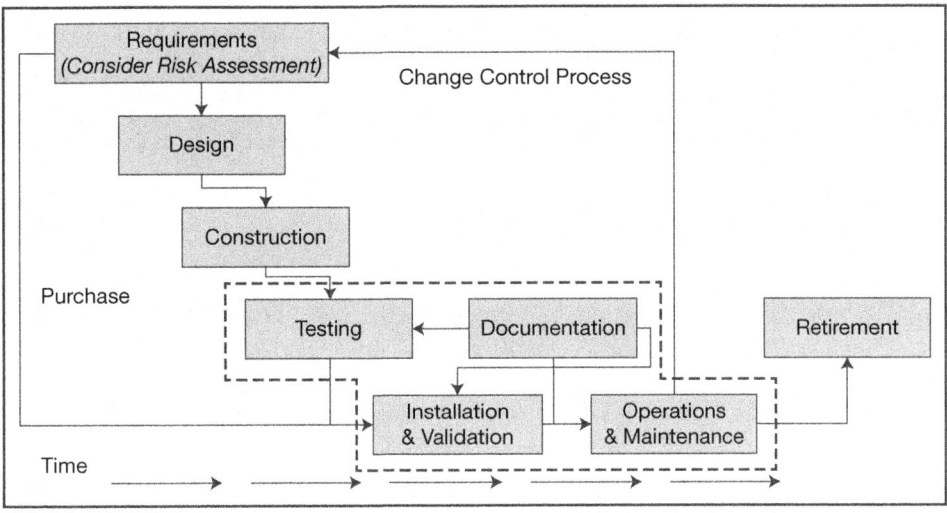

Figure 18.1 NIST GLP 15, Figure 1—Software Life Cycle.[7]

computer software and systems."[8] Not only must the verification and validation process occur at all phases of the software life cycle (as depicted in NIST GLP 15, Figure 1; see Figure 18.1), the knowledge, skills, and attitudes of the metrologist and staff must also be assessed to ensure that, as a result of the software engineering and validation process, no inadvertent measurement errors are introduced into the software.[9]

Commercial off-the-shelf (COTS) software packages commonly used by laboratories are considered sufficiently validated for all use; however, as pointed out in NIST GLP 15, "each laboratory must choose the software appropriately. Some spreadsheet functions, such as rounding or even simple calculations with large numbers having small differences, may not be adequate for the reporting of calibration results and must comply with the accuracy requirements of the procedure as well as other good laboratory practices. The modification and use of COTS spreadsheets in the laboratory are considered software engineering and must be validated."[10]

Even COTS software, while considered sufficiently validated, still has usage risk that should be considered. "The final user (laboratory) must study the information provided by the supplier to properly assess the risk of usage, as the supplier may have a different application in mind, errors may not have been discovered, or software may not be completely validated."[11]

For additional information on Risk Analysis practices and procedures, see Validation of Software in Measurement Systems (Software for Metrology Best Practice Guide No. 1), National Physical Laboratory (NPL), http://www.npl.co.uk/.[12]

ISPE's *GAMP 5 Guide: A Risk-Based Approach to Compliant GxP Computerized Systems* states in section 1.5, "Business Benefits": "There are major business benefits in having a defined process that delivers systems that are fit for intended use, on time, and within budget. Systems that are well defined and specified are easier to support and maintain, resulting in less downtime and lower maintenance costs. Specific benefits to both regulated companies and suppliers include:

- reduction of cost and time taken to achieve and maintain compliance
- early defect identification and resolution leading to reduced impact on cost and schedule
- cost effective operation and maintenance
- effective change management and continuous improvement
- enabling of innovation and adoption of new technology
- providing frameworks for user/supplier co-operation
- assisting suppliers to produce required documentation
- promotion of common systems life cycle, language, and terminology
- providing practical guidelines and examples
- promoting pragmatic interpretation of regulations."[13]

One of the most general and accessible sets of guidelines for software validation is the *General Principles of Software Validation: Guidance for Industry and FDA Staff*[14] from the FDA. This document's guidance is geared toward software life cycle management and risk management, and for application by medical device manufacturers who also develop software for their products as well as software used in implementation of quality systems (e.g., software that records and maintains calibration equipment records). This document is based on recognized software validation principles and thus can be applied to any software.

Translating the FDA medical device terminology to metrology applications, this guidance would cover:

- Software that is part of the inspection, measurement, and test equipment (IM&TE). This includes any built-in firmware.
- Software that is, itself, the IM&TE. An example would be a virtual meter (on a computer system) that responds to data acquisition system inputs.
- Software used in the calibration of workload, such as automated calibration procedures.
- Software used to implement the quality management system. This can include calibration automation systems and laboratory information management systems.

In the software development industry, the words *verification* and *validation*, along with *testing*, are often used as if they were synonyms, e.g., verification, validation, and testing (VV&T). Because ultimately our focus is metrology, we look at the *International Vocabulary of Metrology*'s (the *VIM*; JCGM 200:2012) definition of *verification*: "provision of objective evidence that a given item fulfils specified requirements."[15] The *VIM*'s definition for *validation* is "verification, where the specified requirements are adequate for an intended use."[16] As a separate distinction also made in the *VIM*, "not every verification is a validation."[17] Similarly, the FDA guidance also considers *Verification* as objective evidence that a particular phase of the software development life cycle has met all of the phase requirements.[18] Testing is one of many activities used to develop the objective evidence. *Validation* is "confirmation by examination and provision of objective evidence that software specifications conform to user needs and intended uses, and that the particular requirements implemented through software can be consistently fulfilled."[19] Validation is dependent on, among other things, verification and testing throughout the software development life cycle.

When compared with hardware, software is unique in several ways:[20]

- Most software problems are traceable to errors made during the design and development phase.

- A significant feature of software is the concept of "branching." Branching is the ability to execute different series of commands based on differing input choices. Thus, even short software programs can be complex and hard to understand.

- Testing alone cannot fully verify that the software is complete and correct. Additional verification techniques need to be designed to ensure a comprehensive validation approach.

- Software failures occur without any early indication. This is due to the "branching" aspect, which can hide initial design defects long past implementation.

- Minor changes in software code can sometimes lead to unexpected and significant problems in other areas of the software processes. Regression testing is performed to help minimize this potential. Changes in spreadsheet formulas and codes from access without suitable control and lack of follow-up verification and validation creates significant errors and problems.

The FDA also has specific requirements for electronic records and electronic signatures. The 21 CFR Part 11, Electronic records; electronic signatures[21] regulation includes requirements for user identification (usually meaning password-controlled access); permanent audit trails of records, training and competency; and electronic signatures. Even in unregulated industries, these requirements are a good practice for ensuring that software applications are controlled for data integrity.

CALIBRATION MANAGEMENT SOFTWARE SYSTEMS

Most calibration laboratories use a calibration management software system of some kind to manage calibration activities and record-keeping. These computer systems can be as simple as a spreadsheet or small, local database, such as Microsoft Access™, or can be as complex as an organization's enterprise resource planning system. They can be internally developed or purchased from an outside supplier. No matter the size or origin, the calibration management software system should be validated to ensure that the system performs as intended and designed. There are three main lifecycle phases to calibration management software systems: selection, implementation, and maintenance.

Selection

When we think of software systems, we often focus on the validation and use of the system, but the most important aspect is selecting the right software system for your needs. If you do not select the most appropriate software system first, it will not matter much how well you perform the validation. You just have the wrong tool for the job. One way to prevent this failure is to have a robust selection process. One example is in Figure 18.2.[22]

There are three main workflows that are started together: develop requirements, identify suppliers, and identify project participants. The most important aspect of the selection process is identifying your software system business requirements. One example of a business requirements list comprises more than 100 deliverables broken down into 11 categories: regulatory, workflow, management functions, database administrator functions, logistical functions, external systems interface, implementation, technical, vendor, price, and quality.[23] The requirements list is then used throughout the selection process, not only to identify the best software of choice but also to help develop workflows and testing scripts.

Implementation

Once you have selected the best software system that meets your business requirements, it is time to implement the software. No matter the software's

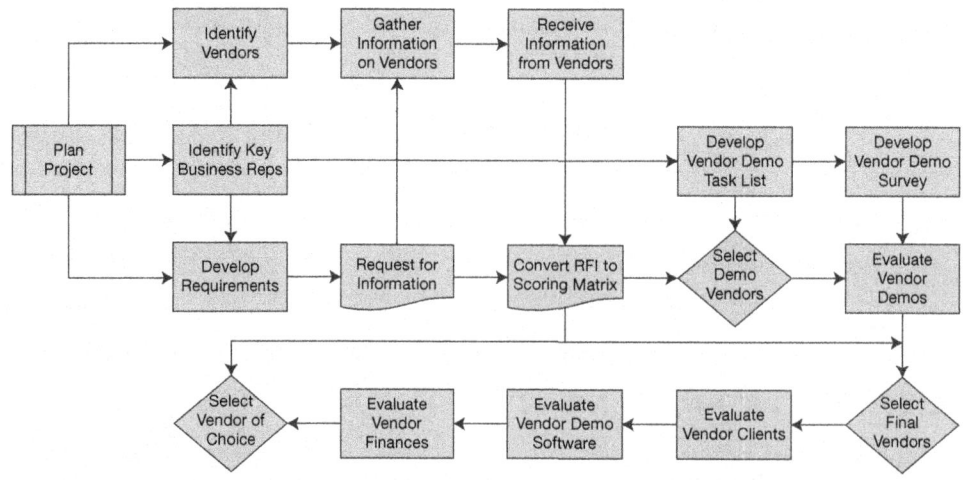

Figure 18.2 Software selection process.

Source: Reproduced by permission from Walter Nowocin, "Selecting a calibration management software system in a regulated environment." *Cal Lab Magazine: The International Journal of Metrology* 28:1 (2021), p. 29.

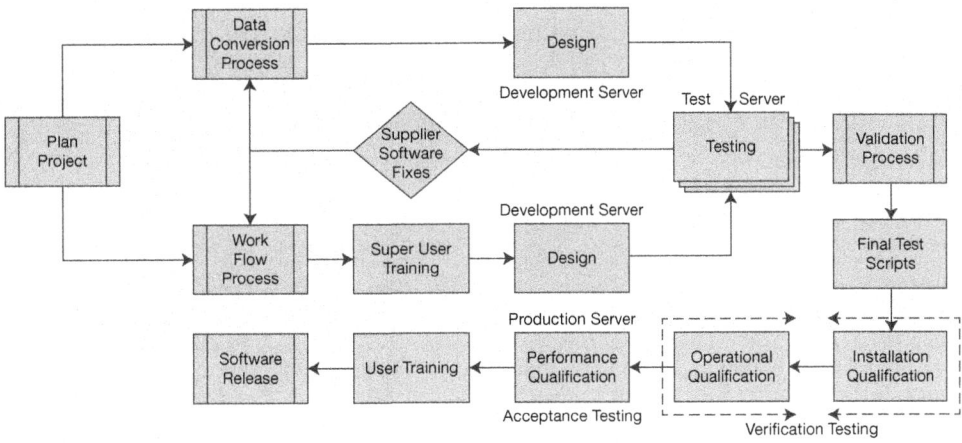

Figure 18.3 Software implementation process.[24]

Source: Reproduced by permission from Walter Nowocin, "Implementing a calibration management software system in a regulated environment." *Cal Lab Magazine: The International Journal of Metrology* 28:2 (2021), p. 27.

size, complexity, or use in a regulated or unregulated business environment, the best practice for implementation is to treat the process as a project that will be sufficiently robust in testing to ensure that the software system will perform to the designed specifications for the intended use. One example for a robust process is in Figure 18.3.

There are two main starting processes for calibration management software system implementation: data conversion and workflows. *Data conversion* is the process of transferring data from the previous software system into the new calibration management software system. The first question to ask is: "Should we convert data into the new database, or should we archive the data and start from scratch?" This is a useful question because the answer can have a significant impact on the design of the implementation project. There are advantages and disadvantages to each choice (see Table 18.1). For regulated industries, it is

Table 18.1 Advantages and disadvantages for converting data.[25]

CONVERT DATA	START NEW
Complex validation	Simple validation
Critical project milestone	Not a project critical path
Higher project costs	Lower project costs
More data for trending	Less data for trending
More historical data to support audits	Less historical data to support audits
Tests software functions more completely	More difficulty in fully testing functions
Workflows can be more fully tested	Workflows are less fully tested
Previous database must be normalized	Previous database does not require clean data

Source: Reproduced by permission from Walter Nowocin, "Implementing a calibration management software system in a regulated environment." *Cal Lab Magazine: The International Journal of Metrology* 28:2 (2021), p. 27.

Table 18.2 Advantages and disadvantages for using existing workflows.[26]

Use Existing Workflows	Use New Workflows
Fewer SOPs to rewrite	New SOPs to develop
Quicker implementation	Longer implementation
Less initial use of new functions	Take advantage of new functions quicker
Less software training time	More software training time
Can identify software flaws	May hide software flaws
Requires mature workflows	Does not require mature workflows

Source: Reproduced by permission from Walter Nowocin, "Implementing a calibration management software system in a regulated environment." *Cal Lab Magazine: The International Journal of Metrology* 28:2 (2021), p. 27.

recommended to convert data into the new software system for all the quality benefits and to decrease quality risk.

Workflow implementation is the next key step in the calibration management software system implementation project. *Workflows* are the individual tasks that together make up the working design of the software system. The first question to ask is: "Do we use the existing workflows for a simpler conversion, or do we create new workflows to take advantage of new software functions?" Again, there are advantages and disadvantages to consider for each choice; see Table 18.2.

It can be tempting to take immediate advantage of the new software features and functionality, but careful planning, testing, and training will be needed to execute the new workflows upon switchover to the new calibration management software system. This direction can easily fall prey to project creep, where a planned project keeps getting larger and taking longer to complete.

The final process for implementing calibration management software systems is the verification, validation, and testing of the software. There are several reliable sources that describe software validation. A more general-purpose software validation process is the System Development Life Cycle (SDLC). SDLC comprises seven phases: planning, analysis and requirements, design, development, testing, implementation, and maintenance.[27] A useful source for software validation for the biomedical, regulated industry is ISPE's *GAMP 5*. *GAMP 5* comprises these sections: key concepts, life cycle approach, life cycle phases, quality risk management, regulated company activities, supplier activities, and efficiency improvements. *GAMP 5* contains document templates that can be used as part of the software validation process, such as forms, checklists, questionnaires, and more.

Software validation terms may differ among organizations and between different business industries, but the underlying details are very much the same. They typically fall within the general SDLC concepts of design and performance verification testing. In Figure 18.3, the terms *installation qualification* (IQ), *operational qualification* (OQ), and *performance qualification* (PQ) are used to describe the verification and acceptance testing phase of the calibration management software system implementation.

IQ ensures the correct installation and configuration of the software and hardware. Typical areas of focus are the data migration, the hardware interfaces and architecture, and the environmental conditions. Some other areas that IQ can cover are data conversion verification, database server architecture and design, database hardware component selection and implementation, printing and email interface functions, and web-based interface functions.

OQ tests the system against specifications to demonstrate correct operation of the software system functions. The OQ is the centerpiece of software validation and will take up the most resources and time. The OQ process is intended to test the software functions and workflows. Therefore, it is a very detailed testing process and the largest component of the validation process.

PQ tests the system to demonstrate fitness for intended use and to accept the system to documented specified requirements. The PQ plan covers these areas: functional testing, user limit testing, data conversion testing, stress testing, performance and load testing, volume testing, failure and recovery testing, and configuration testing.

Test scripts are used to document the software verification, validation, and testing process. Test scripts will describe the test to be performed and the expected results from each test, and then will have an area for the tester to describe the results of the actual test and identify whether the test passed or failed. Some important considerations for test scripts are:

- Document the actual results and provide objective evidence in a manner which demonstrates that the specified expected results were met. Software screen shots are especially useful to meet this objective evidence expectation.

- Document and manage defects that are encountered during testing.

- Provide documented rationale/mitigation for every failed test step.

The FDA describes several software testing principles:[28]

- "The expected test outcome is predefined.

- A good test case has a high probability of exposing an error.

- A successful test is one that finds an error.

- There is independence from coding.

- Both application (user) and software (programming) expertise are employed.

- Testers use different tools from coders.

- Examining only the usual case is insufficient.

- Test documentation permits its reuse and an independent confirmation of the pass/fail status of a test outcome during subsequent review."

The FDA is moving from a traditional computer system validation approach to a new computer software assurance approach.[29] In the new approach, a calibration management software system would be classified as an "indirect system" because it does not directly affect product quality and patient safety. As such, unscripted tests can be used for lower-risk attributes. Unscripted tests do not require either detailed test scripts or a step-by-step test procedure. Instead, the tests can be assigned a test objective and have a "pass or fail" test result. The FDA objective is to lower the documentation burden by 80 % and change from a compliance-centric culture to a quality-centric one. This advances the *GAMP 5* concept of taking a more risk-based approach, increasing the use of suppliers' testing data and not duplicating their validation efforts.

Maintenance

We tend to think that once we have selected and implemented a new calibration management software system, the hardest work is behind us. This may be true in some respects, but because a software system can remain operational for many years, the operational life cycle becomes increasingly more important and critical. Therefore, it is a good practice to have a well-defined plan to manage the new database in an operational state. This plan should comprise the three main components of change control, configuration management, and system operations (see Figure 18.4). Defining roles, providing training, and establishing policies and procedures will ensure that the calibration management software system will remain operational for its intended purpose to meet organizational quality, compliance, and regulatory requirements.

The purpose of the change management process is to ensure that the calibration management software system is maintained in a constant state of operational readiness for the compliant and validated intended use. This activity is performed using a formal change control process (configuration management) that documents all changes made to software, hardware, and infrastructure of the software system. The change management process is comprised of these types of changes:

- System administration
- Updates and patches
- Repairs
- Improvements
- Backup and restore

The most important part of change management is *system administration*, the process of providing dedicated administrative support for the calibration management software system. This role is traditionally performed by a system administrator who is trained in information technology (IT) principles and methods. Some of the standard tasks for a system administrator are identified in the following list and should be included in local standard operating procedures (SOPs) on how to maintain the calibration management software system. Standard system administrative tasks include:

- Setting up new employee accounts and training
- Resetting user passwords

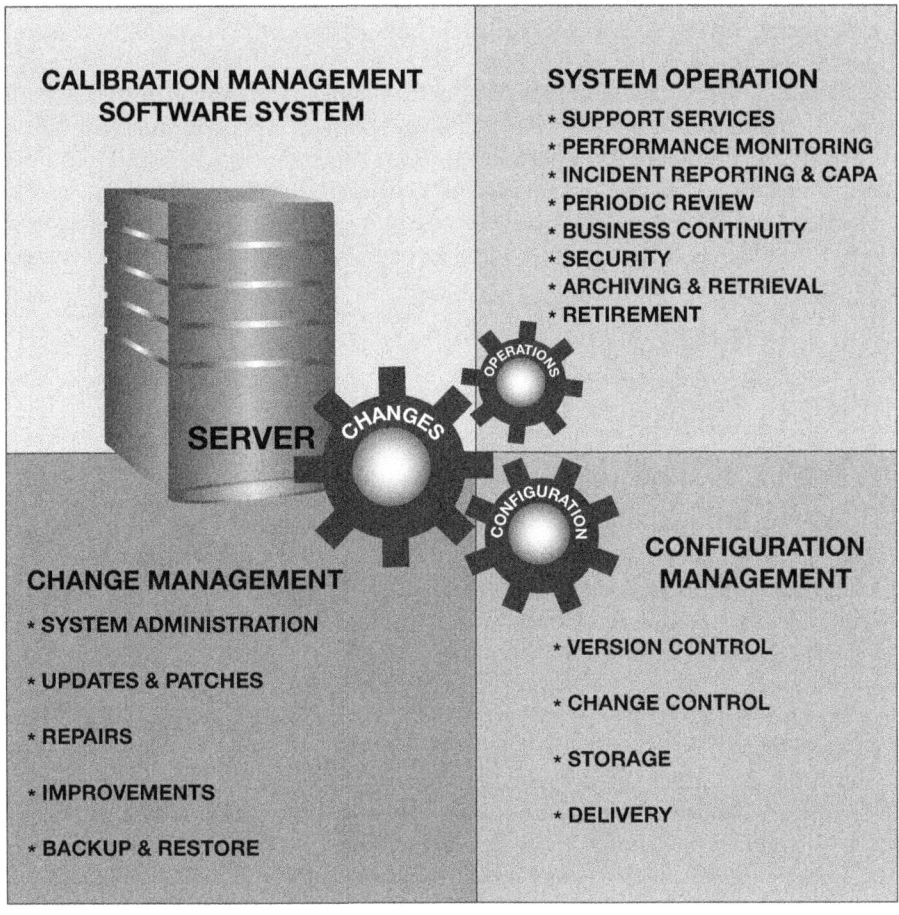

Figure 18.4 Maintenance lifecycle of a software system.[30]
Source: Reproduced by permission from Walter Nowocin, "Maintaining a calibration management software system in a regulated environment." *Cal Lab Magazine: The International Journal of Metrology* 28:3 (2021), p. 35.

- Updating user accounts
- Running and monitoring standard reports
- Responding to incident reports
- Providing software records for audits
- Uploading system administrative database records

Configuration management and change management work together to ensure that any change to the configuration of the calibration management software system is traceable through documentation to determine, at any point in time, the system's requirements state the how, what, where, when, and why. Configuration management begins at the hand-over from implementation to the retirement of the computer system. The four main activities within configuration management are: version control, change control, configuration item storage, and delivery control.

Change control is the most important aspect for managing your calibration management software system. Change control incorporates the aspects of describing the change, documenting and justifying the change, evaluating risks and impact of the change to the configured computer system, accepting or rejecting the change request, developing and verifying the change, and approving and implementing the change. See Figure 18.5. The form example is comprised of four elements:

1. Change request information
2. Change request review
3. Change details
4. Change request completion and approval

In comparison to computer system validation, we tend to think that computer system operation is a simpler process. Yet, in both complexity and time, computer system operation will take more resources to support the system, as the system can stay operational for many years. And as far as complexity, there are many aspects to computer system operation to contend with, such as number of people interfacing with the system and the number of roles and skill sets needed to sustain operations. Some of the operations to be defined are support services, performance of monitoring, incident reporting, periodic review, business continuity, security, archiving and retrieval, and retirement.

Most calibration management software systems are purchased from outside suppliers and thus the buyer will need some level of supplier support during the operations of the software system. A formal supplier agreement should be drawn up that documents the two focus areas for operational support: incident reporting with priority commitment and upgrades. The supplier agreement should detail the process for reporting software deficiencies and the priority level for response commitments. For example, there can be three levels of priority from routine, to urgent, to critical along with a response-level commitment such as routine response in five days, urgent response in three days, and critical response in one day. The supplier agreement should detail what is part of the annual supplier maintenance agreement, which will include at what cost and timing software updates can take place. Additionally, it should set out the time and cost expectations for other types of requests, such as creating reports, customizing workflows, implementing new hardware interfaces, etc. Once a supplier agreement is in place, there should be a periodic review or audit of the supplier to ensure that all contract terms are still in place and committed to and that the supplier is still in a state of compliance with healthy financial and organizational readiness.

SECURITY

As you can readily understand, security management is the most important aspect of operation integrity for calibration management software systems. Therefore, equally important is selecting a calibration management software system that has a robust security management application. This should be one

CHAPTER 18: COMPUTERS, SOFTWARE, AND SOFTWARE VALIDATION 163

CHANGE REQUEST		
Requestor:	Computer System:	Date:
Proposed Change Description:		
Change Reason/Rationale:		
Priority Level:	☐ Critical ☐ Urgent	☐ Routine
CHANGE REQUEST REVIEW		
Change Approval Status:	☐ Accepted	☐ Rejected Reason:
Event Configured:	Current Revision Level:	New Revision Level:
Risk Assessment Classification:	Comments:	Testing Method:
☐ Risk Class 1 – High ☐ Risk Class 2 – Medium ☐ Risk Class 3 – Low		☐ Scripted ☐ Unscripted ☐ Administrative
System Admin Name:	Signature:	Date:
Manager/Supervisor Name:	Signature:	Date:
CHANGE DETAILS		
Summary of Testing Results:		
☐ Fully Passed ☐ Limited Pass – With Accepted Deviation ☐ Failed – Change Not Made		
Testing Comments:		
CHANGE REQUEST COMPLETION AND APPROVAL		
Completed By Name:	Signature:	Date:
Approved By Name:	Signature:	Date:

Figure 18.5 Configuration management change control form example.[31]

Source: Reproduced by permission from Walter Nowocin, "Maintaining a calibration management software system in a regulated environment." *Cal Lab Magazine: The International Journal of Metrology* 28:3 (2021), p. 37.

of the critical requirements for the business when purchasing a new calibration management software system. One effective way to ensure a secure calibration management software system is to establish different levels of security profiles within the database. For example, you could set up the following security level roles (listed from higher-level database access to lower-level, more restricted access):

- **System administrator.** Identify a primary role and secondary, backup role

- **Power user.** Assigned to managers and supervisors

- **Standard user.** Assigned to technicians, engineers, etc.

- **Admin user.** Assigned to those in entry-level administration roles

- **Client user.** Assigned to employees outside of the calibration laboratory for equipment records access

For biomedical companies, an important aspect of security management of calibration management software systems is adherence to 21 CFR Part 11, Electronic Records; Electronic Signatures.[32] A robust security management application will help ensure that electronic records and electronic signatures are controlled, unique, and traceable for any changes. Additionally, the security management application will ensure that passwords are kept up to date and that passwords are uniquely assigned to each individual database user and sufficiently controlled to prevent unauthorized use.

For any digital information, and applicable to all types and sizes of organizations, ISO/IEC 27000-series standards[33] provide an overview of information security management systems (ISMS) vocabulary, practices, requirements, techniques, risk management, and more. Information security is critical to modern business practices and should be considered along with (at the least) confidentiality, data integrity, and risk management.

INSTRUMENTATION SOFTWARE CLASSIFICATIONS

The *GAMP Good Practice Guide* classifies instrumentation software into three categories: simple, medium, and complex.[34]

"Simple" instrumentation software is defined as not being configurable; it indicates that the instrument's firmware produces the numeric output. Typical instruments in this category are analytical balances (not connected to computerized systems), pH meters, and general commercial off-the-shelf electronic instruments. Firmware-based instrumentation can be fully verified by performance testing or calibrating to the instrument's test specifications. It can be argued that the firmware is operating properly if the instrument's test specifications are periodically confirmed.

"Medium" instrumentation software is defined as having configurable components where the firmware generates a numeric output, and the software can have specific configurations whether internal or external to the instrument. Examples of typical instruments in this category are high-performance liquid chromatography (HPLC) and gas chromatography (GC) instruments, sterilizers, and automated electronic test systems. In addition to fully verifying the instrument to test specifications, the software must be verified for its configuration functionality. This verification is to include the security protocols required to ensure that the software is controlled for appropriate user access and data integrity. Software security protocols are a critical element for instrumentation software, especially for older, legacy instrumentation that does not have robust software controls. There are many instances where the FDA has cited a company for regulatory violations for software configured instruments where the user level and password access were not adequately controlled. In these violations, multiple users shared the same password and were able to change the software configuration and stored data without proper authorization. The FDA expectation is clearly stated in 21 CFR Part 211.68(b), Automatic, mechanical, and electronic equipment: "Appropriate controls shall be exercised over computer or related systems to assure that changes in master production and control records or other records are instituted only by authorized personnel. Input to and output from the computer or related system of formulas or other records of data shall be checked for accuracy."[35] A robust software security protocol is appropriate for both regulatory and nonregulatory organizations that use configurable instrumentation software.

The "complex" category of instrumentation software is defined as having multiple configurable, custom components and being connected externally to a computer network. One example would be a sterilizer system connected to an external computerized monitoring and reporting system with multi-site installation and operation. This category of instrumentation is similar to a calibration management software system in its operational complexity and computer architecture. Therefore, "complex" instrumentation can follow the earlier guidelines for calibration management software systems in selection, implementation, and operational requirements with full validation, verification, and testing protocols.

AUTOMATED CALIBRATION PROCEDURE SOFTWARE

With computers continuing to deliver more functionality, faster processing, more storage, and expanding applications, more calibration laboratories are taking advantage of the automation gains. One area that is seeing more acceptance is the automation of the calibrations themselves. From a validation perspective, there are two aspects to automating the calibration of instruments.

The first is the software system platform that has the capability to create automated calibration procedure executables that allow primary calibration instruments to control the automated calibration routines of the unit under test. The validation of the automated calibration software system can follow similar guidelines for the calibration management software system in the areas of selection, implementation, and maintenance. If using a commercial, off-the-shelf software system from an outside supplier, there still is a need to confirm that the software system can perform to the user requirements. This can be considered user acceptance testing of the software system.

The other aspect is the executable calibration procedures themselves: those that perform the actual calibrations. These automated calibration procedures must be individually verified to ensure that they adequately calibrate the instrument under test. This verification requirement applies even if the automated calibration software system supplier provided the automated calibration procedures. The saying "trust but verify" comes into play here. There are several methods that can be used to verify automated calibration procedures: manual comparison, automated comparison, and interlaboratory comparison.

Manual Comparison Method

The manual comparison method is performed by first calibrating the unit under test using a manual calibration procedure process, and then calibrating the unit under test using the automated calibration procedure process. The results are then analyzed, reviewed, and accepted/rejected based on how well the two results align to each other to confirm an acceptable match. One statistical tool that can be used for this analysis is the "paired t-test."[36] This statistical tool will calculate the mean, standard deviation of the mean for each test point. If the difference of the means is less than the critical value of t, which is typically 0.05, then the two samples (manual and automated) are considered to be similar (acceptable) and not different, statistically speaking. This manual comparison method can be also designed to identify the number of tests (repeatability) to perform and by the number of testers (reproducibility). More tests and more testers can be correlated with increased levels of quality risk of the calibration.

Automated Comparison Method

The automated comparison method can be used if no manual calibration capability is readily available or established. This method is similar to the manual comparison method except you are now only comparing the number of tests, different testers, and using just the automated calibration procedure process. Again, the results are analyzed, reviewed, and accepted/rejected based on how well the results align to each other to confirm acceptance. The paired t-test statistical method described previously can be used to help with this analysis.

An interlaboratory comparison method can be used if there are few resources available to perform the manual or automated methods. In this method, the instrument can be sent to an outside supplier and calibrated. Upon return, the calibration certificate can be used to compare to the internal automated calibration results and analyzed for an acceptable match. Again, the statistical paired t-test method can be used to assist with this analysis.

Once the verification has been performed and accepted, the automated calibration procedure executable must be version-controlled and released in such a manner that the executable cannot be further edited or changed without authorization.

SPREADSHEETS

The spreadsheet is a very popular computer application in calibration laboratories, with uses ranging from as a simple, tabular documentation tool to a more complex, standalone database application. The FDA has identified the spreadsheet as one of the most undocumented and uncontrolled computer application tools. An example is an FDA Warning Letter from 3 August 2016: "Failure to validate computer software for its intended use according to an established protocol when computers or automated data processing systems are used as part of production or the quality system, as required by 21 CFR 820.70(i). For example, your firm was utilizing an uncontrolled spreadsheet to track equipment requalification due dates."[37]

Properly controlling a spreadsheet to ensure that it performs as intended is particularly challenging, because the spreadsheet has functional limitations such as lack of audit trails to track data changes and lack of robust password control configuration, to name just two. This lack of control and lack of robust design and documentation contribute to a high rate of spreadsheet errors. One academic study showed that the use of spreadsheets with at least one formula resulted in 24 % of them having a spreadsheet error.[38] Furthermore, another experiment in which participants attempted to identify errors in a spreadsheet found that they only caught about 60 % of the errors.[39] We can conclude that where spreadsheets are used in calibration laboratories, robust design, testing, and documentation ought to be considered to properly control spreadsheets for their intended purpose.

Spreadsheets are not equally created, as they can be used in a variety of ways. *GAMP-5* classifies spreadsheets at several levels of use:[40]

- **No Calculations.** When a spreadsheet is used for tabulation formatting without any calculations, it is to be treated as a document and not as a spreadsheet. Control should be appropriate to the document use condition.

- **Disposable.** When spreadsheets are used solely to make quick calculations in place of a handheld calculator. The spreadsheet is not retained as an electronic record. The results are converted to a paper document which records the data and should have adequate description of the formulas used and the verification of the correct formula to use.

- **Retained as Documents.** When a spreadsheet is used to make calculations and to record the data, it should have adequate design, testing, and verification of all applicable calculations and be subject to adequate control and storage requirements.

- **Used as Templates.** A more effective method over the "retained as a document" method is to use spreadsheet templates that are designed for reuse by different users on a repeated frequency. Robust design, testing, and verification are required with the added level of control regarding revision-level documentation for the template and controlled template access. However, the benefit is realized for each repeat use without requiring additional verifications, as the spreadsheet template has already been verified. Additionally, because of reuse, the quality of the spreadsheet is very high, and the error rate is significantly lower.

- **Used as Databases.** As has been mentioned before, spreadsheets have inherent functional limitations that make them less than ideal for use as standalone databases, especially in regulated industries. It is better to identify a computer application that is designed specifically for database functionality.

Here are a few spreadsheet verification measures from the FDA:[41]

- Lock all cells of a spreadsheet, except those for data input.

- Make spreadsheets read-only, with password protection, so that only authorized users can alter the spreadsheet.

- Automatically reject data outside valid ranges.

- Manually verify spreadsheet calculations by entering data at extreme and expected values.

- Test the spreadsheet by entering nonsensical data.

- Keep a permanent record of all cell formulas when the spreadsheet has been developed. Document all changes made to the spreadsheet and control using a system of version numbers with documentation.

Here are a few spreadsheet design features recommended by the FDA:[42]

- Use drop-down lists whenever possible. This ensures both consistency and reliability of raw data input.

- Ensure that each raw data input has the appropriate measuring unit identified, (e.g., mL, µV, etc.).

- Specify the number of decimal places the measured data input should be rounded off to where appropriate.

- Use validation criteria for a particular cell to ensure that data entry is of the proper type. Restrict data to whole numbers, decimal numbers, or text or set limits on entries.

- Use spreadsheet protection measures. Protecting the spreadsheet prevents new worksheets from being created. Protecting the worksheet prevents formulas from being changed.

Another excellent resource for spreadsheet quality assurance is NIST GLP 15, *Good Laboratory Practice for Software Quality Assurance*. GLP 15, Table 1—*Validation Methods and Example Assessments*, describes a brief overview of assessment methods used to evaluate software and complete the "Software Verification and Validation" form that appears as *GLP 15* Appendix A. Appendix A is a template to assist with documenting the validation testing across the ten areas listed in *GLP 15*, Table 1:[43]

1. Software Inspection;
2. Mathematical Specification;
3. Code Review;
4. Numerical Stability;
5. Component Testing;
6. Numerical Reference Results;
7. Embedded Data Validation;

8. Back-to-Back Testing;

9. Analysis Without Computer Assistance; and

10. Security.

This guide can be used to design compliance into spreadsheets as the spreadsheets are being planned. For each area of *GLP 15*, Table 1, Appendix A[44] lists the "Assessment" activity, whether the validation of that assessment Passed or Failed, and the Result/Observations (including evidence). This quality practice is also known as PDCA (plan–do–check–act) and can be applied to numerous activities, including spreadsheet validation. It is wise to use and include Appendix A in spreadsheet development and demonstrate validation and verification as you design and test, as you implement, and as you fix or improve in-service spreadsheets.

Some of the themes repeated by multiple sources in this chapter emphasize the importance of verifying and validating software, whether it is built in-house, commercial off-the-shelf (COTS), a modification of COTS, instrument software, data-collection and processing software, or spreadsheets. These activities must be planned and executed by personnel qualified for the activity and include users of the software. There are several resources in this chapter, including sample templates and sample processes, which provide useful guidance to the approaches, methodologies, and practices of software verification and verification. A common theme for any validation activity is to 1) plan what you are going to do, 2) do what you planned to do, 3) check what you did and compare the result to what you expected, and 4) act on the difference (PDCA). By practicing the steps in this chapter and using appropriate quality tools, a laboratory can improve its efficiency and processes and reduce risk.

Part III
Metrology Concepts

Chapter 19 A General Understanding of Metrology

Chapter 20 Measurement Methods, Systems, Capabilities, and Data

Chapter 21 Specifications

Chapter 22 Substituting Calibration Standards

Chapter 23 Proficiency Testing, Interlaboratory Comparisons, and Measurement Assurance Programs

Chapter 19

A General Understanding of Metrology

The field of metrology spans a multitude of different disciplines. Metrology incorporates an ensemble of knowledge gathered from multiple diverse fields such as mathematics, statistics, physics, quality, chemistry, mechanical, and computer science; all applied with a liberal sprinkling of common sense. Essential to the field of metrology is understanding the fundamental methods by which objects and phenomena occur and are measured, as well as the means for assigning values to measurements and the uncertainty of these assigned values. Encompassing some of these essentials are the establishment and maintenance of units, measurement methods, measurement systems, measurement capability, measurement data, measurement equipment specifications, measurement standards usage, measurement confidence programs, etc. The adage "A chain is only as strong as its weakest link" also applies to metrology because metrology essentials are interdependent; each relies on an assortment of clearly stated definitions and postulations. The ensemble of metrology essentials lays the foundation for the realization and agreement throughout the world that measurements are accepted and appropriate for their intended purposes. Mathematics is widely considered the universal language; the metrology essentials extend that language into our daily existence as quantifiable, attributed information without which our world, as we know it, could not exist.

The basic concepts and principles of metrology were formulated from the need to measure and compare a known value or quantity to an unknown to define the unknown relative to the known. This may seem like double-talk, but upon further investigation you can see that what is being described is a method for determining the value of an unknown by assigning it a quantity of divisions commonly referred to as units (for example, meters, degrees Celsius, minutes, and so on). Everything we buy, sell, consume, or produce can be compared, measured, and defined in terms of units of a measurement. Without commonly agreed-on units, it would not be possible to accurately quantify the passing of time, the length of an object, or the temperature of one's surroundings. In fact, almost every aspect of our physical world can be related in terms of units of measurement. Units allow us to count things in a building-block type fashion, so they have meaning beyond a simple descriptive comparison such as smaller than, brighter than, longer than, and so on. Determination of measurement units that are deemed acceptable and repeatable, and maintaining them as measurement standards, lies at the heart of fundamental metrology concepts and principles.

Measurement units must be accepted or recognized and agreed on to conduct most commercial transactions. The *VIM* (1.9) defines a *unit of measurement* as a "real scalar *quantity*, defined and adopted by convention, with which any other quantity of the same *kind* can be compared to express the ratio of the two quantities as a number."[1] The *VIM* goes on to note: "Measurement units are designated by conventionally assigned names and symbols."[2] Mutually accepted measurement units for parameters such as mass and length provide the means for fair exchange of commodities. For example, the value of one gram of gold can often be equated to its equivalent worth in local currency throughout the world. The same cannot be said when using a nonaccepted unit of measurement, as its equivalent worth cannot be easily determined. A rectangle of gold has no defined equivalent worth because *rectangle* is not an accepted unit for mass.

In addition to their importance in commerce, consistent and accepted measurement units are also critically important in the sciences and engineering. They serve as a common frame of reference that everyone understands and can relate to. To facilitate the acceptance of units throughout the world, the General Conference on Weights and Measures (CGPM) established the modern International System of Units (SI) in 1960 as an improvement on earlier measurement units. The SI provides a uniform, comprehensive, and coherent system for the establishment and acceptance of units. The SI system comprises seven fundamental units. These seven units are used to derive other units as required to quantify our physical world. Congruent with SI units and their use are values that have been measured or determined for fundamental physical constants. The accepted values of these constants, along with the uncertainty associated with them (if any uncertainty at all since their redefinitions in 2018), are published and updated as needed for use in all areas of science, engineering, and technology. A few of the frequently used constants are listed in Table 19.1.[3]

Table 19.1 Frequently used constants.

Physical Constant	Value	Standard Uncertainty
atomic mass constant	$1.660\ 539\ 066\ 60 \times 10^{-27}$ kg	$0.000\ 000\ 000\ 50 \times 10^{-27}$ kg
Avogadro constant	$6.022\ 140\ 76 \times 10^{23}$ mol^{-1}	exact
Boltzmann constant	$1.380\ 649 \times 10^{-23}$ J K^{-1}	exact
conductance quantum	$7.748\ 091\ 729 \times 10^{-5}$ S	exact
electron mass	$9.109\ 383\ 7015 \times 10^{-31}$ kg	$0.000\ 000\ 0028 \times 10^{-31}$ kg
electron volt	$1.602\ 176\ 634 \times 10^{-19}$ J	exact
elementary charge	$1.602\ 176\ 634 \times 10^{-19}$ C	exact
Faraday constant	$96\ 485.332\ 12$ C mol^{-1}	exact
fine-structure constant	$7.297\ 352\ 5693 \times 10^{-3}$	$0.000\ 000\ 0011 \times 10^{-3}$
hyperfine transition frequency of Cs-133	$9\ 192\ 631\ 770$ Hz	exact
inverse fine-structure constant	$137.035\ 999\ 084$	$0.000\ 000\ 021$
luminous efficacy	6831 m W^{-1}	exact

Continued

Table 19.1 *Continued*

Physical Constant	Value	Standard Uncertainty
magnetic flux quantum	2.067 833 848 × 10⁻¹⁵ Wb	exact
molar gas constant	8.314 462 618 J mol⁻¹ K⁻¹	exact
Newtonian constant of gravitation	6.674 30 × 10⁻¹¹ m³ kg⁻¹ s⁻²	0.000 15 × 10⁻¹¹ m³ kg⁻¹ s⁻²
Planck constant	6.626 070 15 × 10⁻³⁴ J Hz⁻¹	exact
reduced Planck constant	1.054 571 817 × 10⁻³⁴ J s	exact
proton mass	1.672 621 923 69 × 10⁻²⁷ kg	0.000 000 000 51 × 10⁻²⁷ kg
proton-electron mass ratio	1836.152 673 43	0.000 000 11
Rydberg constant	10 973 731.568 160 m⁻¹	0.000 021 m⁻¹
speed of light in vacuum	299 792 458 m s⁻¹	exact
standard acceleration of gravity	9.806 65 m s⁻²	exact
standard atmosphere	101 325 Pa	exact
standard state pressure	100 000 Pa	exact
Stefan-Boltzmann constant	5.670 374 419 × 10⁻⁸ W m⁻² K⁻⁴	exact

SI units, SI-derived units, and fundamental constants form the groundwork for most measurement units that are not specialized to a single industry. These measurement units are generated for a variety of measurement parameters to represent many different measurement technologies employed by a vast array of inspection, measurement, and test equipment (IM&TE) and associated calibration standards. Note that the *VIM* defines a *measuring instrument* as a "device used for making *measurements*, alone or in conjunction with one or more supplementary devices."[4] See Table 19.2 for some common measurement parameters along with their associated measurement units and typical IM&TE.

Table 19.2 Common measurement parameters.

Common measurement parameter	Common units of measurement	Common measurement instruments
Angular	radian, degree, minute, second, gon	clinometer, optical comparator, radius gage, protractor, precision square, cylindrical square
Concentricity (roundness)	meter, inch, angstrom	rotary table with indicator
Current	ampere	amp meter, current shunt, current probe, digital multimeter (DMM)
Flatness/ parallelism	meter, inch, angstrom	optical flats and monochromatic light source, Profilometer

Continued

Table 19.2 Continued

Common measurement parameter	Common units of measurement	Common measurement instruments
Flow	liters per minute (LPM), standard cubic feet per minute (SCFM)	flowmeter, rotameter, mass flowmeter (MFC), anemometer, bell prover
Force (compression and tension)	Newton, dyne, pound-force	force gage, load cell, spring gage, proving ring, dynamometer
Frequency	Hertz (Hz)	counter, time interval analyzer
Hardness	Brinell hardness number (BHN), Rockwell hardness number	Brinell hardness tester, Rockwell hardness tester
Humidity	dew point, relative humidity	hydrometer, psychrometer, chilled mirror
Impedance	impedance (Z)	LCR meter, impedance analyzer, vector network analyzer (VNA)
Length, height (linear displacement)	meter, foot, inch, angstrom	steel rule, tape measure, caliper, micrometer, height/length comparator, laser interferometer, coordinate measuring machine (CMM)
Luminance	candela per square meter, lux, footcandles, Lambert	light meter, radiometer
Mass	kilogram, pound, ounce, gram, dram, grain, slug	balance, weighing scale
Power (RF)	Watt, dBm, dBv	diode power sensor, thermopile power sensor, thermal voltage converter (TVC), scalar network analyzer (SNA), vector network analyzer (VNA)
Power (voltage)	Watt, joule, calorie	wattmeter, power analyzer
Pressure and vacuum	Pascal, pound-force per square inch (psi), bar, inches of water, inches of mercury, atmosphere, torr	pressure gage, manometer (mercury), manometer (capacitance), piranha gage, spinning rotor gage (SRG)
Resistance, conductance	ohm, siemen, mho	ohmmeter, milli ohmmeter, tera ohmmeter, digital multimeter, current shunt, LCR meter, current comparator
Rotation	radians per seconds, revolutions per minute (rpm)	stroboscope, RPM meter
Signal analysis	Hz/dBm (frequency domain)	spectrum analyzer, vector network analyzer (VNA), vector voltmeter, FFT analyzer
Signal analysis (time domain)	Second/volt	oscilloscope, logic analyzer
Temperature	Kelvin, Celsius, Fahrenheit, Rankine	thermocouple, thermometer (liquid-in-glass), thermometer (resistance), infrared pyrometer

Continued

Table 19.2 *Continued*

Common measurement parameter	Common units of measurement	Common measurement instruments
Torque	Newton meter, pound-force foot, pound-force inch, ounce-force inch	torque wrench, torsion bar, torque cells, torque transducer
Vibration/acceleration	meter per second, meters pk-to-pk	accelerometer, velocity pickup, displacement meter, laser interferometry
Voltage	volt	voltmeter, digital voltmeter (DVM), digital multimeter (DMM), millivoltmeter, nanovoltmeter, HV probe

Note: Many of the units of measurement listed, though commonly used in some areas and industries, are not part of the SI and are not generally used in scientific work. The accepted SI unit is listed first in all cases.

Not only are measurement units used in defining the quantity of an unknown measurement parameter, but they are also frequently used when generating known quantities of a parameter. By generating a known quantity of a measurement parameter, equipment used to measure these parameters can be evaluated as to its accuracy (the fundamental principle behind most IM&TE calibrations). See Table 19.3.

Table 19.3 Common measurands and equipment used to source them.

Common units for source	Common source instruments
Angular	angle blocks, sine plate
Current	current source, power supply, multifunction calibrator
Flow	bell prover, flow test stand, flow calibrator
Force (compression and tension)	weights, force test stand
Frequency	timebase, frequency standards
Hardness	Rockwell hardness standards
Humidity	environmental test chamber, saturated salts
Impedance	impedance test artifacts (capacitors, inductors, AC resistors)
Length, height (linear displacement)	gage blocks, parallels
Luminance	calibration light source, laser
Mass	weights
Power (RF)	power meter reference output, generator, synthesizer
Power (voltage)	power supply, multifunction calibrator

Continued

Table 19.3 *Continued*

Common units for source	Common source instruments
Pressure and vacuum	dead weight tester, vacuum test stand, pressure pump, vacuum pump
Resistance	standard resistors, decade resistors, multifunction calibrator
Rotation	AC motor
Temperature	oven, environmental chamber, temperature calibrator, triple point of water, freezing or melting point cells
Torque	weights and torque arm
Vibration/acceleration	shaker table
Voltage	power supply, multifunction calibrator

Most measurement parameters are defined by fundamental principles and concepts. These definitions serve to describe a measurement parameter in terms of the physical world. A simple illustration of this is *force* (F), which is defined in terms of mass (m) and acceleration (a). Most measurement parameters can be represented by mathematical formulas that show the relationship between the physical world factors that comprise them. See Table 19.4.

Table 19.4 Common measurands and some of their associated formulas.

Common measurement parameter	Fundamental formulas	Variable	Variable	Variable
Angular	$SA = O / H$	SA = sine of angle between the hypotenuse and adjacent side	O = opposite	H = hypotenuse
Current	$I = V / R$	I = current	V = voltage	R = resistance
Flatness/parallelism	$D = (n + 1) / 2$	D = deviation measured with optical flats	n = The number of bands between to high point	
Flow	$MF = Qv * DT$	MF = mass flow	Qv = volume flow rate	DT = weight density of gas or liquid at temperature T
Force (compression and tension)	$F = m * a$	F = force	m = mass	a = acceleration
Frequency	$F = C / S$	F = frequency	C = cycle or iterations	S = seconds
Humidity	$\% RH = (Pv / Ps) * 100$	$\% RH$ = percent relative humidity	Pv = pressure of water vapor	Ps = saturation pressure

Continued

Table 19.4 *Continued*

Common measurement parameter	Fundamental formulas	Variable	Variable	Variable
Impedance	$Z = \sqrt{L/C}$	Z = impedance	L = inductance	C = capacitance
Length, height (linear displacement)	Length change = $L * L_{Coef} * t_{Delta}$	L = original length	L_{Coef} = linear expansion coefficient of material	T_{Delta} = change in temperature
Luminance	CF = Pref / Pdut	CF = calibration factor at a specific wavelength or wavelength range	Pref = measured reference standard power	Pdut = measured power from device under test
Mass	p = m / v	p = absolute density	m = mass	v = volume
Power (RF)	$P_{dbm} = 10 \log(P_{Meas}/0.001)$	P_{dbm} = power in dbm	P_{Meas} = measured power	
Power (voltage)	$P = I^2 * R$	P = power	I = current	R = resistance
Pressure and vacuum	P = F / A	P = pressure	F = force	A = area
Resistance	R = V / I	R = resistance	V = voltage	I = current
Rotation	w = a / t	w = angular velocity	a = angular displacement	t = elapsed time
Temperature	$R_t = R_0(1 + At + Bt^2)$	R_t = resistance at some temperature	R_0 = resistance at 0 Celsius	A & B = constants for a particular element that describes its temperature behavior
Torque	T = F × s	T = torque around a point	F = force applied	s = distance through which the force is acting
Vibration/ acceleration	As = E_o / Ai	As = accelerometer sensitivity	E_o = electrical signal output	A_i = acceleration input
Voltage	V = I * R	V = voltage	I = current	R = resistance

Chapter 20

Measurement Methods, Systems, Capabilities, and Data

MEASUREMENT METHODS

Measurements are made by using one or more well-defined measurement methods to obtain quantifiable information about an object or phenomena. These measurement methods include certain fundamental characteristics that allow them to be used to categorize different types of measurements. Usually, a particular situation or application will make obvious the appropriate measurement method(s) needed to achieve the desired measurement results. Knowledge and experience tell us which specific measurement method will typically yield the best results for a particular situation or application. Factors such as phenomena stability, resolution requirements, environmental influences, timing restraints, and so on, must be considered to determine the optimum measurement method(s). Understanding the mechanics and theory behind measurement methods is helpful not only for determining the best method for a particular situation or application, but also for understanding its limitations and the measurement results it produces.

Measurement methods employ various metrology-related terms that are useful when interpreting the method. Table 20.1 lists definitions from the *VIM*, including their respective *VIM* reference number, for some of these frequently used terms:

Table 20.1 Frequently used VIM terms.[1]

Metrology term	VIM definition
Kind of quantity (1.2)	Aspect common to mutually comparable quantities
Measurement (2.1)	Process of experimentally obtaining one or more *quantity values* that can reasonably be attributed to a *quantity*
Measurand (2.3)	*Quantity* intended to be measured
Method of measurement (2.5)	Generic description of a logical organization of operations used in a measurement
Quantity (1.1)	Property of a phenomenon, body, or substance, where the property has a magnitude that can be expressed as a number and reference
Quantity value (1.19)	Number and reference expressing magnitude of a *quantity*
Reference quantity value (5.18)	*Quantity value* used as a basis for comparison with values of *quantities* of the same *kind*

Measurement methods are commonly grouped into one or more of seven categories. The seven measurement method categories are direct, differential, indirect, ratio, reciprocity, substitution, and transfer. Table 20.2 lists the generalized definitions for each of the seven categories of measurement methods.

Examples of the seven measurement methods are given in Table 20.3.

Table 20.2 Seven categories of measurement methods.

Measurement method	Definition
Direct	A measurement that is in direct contact with the measurand and provides a value representative of the measurand as read from an indicating device
Differential	A measurement made by comparing an unknown measurand with a known quantity (standard) such that when their values are equal, a difference indication of zero (null) is given
Indirect	A measurement made of a nontargeted measurand that is used to determine the value of the targeted measurand (measurand of interest)
Ratio	A measurement made by comparing an unknown measurand with a known quantity (standard) in order to determine how many divisions of the unknown measurand can be contained within the known quantity
Reciprocity	A measurement that makes use of a transfer function(s) (relationship) in comparing two or more measurement devices subject to the same measurand
Substitution	A measurement made using a known measurement device or artifact (standard) to establish a measurand value, after which the known measurement device or artifact is removed, and an unknown measurement device (unit under test) is inserted in its place so that its response to the measurand can be determined
Transfer	A measurement employing an intermediate device used for conveying (transferring) a known measurand value to an unknown measurement device or artifact

Table 20.3 Measurement method category examples.

Measurement method	Examples
Direct	A multimeter reading the alternating voltage current (VAC) of a power outlet, using a ruler to measure length, determining temperature by reading a liquid-in-glass thermometer, measuring tire pressure using a pressure gage
Differential	Comparison of two voltages using a null meter, measuring the length of a gage block using a gage block comparator, determining a resistance using a current comparator, measuring a weight using a two-pan balance
Indirect	Calculating a current value by measuring the voltage drop across a shunt, determining a temperature by measuring the resistance of a platinum resistive thermometer, determining unknown impedance by measuring reflected voltage
Ratio	Creating intermediate voltage values using a Kelvin Varley divider and a fixed voltage source

Continued

Table 20.3 *Continued*

Measurement method	Examples
Reciprocity	Determining the sensitivity of a microphone via the response of another microphone
Substitution	Measuring weight using a single-pan scale
Transfer	Determining VAC using an AC/DC transfer device, determining timing deviations via a portable synchronized quartz clock

Measurement methods are the schemes by which measurement data are obtained. The type of measurement method that is appropriate for a particular situation should be thought out and determined before engaging in a measurement. Selection of an inappropriate measurement method will result in wasted time and other resources and produce undesirable and/or unreliable measurement data. Looking before leaping is sound advice when it comes to selecting a measurement method.

MEASUREMENT SYSTEMS

Measurement systems are how measurement data are obtained. A measurement system is an ensemble comprising various elements such as measurement personnel, calibration standards, measurement devices, measurement fixtures, measurement environment, measurement methodology, and so on. These systems are used to obtain quantifiable, attributable data related to an object or phenomenon. You can visualize a measurement system as a process of interactive, interrelated activities by which various objects or phenomena are related to measurement data.

Measurement systems are created and used based on needs for specific measurement data. The makeup of a measurement system is determined by an application or particular situation. The adequacy of a measurement system depends on the accuracy and reliability requirements of the measurement data. Less stringent requirements demand less of a measurement system in terms of sophistication, variability, repeatability, and so on. How the measurement data will be used will drive the selection, composition, and sophistication of a measurement system in order to meet measurement objectives.

Measurement systems are designed to produce measurement data that are assumed to be faithful and representative of the measurand(s) they are intended to measure. Table 20.4 lists *VIM* definitions, including their respective *VIM* reference numbers, that are applicable to this discussion.

For a measurement system to be properly constructed, a comprehensive understanding of applicable measurement application(s) is required. This understanding will direct the development and definition of the measurement system. Also, a thorough understanding of measurement data requirements will provide guidance as to the confidence level and reliability traits required of the measurement system. NASA Reference Publication 1342, *Metrology—Calibration*

Table 20.4 Measurement systems VIM definitions.[2]

Term	VIM Definition
Measuring system (3.2)	Set of one or more *measuring instruments* and often other devices, including any reagent and supply, assembled, and adapted to give information used to generate *measured quantity values* within specified intervals for *quantities* of specified *kinds*
Measuring instrument (3.1)	Device used for making *measurements*, alone or in conjunction with one or more supplementary devices
Measuring chain (3.10)	Series of elements of a *measuring system* constituting a single path of the signal from a *sensor* to an output element
Measurement result (2.9)	Set of *quantity values* being attributed to a *measurand* together with any other available relevant information
Indication (4.1)	*Quantity value* provided by a *measuring instrument* or a *measuring system*

and Measurement Processes Guidelines,[3] notes the following 10-stage sequence in defining measurement requirements, as listed in Table 20.5.

S. K. Kimothi, in his book *The Uncertainty of Measurements*,[4] lists entities that can be considered part of the measurement system:

- Measurement/test method
- Measurement equipment and measurement setup
- Personnel involved in the maintenance and operation of the measuring equipment and the measurement process
- Organizational structure and responsibilities
- Quality control tools and techniques
- Preventive and corrective mechanisms to eliminate defects in operations

The *Measurement Systems Analysis (MSA) Reference Manual*[5] specifies the following as fundamental properties that define a good measurement system:

- Adequate discrimination and sensitivity
- Being in a state of statistical process control
- For product control, exhibiting small variability compared to specification limits
- For process control, exhibiting small variability compared to manufacturing process variations and demonstrating effective resolution

Measurement systems produce data within a window normally associated with a probability or likelihood that the data obtained faithfully represent their intended measurand(s). This likelihood is, as a rule, described in terms of standard deviation regarding a normal frequency distribution curve (see Chapter 27) such

Table 20.5 NASA 10-stage sequence in defining measurement requirements.

Stage	Description
1. Mission profile	Define the objectives of the mission. What is to be accomplished? What reliability is needed and what confidence levels are sought for decisions to be made from the measurement data?
2. System performance profile	Define the needed system capability and performance envelopes needed to accomplish the mission profile. Reliability targets and confidence levels must be defined.
3. System performance attributes	Define the functions and features of the system that describe the system's performance profile. Performance requirements must be stated in terms of acceptable system hardware attribute values and operational reliability.
4. Component performance attributes	Define the functions and features of each component of the system that combine to describe the system's performance attributes. Performance requirements must be stated in terms of acceptable component attribute values and operational reliability.
5. Measurement parameters	Define the measurable characteristics that describe component and/or system performance attributes. Measurement parameter tolerances and measurement risks (confidence levels) must be defined to match system and/or component tolerances and operational reliability.
6. Measurement process requirements	Define the measurement parameter values, ranges, and tolerances; uncertainty limits; confidence levels; and time between measurement limits (test intervals) that match mission, system, and component performance profiles (Stages 2, 3, and 4) and the measurement parameter requirements (Stage 5).
7. Measurement system design	Define the engineering activities to integrate hardware and software components into measurement systems that meet the measurement process requirements. This definition must include design of measurement techniques and processes to assure data integrity.
8. Calibration process requirement	Define the calibration measurement parameter values, ranges, uncertainty limits, confidence levels, and recalibration time limits (calibration intervals) that match measurement system performance requirements, so as to detect and correct for systematic errors and/or to control uncertainty growth.
9. Calibration system design	Define the integration of sensors, transducers, detectors, meters, sources, generators, loads, amplifiers, levers, attenuators, restrictors, filters, switches, valves, etc., into calibration systems that meet the calibration process requirements. This definition must include design of calibration techniques and processes to assure data integrity.
10. Measurement traceability requirements	Define the progressive chain of calibration process requirements and designs that provide continuous reference to national and international systems of measurement from which internationally harmonized systems measurement process control is assured.

184 Part III: Metrology Concepts

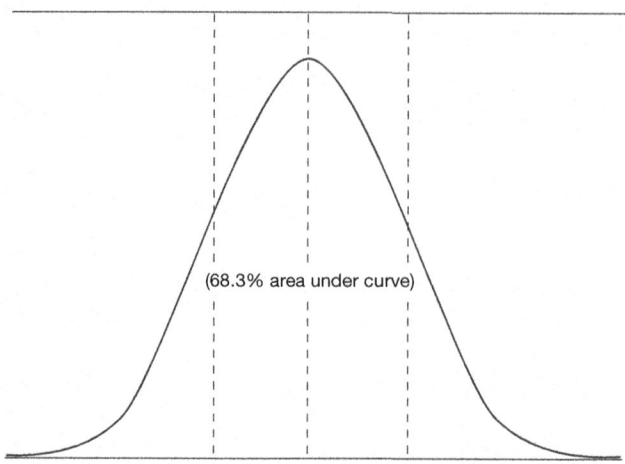

Figure 20.1 Normal frequency distribution curve—1 standard deviation.

that a percentage of probability of occurrence equates to an area under the normal distribution curve. See Figure 20.1 for an illustration of one standard deviation in a normal frequency distribution curve.

Table 20.6 lists the percentages of probability of occurrence for various multiples of standard deviations in regard to a normal frequency distribution curve.

Information such as this may be used to derive reliability targets in terms of risk assessment associated with the probability that inspection, measurement, and test equipment (IM&TE) will drift out of specification during its calibration

Table 20.6 Normal frequency distribution curves and associated standard deviation.

Standard deviation	Area under the normal frequency distribution curve
0.6745	50.00 %
1	68.30 %
1.036	70.00 %
1.282	80.00 %
1.645	90.00 %
1.96	95.00 %
2	95.50 %
2.58	99.00 %
3	99.75 %
3.291	99.90 %
4	99.9940 %
5	99.99994 %

interval (see Chapter 10) or in the estimation that measurements may be expected to lie within a probable range (see Chapter 29).

Calibration considerations regarding measurement systems are of particular interest to metrologists. NASA Reference Publication 1342 notes that "[m]easurement processes are accompanied by errors and uncertainties that cannot be eliminated. However, they can be quantified and limited or controlled to 'acceptable' levels. Calibration is done for this purpose."[6]

Fluke's *Calibration: Philosophy in Practice* notes that calibration is "[a] set of operations, performed in accordance with a definite, documented procedure, that compares the measurements performed by an instrument to those made by a more accurate instrument or standard, for the purpose of detecting and reporting, or eliminating by adjustment, errors in the instrument tested."[7] Calibration can thus be considered as defining attributes about a measurement system in terms of how well they can correlate an unknown (object or phenomenon being measured) to a known (calibration standard). Calibration relates measurement systems to performance indices, thereby providing the means to estimate expected performance.

Measurement systems, like manufacturing processes, transform inputs to desired outputs and are composed of various interactive, interdependent elements that determine the quality and makeup of these outputs. Careful consideration must be given to the development or selection of a measurement system in order for it to correspond with measurement application requirements, thus ensuring the validity and usability of derived measurement data. Measurements are metrological tools used in performing measurement tasks. As the age-old adage says, "The right tool for the right job" cannot be overemphasized.

MEASUREMENT CAPABILITIES

Measurement systems embody a variety of measurement capabilities inherent in their design for their intended purpose(s). *Measurement capabilities* are attributes of a measurement system that determine the extent to which measurements may be made within some qualifying restraints such as measurement range, ambient conditions, required input amplitude, and so on. Measurement system capabilities should be congruent with the requirements of the measurement application they are intended for. Whether a measurement system has the required capabilities to meet a measurement application is not always readily apparent and often must be established through user-assessment activities. It is the responsibility of the measurement system user (or their management) to ascertain whether a measurement system is capable of meeting the requirements of a particular measurement application. Intentional/unintentional use of a measurement system in terms of operation beyond its established capabilities normally results in measurement data with unknown uncertainties at best, or completely erroneous measurement data in worst-case scenarios.

The adequacy of a measurement system to fulfill the requirements of a measurement application is addressed within ISO 9001:2015, Clause 7.1.5, "Monitoring and measuring resources," sub-clause 7.1.5.1, General: "The organization shall determine and provide the resources needed to ensure valid and reliable results when monitoring and measuring is used to verify the conformity of products and services to requirements. The organization shall ensure that the resources provided:

a) Are suitable for the specific type of monitoring and measurement activities being undertaken.

b) Are maintained to ensure their continuing fitness for their purpose."[8]

As continued in ISO 9001:2015, Clause 9.1, "Monitoring, measurement, analysis and evaluation," sub-clause 9.1.1, "the organization shall determine:

a) What needs to be monitored and measured,

b) The methods for monitoring, measurement, analysis, and evaluation needed to ensure valid results,

c) When the monitoring and measuring shall be performed,

d) When the results from monitoring and measurement shall be analysed and evaluated."[9]

Bias Errors

A measurement system's capabilities are often characterized in terms of bias, linearity, repeatability, reproducibility, and stability. *VIM* 2.18 defines *bias* (or *measurement bias*) as "estimate of a *systematic measurement error*."[10] *Systematic measurement error* (*VIM* 2.17) is the "component of *measurement error* that in replicate *measurements* remains constant or varies in a predictable manner."[11] Bias is normally established by averaging the error of indication over an appropriate number of repeated measurements. These measurements are assumed to be of the same measurand using the same measurement system. Bias is frequently referred to as a *systematic offset*. Some possible causes of bias are:

- Measurement system requires calibration or has been improperly calibrated.

- Measurement system is defective, worn, or contaminated.

- Measurement system is inadequate or inappropriate for the measurement application.

- Environmental conditions are excessive.

- Compensation was not applied.

- Operator error.

- Computational error.

Linearity Errors

The NIST/SEMATECH *e-Handbook of Statistical Methods* defines *linearity* (of a gage, section 2.4.5.2) as "gauge response increases in equal increments to equal increments of stimulus, or, if the gauge is biased, that the bias remains constant throughout the course of the measurement process."[12] Some possible causes for linearity errors are:

- Measurement system requires calibration or is improperly calibrated.
- Measurement system is defective, worn, or contaminated.
- Measurement system environment is excessive and/or unstable.
- Measurement system is inadequately maintained.

Precision, Repeatability, and Reproducibility

Continuing with *VIM* definitions, *measurement precision* (*VIM* 2.15) is defined as "closeness of agreement between *indications* or *measured quantity values* obtained by replicate *measurements* on the same or similar objects under specified conditions."[13] *VIM* subsequently notes that measurement precision is used to define measurement repeatability. Repeatability is commonly referred to as *within-system variation* or *equipment variation*. *VIM* defines *measurement repeatability* (*VIM* 2.21) as "*measurement precision* under a set of *repeatability conditions of measurement*."[14] *Repeatability condition of measurement (repeatability condition)* (*VIM* 2.20) is a "condition of *measurement*, out of a set of conditions that include the:

- Same *measurement procedure*
- Same operators
- Same *measuring system*
- Same operating conditions
- Same location
- And replicate measurements on the same or similar objects over a short period of time."[15]

VIM defines *measurement reproducibility* (*VIM* 2.25) as "*measurement precision* under *reproducibility conditions of measurement*."[16] *Reproducibility condition of measurement (reproducibility condition)* (*VIM* 2.24) is the "condition of *measurement*, out of a set of conditions that includes:

- Different locations
- Different operators
- Different *measurement systems*
- And replicate measurements on the same or similar objects."[17]

Additional variations sometimes employed in reproducibility include:

- Different environmental conditions
- Different calibration standard used
- Different time and/or location
- Different measurement technique

Reproducibility is often referred to as the average variations between measurement systems or the average variations between changing conditions of a measurement.

VIM defines *stability of a measuring instrument* (*VIM* 4.19) as "property of a *measuring instrument*, whereby its metrological properties remain constant in time."[18] Stated a little differently, stability is a measure of a measurement system's total variation regarding a specific measure and over some time interval. Quite simply, stability is the change in a measurement system's bias over time. Some possible causes for measurement system bias changes over time are:

- The measurement system requires calibration.
- The measurement system is defective, worn, or contaminated.
- The measurement system is aging.
- The measurement system environment is changing.
- The measurement system is inadequately maintained.

Nonstability of a measuring instrument is closely related to *instrumental drift* (*VIM* 4.21), which *VIM* defines as "continuous or incremental change over time in *indication*, due to changes in metrological properties of a *measuring instrument*."[19]

Statistical Process Control and Control Charts

The variability of a measurement system may be determined via the use of control charts. Walter A. Shewhart, the creator of statistical process control (SPC), pioneered the use of control charts, frequently referred to as Shewhart charts, during the 1920s while working for Bell Telephone Laboratories. Juran's *Quality Control Handbook* defines SPC as "the application of statistical techniques for measuring and analyzing the variation in processes."[20] Shewhart analyzed many different processes and identified two variation components common to all: a steady component inherent to the process and an intermittent component. He referred to the steady variation component as random variations attributable to chance and undiscovered causes such that when averaged its variance is about the same as the parameter being measured. The intermittent component, he concluded, could be attributed to assignable sources (systematic) and as such be removed from a process. He went on to say that a process with only random components could be said to be in a state of statistical control. (Note: A process can be in a state of statistical control and not meet specifications, as statistical control merely means that only random variations are present.) Shewhart envisioned using control charts as a means of applying statistical principles to identify and monitor process variation using intuitive graphics. Control chart limits, based on statistical variations of a process in terms of multiples of standard deviation (one, two, or three standard deviations), are normally included in control charts as a ready means of determining whether a process is in a state of statistical control.[21] The two most popular control charts deal with measurement averages (*X*-bar charts) and measurement ranges (*R*-bar charts). The GOAL/QPC *Memory Jogger 2* recommends the following steps in constructing a control chart:

- Select process to be charted.
- Determine sampling method and plan.

- Initiate data collection.
- Calculate the appropriate statistics.[22]

The GOAL/QPC *Memory Jogger 2* gives the following criteria for determining if your process is out of control by dividing a control chart into different zones (see Table 20.7):

- One or more points fall outside of the control limits.
- Two points, out of three consecutive points, are on the same side of average, in Zone A or beyond.
- Four points, out of five consecutive points, are on the same side of average, in Zone B or beyond.
- Nine consecutive points on one side of average.
- There are six consecutive points increasing or decreasing.
- There are 14 consecutive points that alternate up and down.
- There are 15 consecutive points within Zone C (above and below the average).[23]

Table 20.8 is an example of time interval measurements (in seconds) made of four thermocouples in an environmental chamber. Measurements are made once a day for 25 consecutive days. Table 20.9 lists values used for calculating control limits. The *X*-bar chart in Figure 20.2 and the *R*-bar chart in Figure 20.3 are derived from Table 20.8 calculations.

Gary Griffith, author of *The Quality Technician's Handbook*, recommends that you always annotate the control chart for any of the following:

- Out-of-control conditions and the cause
- Reason the chart was stopped, such as during equipment downtime
- Reason the chart was started again, such as a new setup
- When control limits are recalculated
- When any adjustments are made to the process
- Any other pertinent information about the process[25]

Table 20.7 Control chart zones.

Upper control limit (UCL)
Zone A
Zone B
Zone C
Average
Zone C
Zone B
Zone A
Lower control limit (LCL)

Table 20.8 X-bar and R-bar control chart calculations.

No.	TC 1	TC 2	TC 3	TC 4	Measure Means	Measure Range
1	27.347	27.501	29.944	28.212	28.251	2.597
2	27.797	26.150	31.213	31.333	29.123	5.183
3	33.533	29.330	29.705	31.053	30.905	4.203
4	37.984	32.269	31.917	29.443	32.903	8.541
5	33.827	30.325	28.381	33.701	31.559	5.44
6	29.684	29.567	27.231	34.004	30.121	6.773
7	32.626	26.320	32.079	36.172	31.799	9.852
8	30.296	30.529	24.433	26.852	28.027	6.096
9	33.533	29.330	29.705	31.053	30.905	4.203
10	37.984	32.269	31.917	29.443	32.903	8.541
11	33.827	30.325	28.381	33.701	31.559	5.446
12	29.684	29.567	27.231	34.004	30.121	6.773
13	26.919	27.661	31.469	29.669	28.930	4.551
14	28.465	28.299	28.994	31.145	29.226	2.846
15	32.427	26.104	29.477	37.201	31.302	11.097
16	28.843	30.518	32.236	30.471	30.517	3.393
17	30.751	32.999	28.085	26.200	29.509	6.799
18	31.258	24.295	35.465	28.411	29.857	11.170
19	28.278	33.949	30.474	28.874	30.394	5.671
20	26.919	27.661	31.469	29.669	28.930	4.551
21	28.465	28.299	28.994	31.145	29.226	2.846
22	32.427	26.104	29.477	37.201	31.302	11.097
23	28.843	30.518	32.236	30.471	30.517	3.393
24	30.751	32.999	28.085	26.200	29.509	6.799
25	31.258	24.295	35.465	28.411	29.857	11.170

X-bar control chart					
UCL = Xbar + A2 × Rbar	34.928				
LCL = Xbar – A2 × Rbar	25.653	UCL	Upper control limit		
CL = Xbar	30.290	LCL	Lower control limit		
R-bar control chart		CL	Center line		
UCL = D4 × Rbar	14.517	n	Sample size	4	
LCL = D3 × Rbar	0.000	R-bar	Mean of ranges	6.361	
CL = Rbar	6.361	X-bar	Mean of measured means	30.290	

Table 20.9 Table for calculating the control limits.

n	A2	D3	D4	n	A2	D3	D4
2	1.88	0	3.27	14	0.235	0.328	1.672
3	1.023	0	2.57	15	0.223	0.47	1.653
4	0.729	0	2.28	16	0.212	0.363	1.637
5	0.577	0	2.12	17	0.203	0.378	1.622
6	0.483	0	2	18	0.194	0.391	1.608
7	0.419	0.076	1.92	19	0.187	0.403	1.597
8	0.373	0.136	1.86	20	0.18	0.415	1.585
9	0.337	0.184	1.82	21	0.173	0.425	1.575
10	0.308	0.223	1.78	22	0.167	0.434	1.566
11	0.285	0.256	1.74	23	0.162	0.443	1.557
12	0.266	0.283	1.72	24	0.157	0.451	1.548
13	0.249	0.307	1.69	25	0.153	0.459	1.541

Note: To avoid errors associated with small sample sizes (< 25), the above control chart table values are used for A2, D3, and D4 as condensed from Juran's *Quality Control Handbook,* 4th edition, Table A, Factors for Computing Control Charts.[24]

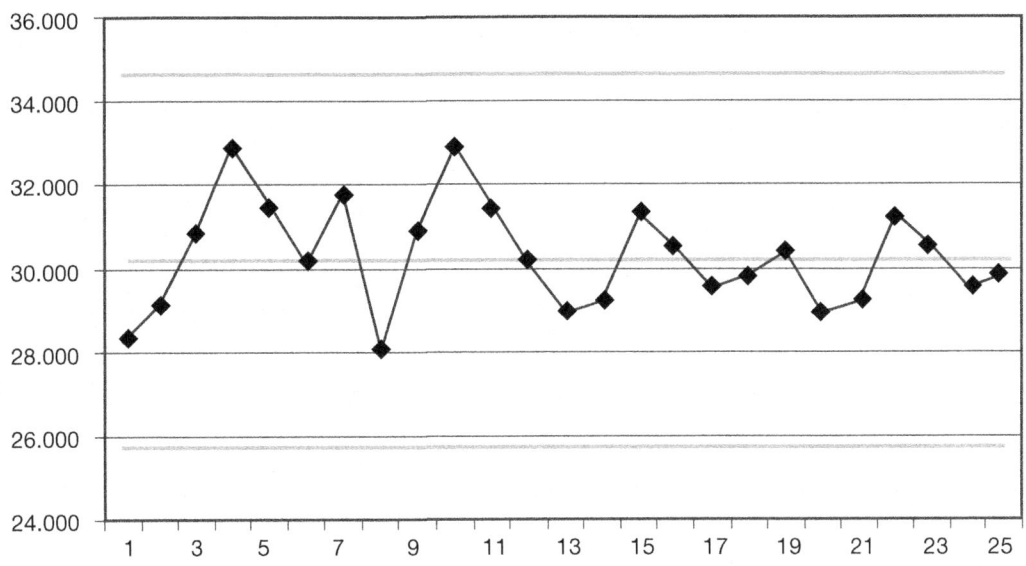

Figure 20.2 *X*-bar chart for example of time interval measurements.

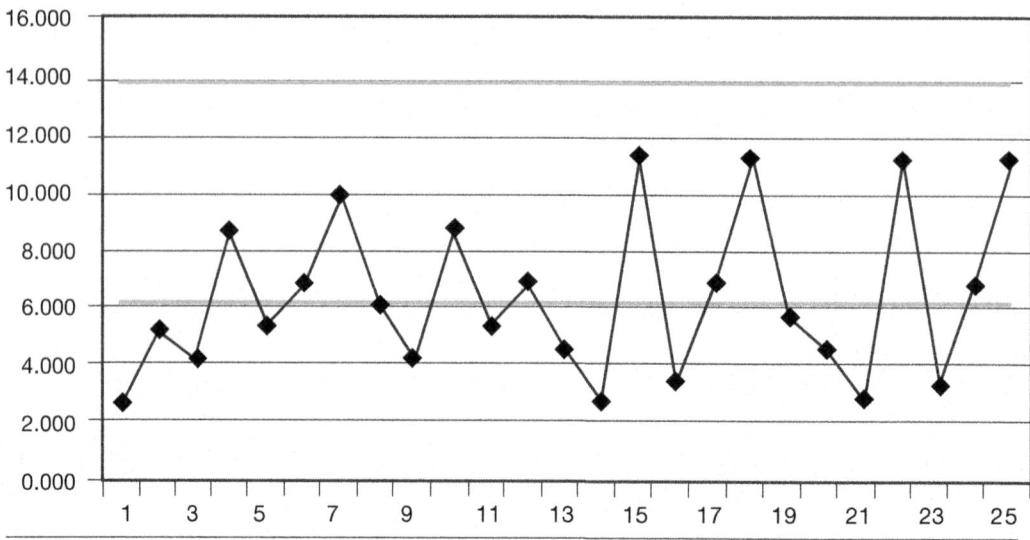

Figure 20.3 *R*-bar chart for the example of time interval measurement.

Gage R&R

A method for assessing the capability of a measurement system is known as a *gage repeatability and reproducibility assessment*, more commonly called a *gage R&R study*. The gage R&R study is designed to measure both the repeatability and reproducibility of a measurement process. The *Quality Technician's Handbook* recommends the following elements of planning prior to performing a gage R&R study:

- Select the proper instrument.
- Make sure the measurement method is appropriate.
- Follow the 10 percent rule of discrimination.
- Look for obvious training/skill problems with observers.
- Make sure all observers use the same gage (or same type of gage).
- Make sure all the measuring equipment is calibrated.[26]

The range method is the most common method for performing a gage R&R study. The range method for a gage R&R study is similar to setting up *X*-bar and *R*-bar control charts in terms of data acquisition and computing the means of observations and the ranges of these observations. Normally, a gage R&R study involves more than one observer (known as an appraiser) performing measurements. The following parameters are typically calculated in a range-method gage R&R study: equipment variation (EV), equipment variation percentage (% EV), appraiser variation (AV), appraiser variation percentage (% AV), repeatability and reproducibility (R&R), R&R percentage (% R&R), part variation (PV), part variation percentage (% PV), and total variation (TV).

Table 20.10 shows an example of a range method gage R&R study involving three appraisers measuring 10 parts.

Figure 20.4 is another way to look at the R&R study involving three appraisers.

Table 20.10 Range R&R study.

Measurements

Part #	1	2	3	4	5	6	7	8	9	10	Average	
Appraiser 1 Trial # 1	0.39	−0.56	1.34	0.47	−0.70	0.03	0.59	−0.31	3.36	−1.36	0.33	
Appraiser 1 Trial # 2	0.41	−0.67	1.17	0.50	−0.93	−0.11	0.75	−0.30	1.99	−1.35	0.15	
Appraiser 1 Trial # 3	0.64	−0.57	1.37	0.64	−0.74	−0.31	0.66	−0.17	3.01	−1.31	0.32	
Average	0.480	−0.600	1.293	0.537	−0.790	−0.130	0.667	−0.260	2.787	−1.340	0.26	Xbar₁
Range	0.25	0.11	0.2	0.17	0.23	0.34	0.16	0.14	1.37	0.05	0.302	Rbar₁
Part #	1	2	3	4	5	6	7	8	9	10	Average	
Appraiser 2 Trial # 1	0.07	−0.47	1.19	0.01	−0.56	−0.3	0.47	−0.63	1.7	−1.67	−0.02	
Appraiser 2 Trial # 2	0.35	−1.33	0.94	1.03	−1.3	0.33	0.55	0.07	3.13	−1.63	0.21	
Appraiser 2 Trial # 3	0.07	−0.67	1.34	0.3	−1.37	0.06	0.73	−0.34	3.19	−1.5	0.18	
Average	0.163	−0.823	1.157	0.447	−1.077	0.030	0.583	−0.300	2.673	−1.600	0.125	Xbar₂
Range	0.28	0.86	0.40	1.02	0.81	0.63	0.26	0.70	1.49	0.17	0.662	Rbar₂
Part #	1	2	3	4	5	6	7	8	9	10	Average	
Appraiser 3 Trial # 1	0.04	−1.37	0.77	0.14	−1.46	−0.39	0.03	−0.46	1.77	−1.49	−0.24	
Appraiser 3 Trial # 2	−0.11	−1.13	1.09	0.3	−1.07	−0.67	0.01	−0.46	1.44	−1.77	−0.24	
Appraiser 3 Trial # 3	−0.14	−0.96	0.67	0.11	−1.44	−0.49	0.31	−0.49	1.77	−3.16	−0.38	
Average	−0.070	−1.153	0.843	0.183	−1.323	−0.517	0.117	−0.470	1.660	−2.140	−0.287	Xbar₃
Range	0.18	0.41	0.42	0.19	0.39	0.28	0.30	0.03	0.33	1.67	0.420	Rbar₃
Part #	1	2	3	4	5	6	7	8	9	10		
Average of trial averages	0.191	−0.859	1.098	0.389	−1.063	−0.206	0.456	−0.343	2.373	−1.693	0.034	A_Xbar

Continued

Table 20.10 *Continued*

Computations

Average of (Rbar₁, Rbar₂, Rbar₃)	A_Rbar	=	0.4613
Range of (Xbar₁, Xbar₂, Xbar₃)	R_Xbar	=	0.5513
Range of averages of trial averages	RA_Xbar	=	4.0667
Number of parts	n	=	10
Number of trial	t	=	3
Equipment variation	EV	=	A_Rbar × K₁ = 0.27256
Appraiser variation	AV	=	Sqrt ((R_Xbar × K₂)^2 – (EV^2 / (n × t))) = 0.28408
Repeatability and reproducibility	RR	=	Sqrt (EV^2 + AV^2) = 0.39368
Part variation	PV	=	RA_Xbar × K₃ = 1.27937
Total variation	TV	=	Sqrt (RR^2 + PV^2) = 1.33857
% Equipment variation	% EV	=	100 × (EV/TV) = 20.36 %
% Appraiser variation	% AV	=	100 × (AV/TV) = 21.22 %
% Repeatability and reproducibility	% RR	=	100 × (RR/TV) = 29.41 %
% Part variation	% PV	=	100 × (PV/TV) = 95.58 %

Constants values	Trial #	2	3	4	5	6	7	8	9	10
	D₄	3.27	2.57	2.28	2.11	2.00	1.92	1.86	1.82	1.78
	K₁	0.8862	0.5908							
	K₂	0.7071	0.5231							
	# of parts	2	3	4	5	6	7	8	9	10
	K₃	0.7071	0.5231	0.4467	0.4030	0.3742	0.3534	0.3375	0.3249	0.3146

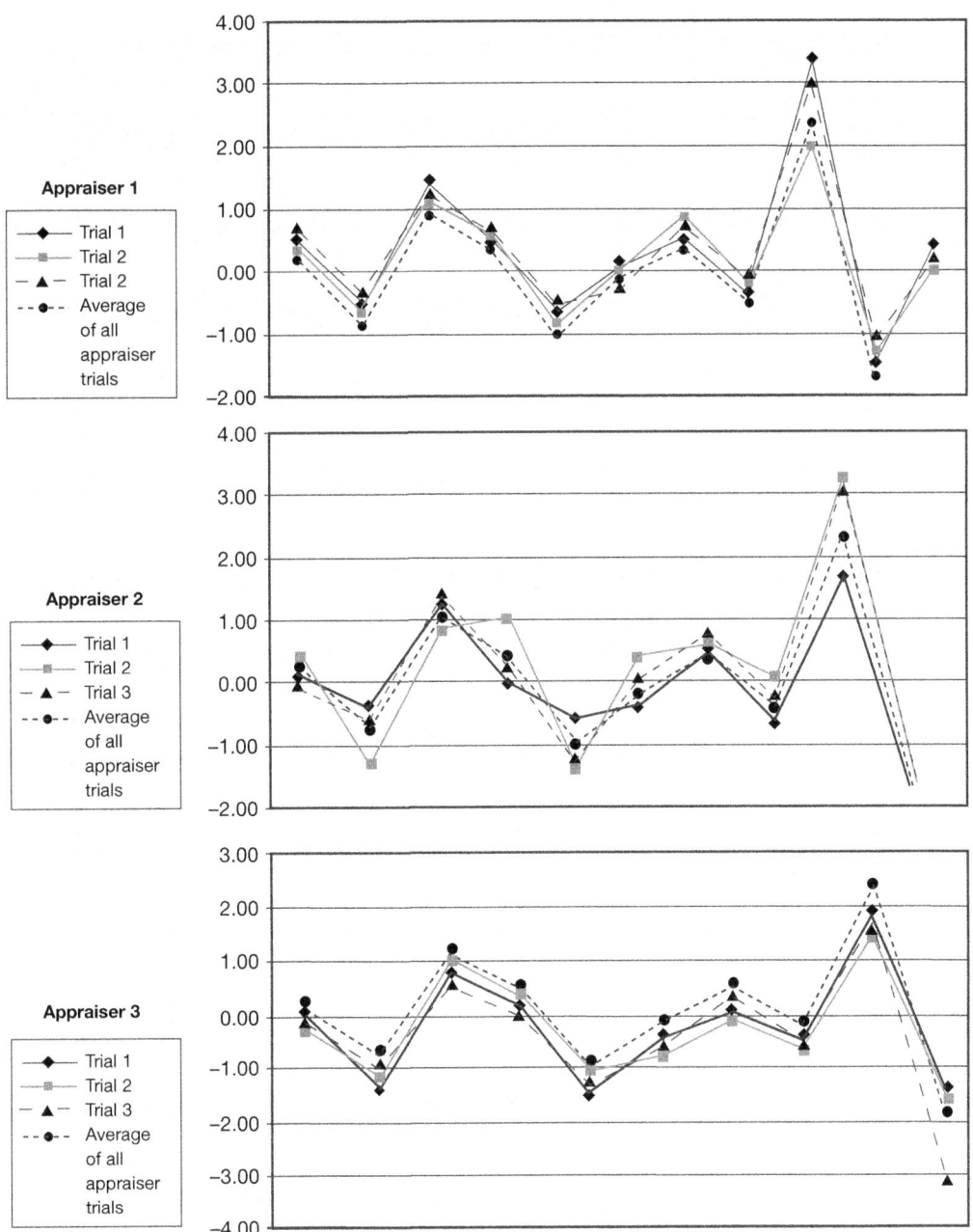

Figure 20.4 Example of appraiser variations.

MEASUREMENT DATA

A measurement system's principal purpose is to generate measurement data. Measurement data come in many varieties, such as alphanumeric characters, plots and graphs, increasing or decreasing audiovisual displays, limit indicators, and so on. Measurement data type should be compatible with the requirements of the intended measurement application. Measurement data are often the only basis for making decisions as to whether a process is in statistical control or a product is in conformance with published specifications. In this context, measurement data must be of a type and quality sufficient to provide adequate information about a measurement application in order to make informed decisions about it. Inappropriate measurement data type, format, or quality can be misleading, resulting in erroneous assumptions about a measurement application. An example of inappropriate measurement data type would be trying to use the graphic display of an oscilloscope to derive numeric data about a process having both very large and very small amplitude changes. In this case, the digital readout of a digital multimeter having both sufficient resolution and acquisition speed would be a better choice. To avoid masking or distorting relevant information, measurement data type, format, and quality should always be considered when evaluating measurements made by measurement systems.

Measurement data, to be useful, must be faithful to the represented measurement application. Measurement data considerations should be addressed to ensure that the data are accurate, credible, and usable for their intended purpose(s). The following are some key measurement data considerations:

- **Format.** This refers to the way measurement data are oriented (layout), type of graphic (bar, pie, and so on), font type and font size, numerical convention, date convention, and so on.

- **Resolution.** This is the smallest or least significant digit (LSD) distinguishable within measurement data. *VIM 4.15* defines *resolution of a displaying device* as the "smallest difference between displayed *indications* that can be meaningfully distinguished."[27] For a digital measuring device, this is the change in the indication when the least significant digit changes by one increment.

- **Readability.** This refers to the ergonomic way measurement data are presented, in terms of how easily they can be read and deciphered by observers. Note that this does not imply comprehension of the data—only that the data are presented in a way that their intent can be readily determined.

- **Suitability.** This refers to how measurement data are presented regarding both the application from which they are derived and the intent as to how the data will be used. An example of this would be in presenting seldom-occurring, slight changes in a large quantity of measurement data. Presenting the measurement data in table format would not readily identify these small, seldom-occurring changes, whereas a log-linear bar chart would allow fairly easy identification.

- **Confidentiality.** This refers to protection and control issues focusing on both measurement data and the source(s) from which they were obtained. Often, measurement data are used in benchmarking and/or proficiency evaluations and as such have the potential for:

    o Competitors to use them to their advantage

    o Publicizing uncomplimentary performance

o Being interpreted outside of their intended context

o Disclosing capabilities or limitations

o Giving insight into programs or products in development

Measurement data confidentiality should be explicitly addressed, and proper safeguards incorporated to prevent unauthorized disclosure.

Measurement data considerations can, if not satisfactorily addressed, make good data unusable for their intended purpose(s). Without a satisfactory understanding of measurement systems output data in relation to the requirements of a measurement application, said data consideration can result in time and effort being wasted and bad decisions being made. Measurement practitioners and those interpreting measurement data would do well to consider how measurement data are to be used before selecting a measurement system for a measurement application to avoid many of the pitfalls.

CALIBRATION METHODS AND TECHNIQUES

IM&TE requires the use of various methods and techniques for determining whether a unit is operating to its published specifications. These methods and techniques are used to establish a relationship between an applied signal and the corresponding IM&TE measurement display, cancel out residuals that can offset the measurand of interest, provide sufficient sensitivity to determine small differences between a unit and a calibration standard, and so on. Often, a calibration method/technique is recommended by the original equipment manufacturer (OEM) as a necessary step to be performed prior to using a unit.

The selection of a particular calibration method/technique is dictated by various factors. These factors include the measurand of interest, the inherent functionality and/or limitations of the IM&TE, the measurement scenario and its associated environment, the operator knowledge and skill, OEM recommendations, and so on. Use of an inappropriate calibration method/technique can often mask an IM&TE measurement response, rendering it inaccurate. Not performing an OEM-recommended calibration method/technique can also degrade a unit's performance. It is essential that calibration practitioners be aware of the calibration methods/techniques appropriate for a particular application in order to ensure that their evaluations are based on reliable measurement data and to avoid misadjusting a unit as a result of misleading measurement data.

Calibration methods and techniques can be adapted to a wide variety of IM&TE types. Some of the most common calibration methods and techniques include:

- **Linearization.** This is a method by which IM&TE is corrected for a linear response such that a step change in an applied signal will result in a corresponding step change in the IM&TE indication. This method may also be used to correct a nonlinear output using a linear-responding measuring device. Linearization is typically used to correct nonlinear measurement sensors such as those used to make high-temperature measurements.

- **Nulling.** This is a method by which two applied signals are algebraically summed whereby the common portion of each signal cancels out, leaving only the difference between the signals. This allows very small differences in signal to be measured that would otherwise be very difficult to detect due to the size of the applied signals relative to these differences. Nulling may also be used to establish a known quantity from an unknown quantity by increasing or decreasing the unknown quantity until the difference between it and the known quantity is sufficiently small. Nulling is frequently used in intercomparing 10V DC calibration standards, allowing for difference measurements at the parts-per-million (ppm) level.

- **Spanning.** This is a method by which an IM&TE-specific range is defined. Spanning typically involves bringing an IM&TE measurement readout scale into agreement with the intended range of the unit via adjustments, changing component values, or firmware correction. Spanning helps ensure that a unit's high, low, and midrange measurement responses correspond to its high, low, and mid-scale measurement readout values. This method is commonly used to set up the range of a pressure gage.

- **Spot frequency.** This is a method by which specific outputs or measurement ranges, commonly referred to as *sweet spots*, are enhanced via correction factors. Outputs and measurement ranges between sweet spots normally include additional interpretation uncertainties as a result of not being directly compared to calibration standards. This method is commonly used to correct IM&TE AC voltage and current response at standardized amplitudes and frequencies as provided directly by AC voltage and current calibration standards.

- **Zeroing.** This is a method by which an IM&TE measurement readout offset present in the absence of an applied signal is excluded from measurements via hardware adjustment or algebraic cancellation. IM&TE numeric indication in the absence of an applied signal after this exclusion is nominally zero. Zeroing is commonly used to establish a datum, or starting or reference point, such as when zeroing a height gage on a surface plate before measuring the height of gage blocks placed on the plate.

Chapter 21

Specifications

Understanding specifications and tolerances is important when applying measurement theory to the real world. The terms *specification* and *tolerance* are closely related and often confused. Although both terms define quantitative limits in relation to a nominal value, they are used in different situations. Unfortunately, some definitions even conflict with one another for the same term.

Specification defines the limits within which an instrument is able to be useful. Properly interpreted and used, specifications are useful in evaluating the performance of inspection, measurement, and test equipment (IM&TE) during calibration and for evaluating the capability of a measurement standard to perform a calibration.

Tolerance is the maximum permissible variations or limits allowed a quantity/parameter from a specified value. Information is described in the form of "± n %," "not less than *n* dB," "maximum," "minimum," etc., in the performance requirement section of the calibration procedures of measuring and test equipment and measurement standards being calibrated. Tolerances are further limits *within* specifications.

Here are other sources of definitions of the words *specification* and *tolerance*:

- *Specifications* define the expected performance limits of a large group of substantially identical finished products (all units of a specific model of digital thermometer, for example). Customers use specifications to determine the suitability of a product for their own applications.

- In JCGM 200-2012, *VIM 3*, *specification* is referenced in the first note for "measurand," where measurand is "quantity intended to be measured."

  o "NOTE 1 The **specification** of a measurand requires knowledge of the *kind of quantity*, description of the state of the phenomenon, body, or substance carrying the quantity, including any relevant component, and the chemical entities involved."[1]

- *VIM* 3 also mentions "specification" in **maximum permissible measurement error** (maximum permissible error, limit of error) (*VIM* 3, 4.26): "extreme value of *measurement error*, with respect to a known *reference quantity value*, permitted by **specification or regulations** for a given *measurement, measuring instrument,* or *measuring system*."[2]

- ISO/IEC 17000:2020(en), Clause 5.1: "specified requirement is a need or expectation that is stated . . . in normative documents such as regulations, standards, and technical specifications."[3]

- ISO 9000:2015, Clause 3.8.7, "specification is a document stating requirements."[4]

- NCSL International Recommended Practice-5, 2016, *Recommended Practice for Measuring and Test Equipment Specifications* (RP-5): "A numerical value or range of values that bound the performance of an MTE parameter or attribute."[5]

- NIST *Handbook 44*:2022: "Specification:—A requirement usually dealing with the design, construction, or marking of a weighing or measuring device. Specifications are directed primarily to the manufacturers of devices."[6]

- NCSL International Laboratory Management-3 1999, *Laboratory Related Terms* (LM-3): "Specification: A quantitative description of the specified characteristics of an instrument, device system, product, or process. A group of statements that define capabilities of product."[7]

Here are various definitions of the word *tolerance*:

- JCGM 200-2012, *VIM* 3, mentions "tolerance" in "**Maximum permissible measurement error**, maximum permissible error, limit of error (4.26): extreme value of *measurement error*, with respect to a known *reference quantity value*, permitted by specifications or regulations for a given *measurement, measuring instrument,* or *measuring system.*

    o Note 1: Usually, the term 'maximum permissible errors' or 'limits of error' is used when there are two extreme values.

    o Note 2: the term 'tolerance' should not be used to designate 'maximum permissible error'."[8]

- NCSL International *RP-5* (2016): "Tolerance Limits: Typically, engineering tolerances that define the maximum and minimum values for a product to work correctly. These tolerances bound a region that contains a certain proportion of the total population with a specified probability or confidence."[9]

- NIST *HB 44*:2022: "Tolerance: A value fixing the limit of allowable error or departure from true performance or value."[10]

- Shewhart: "*Tolerance*" when used regarding manufactured products, is a design feature that defines limits within which a quality characteristic is supposed to be on individual parts. "Whereas tolerance is sometimes defined either as the difference between two limiting sizes as a means of specifying the degree of accuracy or as a specified allowance for variations from a standard, the connect of tolerance as used in this monograph implies not only the concept of tolerance limits *but also that of the percentage of the commercial product that may be expected to have a quality falling within this tolerance range.*"[11]

Application of Specification and Tolerance

The essential result of a calibration (ignoring the data for now) is a pass/fail decision about the item. The decision is based on the results of one or more measurements. Historically, the terminology has been that the result of a measurement is *in tolerance* or *out of tolerance*. This is probably based on the sense that calibration is a process that acts on one item at a time, so the result applies to only one item. It should be understood, however, that the performance specifications of an instrument or tool are the basis of a calibration procedure, as they are what the performance is evaluated against. It is more correct to say that the result of a measurement is within specification or out of specification.

TYPES OF SPECIFICATION LIMITS

There are two general types of specification limits. Which is used depends on the equipment, application, and what the information is used for.

1. A *one-way specification* allows variation in only one direction from the nominal value. The result of a measurement may be either more or less than one of the limits, but not both.[12] The nominal value is either a lower limit or an upper limit of the specification. This type of limit is often found in instruments used for work related to safety. Some companies may refer to a one-way specification as *unilateral* or *single-sided*.[13]

    Here are two examples of one-way specification:

    o An insulation tester may have a specification that the applied test voltage is 500 V, –0 % + 10 % when the test voltage selector is in the 500 V position. The purpose of this is to ensure that at least the specified test voltage is applied to the tested part to ensure that the part meets safety requirements. For the calibration technician, without considering the impact of measurement uncertainty, any measured value from 500 V to 550 V would pass, but 499.9 and 550.1 would both fail.

    o An appliance safety tester may specify that an alarm is to sound if a leakage current is 1 mA, –25 % + 0 %. The purpose of this test is to ensure that the user of the appliance cannot receive a dangerous shock. In this case, the instrument would fail if the applied current exceeded 1 mA and the alarm still had not sounded.

2. A *two-way specification* allows variation in either direction from the nominal value. The nominal value is a target, and the specification defines lower and upper limits of acceptable values.[14] The nominal value is not necessarily centered between the limits. Some companies may refer to a two-way specification as *bilateral*, or as *double-sided*.[15]

    Here are three examples of a two-way specification:

    o The output of a voltage source may have a specification of ± (0.005 % of setting + 10 mV). If the calibration technician tests it at 500 mV, the specification limits will be 499.965 mV to 500.035 mV.

o A signal generator may have an output flatness specified as ± 2 dB with respect to a 1.000 mW reference level. If a meter scaled in dB is used to make the measurement, the limits are –2 to +2 dB. Sometimes it may happen that the only available meter is scaled in watts. The limits must be recalculated using the relationship:

$$\frac{P_2}{P_{ref}} = 10^{(dB/10)}$$

This results in a nominal value of 1.000 mW, and the specification limits are 0.631 mW to 1.585 mW. It is important to note here that the nominal value is not centered between the limits.

o An ANSI/ASME B89.1.9–2002 (R2012) grade AS-1 gage block with a nominal size of 2 inches has a specification tolerance of ± 16 microinch.[16] (AS-1 is generally equivalent to U.S. Federal Specification GGG-G-15C grade 3. U.S. Federal Standard GGG-G-15C was withdrawn and replaced by ANSI/ASME B89.1.9 effective July 2002.) The GGG-G-15C Grade 3 specification for the same nominal 2-inch gage block was +16 microinch, –8 microinch.

Characteristics of a Specification

For an item of IM&TE, a specification is a condensed way of giving information about the probable uncertainty of a measurement made with the instrument. As noted earlier, a specification applies to all instruments of that make and model. As defined by Fluke,[17] there are three groups of uncertainty terms that make up a specification:

- *Baseline* specifications describe the basic performance of the instrument. Terms for the output, scale, and floor are included, although not all three are included every time. Baseline specifications that include one or two of these terms are more common than those with all three.

- *Modifier* specifications apply changes to the baseline, usually to express variation caused by environmental factors.

- *Qualifier* specifications are other important factors that may have to be considered in the application of the instrument.

Specifications are defined in relation to a nominal value. For fixed-value devices such as gage blocks, the nominal value is the specified size or other value. For any variable source or measuring instrument, the specification tables usually state the nominal value in terms of a range. The specification applies to any specific value in that range.

Specification tables for instruments may also contain values that are qualitative specifications. Typical examples include:

- Design parameters and other specifications that cannot be verified by a performance test

- Typical (and other) specifications that are not guaranteed by the manufacturer
- Specifications that are not critical to the quality of the measurements performed by the instrument (determined by the user)

In most cases, calibration procedures do not attempt to verify such qualitative specifications, even if they are shown in quantitative terms.

Baseline

A baseline specification that contains all three of the terms defined by Fluke could be shown as:

$$\text{Nominal value} \pm (\text{output} + \text{scale} + \text{floor})$$

where

Output = A percentage or parts per million (ppm) of the nominal value.
Scale = A percentage or ppm of the range or full-scale value.
Floor = A specific value expressed in the applicable SI units.

Note that the nominal value may be an operating range rather than a specific value. The metrologist may have to apply the specification to each specific value actually used.

Baseline Specification: Output Term

The *output* term is a fixed percentage or ppm ratio of the output (for a source) or the input (for a measuring instrument). Its value is always the input or output value multiplied by the percentage or ppm that applies to the range. For a physical device such as a gage block (or any fixed device), this part of the specification is normally omitted.

Baseline Specification: Scale Term

The *scale* term can be expressed in different ways and can be the most confusing to interpret. On instruments with analog meters, it is usually a percentage of the full-scale value of the meter. For example, a profilometer has an analog meter with a full-scale value of 100 μin and a specification of $\pm$ 2 % of full scale. This means that the possible error of the reading is $\pm$ (100 × 2 %) or $\pm$ 2 μin at any point on the scale. It is apparent that this value becomes more significant as the measured value is smaller, which is why it is common practice to select ranges that allow analog meter readings to be in the upper ⅔ of their scale.

On many digital instruments, the scale term is expressed as a number of digits or counts. This always refers to the least significant digits—the ones at the right of the display. The actual value they represent depends on the meter scale being used. For example, a 4½ digit meter is measuring current on the 20 A range, and the specification is $\pm$ (0.1 % reading + 5 digits). The maximum possible displayed value on this range is 19.999, so the least significant digit represents steps of 0.001 A. In SI units, the scale term of this specification is (5 × 0.001) or 0.005 A.

> ### About Display Size: Digits and Counts
>
> Instruments with a digital display (numbers instead of a moving-pointer meter) are often described as having a display of a certain number of digits (such as 4½) or counts (such as 20,000). These are two different ways of talking about the same thing.
>
> In general, any given position of a digital display can show any of the counting digits, 0 through 9. Historically, the leftmost digit position is often designed to show a 1 only or be blank. So, for example, a typical display could show any value from 0000 through 19999 (ignoring any decimal points or polarity signs). This would be a 4½-digit display—four positions that display 0 through 9, and one position at the left that is either blank or 1. The number of digits of the display is the maximum number of nines that can be displayed, plus ½ (or some other fraction) to represent the incomplete leftmost digit.
>
> In this example, it can also be seen that the range from 0000 through 19999 covers 20,000 distinct values. (Remember that zero is a counting figure.) Therefore, this can be referred to as a 20,000-count display. The number of counts of the display is the total number of discrete values included by the range zero through the maximum displayed value.
>
> Many newer digital instruments do not have the restriction of having the most significant digit being 1 or blank. A popular handheld digital multimeter, for example, has a maximum displayed value of 3999 (ignoring the decimal point location) on most of its ranges. For this reason, it is becoming more practical to refer to a digital meter display in terms of the number of counts instead of terms such as 4½ digits. This meter could be described as having a 4000-count display (0000 through 3999).
>
> Digital displays usually have some of their specifications expressed in terms of digits or counts. Again, these terms are two different ways of talking about the same thing. They always refer to the least significant digits—the ones at the right end of the display. A single count is the smallest interval that can be displayed, the least significant digit, which also defines the resolution of the display. The number of digits or counts in a specification is multiplied by the resolution to get the actual value. For example, assume that a gauge specification is ± (0.05 % indicated value + 12 counts) and the gauge display is 100.000 kPa. Then:
>
> - The digit/count/resolution value is 0.001 kPa.
>
> - The value of the output term of the specification is 100 × 0.0005 = 0.05 kPa.
>
> - The value of the scale term of the specification is 0.001 × 12 = 0.012 kPa.
>
> - The sum—which represents the uncertainty of the gauge at this value—is ± 0.062 kPa.

On some digital meters, especially higher-accuracy models, the scale term is expressed as a *percent of range*. Use caution with this type of specification, and read the instrument manual closely! This usually—but not always—means the range as labeled on the front panel of the meter or as listed in the specification table.

Many digital multimeters have an over range display; that is, the 1 V range may display values larger than 1 V. The over range may go up to a value such as 1.200. There are some models with 100 % over range, which means that the 1.000 V range will read up to 1.999 Volt. In most cases, the percent of range would apply to the name of the range (1 V), but in a few cases the manufacturer intends for it to apply to the maximum displayed value. Check the manual to be sure. If the scale term is listed as percent of scale, then it usually means the maximum possible reading on that range (name) of the meter. Check the manual to be sure. For example, a long-scale digital meter has an accuracy specification of ± (0.0009 % reading + 0.0010 % scale) on the 1.000000 V range. The maximum reading on that range is 1.200000 V. The scale term becomes (1.2 × 0.0010 %) or 0.000012 V.

For an artifact such as a gage block, the scale term is not as obvious. Also, the factors that equate to the scale term are often not stated in a specification sheet because they are contained in other documents that are included by reference and considered to be common knowledge among users of such devices. One part is the dimension specification, which varies by nominal length. This normally does not apply to an individual block, but rather to a full set. For example, a grade 1 one-inch block has a specification of ± 2 microinch (μinch). A 20-inch block has a specification of ± 20 microinch (μinch).[18] The other part is the thermal coefficient of expansion. This is a constant for the particular material, affects every block, and must be considered whenever the measurement is not being done at 20 °C. In the inch/pound system, the expansion coefficient is given in terms of inches per degree per inch of length.

Baseline Specification: Floor Term

The *floor* value is expressed in SI units and is a fixed amount that is added to the output at any value on the applicable range. At low levels, the floor term can be a significant part of the uncertainty. Here's an example. On the 1.000000 V DC range, a thermocouple simulator voltage source specification is ± (0.0005 % output + 5 mV). At an output setting of 0.450000 V, the floor value is only 0.001 % of the output. At 0.080000 V output, the floor value is up to 0.006 % of the output. At 0.001174 V output, the floor is more than 0.4 % of the voltage output and can be very significant. Because this instrument is a thermocouple simulator, the floor on the last value is even more significant. The voltage represents a thermocouple output at 23.0 °C, and, in those terms, the temperature uncertainty due to the floor specification is approximately 0.1 °C.

The floor term is usually not explicitly found in dimensional measurements, but the effect is there. Using a grade 0.5 gage block set specification as an example, all blocks 1 inch and smaller in nominal dimension have a specification of ± 1 microinch (μin).[19] This effectively places that as a limit on the best available measurement uncertainty and is therefore a floor.

The most commonly seen forms of these three terms are scale only (usually on analog meters and fixed devices), output plus scale, or output plus floor.

Forms of Writing Specifications

There are several ways that equipment manufacturers use to print their specifications, with three being the most common. In a strict mathematical sense, the expressions mean different things and can result in different uncertainty intervals for the instrument. The three most common forms are:

1. **(± output ± scale):** This format implies that both terms are random and therefore could be combined using the rss method. Using example values of 0.005 and 0.01, the result would be a range of ± 0.011.

2. **(± output + scale):** This format implies that the first term is random and the second term always acts as a bias in a specific direction. Using example values of 0.005 and 0.01, the result would be a range of 0.005 to 0.015.

3. **± (output + scale):** This format implies that the specification as a whole is random. Using example values of 0.005 and 0.01, the result would be a range of ± 0.015. *This is the most conservative method of showing specifications and is the one we recommend.*

When comparing instruments, determining performance verification limits, writing calibration procedures, or in any other work involving specifications, it is much easier and fewer errors are made if all specifications are expressed in the same format first.

Modifiers

As noted by Fluke, modifiers change the baseline specification. If present, a modifier may indicate differences in performance relative to time, ambient conditions, load, or line power. Other conditions may be mentioned occasionally, but these are the most common.

Modifiers: Time Term

Most specifications of electronic equipment include a time during which the specification is valid. Many bench or system digital multimeters, for example, have performance specifications for 24 hours since calibration, for 30 and 90 days, and for one year. (Other time periods are also common.) Mechanical measuring instruments are not often specified this way because they do not drift in the same way electronic ones do and because their physical wear rate is highly dependent on the type and frequency of use.

Some users have tried to set specifications for their instruments by doing a linear interpolation between the published time intervals. Experiments performed by Fluke indicate that this is not a good practice.[20] Although the long-term drift of a standard may be nearly linear, it is subject to short-term random variation due to noise and other factors. A more conservative approach is to use the specification for the time period that has not yet passed. The most conservative approach is to always use the specification that corresponds to the recalibration interval that has been set. For example, if the instrument is calibrated every 12 months, then use the one-year performance specification all the time.

What if there is no time term and no suggested calibration interval? Then the laboratory and the customer have to work together to select an appropriate calibration interval. See Chapter 10 for more information on this.

If the calibration activity uses data to adjust the calibration interval of instruments, note that changing the interval does not affect which (time-related) specifications are used for calibration. For example, consider a group of instruments with 12-month specifications that have been calibrated every 12 months for a few years. Analysis of the data may indicate that the interval can be extended to 14 months and maintain the same reliability (probability of being in tolerance at the end of that time). Or, the analysis may indicate that the interval must be shortened to 10 months to achieve the desired reliability. In either case, when the instruments are calibrated the 12-month specifications will still be used. Those specification limits are part of the analysis process in that they are the basis for pass/fail decisions. (See Chapter 10 for more information on calibration intervals.)

Time is also a factor in other performance specifications that may be given. Two important ones that may be seen are *drift* and *stability*:

- *Drift* is a long-term characteristic of electronic circuits. In the *VIM* 3, "instrumental drift (*VIM* 4.21) is the continuous or incremental change over time in *indication*, due to changes in the metrological properties of a *measuring instrument*. NOTE—Instrumental drift is related neither to a change in a quantity being measured nor to a change of any recognized influence quantity."[21] Drift is normally specified over a time interval that may range from a month to a year or more.

- *Stability (of a measuring instrument)* as defined in the *VIM* (*VIM* 3, 4.19) is the "property of a measuring instrument, whereby its metrological properties remain constant in time."[22]

- *Stability* can be further defined as "long term" or "short term."

    o *Long-term stability* is normally specified over a time interval that may range from seconds to days, but rarely more than a week.

    o *Short-term stability* (sometimes called *jitter*) is a short-term characteristic of electronic circuits. Short-term stability sees variations that take place over shorter periods of time, typically over periods of less than a second.[23]

Drift and stability are measures of the long-term and short-term change in the IM&TE performance when all other factors are accounted for. These effects are present to some degree in all electronic devices and are commonly specified in oscillators. In other instruments, they may be included as part of the overall performance specification.

Some instruments, such as counters or digital oscilloscopes, have time-related specifications on an even shorter scale. A frequency counter always has a half-count (minimum) uncertainty because the master time base in the counter is not synchronous with the signal being counted. Other timing and pulse characteristics may cause other effects as well.

Modifiers: Temperature Term

Temperature affects all measuring instruments to some extent. A reference temperature is an important part of a performance specification. Performance of dimensional/mechanical/physical measuring instruments is normally specified at 20 °C even if that is not explicitly stated on the specification sheet. If the actual temperature is different at the time of measurement, thermal effects on the measuring instrument and the item being measured must be considered and possibly accounted for. Performance of electronic instruments is normally (but not always) specified at 23 °C. There will often be a band around that temperature where the performance is expected to meet the specification. If the actual calibration temperature is outside that band, a correction may be needed.

Some instruments do not have a temperature band specified, so the temperature correction must be made any time the actual temperature is different from the specified reference temperature.

If the item is mounted in an equipment rack when it is calibrated, the temperature inside the rack should be measured and appropriate corrections made. This may be the case, for example, when performing an in-place calibration of an automated test system.

Modifiers: Line Power Term

In some cases, the output or measuring capability of an electronic instrument may vary with changes in the applied line power. The effect is usually slight, but should be evaluated if the IM&TE is used in an environment where the line power fluctuates or where stability of the output is a critical parameter.

Modifiers: Load Term

In some cases, the output or measuring capability of an electronic instrument may vary with changes in the load applied to the front panel inputs (for a measuring instrument) or outputs (for a source). A current source, for example, may be suitable for use when supplying direct current, but have limitations on the output when supplying alternating current. That is due largely to the compliance voltage generated across reactive components in the load.

Qualifiers

Qualifier specifications are other factors that may affect the usability of the IM&TE for a particular location or for a particular application; however, they rarely affect the general operating specifications and are rarely checked during the performance verification accomplished by a calibration procedure. Many qualifiers fall into the categories of conformance to other standards such as safety or electromagnetic interference. Others are operating or storage environmental conditions; these do not normally affect calibration because the laboratory environment is controlled. There is one type of qualifier, however, that may affect any calibration laboratory (relative versus absolute specifications), and one environmental qualifier that may affect a few (operating altitude).

A manufacturer may describe performance specifications in two ways: in reference to the standards used to calibrate the instruments on the production line (relative specifications) or in reference to the SI values of the particular parameter (absolute or total specifications). The absolute specification is, of course, the most useful to a calibration laboratory. Some manufacturers even list the specifications that way, at least for their standards-grade instruments. Other manufacturers state that the specifications are relative to their calibration standards. The laboratory may have to add uncertainty to them to account for their measurement traceability. Most manufacturers don't explicitly state what type of specification it is. In those cases, if the type cannot be determined, the most conservative path is to assume that the specifications are relative. Always check the IM&TE manual to be sure.

Many electronic devices (IM&TE and other things) are specified for operation up to a maximum altitude (or elevation) of about 3050 m (about 10,000 feet). For most of us, this is not a concern. If the work site happens to be in a place such as Leadville, Colorado (about 3110 m or 10,200 ft), or La Paz, Bolivia (about 3625 m or 11,900 ft), then it is a definite problem. Remember, though, the specifications must be read with care. As an example, a popular 6½-digit digital multimeter has a specified maximum operating altitude of 2000 m (about 6562 ft). If this meter is used in Denver, Colorado (elevation 1600 m or 5260 ft), it is within its altitude specifications. If it is taken 42 km (26 miles) southwest to the town of Evergreen (elevation 2145 m or 7040 ft), it will be over the specification limit. If an organization works in a mountain area, it may have to perform some type of designed experiment to determine how elevation affects the measuring instruments. There are a couple reasons for altitude effects:

- As altitude increases, there is less mass of air to absorb heat generated in the instrument, so cooling efficiency is reduced.[24]

- Altitude is also a problem where high-voltage circuits are concerned, because the dielectric strength of air reduces with altitude.[25]

Note that a user cannot rely on weather-related barometric pressure reports to determine elevation. In order to maintain worldwide consistency, and as an important aviation safety element, those reports are always corrected to what the pressure would be at mean sea level at that location and instant in time.

Specification Tables

Following are some examples of specification tables for various kinds of electronic IM&TE. The format and values are abstracted from actual data sheets but are not complete. Not all functions or ranges are shown. They are not intended to represent any specific instrument or to be examples of anything other than what may be seen in practice.

EXAMPLE 1. In this example, the source instrument is a calibrator.

Direct voltage source:

Voltage output	12-month accuracy ± (% of output + floor)	Max. current	Resolution
0.001 mV to 320.000 mV	0.006 % + 4.16 μV	20 mA	0.001 mV (1 μV)
3.2001 V to 32.0000 V	0.0065 % + 416 μV	20 mA	0.1 mV (100 μV)
320.01 V to 1050.00 V	0.006 % + 19.95 mV	6 mA	10 mV

Note: Specifications apply:

    To positive and negative polarities

    Within ± 5 °C of the temperature at last calibration

    With added load regulation error if load is < 1 MW

Alternating voltage source:

Voltage output	Frequency (Hz)	12-month accuracy ± (% of output + floor)	Max. current	Total harmonic distortion (% of output)	Resolution
0.32001 V to 3.20000 V	10 Hz to 3 kHz	0.04 % + 192 µV	20 mA	0.06 %	10 µV
	3 kHz to 10 kHz	0.04 % + 256 µV	20 mA	0.10 %	10 µV
	10 kHz to 30 kHz	0.06 % + 480 µV	20 mA	0.13 %	10 µV
	30 kHz to 50 kHz	0.09 % + 960 µV	10 mA	0.20 %	10 µV
	50 kHz to 100 kHz	0.20 % + 2.56 µV	10 mA	0.32 %	10 µV

Note: Specifications apply:

    To sine wave only

    Within ± 5 °C of the temperature at last calibration

    With added load regulation error if load is < 1 MΩ

Direct current source:

Current output	12-month accuracy ± (% of output + floor)	Compliance voltage (at lead end)	Resolution
3.2001 mA to 32.0000 mA	0.014 % + 900 nA	4 V	100 nA
0.32001 A to 3.20000 A	0.060 % + 118 µA	2.2 V	10 µA
10.5001 A to 20.0000 A	0.055 % + 4.50 mA	2.2 V	100 µA

Note: Specifications apply:

    To positive and negative polarities

    Within ± 5 °C of the temperature at last calibration

    When using special test cable (instrument accessory)

    With maximum duty cycle of 1:4 on 20 A output range

EXAMPLE 2. Here the measuring instrument is a digital multimeter.

Direct voltage measurement ± (% of reading + % of range):

Range	24 hour 23 ± 1 °C	90 days 23 ± 5 °C	12 months 23 ± 5 °C	Temp. coeff. 0 to 18 °C and 28 to 55 °C
100.0000 mV	0.0030 + 0.0030	0.0040 + 0.0035	0.0050 + 0.0035	0.0005 + 0.0005
10.00000 V	0.0015 + 0.0004	0.0020 + 0.0005	0.0035 + 0.0005	0.0005 + 0.0001
1000.000 V	0.0020 + 0.0006	0.0035 + 0.0010	0.0045 + 0.0010	0.0005 + 0.0001

Note:

> Specifications apply after one-hour warm-up, 6½ digits, AC filter set to slow.
>
> Specifications are relative to calibration standards.
>
> There is a 20 % over-range capability on all ranges except 1000 V DC.

Observe these points:

- The specifications are larger both below and above the 10 V range and as the time interval increases.

- When calculating the values in SI units, the percent of range applies to the value in the range column. The over-range amount is not included.

- The temperatures listed in the 24-hour, 90-day, and 12-month columns can be taken as the required temperature limits in the calibration lab.

- The temperature coefficient is a per-degree term. For example, assume a meter is being used to measure 5.0 V and is mounted in an automated test system rack with an internal temperature of 32 °C.

  o Using the 12-month specifications, the basic uncertainty is ± (5 × 0.0035 % + 10 × 0.0005 %) or ± 0.00022 V.

  o The temperature coefficient is ± (5 × 0.0005 % + 10 × 0.0001 %) per degree. The number of degrees to use is (32 − 28) = 4. The result is 0.000035 × 4, or 0.00014.

  o The uncertainty of the 5 V measurement is the sum of those two values, or ± 0.00036 V.

EXAMPLE 3. In this example, the measuring instrument is a digital oscilloscope.

Sensitivity is 5 mV/division to 50 V/division in 1, 3, 5 sequence. Vertical resolution is 8 bits.

An oscilloscope display is like a graph, almost always with eight major divisions up the vertical scale and 10 major divisions along the horizontal scale. An important specification is the amplitude per division on the vertical scale. The specification above indicates the sensitivity range and indicates that it can be set to 5 mV, 10 mV, 30 mV, 50 mV, 100 mV, 300 mV, 500 mV, 1 V, 3 V, 5 V, 10 V, 30 V, or 50 V per division.

The vertical resolution indicates the available resolution of the vertical scale at any particular sensitivity setting. This indicates that an eight-bit analog-to-digital converter is used, which means that any input voltage will be one of 2^8 or 256 possible output positions. This means that the vertical scale has a resolution of (sensitivity × 8) ÷ 256.

- If the sensitivity is set to 5 mV/division, the resolution is (5 × 8) ÷ 256 or 0.156 mV.

- If the sensitivity is set to 10 V/division, the resolution is (10 × 8) ÷ 256 or 0.312 V.

Unusual Terminology

There are occasions where some work, or contact with the manufacturer, is required to interpret a specification. For example, a laboratory in the southeastern United States received a new handheld insulation resistance tester from a customer. The customer did have the operating manual. The tester was manufactured in Germany. The specification data table in the manual was fairly straightforward, with the expected columns for measuring range, test voltage, and so on. For the performance specifications, however, there were two columns instead of the usual one; they were labeled Intrinsic Error and Measuring Error.

Range	Test voltage	Intrinsic error	Measuring error
200 kW to 10 GW	500 V, +15 % – 0 %	± (5 % reading + 3 digits)	± (7 % reading + 3 digits)

It took several email exchanges with the manufacturer to determine that the Intrinsic Error is the performance specification for calibration and the Measuring Error is what the customer could expect when using the tester in field conditions. The terms used by different manufacturers—especially when translated across languages—may not always mean the same as we think they do. Or we may have no idea what they mean. This is an example of when it is a proper time to contact the manufacturer.

Comparing an Instrument to the Measurement Task

Comparing one instrument to another has a number of practical pitfalls. The most common, of course, is that performance specifications may not be available for one of the instruments. In some regulated environments, an equivalent instrument may have to duplicate every function and range, even the ones that are not used. Yet it is common to be asked to make such a comparison, typically to find a replacement for a meter that is specified in an old procedure but no longer exists.

It is often more practical, and easier, to compare an instrument that is available to the measurement task that is to be performed. This assures that the instrument is capable of making the measurement and helps in quantifying the uncertainty of the measurement. Comparing the specifications of one instrument to another does not provide that assurance, as it is not an unknown occurrence for the original documentation to specify instruments that were not capable of making a measurement with acceptable uncertainty.

An application of specifications continues in Chapter 22, when one must substitute a calibration reference standard in the event the needed reference standard is unavailable or obsolete.

Chapter 22

Substituting Calibration Standards

In an ideal world, a technician would be able to select a calibration procedure, always select the exact measurement standards specified, and always follow the procedure exactly. We must deal with the world as it really is, however, not as we wish it to be.

In the real world there will be cases where the laboratory does not have—and cannot easily obtain—the specified calibration standards. In those cases, a way must be found to make the required measurements using equivalents or substitutes for the originally specified equipment.

MEASUREMENT STANDARD SUBSTITUTION

Why Might a Substitute for a Measurement Standard Be Needed?

Industrial and commercial systems are designed to have an expected useful life cycle on the order of decades. For example, the Rankine Power Station hydroelectric power plant at Niagara Falls (Ontario, Canada) was put in service in 1905 and retired at the end of 1999.[1] Many commercial airliners are up to 30 or more years old,[2] and new ones are expected to last at least that long. When system maintenance requirements are developed during the prime system development phase, inspection, measurement, and test equipment (IM&TE) is selected from what is available at that time. Because that IM&TE must be calibrated, this inevitably means that there will always be equipment to be calibrated that is much older than current measurement standards. That, in itself, is not a problem. The problem is that the documentation is often of the same vintage, if it still exists at all. Quality management systems require that a documented procedure must exist and be followed, but that can be hard to achieve when the specified equipment is not available, and the companies that made it may no longer exist. Therefore, the calibration laboratory needs a method to identify suitable equivalents or substitutes for the unavailable equipment.

How Is a Suitable Substitute Selected?

An instrument that will be used as a substitute for another one must have certain technical characteristics and one practical consideration. When compared to the instrument originally specified:

- It must be capable of measuring the same parameter at the same level.
- The resolution must be equal or better.
- The accuracy and precision (or measurement uncertainty, if stated directly) must be equal or better.
- It should be readily available.

In an ideal case, one would be able to compare the specifications of the old and new to decide on what standard to substitute. That is not always possible and is often not practical. In many cases, the original manufacturers of the test equipment have been absorbed into another company or have disappeared completely. Even if the company still exists, it often does not keep information on equipment that is beyond its support life (or the information may have been lost by fire, flood, effects of aging, or other events). Also, it is a known occurrence to find that the original equipment is not capable of adequately making the measurements when evaluated against modern criteria. In these cases, the documentation of the unit under test will have to be studied to determine what measurements are to be made. Therefore, the best approach might be to begin by comparing the original specifications and performance to the currently needed specifications and performance.

A fundamental requirement is that, in all cases, the calibration standards must be capable of making the required measurement, such as with a specified minimum test uncertainty ratio (TUR). This must be the basis from which any substitution decision is made. Even if the laboratory can determine that the substitute instrument is equivalent to the original based on specifications, the substitute must still be examined against the measurement requirements to see if it is adequate. Once the measurement requirements have been determined, it is typically easy to see what standards are adequate and available.

Comparing Specifications

Sometimes technicians will have to compare specifications of instruments. They may need to compare specifications of two instruments to see if the instruments are equivalent. Other times, the need may be to compare the specifications of a measuring instrument to the requirements of the measurement to be made to determine if the instrument has the required capability. There are well-documented methods for doing these comparisons, and this section is a quick review of them.

Comparing the Specifications of Two Instruments

There are occasions when a calibration laboratory may need to compare specifications of one instrument to another, either to determine if they are equivalent or to determine if one can be used to calibrate the other. A common reason for having to do this is use of a calibration procedure that may have been written years ago and lists equipment models that the lab no longer has. The lab may need to see if its current measurement standards are equal to or better than the ones specified (equivalency) or if the measurement standards have the capability to calibrate the item in question.

To compare specifications, several things must be done.[3] All of these are necessary to ensure that equal values are being compared. The process must:

- Identify the specifications to be compared.

- Convert the specifications to equal units of measure, of the same order of magnitude.

- Apply all modifications required by the specifications, for each value.

- Adjust the uncertainties according to the stated confidence intervals.

For example, say the measurement requirement is to monitor a voltage that is to be maintained at 28.0 V AC, 400 Hz, ± 300 mV. The measurement will be made in an environment where the air temperature is 32 °C (about 90 °F). The meter specified in the procedure (meter A) is no longer available, but specifications are published in an old manufacturer's catalog. The engineer wants to see if the meter currently available (meter B) is equivalent—that is, whether it can be used to make the measurement with equal or better accuracy. Table 22.1 lists the specifications

Table 22.1 Specifications of meters A and B.

Specification	Meter A	Meter B
Range (V)	30.0000	100.0000
Full scale (V)	30.1000	120.0000
Resolution (V)	0.0001	0.0001
Accuracy (% reading)	0.26 %	0.06 %
Accuracy (digits)	102	
Accuracy (% range)		0.03 %
Specification period (months)	12	12
Temperature	$T_{Cal} \pm 5$ °C	23 °C ± 5 °C
T_{Cal} range	20 °C to 30 °C	
Operating temperature	0 °C to 55 °C	0 °C to 55 °C
Temperature coefficient	Included	
TC (% reading)		0.005 %
TC (% range)		0.003 %
Confidence level	Unknown	Unknown
Relative/absolute	Unknown	Relative
Conditions	Auto zero *on* Sine wave > 10 % full scale	Slow AC filter Sine wave > 5 % range Altitude < 2000 m

of each meter as they apply to making this measurement. Note these features as they apply to the description of specifications:

- Accuracy in percent of reading is the output term. (Because this is a measuring instrument, it is really the input, but the term is the same for consistency.)

- Accuracy in number of digits or in percent of range is the scale term.

- These meters do not have a floor term in their specifications.

- For the purposes of this example, both meters are calibrated at 23 °C.

- Meter A does not have a specification for temperature coefficient on AC voltage measurements, so it is assumed to be included in the performance specification.

- For meter B, the temperature coefficient terms are per °C away from the lower or upper limits of the temperature specification. For example, if the meter was being used at 15 °C (or 31 °C), three degrees outside the 23 °C ± 5 °C band, then the two temperature coefficient values would be multiplied by three.

- Meter A gives no indication of whether the performance specifications are absolute or relative to the calibration standards. Therefore, the conservative assumption for this type of meter is that the specifications are relative. Because meter B states that the specs are relative, they can both be treated the same.

- The confidence levels for both meters are unknown, so the most conservative assumption is that they both represent a uniform distribution at about the 99 % level. Because this applies to both, there is no reason to adjust the values. If one of the meters specified the confidence interval or coverage factor, however, that adjustment would have to be made. Methods for this are detailed in the *ISO Guide to the Expression of Uncertainty in Measurement (GUM)* and equivalent documents.

- The conditions listed are examples of specification qualifier terms.

Table 22.2 shows the conversion of the specifications to units of measure (volts, in this case). At the end, each column is added to get a final result. The uncertainty of meter B is lower, so it is suitable for this measurement.

An additional result of this comparison is that it reveals one of the ugly little secrets of performance specifications. The natural tendency of most users is to accept the meter's reading at face value—particularly if it is a digital display. There is a common belief that "if it has more digits, then that means it must be better!" Those of us working in metrology, however, must be more skeptical. The uncertainty numbers calculated in this example make up three of the four digits to the right of the decimal point. So, as a practical matter, the usable resolution of either of these meters is only 0.1 V (or possibly 0.01 V) when making this measurement, not the 0.0001 V that the naïve user might believe. If a measurement

Table 22.2 Measurement unit values at 28 V AC.

Specification	Meter A	Meter B
Accuracy (% reading) volts	0.0728	0.0168
Accuracy (digits) volts	0.0102	
Accuracy (% range) volts		0.0300
Operating temperature (°C)	32.0	32.0
Temperature coefficient	Included	
TC (% reading)		0.0056
TC (% range)		0.0120
TOTAL ± (Volts)	0.083	0.064

result were recorded to the full resolution of the meter's display, the effect would be assigning two or three more significant figures to the result than can be justified by the uncertainty.

Fluke has a detailed comparison using a pair of calibrators in the following case study.[4]

Case Study: Comparing the Ratio Function of Two Digital Multimeters

A calibration laboratory's customer is trying to determine an equivalent meter to replace an obsolete one. The measurement function being evaluated is DC ratio. The metrology engineer is asked to assist, but an opportunity to observe the work is not available.

The meter originally specified for the measurement is a Data Precision 3500. The customer has one that is no longer economical to support. There is a manual dated 1978. The customer also has the instructions for the measurement process being performed.

The observed display is described as 0.0001 and the reference input is 5 V. This implies that the Data Precision 3500 is in DC ratio mode and the 10 range is selected:

- The measurement input and the reference appear to be measured on the same range.
- The 10 range has a full-scale display of 11.9999.
- The displayed value is 10 times the actual ratio, so the true maximum ratio is 1.19999:1.
- The actual ratio is 0.00001, which implies that the test input voltage is 0.00005 V, or 50 mV.
- The ratio performance specifications for the Data Precision 3500 on that range are:

± (0.008 % reading + 0.001 % full scale + 1 digit) × (10 ÷ Reference voltage)

Continued

Case Study: Comparing the Ratio Function of Two Digital Multimeters (*Continued*)

The specification period is within six months of calibration with the temperature at 23 °C ± 5 °C.

0.008 % reading =	(0.00008 × 0.0001) =	0.000000008
0.001 % full scale =	(0.00001 × 11.9999) =	0.000119999
1 least digit =		0.0001
Sum =		0.000220007
10 ÷ Reference voltage =		2
Product =		0.000440014
Rounded to display resolution =		± 0.0004

The results indicate that the display on the Data Precision 3500 would be 0.0001 ± 0.0004. This corresponds to a true ratio of 0.00001 ± 0.00004. This implies that the nominal 50 mV could have a range from +250 mV to –150 mV.

The customer's proposed equivalent is the Fluke 8840A/AF. This model is also very old, but the customer has a dozen of them. It is determined that the /AF modification is required because the standard 8840A does not have the ratio function. The calibration laboratory does have a manual dated 1993 for the basic model and 1991 for the /AF modification.

From the discussion of the original meter, we know that the reference voltage is 5 V, the true ratio is believed to be 0.00001, and therefore the nominal test input voltage is 50 mV. From the information on the Fluke 8840A/AF, it is determined that:

- The measurement range is independent of the reference range.
- The measurement range would be the 200 mV range, with a maximum display of 199.999 mV.
- On that range, the display is the actual ratio × 1000. Therefore, the expected display is (0.00001 × 1000) = 0.010.

The ratio performance specifications of the 8840A/AF for this measurement are: ± (0.01 % reading + 5 counts) × (10 ÷ Reference voltage).

The specification period is within 12 months of calibration with the temperature at 25 ± 5 °C.

0.01 % reading =	(0.0001 × 0.01) =	0.000001
5 least digits =		0.005
SUM =		0.005001
10 Reference voltage =		2
Product =		0.010002
Rounded to display resolution =		± 0.010

Continued

> **Case Study: Comparing the Ratio Function of Two Digital Multimeters (*Continued*)**
>
> The results indicate that the display on the Fluke 8840A/AF would be 0.010 ± 0.010. This corresponds to a true ratio of 0.00001 ± 0.00001. This implies that the nominal 50 mV could have a range from + 100 mV to 0 mV.
>
> This is adequate, but the metrology engineer realizes that a number of assumptions have been made based on the inability to observe the measurement. The results are passed to the customer, along with a recommendation to verify whether the test input voltage really is 50 mV!

Example: Determining Substitutes for Obsolete Calibration Standards

In 2003, a maintenance department of a large company sends a new test set to the corporate electronics calibration laboratory. The lab has never calibrated the test set model before, so the manuals are requested and received. The test set can best be described as new/old: It is a physically new unit, manufactured in 2002 by ABC Company, but the design is old because it was originally built by XYZ Company for about 20 years starting in the mid-1960s. The manual provided by ABC Company is a photocopy of the old XYZ Company manual with the most recent modifications (1989) and revisions (1993). Due to a series of corporate acquisitions in the past 15 years, XYZ Company no longer exists.

The test set manual includes a calibration procedure, but it is really an adjustment/alignment procedure. There is enough information for the metrology engineer to produce an in-house calibration (performance verification) procedure, but the original calibration standards specified are a problem. Most models listed have been discontinued by the manufacturer. In some cases, the manufacturer has changed names or been acquired by another company. It can take a bit of internet sleuthing and industry historical knowledge to put the pieces together for company acquisitions. Many manufacturers, when they buy out another manufacturer, will try to maintain the same model numbering so that customers can have continuity. An acquiring manufacturer also can provide a new historical timeline of updated models and their relation to discontinued models inherited from the purchased manufacturer.

Because the calibration lab does not have any of the listed standards, and only one replacement is readily identifiable, the lab has to attack the problem from the other direction. The actual measurement requirements are collected from the test set manual and analyzed. The identified measurement requirements are listed in Table 22.3.

In Table 22.3, the first four columns identify the parameter, value, frequency, and tolerance as stated in the test set manual. The last column (Minimum required accuracy) lists the minimum performance requirement of a calibration standard based on a 4:1 test accuracy ratio (TAR). The requirement is stated as a percentage of the measured value and in the applicable SI units. That can be compared to the performance specifications of the laboratory's available calibration standards.

No lower-accuracy digital multimeter in the laboratory's inventory can meet the performance requirement for the first two AC voltage measurements in Table 22.3. Because the HP 3458A was specified for AC voltage, the equipment list and the calibration process can be simplified by using the same meter for the DC voltage and resistance measurements as well. Its performance is one or two orders of magnitude better than the minimum requirement for those measurements, but that is not a reason to specify an additional meter.

The phase angle tests in the original XYZ Company manual are based first on the assumption that the synchros have to be re-zeroed (which is not a valid assumption for performance verification) and also on the capabilities of test equipment available in 1965. The purpose of the test is to verify the bearing accuracy of the synchro and resolver, for which the angle position indicator is well suited. (If the synchro or the resolver need adjustment, then a phase angle voltmeter would be needed, but that is adjustment or repair, not calibration.)

Analysis of the requirements determines that AC frequency is not a critical parameter. The signal generator originally specified had a frequency accuracy of ± 3 %. The 3325B frequency accuracy is ± 5 ppm or better, and this would be similar for any modern signal generator selected.

Table 22.3 Example test set manual's measurement requirements.

Parameter	Value	Frequency	Tolerance	Minimum required accuracy
AC voltage measure	0.220 V	400 Hz	± 1 %	± 0.25 % (550 mV)
AC voltage measure	0.440 V	400 Hz	± 1 %	± 0.25 % (1.10 mV)
AC voltage source and measure	1.5 V	400 Hz	± 2 %	± 0.5 % (7.50 mV)
AC voltage source and measure	3.0 V	400 Hz	± 2 %	± 0.5 % (15.0 mV)
AC voltage source and measure	7.07 V	1000 Hz	± 0.5 %	± 0.125 % (8.84 mV)
AC voltage source and measure	26.0 V	400 Hz	± 1 %	± 0.25 % (65.0 mV)
AC voltage source and measure	36.0 V	400 Hz	± 2 %	± 0.5 % (180 mV)
AC voltage/phase measure	26.0 V	400 Hz, 90°	None	Set and used as phase reference
DC voltage source and measure	0.080 V		± 1 mV	± 0.31 % (0.25 mV)
DC voltage source and measure	0.150 V		± 1 %	± 0.25 % (0.375 mV)
DC voltage source and measure	0.150 V		± 3 mV	± 0.5 % (0.75 mV)
DC voltage source and measure	0.45 V		± 1 %	± 0.25 % (1.125 mV)
DC voltage source and measure	16.0 V		± 1 %	± 0.25 % (40.0 mV)
DC voltage source and measure	27.5 V		± 1 %	± 0.25 % (68.75 mV)
Phase angle measure	0 ... 360	Every 10°	± 0.10 °	± 0.025 °
Resistance measure	110 Ω		± 2 %	± 0.5 % (0.55 W)
Resistance measure	500 Ω		± 1 %	± 0.25 % (1.25 W)

At this point, it is possible for the metrology engineer to rewrite the procedure so that it becomes a performance verification using the substituted calibration standards.

EQUIPMENT SUBSTITUTION AND CALIBRATION PROCEDURES

When an equivalent or substitute measurement standard has been identified, the calibration procedure may have to be examined as well. If the procedure is focused on adjustment instead of performance verification, it will have to be revised as discussed in Chapter 5. But other things may have to be examined as well, depending on the procedure's age:

- If the original procedure was written before 1990, measurements involving voltage or resistance should be examined. The conventional values of the SI volt and ohm were adjusted at the beginning of 1990.

- If the original procedure was written before 1990, any temperature measurements will have to be examined. The ITS-90 thermodynamic scale went into effect then, replacing IPTS-68. Although the differences are minor in commonly used ranges, they still have to be evaluated. Also, some items may have been written using the old IPTS-48 (from 1948) temperature scale, and there are significant changes from that version.

- If the original procedure was written before 1968 (or even after) and contains values expressed in terms of the old centimeter, gram, second (CGS) or gravitational systems, or old meter, kilogram, second (MKS)–derived units, they may have to be converted to SI units.

- Finally, any instructions that are written for using or operating a specific measurement standard must be revised to accommodate the new one or to adopt a more generic format.

Chapter 23

Proficiency Testing, Interlaboratory Comparisons, and Measurement Assurance Programs

PROFICIENCY TESTING

The term *proficiency testing* is often misunderstood when laboratories seeking accreditation hear about it the first time. Many think that its purpose is to assess employees' skills. Although indirectly it is related to the employees' skills and other factors, directly it is a means of determining the laboratory's competence compared to other laboratories performing the same kind of work.

ISO/IEC 17043:2010 defines *proficiency testing* as "evaluation of participant performance against pre-established criteria by means of interlaboratory comparisons."[1] Additionally, ISO/IEC 17043:2010 defines *interlaboratory comparisons* as "organization, performance and evaluation of measurements or tests on the same or similar items by two or more laboratories in accordance with predetermined conditions."[2]

Laboratory proficiency testing is an essential element of laboratory quality assurance. In addition to laboratory accreditation and the use of validated methods, it is an important requirement of the accreditation process. Customers increasingly demand independent proof of competence from laboratories. It is also a requirement if the laboratory is involved in any regulated industry such as nuclear, food, drug, and pharmaceuticals. Laboratories can determine how well they perform against other laboratories and how their measurements compare with others.

For participating laboratories, proficiency testing is a comparison of their individual performance against industry performance. It identifies areas for improvement in measurement and techniques. It may identify best practices by comparing other laboratories in the same field.

For the customers of laboratory services, proficiency testing:

- Establishes confidence and is a demonstration of accreditation

- Helps the customer decide if the laboratory meets its measurement, calibration, and testing requirements

- Acts as a measure for ensuring that the laboratory will continuously meet its quality requirements

Ensuring the validity of results both within one's own laboratory *and* in comparison with other laboratories is a requirement of ISO/IEC 17025:2017. ISO/IEC 17025:2017, Clause 7.7, "Ensuring the validity of results," subClause 7.7.2 states that "[t]he laboratory shall monitor its performance by comparison with results of other laboratories, where available and appropriate. This monitoring shall be planned and reviewed and shall include, but not be limited to, either or both of the following:

a) Participation in proficiency testing,
    NOTE—ISO/IEC 17043 contains additional information on proficiency tests and proficiency testing providers. Proficiency testing providers that meet the requirements of ISO/IEC 17043 are considered to be competent.

b) participation in interlaboratory comparison other than proficiency-testing."[3]

Various schemes for organizing proficiency testing and coordination and analysis of data exist and are referenced in the standards. This chapter covers the more popular schemes and their associated statistics.

Measurement Comparison Scheme

In this scheme, an artifact with an assigned value is circulated among participating laboratories. Data are collected and published with the appropriate statistics. A national laboratory usually provides the artifact's reference value. Laboratories follow a predetermined process (procedure) when testing the artifact and providing the measurement data. The measurement data are collected and published with the appropriate statistics.

In Figure 23.1, an artifact with a true value of 10.0000 units is measured by the reference laboratory and assigned a reference value of 9.999 98 units.

The assigned value is provided by the reference, or pivot, laboratory making 10 measurements on the artifact and calculating the mean of the 10 measurements as shown in Table 23.1. The uncertainty of the measurement is also determined.

The artifact with the assigned reference value is then circulated to 10 other laboratories participating in the proficiency testing program, labeled 1 through 10. The laboratories are coded to ensure the confidentiality of the participants. The participating laboratories' data are summarized in Table 23.2.

It is easy to visualize the data if they are graphed. The data are graphed in Figure 23.2 with the mean of each laboratory and the ± 3 standard deviations.

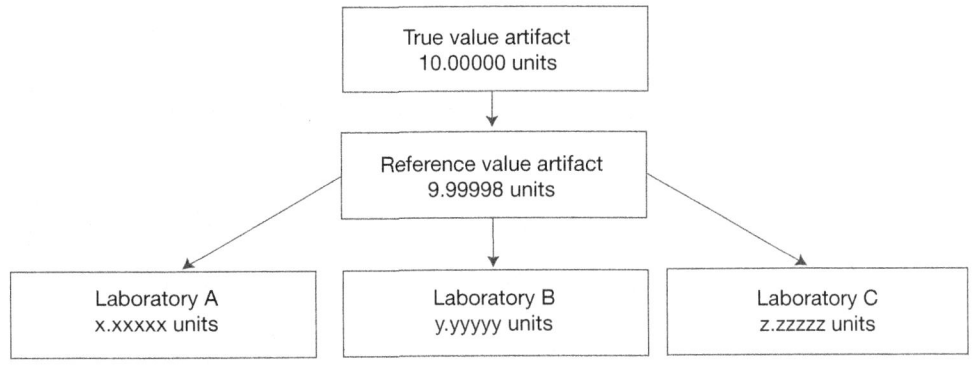

Figure 23.1 Measurement comparison scheme.

Table 23.1 Assigning a reference value.

Specification	± 0.00001
Uncertainty of the calibrator ($k = 1$)	5.7735E-06
Laboratory	Reference
Measurement observation 1	9.99999
Measurement observation 2	10.00000
Measurement observation 3	9.99999
Measurement observation 4	10.00000
Measurement observation 5	10.00000
Measurement observation 6	10.00000
Measurement observation 7	9.99999
Measurement observation 8	9.99999
Measurement observation 9	10.00000
Measurement observation 10	10.00000
Sum	99.99998
Mean	10.00000
Maximum value	10.00000
Minimum value	9.99999
Range	0.00001
Standard deviation	0.00001
Median	10.00000
Uncertainty (type A)	0.00001
Combined uncertainty ($k = 2$)	0.00002
Reference value	10.00000

From the data in Table 23.2 and Figure 23.2, it should be noted that laboratories 2 and 6 have a higher variability in their data. This is based on the specification (or capability of the laboratory) as shown in Table 23.2. It is also important to consider the mean values of the data generated by the laboratories and ensure that they do fall within ± 3 standard deviations of the reference laboratory data. Another way to look at an individual laboratory's data is to look at the individual laboratory mean ± U ($k = 2$) data, as shown in Figure 23.3. Other statistical tests can also be performed using the data generated to test for statistical significance. Most computer spreadsheet software enables this to be done easily. Discussion of the acceptability of data using E_n numbers is covered later in this chapter.

It is also important to note the individual laboratory method used, operator training, test environment, calibration of the measurement system (including measurement uncertainty analysis), and any other important parameter of the process.

Table 23.2 Measurement comparison scheme: raw data and calculations.

Specification of equipment	± 0.00001	± 0.005	± 0.03	± 0.001	± 0.0005	± 0.005	± 0.015	± 0.005	± 0.005	± 0.005	± 0.001
Uncertainty of calibrator ($k = 1$)	5.7735E-06	0.0029	0.0173	0.0006	0.0003	0.0029	0.0087	0.0029	0.0029	0.0029	0.0006
Laboratory	Reference	1	2	3	4	5	6	7	8	9	10
Measurement observation 1	9.99999	10.0011	9.983	9.9968	10.00042	9.9984	10.0054	10.0046	10.0021	10.0045	9.9992
Measurement observation 2	10.00000	10.0002	9.970	9.9966	10.00002	9.9958	10.0148	10.0032	10.0014	10.0027	10.0001
Measurement observation 3	9.99999	9.9958	10.030	9.9953	10.00020	9.9991	9.9929	10.0042	10.0023	10.0048	10.0002
Measurement observation 4	10.00000	10.0004	10.009	9.9961	10.00008	10.0022	9.9929	9.9992	9.9969	9.9982	10.0001
Measurement observation 5	10.00000	9.9999	10.008	9.9955	10.00006	9.9971	10.0081	10.0029	9.9987	10.0031	9.9995
Measurement observation 6	10.00000	10.0041	10.007	9.9966	9.99987	10.0017	9.9979	10.0040	9.9985	9.9961	10.0006
Measurement observation 7	9.99999	9.9951	10.027	9.9965	10.00036	10.0014	9.9961	9.9956	10.0007	9.9976	10.0009
Measurement observation 8	9.99999	10.0031	10.016	9.9957	9.99971	9.9987	10.0121	10.0018	9.9965	10.0024	9.9993
Measurement observation 9	10.00000	9.9998	10.000	9.9963	9.99969	10.0002	10.0090	10.0047	10.0036	10.0009	9.9994
Measurement observation 10	10.00000	9.9964	9.989	9.9966	10.00013	9.9951	9.9890	9.9951	10.0004	9.9990	10.0002
Sum	99.999977	99.99571	100.0398	99.96209	100.000548	99.98958	100.01815	100.01534	100.00112	100.00927	99.99956
Mean	9.999998	9.99957	10.0040	9.99621	10.000055	9.99896	10.00181	10.00153	10.00011	10.00093	9.99996

Continued

Table 23.2 Continued

Maximum value	10.000005	10.00405	10.0300	9.99684	10.000415	10.00216	10.01477	10.00470	10.00361	10.00477	10.00085
Minimum value	9.999990	9.99507	9.9704	9.99530	9.999694	9.99509	9.98899	9.99509	9.99646	9.99608	9.99919
Range	0.000015	0.00898	0.0596	0.00154	0.000721	0.00706	0.02578	0.00961	0.00714	0.00868	0.00167
Standard deviation	0.000006	0.00299	0.0189	0.00054	0.000243	0.00246	0.00912	0.00365	0.00239	0.00304	0.00057
+3 sigma	10.000016	10.00854	10.0608	9.99783	10.000783	10.00634	10.02919	10.01247	10.00728	10.01006	10.00165
−3 sigma	9.999980	9.99060	9.9472	9.99458	9.999326	9.99157	9.97444	9.99060	9.99295	9.99179	9.99826
Median	9.999999	10.00004	10.0078	9.99641	10.000074	9.99890	10.00164	10.00303	10.00056	10.00162	10.00013
Uncertainty (type A)	0.000006	0.00299	0.0189	0.00054	0.000243	0.00246	0.00912	0.00365	0.00239	0.00304	0.00057
Expanded uncertainty ($k = 2$)	0.000017	0.008314	0.051321	0.001583	0.000754	0.007588	0.025159	0.009301	0.007493	0.008391	0.001616
E_n		0.051	0.078	2.393	0.076	0.137	0.072	0.165	0.015	0.111	0.026
Difference		−0.0004	0.0040	−0.0038	0.0001	−0.0010	0.0018	0.0015	0.0001	0.0009	0.0000
Percent difference		−0.0043 %	0.0398 %	−0.0379 %	0.0006 %	−0.0104 %	0.0182 %	0.0154 %	0.0011 %	0.0093 %	−0.0004 %
Z-score		−0.107	0.530	−0.593	−0.037	−0.196	0.217	0.177	−0.029	0.089	−0.051

CHAPTER 23: PROFICIENCY TESTING, INTERLABORATORY COMPARISONS

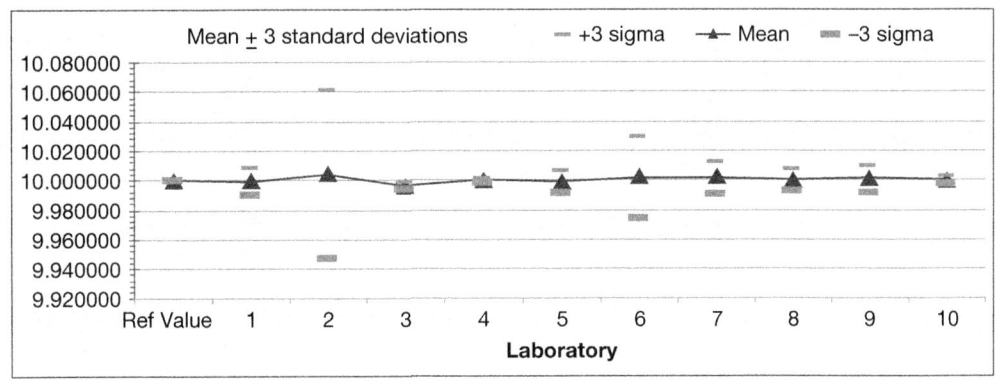

Figure 23.2 Measurement comparison scheme: mean ± 3 standard deviations.

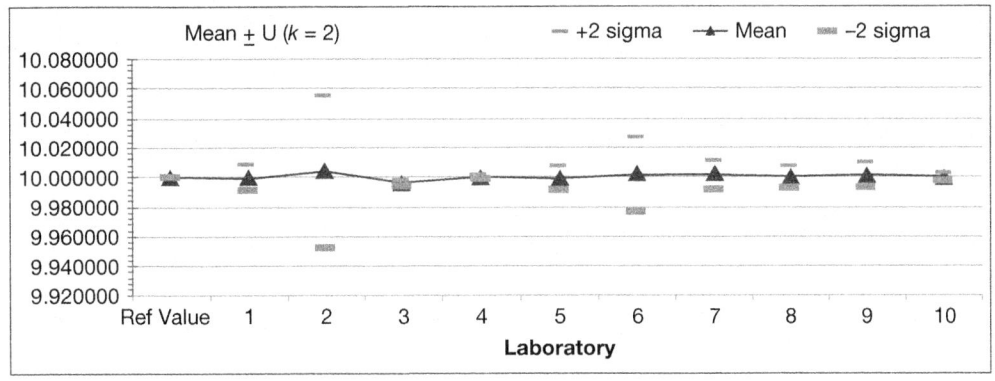

Figure 23.3 Measurement comparison scheme: mean ± U (k = 2).

Another popular graphic technique that gives a good visual presentation of a participant's test data and associated uncertainty compared to a test artifact's assigned value and its associated uncertainty does so by displaying the boundaries created by the participant's and test artifact's uncertainties and evaluating whether there is an overlap between them and to what degree (see Figure 23.4). The following is commonly used to evaluate performance levels for this type of uncertainty overlap graph:

- **In.** Participant's uncertainty overlaps the test artifact's uncertainty, and this overlap encompasses the test artifact's assigned value.

- **Within.** Participant's uncertainty overlaps the test artifact's uncertainty, but this overlap does not encompass the test artifact's assigned value.

- **Out.** Participant's uncertainty does not overlap test artifact's uncertainty proficiency tests.

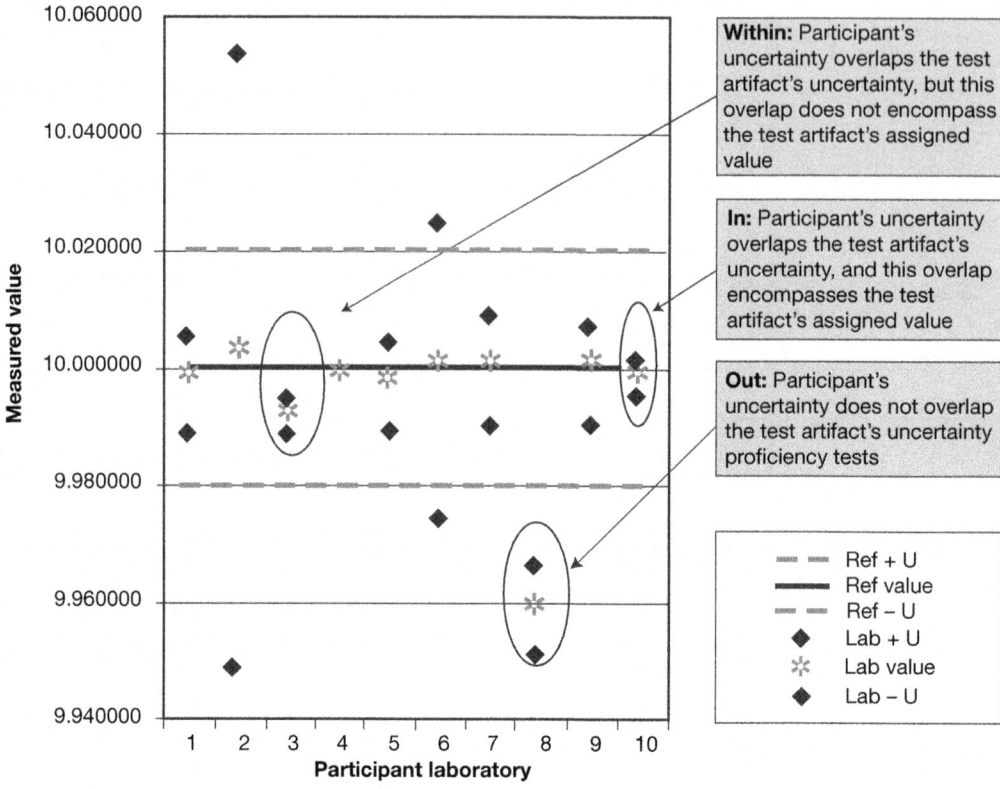

Figure 23.4 Uncertainty overlap plot.

Interlaboratory Testing Scheme

In this scheme, homogeneous material, such as a batch of compounded rubber, is split between the laboratories and tested simultaneously under agreed-on conditions. The test data are sent to the proficiency testing coordinator. The material sometimes has an assigned value. Alternatively, the assigned value for the material can be derived from the results of the tests.

In this example, the assigned value is derived from five laboratories' test data. See Table 23.3. It is always easier to visualize the data if they are presented in graphical form. The laboratory's data are presented in Figure 23.5.

Results from the data are used to calculate summary statistics and assign a reference value to the material being tested or measured. See Table 23.4.

Split-Sample Testing Scheme

When customers of laboratory services wish to compare the performance of a particular laboratory, they may split the homogeneous sample material or circulate a known-value artifact (value unknown to the laboratories) among the laboratories. Data sent by the laboratories are analyzed by the customer to select a particular laboratory's services. In the example shown in Table 23.5, a homogeneous material with a known true value of 100.00 units is distributed to the two laboratories

Table 23.3 Interlaboratory testing comparison data.

Laboratory	A	B	C	D	E
1	99.983	99.9997	99.9851	99.9994	100.00033
2	99.993	100.0013	100.0043	99.9997	100.00043
3	100.027	100.0023	100.0057	100.0007	100.00002
4	99.986	100.0039	99.9850	100.0006	99.99952
5	100.009	99.9965	100.0112	100.0008	99.99985
6	99.982	99.9956	100.0011	100.0000	100.00008
7	100.013	99.9955	100.0049	100.0008	99.99957
8	99.977	100.0021	99.9968	100.0009	100.00010
9	100.024	99.9968	100.0038	99.9990	99.99995
10	99.982	100.0034	100.0144	99.9997	100.00030
Sum 999.97501	999.99705	1000.01230	1000.00166	1000.00015	
Mean 99.99750	99.99970	100.00123	100.00017	100.00001	
Maximum value	100.02661	100.00387	100.01438	100.00087	100.00043
Minimum value	99.97742	99.99552	99.98504	99.99905	99.99952
Range 0.04919	0.00836	0.02934	0.00182	0.00091	
Standard deviation	0.01886	0.00333	0.00979	0.00067	0.00031
Median 99.98946	100.00051	100.00402	100.00031	100.00005	
Established ref. value	99.99972	99.99972	99.99972	99.99972	99.99972
+ 3 Std. dev.	100.02717	100.02717	100.02717	100.02717	100.02717
− 3 Std. dev.	99.97228	99.97228	99.97228	99.97228	99.97228

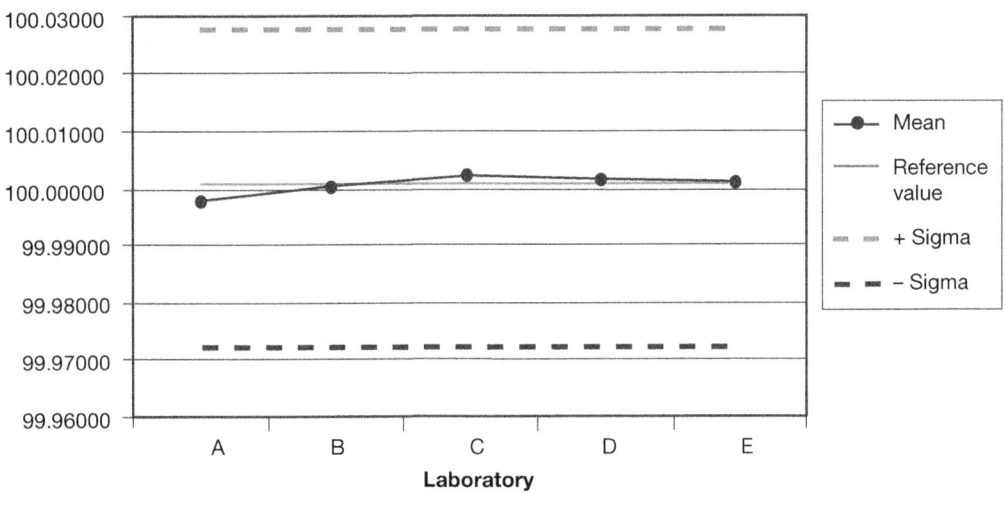

Figure 23.5 Interlaboratory testing comparison data.

Table 23.4 Derivation of consensus or reference value.

Sum	4999.98616
Mean	99.99972
Maximum value	100.02661
Minimum value	99.97742
Range	0.04919
Standard deviation	0.00930
Median	100.00009
+ 3 Std. dev.	100.02763
− 3 Std. dev.	99.971812

competing for the customer's business. Twenty measurements are taken at each laboratory, and the data are sent to the customer for analysis.

Analysis of variance (ANOVA) techniques can be used in the selection of the laboratory by subjecting the split-sample data from Table 23.5 to ANOVA analysis.

Looking at the repeatability data (standard deviation) of the two laboratories by itself (Table 23.5), it would be hard to judge if there were much difference between the two. A case can be made to select laboratory A because its repeatability is better than laboratory B's, and the mean value of the data is closer to the true value of 100.00 units. The ANOVA test (Figure 23.6) also shows that there is not a significant statistical difference between the data provided by the two laboratories (between groups and within groups). A brief explanation on how the ANOVA standard output can be interpreted is given in Figure 23.7.

According to NIST/SEMATECH *Engineering Statistics Handbook*, 1.3.5.9. F-Test for Equality of Two Variances (https://www.itl.nist.gov/div898/handbook/eda/section3/eda359.htm), "An F-test is used to test if the variances of two populations are equal. This test can be a two-tailed test or a one-tailed test. The two-tailed version tests against the alternative that the variances are not equal. The one-tailed version only tests in one direction, which is the variance from the first population is either greater than or less than (but not both) the second population variance. The choice is determined by the problem."[4]

Based on the data, the F value of 3.09971 is less than the F_{Crit} value of 4.098169. Therefore, there is no statistical difference in the mean value of data from laboratory A or B, and either can be chosen based on the one criterion.

Acceptability of Data

Acceptance of proficiency testing (PT) data is based on several factors. Statistical tests are one way to determine compliance and to form the basis for such things as data outliers. ASTM E1323, ASTM E2489, ISO 13528, and ISO/IEC 17043 provide discussion, guidance, and other references on statistical methods. It is important that the proficiency testing coordinator have a good statistical support base to ensure that the correct, unbiased assumption about the data is made and reported.

Table 23.5 Split-sample data analysis.

Laboratory:	A	B
1	100.0025	99.00752
2	99.9977	99.00601
3	99.9992	99.00902
4	100.0047	99.00282
5	100.0020	99.00581
6	100.0043	99.00422
7	99.9957	99.00001
8	99.9955	99.00198
9	100.0003	99.00571
10	100.0036	98.99759
11	99.9952	99.00170
12	100.0036	99.01047
13	100.0034	99.01045
14	99.9953	99.01258
15	100.0030	99.00031
16	99.9996	98.99871
17	100.0018	99.00280
18	100.0031	99.00805
19	100.0028	99.01215
20	99.9982	99.00293
Mean	100.00058	99.005042
Standard deviation	0.00327	0.004461

ANOVA: Single factor Summary

Groups	Count	Sum	Average	Variance
A	20	2000.177	100.0088	7.28E-06
B	20	2000.145	100.0072	8.83E-06

ANOVA

Source of variation	SS	df	MS	F	P-value	F_{Crit}
Between groups	2.5E-05	1	2.5E-05	3.099271	0.086375	4.098169
Within groups	0.000306	38	8.06E-06			
Total	0.000331	39				

Figure 23.6 Analysis of variance.

ANOVA: Single factor Summary

Groups	Count	Sum	Average	Variance
A	5	49	9.8	11.2
B	5	77	15.4	9.8
C	5	88	17.6	4.3
D	5	108	21.6	6.8
E	5	54	10.8	8.2

ANOVA

Source of variation	SS	df*	MS**	F***	P-value****	F_Crit*****
Between groups	475.76	4	118.94	14.75682	9.13E-06	2.866081
Within groups	161.2	20	8.06			
Total	636.96	24				

* df = Number of groups−1
** MS = SS/df (Mean square = Sample variance)
*** F = Ratio of between groups MS/within groups MS
**** P-value: Probability of significance
***** F_{Crit} = From F-Table: $\frac{\text{Degree of freedom (Numerator)}}{\text{Degree of freedom (Denominator)}}$

If $F > F_{Crit}$, then there is a significant difference between the means of data sets.

Figure 23.7 Interpretation of one-way ANOVA data.

At a minimum, the following statistical parameters should be considered when making assumptions about the proficiency testing data:

- Mean
- Standard deviation
- Range (range can be a good estimator of variability)
- Statistical significance using z, t, or F tests

The following information is normally analyzed to compare the performance of the laboratories from the data. There are rules to determine *satisfactory* versus *unsatisfactory* performance of a lab.

Difference. This is an arithmetic difference between the reference value and that of the participant laboratory.

$$x_{\text{Lab value}} - X_{\text{Assigned value (ref)}}$$

x: Participant's result

X: Assigned value

Percent difference. This is an arithmetic difference expressed as a percentage between the reference value and that of the participant laboratory.

$$\frac{\left(x_{\text{Lab value}} - X_{\text{Assigned value (ref)}}\right)}{X_{\text{Assigned value}}} \times 100$$

x: Participant's result

X: Assigned value

z-score. Here the participant laboratory's result is converted to a standardized z-score and the result compared.

$$z = \frac{(x_{\text{Lab value}} - X_{\text{Assigned value (ref)}})}{s}$$

$|z| \leq 2$ = Satisfactory
$2 < |z| < 3$ = Questionable
$|z| \geq 3$ = Unsatisfactory

x: Participant's result

X: Assigned value

s: Standard deviation of the participant's data (unless assigned)

E_n number. The E_n number takes into consideration the expanded measurement uncertainty (usually at $k = 2$) of the measured artifact when comparing the performance of the laboratory. Laboratories participating in formalized proficiency testing programs will see this number reported.

Care should be taken to ensure that laboratories do not give importance to just one pass/fail criterion in the proficiency testing program.

$$E_n = \frac{(x_{\text{Lab value}} - X_{\text{Assigned value (ref)}})}{\sqrt{U^2 \text{Lab} + U^2 \text{Ref}}}$$

$|E_n| \leq 1$ = Satisfactory
$|E_n| > 1$ = Unsatisfactory

x: Participant's result

X: Assigned value

U^2Lab: Uncertainty of participant's result

U^2Ref: Uncertainty of reference laboratory's assigned value

Other Considerations

It is critical that the confidentiality of the laboratories be maintained when the data are reported publicly. The testing coordinator also should ensure and maintain neutrality and report data in an unbiased manner.

Laboratories should ensure that their processes are in statistical control before participating in the proficiency testing program. Use of Shewhart X-bar and R control charts in the laboratory calibration and maintenance program is one way to do this in a preventive manner.

Process control should cover operator training, controlled procedures, and measuring equipment repeatability and reproducibility studies.

Development of measurement uncertainty budgets and measurement uncertainty analysis of the measurement process is another important consideration when reporting the measurement data for proficiency testing. See Chapter 29.

MEASUREMENT ASSURANCE PROGRAMS

It is one thing to compare a laboratory's proficiency against that of another laboratory. But it is equally important in a laboratory's day-to-day operations to ensure that there is a measure of confidence in the testing and calibration work performed.

Some good elements of a measurement assurance program (MAP) include:

- Scheduled, traceable calibration of the laboratory's standards

- Use of check standards

- Comparison of reference standards against check standards and working standards

- Continuous evaluation of the measurement uncertainties associated with the standards and test and calibration processes

- Implementation of a quality system in accordance with recognized international standards such as ISO/IEC 17025:2017

- Controlled, documented test and calibration procedures established under a quality system

- Trained calibration and test technicians (metrologists)

- A formal internal audit program to track the quality of calibration and tests performed

- Control charting and analyzing the results in a timely, proactive manner

Before implementing a MAP, it is important to understand what a laboratory is trying to accomplish. The objectives of a MAP should be defined. The following should be considered as a minimum:

- Select the process to monitor.

- Select the check standard.

- Establish a value for the check standard through traceable calibration or other means (for reference materials).

- Ensure that the check standard is controlled so that its established value is immune to drift due to external factors.

- Monitor and record the process data at a regular, determined interval.

- Chart the data. Control charting is a powerful technique.

- Analyze data using statistical techniques. Observe any trends.
- Make an objective decision based on the data and statistical evidence.

Establishing a MAP

The rest of this chapter is an example of how a laboratory can establish a MAP.

The laboratory has a 10.0 g weight that it is using as a check standard. It also has another 10.0 g weight that is designated as a site reference standard. The check standard is compared against the reference standard on a scheduled basis and the data are recorded and charted. The reference standard is sent to an accredited primary standards laboratory to obtain traceable calibration per its calibration interval.

The laboratory wants to monitor a digital scale that it uses to calibrate customers' weights. The check standard has an established reference value of 10.001 g. To simplify this example, all other factors are assumed constant. In practice, more issues should be considered when making a measurement (environment, operator, training, method, and so on).

The laboratory measures the check standard with a digital scale once every week and records the data on the control chart. This measurement does not take long, so the laboratory makes five measurements. For more than one measurement, the laboratory will utilize an X-bar and R chart.

- An *X-bar chart* is a control chart that monitors the average of observations within a subgroup.[5]
- An *R chart* is a control chart that monitors the range of observations within a subgroup.[6]

If the measurement took a longer time, the laboratory might choose to make a single measurement every week. For a single measurement, the laboratory will utilize an individual X-bar and R chart. Table 23.6 is used to calculate control limits.

Table 23.6 Control chart constants.

| | | Variables data | | |
| | | Control limit constants | | |
n	A2	D3	D4	d2
2	1.88	0	3.267	1.128
3	1.023	0	2.574	1.693
4	0.729	0	2.282	2.059
5	0.577	0	2.115	2.326
6	0.483	0	2.004	2.534
7	0.419	0.076	1.924	2.704
8	0.373	0.136	1.864	2.847
9	0.337	0.184	1.816	2.97
10	0.308	0.223	1.777	3.078

The formula for calculating control limits for the individuals and moving range chart is:

$$UCL_{IX} = \overline{IX} + A_2 \overline{MR}$$
$$LCL_{IX} = \overline{IX} - A_2 \overline{MR}$$
$$UCL_{MR} = D_4 \overline{MR}$$
$$LCL_{MR} = 0$$

The formula for calculating control limits for the X-bar and range chart (R chart) is:

$$UCL_X = \overline{\overline{x}} + A_2 \overline{R}$$
$$LCL_X = \overline{\overline{x}} - A_2 \overline{R}$$
$$CL_X = \overline{\overline{x}}$$
$$UCL_R = \overline{R} D_4$$
$$UCL_R = \overline{R} D_3$$
$$CL_R = \overline{R}$$

The formula for calculating control limits for the R chart is:

UCL = X-bar upper control limit
LCL = X-bar lower control limit
UCL_R = Range upper control limit
LCL_R = Range lower control limit

The data for individual measurements are shown in Table 23.7. Control charts to monitor the measurements are shown in Figures 23.8 and 23.9. The data for

Table 23.7 Data for individual measurement.

Date	Scale	Range	Mean	Range mean	X-bar UCL	X-bar LCL	Range UCL
2022-01-03	10.000		10.001	0.0007	10.0026	9.9991	0.0022
2022-01-10	10.000	0.000	10.001	0.0007	10.0026	9.9991	0.0022
2022-01-17	10.000	0.000	10.001	0.0007	10.0026	9.9991	0.0022
2022-01-24	10.001	0.001	10.001	0.0007	10.0026	9.9991	0.0022
2022-01-31	10.001	0.000	10.001	0.0007	10.0026	9.9991	0.0022
2022-02-07	10.000	0.001	10.001	0.0007	10.0026	9.9991	0.0022
2022-02-14	10.000	0.000	10.001	0.0007	10.0026	9.9991	0.0022
2022-02-21	10.001	0.001	10.001	0.0007	10.0026	9.9991	0.0022

Continued

Table 23.7 *Continued*

Date	Scale	Range	Mean	Range mean	X-bar UCL	X-bar LCL	Range UCL
2022-02-28	10.000	0.001	10.001	0.0007	10.0026	9.9991	0.0022
2022-03-07	10.001	0.001	10.001	0.0007	10.0026	9.9991	0.0022
2022-03-14	10.001	0.000	10.001	0.0007	10.0026	9.9991	0.0022
2022-03-21	10.000	0.001	10.001	0.0007	10.0026	9.9991	0.0022
2022-03-28	10.001	0.001	10.001	0.0007	10.0026	9.9991	0.0022
2022-04-04	10.002	0.000	10.001	0.0007	10.0026	9.9991	0.0022
2022-04-11	10.000	0.001	10.001	0.0007	10.0026	9.9991	0.0022
2022-04-18	10.000	0.000	10.001	0.0007	10.0026	9.9991	0.0022
2022-04-25	10.001	0.000	10.001	0.0007	10.0026	9.9991	0.0022
2022-05-02	10.001	0.001	10.001	0.0007	10.0026	9.9991	0.0022
2022-05-09	10.001	0.000	10.001	0.0007	10.0026	9.9991	0.0022
2022-05-16	10.000	0.000	10.001	0.0007	10.0026	9.9991	0.0022
2022-05-23	10.001	0.001	10.001	0.0007	10.0026	9.9991	0.0022
2022-05-30	10.000	0.001	10.001	0.0007	10.0026	9.9991	0.0022
2022-06-06	10.001	0.000	10.001	0.0007	10.0026	9.9991	0.0022
2022-06-13	10.002	0.001	10.001	0.0007	10.0026	9.9991	0.0022
2022-06-20	10.001	0.001	10.001	0.0007	10.0026	9.9991	0.0022
2022-06-27	10.002	0.001	10.001	0.0007	10.0026	9.9991	0.0022
2022-07-04	10.000	0.001	10.001	0.0007	10.0026	9.9991	0.0022
2022-07-11	10.001	0.000	10.001	0.0007	10.0026	9.9991	0.0022
2022-07-18	10.001	0.001	10.001	0.0007	10.0026	9.9991	0.0022
2022-07-25	10.001	0.001	10.001	0.0007	10.0026	9.9991	0.0022
2022-08-01	10.000	0.000	10.001	0.0007	10.0026	9.9991	0.0022
2022-08-08	10.000	0.000	10.001	0.0007	10.0026	9.9991	0.0022
2022-08-15	10.002	0.002	10.001	0.0007	10.0026	9.9991	0.0022
2022-08-22	10.001	0.000	10.001	0.0007	10.0026	9.9991	0.0022
2022-08-29	10.001	0.000	10.001	0.0007	10.0026	9.9991	0.0022
2022-09-05	10.001	0.000	10.001	0.0007	10.0026	9.9991	0.0022
2022-09-12	10.000	0.001	10.001	0.0007	10.0026	9.9991	0.0022
2022-09-19	10.000	0.000	10.001	0.0007	10.0026	9.9991	0.0022
2022-09-26	10.002	0.001	10.001	0.0007	10.0026	9.9991	0.0022
2022-10-03	10.002	0.000	10.001	0.0007	10.0026	9.9991	0.0022
2022-10-10	10.002	0.000	10.001	0.0007	10.0026	9.9991	0.0022
2022-10-17	10.000	0.002	10.001	0.0007	10.0026	9.9991	0.0022

238 Part III: Metrology Concepts

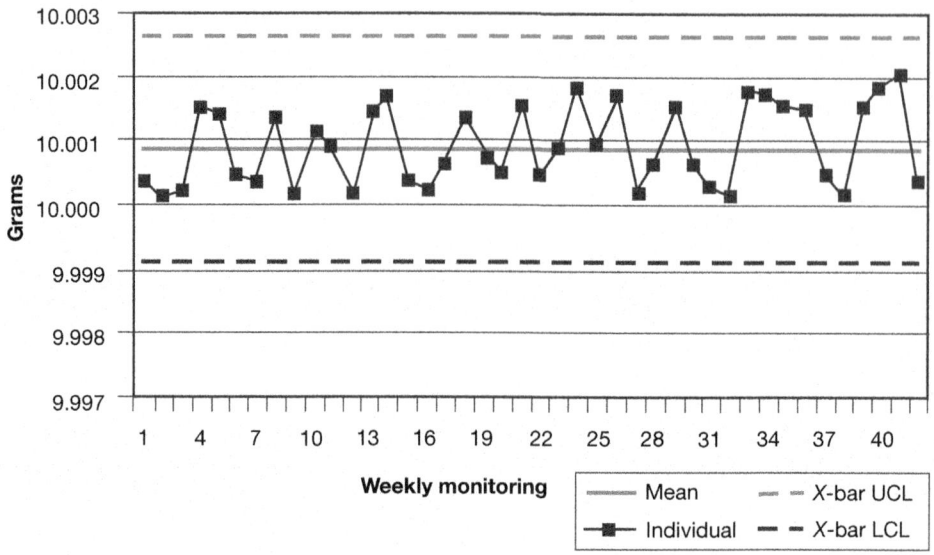

Figure 23.8 Individual measurements chart.

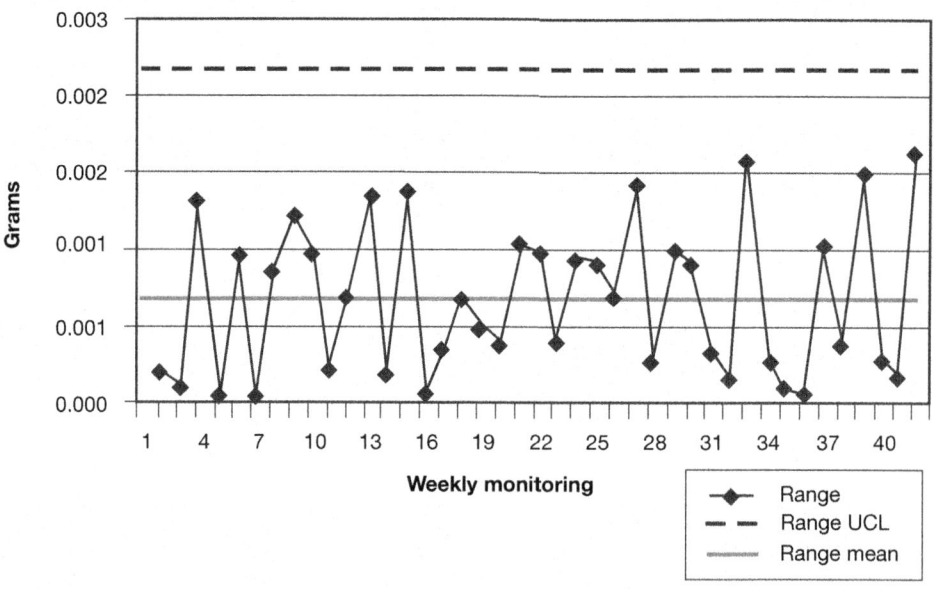

Figure 23.9 Moving range chart.

multiple measurements are shown in Table 23.8. Control charts to monitor the measurements are shown in Figures 23.10 and 23.11.

If the measurements are normally within the control limits, there is no need for concern. By graphically monitoring data, one can observe rising or declining trends, erratic up-and-down fluctuations, and other trends. SPC books describe how to interpret control charts in detail.

Table 23.8 Multiple measurement data.

Date	1	2	3	4	5	X-bar	Range	Mean	Range Mean	X-bar UCL	X-bar LCL	Range UCL
2022-01-03	10.001	10.001	10.000	10.001	10.001	10.0008	0.001	10.0010	0.001	10.0017	10.0003	0.0025
2022-01-10	10.000	10.001	10.001	10.001	10.000	10.0009	0.001	10.0010	0.001	10.0017	10.0003	0.0025
2022-01-17	10.001	10.000	10.002	10.001	10.002	10.0011	0.002	10.0010	0.001	10.0017	10.0003	0.0025
2022-01-24	10.001	10.001	10.002	10.001	10.002	10.0014	0.001	10.0010	0.001	10.0017	10.0003	0.0025
2022-01-31	10.002	10.001	10.000	10.001	10.002	10.0011	0.002	10.0010	0.001	10.0017	10.0003	0.0025
2022-02-07	10.002	10.000	10.001	10.000	10.002	10.0010	0.002	10.0010	0.001	10.0017	10.0003	0.0025
2022-02-14	10.000	10.002	10.001	10.000	10.002	10.0009	0.001	10.0010	0.001	10.0017	10.0003	0.0025
2022-02-21	10.001	10.001	10.001	10.001	10.000	10.0008	0.001	10.0010	0.001	10.0017	10.0003	0.0025
2022-02-28	10.001	10.000	10.002	10.001	10.001	10.0011	0.001	10.0010	0.001	10.0017	10.0003	0.0025
2022-03-07	10.001	10.000	10.001	10.000	10.001	10.0009	0.001	10.0010	0.001	10.0017	10.0003	0.0025
2022-03-14	10.000	10.000	10.002	10.000	10.001	10.0007	0.001	10.0010	0.001	10.0017	10.0003	0.0025
2022-03-21	10.000	10.001	10.001	10.002	10.001	10.0010	0.002	10.0010	0.001	10.0017	10.0003	0.0025
2022-03-28	10.000	10.000	10.001	10.001	10.002	10.0009	0.002	10.0010	0.001	10.0017	10.0003	0.0025
2022-04-04	10.001	10.002	10.001	10.001	10.002	10.0013	0.001	10.0010	0.001	10.0017	10.0003	0.0025
2022-04-11	10.000	10.000	10.001	10.000	10.001	10.0005	0.001	10.0010	0.001	10.0017	10.0003	0.0025
2022-04-18	10.002	10.001	10.002	10.001	10.002	10.0016	0.000	10.0010	0.001	10.0017	10.0003	0.0025
2022-04-25	10.000	10.000	10.001	10.001	10.001	10.0004	0.001	10.0010	0.001	10.0017	10.0003	0.0025
2022-05-02	10.001	10.002	10.001	10.001	10.002	10.0014	0.001	10.0010	0.001	10.0017	10.0003	0.0025
2022-05-09	10.001	10.001	10.000	10.001	10.000	10.0005	0.001	10.0010	0.001	10.0017	10.0003	0.0025
2022-05-16	10.001	10.001	10.000	10.001	10.001	10.0007	0.000	10.0010	0.001	10.0017	10.0003	0.0025
2022-05-23	10.001	10.001	10.001	10.001	10.001	10.0010	0.000	10.0010	0.001	10.0017	10.0003	0.0025

Continued

Table 23.8 Continued

2022-05-30	10.001	10.002	10.001	10.001	10.001	10.0013	0.001	10.0010	0.001	10.0017	10.0003	0.0025
2022-06-06	10.000	10.001	10.000	10.001	10.001	10.0005	0.001	10.0010	0.001	10.0017	10.0003	0.0025
2022-06-13	10.001	10.001	10.001	10.002	10.002	10.0013	0.001	10.0010	0.001	10.0017	10.0003	0.0025
2022-06-20	10.000	10.001	10.001	10.002	10.001	10.0009	0.002	10.0010	0.001	10.0017	10.0003	0.0025
2022-06-27	10.002	10.001	10.001	10.001	10.002	10.0012	0.001	10.0010	0.001	10.0017	10.0003	0.0025
2022-07-04	10.001	10.000	10.001	10.001	10.000	10.0007	0.001	10.0010	0.001	10.0017	10.0003	0.0025
2022-07-11	10.002	10.001	10.000	10.000	10.002	10.0010	0.002	10.0010	0.001	10.0017	10.0003	0.0025
2022-07-18	10.001	10.000	10.001	10.001	10.000	10.0009	0.001	10.0010	0.001	10.0017	10.0003	0.0025
2022-07-25	10.000	10.000	10.000	10.000	10.001	10.0004	0.001	10.0010	0.001	10.0017	10.0003	0.0025
2022-08-01	10.001	10.002	10.001	10.000	10.001	10.0011	0.001	10.0010	0.001	10.0017	10.0003	0.0025
2022-08-08	10.002	10.001	10.001	10.001	10.002	10.0013	0.001	10.0010	0.001	10.0017	10.0003	0.0025
2022-08-15	10.000	10.002	10.001	10.001	10.000	10.0009	0.002	10.0010	0.001	10.0017	10.0003	0.0025
2022-08-22	10.001	10.001	10.002	10.001	10.000	10.0011	0.001	10.0010	0.001	10.0017	10.0003	0.0025
2022-08-29	10.000	10.001	10.001	10.001	10.002	10.0010	0.001	10.0010	0.001	10.0017	10.0003	0.0025
2022-09-05	10.000	10.001	10.000	10.000	10.001	10.0008	0.001	10.0010	0.001	10.0017	10.0003	0.0025
2022-09-12	10.002	10.002	10.001	10.001	10.001	10.0014	0.001	10.0010	0.001	10.0017	10.0003	0.0025
2022-09-19	10.001	10.001	10.001	10.001	10.000	10.0009	0.001	10.0010	0.001	10.0017	10.0003	0.0025
2022-09-26	10.002	10.001	10.001	10.000	10.002	10.0012	0.002	10.0010	0.001	10.0017	10.0003	0.0025
2022-10-03	10.000	10.001	10.001	10.001	10.002	10.0008	0.001	10.0010	0.001	10.0017	10.0003	0.0025
2022-10-10	10.002	10.001	10.001	10.001	10.001	10.0011	0.001	10.0010	0.001	10.0017	10.0003	0.0025
2022-10-17	10.002	10.002	10.002	10.002	10.002	10.0016	0.000	10.0010	0.001	10.0017	10.0003	0.0025

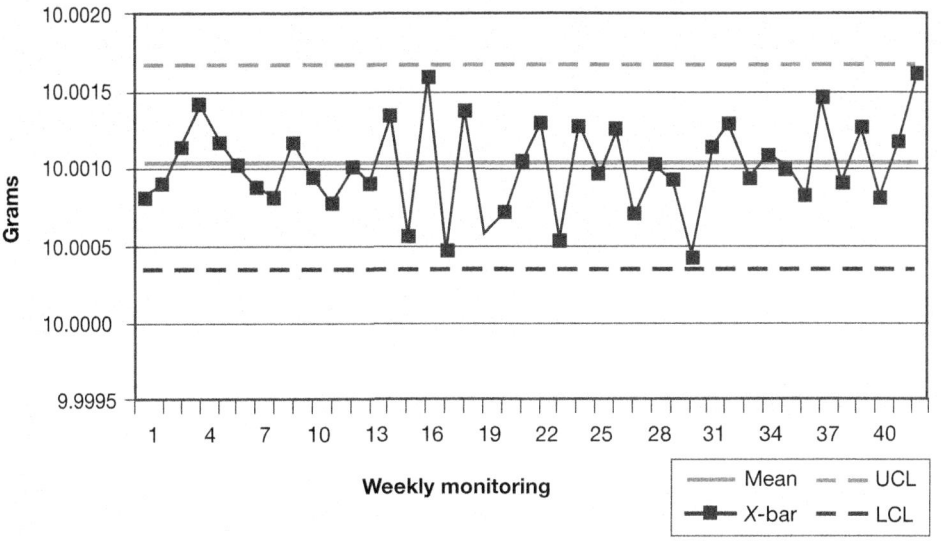

Figure 23.10 X-bar chart.

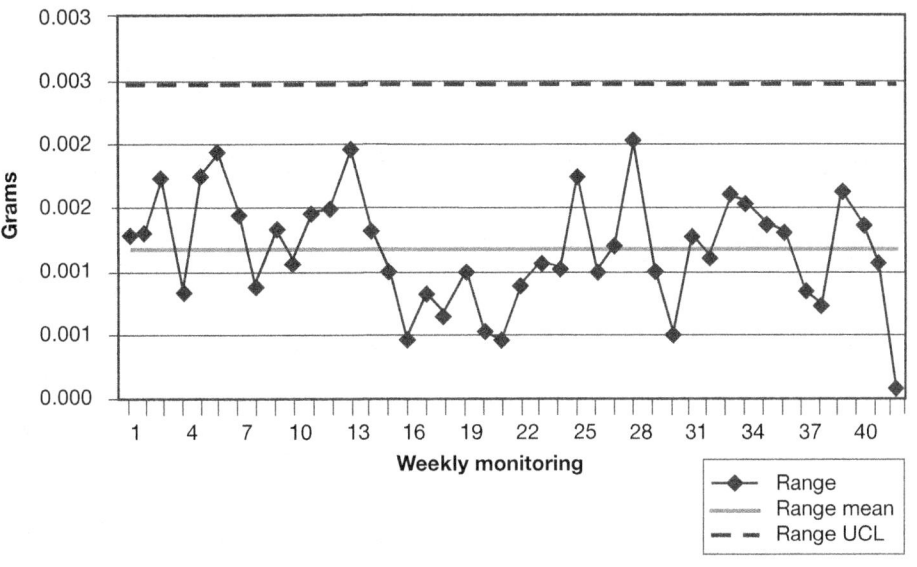

Figure 23.11 Range chart.

Monitoring the performance of equipment with control charts helps the laboratory catch conformance-related problems relating to the measurement in a proactive way before the problem really becomes a problem. It also helps in quantifying drift when used with other statistical tools. In addition, it provides guidelines on determining calibration intervals.

MAPs provide confidence that the accuracy of standards or equipment is maintained when used in a proper manner.

Part IV
Mathematics and Statistics: Their Use in Measurement

Chapter 24   Number Formatting

Chapter 25   Unit Conversions

Chapter 26   Ratios

Chapter 27   Statistics

Chapter 28   Mensuration, Volume, and Surface Areas

Chapter 24

Number Formatting

SIGNIFICANT DIGITS

The definition of a significant digit is "any digit in a number that is necessary to define a numerical value of a quantity." Wherever possible, the number of significant digits must be determined and stated in the formatting of a measurement result or in the accompanying text. Table 24.1 gives examples of numbers with a statement of resolution or precision in the accompanying text (followed in parentheses by the resulting number of significant digits).

Where clarification of intended meaning is not accomplished by an accompanying text or other means, the assumptions about the number of significant digits could change. For instance, for the same data items we could conclude the following, as shown in Table 24.2.

The number of significant digits could change because the determination of significant digits is based on either assumption or knowledge about how each

Table 24.1 Numbers reported and significant digits.

Number	How reported and number of significant digits
130,000,000	Reported in millions (three significant digits)
307,450,000	Measured in thousands (six significant digits)
805,006,700	Resolved in units (nine significant digits)
2.030, 0.0140, 0.68240	Precise to thousandths (four, three, and three significant digits, respectively)

Table 24.2 Numbers and number of significant digits.

Number	Number of significant digits
130,000,000	(Two significant digits)
307,450,000	(Five significant digits)
805,006,700	(Seven significant digits)
2.030, 0.0140, 0.68240	(Three, two, and four significant digits, respectively)

datum (reference, measurement) was taken and the degree of precision intended to be asserted, as presented.

When a value is stated in scientific notation with trailing zeros on the right, the implication is that these digits are significant. Examples of this are:

$$2.8600 \times 10^{-4} \text{ A} = \text{Five significant digits}$$

$$3.780 \times 10^{-8} \text{ A} = \text{Four significant digits}$$

$$1.24400 \times 10^{9} \text{ Hz} = \text{Six significant digits}$$

This implication of resolution/significance is a unique aspect of the scientific notation data format.

As leading zeros are automatically omitted, and magnitude is contained in the power of 10 exponent (for example, 10^{-4}, E—4, and so on), the number of significant digits may be clearly stated and understood by the value in the coefficient. Whichever number format is used, a conservative practice is this: If a number is stated to a given number of digits, without further available basis information, assume that all the stated digits are significant.

In mathematical operations, exact numbers are considered to have an infinite number of significant digits and, therefore, are not considered in determining the number of significant digits in a result. Actual counts, not estimates, are exact numbers. So are defined numbers (1 foot = 12 inches, 1 meter = 1000 millimeters).

Certain irrational number constants (for example, π, ε, and so on), commonly used in mathematical operations, occupy a kind of middle zone of significant digits. These constants may be determined to any number of significant digits—more than sufficient to not be a limitation on the accuracy of the math operation they are used in. When as a shorthand they are used with a low number of significant digits in the calculation (the most common, of course, is π, which is often stated to only three significant digits, 3.14), however, that must be considered in determining the significant digits stated in the calculation result.

There are two approximate rules for determining significant digits in the results of math operations on data. Note that in some cases these rules may yield results that are too small by one or two digits:

1. **Addition and subtraction significant digits rule.** The answer should contain no more significant digits, relative to the decimal marker, further to the right than is present in the least precise number. For example:

    $$134{,}000{,}000 \text{ (in millions)} + 1{,}843{,}991 - 4.600 \times 10^5 = 135{,}383{,}991$$

    This number is adjusted to the precision of the least precise number, 134,000,000 (stated in millions), giving 135,000,000, or 1.35×10^8.

2. **Multiplication and division significant digits rule.** The product or quotient should contain no more significant digits than the number with the fewest significant digits. For example:

$$134{,}000{,}000 \text{ (stated in millions)} / (1{,}843{,}991 \times 4.600 \times 10^{-1}) = 157.9749293$$

This number is adjusted to three significant digits, giving a result of 150, or 1.50×10^2.

Including digits that are not significant would imply resolution not appropriate for the measurement statement. This burdens further uses of the stated measurement result with transcription and computation complexity that has no value. It would be misconstrued as result resolution, or even worse, precision and accuracy that is not warranted.

It is important to note that there is some further degradation in resulting significant digits in exponentiation, transcendental, and other complex math operations. Use of more advanced methods, such as error calculus, is required to adjust for these effects.

NOTATION METHODS

When measurements are made and their values recorded, a normal requirement is to communicate those values to others. Most often, the purpose for communicating these results is to enable either subsequent information gathering or decision making. Nearly always, the recorded or communicated values are expressed as numbers. Thus, there is a need to express these numerical values in a consistent, comparative, and understandable format.

Numbers have been represented in many ways over the years. The simplest is to express decimal numbers as significant digits related to zero by the location of the *decimal marker* (sometimes called *decimal point* or *decimal place*).

There are various notation and formatting methods in use for representing numerical data. Some of these are:

- Standard notation
- Scientific notation
- Engineering notation
- SI prefix system (see Chapter 25, "Unit Conversions")
- Coding (this will not be discussed here)

STANDARD NOTATION

The term *standard notation* is used to describe a numerical value that is expressed simply as a number whose magnitude is conveyed by the placement of the decimal marker. Hence, the magnitude of 1000.1 is 10 times the magnitude of 100.01, and the magnitude of 0.0034 is one-hundredth the magnitude of 0.34.

From this we can see that moving the decimal marker one place to the right has the same effect as multiplying the number by 10; moving it one place to the left has the same effect as dividing it by 10. Moving it two places to the right increases the value by a factor of 100 and moving it to the left decreases the value by a factor of one-hundredth. Moving the decimal marker by three places changes the value by a factor of 1000.

The resulting numbers in standard notation from measurements can vary from quantities that are comparatively small in magnitude to ones that are quite large. All values of fundamental physical constants in this chapter are from the 2018 CODATA International recommended values.[1]

In the following, u = standard uncertainty.

For the definition of *standard uncertainty*, see Chapter 29, "Measurement Uncertainty," and Appendix C, "Glossary of Terms."

Avogadro constant, N_A, is 602214076000000000000000 mol^{-1}

which specifies the number of atoms, molecules, and so on, in a gram mole of any chemical substance. This is an exact number value that no longer has standard uncertainty.

Loschmidt constant, n_0 (given at 273.15 K, 101.325 kPa)

26867801110000000000000 m^{-3},

which specifies the number of molecules in a cubic centimeter of a gas under standard conditions (often confused with Avogadro's number).

The *Andromeda galaxy* (closest to our Milky Way galaxy) is estimated to contain at least 200,000,000,000 stars.

The neutron *magnetic moment* μ_n

−0.00000000000000000000000000096623651 J·T^{-1},

u = 0.00000000000000000000000000000023 J·T^{-1}.

The *neutron mass*, m_n

0.00000000000000000000000000167492749804 kg,

u = 0.00000000000000000000000000000000095 kg.

One thing that is immediately obvious from these measurement numbers is the number of zeros involved in making the standard notation numerical measurement statement. Standard notation clearly is a poor way of representing very large and very small values.

Scientific Notation

Fortunately, these values can be expressed much more efficiently in another numerical format called *scientific notation*. The scientific notation method of formatting numbers is based on the observation that the value of a number is unchanged if you multiply and divide it by the same constant (or power of ten). In scientific notation, numbers are typically written with one integer to the left of the decimal point (in the 10s place) and the following numbers after the decimal place. If the first non-zero number is several places after the decimal, then the first non-zero number becomes the leading value in the scientific notation expression of that number. For instance, for the value 0.00512, we can see that 0.00512 × 1000 / 1000 = 0.00512. Doing this in parts, 0.00512 × 1000 = 5.12, and then 5.12/1000 = 0.00512.

We also know that 1/1000 can be written as 10^{-3} (where 10 is the base and –3 is the exponent of the expression 10^{-3}). Saying 10^{-3} is the exponential form of representing "divide by 1000," or "multiply by 1/1000."

This understanding, and these relationships, allow us to transform 0.00512 (in standard notation) into scientific notation as 5.12×10^{-3}.

In scientific notation we express the above standard notation values as:

Avogadro's number

$$6.022\ 140\ 76 \times 10^{23}\ mol^{-1}$$

Loschmidt constant (273.15 K, 101.325 kPa)

$$2.686\ 7811 \times 10^{25}\ m^{-3}$$

The *Andromeda galaxy* contains at least 2.0E+11 stars.

The neutron *magnetic moment* μ_n

$$-9.662\ 3651 \times 10^{27}\ J\text{-}T^{-1},$$

$$u = 2.3 \times 10^{33}\ J\text{-}T^{-1}$$

The *neutron mass,* m_n

$$1.674\ 927\ 498\ 04 \times 10^{-27}\ kg,$$

$$u = 9.5 \times 10^{-36}\ kg$$

By these comparisons you can immediately see how much more compact scientific notation is. There are several benefits of using scientific notation: It is compact and efficient both for the purposes of reading and comparing values because the exponent concisely gives magnitude as a number (rather than forcing one to count zeros). It reduces typographical errors that could come from erroneously adding or omitting zeros. Also, using values in subsequent math operations is greatly simplified, as will be seen.

There are several frequently used formats for stating a number in scientific notation. In scientific notation, numbers are always expressed in relation to an exponent of the base 10.

In the following, N is the number we wish to express, with the decimal marker placed to the right of the first digit; M is the exponent of the base 10; and E is sometimes used to mean $10^{Exponent}$.

Engineering Notation

Engineering notation or engineering form is a version of scientific notation in which the exponent of 10 must be divisible by three (or the equivalent of the power of a thousand). On most calculators, engineering notation is called "ENG" mode.

In Engineering notation (E-notation) we express the previous standard notation values as:

Avogadro's number

$$602.21476 \text{ E}+21 \text{ mol}^{-1}$$

Loschmidt constant (273.15 K, 101.325 kPa)

$$26.867811 \text{ E}+24 \text{ m}^{-3}$$

The *Andromeda galaxy* contains at least 2.0E+11 stars.

Magnetic moment

$$-9.662\ 3651 \text{ E}-27 \text{ J·T}^{-1},$$

$$u = 2.3\text{E}-33 \text{ J·T}^{-1}$$

Neutron mass

$$1.674\ 927\ 498\ 04 \text{ E}-27 \text{ kg},$$

$$u = 9.5 \text{ E}-36 \text{ kg}$$

The following are the formats in most frequent use.

- **Scientific Notation.** This is used in scientific publications.

$$\pm N.N \times 10^{\pm M}. \text{ For example, } 5.12 \times 10^{-3}.$$

When this is used: in spreadsheet formulas entered in the formula bar (Microsoft Excel™).

$$\pm N.N \times 10^{\wedge} \pm M. \text{ For example, } 5.12 * 10^{\wedge}-3.$$

- **Engineering Notation.** This is used in spreadsheet displays and in most programming languages (MS Excel™, Fortran).

$$\pm N.N \text{ E} \pm M. \text{ For example, } 5.12 \text{ E}-03.$$

The notation of a value expressed in scientific notation has three parts:

± N.N   A numerical value, sometimes called the *coefficient*, having an absolute magnitude greater than or equal to 1 and less than 10. Here, the decimal marker is placed immediately to the right of the leftmost nonzero digit. This value may be positive or negative signed (±).

× 10   A *base*—scientific notation is always base 10.

± M   An *exponent* that, with the base, reflects the magnitude of the numerical value. This value may be positive or negative signed (±).

A number is transformed from standard to scientific notation by recognizing a relationship of powers of 10 as shown here:

$0.000002 = 2.0 \times 10^{-6} = 2.0 \times 10^{-1} \times 10^{-1} \times 10^{-1} \times 10^{-1} \times 10^{-1} \times 10^{-1}$

$0.00002 = 2.0 \times 10^{-5} = 2.0 \times 10^{-1} \times 10^{-1} \times 10^{-1} \times 10^{-1} \times 10^{-1}$

$0.0002 = 2.0 \times 10^{-4} = 2.0 \times 10^{-1} \times 10^{-1} \times 10^{-1} \times 10^{-1}$

$0.002 = 2.0 \times 10^{-3} = 2.0 \times 10^{-1} \times 10^{-1} \times 10^{-1}$

$0.02 = 2.0 \times 10^{-2} = 2.0 \times 10^{-1} \times 10^{-1}$

$0.2 = 2.0 \; 10^{-1} = 2.0 \times 10^{-1}$

$2.0 = 2.0 \times 10^{0}$

$20.0 = 2.0 \times 10^{1} = 2.0 \times 10$

$200.0 = 2.0 \times 10^{2} = 2.0 \times 10 \times 10$

$2000.0 = 2.0 \times 10^{3} = 2.0 \times 10 \times 10 \times 10$

$20000.0 = 2.0 \times 10^{4} = 2.0 \times 10 \times 10 \times 10 \times 10$

$200000.0 = 2.0 \times 10^{5} = 2.0 \times 10 \times 10 \times 10 \times 10 \times 10$

$2000000.0 = 2.0 \times 10^{6} = 2.0 \times 10 \times 10 \times 10 \times 10 \times 10 \times 10$

This obviously can, beneficially, be extended to any magnitude, large or small!

Any standard notation value may be easily converted to scientific notation. This is done by recognizing that a value is unchanged when the decimal marker is moved to behind the first nonzero digit when one adds a $\times 10^{\pm M}$ multiplier that adjusts for this movement, where M is the number of digits that the decimal marker was moved to the left (+) or to the right (–), respectively.

As just mentioned, one more very important advantage of using scientific notation is in the efficiency of doing math (addition, subtraction, multiplication, division, and raising to whole or fractional powers) on numerical results.

Math Operations Using Scientific Notation

When doing math operations on measurement numbers near the same order of magnitude as each other, ordinary math methods can be employed. When the order of magnitude of any of the numbers involved in a math operation or operations is substantially different from other numbers, however, all the numbers should be converted to scientific notation before performing the required operations.

The rules for doing math operations on numbers formatted in scientific notation relate to the operation being performed, as will now be explained.

Addition/Subtraction

The steps in adding or subtracting using scientific notation are:

1. Convert the exponents to the same magnitude before performing the operation. This will often require changing the location of the decimal marker for the numerical value or coefficient.

2. Perform the addition/subtraction.
3. Adjust for required significant digits in the result, as required.
4. Restore to scientific notation format, as required.

For example, add 4.875324×10^6 and 5.02×10^2. Follow the steps.

1. Change the decimal marker of 4.875324×10^6 to result in 48753.24×10^2.
2. $48753.24 \times 10^2 + 5.02 \times 10^2 = 48758.26 \times 10^2$.
3. Restore to scientific notation format, or 4.875826×10^6. There is no adjustment for significant digits.

A more direct method would have been to change 5.02×10^2 to 0.000502×10^6 and add to 4.875324×10^6. For addition, converting to the algebraically more positive exponent (here, 6) often avoids final adjustment for scientific notation formatting.

In another example, subtract 5.02×10^2 from 4.87532448×10^6. Follow the steps.

1. Change the decimal marker of 5.02×10^2 to result in 0.000502×10^6.
2. $4.87532448 \times 10^6 - 0.000502 \times 10^6 = 4.87482248 \times 10^6$, which is already in scientific notation format.
3. Adjust for the significant digits in 0.000502×10^6, giving 4.874822×10^6.

Multiplication

The steps in multiplying using scientific notation are:

1. Algebraically multiply the coefficients and add the values of the exponents.
2. Round to the number of significant digits of the lower significant digit number.
3. Adjust for required significant digits in the result, as required.
4. Restore to scientific notation format, if required.

For example, multiply 4.875324×10^6 and 5.02×10^2. Follow the steps.

1. $(4.875324 \times 10^6)(5.02 \times 10^2) = (4.875324)(5.02) \times (10^6)(10^2) (24.474126) \times (10^{6+2}) = 24.5 \times 10^8$.
2. Note that this number was rounded to the number of significant digits (three) of the lower-resolution number, 5.12×10^2.
3. Restore to scientific notation format, giving 2.45×10^9.

Division

The steps in dividing using scientific notation are:

1. Algebraically divide the coefficients.
2. Subtract the value of the denominator exponent (below the division line) from the numerator exponent. Or, change the signs of the

exponents below the division line (– to +, + to –) and add these to the sum of those above the division line.

3. Round to the number of significant digits of the lower significant digit number.

4. Adjust for required significant digits in the result, as required.

5. Restore to scientific notation format, if required.

For example, divide 4.875324×10^6 by 5.02×10^2. Follow the steps.

1. $(4.875324 \times 10^6) / (5.02 \times 10^2) = [(4.875324) / (5.02)] \times (10^6)(10^{-2}) = (0.97118008) \times (10^{6-2}) = 0.971 \times 10^4$.

2. Note that this number was rounded to the number of significant digits of the lower-resolution number, 5.02×10^2.

3. Restore to scientific notation format, giving 9.71×10^3.

Raising to Powers (Integral and Some Fractional)

Raising to an integral (that is, whole number) power is like multiplying the number by itself the number of times given by the power expressed by the exponent. So, use the multiplication methods. (Though not as intuitive, the same is also true for fractional exponentiation, involving a kind of repeated division and multiplication, not discussed here.)

For example, square 5.02×10^2. Follow the steps for multiplication.

1. Squaring, $(5.02 \times 10^2)^2$ is like multiplying 5.02×10^2 and 5.02×10^2.

2. $(5.02 \times 10^2)(5.02 \times 10^2) = (5.02)(5.02) \times (10^2)(10^2) = (5.02)^2 \times (10^2)^2 = 25.2 \times 10^4 = 2.52 \times 10^5$.

Obviously, we can extend this to any integral power.

Raising to a fractional power is somewhat like raising to integral powers, except the fractional power must be resolved to bring the result into scientific notation format.

The steps in raising to a fractional power are:

1. Algebraically raise the coefficients to the given power (integral or fractional).

2. Raise the exponents to the given power by multiplying (not adding) the power of 10 exponent by the separate power exponent. If the resulting power of (base) 10 exponent is not a whole-number exponent, factor the fractional part to a coefficient and resulting whole-number exponent.

3. Round to the number of significant digits of the lower significant digit number.

4. Adjust for required significant digits in the result, as required.

5. Restore to scientific notation format, if required.

For example, raise 5.02×10^2 to the one-half power (take the square root of 5.02×10^2).

$$(5.02 \times 10^2)^{1/2} = (5.02)^{1/2} \times (10^2)^{1/2} = 2.24053565 \times 10^{2/2} = 2.24 \times 10^1.$$

In another example, raise 5.02×10^2 to the one-third power (take the cube root of 5.02×10^2).

1. $(5.02 \times 10^2)^{1/3} = (5.02)^{1/3} \times (10^2)^{1/3} = 1.712252881 \times 10^{2/3}$.

2. Note that $10^{2/3}$ cannot be reduced to a whole-number exponent. Using a calculator, we see that $10^{2/3}$ has a value of 4.641588834 and make this a factor of the result, giving

3. $1.712252881 \times 4.641588834 = 7.95 \times 10^0$, or 7.95.

Notice that in all cases we do not adjust for the number of significant digits until after all the math operations are performed and we have the result. This is so that we do not lose any available resolution of the result due to the math operations performed. Fortunately, we have computers and spreadsheet programs that make the math portion of this easy. The assignment of significant digits, however, currently requires our intervention in the formatting of the result.

Estimation

Another benefit of using the scientific notation format is the ease and efficiency of estimating approximate results. Using the previous example, estimate the result of 134,000,000 (stated in millions) divided by (1,843,991 multiplied by 4.600×10^{-1}).

Restating these in scientific notation gives 1.34×10^8 / ($1.843991 \times 10^6 \times 4.600 \times 10^{-1}$). Changing these to estimated values yields 1.50×10^8 / ($2.000000 \times 10^6 \times 4.600 \times 10^{-1}$), giving a result of 150, or 1.50×10^2.

In this case the estimated result is the same as that obtained using the full precision of the originating data. Obviously, care must be taken in the judgments made by estimation. In this case, had we estimated the first number to be 1.00×10^8 instead of 1.50×10^8, the result would have been 1.00×10^2 instead of 1.50×10^2, an error of 50 %.

One method for minimizing accumulated estimation errors is to make the estimate values close to the actual data. Another is to alternate the directions of the estimates relative to the data: one higher, the next lower, and so on. Yet another is to alternate the directions of the estimates above and below a divisor line when present.

SI PREFIX SYSTEM

The SI unit prefix system has the advantage of aggregating measurements into easily recognized groups of magnitudes. See Chapter 25 for these units and how they are used. These SI unit prefixes permit compact expression of large and small orders of magnitude and easy comparison of units of similar magnitude. However, it will be seen that it is not as easy to compare relative orders of magnitude, numerically, as compared to the scientific notation method.

Here is a comparison of standard, scientific, and SI prefix notation. One set of measurements is reported in standard and scientific notation (in superscript format) for electrical current measurements in amperes:

Standard notation	Scientific notation	SI prefix notation
0.000286 A	2.86×10^{-4} A	0.286 mA
126.78 A	1.2678×10^{-2} A	0.12678 kA
0.45965 A	4.5965×10^{-1} A	0.45965 A
0.0000000378 A	3.78×10^{-8} A	37.8 nA, or 0.0378 pA

and for frequency measurements in Hertz:

Standard notation	Scientific notation	SI prefix notation
1,244,000,000 Hz	1.244×10^9 Hz	1.244 GHz
3.43 Hz	3.43 Hz	3.43 Hz
60.25 Hz	6.025×10^1 Hz	60.25 Hz, or 6.025 daHz
56,734,200 Hz	5.67342×10^7 Hz	56.7342 MHz

ISO PREFERRED NUMBERS

We are all familiar with products that are manufactured in a series of standard sizes. It is often more efficient for the manufacturer to produce the ideal spacing of sizes so that a range is covered with the fewest possible number of intermediate sizes and with ones that are evenly spread across the range. In the fluid volume series "tablespoon, fluid ounce, quarter-cup, gill, cup, pint, quart, pottle, gallon," each successive unit increases by a factor of two. For the SI system there is a preference for decimal units, and, additionally, a method is given for distributing evenly spaced values between successive powers of 10, when needed.

In 1877 a French military engineer, Col. Charles Renard, was charged with standardizing the size of mooring cables used for balloons that did surveillance during wartime. He reduced the sizes used from 425 to 17 by developing a geometric series basis that resulted in every fifth step increasing by a factor of 10. We now call such a series the *Renard series*. ISO has defined four such basic (Renard) series of preferred numbers, with the designator *R* in honor of Renard. Listed here are values from 10 to 100 for these series:

Table 24.3 ISO (Renard's) series of numbers.

R5:	10, 16, 25, 40, 63, 100
R10:	10, 12.5, 16, 20, 25, 31.5, 40, 50, 63, 80, 100
R20:	10, 11.2, 12.5, 14, 16, 18, 20, 22.4, 25, 28, 31.5, 35.5, 40, 45, 50, 56, 63, 71, 80, 90, 100
R40:	10, 10.6, 11.2, 11.8, 12.5, 13.2, 14, 15, 16, 17, 18, 19, 20, 21.2, 22.4, 23.6, 25, 26.5, 28, 30, 31.5, 33.5, 35.5, 37.5, 40, 42.5, 45, 47.5, 50, 53, 56, 60, 63, 67, 71, 75, 80, 85, 90, 95, 100

By moving the decimal marker to the left or right, these same series may be extended to any magnitude number. For instance, the R5 series could be used to define a preferred-number sequence between 0.1 and 1.0, as 0.10, 0.16, 0.25, 0.40, 0.63, or 1.00. One would use this method where it is desired to have a range of produced results restated as values conforming to the spacing of a preferred-number series.

As an example, say a series of measurements has been made with the results being 0.4812, 0.0125, 0.9823, 0.5479, 0.2258, 0.7601, 0.4271, 0.15812, and 0.7013. Report these in a series conforming to the ISO R10 preferred number series.

The adjusted R10 equivalent series would be:

R10: 0.01, 0.125, 0.16, 0.20, 0.25, 0.315, 0.40, 0.50, 0.63, 0.80, 1.00

The R10 preferred number transformation of the results would then be:

0.50, 0.01, 1.00, 0.50, 0.20, 0.80, 0.40, 0.16, 0.63

Note: If there are paired results, adjusting one set of the paired values to a preferred number spacing will require interpolating the results of the other set in the pair for each value in the series (see Chapter 27, "Statistics").

NUMBER ROUNDING METHODS

The purpose for rounding numbers is to present data in a more concise format that is appropriate to the purpose at hand. For instance, a bar of steel has a diameter of 9/16 in (0.5625 in). If the purpose at hand is to compare it to other choices of ½ in (0.50 in) and ¾ in (0.75 in), it is more convenient (and may be more appropriate) to report the 9/16-in measurement decimally as 0.56 in.

Caution must be exercised in employing rounding methods to ensure that the necessary precision in numbers or measurements is not sacrificed by rounding merely for the sake of brevity.

There are some uses of data where reporting data to a larger number of significant digits is advantageous. Some such applications include, but are not limited to, comparisons involving minor differences in measurements, and some statistical calculations (paired *t*-tests, ANOVA, design of experiments, correlations, autocorrelations, nonparametric tests).

There are several number rounding methods in common use. Prominent among these are:

1. **ISO rounding[2] (preferred method).** This method is similar to the "round up for 5 or over" rule, but it balances what is done when the discarded digit is 5 followed by all zeros. In the special case where the discarded digit is 5 followed by all zeros, the rightmost retained digit, before rounding, determines if it is kept unchanged or increased by one. If before rounding this digit is even, it is left unchanged. If before rounding this digit is odd, it is increased by one to make it even. This method is preferred for most scientific measurements. It is required for most weights and measures reporting.

2. **Rounding up for 5 or over.** Here, the rightmost retained digit is unchanged if the discarded digit is less than 5 and increased by one if the discarded digit is 5 or over. This is the method most often taught.

Be aware that this is the method implemented in the displayed result as well as the ROUND() function for MS Excel™ and some other spreadsheet programs.

3. **Truncation.** Here, deleted digits on the right side of a number are simply omitted with no accommodation to remaining digits. The obvious problem with this method is that the rightmost retained digit may not reflect the precision implied from data in the deleted digits: 0.343×10^3 could be the truncated result of either 0.343999×10^3 or 0.343001×10^3 actual measurements.

4. **Rounding to closest ISO-preferred number.** Use one of the four ISO basic series of preferred numbers (ISO R5, R10, R20, R40). This method ensures that reported values aggregate into preferred values. This method is recommended for use in design and both eases comparison and increases standardization in reporting of results.

5. **Rounding to nearest stated multiple of significance.** In spreadsheet programs, such as MS Excel™, other methods are available:

    o Floor(), rounds toward zero, to the nearest stated multiple of significance.

    o Ceiling(), rounds away from zero, to the nearest stated multiple of significance.

    Such methods are useful in cases where a governing standard or practice specifies a multiple of significance that a result must be rounded to (for example, ASTM standards that specify rounding a measurement to a stated increment, such as to the nearest 1 %, 100 kPa, and so on).

As previously emphasized, do not round intermediate results of math operations. Round only the final result.

Here are examples of rounding by various methods. The measurement is 1⅞ in (1.875 in or 47.625 mm, exactly). Referring to the five most prominent methods:

1. ISO rounding to

    3 significant digits: 1.88 in, 47.6 mm

    2 decimal places: 1.88 in, 47.62 mm

2. Round up for 5 or over to

    3 significant digits: 1.88 in, 47.6 mm

    2 decimal places: 1.88 in, 47.63 mm

    1 decimal place: 1.9 in, 47.6 mm

3. Truncate to

    3 significant digits: 1.87 in, 47.6 mm

    1 decimal place: 1.8 in, 47.6 mm

4. Round to closest ISO preferred number:

   R5:   1.6 in, 40 mm

   R10: 2.0 in, 50 mm

   R20: 1.8 in, 50 mm

   R40: 1.9 in, 47.5 mm

5. Round the value 126.34 to the nearest stated multiple of significance (for example, with MS Excel™ spreadsheet or similar built-in functions):

   —Floor (126.34,25) = 125, Floor (0.623,0.25) = 1.0

   —Ceiling (126.34,25) = 150, Ceiling (0.123,0.25) = 0.25

OTHER NUMBER FORMATTING ISSUES

The symbol for:

- Degrees (often written as the abbreviation, deg) is ° (small superscript circle).

- Minutes (often written as the abbreviation, min) is ' (single quotation mark).

- Seconds (often written as the abbreviation, sec) is " (double quotation mark).

- Angles with a magnitude less than one degree are written beginning with 0°, as follows:

    o 0° 43'

    o 0° 0.35', or 0° 0' 30" (preferred)

Leading Zeros

Including a zero preceding the decimal marker is an essential writing format practice to represent and accurately interpret place value for *both SI and non-SI units*. NIST SP 811 addresses this leading-zeros matter in section NIST SP 811:2008, 10.5.2: "Decimal sign or marker: The recommended decimal sign or marker for use in the United States is the dot on the line [3, 6]. For numbers less than one, a zero is written before the decimal marker. For example, 0.25 s is the correct form, not .25 s."[3]

U.S. Customary Number Formatting and Decimal Markers

In United States customary units, numbers with four or more digits to the left of the decimal marker have typically been represented with commas separating each group of three digits, and a decimal point as a decimal marker. However, in most other countries the comma has been widely used as a decimal marker. Resolution 10 of the 22nd CGPM (2003) *"declares* that the symbol for the decimal marker shall be either the point on the line or the comma on the line,"[4] but it has become international practice to use the dot (point on the line) to represent the decimal marker in English-language documentary standards because use of the comma conflicts with customary practice (of the comma [on the line] remaining the decimal marker in all of its French-language publications).[5] Resolution 10 of the 22nd CGPM (2003) also *reaffirms* that "[n]umbers may be divided in groups of three in order to facilitate reading; neither dots nor commas are ever inserted in the spaces between groups," as stated in Resolution 7 of the 9th CGPM, 1948.[6] NIST SP 330:2019, 5.4.4 offers the guidance that "the decimal marker chosen should be that which is customary in the context concerned."[7]

NIST SP 811:2008, 10.5.3, "Grouping digits," offers uniform guidance on use of commas when expressing numbers: "Because the comma is widely used as the decimal marker outside the United States, it [the comma] should not be used to separate digits into groups of three. Instead, digits should be separated into groups of three, counting from the decimal marker towards the left and right, by the use of a thin, fixed space. However, this practice is not usually followed for numbers having only four digits on either side of the decimal marker except when uniformity in a table is desired."[8] When there are more than four digits to the right of the decimal marker, also adding a thin, fixed space between groups of three numbers is also recommended.

Examples of these representations appear in Table 24.4.

Table 24.4 Recommended spacing and use of decimal marker.

SI recommended	Not recommended
0.284 5 or 0.2845	
4 683 or 4683	4,683
2 352 491	2,352,491
44 231.112 34	44,231.11234
1403.2597 or 1 403.259 7	1,403.2597
0.491 722 3	0.4917223

DATE AND TIME FORMATS

The concept of time has evolved over the millennia. For cultures in which time was an important agreement, time was defined and kept locally. In the United States alone, there were hundreds of local time standards by the mid-1800s.[9] This was not a big deal when it took days to travel from place to place.[10] The expansion of railroads and rail travel provided faster travel. Because of the lack of time standardization, schedules on the same tracks often could not be coordinated, resulting in collisions.[11] Imagine the challenges to railroads to coordinate departures and arrivals among their routes. The expansion of transportation and communication worldwide in the 1800s highlighted the need for a unified timekeeping system. In 1878, Sir Sanford Fleming, Canada's foremost railway construction engineer, proposed the system of 24 worldwide time zones (one hour apart) that are still used today.[12] In the United States, "on November 18, 1883, precisely at noon, North American railroads switched to a new standard time system for rail operations, which they called Standard Railway Time (SRT). Almost immediately after being implemented, many American cities enacted ordinances, thus resulting in the creation of time "zones." The four standard time zones adopted were Eastern, Central, Mountain, and Pacific."[13] Then, in 1884, at the International Meridian Conference, the establishment of timekeeping with Greenwich Mean Time (GMT) as the world's time standard was adopted. From this, the worldwide 24-hour time-zone system was adopted.

It was in 1954 that the second was adopted as a base unit (of measurement) for time at the 10th CGPM. The CIPM in 1956 defined the second as "the fraction of 1/31 556 925.9747 of the tropical year 1900 January 0 at 12 hours ephemeris time."[14] Then, in 1960, at the 11th CGPM, the 1956 CIPM definition of the second was ratified.[15] In 1972, Coordinated Universal Time (UTC, for Universal Time Coordinated) replaced Greenwich Mean Time (GMT) as the world's time standard.[16] UTC is not adjusted for daylight saving time. This makes it particularly useful for reducing confusion across time zones or during changes to and from daylight saving time.

Since the 1970s, ISO 8601 and its predecessors have been the gold standard for representing date and time. ISO 8601 is the International Standard for representing date and time for interchange, standardization, and agreement of date and time formats. ISO 8601-1:2019, Date and time—Representations for information interchange—Part 1: Basic rules is the direct successor to ISO 8601:2004. ISO 8601-2:2019 provides extensions on top of ISO 8601-1:2019. See the introduction to the new ISO 8601-1 and ISO 8601-2—ISO/TC 154: Processes, data elements and documents in commerce, industry and administration.[17] Businesses based in different countries or economies may still use local date formats, which may conflict with date formats in other countries or economies. The United States official date format is for dates to be expressed in numbers as MM-DD-YYYY, where MM = one to two digits reflecting the month, DD = one to two digits reflecting the day of the month, and YYYY = four digits reflecting the year. This means that 3-10-2023 is March 10, 2023, in the United States. However, in most of Europe, it is common for dates to be expressed in numbers as DD-MM-YYYY, where DD = two digits reflecting the day of the month, MM = two digits reflecting the month, and YYYY = four digits reflecting the year. This means that 3-10-2023 means October 3, 2023, in Europe. If we further shorten a date to 3-10-23, we may not know which one is the day, month, or year! It is easy to imagine how these different date formats might create confusion to the unwitting reader. When presented with the date of 03-10, the reader cannot be confident if the date is March 10 or October 3! This is where ISO 8601-1:2019 is helpful. In it, a common representation of dates as [YYYY]["-"][MM] ["-"][DD] would give the expression of 2023-03-10 which identifies March 10, 2023. Dates can also be expressed as YYYYMMDD, or YYYY-MM. ISO 8601-1:2019 also further clarifies week numbers and time formats.[18]

Weeks begin with Monday and end with Sunday. Week numbering, such as Week 01, begins with starting the week with the year's first Thursday in it (or the first week with four or more of its days in the starting year). Another way to determine Week 01 is the week starting with the Monday in the period of December 29 to January 4. December 28 is always in the last week of its year.[19]

Time is based on a 24-hour clock system. The *basic format* for time is T[hh][mm][ss] and the *extended format* is T[hh]:[mm]:[ss]. T represents that the numbers that follow are for "time." [hh] refers to an hour between 00 and 23. [mm] refers to minutes between 00 and 59. [ss] refers to seconds between 0 to 60 (where 60 is only used to denote the added leap second used every four years). Time can be expressed without the T using the *extended format* as [hh]:[mm]:[ss].

Time zones are represented as local time (based on coordinated Universal Time [UTC]) and time shifts are represented as an offset from coordinated Universal Time.[20]

Chapter 25

Unit Conversions

The SI, also known as the metric system, is in use throughout the world and is the most widely recognized and accepted system of measurement units. The benefits of use of the SI in industry, trade, and the sciences are numerous. SI units enable greatly improved comparability of measurements of both similar and widely varying magnitudes. Products and components dimensioned in SI units afford more direct engineering and scientific analysis, as the units are often directly associated with underlying natural physical quantities. The SI is extensively covered in Chapter 11, "The International System of Units (SI) and Measurement Standards."

Another frequently used system is often called the *English, Imperial, U.S. Customary, conventional,* or *inch-pound* system of units of measure. These units, based on inch/pound/horsepower/etc. units of measure, are still in wide use today in the United States even though SI is the legal basis for all measurements. Although their use is found mostly in the United States, there are also some countries that have had significant trade with and/or been influenced by both British and U.S. commerce and industry and will at least dual-label products with metric and the secondary unit systems.

Because of this continuing divergence from the practical use of a single system of units in the world, it continues to be important to accurately and easily convert from measurements made in one system, say, SI units, to those of the other, U.S. customary units or vice versa. To be able to compare measurements it is necessary to convert to and from each unit with confidence.

This chapter provides the tools for correct and adequate conversion of units between SI and customary units. Additional conversions can also be found in Appendix D.

SI UNITS

SI has currently adopted seven base units (see Table 25.1). The names of SI units are not capitalized, and the symbol is a capital letter only if the name is derived from a person's proper name.

Additionally defined are the following dimensionless units:

- A *radian*, symbol rad, is the plane angle between two radii of a circular arc that has a length equal to the radius.

- A *steradian*, symbol sr, is the solid angle such that where its vertex is at the center of a sphere, cuts off a surface area on the spherical surface equal to that of a square with sides of arc length equal to the radius of the sphere.

Table 25.1 Seven base units in the International System of Units (SI).[1]

Base quantity name	Typical symbol	Base unit	Symbol
length	l, x, r, etc.	meter (metre)	m
mass	m	kilogram	kg
time	t	second	s
electric current	I, i	ampere	A
thermodynamic temperature	T	kelvin	K
amount of substance	n	mole	mol
luminous intensity	I_v	candela	cd

In a metrology laboratory, the measurement standards are representations of the SI units. This is true whether you are using an intrinsic standard such as a Josephson junction array or a triple point of water cell, or if you are using a standard such as a multifunction calibrator or gage block. However, National Metrology Institutes around the world are creating the next level of measurement standards based on quantum physics. This may restructure the metrological traceability pyramid into a flattened layer where a typical metrology lab may have its own quantum measurement standards for providing exceptionally accurate measurements.

COHERENCE OF SI UNITS

The SI system of units is often referred to as *coherent* because all other units are derived from these base units by the rules of multiplication and division, with no numerical factor other than unity. Hence, all SI-recognized units are direct (by multiplication and/or division) combinations of base units. *Power of ten* SI-prefixes are added for representing larger and smaller magnitudes, making the entire SI system decimally based. One exception is time, which still has recognized derived units from the second of minute, hour, day, etc. Note that chronological time, which still has recognized derived units from the second of minute, hour, day, and year, is a set of non-SI units that are acceptable for use along with the SI. (See Chapter 32 for more information on the second and customary time scales such as universal coordinated time [UTC]). Time units are part of non-SI units accepted for use with the SI units. See Table 25.2 for these and other non-SI units accepted for use with SI units.

This decimal basis is a quite distinctive feature of the SI system of units. For comparison, U.S. customary units often are products of doubles (for example, pint versus quart), triples (for example, foot versus yard), combinations of these (as in inch versus foot), sexagesimal (degree, minute, second—with origins as early as ancient Babylonian usage), and other integral and nonintegral multiples. In fact, the decimal system's attractiveness is based on the great simplification that it provides over competing unit systems, providing for this coherence that, with SI

Table 25.2 Some non-SI units accepted for use with SI units.[2]

Quantity	Unit name	Unit symbol	Value in SI units
time	Minute	min	1 min = 60 s
	hour	h	1 h = 60 min = 3600 s
	day	d	1 d = 24 h = 86 400 s
length	Astronomical unit[a]	au	1 au = 149 597 870 700 m
plane and phase angle	degree	°	$1° = (\pi/180)$ rad
	minute	′	$1' = (1/60)° = (\pi/10\,800)$ rad
	second	″	$1'' = (1/60)' = (\pi/648\,000)$ rad
volume	liter (litre)	l, L	1 l = 1 L = 1 dm^3 = 10^3 cm^3 = 10^{-3}m^3
energy	electronvolt	eV	1 eV = 1.602 176 634 × 10^{-19} J

prefixes and scientific notation (see Chapter 24), yields expression of wide ranges of magnitudes in a way that is easily understood and compared.

SI-DERIVED UNITS

A significant number of units have been developed to satisfy specific application needs that are derived from these base SI units. Some of these derived units have names expressed in terms of the base units from which they are formed.

Table 25.3 lists some of these units and how they are derived from base units. Table 25.4 lists other derived units with specialized names and symbols. Other derived units are themselves expressed in terms of derived units, some with special names, as in the examples listed in Table 25.5.

Table 25.3 SI units derived from base units.[3]

	SI Derived Unit	
Quantity	Name	Symbol
Area	Square meter	m^2
Volume	Cubic meter	m^3
Speed, velocity	Meter per second	m/s
Acceleration	Meter per second squared	m/s^2
Wave number	Reciprocal meter	m^{-1}
Mass density (density)	Kilogram per cubic meter	kg/m^3
Specific volume	Cubic meter per kilogram	m^3/kg
Current density	Ampere per square meter	A/m^2
Magnetic field strength	Ampere per meter	A/m
Amount-of-substance concentration (concentration)	Mole per cubic meter	mol/m^3
Luminance	Candela per square meter	cd/m^2

Table 25.4 The 22 SI units with special names and symbols.[4]

Derived quantity	Unit (symbol)	Unit in terms of base units	Unit expressed in other SI units
plane angle	radian (rad)	m/m	
solid angle	steradian (sr)	m^2/m^2	
frequency	hertz (Hz)	s^{-1}	
force	newton (N)	$kg\ m\ s^{-2}$	
pressure, stress	pascal (Pa)	$kg\ m\ s^{-2}$	
energy, work, amount of heat	joule (J)	$kg\ m^2\ s^{-2}$	N m
power, radiant flux	watt (W)	$kg\ m^2\ s^{-3}$	$J\ s^{-1}$
electric charge	coulomb (C)	A s	
electric potential difference	volt (V)	$kg\ m^2\ s^{-3}\ A^{-1}$	W/A
capacitance	farad (F)	$kg^{-1}\ m^{-2}\ s^4\ A^2$	C/V
electric resistance	ohm (Ω)	$kg\ m^2\ s^{-3}\ A^{-2}$	V/A
electric conductance	siemens (S)	$kg^{-1}\ m^{-2}\ s^3\ A^2$	A/V
magnetic flux	weber (Wb)	$kg\ m^2\ s^{-2}\ A^{-1}$	V S
magnetic flux density	tesla (T)	$kg\ s^{-2}\ A^{-1}$	Wb/m^2
inductance	henry (H)	$kg\ m^2\ s^{-2}\ A^{-2}$	Wb/A
Celsius temperature	degree (°C)	°C = K	
luminous flux	lumen (lm)	cd sr	cd sr
illuminance	lux (lx)	$cd\ sr\ m^{-2}$	lm/m^2
activity (of a radionuclide)	becquerel (Bq)	s^{-1}	
absorbed dose, kerma	gray (Gy)	$m^2\ s^{-2}$	J/kg
dose equivalent	sievert (Sv)	$m^2\ s^{-2}$	J/kg
catalytic activity	katal (kat)	$mol\ s^{-1}$	

Table 25.5 Other derived units.[5]

Derived quantity	SI-derived unit Name	Symbol	Expression in terms of SI base units
Absorbed dose rate	gray per second	Gy s–1	$m^2\ s^{-3}$
Angular acceleration	radian per second squared	rad/s^2	$m\ m^{-1}\ s^{-2} = s^{-2}$
Angular velocity	radian per second	rad/s	$m\ m^{-1}\ s^{-1} = s^{-1}$
Electric charge density	coulombs per cubic meter	C/m^3	$m^{-3}\ s\ A$
Electric field strength	volt per meter	V/m	$m\ kg\ s^{-3}\ A^{-1}$
Electric field strength	newtons per coulomb	N/C	$m\ kg\ s^{-3}\ A^{-1}$
Electric flux density	coulomb per square meter	C/m^2	$m^{-2}\ s\ A$

Continued

Table 25.5 *Continued*

Derived quantity	SI-derived unit Name	Symbol	Expression in terms of SI base units
Energy density	joule per cubic meter	J/m³	m⁻¹ kg s⁻²
Entropy	joule per kelvin	J/K	m² kg s⁻² K⁻¹
Exposure (x and g rays)	coulomb per kilogram	C/kg	kg⁻¹ s A
Heat capacity	joule per kelvin	J/K	m² kg s⁻² K⁻¹
Heat flux density, irradiance	watt per square meter	W/m²	kg s⁻³
Molar energy	joule per mole	J/mol	m kg s⁻² mol⁻¹
Molar entropy molar heat capacity	joule per mole kelvin	J/(mol K)	m kg s⁻² K⁻¹ mol⁻¹
Moment of force	newton meter	N m	m² kg s⁻²
Permeability (magnetic)	henry per meter	H/m	m kg s⁻² A⁻²
Permittivity	farad per meter	F/m	m⁻³ kg⁻¹ s⁴ A⁻²
Power density	watt per square meter	W/m²	kg s⁻³
Radiance	watt per square meter steradian	W/(m² sr)	kg s⁻³
Radiant intensity	watt per steradian	W / sr	m² kg s⁻³
Specific energy	joule per kilogram	J/kg	m² s⁻²
Specific entropy	joule per kilogram kelvin	J/(kg K)	m² s⁻² K⁻¹
Specific heat capacity, specific entropy	joule per kilogram kelvin	J/(kg K)	m² s⁻² K⁻¹
Surface tension	newton per meter	N/m	kg s⁻²
Surface tension	joule per square meter	J/m²	kg s⁻²
Thermal conductivity	watt per meter kelvin	W/(m K)	m kg s⁻³ K⁻¹
Viscosity, dynamic	pascal second	Pa s	m⁻¹ kg s⁻¹
Viscosity, kinematic	square meter per second	m²/s	m² s⁻¹

SI Prefixes

Table 25.6 lists currently recognized SI unit prefixes. Some common examples of use of these SI prefixes are:

- A ruler that is graduated in 0.5 centimeter units, or 0.5 cm graduations
- A microprocessor operating at 4 gigahertz, or 4 GHz
- A voltmeter that has a resolution to 10 nanovolts direct current, or 10 nV DC
- A power meter that measures in 5 kilowatt units, or 5 kW units

Table 25.6 The SI prefixes for multiples and submultiples of units.[6]

Multiplication factor	Prefix	Symbol
10^{24}	yotta	Y
10^{21}	zetta	Z
10^{18}	exa	E
10^{15}	peta	P
10^{12}	tera	T
10^{9}	giga	G
10^{6}	mega	M
10^{3}	kilo	k
10^{2}	hecto	h
10^{1}	deka	da
10^{-1}	deci	d
10^{-2}	centi	c
10^{-3}	milli	m
10^{-6}	micro	μ
10^{-9}	nano	n
10^{-12}	pico	p
10^{-15}	femto	f
10^{-18}	atto	a
10^{-21}	zepto	z
10^{-24}	yocto	y

Note that there is a major difference between the meaning of an uppercase and a lowercase letter. For example, M is 10^6, but m is 10^{-3}. The correct case must be used or the magnitude of the number will be incorrect.

- A current meter that is sensitive to 2.0 femtoamperes, or 2.0 fA
- A pressure gage that is accurate to 5 kilopascal, or 5 kPa

These SI prefixes may be used singly as prefixes to base or named derived units. In other words, compound prefixes are not permitted. Instead of 1.45 mmm, 1.45 nm should be used, and 4.65 pF should be used instead of 4.65 mmF. See Table 25.7 for a sampling of units that are multiples of SI base or derived units.

UNITS NOT TO BE USED WITHIN THE SI SYSTEM OF UNITS

SI has designated certain units as not to be used. These units, many of which are often found in the literature, are discouraged because equivalent SI units are available and are now preferred. Some of these prior units to be avoided are CGS-based (centimeter-gram-second) special units, such as lambert, emu, esu, gilbert, biot, franklin, and names prefixed with *ab-* or *stat-*. Table 25.8 shows all of the units to be avoided according to SI usage. See NIST Special Publication 811:2008 edition or the most current edition for the most up-to-date list.

Table 25.7 SI-named units that are multiples of SI base or derived units.

Quantity	SI derived unit Name	Symbol	Value in SI units
Angle, plane	degree	°	$\pi/180$ rad
Angle, plane	minute	'	$(1/60)° = \pi/10\,800$ rad
Angle, plane	second	"	$(1/60)' = \pi/648\,000$ rad
Angle, plane	revolution, turn	r	2π rad
Area	hectare	ha	hm^2 = 10^4 m^2
Mass	metric ton	t	Mg = 10^3 kg
Time	minute	min	60 s
Time	hour	h	60 min = 3 600 s
Time	day	d	24 h = 86 400 s
Volume	liter	L	dm^3 = 10^{-3} m^3

Table 25.8 Units not to be used within the SI system of units.[7]

Unit name	Symbol	Value in SI units
	metric carat	
	metric horsepower	735.4988 W
angstrom	A°	10^{-10} m
are	a	100 m^2
atmosphere, standard	atm	101.325 kPa
atmosphere, technical	At	98.0665 kPa
b	barn	10^{-28} m^2
bar	bar	100 kPa
calorie, nutrition candle	Cal	4.184 kJ cd
calorie, physics	cal	4.184 J
candlepower	cp	cd
Cu × unit	xu (Cu Ka1)	1.002 077 03 × 10^{-13} m
dyne	dyn	10^{-5} N
erg	erg	10^{-7} J
fermi	fermi	10^{-15} m
gal	Gal	10^{-2} m/s^2
gamma	γ	nT = 10^{-9} T

Continued

Table 25.8 Continued

Unit name	Symbol	Value in SI units
gauss	G	10^{-4} T
gon, grad, grade	gon	$(\pi/200)$ rad
gravity, standard acceleration due to	g_n (G, g)	9.806 65 m/s^2
kilocalorie	kcal	4.184 kJ
kilogram force	kgf	9.806 65 N
kiloliter	1000 L	m^3
langley	cal/cm^2	4.184×10^4 J/m^3
maxwell	Mx	10^{-8} Wb
mho	mho	S
micron	μ	10^{-6} m
millimeter of mercury	mmHg	133.3 Pa
millimicron	mμ	10^{-9} m
oersted	Oe	$(1000/4p)$ A/m
phot	ph	10^4 lx
poise	P	0.1 Pa s
stilb	sb	10^4 cd/m^2
stere	st	m^3
stokes	St	cm^2/s
torr	Torr	$(101\ 325/760)$ Pa

SI UNITS ARRANGED BY UNIT CATEGORY

SI units, in particular, relate to the physical world we live in. They represent quantities that may be categorized by the general kind of unit, or the branch of science, that they are related to. Table 25.9 shows the current base and derived units arranged by unit category.

Table 25.9 SI units arranged by unit category.

Unit category	Quantity	Symbol	Name
Electricity and magnetism	charge, electric, electrostatic	A·h	ampere-hour
	charge, electric, electrostatic, quantity of electricity	C	coulomb
	current density	A/m^2	ampere per square meter
	electric capacitance	F	farad

Continued

Table 25.9 Continued

Unit category	Quantity	Symbol	Name
Electricity and magnetism	electric charge density	C/m³	coulomb per cubic meter
	electric current	A	ampere
	electric dipole moment	C·m	coulomb meter
	electric field strength	N/C	newton per coulomb
	electric field strength	V/m	volt per meter
	electric flux density	C/m²	coulomb per square meter
	electric inductance	H	henry
	electric potential difference, electromotive force	V	volt
	electric resistance	W	ohm
	electrical conductance	S	siemens
	inductance, electrical	H	henry
	magnetic constant	N/A²	newton per square ampere
	magnetic field strength	A/m	ampere per meter
	magnetic flux	Wb	weber
	magnetic flux density, induction	T	tesla
	magnetic flux density, induction	Wb/m²	weber per square meter
	magnetomotive force	A	ampere
	magnetomotive force	A	ampere-turn
	magnetomotive force	Oe·cm	oersted-centimeter
	permeability (magnetic)	H/m	henry/meter
	permittivity	F/m	F/m
	resistance, electrical	Ω	ohm
	resistance, length	Ω·m	ohm meter
Light	illuminance	lm/m²	lumen per square meter
	illuminance	lx	lux
	irradiance, heat flux density, heat flow rate/area	W/m²	watt per square meter
	luminance	cd/m²	candela per square meter
	luminous flux	lm	lumen
	luminous intensity	cd	candela
	radiance	W/(m²·sr)	watt per square meter steradian
	radiant intensity	W/sr	watt per steradian

Continued

Table 25.9 *Continued*

Unit category	Quantity	Symbol	Name
Mechanics	angular momentum	kg·m²/s	kilogram meter squared per second
	catalytic activity	kat	katal
	coefficient of heat transfer	W/(m² K)	watt per square meter-kelvin
	concentration (of amount of substance)	mol/m³	mole per cubic meter
	density, mass/volume	kg/m³	kilogram per cubic meter
	density, mass/volume	kg/L	kilogram per liter
	density, specific gravity	kg/m³	kilogram per cubic meter
	energy	eV	electron volt
	energy	J	joule
	energy	W s	watt second
	energy density	J/m³	joule per cubic meter
	energy per area	kJ/m²	kilojoule per square meter
	energy per mass	J/kg	joule per kilogram
	energy per mass, specific energy	J/kg	joule per kilogram
	energy per mole	J/mol	joule per mole
	energy, work	N·m	newton-meter
	force	N	newton
	force per length	N/m	newton per meter
	frequency (periodic phenomena)	Hz	cycles per second
	fuel economy, efficiency	L/(100 km)	liter per hundred kilometer
	heat capacity, enthalpy	J/K	joule per kelvin
	heat capacity, entropy	J/K	joule per kelvin
	heat flow rate	W	watt
	insulance, thermal	K m²/W	kelvin square meter per watt
	linear momentum	kg·m/s	kilogram meter per second
	mass	kg	kilogram
	mass	t	ton (metric), tonne
	mass	u	unified atomic mass unit
	mass/area	kg/m²	kilogram per square meter
	mass/energy	kg/J	kilogram per joule
	mass/length	kg/m	kilogram per meter
	mass/time	kg/s	kilogram per second
	mass/volume	kg/m³	kilogram per cubic meter

Continued

Table 25.9 *Continued*

Unit category	Quantity	Symbol	Name
Mechanics	mass per mole	kg/mol	kilogram per mole
	molar entropy, molar heat capacity	J/(mol·K)	joule per mole kelvin
	molar gas constant	R	J mol^{-1} K^{-1}
	molar heat capacity	J/(mol·K)	joule per mole kelvin
	moment of force, torque, bending moment	N·m	newton meter
	moment of inertia	kg·m^2	kilogram meter squared
	moment of section	m^4	meter to the fourth power
	per viscosity, dynamic	1/(Pa·s)	per pascal second
	permeability	m^2	square meter
	power	W	watt
	power density, power/area	W/m^2	watt per square meter
	power density, power/area	W/sr	watt per steradian
	pressure, stress	kg/m^3	kilogram per cubic meter
	pressure, stress	Pa	pascal
	quantity of heat	J	joule
	section modulus	m^3	meter cubed
	specific heat capacity, specific entropy	J/(kg·K)	joule per kilogram kelvin
	specific volume	m^3/kg	cubic meter per kilogram
	surface tension	J/m^2	joule per square meter
	temperature	K	kelvin
	thermal conductance	W/(m^2·K)	watt per square meter-kelvin
	thermal conductivity	W/(m·K)	watt per meter-kelvin
	thermal diffusivity	m^2/s	square meter per second
	thermal insulance	K m^2/W	kelvin square meter per watt
	thermal resistance	K/W	kelvin per watt
	thermal resistivity	K m/W	kelvin meter per watt
	thrust/mass	N/kg	newton per kilogram
	viscosity, dynamic	kg/m/s	kilogram per meter-second
	viscosity, dynamic	N·s/m^2	newton-second per meter squared
	viscosity, dynamic	Pa·s	pascal second
	viscosity, kinematic	m^2/s	meter squared per second
	volume/energy	m^3/J	cubic meter per joule

Continued

Table 25.9 *Continued*

Unit category	Quantity	Symbol	Name
Other	count	n/a	n/a
Radiology	absorbed dose	Gy	gray
	absorbed dose	J/kg	joule per kilogram
	absorbed dose rate	Gy/s	gray per second
	activity	Bq	becquerel
	activity	s^{-1}	per second (disintegration)
	dose equivalent	Sv	sievert
	exposure (x and gamma rays)	C/kg	coulomb per kilogram
Space and time	acceleration, angular	rad/s^2	radian per second squared
	acceleration, linear	m/s^2	meter per second squared
	angle, plane	°	degree
	angle, plane	'	minute (arc)
	angle, plane	rad	radian
	angle, plane	r	revolution
	angle, plane	"	second
	angle, solid	sr	steradian
	area, plane	ha	hectare
	area, plane	hm^2	square hectometer
	area, plane	m^2	square meter
	length	m	meter
	time	d	day (24h)
	time	h	hour
	time	min	minute
	time	s	second
	velocity/speed	m/s	meter per second
	velocity, angular	rad/s	radian per second
	velocity, angular	r/s	revolution per second
	volume/time, (flow rate)	m^3/s	cubic meter per second
	volume/time, (flow rate)	L/s	liter per second
	volume/time, (leakage)	slpm	standard liter per minute
	volume, capacity	m^3	cubic meter
	volume, capacity	L	liter
	wavenumber	m^{-1}	per meter

CONVERSION FACTORS AND THEIR USES

Equivalent quantities that are stated in different units are related to each other by a conversion factor that is the ratio of the differing units. Because of the need to state quantities in the units required by a user or specifying organization, the conversion of a quantity from one unit to another occurs often.

A conversion factor is based on the ratio of the units between which a conversion is needed. For instance, the often-required conversion from inches to millimeters involves multiplying a number in inches by the conversion factor 25.4 to obtain the equivalent number in millimeters.

Appendix D lists a number of frequently required unit conversions between various customary units, SI units, and also between many customary and SI units. Also see NIST Special Publication 1038:2006 and NIST Special Publication 811:2008 edition or the most current edition for the most up-to-date list. The conversion factors given are accurate to the number of decimal places shown or, where bold-faced, are exact conversion factors with zero.

The format for conversion factor listings most often follows one of two formats: the from/to list and from/to matrix table. In Appendix D, the from/to list is given for a wide range of commonly required unit conversions.

From/To Lists

The following example shows how from/to lists are arranged:

To convert from	To	Multiply by:
inch	millimeter	2.54 E+01
circular mil	square meter (m^2)	5.067 075 E–10

From/To Matrix Table

In these tables, a matrix of *from* and *to* values are listed in column and row headings, with the conversion factors for each combination given at the intersection of the column and row. Table 25.10 is an example of how these tables are arranged.

UNCERTAINTIES FOR SOME FUNDAMENTAL UNITS (STANDARD AND RELATIVE)

Table 25.11 is compiled from the CODATA Internationally recommended 2018 values of the Fundamental Physical Constants. This list can be accessed via NIST at https://physics.nist.gov/cuu/Constants/index.html, where one can search any category of fundamental physical constants.

Table 25.10 Conversion matrix: plane angle units.

From

Symbol	Name	deg	grad grade gon	mil	min	quad	rad	rev	sec
deg	degree	**1**	1.1111	17.7778	60	0.0111	0.0175	0.0028	3600
grad, grade, gon	grad, grade, gon	0.9000	**1**	16	0.015	0.01	0.0157	0.0025	0.0003
mil	mil	0.0563	0.0625	**1**	0.0009	0.0006	0.001	0.0002	2E–05
min	minute	0.0167	66.667	1066.67	**1**	0.0002	0.0003	5E–05	60
quad	quadrant	90	100	1600	5400	**1**	1.5708	0.25	324000
rad	radian	57.2958	63.662	1018.59	3437.7	0.6366	**1**	0.1592	206265
rev	revolution	360	400	6400	21600	4	6.2832	**1**	1E+06
sec	second	0.0003	4000	64000	0.0167	3E–06	5E–06	8E–07	**1**

NUMBER BASES

A *number base* is the exponent base that defines the magnitude of each digit in a number, where the position of each digit determines its value by the relationship:

$$\text{Digit value} = \text{Number} \times \text{Base}^N$$

where N is the number of positions to the left or right of the decimal marker, with those on the left being positive N and those on the right being negative N.
For example, for the decimal (base 10) number 453.0, the 5 has a value of 50 because:

$$5 \times 10^2 = 50.$$

Without thinking, we interpret the value of numbers by using this almost unconscious rule for decimal numbers. Some, to clarify the base that a number is related to, will list that base as a trailing subscript, as in 453.0_{10}. 453.0 is interpreted as four hundred, fifty, and three units.

Other number bases that are used quite often in computer programming are *binary, octal,* and *hexadecimal.* Whereas in decimal numbers each digit can take on values from 0 to 9 (10 values), binary digits can take on only 0 and 1 (two values), octal 0 to 7 (eight values), and hexadecimal 0 to F (zero to nine plus A, B, C, D, E, F, or 16 values).

All SI units and the vast majority of customary unit measurements are made in decimal (base 10) units and, therefore, have a base of 10. Digital computers and calculators use binary numbers internally, however, and many computer programming languages use hexadecimal numbers.

Table 25.11 Frequently used constants.[8]

NIST Category	Name	Symbol	Value	Units	Standard uncertainty	Relative standard uncertainty	Equivalent Equation
PC	Avogadro constant	"NA, L"	6.022 140 76 E+23	mol^{-1}	Exact	Exact	
PC	Boltzmann constant	k	1.380 649 E-23	$J K^{-1}$	Exact	Exact	$k = R/NA$
EM	conductance quantum	G_0	7.748 091 729 E-05	S	Exact	Exact	$G_0 = 2e^2/h$
AN	electron mass	me	9.109 383 7015 E-31	kg	28.0 E-39	30.0 E-9	
NS	electron volt	eV	1.602 176 634 E-19	J	Exact	Exact	$1 eV = (e/C) J$
EM	elementary charge	e	1.602 176 634 E-19	C	Exact	Exact	
PC	Faraday constant	F	96 485.332 12	$C\, mol^{-1}$	Exact	Exact	
AN	fine-structure constant	a	7.297 352 569 3 E-03		1.1 E-12	15.0 E-9	$a = e^2/4pe0hc$
EM	magnetic flux quantum	F0	2.067 833 848 E-15	Wb	Exact	Exact	$F0 = h/2e$
PC	molar gas constant	R	8.314 462 618	$J\, mol^{-1}\, K^{-1}$	Exact	Exact	
UN	Newtonian constant of gravitation	G	6.674 30 E-11	$m^3\, kg^{-1}\, s^{-2}$	1.5 E-15	120 E-6	
UN	Planck constant	h	6.626 070 15 E-34	J s	Exact	Exact	
UN	Planck constant over 2 pi	h	1.054 571 817 E-34	J s	Exact	Exact	
AN	proton mass	m_p	1.672 621 923 69 E-27	kg	51.0 E-36	44.0 E-9	
AN	proton-electron mass ratio	m_p / m_e	1836.152 673 43		11.0 E-6	410 E-12	
AN	Rydberg constant	Rx	10 973 731.568 160	m^{-1}	20.0 E-6	5.0 E-12	$Rx = a^2 m_e c/2h$
NS	speed of light in vacuum	c, c0	299 792 458	$m\, s^{-1}$	Exact	Exact	
PC	Stefan-Boltzmann constant	s	5.670 374 419 E-08	$W\, m^{-2}\, K^{-4}$	Exact	Exact	$s = (p^2/60) k^4 / h^3 c^2$
NS	unified atomic mass unit	1 u	1.660 539 066 60 E-27	kg	0.50 E-36	30.0 E-9	$1 u = m_u = (1/12) m(12C) = (10^{-3} kg\, Mol^{-1})$
UN	Vacuum electric permittivity	e0	8.854 187 8128 E-12	$F\, m^{-1}$	1.3 E-21	15 E-09	

CONVERSION FACTORS FOR CONVERTING BETWEEN EQUIVALENT MEASUREMENT UNITS

Unit conversion factors, otherwise known as *conversion factors*, are made widely available in various forms and extents in texts, publications, and standards, and are often supplied as appendices in catalogs of manufactured products, as well as being provided in specific measurement instrumentation catalogs.

Conversion factors typically follow the format:

To convert from (first) given unit to (second) desired unit, multiply (first unit) measurement by conversion factor given.

To convert from (second) given unit to (first) desired unit, divide the relationship of the second unit by the first unit.

Appendix D contains conversion factors for many common sets of units. Those listed are also found in NIST Special Publication 1038:2006, *The International System of Units (SI)—Conversion Factors for General Use*, and in NIST Special Publication 811:2008, *Guide for the Use of the International System of Units (SI)*; there are more than 500 factors in all. Two lists are provided. The first is conversion factors sorted by category and then by alphabetical unit. This table is useful if you know the category of physical unit and are looking for options within that category. The other list, where conversion factors are sorted by alphabetical unit only, is useful if you know the units you want to convert from and to, irrespective of the physical unit's category. An example of the second table is as follows:

To convert	From		To		Multiply by
	Unit A		Unit B		Factor = Unit B/ Unit A
Category	Symbol	Name	Symbol	Name	
electric current	abA	abampere	A	ampere	10
electric capacitance	abF	abfarad	F	farad	1000 000 000
electric inductance	abH	abhenry	H	henry	1000 000 000

As an example, using this table to convert a 25-abampere value to amperes, multiply by the conversion factor shown (10), giving 250 amperes. These tabled values are computed from fundamental units and are accurate to the number of digits shown in the table. Generally, these factors are accurate to at least 10 decimal places, with many to 15 decimal places.

Another useful website for unit conversions is the "Unit Conversion" section of NIST's Physical Measurement Laboratory/Weights and Measures—Metric Program: https://www.nist.gov/pml/weights-and-measures/metric-si/unit-conversion.

Chapter 26

Ratios

DECIBEL MEASURES

All measurements given in decibels are statements of a ratio between either a measurement (a) and a reference value (b) or between two measurements (a / b).

Decibels are measures stated using logarithms. A few simple rules of logarithms should be reviewed first to enable better understanding of calculations involving decibels.

A *logarithm* is an exponential function of an input variable (the number we supply) and an exponential base. The logarithm of a variable x is defined as the following function:

$$f(x) = \log c\,(x), \text{ or conversely, } x = cf(x)$$

Said another way, let $y = f(x)$. Then

$$y = \log c\,(x), \text{ or conversely, } x = cy$$

Here, it is required that $x > 0$, $c > 0$ and not $= 1$.

One of the nice things about using logarithms is that they make multiplication and division transform into simple addition and subtraction. Anyone who has used a slide rule knows this, as the scales on a slide rule are proportioned logarithmically. So,

$\log c\,(f) + \log_c\,(g) = \log_c\,(fg)$, or conversely, for $f = c^{y1}$ and $g = c^{y2}$,

we get $fg = c^{y1+y2}$

and

$\log_c\,(f) - \log_c\,(g) = \log_c\,(f/g)$, or conversely, for $f = c^{y1}$ and $g = c^{y2}$,

we get $f / g = c^{y1-y2}$.

Similarly, we get

$\log_c\,(f^n) = n \log_c\,(f)$, or conversely, for $f^n = (c^y)^n$, we get $f^n = c^{n\,(y)} = c^{ny}$.

One thing we note is that the argument for the log() function cannot be negative or zero.

The bel is named after Alexander Graham Bell, hence the capital B used as the symbol. To express a value in bels, we use the logarithm function and the base of $c = 10$, giving:

$$B = \log_{10}(x), \text{ or conversely, } x = 10B.$$

The general form of the equation for determining a value in decibels is

$$dB = 10 \log_{10}(a/b), \text{ or conversely, } a/b = 10^{dB/10}.$$

Where (often) the ratio being examined is a squared ratio (as in inputs to power calculations), this equation takes a slightly different form:

$$dB = 10 \log_{10}(a^2/b^2) = 20 \log_{10}(a/b), \text{ or conversely, } a/b = 10^{dB/20}.$$

With these two expressions we have the basis for calculating all decibel results. Where one of the input values (b) is a reference value, we merely supply that in the denominator.

Logarithms are useful because they are a shorthand for obtaining products and quotients of numbers. The following are useful decibel relationships:

An increase/decrease* in dB	Is equivalent to an increase/reduction by a factor (ratio) of
3	2
6	4
10	10
2	100
30	1000
n × 10	10^n

* Increasing dBs are positive, decreasing dBs are negative.

From this we immediately conclude that:

- A 56 dB measurement describes a value of a parameter that is approximately twice the magnitude of a 53 dB measurement.

- A 46 dB measurement describes a value of a parameter that is approximately one-tenth the magnitude of a 56 dB measurement.

- A 46 dB measurement describes a value of a parameter that is approximately 1000 times the magnitude of a 16 dB measurement.

An Example of Decibel Use

A sound source radiates power, and this results in sound pressure. The power in a sound wave is associated with the square of the peak pressure of the sound wave and is expressed in the SI unit of watts (W). Sound pressure level measurements are generally based on a reference sound pressure level of 20 micro-pascals, 20 mPa, 0.02 mPa, or 2.0^{-5} Pa and expressed in the SI unit of Pascal (Pa). Sound pressure is also commonly expressed in decibels (dBs).

Note: Alternatively, 1 microbar has also gained wide acceptance for calibration of certain transducers and sound measurements in liquids. Also, unless otherwise specifically given, sound pressure is taken to be the effective (RMS) pressure.

Sound power levels, distinct from sound pressure levels, are expressed in relation to a reference power level of one picowatt (1.0×10^{-12} W) exactly.

Unless otherwise explicitly stated, it is to be understood that the sound pressure value used is the effective (root mean square) of the measured sound pressure. Also, in many specific sound fields the sound pressure ratio is not the square root of the corresponding power ratio.

If a sound has a pressure level of $p_2 = 1.00$ Pa, to express this sound pressure level in decibels the result is obtained by use of the following formula:

$$\text{Sound pressure level, dBspl} = 20 \log (p_2 / p_1)$$

or, sound pressure level, dBspl = 20 log (1.0 / 0.00002) = 94.0 dB

Various Decibel Scales in Use

Decibel scales have been specialized for use in the fields of acoustics, electricity, and electromagnetics, differing from each other by chosen reference values and units employed.

dB (Acoustics, Sound Power Level)

Here, the reference level is 1 picowatt (1.0 pW). The corresponding level of sound intensity, 1.0×10^{-12} W / m², corresponds to a sound power level of 0 dB, which is the lower limit threshold of normal hearing.

dBA and dBC Scales (Acoustics)

The most widely used sound level filter is the A scale, which roughly corresponds to the inverse of the 40 dB (at 1 kHz) equal loudness curve. Measurements are expressed in dBA, and sound meter readings on this scale are less sensitive to very high and low frequencies.

The C scale is nearly linear from 80 to 2.5 kHz, becoming less sensitive below and above this range. Another scale, B, is rarely used, and is midway between the A and C scales.

dB (General, Electrical Signals)

When impedances are equal:

$$\text{dB} = 10 \log (P_2 / P_1) = 20 \log (E_2 / E_1) = 20 \log (I_2 / I_1)$$

When impedances are unequal:

$$\text{dB} = 10 \log (P_2 / P_1) = 20 \log [(E_2 \sqrt{Z_1}) / (E_1 \sqrt{Z_2})]$$

$$= 20 \log [(I_2 \sqrt{Z_2}) / (I_1 \sqrt{Z_1})]$$

Here, *P* refers to power in watts, rms; *E* refers to voltage, RMS; *I* refers to current, RMS; and *Z* refers to impedance in the general form, including inductance and capacitance.

dB/bit (Electrical Signals)

$$dB/bit = 20 \log(2)/bit = \text{approximately } 6.02 \text{ dB/bit}$$

This is commonly used for specifying the dynamic range or resolution for pulse coded modulation (PCM) systems.

dB/Hz (Electrical Signals)

This refers to dB measurements of relative noise power in a 1 Hz bandwidth and is used in determining a laser's *relative intensity noise* (RIN).

dBi (Electrical Signals)

This refers to dB isotropic, with reference to defining antenna gain.

dBm Scale (Electric Signals, also dBmW)

$$dBm = 10 \log (W_2 / 1.0 \text{ mW}_{RMS})$$

Here, 1 milliwatt (1 mW) across a specified impedance is the reference level. For example, a 0 dBm signal in a circuit with an impedance of 600 Ω corresponds to approximately 0.7746 V RMS. In most cases, the specified reference impedance is assumed from the nature of the circuit:

- Audio and communications circuits and inspection, measurement, and test equipment (IM&TE) typically use 600 Ω. In audio circuits, this is also the same as indications on a VU (volume unit) meter.
- RF circuits and IM&TE typically use 50 Ω.
- Cable television systems (and some other systems) and IM&TE typically use 75 Ω.

If the impedance is different from one of these customary values, it must be explicitly stated.

dBm Scale (Electric Signals)

$$dBm = 10 \log (W_2 / 1.0 \text{ mW}_{RMS})$$

Here, 1 microwatt (1 mW) is the reference level.

dBr (Electrical Signals)

$$dBr = 20 \log (V_2 / V_1)$$

Here, the reference level is specified in the immediate context of the value.

dBu (Preferred) or dBv (Electrical Signals)

$$dBu = 20 \log (V_2 / 0.775 \, V_{RMS})$$

Here, the reference level is defined as 0.775 volt, RMS (0.775 V, RMS), across any impedance. Compare to dBm.

dBuV (Electrical Signals)

$$dBuV = 20 \log (V_2 / mV_{RMS})$$

This is commonly used for specifying RF levels to a communications receiver. Here, the reference level is defined as 1.0 microvolt, RMS (1.0 mV, RMS), across any impedance. Compare to dBm.

dBV (Electrical Signals)

$$dBV = 20 \log (V_2 / V_1)$$

Here, the reference level is 1 volt, RMS (1.0 V, RMS), across any impedance. The following is a useful table of dBm equivalents:

$$dBV = dBm - 13.0$$
$$dBmV = dBm + 47.0$$
$$dBuV = dBm + 107.0$$
$$V = 10^{(dBm - 13.0) / 20}$$
$$mV = 10^{(dBm + 47.0) / 20}$$
$$uV = 10^{(dBm + 107.0) / 20}$$

dBW Scale (Electric Signals)

$$dBW = 10 \log (W_2 / 1.0 \, W_1)$$

Here, the reference level is 1 watt (1.0 W). Usually, the impedance is 50 W.

DBW/KHz Scale (Electric Signals)

$$DBW/KHz = 10 \log (k) / KHz = 10 \log (1.38065^{-23})$$
$$= -228.5991631 \, DBW/KHz$$

Here, the reference level is defined in relation to Boltzmann's constant. This is commonly used when analyzing carrier-to-noise (C/N) ratios in communication links.

LOGARITHMS IN MICROSOFT EXCEL

Microsoft Excel™ makes three logarithmic functions available: LOG(number, base), LOG10(number), and LN(number).

In the LOG(number, base) function, "number" is the value you provide, and "base" is the base of the logarithm. If the base is omitted, then Excel™ assumes it to be 10.

In the LOG10(number) function, the base is 10. Actually, if you omit the base argument in LOG()—that is, provide only LOG(number)—the calculation assumes that the base is 10 and computes identically to LOG10(number).

In the LN(number) function, the base is the Naperian constant, e, an irrational number that to nine decimal places is 2.718 281 83. We can obtain this constant within Excel™, where needed, by using the Microsoft Excel™ formula "= EXP(1)." We will not consider the uses of this function.

TYPES OF LINEARIZING TRANSFORMATIONS

In general, linearizing transformations convert one or more data variables in a way that the plot of a variable against another results in a graphical straight-line relationship between pairs of the data sets. Exponential transforms raise or lower one or more data variables by an exponential power or fractional power such that the transformed data sets plot as a straight line when graphed against one another. In log transformations one or more data variables are transformed by taking the logarithm of the data variable, and when the log of the data is plotted against other variables, the result is a linear (straight-line) relationship.

GRAPHS

Graphs are a means of visualizing the shape of data sets when compared by plotting one data variable against another. Examples of kinds of plots are:

Scatter plots. These graphs plot one or more variables against a common reference variable. Examples of these are $X - Y$ charts.

Histograms. These graphs show the aggregation of data in bins, which indicate the distribution of data items over a range of a given variable.

Bar (or column) graphs. These graphs show the relative magnitude of categories of variables, compared side by side. Bar graphs have the magnitudes compared by lengths of horizontal bars, one below the other. Column graphs, similar to bar graphs, show the same side-by-side comparison of categories, but the magnitudes are in a vertical direction.

Pie (or doughnut) charts. These graphs compare relative magnitudes of categories by their portion of a circle (360° being equal to the sum, 100 percent, of the magnitudes).

Radar (or spider) graphs. These graphs show comparative magnitudes of categories of data by the radial distance from a center point, arranged around a circle, with the adjacent endpoints connected.

Bubble graphs. These graphs compare three separate data variables: two variable ones and one categorical one. Categories of data are circles, whose location in an $X - Y$ plane and the size of the circle depicts the three magnitudes.

3-D (or surface) graphs. These graphs plot three continuous variables as a surface in 2-D space.

Line (or run) charts. These graphs show the running relationship of one variable versus another by data item number or sequence (one form of which is by time).

Microsoft Excel™ has a rich variety of graphing wizards and tools for achieving a wide range of data graphing results. All of the graph types mentioned here can be easily constructed from tabular data in a few easy steps in Excel™.

Chapter 27
Statistics

FUNDAMENTAL MEASUREMENT ASSUMPTIONS

The four basic assumptions on which the validity of all measurements depends are *random distribution, fixed model, fixed variation,* and *fixed distribution*. The majority of statistical tests depend very heavily on the assumption that the data were collected in a random manner. This is an important assumption because of the significant probability during measurement of the occurrence of aggregations, changes in condition, assignable causes, trends, cyclical effects, and progressive learning, and drift in natural, industrial, and human processes. The random distribution assumption is particularly important when calculating statistics that are the basis of an inference or comparison; for example, taking a sample that will be the basis for making an inference as to the population it is drawn from. Another example of the random distribution assumption is in experiments, where the results of one treatment are compared to those of another factor setting.

A commonly used definition for normality of data distributions is IID (independently and identically distributed), which is defined in NIST SP 800-90B as "[a] quality of a sequence of random variables for which each element of the sequence has the same probability distribution as the other values, and all values are mutually independent."[1] This assumption is often made regarding data that have been randomly selected for use in statistical calculations. This is in a Normal distribution with mean zero (0) and variance (s^2).

DEGREES OF FREEDOM

Degrees of freedom (often represented by n or its Greek letter equivalent ν, nu), in relation to the count of data items on which a statistic is based, refers to the number of ways a statistic's value can vary with a variation in each of the source data items. It is one of the first items determined in the calculation of many statistics. When the number of data items is very large or includes the total population of items on which an inference is being made, the symbol *N* is often used. Conversely, when the degrees of freedom (DoF) relate to a statistic based on a sample from a larger population that is being described, the symbols *n* or ν are most often used. This is to help identify the meaning to the user.

Because determining DoF involves summation of the number of data items being evaluated (minus a constant in some cases), the resulting DoF, being a count, is dimensionless (not associated with a unit). The determination of DoF depends on the net ways that a statistic may vary with variation in the data elements going into it. For instance, in the calculation of the mean statistic based on 20 values, the resulting statistic can vary 20 different ways with the change of any of the source data values. Hence, the DoF for mean is merely the count, N, of the data items.

In contrast, as mentioned, some statistics have a DoF that is the count of data items used in the statistic calculation minus a constant. For instance, in the calculation of the sample standard deviation, part of the calculation involves first calculating the mean. Because the mean statistic is part of the sample standard deviation calculation, one degree of freedom is lost due to the use of the mean, and the resulting DoF is the data count minus one.

This is not a book on statistical theory, so the formulas for various statistics will be given and their DoF basis merely listed without discussion as to the mathematical basis for them.

$$df = N - 1$$

RESIDUALS

A *residual* is the difference between the value of a data item and a statistic describing it. For instance, if the average of a set of measurements is x-bar = 4.583, and the first two values are $x_1 = 4.282$ and $x_2 = 4.632$, the residuals of these values is the difference between the measurement value and the average, or $r_1 = (4.282-4.583) = -0.301$ and $r_2 = (4.632-4.583) = 0.049$.

Because the operation involved in calculating residuals is subtraction, the resulting statistic has the same units as the data from which it is calculated. For instance, if measurements of voltage are used to calculate the mean and residuals for some or all of the data, the units for the residuals are volts.

CENTRAL TENDENCY

This is obviously the case because quite often the objective is to understand where the central value is. In mathematical terms, such statistics are said to determine central tendency. The primary statistics for describing central tendency are the mean, median, and mode.

Mean

One of the most common and often used or referred-to statistics is the *mean* or *average*. Because the operations involved in calculating the mean are addition and division, the resulting statistic has the same units as the data from which it is calculated. The mean or average is given by the following equation:

$$\bar{x} = \frac{\sum_{i=1}^{n} x_i}{n}$$

For example, in the data series of amperage measurements:

13, 14, 23, 23, 32, 33, 45, 99, and 105 A,

the mean (or average) value is 44.375 A.

Use Excel's AVERAGE() function. For example, for the data named range DATA, use the Excel function expression "=AVERAGE(DATA)".

Median

The *median* statistic is the middle value in a set of data items arranged in order of increasing value. That is, the median is the value where half the data items are less in value and half are greater in value. Increasingly, we are hearing this statistic being used for economic and demographic purposes, as in median income, median house value, median age.

To determine the median of a set of measurements, arrange the values in order of increasing value. If the number of values in the data set is odd, the median is the middle value in this ordered set. If the number of values in the data set is even, the median is the average of the middle two values.

For example, in the data series of amperage measurements:

13, 14, 23, 23, 32, 33, 45, 99, and 105 A,

the median (or middle) value is 32.

Use Excel's MEDIAN() function. For example, for the data named range DATA, use the Excel function expression "=MEDIAN(DATA)".

Mode

The *mode* statistic describes the most often (in statistics, described as most frequently) occurring value.

For example, in the data series:

13, 14, 23, 23, 32, 33, 45, 99, and 105 A,

the mode (or most often occurring) value is 23 A.

Use Excel's MODE() function. For example, for the data named range DATA, use the Excel function expression "=MODE(DATA)".

BIMODAL DISTRIBUTIONS

Some distributions either have more than one mode or have frequencies of occurrence of values where the histograms indicate more than one peak.

Such distributions are not normal distributions, that is, distributions that likely fit a normal distribution model, and in these cases statistics based on the assumption of a normal distribution may not be valid. Figure 27.1 is a sample graph of a bimodal distribution.

CENTRAL LIMIT THEOREM

The central limit theorem states that, for large enough sample size n, the distribution of the sample mean $\bar{x}$ will approach a normal distribution (see Figure 27.2). This is true for a sample of independent random variables from *any* population distribution, as long as the population has a finite standard deviation $\sigma.^2$

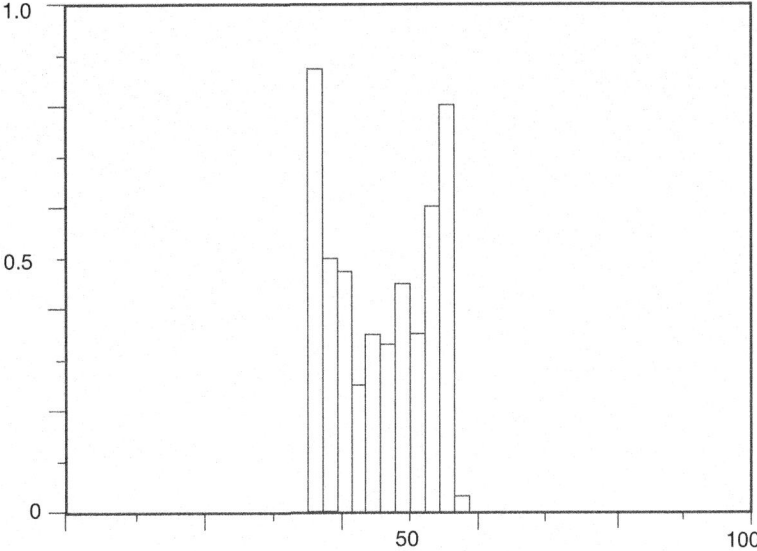

Figure 27.1 Bimodal distribution.

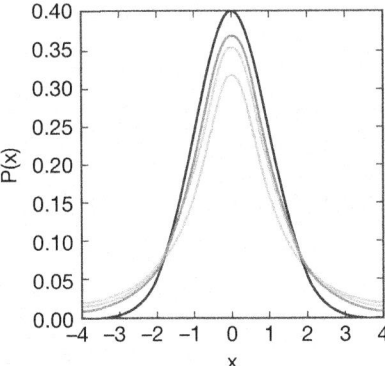

Figure 27.2 Normal distributions.

The sample mean is defined to be

$$\bar{x} = 1/n(x_1 + x_2 + \cdots + x_n)$$

The central limit theorem consists of three statements:[3]

1. The mean of the sampling distribution of means is equal to the mean of the population from which the samples were drawn.

2. The variance of the sampling distribution of means is equal to the variance of the population from which the samples were drawn divided by the size of the samples.

3. If the original population is distributed normally (in other words, bell shaped), the sampling distribution of means will also be normal. If the original population is not normally distributed, the sampling distribution of means will increasingly approximate a normal distribution as sample size increases (in other words, when increasingly large samples are drawn).

CONFIDENCE INTERVAL

A *confidence interval* gives an estimated range of values, which is likely to include an unknown population parameter, the estimated range being calculated from a given set of sample data.

Confidence Intervals for Unknown Mean and Known Standard Deviation[4]

For a population with unknown mean μ and known standard deviation σ, a confidence interval for the population mean, based on a simple random sample (SRS) of size n, is

$$\bar{x} \pm z^* \, \sigma/\sqrt{n}$$

where z^* is the upper (1-C)/2 critical value for the standard normal distribution (see Figure 27.3).

Note: This interval is only exact when the population distribution is normal. For large samples from other population distributions, the interval is approximately correct by the central limit theorem.[5]

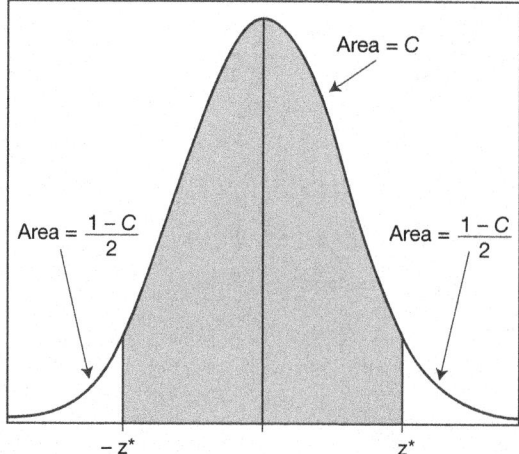

Figure 27.3 Confidence intervals.

ROOT MEAN SQUARE

The *root mean square* (RMS) statistic is most often associated with continuous variables such as time. There are occasions where RMS is associated with discrete variables as well, such as specific measurements of a variable, for example, weight. As the name implies, RMS involves squaring the values of data, finding the mean (average) of these values, followed by taking a (square) root of this mean. As such, the unit of RMS is the same as the measurement unit. For instance, if the RMS value of a set of data is taken in "meter," the RMS result also has the unit meter.

Root mean square (RMS) is given by the following equation:

$$x_{rms} = \sqrt{\frac{\sum_{i=1}^{n} Y^2}{n}}$$

For instance, in the data series of amperage measurements:

13, 14, 23, 23, 32, 33, 45, 99, and 105 A,

the RMS value is 54.15 A, rounded to two decimal places.

SUM OF SQUARES

Sum of squares (SS) is often used as one step in ANOVA calculations. The unit of SS is the measurement unit squared. The sum of squares is given by the following equation:

$$SS_{Total} = \sum(x_i - \bar{x})^2$$

SS_{Total} = Sum Squared Total Error

Σ = sum of data points

x_i = each data point

$\bar{x}$ = mean value

For instance, in the data series of amperage measurements:

13, 14, 23, 23, 32, 33, 45, 99, and 105 A,

the SS value is 26,387 A.

ROOT SUM OF SQUARES

The unit of *root sum of squares* (RSS) is the same as the measurement unit. The RSS is given by the following equation(s):

$$RSS = \sqrt{X_1^2 + X_2^2 + X_3^2 + \ldots + X_n^2}$$

For instance, in the data series of amperage measurements

13, 14, 23, 23, 32, 33, 45, 99, and 105 A

the RSS value is 162.44 A, rounded to two decimal places.

POPULATION VARIANCE

Unlike standard deviation, variances may be combined by addition. The **population variance**, V, is given by the following equation:

$$V = \sigma^2 = \frac{\sum_{i=1}^{n}(x_i - \mu)^2}{N}$$

σ^2 = population variance

x_i = value of the ith element

μ = population mean

N = population size

For example, in the data series of amperage measurements:

13, 14, 23, 23, 32, 33, 45, 99, and POP105 A,

the variance, V, value is 1201.23 A, rounded to two decimal places.

Use Microsoft Excel™ VARP() function. For example, for the data range named DATA, use the Excel™ function expression "=VARP(DATA)".

POPULATION STANDARD DEVIATION

The **population standard deviation** is used in the determination of several other statistics, for example, in determining a confidence interval and for hypothesis testing.

The population standard deviation, s, is given by the following equation:

$$\sigma = \sqrt{\frac{\sum_{i=1}^{n}(x_i - \mu)^2}{N}} = \sqrt{\frac{\sum_{i=1}^{n}x_i^2}{N} - \mu}$$

For example, in the data series of amperage measurements:

13, 14, 23, 23, 32, 33, 45, 99, and 105 A,

the population standard deviation, s, value is 34.66 A, rounded to two decimal places.

Use Excel's STDEVP() function. For example, for the data range named DATA, use the Excel function expression "=STDEVP(DATA)".

SAMPLE VARIANCE

As noted, variances may be combined by addition. The **sample variance,** v, is given by the following equation:

$$v = s^2 = \frac{\sum_{i=1}^{n}(X_i - \bar{X})^2}{n-1}$$

s^2 = sample variance

X_i = value of the ith element

$\bar{X}$ = sample mean

n = number of observations

For example, in the data series of amperage measurements

13, 14, 23, 23, 32, 33, 45, 99, and 105 A

the sample variance, v, value is 1372.84 A, rounded to two decimal places.
 Use Excel's VAR() function. For example, for the data range named DATA, use the Excel function expression "=VAR(DATA)".

SAMPLE STANDARD DEVIATION

The **sample standard deviation** is used in the determination of several other statistics; for example, in determining a confidence interval and for hypothesis testing. The **sample standard deviation**, *s*, is given by the following equations:

$$s = \sqrt{\frac{\sum_{i=1}^{n}(X_i - \bar{X})^2}{n-1}} \quad s = \sqrt{\frac{\sum_{i=1}^{n}\left(X_i - \frac{\sum X_i}{n}\right)^2}{n-1}} \quad s = \sqrt{\frac{n\sum_{i=1}^{n}X_i^2 - \left(\sum_{i=1}^{n}X_i\right)^2}{n(n-1)}}$$

For example, in the data series of amperage measurements:

13, 14, 23, 23, 32, 33, 45, 99, and 105 A,

the sample standard deviation, *s*, value is 37.05 A, rounded to two decimal places
 Use Excel's STDEV() function. For example, for the data range named DATA, use the Excel function expression "=STDEV(DATA)".

STANDARD ERROR OF THE MEAN

Standard error of the mean (SEM) does not assume a normal distribution. Many applications of SEM do assume a normal distribution. For SEM, the larger the sample size, the smaller the standard error of the mean. In other words, the size of SEM is inversely proportional to the square root of the sample size.
 SEM, or *s-y-bar*, is given by the following formula:

$$s_{\bar{y}} = \frac{s}{\sqrt{n}}$$

where s = sample standard deviation and n = number of items in the sample. For example, in the data series of amperage measurements,

13, 14, 23, 23, 32, 33, 45, 99, and 105 A

the sample standard deviation, *s*, was 37.05 and the number of items, *n*, was 8, making the SEM 13.10, rounded to two decimal places.

MEASURES OF SKEWNESS AND KURTOSIS[6]

A fundamental task in many statistical analyses is to characterize the location and variability of a data set. A further characterization of the data includes skewness and kurtosis.

SKEWNESS

Skewness is a measure of symmetry (or lack of symmetry). Data sets generally are not completely normally distributed. There may be more low values than high values, resulting in histograms of the data that are higher on one side or the other of the mean (see Figure 27.4).

There are data sets where more values are below the mean value, and the graphed shape of the distribution leans toward higher, more positive values. Such distributions have a *positive skewness* (the value of the distribution's skew is positive, or greater than zero).

There are other data sets where more values are above (at a more positive value than) the mean, and the distribution appears to lean toward the more negative values. Such distributions have a *negative skewness* (the value of the distribution's skew is negative, or less than zero).

Distribution skewness is determined by the formula shown in Figure 27.4.[7] For example, in the data series of amperage measurements:

13, 14, 23, 23, 32, 33, 45, 99, and 105 A,

the skewness value is 1.15, dimensionless, rounded to two decimal places.

Use Excel's SKEW() function. For example, for the data range named DATA, use the Excel function expression "=SKEW(xx:yy)" where xx:yy is the cell location of your data (for example, B3:B18).

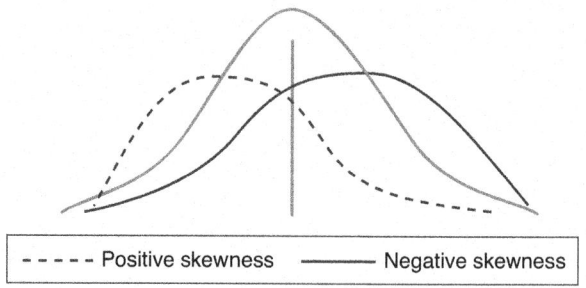

Figure 27.4 Example of positive and negative skewness.

KURTOSIS

Kurtosis is a measure of whether the data are heavy-tailed or light-tailed relative to a normal distribution. That is, data sets with high kurtosis tend to have heavy tails, or outliers. Data sets with low kurtosis tend to have light tails, or lack of

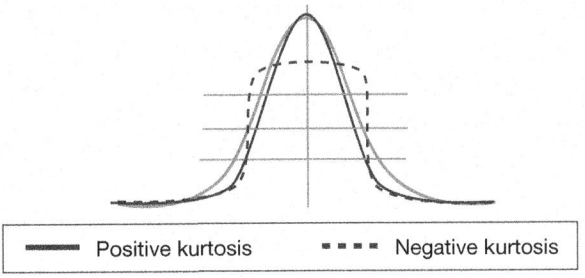

Figure 27.5 Example of positive and negative kurtosis.

outliers. A uniform distribution would be the extreme case.[8] Other data sets, not completely normally distributed, may have more values toward the tails than around the mean, and so on, resulting in histograms of the data that are more rectangular-shaped than bell curve (normal distribution)–shaped.

In the data sets where more values are toward the tails, and the distribution appears to be more rectangular or flat, the distribution is said to have negative kurtosis. Alternately, if more data are located closer to the mean, the distribution appears to be more triangular or peaked, and the distribution is said to have positive kurtosis (see Figure 27.5).

Distribution kurtosis is determined by the following formula:

$$\text{kurtosis} = \frac{n(n+1)}{(n-1)(n-2)(n-3)} \sum_{i=1}^{n} \left(\frac{x_i - \bar{x}}{s}\right)^4 - \frac{3(n-1)^4}{(n-2)(n-3)}$$

For example, in the data series of amperage measurements

13, 14, 23, 23, 32, 33, 45, 99, and 105 A

the kurtosis value is –0.41, dimensionless, rounded to two decimal places.

Use Excel's KURT() function. For example, for the data range named DATA, use the Excel function expression "=KURT(DATA)".

CORRELATION

When a plot of data is made between one variable and another, with the presence of some amount of random variation in one of the variables, the resulting graph can look like one of those in Figure 27.6.

When the general shape of the data is upward-sloping to the right, the data are said to be positively correlated. When the shape is downward-sloping to the right, they are negatively correlated.

Mathematically, data are described by the statistic r, or Pearson's correlation coefficient. Pearson's correlation coefficient is used to measure the strength of a linear association between two variables, where the value $r = 1$ means a perfect positive correlation and the value $r = -1$ means a perfect negative correlation.

294 Part IV: Mathematics and Statistics: Their Use in Measurement

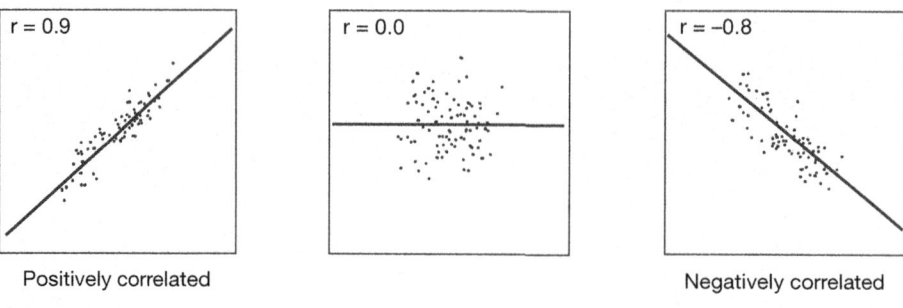

Figure 27.6 Examples of correlation graphs.

$$r = \frac{\sum_{i=1}^{n}\left((x_i - \bar{x})(y_i - \bar{y})\right)}{\sqrt{\sum_{i=1}^{n}(x_i - \bar{x})^2 \sum_{i=1}^{n}(y_i - \bar{y})^2}}$$

Figure 27.7 Pearson's correlation coefficient formula.

Figure 27.7 is the Pearson correlation coefficient formula for a given set of observations $(x_1, y_1), (x_2, y_2), \ldots (x_n, y_n)$ and where $s() =$ sample standard deviation:

$$r = \frac{n\sum_{i=1}^{n} x_i y_i - \left(\sum_{i=1}^{n} x_i\right)\left(\sum_{i=1}^{n} y_i\right)}{\sqrt{\left[n\sum_{i=1}^{n} x_i^2 - \left(\sum_{i=1}^{n} x_i\right)^2\right]\left[n\sum_{i=1}^{n} y_i^2 - \left(\sum_{i=1}^{n} y_i\right)^2\right]}}$$

$$= \frac{\sum_{i=1}^{n}(x_i - \bar{x})(y_i - \bar{y})}{\sqrt{\sum_{i=1}^{n}(x_i - \bar{x})^2 (y_i - \bar{y})^2}} = \frac{\sum_{i=1}^{n} x_i y_i - \overline{xy}}{(n-1)s(x)s(y)}$$

Positively correlated data are seen when there is a trend in the data where an increase in one variable relates to increasing values in the other variable. For example, in the pair of data series of amperage measurements

13, 14, 23, 23, 32, 33, 45, 99, and 105 A

and induced amperage measurements

34.60, 36.43, 59.23, 59.56, 82.03, 111.91, 240.50, and 255.19, respectively

the Pearson coefficient of variation, r, value is 0.99998, dimensionless.

Use Excel's PEARSON() function. For example, for the data ranges named DATA1 and DATA2, use the Excel function expression "=PEARSON(DATA1, DATA2)".

LINEAR RELATIONSHIPS

Many physical relationships are known to be generally linear relationships. As an example, many elastic relationships are assumed to follow this rule. The relationship between the deflected length of a helical spring and the resulting force it produces is considered linear over a portion of its available deflection range. The current flowing in a purely resistive circuit is linearly related to voltage drop over a range of current values.

Departures from a linear relationship occur at the limits of such linear ranges. For example, compression springs are known to be nonlinear near the end limits of their available stroke (near their free length and also near their solid height). The relationship of current to voltage in an electrical circuit can become nonlinear when the value of resistance (or impedance) varies with the magnitude of current through it (for example, when heating effects change the magnitude of resistance).

A linear relationship is one where, if two or more pairs of data values (for example, two or more points in an $x-y$ relationship) are plotted against one another, the resulting graph produces a straight line.

Two-Point Slope-Intercept Relationship in Linear Data Sets

When there are only two points, the line connecting them is simply a straight line. Such a linear graphical relationship can be described by a formula with two constants: one for the slope of the line and another for the intercept of the line with the y-axis.

This equation is the slope-intercept formula:

$$y = mx + b$$

where the formula for the slope is

$$m = \frac{y_2 - y_1}{x_2 - x_1}$$

The units of slope are a ratio: unit of y divided by the unit of x.

The formula for the intercept is:

$$b = y_1 - mx_1$$

The units of intercept are the same unit as y.

For example, in the x, y data pairs $x1 = 12.5$ mm, $y1 = 4.2$ N, $x2 = 22.1$ mm, $y2 = 8.4$ N, the slope is $m = 0.4$ N / mm, and the intercept is $b = -1.3$ N, rounded to one decimal place.

Use Excel's INTERCEPT() function. For example, for the data ranges named DATA1 and DATA2, use the Excel function expression "=INTERCEPT(DATA1, DATA2)".

Figure 27.8 illustrates a graph with slope and intercept.

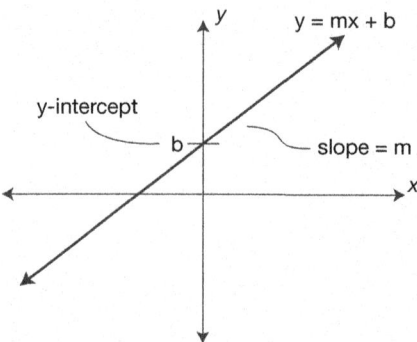

Figure 27.8 Slope and intercept graph.

Linear Regression, Best-Fit Line Through a Data Set

Linear regression is a way to perform predictive analysis. Often, data that are well-correlated can be described by a best-fit line through the set of measurements. The method by which the best-fit line is determined is through use of the least-squares method.

Data sets of linear data containing some random variation result in deviations from a straight-line relationship. For such data, the best-fit line through the data may need to be determined. The least-squares regression line is the line that minimizes the sum of the residuals squared.

The least-squares regression line must pass through $(\bar{x}, \bar{y})$. The general equation for the **least-squares regression line (best-fit line)** through a set of input–output data values is as follows:

$$\hat{y} = mx + b$$

$\hat{y}$ = the predicted y-value given x

m = slope

x = independent variable

b = y-intercept

where for the slope (m):

$$m = r * S_y/S_x$$

residual, $r = y - \hat{y}$

y = observed point

$\hat{y}$ = the predicted y-value given x

Standard deviation $S_y = \sqrt{\dfrac{\sum y - \text{residuals}^2}{n-2}}$

Standard deviation $S_x = \sqrt{\dfrac{\sum x - \text{residuals}^2}{n-2}}$

and for the intercept (b):

$$b = \bar{y} - m\bar{x}$$

For example, in the following pair of data series of amperage measurements,

13, 14, 23, 23, 32, 33, 45, 99, and 105 A,

and induced amperage measurements,

34.60, 36.43, 59.23, 59.56, 82.03, 111.91, 240.50, and 255.19 A, respectively,

the slope of the best-fit line through the data pairs is 2.39, dimensionless, and the intercept is 3.66 A, rounded to two decimal places.

Use Excel's SLOPE() function. For example, for the data ranges named DATA1 and DATA2, use the Excel function expression "=SLOPE(DATA1, DATA2)".

Also, the intercept of the best-fit line through the data pairs is 3.66 A, rounded to two decimal places.

Use Excel's INTERCEPT() function. For example, for the data ranges named DATA1 and DATA2, use the Excel function expression "=INTERCEPT(DATA1, DATA2)".

ZERO AND SPAN RELATIONSHIPS

In most measuring and process control instruments, it is desired to have the displayed value, or value sent to a control final element (valve, motor, and so on), scaled such that a given lower output is used as a zero value, and the difference between an upper output and the lower zero value is termed as the *span*.

INTERPOLATION

Many numerical relationships are given in table (tabular) form. Often, acceptance criteria, standard values, and empirical data are transmitted in such format. The user of these data formats often needs to obtain intermediate values, by interpolation.

Interpolation is determining an intermediate output value in relation to one or more intermediate input values.

A related term, *extrapolation*, refers to estimating a value that is outside the range of given data, at a higher or lower value. Generally, extrapolation is to be discouraged unless specifically allowed by written guidance from the governing source of the data.

FORMATS OF TABULAR DATA

One-Way Tabulations. Tabular data are most often provided in input–output format. Each input data item value is related in a one-to-one relationship to an output value.

Two-Way Tabulations. Some data, such as in published steam tables, relate more than one input variable to an output variable. For example, in two-way tabulations, the first input is listed down a side column and the second across a row (usually at the top of the table), with the output variable listed as an array on a page. An example of this would be relating gas volume (output) to pressure and temperature (input variables).

Three-Way Tabulations. This is where three input variables are associated with an output variable. Extending the two-way thought, such tabulations have separate pages or tables for each value of the third input variable. An example of this is the *F*-distribution table.

LINEAR INTERPOLATION METHODS

Fortunately, much available data are either strictly linearly related or may be approximated satisfactorily with linear interpolation methods, over small known increments of the available tabular data. Depending on the type of data and the degree of accuracy required, the following interpolation methods are used.

One-Way, Two-Point Interpolation, Linear

One-way interpolation is used for graphical or tabular data, where the objective is to determine an intermediate output value in relation to a given intermediate input value.

A special case of one-way interpolation is where the input–output relationship is either known or assumed to be linear. This form of interpolation is most often used in practice, where the desired intermediate values are taken as linearly related.

The method for one-way, two-point linear interpolation involves locating known higher and lower input values, their related higher and lower output values, and use of the intermediate input value to compute the desired intermediate output value by the following formula:

$$y(x) = y_1 + \frac{x - x_1}{x_2 - x_1}(y_2 - y_1)$$

Here, $y(x)$ is the desired interpolated output value. x_1, x_2, y_1, and y_2 are the tabulated high and low input and output values, respectively.

For example, for the following source data voltage values, for points x and y,

x	y
1	8
2	1

the interpolated value of $y(x)$ at $x = 1.6$ V is $y(x) = 3.8$ V.

Two-Way, Two-Point Interpolation, Linear

For this requirement, the preceding formula is used incrementally, as follows. Using the formula and the intermediate value of the first input variable, determine two intermediate output values for the second input variable at both the high and low value of the first variable. Then repeat this process and, using the intermediate value of the second input variable, now having the values of the output variable at the intermediate value of the first input variable, compute the intermediate value of the output variable (which is the interpolated output value related to both intermediate input values).

For example, we have the following table of source data voltage values, for voltages at points w and x, related to a temperature, y °C, the interpolated value of $y(x)$ at $w_i = 77$ V, and $x_i = 86$ V is $y(w, x) =$ **0.507 V**.

Source data table for output temperature, y °C

		x_1	x_2
		80	90
w_1	70	0.483	0.534 y
w_2	80	0.474	0.524 y

Three-Way Interpolation, Linear

Three-way interpolation is merely an extension of two-way interpolation, following the same logic. Obviously, the process becomes more and more repetitive and time-consuming.

In the CD-ROM the calculation is done automatically by providing the desired intermediate and tabulated input and output values from known tabular data.

For example, we have the following table of source data voltage values, for voltages at points w and x, related to a temperature, y °C, the interpolated value of $z(y_i)$ at $w_i = 77$ V, $x_i = 86$ V and $y_i = 0.50$, is $y(w, x, y) =$ **0.517 V**.

Source data tables for output temperature, y °C

$Y_1 = 0.25$	x_1	x_2		$Y_1 = 0.75$	x_1	x_2	
	80	90			80	90	
w_1	70	0.483	0.534	$z(y_1)$	w_1	70	0.503	0.554	$z(y_2)$
w_2	80	0.474	0.524	$z(y_1)$	w_2	80	0.494	0.544	$z(y_2)$

INTERPOLATION METHODS FOR NONLINEAR DATA

Data that are slightly nonlinear, where higher-accuracy interpolated values are required, or substantially nonlinear data where known values between input and output values are more widely separated, require use of nonlinear interpolation methods. Generally, interpolation methods for such data are restricted to one-way (one input variable associated with each output variable) data, as the complexity and uncertainty can increase in nonlinear relationships.

One-Way, Three-Point Interpolation—Quadratic

When the data relationship is slightly nonlinear, a better and more accurate interpolation method is to use three-point, quadratic interpolation, with the following formula:

$$y(x) = \frac{(x-x_2)(x-x_3)}{(x_1-x_2)(x_1-x_3)}y_1 + \frac{(x-x_1)(x-x_3)}{(x_2-x_1)(x_2-x_3)}y_2 + \frac{(x-x_1)(x-x_2)}{(x_3-x_1)(x_3-x_2)}y_3$$

Here,

$y(x)$ is the desired interpolated output value.

$x_1, x_2, x_3, y_1, y_2,$ and y_3 are the tabulated high, intermediate, and low input and output values, respectively.

For example, for the following source data voltage values, for points x and y,

x	y
1	8
2	1
4	5

the interpolated value of $y(x)$ at $x = 1.6$ V is $y(x) = 3.08$ V.

One-Way, n-Point Interpolation—Nonlinear (Lagrangian)

Both the linear and quadratic interpolation equations just mentioned are special cases of the general Lagrangian interpolation equation. This equation may be extended to any number of input/output pairs for consideration in producing even more accurate interpolation results.

For example, for the following source data voltage values, for points x and y,

x	y
1	12
2	4
4	16
5	25

the four-point interpolation, cubic, third-order exponential interpolated value of $y(x)$ at $x = 1.5$ V is $y(x) = 6.3$ V, rounded to one decimal place.

TYPES OF DISTRIBUTIONS AND THEIR PROPERTIES

Normal (Gaussian) Probability Distribution

The *normal distribution*, or *gaussian distribution*, describes randomly occurring events and their frequency relative to the magnitude of their comparative measurement values. The density function of a normal probability distribution is bell shaped and symmetric about the mean.

The normal (gaussian) distribution is given by the normal probability density function formula in Figure 27.9 where x is normally distributed with a population mean μ and population variance σ^2.

$$f(x) = \frac{e^{-(x-\mu)^2/(2\sigma^2)}}{\sigma\sqrt{2\pi}}$$

Figure 27.9 Normal probability density function.

STANDARD NORMAL (GAUSSIAN) Z-DISTRIBUTION

The *standard normal (gaussian) distribution*, or *z-distribution*, is normalized to produce an area under the normal curve of 1.0 and place the peak of the curve at $z = 0$. This also means that the mean is 0 and the standard deviation is 1. This allows the convenient use of the z-statistic, also called the z-score, to locate the point where the area under the curve, at greater or lesser values than the z-statistic, is associated with the probability of occurrence of events having any of those z-statistic values. In other words, the z-score can tell you how many standard deviations from the mean each value lies.

The standard normal (gaussian) z-distribution is given by the formula in Figure 27.10.

$$z = \frac{x - \mu}{\sigma}$$

Figure 27.10 Calculating z-score.

Where x = individual value, μ = mean, σ = standard deviation.

To standardize a value from a normal distribution, convert the individual value into a z-score by:

1. Subtracting the mean from your individual value.
2. Dividing the difference by the standard deviation.

t-Distribution

The t-*distribution* relates the distribution of a sample to the z-distribution's z-statistic and probabilities. As the sample size is increased to larger and larger values, the t-distribution approaches and, in the limit of infinite sample size, becomes identical to the z-distribution.

F-Distribution

The F-*distribution* is used to compare the variances of two sample or population statistics. It is used in ANOVA and other variance-based statistic calculations. The F-distribution is the ratio of the two chi-square distributions with degrees of freedom v_1 and v_2, respectively, where each chi-square has first been divided by its degrees of freedom. See Figure 27.11 for the formula.

$$f(x) = \frac{\Gamma(\frac{v_1+v_2}{2})(\frac{v_1}{v_2})^{\frac{v_1}{2}} x^{\frac{v_1}{2}-1}}{\Gamma(\frac{v_1}{2})\Gamma(\frac{v_2}{2})(1+\frac{v_1 x}{v_2})^{\frac{v_1+v_2}{2}}}$$

Figure 27.11 F-distribution formula.

χ^2 (Chi-Squared)-Distribution

The *chi-squared distribution* is used to compare two distributions to test if they may be assumed to represent the same population, or for samples if the samples may be considered to be drawn from a population having a common distribution. The chi-squared (n) random variable is the sum of the squares of n standard normal random variables $\left(x_n^2 = Z_1^2 + \cdots + Z_n^2\right)$.

See Figure 27.12 for the formula for the probability density function of the chi-square distribution.

$$f(x) = \frac{1}{2^{n/2}\,\Gamma(n/2)} x^{n/2-1} e^{-x/2}, \quad x > 0$$

Figure 27.12 Chi-square distribution formula.

Weibull Distribution

Weibull distributions are used in reliability statistics for estimating expected failure rates over the life of an item under consideration. We will not consider this distribution in this text.

Hypergeometric Distribution

The hypergeometric distribution is used to determine the number of successes or failures in a population of events, based on a given sample from that population. It is often the basis for determining an acceptance sampling plan and its performance in correctly identifying the presence of a condition.

The hypergeometric distribution gives the probability of the number of successes, given the number of population successes, sample size, and population size. The hypergeometric distribution is given in the formula in Figure 27.13 where the probability of selecting x marked items when a random sample of size KK is taken without replacement from a population of MM items, NN of which are marked. Marked and unmarked items can also be thought of as successes and failures.[9]

$$p(x) = \frac{\binom{NN}{x}\binom{MM-NN}{KK-x}}{\binom{MM}{KK}}$$

Figure 27.13 Hypergeometric distribution formula.

Binomial Distribution

The *binomial distribution* is often used in situations where there is a fixed number of trials or tests, when the outcomes are pass/fail, trials are statistically independent of one another, and the probability of success is constant. There are exactly two mutually exclusive outcomes of a trial. These outcomes are appropriately labeled "success" and "failure." The binomial distribution is used to obtain the probability of observing x successes in n trials, with the probability of success on a single trial denoted by p. The binomial distribution assumes that p is fixed for all trials.[10] See Figure 27.14. When the probability p is small, and the number of samples n is large, for a fixed np in the limit, the binomial distribution approaches the Poisson distribution.

$$P(x; p, n) = \binom{n}{x}(p)^x(1-p)^{(n-x)} \quad \text{for } x = 0, 1, 2, \cdots, n$$

where

$$\binom{n}{x} = \frac{n!}{x!(n-x)!}$$

$n! = n * (n - 1) * \ldots * 3 * 2 * 1$

Figure 27.14 Binomial distribution formula.

Where:

n = Number of trials

x = Desired number of successes

p = Probability of getting success

$1 - p$ = Probability of getting failure

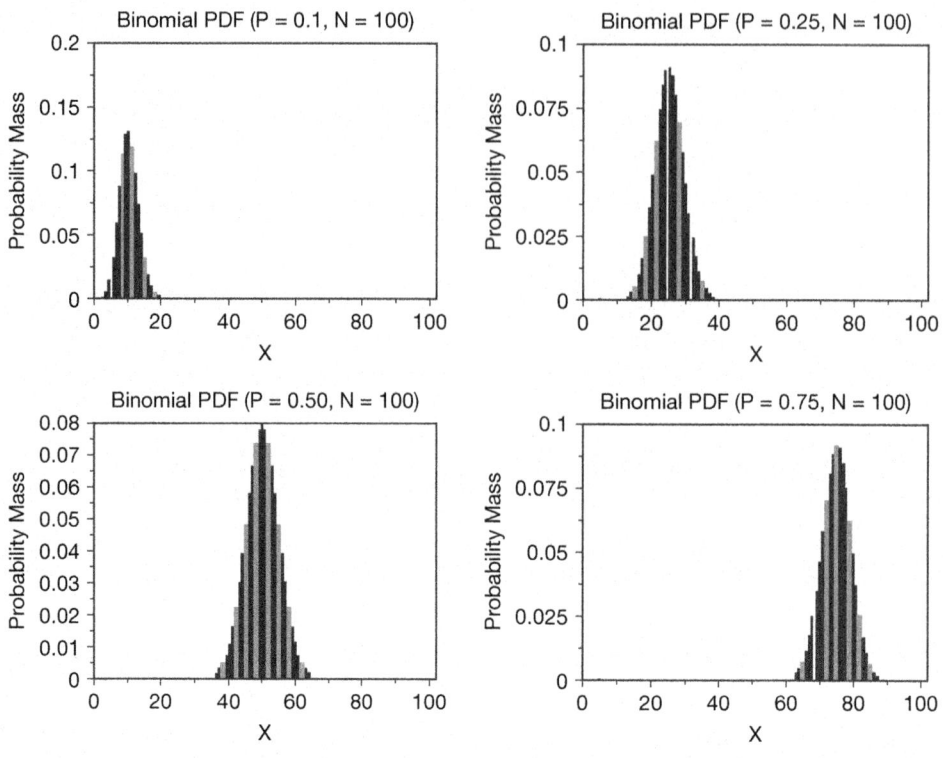

Figure 27.15 Binomial distribution plots.[11]

Figure 27.15 is a plot of the binomial probability density function for four values of *p* and *n* = 100.

Poisson Distribution

A common application of the *Poisson distribution* is predicting the number of events that will occur over a specified time interval. The Poisson distribution has often been called the *small probability distribution* because it is usually applied in situations where there is a small, per-unit, probability *p* of an occurrence of an event in a given standard sample unit size. Multiples of the stated sample unit on which the probability *p* is based are labeled *n* for the number or size of the sample being considered. The probability for the sample size *n* is therefore the product, or *np*. The Poisson distribution is given by the formula in Figure 27.16.

$$P(\lambda, x) = \left(\frac{\lambda^x e^{-\lambda}}{x!}\right)$$

Figure 27.16 Poisson distribution formula.

Where:

$$x = 0, 1, 2, \ldots$$

$$e = \text{Euler's constant} \approx 2.71828$$

$$\lambda = \text{mean number of occurrences in the interval}$$

Uncertainty, Normal Distribution

Many measurements are made under the assumption that the underlying uncertainty is normally distributed. For an assumed normal distribution, the Type B standard uncertainty u_j, where 2σ is the range including approximately 67 % (2s) probability that the value lies in the interval $-\sigma$ to $+\sigma$, is given by the following formula:

$$u_i = \sigma$$

For example, if an ammeter is assumed to have an approximately 100 % probability that the reading lies in the interval of ± 0.02 mA, with a normal distribution, then the standard uncertainty u_j is 0.02 mA.

Uncertainty, Rectangular Distribution

Rectangular (or *uniform*) *distributions*, as their name indicates, have distributions that are rectangular-shaped, with a constant frequency of occurrence over the range of the distribution's measurement values.

For an assumed rectangular distribution, the Type B standard uncertainty u_j, where $2a$ is the range including approximately 100 % probability that the value lies in the interval $-a$ to $+a$, is given by the following formula:

$$u_i = a/\sqrt{3}$$

For example, if a volumetric meter is assumed to have an approximately 100 % probability that the reading lies in the interval of ± 0.15 m^3, with a rectangular distribution, then the standard uncertainty u_i is 0.087 m^3. This distribution is used when referencing calibration certificates or manufacturer's specifications.

Uncertainty, Triangular Distribution

Triangular distributions, as the name indicates, have distributions that are isosceles triangular–shaped, with frequencies and probabilities linearly decreasing in magnitude for values above or below the distribution mean (which is at the peak of the triangle).

For an assumed triangular distribution, the Type B standard uncertainty u_j, where $2a$ is the range including approximately 100 % probability that the value lies in the interval $-a$ to $+a$, is given by the following formula:

$$u_i = a/\sqrt{6}$$

For example, if a weigh scale is assumed to have an approximately 100 % probability that the reading lies in the interval of ± 0.015 N, with a triangular distribution, then the standard uncertainty u_i is 0.0061 N. These types of uncertainty contributors are often used for noise and vibration.

Uncertainty, U-Shaped Distribution

U-shaped distributions generally apply to distributions where the frequency of occurrence of the measurement is lowest at the mean of the distribution, increasing in some manner for values above or below the distribution mean.

For an assumed U-shaped distribution, the Type B standard uncertainty u_j, where 2a is the range including approximately 100 % probability that the value lies in the interval—a to +a, is given by the following formula:

$$u_i = a/\sqrt{2}$$

For example, if a thermometer is assumed to have an approximately 100% probability that the reading lies in the interval of ± 0.1 K, with a U-shaped distribution, the standard uncertainty u_j then is 0.07 K. This distribution is used when referencing cyclic influences, such as temperature variation.

Relative Uncertainty

Relative uncertainty, U_r, is a dimensionless value that expresses a unit's uncertainty independent of the unit's size. It is the measurement uncertainty, U, divided by the magnitude of the unit, y, given by the following formula:

$$U_r = u/y$$

For instance, if an ammeter measures (y = 0.75 A), and the uncertainty, u, of the measurement is known to be 0.005 A, then the relative uncertainty U_r is 0.00667, dimensionless.

AUTOCORRELATION

One important test of the assumption that source data are normally distributed is to determine the *autocorrelation coefficient* for the data set, a value between 0 and 1. If the autocorrelation coefficient is near zero, the data set may be more confidently assumed to be normally distributed. If it is larger, this indicates the presence of cyclic variation in the data, and the presence of a sequence- or time-based factor in the data. The autocorrelation function can be used for the following two purposes:

1. To detect nonrandomness in data

2. To identify an appropriate time series model if the data are not random

The general formula for calculating the autocorrelation for lag_k, denoted r_k, given measurement, $Y_1, Y_2, \ldots, Y_N$ at time $X_1, X_2, \ldots, X_N$ is seen in Figure 27.17.

$$r_k = \frac{\sum_{i=1}^{N-k}(Y_i - \bar{Y})(Y_{i+k} - \bar{Y})}{\sum_{i=1}^{N}(Y_i - \bar{Y})^2}$$

Figure 27.17 Autocorrelation function.

If the autocorrelation coefficient is calculated for all lags, $k = 0 \ldots N-1$, the resulting series is called the *autocorrelation series,* or *correlogram.*

Chapter 28

Mensuration, Volume, and Surface Areas

In this chapter, we cover the basic formulas for common calculations used in measurement of quantities dependent on geometry and known dimensions. This section enumerates often-encountered methods for calculating lengths, angles, arcs, and the more common plane areas, volumes, and surface areas. For other formulas and methods, consult general references containing formulas for specific areas of interest.

MENSURATION, LENGTHS, AND ANGLES

Mensuration is defined as the act or process of measuring. More often, the term *mensuration* is associated with dimensional measurement and is often associated with the determination, by computation from known characteristics, of the lengths of lines, arcs, angles, and distances.

The following are formulas that are often encountered in determining both length and graphical relationships.

LENGTH

y-Intercept of a Line

This equation is used to determine the point where a line crosses the *y*-axis at $x = 0$.

Formula:

$$b = y_i - mx_i$$

where

b, *y*-intercept of line (intercept at $x = 0$)

m, slope of line, dimensionless (dmls)

x_i, value of individual measurement

y_i, value of individual measurement

An example of use of this formula is:

y_1	x_1	m	b
M	m	dmls	m
6.1	3.8	1.2	1.54

Slope of a Line

Formula:

$$m = \frac{y_2 - y_1}{x_2 - x_1}$$

where

m, slope of line, dimensionless

x_1, x_2, values of individual measurements

y_1, y_2, values of individual measurements

An example of use of this formula is:

y_1	y_2	x_1	x_2	m
m	m	m	dmls	m
6.1	10.6	3.8	6.8	1.5

Slope-Intercept Equation of a Line

Formula:

$$y = mx + b$$

where

m, slope of line, dimensionless

x, value of measurement-independent variable

y, value of measurement-dependent variable

b, y-intercept of line (intercept at $x = 0$)

An example of use of this formula is:

x_1	m	b	y_1
m	m	dmls	m
6.8	1.5	2.1	12.3

Point-Slope Equation of a Line

Formula:

$$(y - y_i) = m(x - x_i)$$

where

m, slope of line

x_i, value of individual measurement

y_i, value of individual measurement

Distance Between Two x–y Points

Formula:

$$d = \pm\sqrt{(x_2 - x_1)^2 + (y_2 - y_1)^2}$$

where

d, distance between two points in a plane

x_i, value of individual measurements

y_i, value of individual measurements

An example of use of this formula is:

y_1	y_2	x_1	x_2	d
m	m	m	dmls	m
6.1	10.6	3.8	6.8	5.40833

Distance in a General Plane Between Two Points

Formula:

$$d = \pm\sqrt{(x_2 - x_1)^2 + (y_2 - y_1)^2 + (z_2 - z_1)^2}$$

where

d, distance between two general points

x_i, value of individual measurements

y_i, value of individual measurements

An example of use of this formula is:

y_1	y_2	x_1	x_2	z_1	z_2	d
m	m	m	m	m	m	m
6.1	10.6	3.8	10.6	3.8	6.8	8.6885

Perpendicular Lines, Slope Relationship

Formula:

$$m_1 = -\frac{1}{m_2}$$

where

m, slope of line, dimensionless

An example of use of this formula is:

m_1	m_2
dmls	dmls
3.4	−0.2941

CIRCLE

Circle, General

Formula:

$$r = \sqrt{(x-h)^2 + (y-k)^2}$$

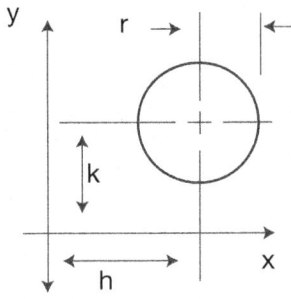

Figure 28.1 Circle general equation.

where

    r, radius of circle

    x, value of x-coordinate

    h, x-offset

    y, value of y-coordinate

    k, y-offset

An example of use of this formula is:

x	h	y	k	r
m	m	m	m	m
4.3	2.1	5.1	7.3	3.11127

ELLIPSE

Ellipse, Major Axis Horizontal

Formula:

$$y = k \pm b\sqrt{1 - \frac{(x-h)^2}{a^2}} \quad 1 = \frac{(x-h)^2}{a^2} + \frac{(y-k)^2}{b^2}$$

where

    major axis length: $2a$

    minor axis length: $2b$

    x, value of x-coordinate

    h, x-offset, center of ellipse from $x = 0$

    y, value of y-coordinate

    k, y-offset, center of ellipse from $y = 0$

An example of use of this formula is:

x	h	a	b	k	y_1	y_2
m	m	m	m	m	m	m
4.1	2.1	4.3	2.4	2.1	−0.0246	4.2246

Ellipse, Major Axis Vertical

Formula:

$$y = k \pm a\sqrt{1 - \frac{(x-h)^2}{b^2}} \quad\quad 1 = \frac{(x-h)^2}{b^2} + \frac{(y-k)^2}{a^2}$$

where

major axis length: $2a$

minor axis length: $2b$

x, value of x-coordinate

h, x-offset, center of ellipse from $x = 0$

y, value of y-coordinate

k, y-offset, center of ellipse from $y = 0$

An example of use of this formula is:

x	h	a	b	k	y_1	y_2
m	m	m	m	m	m	m
4.1	2.1	4.3	2.4	2.1	−0.2769	4.4769

ANGLE

Degree to Radian

Formula:

$$\text{deg} = \frac{\pi}{180}\text{rad}$$

An example of use of this formula is:

Angle		Angle	
Degree	Radian	Radian	Degree
3.4	0.05934	2.4	137.51

General Plane Triangle Relationships—Right Triangle

$$\sin\theta = \frac{\text{Side opposite}}{\text{Hypotenuse}}$$

$$\cos\theta = \frac{\text{Side adjacent}}{\text{Hypotenuse}}$$

$$\tan\theta = \frac{\text{Side opposite}}{\text{Side adjacent}}$$

$$\csc\theta = \frac{\text{Hypotenuse}}{\text{Side opposite}} = \frac{1}{\sin\theta}$$

$$\sec\theta = \frac{\text{Hypotenuse}}{\text{Side adjacent}} = \frac{1}{\cos\theta}$$

$$\cot\theta = \frac{\text{Side adjacent}}{\text{Side opposite}} = \frac{1}{\tan\theta}$$

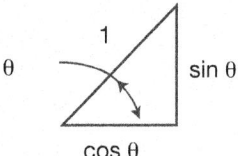

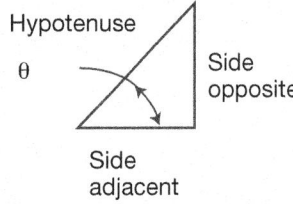

Figure 28.2 Right triangle.

Pythagorean (Right-Angle Triangle) Theorem

Formulas:

$$c^2 = a^2 + b^2$$

where

$$a = \sqrt{c^2 - b^2}$$
$$b = \sqrt{c^2 - a^2}$$
$$c = \sqrt{a^2 + b^2}$$

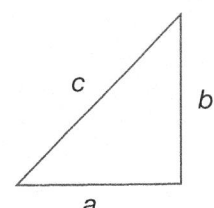

An example of use of this formula is:

a	b	c
m	m	m
6.1	3.8	7.18679

Law of Sines

Formula:

$$\frac{a}{\sin\alpha} = \frac{b}{\sin\beta} = \frac{c}{\sin\gamma}$$

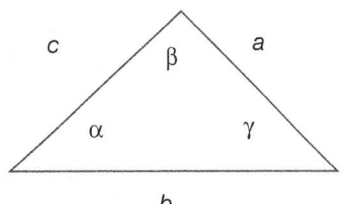

where

a, side opposite angle α

b, side opposite angle $\phi3$

c, side opposite angle ψ

An example of use of this formula is:

b	*a*	*b*	sin *a*	sin *b*	*a*
m	deg	deg	dmls	dmls	m
6.1	60	30	0.86603	0.5	10.5655

Law of Cosines

Formula:

$$c = \sqrt{a^2 + b^2 - 2ab\cos(\gamma)}$$

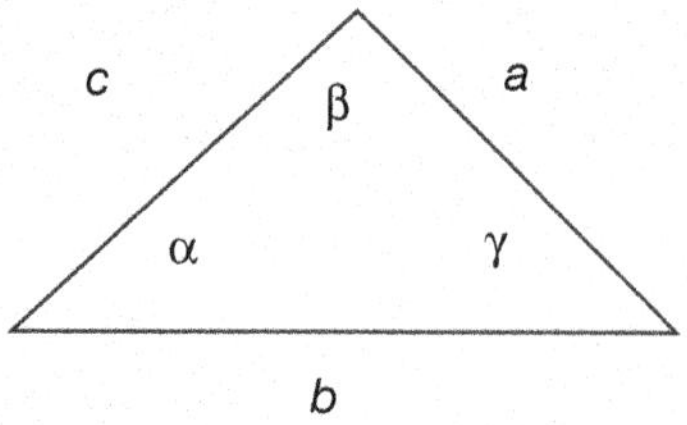

where

a, side opposite angle α

b, side opposite angle ϕ3

c, side opposite angle ψ

An example of use of this formula is:

a	*b*	*g*	*cos b*	*c*
m	deg	deg	dmls	m
1	1.41421	45	0.70711	1

PLANE AREA

Rectangle Area

Formula:

$$A = bh$$

where

A, plane area

b, length of base

h, length of height

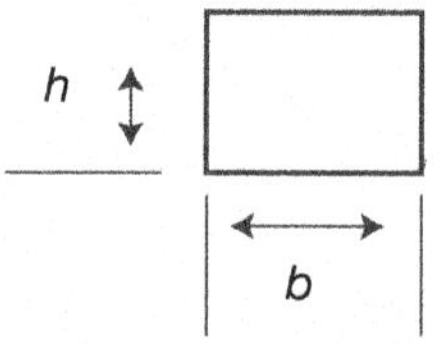

Figure 28.3 Rectangle area.

An example of use of this formula is:

b	*h*	*A*
m	m	m ^2
1	2	2

Parallelogram Area

Formula:

$$A = bh$$

where

    A, plane area

    b, length of base

    h, perpendicular distance from base to top

Figure 28.4 Parallelogram area.

An example of use of this formula is:

b	h	A
m	m	m ^2
1	2	2

Trapezoid Area

Formula:

$$A = \frac{h(b_1 + b_2)}{2}$$

where

    A, plane area h

    b_2, length of base, top

    b_1, length of base, bottom

    h, perpendicular distance from base to top

An example of use of this formula is:

b_1	b_2	h	A
m	m	m	m ^2
1	2	3	4.5

Right Triangle Area

Formula:

$$A \rightarrow b\frac{h}{2}$$

where

    A, plane area b

    b, length of base

    h, perpendicular distance from base to apex

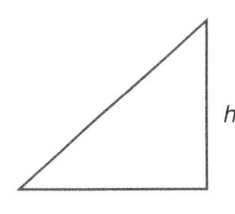

An example of use of this formula is:

b	*h*	*A*
m	m	m ^2
3.45	2.2	3.795

Oblique Triangle Area

Formula:

$$A = b\frac{c\sin\alpha}{2}$$

where

A, plane area

b, length of base

c, length of side adjacent to *b*

α, angle between sides *b* and *c*

h, perpendicular distance from base to apex

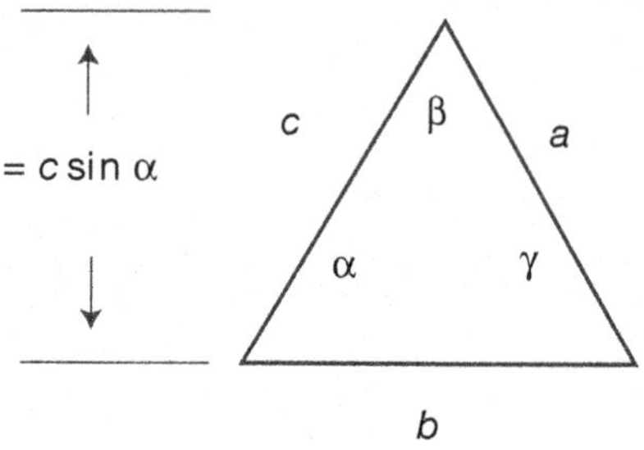

Figure 28.5 Oblique triangle area.

An example of use of this formula is:

b	*a*	*α*	*sin α*	*A*
m	m	deg	dmls	m²
1	1.41421	45	0.70711	0.683

Circle Area

Formula:

$$A = \pi r^2 = \pi\frac{d^2}{4}$$

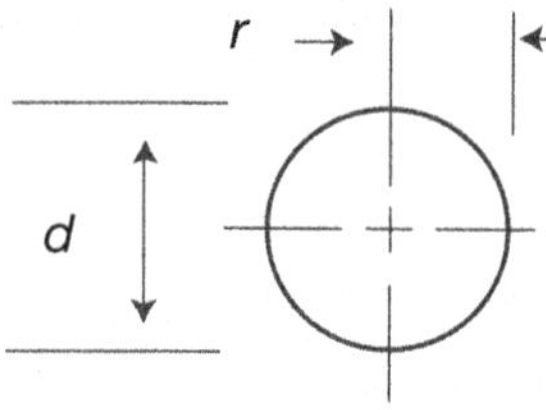

where

A, plane area

d, diameter

r, radius

An example of use of this formula is:

r	*A*
m	m ^2
0.5	0.7854

d	*A*
m	m ^2
1	0.7854

Ellipse Area

Formula:

$$A = \pi ab = \pi \frac{CD}{4}$$

where

A, plane area

a, major axis radius length

b, minor axis radius length

C, major axis length: 2a

D, minor axis length: 2b

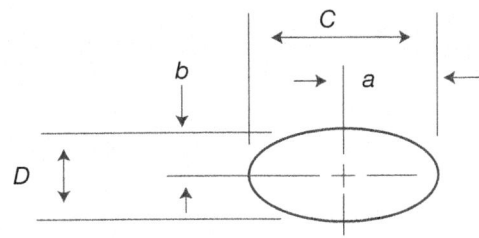

An example of use of this formula is:

a	b	A
M	m	m^2
0.5	0.5	0.7854

C	D	A
m	m	m^2
1	1	0.7854

PERIMETER

Rectangular Perimeter

Formula:

$$P = 2(b + h)$$

where

P, length of perimeter

b, length of base

h, length of height

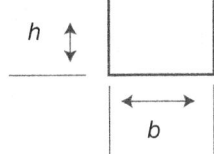

An example of use of this formula is:

b	h	P
m	m	m
0.78	1.32	4.2

Right Triangle Perimeter

Formula:

$$P = a + b + \sqrt{a^2 + b^2}$$

where

P, length of perimeter

a, length of side

b, length of base

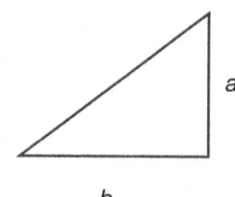

An example of use of this formula is:

a	b	P
m	m	m
1.8	3.4	9.04708

Circle Perimeter (Circumference)

Formula:

$$P = 2\pi r = \pi d$$

where

P, length of perimeter

d, diameter

r, radius

An example of use of this formula is:

r	P
m	M
0.5	3.14159

d	P
m	m
1	3.14159

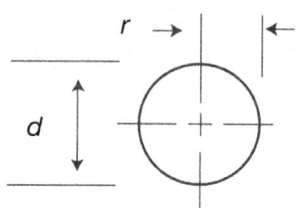

ELLIPSE PERIMETER (CIRCUMFERENCE)

Formula:

$$P = \pi(C + D) = 2\pi\sqrt{\frac{a^2 + b^2}{2}}$$

where

P, length of perimeter

a, major axis radius length

b, minor axis radius length

C, major axis length: 2a

D, minor axis length: 2b

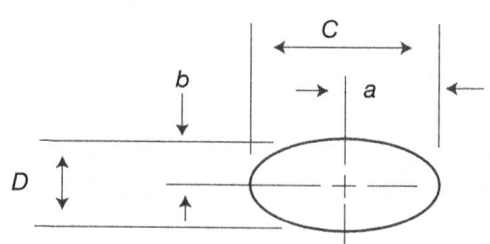

VOLUME

Rectangular Prism Volume

Formula:

$$V = lwh$$

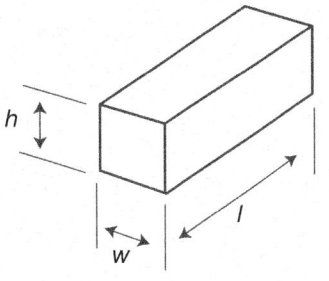

where

    V, volume

    l, length

    w, width

    h, height

Figure 28.6 Rectangular prism volume.

An example of use of this formula is:

l	w	h	V
m	m	m	m ^3
1.24	3.8	2.76	13.005

Sphere Volume

Formula:

$$V = \frac{4}{3}\pi r^3 = \frac{1}{6}\pi d^3$$

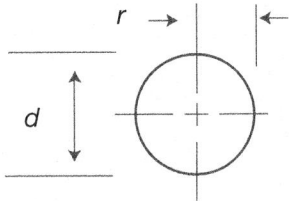

where

    V, volume

    d, length, spherical diameter

    r, length, spherical redius

Figure 28.7 Sphere volume.

An example of use of this formula is:

r	V
m	m ^3
0.5	0.5236

d	V
m	m ^3
1	0.5236

Ellipsoid Volume

Formula:

$$V = \frac{4}{3}\pi abc = \frac{1}{6}\pi ABC$$

where

    V, volume

    a, length, major axis radius

    b, length, first minor axis radius

    c, length, second minor axis radius

    A, lengths, first through third axes

    B, lengths, first through third axes

    C, lengths, first through third axes

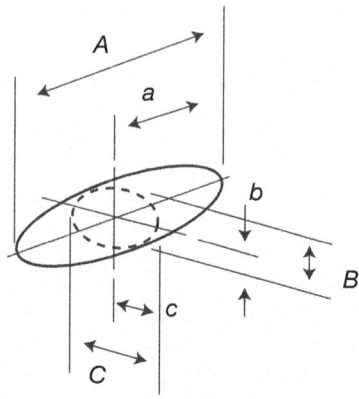

Figure 28.8 Ellipsoid volume.

An example of use of this formula is:

a	b	c	V
m	m	m	m ^3
0.5	0.5	0.5	0.5236

A	B	C	V
m	m	m	m ^3
1	1	1	0.5236

Pyramid Volume

Formula:

$$V = \frac{1}{3} A_{Base} h$$

where

    V, volume

    A_{Base}, plane area of base

    h, height from base

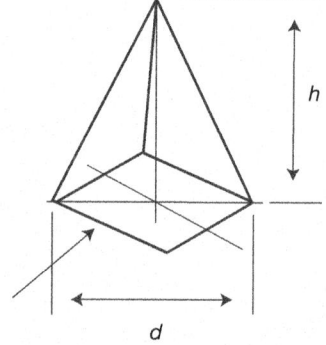

Figure 28.9 Pyramid volume.

An example of use of this formula is:

A_{Base}	h	V
m ^2	m	m ^3
0.5	1	0.16667

Truncated Pyramid Volume

Formula:

$$V = \frac{1}{3}h(A_1 + \sqrt{A_1 A_2} + A_2)$$

where

    V, volume

    A_1, plane area of base 1

    A_2, plane area of base 2

    h, height from base

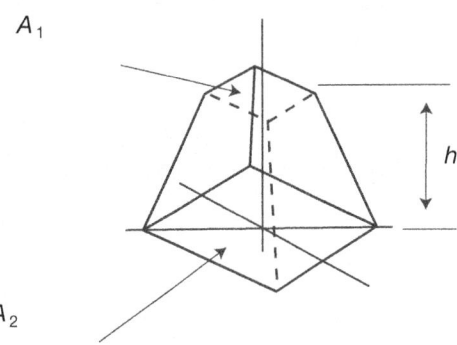

Figure 28.10   Truncated pyramid volume.

An example of use of this formula is:

A_1 base	A_2 base	h	V
m ^2	m	m	m ^3
0.5	1	1	0.7357

Cone Volume

Formula:

$$V = \frac{1}{3}\pi r^2 h = \frac{1}{12}\pi d^2 h$$

where

    V, volume

    r, base radius

    d, base diameter

    h, height from base

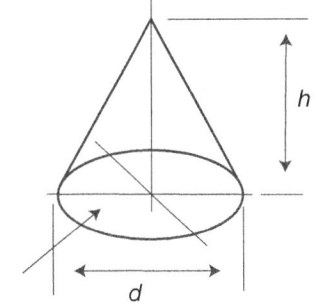

Figure 28.11   Cone volume.

An example of use of this formula is:

r	h	V
m	m	m ^3
0.5	1	0.2618

D	h	V
m	m	m ^3
1	1	0.2618

Truncated Cone Volume

Formula:

$$V = \frac{\pi}{12} h \left(d_1^2 + d_1 d_2 + d_2^2 \right) = \frac{1}{3} h \left(A_1 + \sqrt{A_1 A_2} + A_2 \right)$$

where

V, volume

A_1, plane area of base 1

A_2, plane area of base 2

d_1, diameter of base 1

d_2, diameter of base 2

h, height from base

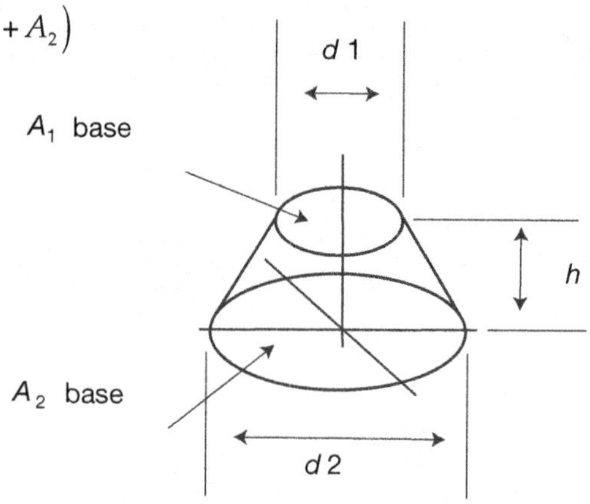

Figure 28.12 Truncated cone volume.

An example of use of this formula is:

d_1	d_2	h	V
M	M	m	m ^3
0.25	0.25	0.5	0.02454

A_1	A_2	H	V
m ^2	m ^2	M	m ^3
0.04909	0.04909	0.5	0.02454

SURFACE AREA

Rectangular Prism Surface Area

Formula:

$$A_S = 2(lw + lh + wh)$$

where

A_S, surface area

l, length

w, width

h, height

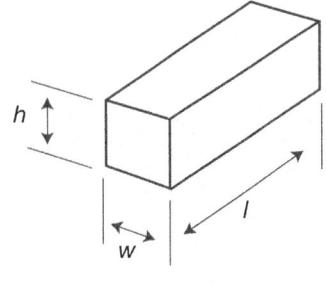

Figure 28.13 Rectangular prism surface area.

An example of use of this formula is:

l	w	h	AS
m	m	m	m ^2
1	2	3	22

Sphere Surface Area

Formula:

$$A_S = 4\pi r^2 = \pi d^2$$

where

A_S, surface area

d, length, spherical diameter

r, length, spherical radius

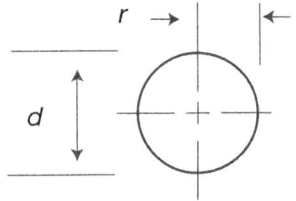

Figure 28.14 Sphere surface area.

An example of use of this formula is:

r	V
m	m ^2
0.5	3.14159

d	V
m	m ^2
1	3.14159

Prolate Ellipsoid of Revolution Surface Area

Physical examples of this shape are eggs.

Formula:

$$A_S = 2\pi \left[b^2 + \frac{a^2 b}{\sqrt{a^2 - b^2}} \sin^{-1}\left(\sqrt{\frac{a^2 - b^2}{a^2}}\right) \right]$$

where

A_S, surface area

a, length, major axis radius

b, length, minor axis radius

Note: a not equal to b

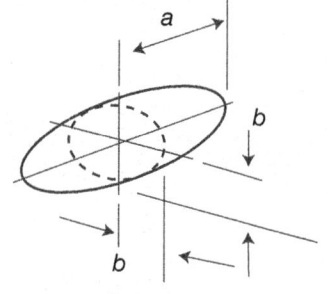

Figure 28.15 Prolate ellipsoid of revolution surface area.

An example of use of this formula is:

a	b	V
m	m	Unit ^2
0.503	0.497	3.12902

Oblate Ellipsoid of Revolution Surface Area

Physical examples of this shape are large rising bubbles.

Formula:

$$A_S = 2\pi \left[a^2 + \frac{ab^2}{\sqrt{a^2 - b^2}} \ln\left[\frac{a + \sqrt{a^2 - b^2}}{b}\right] \right]$$

where

A_S, surface area

a, length, major axis radius

b, length, minor axis radius

Note: a not equal to b

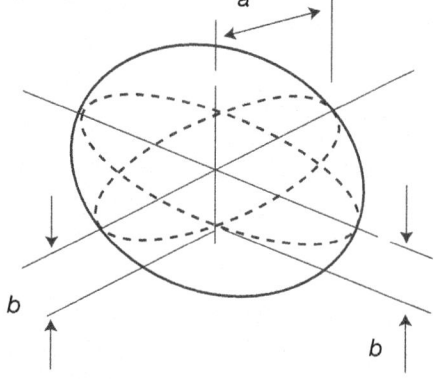

Figure 28.16 Oblate ellipsoid of revolution surface area.

An example of use of this formula is:

a	b	V
m	m	Unit ^2
0.503	0.497	3.15415

Part V
Uncertainty in Measurement

Chapter 29   Measurement Uncertainty

Chapter 30   Measurement Risk and Decision Rules

Chapter 29
Measurement Uncertainty

The natural physical constants that define International System of Units (SI) base units of measurement have no associated measurement uncertainty (because they are, by definition, *natural physical constants*). Once humans become involved, these realized base units of measurement now have measurement uncertainty. The amount of measurement uncertainty can vary depending on the factors (components) discussed in this chapter. *Measurement uncertainty*, as defined in the *International Vocabulary of Metrology* (*VIM*) 2.26, is a "non-negative parameter characterizing the dispersion of the *quantity values* being attributed to a *measurand*, based on the information used."[1] In other words, the "true value" (i.e., "*quantity value* consistent with the definition of a *quantity*"[2]) is somewhere within its measurement uncertainty. We see this likelihood presented as measurement uncertainty $\pm x$ units at an approximately 95 % level of confidence where $k = 2$. In this chapter, we will be breaking down how this statement was formulated and what it means. We will begin by using definitions from the *VIM*.

It is important to begin by differentiating "measurement uncertainty" from "measurement error." They are not the same thing, nor are the terms interchangeable. Measurement error (error of measurement / error [*VIM* 2.16]) is defined as "*measured quantity value* minus a *reference quantity value*."[3]

- *Measured quantity value* / value of a measured quantity / measured value (*VIM* 2.10): **quantity value** representing a **measurement result**[4]

- *Reference quantity value* / *reference value* (*VIM* 5.18): **quantity value** used as a basis for comparison with values of **quantities** of the same **kind**[5]

- *Quantity value* / *value of a quantity* / *value* (*VIM* 1.19): number and reference together expressing magnitude of a **quantity**[6]

- *Measurement result* (*VIM* 2.9): set of **quantity values** being attributed to a **measurand** together with any other available relevant information[7]

- *Measurand* (*VIM* 2.3): **quantity** intended to be measured[8]

- *Quantity* (*VIM* 1.1): property of a phenomenon, body, or substance, where the property has a magnitude that can be expressed as a number and a reference[9]

- *Kind of quantity* / *kind* (*VIM* 1.2): aspect common to mutually comparable quantities[10]

- *Metrological traceability* (*VIM* 2.41): "property of a **measurement result** whereby the result can be related to a reference through a documented unbroken chain of **calibrations**, each contributing to the **measurement uncertainty**."[11]

- *Calibration* (*VIM* 2.39): "operation that, under specified conditions, in a first step, establishes a relation between the **quantity values** with **measurement uncertainties** provided by **measurement standards** and corresponding **indications** with associated measurement uncertainties and, in a second step, uses this information to establish a relation for obtaining a **measurement result** from an indication."[12]

- *Measurement standard* (*VIM* 5.1): "realization of the definition of a given **quantity**, with stated **quantity value** and associated **measurement uncertainty**, used as a reference."[13]

- *Indications* (*VIM* 4.1): "**Quantity value** provided by a **measuring instrument** or a **measuring system**."[14]

- *Measurement repeatability* (*VIM* 2.21): Measurement precision under a set of repeatability conditions of measurement (s_r)[15]

- *Repeatability condition of measurement* (*VIM* 2.20): condition of **measurement**, out of a set of conditions that includes the same **measurement procedure**, same operators, same **measuring system**, same operating conditions and same location, and replicate measurements on the same or similar objects over a short period of time[16]

- *Measurement reproducibility* (*VIM* 2.25): Measurement precision under reproducibility conditions of measurement; (s_R) or (s_L) — depending on test (s_L) or calibration laboratory scenario (s_R)[17]

- *Reproducibility condition of measurement* (*VIM* 2.24): Condition of **measurement**, out of a set of conditions that includes different locations, operators, measuring systems, and replicate measurements on the same or similar objects[18]

- *Intermediate measurement precision* (*VIM* 2.23): measurement precision under a set of intermediate precision conditions of measurement (s_L)

Note: Relevant statistical terms are given in ISO 5725-3:1994.[19]

- *Intermediate precision condition of measurement* (*VIM* 2.22): condition of **measurement**, out of a set of conditions that includes the same measurement procedure, same location, and replicate measurements on the same or similar objects over an extended period of time but may include other conditions involving changes[20]

Now that we have defined some words critical to understanding the relationships for measurement uncertainty, let us look at why measurement uncertainty is not just important, but necessary to measurements (where necessary, refer to the *VIM* definitions listed earlier):

- It estimates the dispersion of values associated with the measurement as a quantifiable numerical value.
- It provides a level of confidence in one's measurement.
- It is required to ensure metrological traceability.
- It must be accounted for in decision rules when making statements of conformity.
- It is a tool to help evaluate risk.
- It is used in validation and verification of methods.
- It is required by ISO/IEC 17025:2017:
    - Clause 6.4.5, "The equipment used for measurement shall be capable of achieving the measurement accuracy and/or measurement uncertainty required to provide a valid result."[21]
    - Clause 6.4.6, "Measuring equipment shall be calibrated when the measurement accuracy or measurement uncertainty affects the validity of the reported results."[22]
- It is required for laboratory accreditation by internationally recognized accreditation bodies.
- It is a good practice to identify and evaluate the potential impact of the many sources of error encountered when one performs a measurement.

MEASUREMENT UNCERTAINTY

How does measurement uncertainty happen? What are the contributing components? The "Notes" of the *VIM* definition of *measurement uncertainty* say: "Measurement uncertainty comprises, in general, many components. Some of these may be evaluated by **Type A evaluation of measurement uncertainty** from the statistical distribution of the quantity values from a series of **measurements** and can be characterized by standard deviations. The other components, which may be evaluated by **Type B evaluation of measurement uncertainty**, can also be characterized by standard deviations, evaluated from probability density functions based on experience or other information. Measurement uncertainty includes components arising from systematic effects, such as components associated with **corrections** and the assigned quantity values of **measurement standards**, as well as the **definitional uncertainty**. Sometimes estimated systematic effects are not corrected for but, instead, associated measurement uncertainty components are incorporated."[23]

Once measurement uncertainty is calculated, it must be evaluated. Measurement uncertainty is not static. For a given measurement or set of measurements, the measurement uncertainty is associated with a stated quantity value attributed to the measurand. When this value is changed, the associated measurement uncertainty also changes.[24]

EVALUATING MEASUREMENT UNCERTAINTY

When one makes any measurement, there is a measurement uncertainty associated with it. In an ideal world, the measurement that one makes would be absolute and have a true value associated with it; we would not have to worry about it because no measurement error would be present. This is described as the "one measurement bliss or one measurement quandary depending on whether one is an optimist or a pessimist."[25] Unfortunately, we do not live in an ideal world, and no measurement is perfect. These imperfections give rise to an error in the measurement result. An error has two components: random and systematic.[26] In this chapter we discuss what these components are and how to quantify them. Granted, there are situations in which one measurement is all that it is possible to make, even though multiple factors contribute to the uncertainty of measurement. That is why one should always consider estimating the uncertainty of measurement.

This section describes the process of estimating measurement uncertainty in a test and calibration environment. Practical examples using various parameters are used to illustrate the process of determining estimated measurement uncertainty.

MEASUREMENT UNCERTAINTY EVALUATION PROCESS

The process of evaluating measurement uncertainty can be done by following the *Guide to Uncertainty in Measurement* (*GUM*)[27] approach, as further clarified in NIST SOP 29 (*SOP for the Assignment of Uncertainty*)[28]:

1. Specify the measurand by specifying the measurement process.

2. Identify all possible sources of measurement uncertainty and characterize the measurement uncertainty components.

3. Quantify each measurement uncertainty component/contributor in applicable measurement units.

4. Convert measurement uncertainty components to standard uncertainties in the units of the measurement result.

5. Calculate the combined uncertainty by combining the contributors (often by the root sum square [RSS] method).

6. Expand the combined uncertainty using an appropriate coverage factor (k factor).

    - Note: There is a common misconception that the coverage factor is $k = 2$. This is only true if the overall effective degrees of freedom is infinite at approximately 95 % confidence interval. At other effective degrees of freedom, the k coverage factor may be different.

7. Review the entries and evaluate the expanded measurement uncertainty against appropriate tolerances, user requirements, and laboratory capabilities.

8. Report correctly rounded (two significant digits) measurement uncertainties with associated measurement results in a measurement

uncertainty report, with the appropriate information at least including coverage factor and confidence interval (add notes and comments for future reference).

It is important to understand that before any measurement uncertainty determination can be made, the process (calibration or test) must be in a state of statistical control. Unfortunately, some practitioners ignore this fact and find that the measurement uncertainty budgets for their measurement parameters cannot be validated or verified later (unless the measurement uncertainty determination is applicable to the environment where the measurement system is being used). For more information on statistical process control and control charts, refer to Chapter 20. A detailed process for determining measurement uncertainty using the seven steps just outlined follows.

Identify the Sources of Measurement Uncertainty in the Measurement Process

This is a brainstorming exercise in which technicians and engineers familiar with the process (test or calibration) determine the factors affecting the measurement. Typical examples of some of the factors affecting the measurement are:[29]

- Environment (temperature, humidity, vibration, air pressure, gravity, etc.)
- Accuracy of measurement standards
- Stability of the measurement standards
- Instrument resolution
- Reference standard(s) calibration measurement uncertainty
- Repeatability
- Reproducibility
- Operator
- Measurement setup
- Method (procedure)
- Software

Evaluate and Classify Type of Uncertainty (Type A or B)

Once the factors affecting the measurement uncertainty are identified, it is important to classify the type of uncertainty. The *GUM* classification of types of uncertainty follows.

Type A evaluation method:[30] The method of evaluation of uncertainty of measurement by the statistical analysis of a series of observations. Three examples are standard deviation of a series of measurements, analysis of variances (ANOVA), and design of experiments (DOE).

Table 29.1 Type A uncertainty example.

Replicate number (n)	Measurement
1	10.05
2	9.98
3	9.97
4	9.98
5	10.01
6	10.02
7	10.03
8	10.01
9	10.05
10	10.00
Sum	100.100
Mean	10.010
Sample Standard Deviation	0.0283

A series of measurements is taken to assist in the evaluation of the estimated uncertainty of measurement. When this series of measurements is taken in a short period of time using the same instrument, the same operator, the same method, in the same conditions, this is known as *repeatability*. When there is a variation in one or more of these factors, it is known as *reproducibility*.

1. A series of measurements is made, and the results are recorded. See Table 29.1.

2. The standard uncertainty of measurement is the standard deviation of the 10 individual measurements. This is derived using a statistical method (sample standard deviation—see Figure 29.2) and is considered Type A uncertainty.

Note that in Table 29.1, the calculated sample standard deviation value is one digit more to the right. This is because one should always carry one extra digit of resolution in one's calculations from the original data.

It is important to note the following standard deviation calculations:

1. Population standard deviation (Figure 29.1)

2. Sample standard deviation (Figure 29.2)—relating to calculation in Table 29.1

3. Standard deviation of the mean (sometimes referred to as standard error of the mean or experimental standard deviation of the mean): a measure of how dispersed the data are in relation to the mean. A low standard deviation indicates that data are clustered around the mean and a high standard deviation indicates that the data are more spread out. See Figure 29.3 and Figure 29.4.

$$\sigma = \sqrt{\frac{\Sigma(x_i - \mu)^2}{N}}$$

Figure 29.1 Population standard deviation.

$$S = \sqrt{\frac{\Sigma(x_i - \bar{x})^2}{n-1}}$$

Figure 29.2 Sample standard deviation.

$$\sigma_{\bar{x}} = \frac{\sigma}{\sqrt{n}}$$

Figure 29.3 Population standard deviation of the mean.

$$S_{\bar{x}} = \frac{S}{\sqrt{n}}$$

Figure 29.4 Sample standard deviation of the mean.

Where:

σ = Population standard deviation
s = Sample standard deviation
$S_{\bar{x}}$ = Standard deviation of the mean
n = Number of measurements
$\bar{x}$ = Average of data
i = Index

In most measurement uncertainty calculations, the *sample standard deviation* and *standard deviation of the mean* are normally encountered.

If the data in Table 29.1 are for individual data measurements, the sample standard deviation is used. If, however, each line of the data in Table 29.1 was an average of five individual measurements, as shown in Table 29.2, then calculating the standard deviation of the mean to determine measurement uncertainty would be more appropriate.

Table 29.2 Examples of Type A uncertainty evaluation calculation.[31]

Calculation of Repeatability and Reproducibility Data (Type A Evaluation)						
	<< Reproducibility (Between Groups) >>					
n	Group 1	Group 2	Group 3	Group 4	Group 5	
1	10.32	9.75	9.71	9.81	9.82	
2	10.35	10.30	10.07	10.31	9.90	
3	9.85	9.71	9.79	9.67	9.90	
4	10.07	10.09	10.27	10.39	10.18	
5	10.03	10.15	10.05	10.26	9.76	
6	10.30	10.09	9.91	9.76	9.91	
7	10.11	9.68	9.94	9.76	9.73	
8	10.05	10.29	9.85	10.01	9.79	
9	9.87	10.26	9.74	9.78	9.85	
10	9.86	10.09	10.31	10.32	9.82	
Sum	50.190	50.410	49.750	49.630	49.100	
Mean (x-Bar)	10.038	10.082	9.950	9.926	9.820	
Range	0.440	0.610	0.570	0.560	0.180	
Sample Standard Deviation	0.183	0.243	0.215	0.244	0.067	
Variance	0.033	0.059	0.046	0.060	0.005	
Repeatability (s_r)	0.2016	=SQRT(AVERAGE(C18:G18))				
Reproducibility ($s_R \gg s_{X-Bar}$)	0.1022	=STDEV(C15:G15)				
Bar	0.0511	$SQRT(s_r^2 + s_R^2) =$			0.2260	
$s_L^2 =$	0.0084	$s_L^2 = s_{X-Bar}^2 - s_r^2/n$				
Intermediate Laboratory Precision (s_{LI}) =	0.0914	if s_L^2 is negative, set $s_L^2 = 0$ and $s_L = 0$				
$s_R = SQRT(s_r^2 + s_L^2) =$	0.2213					

If only one group of data is available, then Repeatability is the Sample Std. Deviation of that sample (n = 10)

Treatment of Std. Dev. of Mean	
	Mean
1	9.882
2	10.186
3	9.784
4	10.200
5	10.050
6	9.994
7	9.844
8	9.998
9	9.900
10	10.080
Mean	9.9632
If these 10 averages are made up of some unknown individual measurements ($m = 10, n = 10$) s_{X-Bar}=	0.0444
If these 10 averages are made up of 5 individual measurements ($m = 5, n = 10$) s_{X-Bar} =	0.0629

$$s_{\bar{x}} = \sqrt{\frac{\sum_{i=1}^{i=n}(\bar{x}-x_i)^2}{n-1}} \Big/ \sqrt{m}$$

Copyright © E = mc³ Solutions

Source: Reproduced by permission from E = mc3 Solutions.

Type B evaluation method:[32] A method of evaluation of uncertainty of measurement by means other than the statistical analysis of a series of observations. Some examples are the reference standard calibration measurement uncertainty; reference standard stability or drift; history of parameter; or other knowledge of the process parameter, based on specification and reference data (for example, a physics handbook).

Distributions Associated With Measurement Uncertainty[33]

There is usually another piece of information that is required to determine standard uncertainty. One must determine the type of distribution that the Type B uncertainty falls under. Usually, there are five distributions to which one can classify individual uncertainty components. They are:

- Normal distribution (also known as bell curve or Gaussian distribution)
- Rectangular distribution
- Triangular distribution
- U-shaped distribution
- Resolution distribution

Normal Distribution

The *normal distribution* is usually associated with Type A uncertainty. Examples of a normal distribution are the Type A uncertainty data obtained when a series of measurements is recorded and the measurement uncertainty is calculated using the standard deviation. It should be recognized that Type A uncertainty may be obtained from external sources, as in the instance of manufacturer specifications that are provided with a stated level of confidence.

Rectangular Distribution

The *rectangular distribution* is one where there is an equal probability of a measurement occurring within the bound limits. An example of a rectangular measurement is the specification data normally supplied by a manufacturer of an instrument, without a stated level of confidence.

Note: When the frequency distribution of a particular component of measurement uncertainty cannot be determined, the *GUM* suggests assuming the rectangular distribution and thereby erring on the conservative side.

Here is an example of determining measurement uncertainty. Say the accuracy specification for a voltmeter at 20-volt scale is ± 0.02 volt. The measurement uncertainty associated with this statement for the voltmeter at this scale is determined in the following manner.

It is classified as Type B uncertainty because the information provided does not state how the accuracy specification was derived. Information like this is usually found in the manufacturer's manual or data sheets. Therefore, it is classified as Type B.

The *distribution* that this specification falls under is rectangular. This is because the specification states that the measurement made has an equal chance of being anywhere within ± 0.02 volt.

In another example, the accuracy specification for the voltmeter at 20-volt scale is ± 0.02 volt.

This information would be considered as a rectangular distribution. To convert it to standard uncertainty, divide 0.02 volt by the square root of 3 as shown here:

$$\frac{0.02}{\sqrt{3}} = 0.01155 \text{ (standard uncertainty attributed to the voltage specification)}$$

The fourth item in a correction factor table is not necessarily a distribution but is mentioned as a special case. It is attributed to the rectangular distribution and shown best by the next example.

Say the resolution of a digital multimeter (DMM) is 0.001 volt. This is referred to as a **3.5-digit multimeter**. The standard uncertainty contribution due to the resolution of the multimeter is determined this way:

The least significant digit (LSD) of the DMM will commonly be displayed as any number from 0 to 9, depending on the fourth invisible digit resolving the meter third decimal digit. In this instance the resolution interval is 0.001 volt.

Note: Most digital hand tools have displays where the least significant digit only displays a 0 or 5. The resolution interval for this situation is 5 units.

Taking the rectangular distribution correction approach, the standard uncertainty associated with the resolution of the DMM is determined by dividing half of the resolution by the square root of 3:

$$\frac{0.0005}{\sqrt{3}} = 0.000288675$$

The resolution interval is divided in half because it is presumed that the displayed value will switch to show the next higher (or lower) digit halfway between the digits.

Another way to perform the conversion is to divide the resolution interval by the square root of 12:

$$\frac{0.001}{\sqrt{12}} = \frac{0.001}{2\sqrt{3}} = 0.000288675$$

In another example, a manufacturer specifies that the Grade 0 gage block has a specification of ± 0.1 mm. The standard uncertainty for this rectangular distribution is:

$$\frac{0.1}{\sqrt{3}} = 0.057735$$

Triangular Distribution

In a *triangular distribution*, there is a central tendency for a measurement to occur with a few dispersing values of a measurement. An example of a triangular distribution is that of a frequency measurement where there is a fixed frequency value with its associated harmonics.

Here is an example of a triangular distribution. Say a series of measurements taken indicates that most of the measurements fall at the center with a few spreading equally (±) 0.5 units away from the mean. The variance associated with this triangular distribution is:

$$u^2 = \frac{(0.5)^2}{6} = 41.667\ E-03$$

The standard uncertainty for this triangular distribution is:

$$u = \sqrt{\frac{(0.5)^2}{6}}\ 204.124\ E-03$$

U-shaped Distribution

The *U-shaped distribution* is one where there is less chance of a measurement occurring at the mean value and more chance of a value occurring at the outer bound limits. A cyclical or sinusoidal measurement usually falls under the U-shaped distribution label.

Here is an example of a U-shaped distribution. Say the temperature of the oil bath stated by the manufacturer is 100.00 ± 0.2 °C.

The variance associated with this U-shaped (trough) distribution is:

$$u^2 = \frac{(0.2)^2}{2} = 20.0\ E-03$$

The standard uncertainty for this U-shaped (trough) distribution is:

$$u^2 = \frac{(0.2)^2}{\sqrt{2}} = 141.421\ E-03$$

Quantify (evaluate and calculate) individual uncertainty by various methods.

This process sometimes works in conjunction with the evaluation process. Sometimes calculations are made in a separate process, depending on the evaluation selected.

Resolution Distribution

Another non-normal distribution, *resolution*, is the rectangular distribution for a digital resolution where the resolution value is stated in the budget as it appears on the device. The *GUM* refers to this in F.2.2.1, The resolution of a digital indication.[34] The resolution distribution contains the math of dividing the resolution in half and dividing by the square root of 3. This calculation results in dividing the actual resolution value by the square root of 12. If one were to use rectangular distribution for the digital resolution contribution, one would enter the value of the resolution divided in half, and then divide by the square root of 3. Both scenarios result in the same standard deviation.

Documenting a Measurement Uncertainty Budget

The measurement uncertainty budget lists the uncertainty contributors of the measurement process in a list with its individual uncertainties. See Figure 29.5 for

Measurement Uncertainty Budget for Calibration of Digital Balance Model DB-110 at 10 g
Created by Chris Smartypants on 2021-12-02, Rev 0

Contributor	Type	Magnitude g	Probability Distribution	Divisor	u_i g	Variance (u_i^2)	Degrees of Freedom	u_i^4/df	Percent Contribution
Repeatability	A	0.0023	Normal 1s	1.00	0.0023	5.3E-06	15	1.9E-12	0.62%
Reproducibility	A	0.00333	Normal 1s	1.00	0.00333	1.1E-05	30	4.1E-12	1.3%
Resolution (UUT)	B	0.01	Digital Res.	3.46	0.00289	8.4E-06	100	7.0E-13	1.0%
Reference Standard Uncert	B	0.0045	Normal 2s	2.00	0.00225	5.1E-06	100	2.6E-13	0.6%
Reference Standard Stability	B	0.05	Normal 2s	2.00	0.025	6.3E-04	100	3.9E-09	73.2%
Environmental Factors (Thermal stability)	B	0.02	U-shaped	1.41	0.0141	2.0E-04	100	4.0E-10	23.3%
Other contributions....					0	0.0E+00		0.0E+00	0.0%
					SUM	8.5E-04		4.3E-09	100.0%
					u_c	0.0292 g			
					veff	168			
					k	1.97			
					U	0.058 g			

Comments: Include information here regarding the identified contributors above to ensure the user of the record understands where the numbers for each contribution were derived. Also include any other relevant information needed for the interpretation of the uncertainty budget.

Definitions:
u_i = standard uncertainty of the ith contributor
df = degrees of freedom
u_c = total standard uncertainty
veff = effective degrees of freedom
k = coverage factor for 95 % coverage probability
U = total expanded measurement uncertainty

Figure 29.5 Example measurement uncertainty budget.[35]

G129—Measurement Uncertainty Budget Template used with the permission and courtesy of A2LA.

Table 29.3 Correction factors for non-normal distributions.

Distribution	Divide by	Divisor	1/Divisor
Rectangular	Square-root 3	1.7321	0.5774
Triangular	Square-root 6	2.4495	0.4082
U-shaped	Square-root 2	1.4142	0.7071
Resolution	Square-root 12	3.4641	0.2887

an example measurement uncertainty budget using the A2LA G129 Measurement Uncertainty Budget Template (https://www.a2la.org).

Note that before the measurement uncertainty contributors are combined, they should be normalized to standard uncertainty. One cannot combine rectangular distribution with triangular or normal distributions. The *GUM* provides the following correction factors for the other three non-normal distributions, rectangular, triangular, u-shaped, and resolution.[36]

All four non-normal distribution types are listed in Table 29.3.

Determining the Combined Standard Uncertainty (RSS Method)[37]

Just as one does not simply walk into Mordor,[38] one cannot simply combine individual standard uncertainty components together algebraically to calculate the combined standard uncertainty. To combine the standard uncertainty components, the root sum square (RSS) method is used. This assumes that the standard uncertainty components are random and independent.

Type A uncertainty components are added together using the RSS method depicted in Figure 29.6.

Type B uncertainty components are added together using the RSS method (see Figure 29.7).

The combined Type A and Type B components are then added to obtain the total combined uncertainty using the RSS method (see Figure 29.8).

$$u_{C_a} = \sqrt{u_{1a}^2 + u_{2a}^2 + u_{3a}^2 + \ldots + u_{na}^2}$$

Figure 29.6 Root sum square for Type A contributors.

$$u_{C_b} = \sqrt{u_{1b}^2 + u_{2b}^2 + u_{3b}^2 + \ldots + u_{nb}^2}$$

Figure 29.7 Root sum square for Type B contributors.

$$u_{C_{ab}} = \sqrt{u_{C_a}^2 + u_{C_b}^2}$$

Figure 29.8 Root sum square for combined Type A and B contributors.

An argument can be made that all Type A and Type B components can be combined at one time. If the components are combined separately in a methodical approach as shown, however, it is easier to troubleshoot calculation-related errors later.

Here is an example of using the RSS method for Type A uncertainty components:

Parameter	Standard uncertainty
Repeatability	0.0152 units
Reproducibility	0.005 units

The combined Type A uncertainty is depicted in Figure 29.9.

Here is an example of using the RSS method for Type B uncertainty components.

Parameter	Standard uncertainty
Resolution	0.001 units
Calibration	0.0002 units
Temperature	0.005 units

The combined Type B uncertainty is depicted in Figure 29.10.
The combined Type A and B uncertainty is depicted in Figure 29.11.

$$uC_a = \sqrt{0.0152^2 + 0.005^2} = 0.015811$$

Figure 29.9 Combined Type A uncertainty example.

$$uC_b = \sqrt{0.001^2 + 0.00023^2 + 0.005^2} = 0.001503$$

Figure 29.10 Combined Type B uncertainty example.

$$uC_{ab} = \sqrt{0.015811^2 + 0.001503^2} = 0.016614$$

Figure 29.11 Combined Type A and B uncertainty example.

Correlation Coefficients[39]

Some measurement uncertainty components may be correlated in their effect on the measurand. In that case, the correlation coefficients and covariance determination may be required. While it is not within the scope of this book to describe in detail how to determine correlation coefficients and covariances, it is important to point this out. Determination of correlated coefficients may be avoided by giving some thought to how the uncertainties operate.

Here is an example. If a voltmeter is used to measure the voltage of a standard thermocouple and of a thermocouple being measured by the standard, the measurement uncertainties contributed by the voltmeter are correlated if made on the same voltmeter range. The uncertainties contributed by the voltmeter for both the thermocouples (standard and measured) will be almost identical and cancel out. Thus, no determination of correlation coefficients is required.

If the measurements are made on a different range of the voltmeter, the measurement uncertainty components are partially correlated. The technician has three options:

1. If the uncertainties are relatively small, use the uncertainty for one of the measurements.

2. Calculate correlation coefficients.

3. Use the voltmeter on the same range for both the measurements, thereby canceling out the correlation effects.

For further information on correlated input quantities, refer to *JCGM 100*, Section 5.2.

Expanded Uncertainty and Choosing a Coverage Factor (*k*)[40]

Combined uncertainty is assigned a coverage factor *k* that denotes a degree of confidence interval associated with it. The *GUM* recommends that expanded uncertainty be reported at approximately a 95 % level of confidence. The *k* coverage factor may vary depending on the effective degrees of freedom, from 12.71 to 1.96 at exactly 95 % level of confidence and 13.97 to 2.00 at exactly 95.45 % level of confidence (for an exact *k* coverage value, the t-distribution table should be used based on the effective degrees of freedom and the appropriate confidence level). The effective degrees of freedom is an important consideration, especially for measurement uncertainty related to testing laboratories (and calibration laboratories, too), where Type A sample sizes for repeatability and reproducibility tests are quite small, thus resulting in smaller degrees of freedom. When this happens, the overall contribution of smaller effective degrees of freedom to measurement uncertainty is significantly larger.

Combined uncertainty multiplied by the coverage factor *k* is known as *expanded measurement uncertainty* and denoted *U*:

$$U = k * u_c$$

Various confidence interval values and their associated *k* values are shown in Table 29.4.

Table 29.4 Confidence level with associated coverage factor (k).[41]

Coverage factor (k) at ∞ degrees of freedom	Confidence level (%)
1.000	68.27
1.645	90.00
1.960	95.00
2.000	95.45
2.576	99.00
3.000	99.73

Using the combined uncertainty sum from the previous example, the expanded measurement uncertainty at $k = 2$ with an approximately 95% confidence interval is:

$$U = 2(0.016614) = 0.033228$$

Measurement Uncertainty Report[42]

All the results of the measurement uncertainty determination are documented in a measurement uncertainty report. It is important to ensure that this document contains not only the calculations (the measurement uncertainty budget), but also the reasoning used to derive and justify each of the calculations. If software is used to determine measurement uncertainty, the software must be validated by an alternate calculation method. (This includes spreadsheet templates too! See Chapter 18 for more information on validating and verifying software, including spreadsheets.) The measurement uncertainty report has to be well documented with appropriate comments where necessary so that someone other than the originator of the report can understand the reasoning behind the measurement uncertainty analysis.

The measurement uncertainty report is a living document such that, from time to time, it should be reevaluated. Examples prompting reevaluation are:

- The method (procedure) changes the measurement process.
- Operator changes.
- New or alternate equipment is used in the measurement process.
- Reference standards are recalibrated, and their reported uncertainty is updated.

When these updates occur, the respective contributors in the measurement uncertainty budget(s) also have to be updated. For changes to the process (method/operator/equipment), new repeatability and new/updated reproducibility must be updated in the uncertainty budget. When the reference standard(s) are recalibrated and there is a change to the reported reference standard(s) uncertainty, the reference standard uncertainty has to be updated in the uncertainty budget. If the laboratory calculates the stability of the reference standard, then the reference standard stability contributor might also have to be updated in the uncertainty budget. It is also a helpful practice to document these changes, such as by a revision or change control for the budget that includes who made the changes, when the changes were made, and which specific changes were made (including from what to what).

Measurement Uncertainty Considerations

The following are guidelines of factors to consider when determining measurement uncertainty. It is by no means complete, but it should give the reader some guidance when estimating measurement uncertainty of the parameters considered.

- **Dimensional (with a caliper).** In a straightforward process of measuring a dimension with a caliper, the following factors may contribute to the measurement uncertainty:
  - Caliper resolution (if not already included in its calibration measurement uncertainty)
  - Caliper calibration measurement uncertainty
  - Caliper accuracy (specification)
  - Operator
  - Method
  - Repeatability and reproducibility
  - Environment

- **Electrical (with a voltmeter).** In a straightforward process of measuring voltage with a DMM, the following factors may contribute to the measurement uncertainty:
  - DMM resolution (if not already included in its calibration measurement uncertainty)
  - DMM calibration measurement uncertainty
  - DMM accuracy (specification)
  - Operator
  - Method
  - Repeatability and reproducibility
  - Environment

- **Pressure (with a digital pressure indicator).** In a straightforward process of measuring pressure with a digital pressure indicator, the following factors may contribute to the measurement uncertainty:
  - Indicator resolution (if not already included in its calibration measurement uncertainty)

- o Indicator calibration measurement uncertainty
- o Indicator accuracy (specification)
- o Sensor specifications
- o Operator
- o Method
- o Repeatability and reproducibility
- o Environment
- o Location

- **Temperature.** In a straightforward process of measuring the temperature of a stirred temperature bath with a digital thermometer and a Type-K thermocouple, the following factors may contribute to the measurement uncertainty:
  - o Digital thermometer resolution (if not already included in its calibration measurement uncertainty)
  - o Digital thermometer calibration measurement uncertainty
  - o Digital thermometer accuracy (specification)
  - o Thermocouple calibration measurement uncertainty
  - o Thermocouple accuracy (specification)
  - o Bath stability
  - o Bath uniformity
  - o Operator (repeatability)
  - o Method
  - o Variation due to different operators (reproducibility)
  - o Environment
  - o Location of the thermocouple probe

- **Mass (with a scale).** In a straightforward process of measuring the weight of an unknown mass (also see Chapter 34, "Mass and Weighing," for more information), the following factors may contribute to the measurement uncertainty:
  - o Scale resolution (if not already included in its calibration measurement uncertainty)
  - o Scale calibration measurement uncertainty
  - o Scale accuracy (specification)
  - o Operator

- o Method (placement of the unknown mass)
- o Repeatability and reproducibility
- o Environment (vibration, local gravity)
- o Location (stability of the scale)

- **Force.** In a straightforward process of measuring force (also see Chapter 35, "Force"), the following factors may contribute to the measurement uncertainty:
  - o Force-measuring instrument resolution (if not already included in its calibration measurement uncertainty)
  - o Force-measuring instrument calibration measurement uncertainty
  - o Force-measuring instrument accuracy (specification)
  - o Mechanical interactions (hardness of material, bending, length of thread engagement, sideloads, level, plumbness, torsion)
  - o Speed of applied force (dynamic versus static, time loading profiles)
  - o Method
  - o Repeatability and reproducibility
  - o Environment
  - o Electrical—if applicable (excitation, waveform, wiring)

- **Torque**
  - o Torque calibrator resolution (if not already included in its calibration measurement uncertainty)
  - o Torque calibrator calibration measurement uncertainty
  - o Torque calibrator accuracy (specification)
  - o Operator (speed of measurement, angle of applied torque)
  - o Method
  - o Cosine error, alignment
  - o Setup (bolting, rigidity of fit)
  - o Overhung moment
  - o Correcting mass weights for force at the location of use
  - o Electrical—if applicable (excitation, waveform, wiring)
  - o Repeatability and reproducibility
  - o Environment
  - o Location

OTHER MEASUREMENT UNCERTAINTY CONSIDERATIONS

The Effect of Resolution (Readability, Sensitivity) on Measurement Uncertainty

ILAC P14:09/2020 addresses when the resolution of the inspection, measurement, and test equipment (IM&TE) being calibrated (UUT) should be included by stating, "When it is possible that the best existing device can have a contribution to [measurement] uncertainty from repeatability equal to zero, this value may be used in the Evaluation of the CMC. However, other fixed uncertainties associated with the best existing device shall be included."[43]

One cannot report the calibration measurement uncertainty for any UUT at the same value as the Calibration and Measurement Capability (CMC) uncertainty component, unless the UUT has the same performance and specifications as the "best existing device." "Contributions to the [measurement] uncertainty stated [on the calibration certificate] shall include relevant short-term contributions during calibration and contributions [such as UUT resolution] that can reasonably be attributed to the customer's device. Where applicable the [measurement] uncertainty shall cover the same contributions to uncertainty that were included in evaluation of the CMC uncertainty component, except that [measurement] uncertainty components evaluated for the best existing device shall be replaced with those of the customer's device. Therefore, reported [measurement] uncertainties tend to be larger than the [measurement] uncertainty covered by the CMC."[44]

At a minimum, the expanded measurement uncertainty should include the uncertainty of the measurement process (labeled as CMC, though it is the CMC uncertainty component), as well as the UUT's resolution. There are some instances in which the UUT's repeatability is substituted with that of the best existing device used for calibration, as referenced in *ILAC P14:09/2020*. As the resolution of the device increases, so does the overall measurement uncertainty. Not accounting for the UUT's resolution can result in an increased risk to the end user unless the same resolution was accounted for in the CMC uncertainty component.[45]

Sensitivity Coefficients[46]

Sensitivity coefficients are defined as "the differential change in the output estimate generated by a differential change in the input estimate divided by the change in that input estimate." These derivatives describe how the output estimate (y) varies with changes in the values of the input estimates ($x_1, x_2, \ldots, x_n$).

Sensitivity coefficients show how components are related to results. The sensitivity coefficient shows the relationship of the individual (measurement) uncertainty component to the standard deviation of the reported value for a test item. Most sensitivity coefficient values for Type B evaluations will be "1," with a few exceptions.

Effective Degrees of Freedom[47]

According to K. A. Brownlee, "Degrees of freedom for the standard uncertainty, u, which may be a combination of many standard deviations, is not generally known.

$$v_{\text{eff}} = \frac{u_c^4(y)}{\sum_{i=1}^{N} \frac{u_i^4(y)}{v_i}}$$

Figure 29.12   Welch-Satterthwaite formula.

This is particularly troublesome if there are large components of uncertainty with small degrees of freedom. In this case, the degrees of freedom are approximated by the Welch-Satterthwaite formula,"[48] as seen in Figure 29.12.

where

v_i = Degrees of freedom
u_c = Combined [measurement] uncertainty
u_i = Individual uncertainties
V_{eff} = Effective degrees of freedom

For further information, refer to *JCGM 100*, Annex G, Section 4.

The *t*-Distribution and Degrees of Freedom

For instances where the effective degrees of freedom are small, one needs to use the *t*-distribution (Student's distribution) to determine the *k* value for the coverage factor in the expanded combined (measurement) uncertainty. Both the *GUM (JCGM 100:2008)*[49] and the NIST/SEMATECH *e-Handbook of Statistical Methods*, 1.3.6.7.2. Critical Values of the Student's *t*-distribution,[50] further explain how and when to use Student's *t*-distributions. See Figure 29.13 for the Student's *t*-distribution table.

MANAGING MEASUREMENT UNCERTAINTY

It is important to ensure that once the measurement uncertainty for a process (calibration or test) has been determined, the data are not simply filed away and forgotten. The work is not finished. Managing measurement uncertainty is a continuous process, not a one-time exercise. When a manufacturer provides measurement uncertainty data for its product, they are specified under stated operating conditions. The manufacturer may specify 30-day, 90-day, one-year, or other periods for use under defined humidity and temperature conditions.

It is important to estimate measurement uncertainty under the laboratory's operating environment and not the manufacturer's stated conditions. From time to time, the measurement uncertainty estimates should be reevaluated to ensure that estimates have not significantly changed. At a minimum, whenever equipment is recalibrated or readjusted, an evaluation is necessary to ensure that its measurement uncertainty has not changed.

Various tools exist for managing measurement uncertainty data. Use of software minimizes calculation errors and helps in managing data from a computer workstation. Computer spreadsheet packages also help in easing data calculations.

Table G.2 — Value of $t_p(v)$ from the t-distribution for degrees of freedom v that defines an interval $-t_p(v)$ to $+t_p(v)$ that encompasses the fraction p of the distribution.

Degrees of freedom v	\multicolumn{6}{c}{Fraction p in percent}					
	68,27 [a]	90	95	95,45 [a]	99	99,73 [a]
1	1,84	6,31	12,71	13,97	63,66	235,80
2	1,32	2,92	4,30	4,53	9,92	19,21
3	1,20	2,35	3,18	3,31	5,84	9,22
4	1,14	2,13	2,78	2,87	4,60	6,62
5	1,11	2,02	2,57	2,65	4,03	5,51
6	1,09	1,94	2,45	2,52	3,71	4,90
7	1,08	1,89	2,36	2,43	3,50	4,53
8	1,07	1,86	2,31	2,37	3,36	4,28
9	1,06	1,83	2,26	2,32	3,25	4,09
10	1,05	1,81	2,23	2,28	3,17	3,96
11	1,05	1,80	2,20	2,25	3,11	3,85
12	1,04	1,78	2,18	2,23	3,05	3,76
13	1,04	1,77	2,16	2,21	3,01	3,69
14	1,04	1,76	2,14	2,20	2,98	3,64
15	1,03	1,75	2,13	2,18	2,95	3,59
16	1,03	1,75	2,12	2,17	2,92	3,54
17	1,03	1,74	2,11	2,16	2,90	3,51
18	1,03	1,73	2,10	2,15	2,88	3,48
19	1,03	1,73	2,09	2,14	2,86	3,45
20	1,03	1,72	2,09	2,13	2,85	3,42
25	1,02	1,71	2,06	2,11	2,79	3,33
30	1,02	1,70	2,04	2,09	2,75	3,27
35	1,01	1,70	2,03	2,07	2,72	3,23
40	1,01	1,68	2,02	2,06	2,70	3,20
45	1,01	1,68	2,01	2,06	2,69	3,18
50	1,01	1,68	2,01	2,05	2,68	3,16
100	1,005	1,660	1,984	2,025	2,626	3,077
∞	1,000	1,645	1,960	2,000	2,576	3,000

a) For a quantity z described by a normal distribution with expectation μ_z and standard deviation σ, the interval $\mu_z \pm k\sigma$ encompasses $p = 68,27$ percent, $95,45$ percent and $99,73$ percent of the distribution for $k = 1$, 2 and 3, respectively.

Figure 29.13 JCGM 100:2008 Table G.2, Student's t-Distribution Table.[51]

When using automated tools and utilities for managing measurement uncertainty, it is important to note that the data are only as good as the user determined them to be. The user must make responsible decisions about the quality of the data and the measurement uncertainty analysis, even when a computer is used to perform the calculations. Remember the adage from the 1960s: garbage in, garbage out. Although software tools help in automating calculations, they do not make sound decisions. That is the responsibility of the technician.

Follow good math practices. Ensure that consistent rounding number conventions are followed. Do not mix units in measurement uncertainty budgets. State all parameters in one unit. If that is not possible, state the data in percentage or express the variable value in relative terms (e.g., microvolts per volt).

Chapter 30

Measurement Risk and Decision Rules

MEASUREMENT RISK

Measurements are frequently utilized in our daily activities. We make decisions on measurements that are made to accept/reject or buy/sell products and services. A company might budget research and development (R&D) costs based on specifications required or mandated by the customer or regulations. Medical professionals use measurement results to assess the outcome of diagnostic tests. Different industries utilize test method results by conducting defined or random sampling methods on individual parts, sub-assemblies, and assemblies to make conformance decisions. For example, a satellite to be deployed in Earth's orbit will undergo various environmental, structural, reliability, and other tests.

Imagine that a satellite that has been launched into space behaves in a way in which it is not designed to function. Suppose the satellite experiences a wobble, which causes connectivity problems at the receiving station on Earth. The cause of the wobbling is identified as use of a calibration provided with an incorrect measurement uncertainty, which thus resulted in an incorrect measurement conformance decision. For example, each of the load cells used to measure the fuel load stored in the satellite must include the confidence probability limit in its calibration. However, if a calibration provider does not have the measurement confidence due to incorrect assumptions, the load cell calibration may not be fit for the purpose. In this case, the result is a wobbling satellite requiring significantly more resources to fix the problem. A wobbly satellite is one example of many where measurement uncertainty, when not adequately accounted for, leads to significant engineering disasters, trillions of dollars in losses, and potentially the loss of human life. Fixing the faulty satellite is not as easy as hopping on a commercial flight and renting a car to get to it! These scenarios are played out daily in all fields of measurement, including the treatment of diseases and medical advances to save or prolong life.

Measurements made and reported for any application *must exhibit measurement confidence*, and the *decision made must be fit for the purpose*.

Measurement confidence comes from these "three pillars of measurement":

1. Measurement Uncertainty

2. Metrological Traceability

3. Measurement Decision Rules fit for the purpose, taking into account the acceptable risk

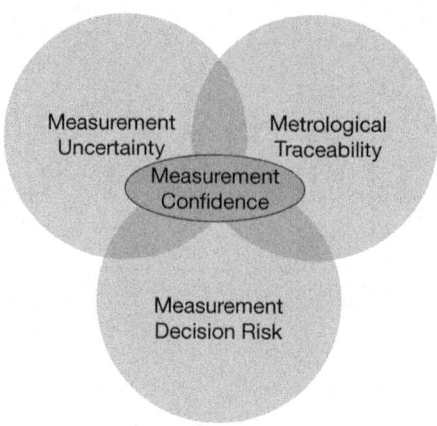

Figure 30.1 Three pillars of measurement.[1]
Source: Reproduced by permission from E = mc³ Solutions.

Accounting for these three considerations shall result in reporting statements of conformity with a high degree of confidence. Consider manufacturers and calibration laboratories that fail to correctly calculate measurement uncertainty and assess risk on equipment that make a statement of conformity (pass or fail) on products and services. Would this make the end users feel confident when they schedule their subsequent surgery, sit in a traffic jam on a bridge, or experience mechanical problems on their next flight?

This chapter examines **measurement risk** and **decision rules**, and how they instill **measurement confidence**. It provides guidance, citing several published standards, to help readers become better consumers of test and calibration services by making more informed decisions to reduce risk to their customers. It provides guidance for suppliers to make better conformance decisions on the products and services they provide.

Understanding Measurement Risk

AS9100D defines *risk* as "[a]n undesirable situation or circumstance that has both a likelihood of occurring and a potentially negative consequence."[2] The definition can be simplified to describe measurement risk as the probability that a measurement result can lead to an incorrect decision. In metrology, there is always uncertainty associated with any measurement value reported. To quote *Introduction to Statistics in Metrology*, "Since the actual value of a measurand can never be known, there is always some probability that a measured value falls within an acceptable region when the actual value lies outside that region. This results in a False Accept (FA) condition. There is also some probability that a measured value falls outside an acceptable region when the actual value lies inside that region. This results in a False Reject (FR) condition. National standard ANSI/NCSL Z540.3 2006 (see Clause 3.7.4) recommends that compliance testing should control the false accept risk to be no more than 2 %. False accept and false reject risks are only one part of a full risk assessment."[3]

False accept and false reject concepts can also be simplified. In short: All measurements have a percentage of likelihood of calling something good when it is bad, and something bad when it is good. Decision rules rely on meeting tolerance or specification limits.

Decision Rules

ISO/IEC 17025:2017 defines a *decision rule* as a "rule that describes how measurement uncertainty is accounted for when stating conformity with a specified requirement."[4] ISO/IEC 17025 does not specify what the decision rule must be. However, there are other documents, such as ILAC G8/2019, *Guidelines on Decision Rules and Statements of Conformity*; JCGM 106:2012, *Evaluation of measurement data— The role of measurement uncertainty in conformity assessment*; UKAS LAB 48, *Decision Rules and Statements of Conformity*; *Handbook for the Application of ANSI Z540.3-2006: Requirements for the Calibration of Measuring and Test Equipment*, and standards that adhere to an approach based on acceptable risk. ISO/IEC 17025:2017 requires calibration laboratories to identify all significant contributions to the uncertainty and then apply that combined uncertainty, along with a "decision rule," when determining pass/fail, in-tolerance/out-of-tolerance condition (conformance). The standard goes into greater detail about a risk-based approach; however, it does not call out what level of risk is acceptable.

ILAC G8:09/2019, *Guidelines on Decision Rules and Statements of Conformity*, has several examples of decision rules and the corresponding risk associated with those rules. The standard gives an excellent summary, stating: "In all cases, decision rules need to be compatible with the customer, regulation or standard requirements. They need to be agreed upon and documented before the work starts. It must be clear that the tolerance limits are consistent with the requirements and that all measurement uncertainty and other calculations are performed consistent to the ISO/IEC 17025:2017 requirements. The agreed decision rule employed for statements of conformance must be clearly documented in the measurement report."[5]

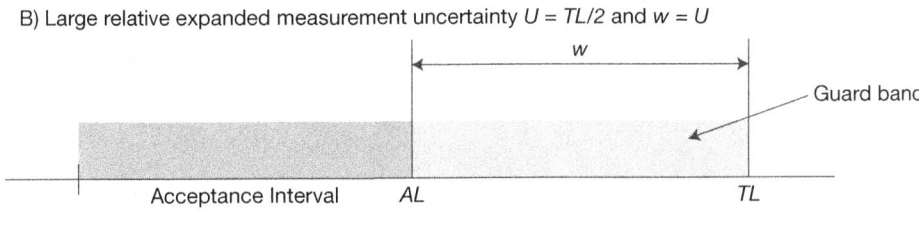

Figure 30.2 Guard band and acceptance interval illustration found in *ILAC G8:09/2019*.
Source: Reproduced with permission from ILAC.

JCGM 106:2012 states, "A decision to accept an item as conforming, or reject it as non-conforming, to specification is based on a measured value of a property of the item in relation to a stated decision rule that specifies the role of measurement uncertainty in formulating acceptance criteria. An interval of measured values of a property that results in acceptance of the item is called an acceptance interval (see 3.3.9), defined by one or two acceptance limits (see 3.3.8)."[6]

UKAS's *LAB 48: Decision Rules and Statements of Conformity*, lists 14 different examples from a validated method to inspection levels (conformity decisions for discrete measurements). Example 3 is a test scenario in which a customer asks a laboratory to ignore uncertainty—something that happens throughout industry. This is dealt with by reaffirming that "accredited reporting of the outcome for such a decision is not permitted by ISO/IEC 17025:2017 nor by ILAC-G8:09/2019 which require that uncertainty should be taken into account (directly or indirectly) when conformity decisions are made."[7]

NCSL International's *Handbook for the Application of ANSI Z540.3-2006: Requirements for the Calibration of Measuring and Test Equipment* provides four methods for estimating probability of false accept (PFA) and two commonly accepted guard-band methods. The guard-band methods are known as Method 5 and Method 6. Method 5 uses guard bands based on the expanded calibration process uncertainty. Method 6 uses guard bands based on the Test Uncertainty Ratio (see Appendix A).[8]

Tolerance/Specification Limits

Most accept or reject decisions are based on specification limits. It should be noted that the terms *specifications* and *tolerance limits* are sometimes used interchangeably. Specifications may be qualitative or quantitative. Tolerances are usually quantitative in nature and are based on scientific evaluation. These can also be referred to as *acceptance limits*. The *International Vocabulary of Metrology* (*VIM*) uses the term "tolerance limits" in place of *specification* or *acceptance limits*. *VIM* defines the tolerance interval to be the range of acceptable values between the limits. *VIM* defines *maximum permissible measurement error* as "extreme value of measurement error, with respect to a known reference quantity value, permitted by specifications or regulations for a given measurement, measuring instrument, or measuring system.

NOTE 1: Usually, the term 'maximum permissible errors' or 'limits of error' is used where there are two extreme values.

NOTE 2: The term 'tolerance' should not be used to designate 'maximum permissible error.'"[9]

This is an important distinction because many wrongly assume that a specification limit indicates the overall error. Many manufacturers do not understand measurement uncertainty and fail to consider it when developing their tolerance or specification limits.

For example, a manufacturer may sell a system with a specification of:

± 0.1 % of full scale

The end user has determined that this is suitable for their application. However, in many cases, the maximum permissible error of the system is far larger than the specification, and the actual uncertainty when using the system (Calibration Process Uncertainty) could be much more significant.

Risk Types

The reader might be familiar with the terms:

- **Consumer Risk (β or Type II error)**
- **Producer Risk (α or Type I error)**

Consumer risk refers to the supplier of a product assuming that a product meets all applicable requirements (tolerance) and passing the risk on to the consumer who assumes the risk. If the product does not conform to the stated requirements, the consumer or the end-user assumes the risk. An example is the under-evaluation of measurement uncertainty utilized in the decision rule, resulting in a False Accept decision.

Another example of this is the case of a consumer smartphone in which the installed batteries could overheat and catch fire. Because the batteries were lithium-based, they generated oxygen upon combustion, which fed the fire and could not be easily extinguished. In this case, the phone device's batteries were tested in an isolated test that was not appropriate for the purpose, rather than in a compact design smartphone. If the batteries had been tested in the compact smartphone packaging—the application for which the item was intended—then the problem might have been caught and prevented, resulting in an alternate design that could have averted the risk posed by self-feeding combustion in an unsuitable environment.

		\multicolumn{2}{c	}{Type I - Type II Error}
		\multicolumn{2}{c	}{Calibration}
		Conforming (GOOD)	Non-Conforming (BAD)
Decision Made	Conforming (ACCEPT)	(1-α) Calibration Lab's Confidence (Probability of Correct Accept – PCA)	Consumer's Risk β or Type II Error (Probability of False Accept – PFA)
	Non-Conforming (REJECT)	Producer's Risk α or Type I Error (Probability of False Reject – PFR)	(1-β) End User's Confidence (Probability of Correct Reject – PCR)

Figure 30.3 Error types.[10]
Source: Reproduced by permission from E = mc³ Solutions.

Producer risk refers to the supplier of a product and service determining that the product does not conform to the requirements (tolerance) and spending extra resources to determine why the nonconformance does not (or does) lead to possible further incorrect assumptions. This may be because the producer does not have the right decision tools to make a conformance decision correctly. An example would be an over-evaluation of measurement uncertainty utilized in the decision rule, resulting in a False Reject decision.

As was noted, measurement confidence comes from measurement uncertainty, metrological traceability, and measurement decision rules based on the risk assumed.

The examples given in the following scenarios are assumed to be based on discrete measurement at the bench level (specific risk).

Scenario 1:
Let us assume the following measurement requirements and the collection of data:

- Making a measurement on a nominal quantity value of 10 units

- With a requirement that it conforms to a value of ± 2 units around the nominal (between 8 to 12 units).

- A first, single measurement of 10.5 units is made on this measurand.

  o It is assumed to be within the conformance, making no other assumptions.

  o This is the measurement decision, within conformance, also called *simple acceptance decision rule*.

- A second measurement is made on the measurand for verification, and it is 11.0 units.

The requirements have not changed: Nominal value is 10 units and the conformance requirement is within 8 to 12 units.

Now there is doubt introduced about the first measurement of 10.5 units.

- A third measurement is made, and it is 10.8 units.

Further doubt is introduced, so it is decided to make two additional measurements. The resulting measurements are:

- 10.9 and 11.1 units, for a total of five measurements (sample size: $n = 5$).

If only one measurement had been made and a conformance decision was made on that basis, the measurement would have been in conformance! This scenario is often referred to as *single measurement bliss*.[11] However, **bliss can be ignorance in disguise.**

As additional measurements are made, more information about the measurement process becomes known, providing further confidence in the reported result. In simple terms, the measurement uncertainty of the measurement process was evaluated considering the sample standard deviation of the five repeated measurements.

Scenario 1:
The resulting data are summarized:

Sample	Measurement
1	10.5
2	11.0
3	10.8
4	10.9
5	11.1
Sample Mean	10.86
Sample Standard Deviation	0.23

Figure 30.4 Scenario 1 data.

The measurement is within the conformance requirements shown graphically:

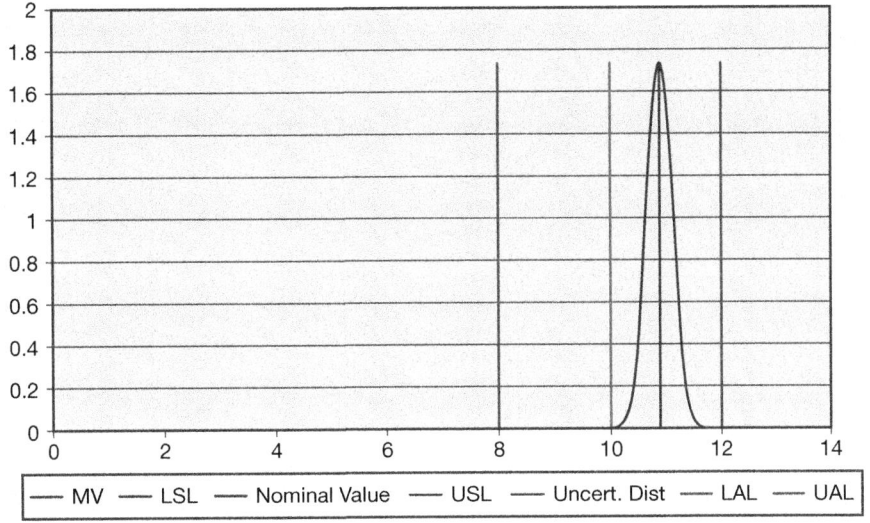

Figure 30.5 Scenario 1 graph.

Source: Reproduced by permission from Morehouse Instruments and E = mc³ Solutions.

Note: Figures 30.5, 30.7, 30.9, 30.11, and 30.27 use the following acronyms that correspond with the information found in the corresponding data figures:

MV = Measured Value
LSL = Lower Specification Limit
USL = Upper Specification Limit
Uncert. Dist = Uncertainty Distribution
LAL = Lower Acceptance Limit
UAL = Upper Acceptance Limit

The decision rule employed is that the measurement result of 10.86 units (average of five measurements) is within the conformance/tolerance requirement of 10 ± 2 units. This is also known as the *simple acceptance decision rule* (shared risk). Nothing else is implied.

How much measurement confidence does this reported result provide if no consideration of measurement uncertainty is made?

Scenario 2:
The same simple acceptance decision rule is employed. The location of the measurement has changed. The measurement result is equal to the upper specification limit of 12 units.

	Reported Result
Nominal Value	10
Lower Specification Limit	8
Upper Specification Limit	12
Measured Value	12
Std. Uncert. (k = 1)	0.230
Total Risk	50.000%
Upper Limit Risk	50.000%
Lower Limit Risk	0.000%

Figure 30.6 Scenario 2 data.

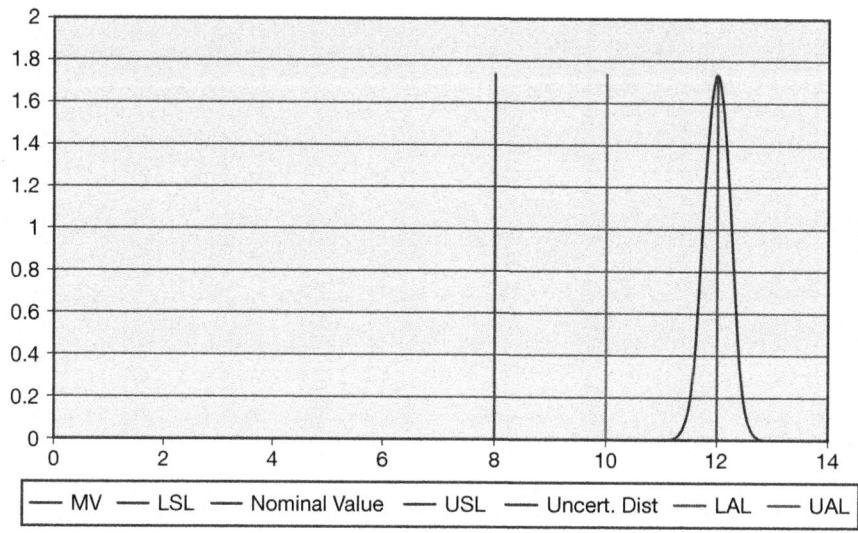

Figure 30.7 Scenario 2 graph.[12]
Source: Reproduced by permission from Morehouse and E = mc³ Solutions.

There is a probability of false accept (PFA) of 50.000 % in this scenario. The calibration laboratory has not taken measurement uncertainty into account and has issued a conformity decision of a pass condition based on simple acceptance. The difference between Scenario 1 and 2 is that the PFA has increased based on the measurement location.

It is always important to consider the location of the measurement and the associated measurement uncertainty when evaluating the measurement risk.

ILAC-G8 defines *simple acceptance* as "a decision rule in which the acceptance limit is the same as the tolerance limit, i.e., $AL = TL$."[13] However, defining a decision rule that does not take measurement uncertainty into account contradicts the very definition of a decision rule in ISO/IEC 17025:2017!

ILAC-G8 continues, "a guard band which has a length equal to zero, $w = 0$, infers that acceptance is when a measurement result is below a tolerance limit. . . . Simple acceptance is also called 'shared risk' because the probability to be outside the tolerance limit may be as high as 50 % in the case when a measurement result is exactly on the tolerance limit (assuming a symmetric normal distribution of the measurements)."[14]

> *If the uncertainty evaluation does not consider known contributors to* ***measurement uncertainty****,*
> *then there is a question of whether the calibration is* ***metrologically traceable****.*

The United Kingdom Accreditation Service (UKAS) elaborates on this point: "Conformity statements under ISO/IEC 17025:2017 require a Decision Rule (3.7) that takes account of measurement uncertainty. Some people argue that it is possible to 'take account' by ignoring it if that is what the customer requests; however, this seems to require a rather contradictory belief that you can be 'doing something' by 'not doing something' (is it possible to 'obey a red stoplight' by 'not obeying a red stoplight'?)."[15]

JCGM 106:2012 further explains the intent of simple acceptance by stating: "In practice, in order to keep the chances of incorrect decisions to levels acceptable to both producer and user, there is usually a requirement that the measurement uncertainty has been considered and judged to be acceptable for the intended purpose."[16]

Scenario 3:
ISO/IEC 17025:2017 states the following about reporting statements of conformity:

> 7.8.6.1 When a statement of conformity to a specification or standard is provided, the laboratory shall document the decision rule employed, taking into account the level of risk (such as false accept and false reject and statistical assumptions) associated with the decision rule employed, and apply the decision rule.
>
> NOTE Where the decision rule is prescribed by the customer, regulations or normative documents, a further consideration of the level of risk is not necessary.
>
> 7.8.6.2 The laboratory shall report on the statement of conformity, such that the statement clearly identifies:
>
> a) to which results the statement of conformity applies;
>
> b) which specifications, standards or parts thereof are met or not met;
>
> c) the decision rule applied (unless it is inherent in the requested specification or standard).

NOTE: For further information, see ISO/IEC Guide 98-4.[17]

This requires that a decision rule be employed with the associated risks which is acceptable to the customer, regulations, or normative documents as required. Quantitative decision rules involve taking the *measurement uncertainty* into account to calculate and assess the risk. To illustrate this, the ILAC G8:03/2009 decision rule referenced in ILAC-G8:09/2019 is used; this is also referenced in the ANSI/NCSL Z540.3 *Handbook* as Method 5. The rule establishes an acceptance level (AL) that is determined by subtracting half of the expanded measurement uncertainty, at 95 % confidence interval, from both the upper and lower tolerance limits and thus minimizing the acceptable risk (PFA) to less than 2.5 %.[18]

Upper acceptance limit (UAL) = Upper specification limit (USL) − $U_{95\%}/2$
Lower acceptance limit (LAL) = Lower specification limit (LSL) + $U_{95\%}/2$

The region between the specification limits (SLs) and the acceptance limits (ALs) is known as the *guard band* (GB).

Using the decision rule from ILAC-G8, limiting the risk to less than 2.5 % PFA, and the measured value of 10.86 Units, the PFA is 0.000 %.[19] This is because the measurement uncertainty distribution area does not go beyond the upper and lower specification limits.

	Reported Result
Nominal Value	10
Lower Specification Limit	8
Upper Specification Limit	12
Measured Value	10.86
Std. Uncert. (k=1)	0.230
Total Risk	0.000%
Upper Limit Risk	0.000%
Lower Limit Risk	0.000%

Figure 30.8 Scenario 3 data where PFA is 0.000 %.

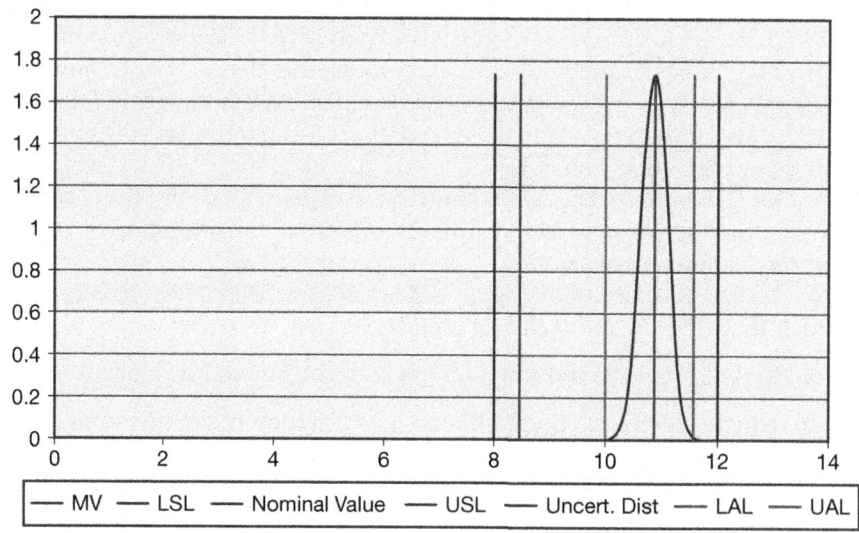

Figure 30.9 Scenario 3 graph where PFA is 0.000 %.[20]

Source: Reproduced by permission from Morehouse Instruments and E = mc³ Solutions.

The actual reported measurement must be less than the upper acceptance limit (11.55 units) before the measurement result is found to be out of conformance (PFA > 2.500 %).

Using the ILAC G8:03/2009 decision rule, the conformance decision limiting the risk (PFA) to less than 2.500 % is illustrated in Figure 30.10 and Figure 30.11.

	Reported Result
Nominal Value	10
Lower Specification Limit	8
Upper Specification Limit	12
Measured Value	11.549
Std. Uncert. (k=1)	0.230
Total Risk	2.497%
Upper Limit Risk	2.497%
Lower Limit Risk	0.000%

Figure 30.10 Scenario 3 data where PFA is larger than 2.5 %.

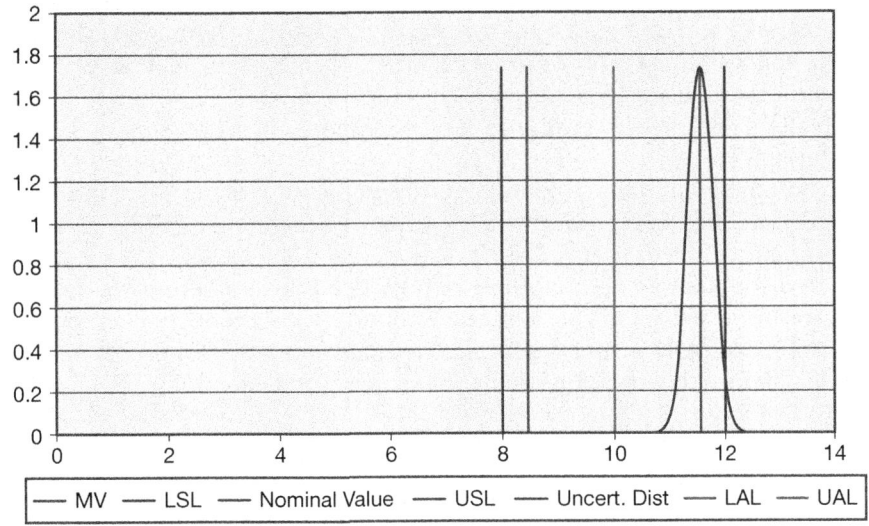

Figure 30.11 Scenario 3 graph where PFA is larger than 2.5 %.[21]
Source: Reproduced by permission from Morehouse and E = mc³ Solutions.

Many different decision rules have been published, and one can determine one's own decision rules based on the risk acceptance level to make conformance decisions. One can use the same equipment in different applications—but the risk associated with the application may be sufficient with one application and not sufficient with another application.

An example would be a simple tire pressure gage whose conformance criteria may be sufficient for checking the pressure of bicycle tires, which is considered low risk, yet not sufficient for checking the pressure of tires on commercial aircraft carrying passengers, which is high risk.

Another example is that of two commercial accredited calibration providers who list different measurement uncertainties for calibrating the same equipment on their scope of accreditation (SoA). This could be for a variety of reasons:

- Use of different standards with different measurement uncertainties (Calibration and Measurement Capability, CMC) utilized by the laboratories

- Incorrect evaluation of measurement uncertainty (CMC) listed on their Scope of Accreditation (SoA)
- Incorrect application of decision rules to make conformance decisions
- Not understanding the conformance requirements of the customer, regulatory standards, or normative standards

Most importantly, making a conformance decision requires the engagement of the customer or end user. Traditionally, the consumer (customer) of calibration and test services relies on the expertise of the calibration and test provider, and the conformance decision relies on the manufacturer's specifications without taking the associated measurement uncertainty into account. Therefore, it becomes important to note the requirements of ISO/IEC 17025:2017: "When the customer requests a statement of conformity to a specification or standard for the test or calibration (e.g., pass/fail, in-tolerance/out-of-tolerance), the specification or standard and the decision rule shall be clearly defined. Unless inherent in the requested specification or standard, the decision rule selected shall be communicated to, and agreed with, the customer.

Note: For further guidance on statements of conformity, see ISO/IEC Guide 98-4."[22]

The customer must understand the capability of its test and calibration provider and its own conformance requirements before it can specify the test and calibration service to reduce their risk appetite. It must specify the decision rules required for the test and calibration provider that is making the conformance decision on a product. The test and calibration provider must understand the customer requirements for conformance decisions so that when the measurement result is reported with a confidence statement it considers:

- The proper evaluation measurement uncertainty
- Proof of metrological traceability
- Proper application of measurement decision rules to address the risk required in reporting any statements of conformity

TEST UNCERTAINTY RATIO (TUR)

Another decision rule that is quite frequently employed is the application of the *Test Uncertainty Ratio (TUR)*. This rule has its roots in the U.S. military specifications, and its application was listed in the MIL STD 45662 as early as the 1960s. The TUR has its limitations regarding conformance decisions and risk mitigation. In addition, there are different interpretations of how the TUR is calculated, which leads to further confusion.

Calculation of Test Uncertainty Ratio

Correctly calculating TUR is crucial because it is a commonly accepted practice when making a statement of conformity. It helps you analyze measurement risk and is often used in decision rules. Although TUR is defined in the ANSI/NCSL Z540.3 standard (NCSL International's *Handbook for the*

$$TUR = \frac{Span\ of\ the\ \pm\ UUT\ Tolerance}{2\ x\ k_{95\%}(Calibration\ Process\ Uncertainty)}$$

Figure 30.12 TUR formula (*UUT*, unit under test).[23]

Interpretation of ANSI/NCSL Z540.3), and is referenced in ILAC-G8:09/2019, these universal guidance documents are not always followed. Furthermore, other documents have different contributors for determining the denominator in the TUR calculation.

Many decision rules require a TUR calculation. The formula's ratio includes a numerator and a denominator. In "The New Dimension to Resolution: Can It Be Resolved?," Zumbrun and Shah state that understanding TUR is the first step when using TUR to make a conformity decision.[24] TUR is defined as:

- The ratio of the span of the tolerance of a measurement quantity subject to calibration, to twice the 95 % expanded uncertainty of the measurement process used for calibration[25] (also see Figure 30.12)

- The ratio of the tolerance, TL, of a measurement quantity, divided by the 95 % expanded measurement uncertainty of the measurement process where TUR = TL/U [U = unit][26]

Zumbrun and Shah describe how these definitions are similar. If the tolerance is not symmetrical, then the ANSI/NCSL Z540.3 definition is more concise. The calculation of TUR is one of the decision rules utilized in making a statement of conformity. "When used in combination with the measurement location, one can calculate measurement risk at the time of calibration."[27]

The ANSI/NCSL Z540.3 *Handbook* states: "For the numerator, the tolerance used in the calibration procedure should be used in the calculation of the TUR. This tolerance is to reflect the organization's performance requirements for the M&TE which are, in turn, derived from the intended application of the M&TE. In many cases, these performance requirements are those described by the manufacturer's tolerances and specifications for the M&TE and are therefore included in the numerator."[28]

In most cases, the numerator is the UUT accuracy tolerance. The denominator is slightly more complicated. Per the ANSI/NCSL Z540.3 *Handbook*, "For the denominator, the 95 % expanded uncertainty of the measurement process used for calibration in accordance with the calibration procedure is to be used to calculate TUR. The value of this uncertainty estimate should reflect the results that are reasonably expected from the use of the approved procedure to calibrate the particular type of M&TE. Therefore, the estimate includes all components of error that have an influence on the measurement results of the calibration which would also include the influences of the item being calibrated with the exception of the bias of the M&TE. The calibration process error, therefore, includes temporary and non-correctable influences incurred during the calibration such as repeatability, resolution, error in the measurement source, operator error, error in correction factors, environmental influences, etc."[29]

$$TUR = \frac{Span\ of\ the\ \pm Tolerance}{2 \times k_{95\%} \left(\sqrt[2]{\left(\frac{CMC}{k_{CMC}}\right)^2 + \left(\frac{Resolution_{UUT}}{\sqrt[2]{12}}\right)^2 + \left(\frac{Repeatability_{UUT}}{1}\right)^2 + \cdots (u_{Other})^2} \right)}$$

Figure 30.13 Example of a TUR formula (adapted from the ANSI/NCSL Z540.3 *Handbook*).
Source: Reproduced by permission from Morehouse and E = mc³ Solutions.

This definition of the TUR denominator aligns very closely with that of ILAC P14:09/2020, which states: "Contributions to the uncertainty stated on the calibration certificate shall include relevant short-term contributions during calibration and contributions that can reasonably be attributed to the customer's device. Where applicable, the uncertainty shall cover the same contributions to uncertainty that were included in evaluation of the CMC uncertainty component, except that uncertainty components evaluated for the best existing device shall be replaced with those of the customer's device. Therefore, reported uncertainties tend to be larger than the uncertainty covered by the CMC."[30]

As seen in Figure 30.13, *TUR* comprises the minimum contributors that should be included in the denominator for Calibration Process Uncertainty (CPU) to correctly calculate TUR. The formula includes the ratio of UUT accuracy tolerance, which manufacturers often request as the accuracy specification, compared against the expanded uncertainty of the calibration process.[31]

"At a minimum, the expanded uncertainty should include the uncertainty of the measurement process (labeled as CMC, though it is the CMC Uncertainty Component), as well as the UUT's resolution. There are some instances in which the UUT's repeatability is substituted with that of the best existing device used for calibration, as referenced in ILAC P-14:09/2020.

Not accounting for the UUT's resolution can result in an increased risk to the end-user unless the same resolution was accounted for in the CMC uncertainty component."[32]

The Calibration Process Uncertainty definition establishes whether relevant uncertainty contributors of the customer's device will be considered at the time of calibration. If a calibration laboratory does not include the appropriate uncertainty contributors, it passes the risk on to the customer or consumer.[33]

Test Uncertainty Ratio Calculation Example

The following example (from "Why a 4:1 TUR is not Enough"[34]) illustrates a TUR calculation.

- A customer sends a **10,000 lbf** load cell for calibration with an accuracy specification of **± 0.05 %** **of full scale**.
- The calibration provider uses a universal calibrating machine to perform the calibration.
- When **10,000 lbf** is applied, the unit reads **10,001 lbf**.
- The display resolution is **1 lbf**.

Step 1: Calculate the numerator.

$$TUR = \frac{\text{Span of the} \pm \text{Tolerance}}{2 \times k_{95\%} \left(\sqrt[2]{\left(\frac{CMC}{k_{CMC}}\right)^2 + \left(\frac{Resolution_{UUT}}{\sqrt[2]{12}}\right)^2 + \left(\frac{Repeatability_{UUT}}{1}\right)^2 + \cdots (u_{Other})^2} \right)}$$

Figure 30.14 TUR formula numerator.

The device is a 10,000-lbf load cell with an accuracy specification of ± 0.05 %. Therefore, **10,000 × 0.0005 = ± 5 lbf**.

- The upper specification limit is 10,000 + 5 = 10,005 lbf.
- The lower specification limit is 10,000 – 5 = 9,995 lbf.

Therefore, the span of the ± Tolerance is 10,005 – 9,995 = 10 lbf.

See Figure 30.15 for the inclusion of 10 lbf as the numerator.

$$TUR = \frac{10 \; lbf}{2 \times k_{95\%} \left(\sqrt[2]{\left(\frac{CMC}{k_{CMC}}\right)^2 + \left(\frac{Resolution_{UUT}}{\sqrt[2]{12}}\right)^2 + \left(\frac{Repeatability_{UUT}}{1}\right)^2 + \cdots (u_{Other})^2} \right)}$$

Figure 30.15 TUR formula with the numerator value included.

Step 2: Calculate the denominator

Everything is calculated to **1 standard deviation (Standard Uncertainty)** for this calculation.

Step 2-a: Calibration and Measurement Capability (CMC). See Figure 30.16 for inclusion of the CMC in the denominator.

$$TUR = \frac{\text{Span of the} \pm \text{Tolerance}}{2 \times k_{95\%} \left(\sqrt[2]{\left(\frac{CMC}{k_{CMC}}\right)^2 + \left(\frac{Resolution_{UUT}}{\sqrt[2]{12}}\right)^2 + \left(\frac{Repeatability_{UUT}}{1}\right)^2 + \cdots (u_{Other})^2} \right)}$$

Figure 30.16 CMC portion of the denominator.

- CMC is the uncertainty at the calibrated force.
- The universal calibrating machine has an uncertainty of **0.02 %** at **10,000 lbf**, which means the CMC is **10,000 × 0.0002 = 2 lbf**.
    - The coverage factor, k_{CMC}, is **2**, which was listed on the calibration provider's certificate.

Dividing the CMC by 2, the standard uncertainty is reported at **one standard deviation**. In most cases, the **CMC uncertainty component is reported at approximately 95 %**, and a coverage factor of $k = 2$ is used. See Figure 30.17 for the inclusion of CMC value in the denominator.

$$TUR = \frac{10\ lbf}{2 \times k_{95\%} \left(\sqrt[2]{\left(\frac{2\ lbf}{2}\right)^2 + \left(\frac{Resolution_{UUT}}{\sqrt[2]{12}}\right)^2 + \left(\frac{Repeatability_{UUT}}{1}\right)^2 + \cdots (u_{Other})^2} \right)}$$

Figure 30.17 TUR formula with CMC value included.

Step 2-b: UUT resolution. See Figure 30.18 for the inclusion of resolution in the denominator.

$$TUR = \frac{Span\ of\ the\ \pm Tolerance}{2 \times k_{95\%} \left(\sqrt[2]{\left(\frac{CMC}{k_{CMC}}\right)^2 + \left(\frac{Resolution_{UUT}}{\sqrt[2]{12}}\right)^2 + \left(\frac{Repeatability_{UUT}}{1}\right)^2 + \cdots (u_{Other})^2} \right)}$$

Figure 30.18 Resolution portion of the denominator.

Note: **Resolution**$_{UUT}$ for force instrument is calculated by dividing the force applied by the output at applied force, then multiplying this value by the instrument's readability.
The **Resolution**$_{UUT}$ is **(10,000 lbf / 10,000 lbf) × 1 = 1 lbf**.

To convert 1 lbf resolution to standard uncertainty, it is either divided by the square root of 12, or the square root of 3 depending on the type of resolution. See Figure 30.19 for inclusion of resolution in the denominator.

$$TUR = \frac{10\ lbf}{2 \times k_{95\%} \left(\sqrt[2]{\left(\frac{2\ lbf}{2}\right)^2 + \left(\frac{1\ lbf}{\sqrt[2]{12}}\right)^2 + \left(\frac{Repeatability_{UUT}}{1}\right)^2 + \cdots (u_{Other})^2} \right)}$$

Figure 30.19 TUR formula with the resolution value included.

Step 2-c: Repeatability. See Figure 30.20 for inclusion of repeatability in the denominator.

$$TUR = \frac{Span\ of\ the\ \pm Tolerance}{2 \times k_{95\%} \left(\sqrt[2]{\left(\frac{CMC}{k_{CMC}}\right)^2 + \left(\frac{Resolution_{UUT}}{\sqrt[2]{12}}\right)^2 + \left(\frac{Repeatability_{UUT}}{1}\right)^2 + \cdots (u_{Other})^2} \right)}$$

Figure 30.20 Repeatability portion of the denominator.

For this example, five replicate readings are taken.

Repeatability is obtained by applying a force of **10,000 lbf** to the **unit under test (UUT)** five times, and the sample standard deviation of five replicated measurements is calculated.

Repeatability of sample size five: (10,000, 10,001, 10,000, 10,001, 10,001) = 0.54772. (Because the repeatability is already expressed as one standard deviation, the divisor is 1.) See Figure 30.21 for inclusion of the repeatability value in the denominator.

$$TUR = \frac{10\ lbf}{2 \times k_{95\%} \left(\sqrt[2]{\left(\frac{2\ lbf}{2}\right)^2 + \left(\frac{1\ lbf}{\sqrt[2]{12}}\right)^2 + \left(\frac{0.54772}{1}\right)^2 + \cdots (u_{Other})^2} \right)}$$

Figure 30.21 TUR formula with repeatability value added.

Step 2-d: Other uncertainty contribution sources. See Figure 30.22 for inclusion of other error sources in the denominator.

$$TUR = \frac{\text{Span of the} \pm \text{Tolerance}}{2 \times k_{95\%} \left(\sqrt[2]{\left(\frac{CMC}{k_{CMC}}\right)^2 + \left(\frac{Resolution_{UUT}}{\sqrt[2]{12}}\right)^2 + \left(\frac{Repeatability_{UUT}}{1}\right)^2 + \cdots (u_{Other})^2} \right)}$$

Figure 30.22 Other uncertainty contribution sources in the denominator.

Other measurement uncertainty contribution sources attributed to the calibration process uncertainty can be considered for the UUT. Some examples are environmental influences, error in correction factors, and so on. For this example, other measurement uncertainty contribution sources are inherent in repeatability and in the CMC.

$$TUR = \frac{10\ lbf}{2 \times k_{95\%} \left(\sqrt[2]{\left(\frac{2\ lbf}{2}\right)^2 + \left(\frac{1\ lbf}{\sqrt[2]{12}}\right)^2 + \left(\frac{0.54772}{1}\right)^2} \right)}$$

Figure 30.23 TUR formula with all uncertainty contribution sources included.

Step 3: Calculate the Denominator

Sum of all the contributors = $\sqrt{((2/2)^2 + (1/3.464)^2 + (0.54772/1)^2)} = 1.1762$

$$\mathbf{TUR} = \frac{10\ lbf}{2 \times k_{95\%}\ (1.1762)}$$

Figure 30.24 TUR calculated-formula.

366 Part V: Uncertainty in Measurement

The specification of **10 lbf** is divided by: **2 × k at 95 %** CPU ($k = 2$ for this example).

$$\text{TUR} = \frac{10\,lbf}{2 \times 2.35231} = \frac{10\,lbf}{4.70462}$$

Figure 30.25 TUR calculated value.

Therefore, **TUR = 2.1256**.

In this example, TUR is calculated for the calibration. Combining this with the location of the measurement, the False Accept may be calculated as shown using the ILAC G8:03/2009 decision rule.

	Reported Result
Nominal Value	10000
Lower Specification Limit	9995
Upper Specification Limit	10005
Measured Value	10002
Std. Uncert. (k=1)	1.176
Total Risk	0.538%
Upper Limit Risk	0.538%
Lower Limit Risk	0.000%
Test Uncertainty Ratio (TUR) =	2.1256

Figure 30.26 TUR data.

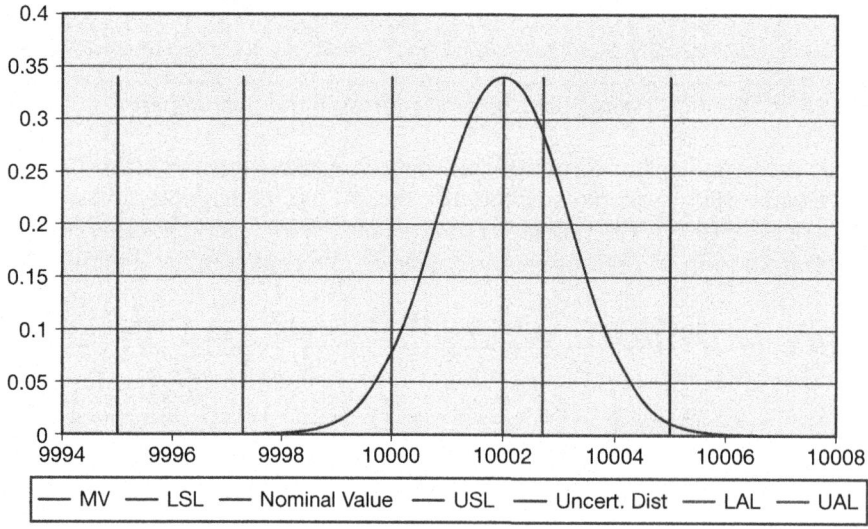

Figure 30.27 TUR graph.[35]

Source: Reproduced with permission from Morehouse Instruments and E = mc³ Solutions.

Note: In Figure 30.27, the following acronyms correspond with the information found in the corresponding data figures:

MV = Measured Value
LSL = Lower Specification Limit

USL = Upper Specification Limit
Uncert. Dist = Uncertainty Distribution
LAL = Lower Acceptance Limit
UAL = Upper Acceptance Limit

A 4:1 TUR

In "Why a 4:1 TUR is not Enough," Zumbrun notes that several organizations and publications reference or insist on maintaining a 4:1 Test Uncertainty Ratio without understanding the level of measurement risk. The assumption is that if the lab performing the calibrations has standards at least four times better than what they are calibrating, everything is good. In Clause 5.3 b), ANSI/NCSL Z540.3, *Requirements for Calibration of Measuring and Test Equipment*, allows for use of a TUR equal to or greater than 4:1 when it is not practical to estimate the false accept risk of less than 2 %. That Clause then goes on to say that objective evidence of nonpracticability of this determination is expected, as in an agreement with the customer TUR use. The 4:1 TUR seems to be a fallback position that many industries have adopted, maybe because they did not understand or want to deal with guard bands. The assumption is that the higher the TUR, the higher the probability that the measuring equipment will have a probability of false accept (PFA) of less than 2 %, assuming that the measured reading is closer to the nominal value.[36]

Many fail to realize what is described in *Introduction to Statistics in Metrology*: Using a TUR assumes that all measurement biases have been removed from the measurement process and the measurements involved follow a normal distribution. If there are significant biases that cannot be removed, then the TUR will not account for the increased risk.[37]

CONCLUSION

Any decision rule used in making a conformance decision can affect the outcome of the application of the unit under test.

ISO/IEC 17025:2017 dictates that the decision rule must be agreed upon by the customer when a statement of conformity is to be made. The customer shall specify the tolerance requirements for conformity. If no conformance requirement is requested, then the requirement for decision rules is not necessary. Simply put, regarding decision rules, if you choose not to decide, you still have made a choice (shared risk). In this scenario, the consumer can bear as much as 50 % of the measurement risk. If the UUT is used to propagate measurement uncertainty further, Type II error is more likely to occur. However, the test and calibration providers are still obligated to report the measurement result and its associated measurement uncertainty.

Part of any risk mitigation strategy should consider the following:

- Selecting a calibration supplier that offers the smallest measurement uncertainty

- Utilizing the appropriate reference standards
- Improving reliability by managing calibration intervals
- Monitoring of standards using control charts
- Continual improvement of the calibration processes
- Using the appropriate decision rules

This chapter has provided guidance on applying and interpreting decision rules to meet customer, regulatory, and normative requirements for reporting statements of conformity and the associated measurement risk. Annex A gives an example of application of both methods 5 and 6 guard bands.

ANNEX A

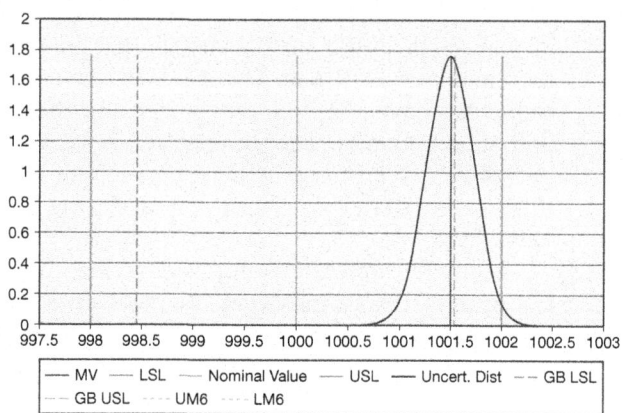

Figure 30.28 *ANSI/NCSL Z540.3 Handbook* Methods 5 and 6 guard bands.
Source: Reproduced by permission from Morehouse and E = mc³ Solutions.[38]

Figure 30.28 shows application of both methods 5 and 6 guard bands per NCSL International's ANSI/Z540.3 *Handbook*. The example uses a 1,000 lbf load cell with a resolution of 0.1 lbf and a 0.2 % of full-scale tolerance requirement. The standard uncertainty, which includes the Calibration and Measurement Capability uncertainty parameter and contributions from the unit under test from resolution, repeatability, environmental, and other error sources, is 0.23 lbf.

When calculated correctly, the TUR is 4.44:1. The methods shown in Figure 30.28 take into account the location of the measurement of 1001.5. If either of these methods were to be used for a conformity assessment, the instrument would pass, assuming the requirement was a PFA of less than 2 %; however, the acceptance limits are different. Method 5 is much more conservative, meaning a higher probability of false reject (PFR) and less risk for the end-user. If the value observed for the unit under test was read anywhere from 1001.6–1001.9, it would fail using method 5 and pass using method 6.

Part VI

Measurement Parameters

Chapter 31 Introduction to Measurement Parameters

Chapter 32 DC and Low Frequency

Chapter 33 Radio Frequency and Microwave

Chapter 34 Mass and Weighing

Chapter 35 Force

Chapter 36 Dimensional and Mechanical Parameters

Chapter 37 Other Parameters: Electro-optical and Radiation

Chapter 38 Chemical and Biological Measurements and Uncertainties

Chapter 31

Introduction to Measurement Parameters

> *When you can measure what you are speaking about, and express it in numbers, you know something about it; but when you cannot measure it, when you cannot express it in numbers, your knowledge is of a meager and unsatisfactory kind: it may be the beginning of knowledge, but you have scarcely, in your mind, advanced to the stage of science . . .*
>
> —William Thomson, Baron Kelvin of Largs, 1883

Part VI is about the process of obtaining the knowledge described by Lord Kelvin: the numbers. Measurements are what transform opinion about something into the facts about that thing.

This part is an overview of the major parameters measured by calibration laboratories, and provides some information on how the measurements of each are made. This part is not intended to be a detailed or complete review of each parameter. The measurement parameters are grouped into six major types, based on the divisions commonly found in the industry:

- Direct current (DC) and low frequency
- Radio frequency (RF) and microwave
- Physical measurements: Mass and weight, force
- Dimensional and mechanical
- Analytical, electro-optical, qualitative, and radiation
- Life science: biological and chemical

The chapters in this part discuss some important measurements that are made in each division. For each measurement, there is information about the relevant SI units, typical measurement and transfer standards, typical workload items, and some information about the parameter and how it is measured. Remember, though, the information here can only supplement—not replace—the information in other more detailed references that apply to specific areas. Table 31.1 lists the principal parameters that are measured in each measurement area, in no particular order.[1] Not all of these parameters are discussed in this book.

Table 31.1 The six major measurement parameters.

DC and low frequency	RF and microwave	Physical	Dimensional and mechanical	Analytical, electro-optical, qualitative, radiation	Life science: Biological, chemical
Capacitance	Antenna gain	Fluid density	Acceleration	Acidity (pH)	Biological properties
Current	Attenuation	Mass	Acoustics	Color	Chemical properties
Electrical phase angle	Electromagnetic field strength	Pressure	Angle	Conductivity	
Energy	Frequency	Relative humidity	Flatness	Ionizing radiation	
Frequency and time interval	Impedance	Temperature	Force	Light intensity	
Impedance	Noise figure	Vacuum	Hardness	Light power	
Inductance	Phase	Viscosity	Length	Light spectral analysis	
Magnetics	Power		Optical alignment	Nuclear activity	
Power	Pulse rise time		Roundness	Optical density	
Resistance	Reflection coefficient (VSWR)		Surface finish	Radiation dosimetry	
Voltage	Voltage		Volume		

Chapter 32
DC and Low Frequency

It is safe to assume that everyone reading this book has had some interaction with devices that use direct current or alternating current in the low-frequency range. Batteries provide direct current to things such as automobiles, radios, and toys. Alternating current is provided by national power distribution systems to industrial plants, businesses, and individual homes. The spread of technology during the twentieth century carried these things to every place on Earth.

DC AND AC

When a voltage or current is at a constant level over time (ignoring random noise), it is said to be direct. When measuring or referring to voltage, it is correct to use the term *direct voltage,* or DV. When measuring or referring to current, it is correct to use the term *direct current,* or DC. It is common, though, for people to use DC to refer to both current and voltage.

When a voltage or current changes magnitude in a periodic manner (ignoring random noise), it is said to be alternating. When measuring or referring to voltage, it is correct to use the term *alternating voltage,* or AV. When measuring or referring to current, it is correct to use the term *alternating current,* or AC. It is common, though, for people to use AC to refer to both current and voltage.

When using measuring instruments, it is often important to consider the input impedance of the measuring instrument. For voltage and current measurements, the input impedance appears as resistance in parallel with the measurement. It can become a significant factor when measuring low voltages or high resistance. For current measurements, the shunt impedance is in series with the measurement and can become a significant factor when measuring low currents.

LOW FREQUENCY

Direct current is the absence of any periodic variation, or 0 Hz. For electronic metrology, low-frequency AC is from a frequency greater than zero up to 100 kHz. There is some overlap with the resistance-factor (RF) area, because a number of AC meters can measure up to one MHz, and some up to 10 MHz or more. Traditionally, this area also includes frequency standards; they commonly operate at frequencies of 1, 5, or 10 MHz (as well as others).

Frequency ranges of particular interest are:

- *Power frequencies* (mostly 45 to 65 Hz and 400 Hz) are used for distribution of energy. The most common are 50 Hz and 60 Hz. 400 Hz is common in aircraft, submarines, and some other applications because the higher frequency allows smaller—and therefore lighter—motors, transformers, and other components.

- *Audio frequencies* (20 Hz to 20 kHz) are defined by the average limits of human hearing. Everything below 20 Hz is called *infrasonic*, and everything above 20 kHz is called *ultrasonic*.

- *Ultrasonics* (above 20 kHz). Ultrasonic frequencies up to several megahertz are used in applications such as nondestructive testing, medical imagery, motion detection, and short-distance range finding.

It is useful to note two important distinctions here. *Electronics* is concerned with variations of voltage or current (the movement of electrons) in a conductor or semiconductor, and with propagation of electromagnetic waves through free space. *Acoustics* is specifically concerned with the propagation of pressure waves (sound) through a physical medium. The medium most referred to is air, which gives rise to the common perception of acoustics as having to do only with hearing. Acoustic energy is also carried through solid materials, and at much higher frequencies. Applications of ultrasonics were mentioned. There is also overlap from ultrasonics into electronics: surface acoustic wave (SAW) devices use mechanical vibration within a structure to control electronic properties. SAW devices are used as filters and for other applications. Without them, many modern electronic devices (such as mobile telephones) would be larger, heavier, and more complex than they are now. SAW devices are well above the frequency range of this section, though.

MEASUREMENT PARAMETERS

Following is a discussion of the most important measurement parameters in the DC and low-frequency area. This book does not include all parameters. Also, it is not a complete discussion of them, especially considering that, in many cases, complete books have been written about each. The parameters are:

- Direct current
- Direct voltage
- Resistance
- Alternating voltage
- Alternating current
- Capacitance
- Inductance
- Time interval and frequency
- Phase angle
- Electrical power

Direct Current

The ampere, symbol A, is the SI unit of electric current (and one of the seven base SI units). It is defined by taking the fixed numerical value of the elementary charge, e, to be $1.602\,176\,634 \times 10^{-19}$ when expressed in the unit C, which is equal to A s, and where the second is defined in terms of Δv_{Cs}.[1]

The effect of this definition is that one ampere is the electric current corresponding to the flow of $1/(1.602\,176\,634 \times 10^{-19})$ elementary charges per second.[2]

The previous definition of the ampere was based on the force between two current-carrying conductors and was related to the value of the vacuum magnetic permeability μ_0. With the redefinition of SI units effective May 2019, the new definition of the ampere fixes the value of e instead of μ_0.[3]

Realization of the Ampere

"In practice, the ampere A can be realized by:

(a) by using Ohm's law, the unit relation $A = V/\Omega$, and using practical realizations of the SI derived units the volt V and the ohm Ω, based on the Josephson and quantum Hall effects, respectively, or

(b) by using a single electron transport (SET) or similar device, the unit relation $A = C/s$, the value of e given in the definition of the ampere and a practical realization of the SI base unit the second s; or

(c) by using the relation, $I = C \cdot dU/dt$, the unit relation $A = F \cdot V/s$, and practical realizations of the SI derived units the volt V and the farad F and of the SI base unit second s."[4]

For general calibration purposes, direct current is measured using the indirect technique. The current is passed through a known resistance (called a *shunt*), and the voltage developed across the resistance is measured. The relationship is one of the arrangements of Ohm's law:

$$I = V / R$$

where I is the current in amperes, V is the voltage in volts, and R is the resistance in ohms. Measurement problems arise when dealing with very small or very high currents. Very small currents, 100 pA or less, are difficult to measure because the input impedance of a typical laboratory digital multimeter is equal to or less than the resistance needed to generate a measurable voltage. Measurements in this area require a pico-ammeter. Very high currents are difficult to measure with a shunt and voltmeter because a very low resistance is necessary to avoid overheating, but the voltage developed may be too low to measure accurately with a normal digital multimeter. Measurements in this area require an electrometer or a nanovoltmeter.[5]

There are some meters that measure direct current by passing it through coils in the meter movement to deflect an indicating needle, but these are normally calibration workload items rather than measurement standards. See Table 32.1 for details about direct current parameters.

Table 32.1 Direct current parameters.

Parameter name	Direct current (DC)
SI unit	ampere
Typical primary standards	Calculable capacitor, current balance
Typical transfer standards	Resistance and voltage standards, using Ohm's law
Typical calibration standards	DC calibrator, multifunction calibrator, transconductance amplifier
Typical IM&TE workload	Direct current function of multimeters, multifunction calibrators, current shunts, current sources
Typical measurement method	Indirect
Measurement considerations	Thermal effects, excessive source loading, very low or very high currents

Direct Voltage

Voltage is the difference in electric potential between two points in a circuit. Direct voltage has been known since ancient times as static electricity. While referred to as "voltage" in many countries, it is also called "electric tension" or simply "tension" in some countries.[6] It is inherent in the structure of matter, as the basic unit is the charge or potential of the electron. Now, the measurement of direct voltage is one of the most common tasks in metrology and can be done with very high accuracy. See Table 32.2 for details about direct voltage. Measurements of direct voltage can be made to uncertainties of a few parts per million or better. Most measuring instruments measure voltage using the direct method, but differential and ratio methods are also used.

When using transfer standards such as electronic volt standards or saturated standard cells, the usual practice is to use them in groups and use the group average as the assigned value. All these devices drift. Using a large number of

Table 32.2 Direct voltage parameters.

Parameter name	Direct voltage (DV or, most commonly, DC)
SI unit	volt (watt/ampere)
Typical primary standards	Josephson junction array Saturated standard cells
Typical transfer standards	Electronic volt standards Saturated standard cells
Typical calibration standards	DV calibrator, multifunction calibrator
Typical IM&TE workload	Direct voltage function of multimeters, thermocouple meters, galvanometers, voltage dividers, multifunction calibrators
Typical measurement method	Direct, ratio, differential
Measurement considerations	Temperature (thermal electromotive force [EMF]) Excessive source loading

standards allows the drift rates and directions to be averaged, which results in a more stable and predictable value. Many laboratories use automated systems to run daily intercomparisons between the cells or electronic standards in the group. Specialized switch matrices and software collect measurements and report on the state of individual units in the group, the group average and standard deviation, and the drift rate. When saturated standard cells are used, it is usual to have a group of up to twelve used as the working standard, a group of four as a transfer standard, and another group of four as a check standard.[7] Electronic volt standards are commonly used in groups ranging from four to six. At least four are needed to maintain the laboratory's local value of the volt and provide redundancy.[8]

The volt is an SI-derived unit, where it is defined as the electric potential along a wire when an electric current of one ampere dissipates one watt (W) of power (W = J/s), meaning it is expressed in terms of the watt and the ampere.[9]

When expressed in terms of base SI units:

$$V = kg\ m^2\ s^{-3}\ A^{-1}$$

When expressed in terms of other SI units:

$$V = Watt/Ampere$$

Realization of the volt measurement uses a property of superconductivity (the Josephson effect) in terms of two constants of quantum physics: the elementary charge of the electron, e, and Planck's constant, h. The volt V can be realized using this Josephson effect. This value follows the accepted and supported equation of the Josephson constant:[10]

$$K_J = 2e/h$$

Effective May 20, 2019, the natural physical constants in the SI were officially redefined:

Elementary charge, e, was redefined as $1.602\ 176\ 634 \times 10^{-19}$ C

Planck's constant, h, was redefined as $6.626\ 070\ 15 \times 10^{-34}$ J/s

As a result, effective May 20, 2019, the CIPM redefined the value of the Josephson constant K_J (calculated to 15 significant digits):[11]

$$K_J = 483\ 597.848\ 416\ 984\ GHz/V$$

This new value of K_J replaces the conventional value of K_{J-90}, which was 483 597.9 GHz/Volt. A plot of voltage versus current shows distinct steps; with an applied frequency of 10 GHz, each step is about 20 mV high.

Zener Diode

An electronic volt standard uses a Zener diode in a temperature-controlled oven. The important property of a Zener diode is that when reverse-biased, it has a nearly constant voltage drop over a very wide current range. The voltage developed across the Zener diode is filtered to remove noise and is usually amplified to the desired level. Most of these standards have a primary output of 10 V nominal, with the actual output known to a resolution of 0.1 mV. They often also have a secondary output that divides this down to 1 V or to 1.018 V for compatibility with older saturated standard cells. Zener-based direct voltage references are normally used in groups of at least four. The members of the group are intercompared regularly (weekly or daily) to provide a group average value. Although these standards can usually source a current of several milliamps, if necessary, they are not power supplies. Measurements are normally made using voltmeters with high input impedance or by using the differential technique.

Important performance characteristics of a Zener-based electronic volt reference are stability, noise, predictability, and hysteresis. *Stability* and *noise* are short-term performance characteristics. *Predictability* is how well the future performance can be predicted from linear regression of historical data. *Hysteresis* is a measure of how closely the standard will return to its previous value after its power has been turned off.

Saturated standard cells were the first widely used stable and predictable direct voltage standards. Over the past 60 years, the most common type has been the Weston saturated mercury-cadmium standard cell that was widely used in all levels of metrology laboratories. (There also was an unsaturated standard cell that was used in portable inspection, measurement, and test equipment [IM&TE] such as thermocouple potentiometers and differential voltmeters.) Because saturated standard cells have a temperature coefficient of about 50 ppm/°C, they are kept in controlled-temperature enclosures or oil baths to minimize this effect. The voltage available from a standard cell is approximately 1.018 V, and measurements are often made with a resolution of 0.1 mV. Because of the high internal resistance of a standard cell, the voltage cannot be measured by any method that draws significant current from it. Voltage measurements and comparisons made using standard cells are always performed using the differential technique with a null meter. At the null point, where opposing voltages are balanced, the current is virtually zero and therefore the load impedance is effectively infinite. Standard cells are now largely being replaced by Zener-based direct voltage references in many general calibration laboratories.

Care must be taken to avoid the effects of thermoelectric potentials when measuring low voltage (10 V and less). A thermal EMF will be produced at every connection and can affect the voltage measurement. In effect, each connection is a thermocouple. Table 32.3 lists some common connector material pairs and their thermoelectric potential.

Table 32.3 Seebeck coefficients: Thermoelectric effects from connector materials.[12]

Materials	Thermoelectric potential
Copper–Copper	<0.2 $\mu V/°C$
Copper–Silver	0.3 $\mu V/°C$
Copper–Gold	0.3 $\mu V/°C$
Copper–Lead/tin solder	1 to 3 $\mu V/°C$
Copper–Nickel	21 $\mu V/°C$
Copper–Copper oxide	>1000 $\mu V/°C$

In Table 32.3, all the metals are assumed to be clean, bright, and untarnished, except one. Copper oxide is the tarnish that quickly forms on copper after its protective insulation is removed. From this, it should be obvious that it is especially important for all copper wires and connection points to be regularly cleaned to keep them bright and shiny.

Ideally, each connection should be at the same temperature, which means allowing time for them to stabilize after handling. Connections must be clean and tight. Connections should be made with low thermal EMF alloys such as gold-plated tellurium copper or bright clean copper. Tarnished copper and nickel-plated connectors should be avoided because of the very high thermal EMF potentials.[13] Residual thermal effects can be evaluated by making two sets of measurements. Measure the voltage, reverse the voltage sense leads at the meter and repeat the measurement, then average the values.[14]

Resistance

Resistance is the opposition to electric current. In most situations it is a property of all materials in varying degrees. Some materials exhibit no measurable resistance (superconductivity) at cryogenic temperatures. See Table 32.4 for detail about resistance.

The ohm is a derived unit, equal to the voltage divided by current. This relationship, (R = E/I), is known as Ohm's law for direct current circuits.

The *quantum Hall effect*, discovered in 1980 by Klaus von Klitzing, is used to make a representation of the ohm. The Hall effect, which was discovered in 1880, is exhibited when a semiconductor with a current passing through it and exposed to a magnetic field produces a voltage that is proportional to the strength and polarity of the field. (A Hall device is used in many current sensors and electronic compasses.) A quantum Hall effect (QHE) device is made so that electron flow is confined to an extremely thin layer, and it is operated at a temperature of less than 4 kelvin. Under these conditions, the voltage changes in steps instead of continuously as the magnetic field is varied. Von Klitzing discovered that the steps are multiples of a ratio of two fundamental values of physics, Planck's constant

Table 32.4 Resistance parameters.

Parameter name	Resistance
SI unit	ohm
Typical primary standards	Quantum Hall effect apparatus, calculable capacitor, standard resistors
Typical transfer standards	Standard resistors
Typical calibration standards	Digital multimeter, multifunction calibrator, standard resistors, resistance bridge
Typical IM&TE workload	Resistance function of multimeters, multifunction calibrators, resistors, current shunts
Typical measurement method	Ratio, transfer, indirect
Measurement considerations	Thermal effects, excessive current

(h), and the elementary charge of the electron (e), where R_k is the von Klitzing constant:[15]

$$R_k = h/e^2$$

Effective May 20, 2019, the natural physical constants in the SI were officially redefined:

Elementary charge, e, was redefined as $1.602\ 176\ 634 \times 10^{-19}$ C.

Planck's constant, h, was redefined as $6.626\ 070\ 15 \times 10^{-34}$ J/s.

Like voltage, effective May 20, 2019, the CIPM redefined the value of R_k, the von Klitzing constant K_J (calculated to 15 significant digits):[16]

$$R_k = 25\ 812.807\ 459\ 3045\ \Omega$$

This new value of R_k replaces the conventional value of $R_{k\text{-}90}$, which was $25\ 812.807\ \Omega$.[17]

Calculating Resistance in Practice

The value of standard resistors is normally transferred by ratio methods. One method is the use of any of several types of resistance bridge circuits. The other method compares the voltages across two resistors when they are connected in series and therefore have the same current flowing through them. In either case, one of the resistors is a known standard used as the reference for the ratio.

Multimeters typically use one of two different indirect measurement methods. One method passes a known current through the resistor and measures the voltage developed across it. This method is implemented in most digital multimeters. The other method places an ammeter and adjustable resistor in series with the unknown resistor and a fixed voltage across the combination. The ammeter is set to a reference with zero resistance and then used to measure the unknown. This method is implemented in most analog meters and some digital multimeters. With either method, the effect is using Ohm's law to determine resistance from values of voltage and current.

High-accuracy resistance measurement is performed using the four-wire measurement method. One pair of wires is used to connect the current that is passed through the resistor. The other pair of wires is used to measure the voltage across it. Advantages of the four-wire method are elimination of the test lead resistance, higher-accuracy voltage measurements, and the possibility of measuring both voltage and current simultaneously. When measuring low resistances (less than 1 ohm), particular care must be taken to avoid the effects of thermoelectric potentials. A thermal EMF will be produced at every connection and can affect the voltage measurement. Each connection should ideally be at the same temperature, which means allowing time for them to stabilize after handling. Connections must be clean and tight. Connections should be made with low thermal electromotive force (EMF) alloys such as gold-plated tellurium copper or bright clean copper. Tarnished copper and nickel-plated connectors should be avoided because of the very high thermal EMF potentials.[18] Residual thermal effects can be evaluated with the meter in DC volt mode. One method is to measure the voltage, reverse the voltage sense leads at the meter and repeat the measurement, then average the values.[19] Another method is to reverse the polarity of the current and repeat the measurement.[20] Some laboratory digital multimeters have measurement modes that use one of these methods automatically.

Alternating Voltage

Alternating voltage changes value in a periodic manner. This makes it useful in many ways that direct voltage cannot be. Alternating voltage can be stepped up or down using transformers, which makes it practical for long-distance energy transmission. See Table 32.5 for details about alternating voltage.

Alternating voltage metrology is based on the conversion of electrical power to heat. When a voltage is across a resistor, it converts the power to a proportional amount of heat. The relationship is a version of Ohm's law:

$$W = V^2/R$$

where W is the power in watts, V is the voltage, and R is the resistance. A resistor produces a certain amount of heat when an unknown alternating voltage is applied. If a direct voltage is then applied and adjusted so that the same amount of heat is produced, then its value must be equal to the alternating voltage. This is the basic principle of a thermal voltage converter.

The thermoelement in a thermal voltage converter is a simple device. It consists of a resistive heater element and a thermocouple that is physically close but electrically isolated. The assembly is enclosed in a glass envelope, and a vacuum is created inside. There are different models to accommodate various voltage ranges. An ideal thermal voltage converter will have square-law response, so the direct voltage output represents the RMS (root-mean-square) value of the alternating voltage. Deviations from the ideal response are on the device's calibration report as an AC–DC difference. Another error, reversal error, is present due to different responses to +DC and –DC in the thermoelement. There are some disadvantages to this type of thermal converter. There is some unit-to-unit variation, so a converter must be recalibrated if the thermoelement is replaced. They are delicate, and their

Table 32.5   Alternating voltage parameters.

Parameter name	Alternating voltage (AV or most commonly AC)
SI unit	volt (watt/ampere)
Typical primary standards	DC Volt, thermal voltage converter, micro-potentiometer
Typical transfer standards	AC/DC thermal voltage converter
Typical calibration standards	AV calibrator, multifunction calibrator, DV reference, thermal converter
Typical IM&TE workload	Alternating voltage function of multimeters, multifunction calibrators, ratio transformers
Typical measurement method	Indirect, transfer, direct
Measurement considerations	Transfer error, frequency

output varies with increasing frequency. In addition, the thermocouple output is in millivolts, which can cause measurement difficulties. The devices also respond slowly, which can be a disadvantage.[21]

A multi-junction thermal converter of this type does exist but is not widely used. This device uses multiple thermocouples in series to increase the DC output to around 100 mV. It is limited in the amount of applied voltage and in frequency.[22] Some work is also being done on a thin film multi-junction thermal converter fabricated using microcircuit techniques, but this is not known to be in commercial use yet.

In the 1970s, Fluke developed another type of thermal AC/DC converter.[23] This sensor uses two resistor/transistor sets on a microcircuit. Each set is thermally isolated from the other. One resistor is the input; applied voltage heats the transistor next to it. That unbalances an amplifier that then produces a direct voltage to the other resistor. When the second transistor is at the same temperature as the first, the circuit is balanced, and the amplifier is producing a direct voltage proportional to the input voltage. Major advantages of this sensor are a 2 V output, shorter response time, and lower DC reversal error.[24]

The thermal devices are used for the lowest uncertainty measurements in their range (50 ppm or less) and as transfer standards. Most meters do not use these methods. Analog and digital alternating voltage meters use one of the following several methods to make direct measurements.

- **High-speed sampling of the input and direct computation of the RMS value.** This method can be very accurate, but it requires an unvarying input at a frequency no higher than half the sampling rate. This method has minimum input frequency and amplitude limits.

- **Logarithmic amplifier.** This method uses a logarithmic amplifier and other circuitry to produce a converter output proportional to the RMS value of the input in real time. Many true RMS digital multimeters use this method. This method has minimum input frequency and amplitude limits.

- **Direct measurement of the average voltage and multiplying by a scale factor.** Unless the meter specification says it does true RMS measurement, this is what it does. The meter measures the average alternating voltage and multiplies it by 1.414. For a pure sine wave, this will make the meter read the same as an RMS responding meter. Any deviation from a sine wave will result in increased error. (The specifications for this type of meter will often say something like "average responding, RMS calibrated.")

At voltages over 1000 V, resistive voltage dividers or electrostatic voltmeters are commonly used.

Alternating Current

Alternating current occurs when the direction of electron flow in a conductor changes direction in a periodic manner. Like alternating voltage, this makes it useful in many ways that direct current cannot be. In particular, alternating

Table 32.6 Alternating current parameters.

Parameter name	Alternating current (AC)
SI unit	ampere
Typical primary standards	Direct current, thermal current converter
Typical transfer standards	AC/DC thermal current converter
Typical calibration standards	AC calibrator, multifunction calibrator, DC reference, thermal converter
Typical IM&TE workload	Alternating current function of multimeters, multifunction calibrators, current transformers, current and power meters
Typical measurement method	Indirect, transfer
Measurement considerations	Transfer error, frequency

current can be stepped up or down using transformers, which makes it practical for long-distance energy transmission. When passing through a transformer, the resulting current ratio has an inverse relationship to the voltage ratio. If the voltage is doubled, the current is halved. The total energy remains the same, but transmission is more efficient because the power loss in a conductor with resistance is proportional to the square of the current. See Table 32.6 for details on alternating current.

Alternating current metrology is based on the conversion of electrical power to heat. When a current is passed through a resistor, it converts the power to a proportional amount of heat. The relationship is a version of Ohm's law:

$$W = I^2/R$$

where W is the power in watts, I is the current, and R is the resistance. Except for being used to measure current, this is the same as a thermal voltage converter.

The thermoelement in a thermal current converter is identical to the one used in a thermal voltage converter. The difference is that a current shunt that has a very small AC/DC difference is connected in parallel with the thermoelement. This forms a current divider, limiting the current throughout the thermoelement to its full-scale current when the shunt's rated current is applied. Instruments using Fluke's solid-state thermal converter are used with current shunts in the same manner. Note that for best results in either case, the shunts and AC/DC transfer standard should be calibrated together.[25]

The thermal devices are used for the lowest uncertainty measurements in their range (0.05 % or less) and as transfer standards. Most meters do not use these methods. Analog and digital alternating current meters typically measure the voltage developed across an internal shunt, using one of the AV measurement methods described earlier. To measure currents over 20 A, current transformer coils are used. The coil itself is the secondary coil of a transformer, sized to provide 5 amperes when the full rated current is passed through the primary. The primary is a cable loop connected to both sides of the current source and passing once through the center of the current transformer.[26]

Capacitance

Capacitance is a property of a circuit or device that opposes a change in voltage. A capacitor can store a charge in the electric field of the dielectric (insulation) between its conductors. See Table 32.7 for details on capacitance.

The unit of capacitance, a *farad*, is equal to one coulomb of electric charge divided by one volt. A *coulomb* is the quantity of electricity moved in one second by a current of one ampere. If a current of one ampere is applied to a one-farad capacitor, the charge stored in it will change at the rate of one volt per second. For practical use, one farad is an extremely large value. Most capacitance measurements are in microfarads to picofarads.

A property of a capacitor in an AC circuit is its reactance X_C:

$$XC = \frac{1}{2\pi f C}$$

where f is the frequency in hertz and C is the capacitance in farads.

The reactance of a capacitor can be measured in the same way as DC resistance: voltage across the capacitor divided by the current in the circuit, with the result expressed as ohms. If the AC frequency is known and the reactance is measured, the value of an unknown capacitor can be found by rearrangement of the defining equation for reactance.

At a standards laboratory, the farad may be realized by using a calculable capacitor. This device is made from a set of four long, parallel metal rods arranged so that when viewed from one end they are at the corners of a square. A short ground rod is fixed in the center at one end, and a movable ground rod is inserted from the other end. The arrangement of the parallel rods provides a constant value of capacitance per meter of length. Changing the position of the movable ground changes the effective length and therefore changes the measured capacitance. The theoretical value can be calculated from the length and the speed of light in air.[27]

Table 32.7 Capacitance parameters.

Parameter name	Capacitance
SI unit	farad (derived from ampere and second)
Typical primary standards	Calculable capacitor, standard capacitors, current comparator
Typical transfer standards	Standard capacitors
Typical calibration standards	Standard capacitors, capacitance bridge, electronic impedance bridge
Typical IM&TE workload	Capacitance current function of multimeters, decade capacitors, tank level indicators or test sets
Typical measurement method	Ratio
Measurement considerations	Transfer error, temperature, frequency, shielding

Capacitance measurements are made by the ratio method. A capacitance bridge is a frequently used item in many calibration labs. The capacitance bridge includes high-stability reference capacitors in the ratio arm, and the range can be extended if necessary by substituting an external standard capacitor. Some types of ratio bridges can be used to compare a standard capacitor to a standard resistor, thereby comparing capacitive reactance to resistance.

Some electronic impedance bridges place the unknown capacitor in series with a known resistor and make a ratio measurement of the voltages across them. This ratio is also the ratio of the capacitive reactance to the known resistance, so the value of the capacitor can be both calculated and compared to a standard.

Inductance

Inductance is a property of a circuit or device that opposes a change in current. An inductor can store energy in the magnetic field around its conductors. See Table 32.8 for details about inductance.

The unit of inductance, a *henry*, is equal to one weber of magnetic flux divided by one volt. A *weber* is the amount of magnetic flux produced by a current that is changing amplitude at the rate of one volt per second.

A property of an inductor in an AC circuit is its reactance, X_L:

$$X_L = 2pfL$$

where f is the frequency in hertz and L is the inductance in henrys.

The reactance of an inductor can be measured in the same way as DC resistance: voltage developed across the inductor divided by the current through it, with the result expressed as ohms. If the AC frequency is known and the reactance is measured, the value of an unknown inductor can be found by rearrangement of the defining equation for reactance.

In principle, a calculable inductance standard can be made by constructing an extremely uniform solenoid. (A *solenoid* is a long wire coil, where the length of the coil is much greater than the radius of each turn.) The inductance is calculated from the length, radius, and number of turns. In practice, various problems make

Table 32.8 Inductance parameters.

Parameter name	Inductance
SI unit	henry (derived from ampere and second)
Typical primary standards	Standard inductor, inductance bridge, standard capacitor
Typical transfer standards	Standard inductors
Typical calibration standards	Standard inductors, inductance bridge, electronic impedance bridge
Typical IM&TE workload	Decade inductors, LCR meters, impedance bridges
Typical measurement method	Ratio
Measurement considerations	Transfer error, temperature, current limits, frequency, magnetic field shielding

this not very practical, although a few have been made. Very precise standard capacitors are easier to make, so they are used with a ratio bridge to determine the values of standard inductors.[28] Most practical standard inductors are wound as toroids to minimize their physical size and their external magnetic field.

Inductance measurements are usually made by the ratio method. Inductors can be compared with each other, standard capacitors, or standard resistors.

Some electronic impedance bridges place the unknown inductor in series with a known resistor and make a ratio measurement of the voltages across them. This ratio is also the ratio of the inductive reactance to the known resistance, so the value of the inductor can be both calculated and compared to a standard.

Time Interval and Frequency

The *second* (symbol, s) is the SI base unit of time interval. The *hertz* is the derived unit of frequency, with one hertz being one complete cycle per second. This makes the two unique in that by knowing one, the other is known automatically. Another unique attribute is the ability to measure these values with extremely high precision. See Table 32.9 for details about time interval and frequency.

With the redefinition of SI units effective in May 2019, the SI definition of *second* was also redefined. The new SI definition of the second, s, is the unperturbed ground state hyperfine transition frequency of the caesium133 atom, Δv_{Cs}, which is 9 192 631 770 Hz. The BIPM *SI Brochure 9* (2019) continues, "The reference to an unperturbed atom is intended to make it clear that the definition of the SI second is based on an isolated caesium atom that is unperturbed by any external field, such as ambient black-body radiation."[29]

Table 32.9 Time interval and frequency parameters.

Parameter name	Time interval/frequency
SI unit	second/hertz
Typical primary standards	Atomic frequency standards (caesium beam, caesium fountain, hydrogen MASER)
Typical transfer standards	Caesium beam frequency standards Terrestrial or satellite RF signals
Typical calibration standards	Caesium beam frequency standards Quartz or rubidium oscillators disciplined to LORAN or GPS transmissions Frequency and time interval counters, signal generators (low frequency, RF, and microwave), function generators
Typical IM&TE workload	Frequency and time interval counters, signal generators (low frequency, RF, and microwave), function generators, spectrum analyzers, and navigation and communication equipment
Typical measurement method	Ratio (phase comparison) Direct (time interval or frequency counter)
Measurement considerations	Phase noise

"The second, so defined, is the unit of proper time in the sense of the general theory of relativity. To allow the provision of a coordinated time scale, the signals of different primary clocks in different locations are combined, which must be corrected for relativistic caesium frequency shifts (see *SI Brochure 9*, section 2.3.6, SI units in the framework of the general theory of relativity).

The CIPM has adopted various secondary representations of the second, based on a selected number of spectral lines of atoms, ions, or molecules. The unperturbed frequencies of these lines can be determined with a relative uncertainty not lower than that of the realization of the second based on the 133Cs hyperfine transition frequency, but some can be reproduced with superior stability."[30]

It is probable that every electronic calibration laboratory has some type of frequency standard. Frequency standards have a fixed output of at least one frequency, and units with several different outputs are common. Common frequencies are 1, 5, and 10 MHz, as well as 1 kHz and 1 Hz. A laboratory frequency standard can readily be compared to a national standard by any of several means. All of these use a radio signal as a transfer standard between the laboratory's frequency standard and the National Metrology Institute's.

Global Positioning System (GPS)

The GPS system has a remarkably high level of performance. Each of the operational satellites has at least four atomic frequency standards (two caesium and either two or three rubidium). GPS receivers intended for time and frequency measurements must be simultaneously tracking at least four satellites to obtain valid signals, and most are capable of tracking as many as eight to 12. The receiver output is an average from all the satellites in view, so in effect the signals of at least 16 atomic standards are being averaged. The receiver output is used to phase lock, or discipline, a frequency standard. A large number of companies manufacture GPS-disciplined quartz or rubidium oscillators, although not all are suitable for frequency metrology. The main risk of using a satellite-based system is that position and timing accuracy for nonmilitary users can be intentionally degraded at any time by the U.S. Department of Defense. This degradation, called *selective availability* or SA, was turned off in May 2000, but can be turned on again any time it is needed for military requirements. When GPS is used as a transfer standard, SA appears as additional phase noise.

The subject of calibration and metrological traceability of time signals could fill an entire book. Included in the following sections are some well-regarded (and free internet download) references for more information on each subject:

- Michael A. Lombardi. 2014. "Time Measurement." In *Measurement, Instrumentation, and Sensors Handbook*, edited by John G. Webster and Halit Eren, ch. 41. Boca Raton, FL: CRC Press/Taylor and Francis Group, LLC.[31] https://tf.nist.gov/general/pdf/2488.pdf

Calibrating a Frequency Standard

When calibrating a frequency standard, three methods are used. One method is to use a phase comparator to compare the unit under test (UUT) to the standard.

(Phase comparison is a ratio method.) The frequency of the UUT is adjusted to minimize the phase shift over a given time period. The accuracy improves as the time period is increased.

Another method is using a time interval counter. The counter is triggered by the start of a pulse from the frequency standard and stopped by the start of a pulse from the time base under test. The time interval between them is measured, and the rate at which it changes is computed and used to adjust the UUT. A limitation is the time resolution of the time interval counter, but units with picosecond resolutions are available.

The third method is direct measurement with a frequency counter. This is limited by the resolution and uncertainty of the counter, but counters with resolution of 0.001 Hz or better at 10 MHz are readily available.

Frequency calibrations can take considerable time, especially when calibrating a high-quality oscillator and using GPS as the transfer standard. The problem is noise. Atmospheric propagation effects add noise and phase uncertainty to the signals. The GPS signal will have significantly greater uncertainty if SA is turned on. Although the long-term stability of both systems is high, the noise creates a great deal of short-term instability. It is recommended that the measurement period for most oscillators should be at least 24 hours. Caesium beam frequency standards and rubidium oscillators may require a measurement period of several days to a week to verify their performance.

More information can also be found in the following publications:

- Michael A. Lombardi. 2014. "Time Measurement." In *Measurement, Instrumentation, and Sensors Handbook*, edited by John G. Webster and Halit Eren, ch. 42. Boca Raton, FL: CRC Press/Taylor and Francis Group, LLC.[32] https://tf.nist.gov/general/pdf/2496.pdf

- Michael A. Lombardi. 2016. *Evaluating the Frequency and Time Uncertainty of GPS Disciplined Oscillators and Clocks*. NCSL International *Measure*, no. 3–4 (October): 30–44. https://doi.org/10.1080/19315775.2017.1316696

The abstract of this paper states, "Global Positioning System (GPS) disciplined oscillators and clocks serve as standards of frequency and time in numerous calibration and metrology laboratories. They also serve as frequency and time references in many industries, perhaps most notably in the telecommunication, electric power, transportation, and financial sectors. These devices are inherently accurate sources of both frequency and time because they are adjusted via the GPS satellites to agree with the Coordinated Universal Time (UTC) time scale maintained by the United States Naval Observatory (USNO). Despite their excellent performance, it can be difficult to evaluate their uncertainty, and even more difficult for metrologists to prove their claims of uncertainty and traceability to skeptical laboratory assessors. This article is written for metrologists and laboratory assessors who work with GPS disciplined oscillators (GPSDOs) or GPS disciplined clocks (GPSDCs) and need to assess their uncertainty. It describes the relationship between GPS time and Coordinated Universal Time (UTC), explains why GPS time is traceable to the International System (SI), and provides methods for evaluating the frequency and time uncertainty of signals produced by a GPSDO or GPSDC."[33]

- D. Matsakis, J. Levine, and M. Lombardi. 2018. Metrological and legal traceability of time signals. 2018 Precise Time and Time Interval (PTTI) Meeting, Reston, Virginia, https://tsapps.nist.gov/publication/get_pdf.cfm?pub_id=924939

The abstract of this paper states, "Metrological traceability requires an unbroken chain of calibrations that relate to a reference, with each calibration having a documented measurement uncertainty. In the field of time and frequency metrology, the desired reference is usually Coordinated Universal Time (UTC), or one or more of its official realizations, termed UTC(k), and traceability to UTC is a legal requirement for many entities. Traceability to UTC can be established in three areas—frequency, time interval, and time-of-day synchronization—but this paper focuses solely on the traceability of time signals used for synchronization. We first examine the definition of traceability, then discuss how traceability can be established via the reception of time signals transmitted by satellites and network time servers, followed by a discussion of how these signals can meet the synchronization and traceability requirements of the financial and electric power industries. Not all the available UTC time signals are considered in this paper, as we primarily focus on direct broadcast and common-view Global Positioning System (GPS) signals, with uncertainties measured in nanoseconds, and Network Time Protocol (NTP) signals, with uncertainties measured in microseconds and milliseconds."[34]

Work on furthering and coordinating time and frequency metrology is conducted worldwide. One working group is the SIM Time and Frequency Metrology Working Group (TFMWG) (https://tf.nist.gov/sim), which covers North, Central, and South America, and the Caribbean Islands. The members of the SIM TFMWG are timing laboratories located at National Metrology Institutes and other designated institutions in the 34 member nations of the Organization of American States (OAS). In most cases, these timing laboratories are responsible for keeping the official time in their respective nations. SIMT complements the world's official time scale, Coordinated Universal Time (UTC), by providing real-time support to operational timing and calibration systems in the SIM region.[35] One can search and link to related websites of National Metrology Institute time and frequency groups as well as national web clocks both within and outside SIM.

Calibration of Stopwatches and Timing Devices

NIST published SOP 24 (2019), Calibration of Stopwatches and Timing Devices, which is typically used for legal metrology and tying in with NIST *Handbook 44*. "This procedure is used to calibrate Type I or Type II stopwatches and timing devices used by Weights and Measures officials, industrial, technical and other interests concerned with traceable time measurements.

Type I: These stopwatches utilize digital, electronic circuitry to measure time intervals.

Type II: These stopwatches utilize analog, mechanical mechanisms to measure time intervals."[36]

NIST SOP 24 (2019) also includes useful guidance on calibration methods, calculations, estimates of uncertainty, and uncertainty budgets. SOP 24 also lists standards with suitable traceability and resolution for calibrating stopwatches and timing devices, which include one of the following:

1. "An operational shortwave receiver to receive the broadcast timing signal on one of the frequencies listed in Table 2, or;

2. A land line telephone to call one of the numbers listed in Table 2, to receive the broadcast timing signal; or

3. A GPS master clock. Verify that the GPS Master Clock is locked to GPS signals. In order for a GPS display to be used as a reference, there must be an indicator on the unit that shows whether the display is currently locked to the GPS signal or is in 'coast' mode. If the receiver is in 'coast' mode, it should not be used as a calibration reference."[37]

For laboratories trying to meet organizational metrology requirements and not just U.S. legal requirements, NIST has also published a Recommended Practice Guide, *Special Publication 960-12, Stopwatch and Timer Calibrations* (2009 edition); https://tf.nist.gov/general/pdf/1930.pdf. According to the guide's foreword, "This document is a recommended practice guide for stopwatch and timer calibrations. It discusses the types of stopwatches and timers that require calibration, their specifications and tolerances, and the methods used to calibrate them. It also discusses measurement uncertainties and the process of establishing measurement traceability back to national and international standards. This guide is intended to serve as a reference for the metrologist or calibration technician. It provides a complete technical discussion of stopwatch and timer calibrations by presenting practical, real-world examples of how these calibrations are performed."[38]

Time of Day

It is important not to confuse the preceding information on time interval with the common use of time of day. "Coordinated Universal Time (UTC) is the international reference time scale that forms the basis for the coordinated dissemination of standard frequencies and time signals; UTC is obtained from International Atomic Time (TAI) by the insertion of leap seconds according to the advice of the International Earth Rotation and Reference Systems Service (IERS) to ensure approximate agreement with the time derived from the rotation of the Earth. The formal definition of UTC and TAI was adopted by the CGPM in 2018."[39]

Time of day has multiple time scales, most of which are based on astronomical observations. The second is used as a unit in time of day, but different time scales use seconds of different duration (or size).

Many customary time scales are based on the mean solar day, which has 86,400 seconds. For example, this is the basis of legal time in the United States. The duration of a day is determined by observations of the sun as it passes through the zenith. This means that the duration of the mean solar second is not a constant.

It slowly increases because of short-term variations and a gradual decrease in the Earth's rotational speed.

Universal coordinated time (UTC) is a uniform atomic time scale, and the UTC second has the same duration as the TAI second. UTC is synchronized with UT1, but this means that UTC must be adjusted occasionally to keep the synchronization. Synchronizing the scales is accomplished by adding a leap second as needed to keep UTC within 0.9 s of UT1. (Therefore, a given year may be one or two seconds longer than the year before or after.) The agencies that provide time-of-day standards around the world have agreed to keep their master clocks coordinated with UTC as determined by the BIPM. The UTC time scale was formerly called Greenwich Mean Time (GMT).

Phase Angle

Accurate measurement of electrical phase is important in areas as diverse as metering electrical energy, precise positioning of control devices, and safe navigation of aircraft. See Table 32.10 for details on phase angle parameters.

The SI unit that applies to phase measurements is the *radian*, but the commonly used unit of measure is the *degree* (°).[40] The division of the circle into 360°, with 60 minutes per degree and 60 seconds to the minute, originated in ancient Babylon. Now it is also common for subdivisions of the degree to be expressed in decimal form: 12.34°, for example. The relationship to radians is:

$$\text{Degree} = 1/360 \text{ circle} = \pi/180 \text{ rad}$$
$$\text{Minute} = 1/60 \text{ degree} = \pi/10\,800 \text{ rad}$$
$$\text{Second} = 1/60 \text{ minute} = \pi/648\,000 \text{ rad}$$
$$360° = 2\pi \text{ rad}$$

A 1978 paper by Turgel and Oldham opens with the statement, "Phase measurements are essentially a determination of the ratio of two time intervals,

Table 32.10 Phase angle parameters.

Parameter name	Phase
SI unit	radian
Typical primary standards	High-precision phase calibration standard, resistor bridges, capacitive bridges
Typical transfer standards	Resistor bridges, capacitive bridges
Typical calibration standards synchro/ servo simulator, ratio transformer	Phase calibration standard, phase angle voltmeter
Typical IM&TE workload	Phase angle voltmeter, angle position indicator, phase meters, power analyzers, radio navigation equipment
Typical measurement method	Ratio
Measurement considerations	Phase noise, RMS voltmeter limits

and as such are not dependent on any system of measurements."[41] There are two things that can be drawn from this:

1. Phase measurements are made by the ratio method.

2. There is no realizable standard radian, at least not in the same sense as we have realizable standards for units such as the meter or the volt.

Electrical phase angles can, of course, be generated and measured. There are also standards that generate phase relationships, and methods for verifying those standards. Signals used in phase standards are generated digitally and use digital timing to shift the phase. Display resolution is usually millidegrees (0.001°), but digital quantization may limit the resolution to something less than this. One commercial standard, for example, has an output that only increments in steps of $360/2^{18}$, or 0.001373°, due to binary counting in the digital circuits.[42] This must be accounted for in the uncertainty. IM&TE that is calibrated often has resolution of 1 or 10 millidegrees.

Phase standards are verified using two-arm resistive and capacitive bridges. One arm is connected to the reference output, the second arm is connected to the variable output, and the center of the bridge is connected to a voltmeter or oscilloscope. If the two arms are equal (a 1:1 ratio bridge) and the phase is set to 180.000°, the signal should cancel exactly. As the phase is varied in 1 m° steps above and below 180.000°, each of the output steps can be observed. Verification of phase performance is simply observing that all the digital steps are present. When a 1:1 bridge is used and the signal amplitudes are equal, it is possible to separate the errors from the phase standard and those from the bridge. After the first set of readings is made, the inputs to the bridge are swapped and the measurement is repeated. The phase errors are calculated for each set. It has been shown that at any step, half the sum of the phase errors is the error of the phase standard, and half the difference is the error in the bridge. This provides an absolute verification of the phase standard and the bridge at the same time.[43] Note that bridges made with other ratios (10:1 or 100:1, for example) are themselves verified by comparison to other well-characterized bridges.

A measurement problem may arise because the theoretical output of the bridge is zero or close to it. This can be a problem for some RMS voltmeters. Some methods to work around this include using angle pairs farther away from the null point, using unequal-signal amplitudes, or using a high-gain preamplifier.

Electrical Power

Measurement of electric power (or energy, when integrated over time) is done so frequently that it is virtually invisible to most people—except when the electricity bill arrives at home or business for payment. See Table 32.11 for details on electrical power.

From Ohm's law we know that DC power in watts is $P = E \times I$ where E is the voltage, and I is the current. If an AC circuit had only pure resistance, the same relationship would be true; however, there is always at least a small amount of inductance or capacitance. This leads to a phase difference between voltage and current and therefore a difference between the apparent power (see the preceding equation) and the true power.

Table 32.11 Electrical power parameters.

Parameter name	Power
SI unit	watt
Typical primary standards	Voltage, current, resistance, time interval, phase sampling wattmeter, synthesized power source
Typical transfer standards	Voltage, current, resistance, time interval, phase, characterized watt-hour meter
Typical calibration standards	Voltage source, current source, AC/DC shunts, voltmeters, phase standard
Typical IM&TE workload	Watt-hour meters, power meters, power analyzers, power supplies, load banks
Typical measurement method	Indirect
Measurement considerations	Safety, AC/DC transfer, measurement uncertainty—especially of small shunt voltages

True power is $P = EI \cos q$, where q is the phase angle between voltage and current.

One watt is one joule (J) of energy delivered in one second. If this is continued for an hour, 3600 J or one watt-hour of energy is delivered. In actual practice, the customary practical units are kilowatt-hour for selling energy and megawatt-hour for generating it. In any case, the accuracy of metering is important to an electric utility company.

There are several methods of measuring power or energy. All are indirect methods, although the first does make a direct measurement of voltage:

- Power alone (without considering time) can be found by placing an appropriate current shunt in a circuit and then measuring the voltages across the load and the shunt. The values are multiplied to find the power. If either the impedance or phase can be measured, then true power can be found; otherwise only apparent power can be found.

- Electrodynamic and electrostatic methods use mechanical meters to multiply the voltage and current. An example of an electrodynamic meter is the familiar electric meter outside a home or business.

- Thermal methods, using either a thermoelement or a calorimeter.

- Electronic methods using a Hall effect sensor.

- Digital sampling of the current and voltage waveforms and integration of the measured values. This method is used by a wattmeter developed at NIST.[44] An advantage of this method is the capability to accurately measure distorted and noisy waveforms.

Multiple aspects of DC and low frequency measurements were discussed in this chapter. Entire books can be, and have been, written on each of these subjects. We hope that the information and supplemental references in this chapter provide a solid foundation upon which to build and grow.

Chapter 33
Radio Frequency and Microwave

The radio frequency (RF) and microwave measurement area is concerned with the part of the electromagnetic spectrum where information is transmitted through free space and received using the electrical or magnetic waves of the signal. Until the last quarter of the twentieth century, that would have been sufficient as a definition and would have covered the vast majority of applications. Since then, the explosive growth of communications systems and digital information processing has greatly expanded the number and variety of products for which RF and microwave engineering and measurement are important. Much of this is due to ever-increasing processing speed. This requires faster processors and logic and thus higher frequencies. This creates new challenges for circuit engineers because, for example, the printed circuit board traces and many electronic components are large compared to the signal wavelength and, therefore, must be treated much differently than a wire with direct current (DC) flowing in it.

Consider what has happened in the United States since 1975:

- In 1975, telephone service was from the phone company—there was only one company. The telephone was attached to the wall with a wire, and the concept of being able to buy (instead of lease) a telephone set was less than 10 years old. Now there are multiple phone companies that can provide hard-lined phone service as well as Voice-over-Internet Protocol (VoIP).

- If you were away from home and wanted to call someone, you had to find a coin payphone and have spare change or a calling card. Now you just pick up your personal cellular phone and place a call or send a text to anywhere in the world. It is rare to find a working payphone anymore.

- Handheld calculators were still a new spin-off from the Apollo space program, and they were expensive. Now we have smartphones that have built-in and optional calculator apps that allow us to perform high-end, complex calculations.

- In 1975, spread-spectrum communication technology and global positioning system(s) (GPS) were classified military technology, and frequencies in the gigahertz range were used almost exclusively by space systems, military, and law enforcement. Now applications such as home and office wireless telephones, wireless internet ports, and wireless links between computers and their accessories use spread-spectrum technology at gigahertz frequencies; and you have a GPS on your smartphone and in many automobiles' built-in navigation systems.

- The January 1975 issue of *Popular Electronics* featured the first of a series of articles about the MITS Altair 8800 microcomputer. This kit, considered to be the first home computer, used the Intel 8080 microprocessor that had been invented the year before. (By late 1977, you could walk into an electronics store and buy a ready-to-use personal computer, such as a Commodore PET 2001, Apple II, or Radio Shack TRS-80. But the internet would be still under the control of the Department of Defense and the National Science Foundation for another 10 years.) Now you can buy a handheld computer with as much power as room-filling business systems of 1975 and connect it wirelessly to the full range of offerings on the internet.

- Not only are computers connected to the internet, but we also now have "smart" devices in our homes, offices, and beyond that connect and communicate wirelessly. These members of the "Internet of Things" (IoT) are prolific. In 2019, there were approximately 7.74 billion connected devices worldwide. It is estimated that by 2030, there will be 500 billion devices connected with the IoT.[1]

All these new systems and things, and their continually increasing operating speeds, demand more of RF and microwave measurement systems and have required advances in related measurement sciences.

The International Telecommunications Union (ITU) Radiocommunication Sector manages the radio frequency part of the spectrum on a worldwide level. Within individual countries, the national government manages the radio spectrum, usually in conformance to the ITU regulations and recommendations. In the United States, the responsible agencies are the National Telecommunications and Information Administration (NTIA), part of the U.S. Department of Commerce, and the Federal Communications Commission (FCC), which is an independent government agency.

RF AND MICROWAVE FREQUENCIES

The general definitions of the terms used in this section are found in Federal Standard 1037C, *Telecommunications: Glossary of Telecommunication Terms*.[2]

- *Radio waves* are electromagnetic waves that are below an arbitrary limit of 3000 GHz (wavelength of 100 mm, in the infrared region).

- *Microwaves* are electromagnetic waves with frequencies from 1 GHz up to 300 GHz (wavelength from 30 cm to 1 mm).

- *Infrared* is the portion of the electromagnetic spectrum from 300 GHz up to the long-wavelength end of the visible light spectrum, approximately 0.7 mm.

These frequency boundaries are fuzzy, however. For example:

- Radio frequencies are not allocated below 9 kHz, but the U.S. Department of Defense has some transmitters that operate below 300 Hz.

- The NTIA frequency allocation chart shows the microwave region starting at 100 MHz.[3]

- There may be other names specific to an industry or company for various portions of the spectrum.

For calibration purposes, the bottom edge of the RF area is 100 kHz. The top edge is a moving target, regularly redefined as new or improved technology makes another advance. Optical fiber metrology (see Chapter 36, "Dimensional and Mechanical Parameters") shares many of the same principles. An optical fiber is essentially a waveguide for photons and may be considered an analog to a metallic waveguide for lower-frequency electromagnetic waves.

For convenience, the RF spectrum is often divided into bands or groupings of frequencies. The divisions most commonly known to the public (in North America) are the AM and FM broadcast radio bands, 535 to 1605 kHz and 88 to 108 MHz, respectively. The most commonly used general divisions are shown in Table 33.1.[4]

The ITU defines 21 specific bands in the electromagnetic spectrum: The lowest starts at 3 Hz and the highest ends at 3000 EHz (Exahertz, 3000 × 10^{18} Hz, X-ray frequencies). Other common methods of referring to groups of frequencies are the bands used by amateur radio operators and the bands used by shortwave radio listeners. An example is the amateur radio two-meter band, 144 to 148 MHz.

Waveguides

In the microwave area, frequencies are grouped in bands that are associated with waveguide sizes. The most used waveguide bands are shown in Table 33.2.[5]

Table 33.1 Common frequency bands and names.

30 to 300 Hz	Extremely low frequency	ELF
0.3 to 3 kHz	Super low frequency	SLF
3 to 30 kHz	Very low frequency	VLF
30 to 300 kHz	Low frequency	LF
0.3 to 3 MHz	Medium frequency	MF
3 to 30 MHz	High frequency	HF
30 to 300 MHz	Very high frequency	VHF
0.3 to 3 GHz	Ultra-high frequency	UHF
3 to 30 GHz	Super high frequency	SHF
30 to 300 GHz	Extremely high frequency	EHF

Table 33.2 Common waveguide bands.

Frequency (GHz)	Common name
8.2 to 12.4	X
12.4 to 18.0	P, KU
18.0 to 26.5	K
26.5 to 40.0	R, KA

Of course, there are many more waveguide sizes than this, and many different names for each size. Waveguide systems are (or have been) used from 320 MHz to 325 GHz.[6] Waveguide band names are created by industry consensus or government action and have changed several times over the past 50 years. There are several overlapping systems for the bands, the waveguide pipe, and even the flanges at waveguide ends. Common designations in North America and the NATO area include consensus names such as in Table 33.2, the EIA WR numbers, the IEC R numbers, and British WG numbers. The U.S. military has both the JAN RG numbers and the MILW numbers, and, in many cases, each system has two different numbers for the same frequency range.[7] The actual interior dimensions of waveguides, however, do have a rational basis. The wide side of plain rectangular waveguide is equal to the half-wavelength of the lowest frequency the waveguide is designed for; the narrow side is about half that dimension. This applies to rectangular waveguide only; the physics are more complicated for ridged and circular waveguide. (See chapter 2 of Adam's *Microwave Theory and Applications* for more details.[8])

Now that coaxial cable is being used to carry signals up to 65 GHz or more, the waveguide name system is becoming less well known. Waveguide is still used, however, in radar and satellite transmitters—and other applications where there is a requirement to deliver high power to the antenna. Waveguide is also still used in metrology because some of the best measurement standards are waveguide types (rotary vane variable attenuators, for example).

MEASUREMENT PARAMETERS

Following is a discussion of the most important measurement parameters in RF and microwave. This does not include all parameters. Also, it is not a complete discussion of them, especially considering that, in many cases, complete books or chapters of books have been written about each. The parameters are:

- RF power
- Attenuation or insertion loss
- Reflection coefficient or standing wave ratio
- RF voltage
- Modulation
- Noise figure and excess noise ratio

RF Power

Power is one of the fundamental measurements that are commonly made in any radio frequency system. It is usually one of the most important parameters specified by customers and designers. Power measurement is also fundamental to other measurements, such as attenuation and reflection coefficient. See Table 33.3 for more details on RF power.

The SI unit of RF and microwave power is the *watt*, a derived unit defined as joules/second. Because power measurements can cover several orders of magnitude, it is usual to express absolute power measurement as decibels relative to one milliwatt (dBm); the logarithmic scale makes the numbers easier to handle. Reference levels other than one milliwatt may also be used.

In DC and in low-frequency AC measurements, power is determined by measuring some combination of voltage, current, and phase angle. For example, when determining the power available in a distribution system, an electrician will use instruments that measure the voltage, current, and power factor (cosine of the phase) simultaneously. (See Chapter 3, "Quality Standards and Their Evolution.")

Table 33.3 RF power parameters.

Parameter name	RF power
SI units	Watt (J/s)
Typical primary standards	Calorimeter, bolometer, direct voltage, AC/DC transfer, dual six-port network analyzer
Typical transfer standards	Thermistor mount (coaxial, waveguide)
Typical calibration standards	Thermistor mount (coaxial, waveguide)
Typical IM&TE workload	Thermistor mounts, thermocouple, and diode-detector power sensors
Typical measurement method	*Calorimeter*. Direct measurement of energy absorbed as heat *Dual six-port*. S-parameter network analysis *Bolometer*. Ratio (resistance and/or AC/DC power transfer) *Thermocouple*. Comparison to bolometer *Diode-detector power sensors*. Comparison to bolometer
Measurement considerations	Excessive power, electrostatic discharge, connector dimensions, connector torque
Major uncertainty sources	*Coaxial connection repeatability*. Always use a torque wrench if the connector design allows it. *Reflection coefficient*. Affected by connector cleanliness, pin recession or protrusion dimensions, and overall connector wear. *Ambient temperature*. Standards and workload must be protected from drafts and temperature changes during the calibration process. *Electromagnetic interference*. In some cases, the calibration may have to be done inside a screen room.

As the frequency increases to the RF region, these parameters become both more difficult to measure and less relevant to the average user. There are several considerations:

- The transmission line characteristics have an increasing effect on the signal and its measurement as frequency increases.

- Impedance becomes a complex vector value that cannot be easily determined and that varies with position on the transmission line.

- At RF and microwave frequencies, the voltage and current parameters are meaningless to the practical user.

In an ideal transmission line with no loss, power is constant along the length because it is a product of voltage and current that is independent of position along the line.[9] Voltage measurements are still used (with a diode detector), but many RF power measurements use a thermal sensor, either a thermistor or a thermocouple. In either case, the measurement system usually displays the result as power or as a power-related computed value.

An RF power measurement system is composed of a sensor and a meter. Some low-level sensors include a 30-dB reference attenuator, and some high-power sensors include attenuators. All of these items must be calibrated before valid measurements can be made with them. Also, some types of power sensors must be used only with a specific power meter (by serial number)—watch out for those.

Sensors and Their Meters

In most cases, what one needs to measure is the RMS (root-mean-square) or average power level. One way to quantify RMS power is to equate it to the heating effect produced by direct current into the same impedance. Calorimeters, thermistors, and thermocouples for RF and microwave power measurement use heat in some form and are inherently average responding. Diode sensors, in contrast, sense either average or peak voltage and use computation to derive the RMS equivalent.

Thermistor Sensors

A thermistor sensor (or thermistor mount) is a type of bolometer. *Bolometer* is a term that is common in the (older) fundamental literature but is seldom used now. It refers to two types of sensors: a barretter or a thermistor. In general, a bolometer is a type of sensor that absorbs the RF power and changes the resistance of the sensing element. The bolometer element is usually in one arm of a bridge circuit, which allows the resistance to be measured by conventional DC or low frequency AC methods. A bolometer is inherently average responding. A *barretter* is a metallic resistive element, often platinum, which increases in resistance as temperature goes up. The temperature change is determined by measuring the resistance.[10] Barretters are no longer used, so this term is rapidly becoming archaic.[11] A *thermistor* is a small bead of semiconductor material, with wires for connecting to a measuring circuit. The temperature change is determined by measuring the resistance. The resistance of a thermistor element decreases as the temperature increases.

A thermistor is a temperature-sensitive resistor and, as such, its resistance can be measured using a DC bridge. (See the discussion of resistance in Chapter 32, "DC and Low Frequency.") Alternatively, a bias current in the bridge can warm the thermistor to a preselected resistance value and then the bridge can be balanced for a zero indication. If RF power is added to the thermistor through the sensor input, the bridge will become unbalanced. If the bias current is reduced to bring the bridge back into balance, the amount of power (I^2R) removed from the bridge by reducing the current is equal to the RF power absorbed by the thermistor. This is the DC substitution method; it is traceable to the SI base units of the ampere and second and the derived unit of the watt. Power meters that use thermistor mounts use this measurement method.

The principal advantage of a thermistor sensor is absolute RMS power measurement by the DC substitution technique. The main disadvantages (for some applications) are limited dynamic range and slow response time. In a calibration lab, the principal uses of thermistor sensors are transfer standards from higher echelons, calibration of power sensors, measurement of absolute RF power, and calibration of the 50 MHz 1 mW reference output used on meters for thermocouple and diode sensors. This last application requires a thermistor sensor selected for low standing wave ratio (SWR) and optimized for low-frequency operation.

Thermistor Sensor Meters

There are two general types (portable and laboratory) of RF power meters that use thermistor sensors. They both use versions of the self-balancing DC substitution bridge that was developed by the U.S. National Bureau of Standards (NBS, now NIST) in 1957 and refined in the 1970s.[12] As discussed earlier, the amount of DC power in the bridge is adjusted to keep the resistance of the thermistor constant, and the amount of change is the measure of the absorbed RF power. This type of meter and sensor is the only one that can give an absolute power measurement.

Thermocouple Sensors

A thermocouple sensor measures power by sensing the heating of a thermocouple junction. A thermocouple produces a voltage that has a known relationship to temperature difference between the junction and the measuring end. The meter measures the voltage and displays the result as power. Like thermistors, thermocouples are inherently average responding.

Thermocouple sensors have several advantages over thermistor types. Thermocouples can measure across larger temperature ranges than thermistors: –200 °C to 2500 °C for thermocouples vs. –100 °C to 260 °C for thermistors. These wider temperature ranges allow for their use in a larger range of applications, such as in ovens, kilns, autoclaves, furnaces, chambers, manufacturing presses and molds, aircraft, satellites, etc. Thermocouples do not require any external power to operate and therefore are not prone to self-heating.

The principal disadvantages of thermocouple sensors are that they are nonlinear across their measurement ranges, and their accuracies tend to be poor. System errors of less than 1 °C are difficult to achieve and maintain during use. Thermocouples wires are very delicate and can easily be damaged from bending of the wires. The bending of the wires causes a change in resistance of the wires, and therefore a change in voltage which translates into changes in its calibration and accuracy. Because thermocouples are made of two dissimilar metals, they are subject to corrosion, which can also affect their calibration and accuracy. It is important to use the correct thermocouple type for an application and heed the temperature limits of its range.

Diode Sensors

A diode power sensor rectifies the RF and produces an output voltage. The meter measures the voltage, references it to current or impedance (depending on the design), and displays the result as power. A meter that uses a diode sensor may respond to the average voltage and be calibrated to display an equivalent RMS value, or it may respond to the peak waveform. A peak-responding meter may be calibrated to display an equivalent RMS value (the most common type), or it may display the actual peak power. The measurements are based on the AC volt and either the ampere or the ohm. Depending on the design, they use the relationship $P = E^2/R$ or $P = I^2/R$ to display the power up to –20 dBm, the square law region.[13] At higher levels, the performance transitions to a linear output region. The degree of linearity in this region is an important specification, particularly at higher power levels. Diodes used to be limited to a maximum input of about –20 dBm due to use in the square law region, but modern diode sensors are usable up to about +20 dBm.[14] The expanded range is possible because of digital storage of correction factors for linearity errors. Depending on the design, the corrections may be stored in the sensor (which makes them interchangeable between compatible meters) or in the meter. If linearity correction factors are stored only in the meter, then the sensor is not interchangeable, and the sensor and meter must be calibrated together.

Peak power measurements are common in pulse modulation applications such as radar, aircraft transponders, and digital communications. Diode sensors are commonly found in peak power meters and network analyzers. Limits are defined by low-level noise from the diode and meter, and damage due to excessive power input. Advantages of a diode sensor are high sensitivity, a simple circuit design, and (in newer models) up to 90 dB dynamic range.

The principal disadvantage of broadband diode sensors is that in most cases they only make relative power measurements, not absolute. The measurements are all made relative to a reference. The power meters for this type of sensor have a 50 MHz oscillator output on the front panel; the output power level is usually set to 1 mW ± 0.4 %. That output is used to set the sensor and meter to a reference level, and measurements are made relative to that setting. This type of system cannot be used for absolute power measurements. Accurate measurement of the power output (using a thermistor mount) is an important part of calibrating such as meter. Naturally, this adds an uncertainty element to the measurements.

Thermocouple and Diode Sensor Meters

Most common RF power meters use either thermocouple or diode sensors. All of them make relative power measurements. With the exception of one type (to be discussed), the reference is a reference oscillator in the power meter. The reference oscillator is usually designed to produce a 1.00 mW output at 50 MHz. The user of the meter connects the sensor and checks or sets this as the reference level before every use. Therefore, calibration of the reference output is a critical part of calibrating the power measurement system.

The other type of meter that uses a diode sensor does not require a reference power setting. This type of meter uses the diode as a voltage detector on the coupled output arm of a directional coupler. The diode, coupler arm, and reference termination are housed in a plug-in element. Each element is designed for a specific frequency and power range, and for either average or peak power. They are not broadband devices (other power meters are), but they are often more suitable for field service work. An advantage of this type of meter is that simply rotating the plug-in element 180° allows measurement of reverse power, which can be used to measure the SWR on the transmission line. This type of meter has lower overall accuracy and makes measurements relative to the calibration standards.

Peak Sensor Meters

Older peak power meters use an average-responding sensor and divide the result by the pulse duty cycle to estimate the peak power. This is suitable only if the modulation waveform is a rectangular pulse with a known and constant duty cycle.[15] Recent meter designs have started to use advanced signal processing to detect and analyze the actual modulation envelope of the signal, and to use time-gated systems that measure the power only during the on portion of a pulse.[16]

Calibration of RF Power Sensors and Meters

Power Sensors

The most important result of calibrating a power sensor is determination of the calibration factor. This value is essential for the sensor and meter to be properly used as a system. The calibration factor is the ratio of power absorbed by the sensing element to power available at the input plane of the sensor. It includes the two primary error sources—mismatch error and sensor efficiency. Mathematically,

$$K_b = \eta_e (1 - \rho_l^2)$$

where

K_b = Calibration factor
η_e = Effective efficiency, ratio of power absorbed by the sensing element to total incident power
ρ_l = Reflection coefficient magnitude of the sensor[17]

RF power sensors are calibrated by comparison to a transfer standard. There are two types of systems in common use. One is based on a DC or low-frequency AC bridge, and the other uses a well-matched RF power splitter. The purpose of calibration is to verify the RF performance by comparison to a standard with known characteristics and to determine a value for a calibration factor.

The bridge system is capable of the lowest uncertainty and direct traceability to national standards. A popular portable meter based on an automatic bridge design is the Keysight (formerly Agilent) 432A, which has been in production for

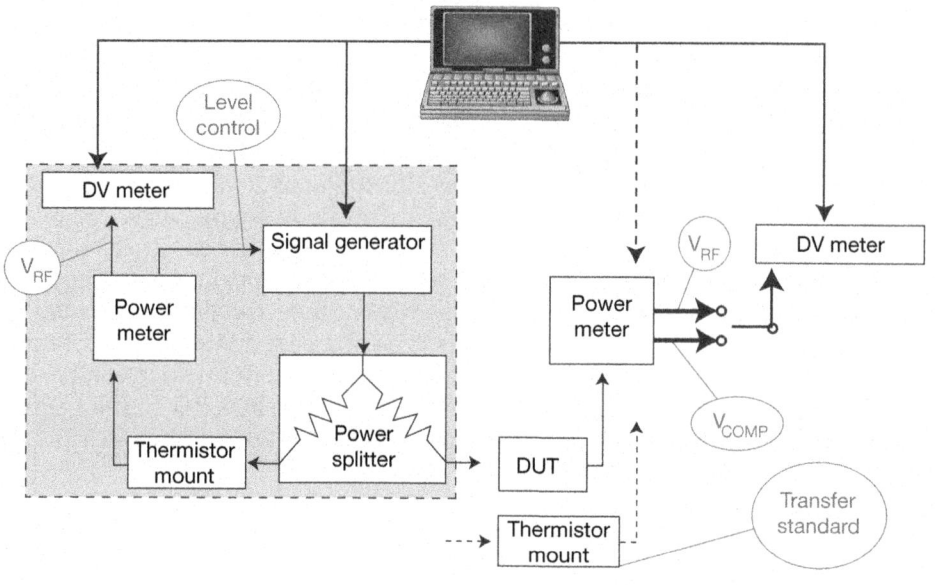

Figure 33.1 Power sensor calibration system—block diagram.

more than 30 years. As a standalone meter, its instrumentation uncertainty is ± 1 % of full scale in addition to the sensor uncertainty. But because the bridge voltages are available on the rear panel, higher-accuracy DC voltmeters can be used for the measurements to achieve instrumentation uncertainty closer to ± 0.2 % of reading.[18] A popular laboratory system is based on the NIST Type IV bridge design developed in the early 1970s. This system is more complex and expensive but can be used (with purchase of appropriate transfer standards) to obtain direct traceability to national standards. In addition to this system, the laboratory needs a signal generator, power meters compatible with the workload, and a means of measuring reflection coefficient.

The power splitter system is based on a two-resistor power splitter (see Figure 33.1). It is simple, easy to use, broadband, and requires little equipment. It is restricted to use with coaxial power sensors, but the majority of them are that type now. The basic components needed are a signal generator, a power splitter, a power sensor and meter for measuring the applied power and leveling the generator, a transfer standard sensor and meter, the sensor to be tested, and a means of measuring reflection coefficient. (The sensor to be tested can be used with its own meter provided the meter has been calibrated first.) A waveguide power sensor could be calibrated with the same type of system by replacing the two-resistor power splitter with a waveguide directional coupler and adjusting the readings to account for the coupling factor.

Power Meters

The power meters must also be calibrated. Functionally, power meters are essentially DC or low-frequency AC voltmeters with a few additional special circuits. The most important parameters are range accuracy and linearity, and calibration factor accuracy. For meters that use thermocouple or diode sensors, the output level of the reference oscillator is a critical parameter.

Attenuation or Insertion Loss Parameters

Attenuation is an important parameter of RF and microwave systems. It is double-edged: In some cases, it is a desirable characteristic, and in others it is something to be minimized. Attenuation is normally measured using RF power and ratio methods. See Table 33.4 for details on attenuation or insertion loss.

Attenuation is a ratio. It is a decrease in power as a signal passes through a transmission line or other passive device. Insertion loss is an equivalent term. Attenuation is the opposite of the power gain provided by an amplifier. An attenuator is a device that is intended to reduce the input power by a predetermined ratio. Attenuation measurement values are typically reported in decibels (dB) due to the wide range of the numbers.[19]

Attenuation can be measured by either voltage or power measuring instruments. Attenuation is normally expressed in decibels, as the ratio of the output to the input voltage or power:

$$dB = 10 \log \frac{P_{out}}{P_{in}}$$

or

$$dB = 20 \log \frac{V_{out}}{V_{in}}$$

Table 33.4 Attenuation or insertion loss parameters.

Parameter name	Attenuation or insertion loss
SI units	Ratio
Typical primary standards	RF power standards, waveguide-beyond-cutoff (piston) attenuator, dual six-port network analyzer
Typical transfer standards	Waveguide-beyond-cutoff (piston) attenuator, waveguide rotary vane attenuator
Typical calibration standards	Waveguide-beyond-cutoff (piston) attenuator, waveguide rotary vane attenuator, precision coaxial attenuators, tuned RF attenuation measurement system, scalar network analyzer, vector network analyzer
Typical IM&TE workload	Coaxial and waveguide attenuators (fixed or variable), tuned RF attenuation measurement system, signal generator output attenuators, receiver input attenuators
Typical measurement method	Audio substitution attenuation measurement system, IF substitution attenuation measurement system, tuned RF attenuation measurement system, scalar network analyzer, vector network analyzer, RF power ratio
Measurement considerations	Excessive power, electrostatic discharge, connector dimensions, connector torque
Major uncertainty sources	*Coaxial connection repeatability.* Always use a torque wrench if the connector design allows it. *Reflection coefficient.* Affected by connector cleanliness, pin recession or protrusion dimensions, and overall connector wear. *Electromagnetic interference.* In some cases, the calibration may have to be done inside a screen room.

If the output is less than the input, solving either of these relationships will result in a negative number, which is mathematically correct because attenuation is a negative gain. By convention, however, the negative sign is commonly dropped as long as the context makes it clear that attenuation is being discussed instead of gain.

Measurement of power is important to attenuation measurement, as the power ratio is often the most convenient to measure and use. Several different measurement methods are used, each with their own set of benefits. At low attenuation ratios (to about 40 dB), systems that use an audio substitution method are commonly used. A typical swept-frequency system that uses audio frequency (AF) substitution is the Keysight (formerly Agilent) 8757E scalar network analyzer. Intermediate frequency (IF) substitution systems can measure attenuation ratios of 90 dB or more. A typical system that uses IF substitution is the TEGAM 8852, which is based on the model VM-7 30 MHz receiver.

Reflection Coefficient, Standing Wave Ratio Parameters

The *reflection coefficient* or *standing wave ratio* (SWR) is another important parameter in RF systems. This must be minimized to improve power transfer and reduce the likelihood of equipment damage. SWR is a ratio measurement. See Table 33.5 for more details on reflection coefficient.

Reflection coefficient, SWR, and return loss are different ways of expressing the same parameter. Their relationships are as follows:[20]

Complex reflection coefficient = $\Gamma = \dfrac{V_{reflected}}{V_{incident}}$ (has magnitude and phase)
Range 0 (no reflection) to 1 (total reflection) with phase 0 to $\pm 180°$)

where

$V_{reflected}$ = voltage of the reflected signal

$V_{incident}$ = voltage of the incident signal

Scalar reflection coefficient = $\rho = |\Gamma|$ (has magnitude only)
Range 0 (no reflection) to 1 (total reflection)

$SWR = \sigma = \dfrac{1+\rho}{1-\rho} = \dfrac{SWR-1}{SWR+1}$ (A unitless ratio)
Range 1 (no reflection) to (total reflection)

SWR (in dB) = 20 log σ

Return loss (dB) = $-20 \log \rho$ Range ∞ (no reflection) to 0 (total reflection)

Mismatch loss (dB) = $-10 \log (1 - \rho^2)$ Range 0 (no loss) to more than 50 dB loss

Impedance of a component being measured Range 0 (short circuit) to ∞ (open circuit)

$Z_{DUT} = Z_0 \dfrac{1+|\rho}{1-\rho}$ Where Z_0 is the characteristic impedance of the system (normally 50 W)

Note: The conversion from reflection coefficient to other values changes it from a vector with magnitude and phase to a scalar value with magnitude only. The loss of phase information increases measurement system uncertainty.

Table 33.5 Reflection coefficient, standing wave ratio parameters.

Parameter name	Reflection coefficient
SI units	Ratio
Typical primary standards	Dual six-port network analyzer
Typical transfer standards	Fixed reflection coefficient standards, sliding loads, sliding shorts, network analyzer calibration standards
Typical calibration standards	Scalar and vector network analyzers, S-parameter test sets, SWR bridges
Typical IM&TE workload	All coaxial or waveguide RF and microwave devices
Typical measurement method	Audio substitution attenuation measurement system, IF substitution attenuation measurement system, tuned RF attenuation measurement system, scalar network analyzer, vector network analyzer, RF power ratio
Measurement considerations	——
Major uncertainty sources	*Coaxial connection repeatability.* Always use a torque wrench if the connector design allows it.

SWR is normally measured as a voltage ratio. SWR can also be measured as a power ratio, and it can be derived in terms of current or impedance ratios.[21]

Measurement of SWR is dependent on the ability to measure RF voltage or power, and attenuation. The ideal case is where the load absorbs all of the forward power, and none is reflected. Imperfect impedance match, connector variations, transmission line discontinuities, or other problems cause reflected power, which is measured as reflection coefficient or SWR. Note: SWR may be measured in terms of either voltage (VSWR) or power (PSWR). The power ratio method is so rarely used now that it is increasingly common usage in the United States to drop the initial letter of the acronym. It is understood that SWR refers to the voltage ratio.[22]

SWR measurement uses attenuation or power measurement ratios and the theory of transmission lines. The measurement systems are initialized and verified with standard terminations: short-circuit, open-circuit, and standard mismatches with known reflection coefficients. (Open-circuit standards are generally not used in waveguide systems.) In a swept-frequency measurement system, attenuation measurements may be presented as a line graph of magnitude versus frequency or on a polar graph or a Smith chart.

Because the Smith chart is frequently used with measurements of reflection coefficient and other RF and microwave parameters, a few comments about it are appropriate. The Smith chart, invented in 1937 by Phillip H. Smith, is a useful tool for graphical analysis of and transformation between impedance and SWR. Full discussion of the Smith chart is beyond the scope of this book (it would take at least another chapter), but some useful references are listed in the references and bibliography. Figure 33.2 is an example of a Smith chart. A printed chart can be plotted and evaluated manually, but more commonly now the chart is a display feature of measurement systems.

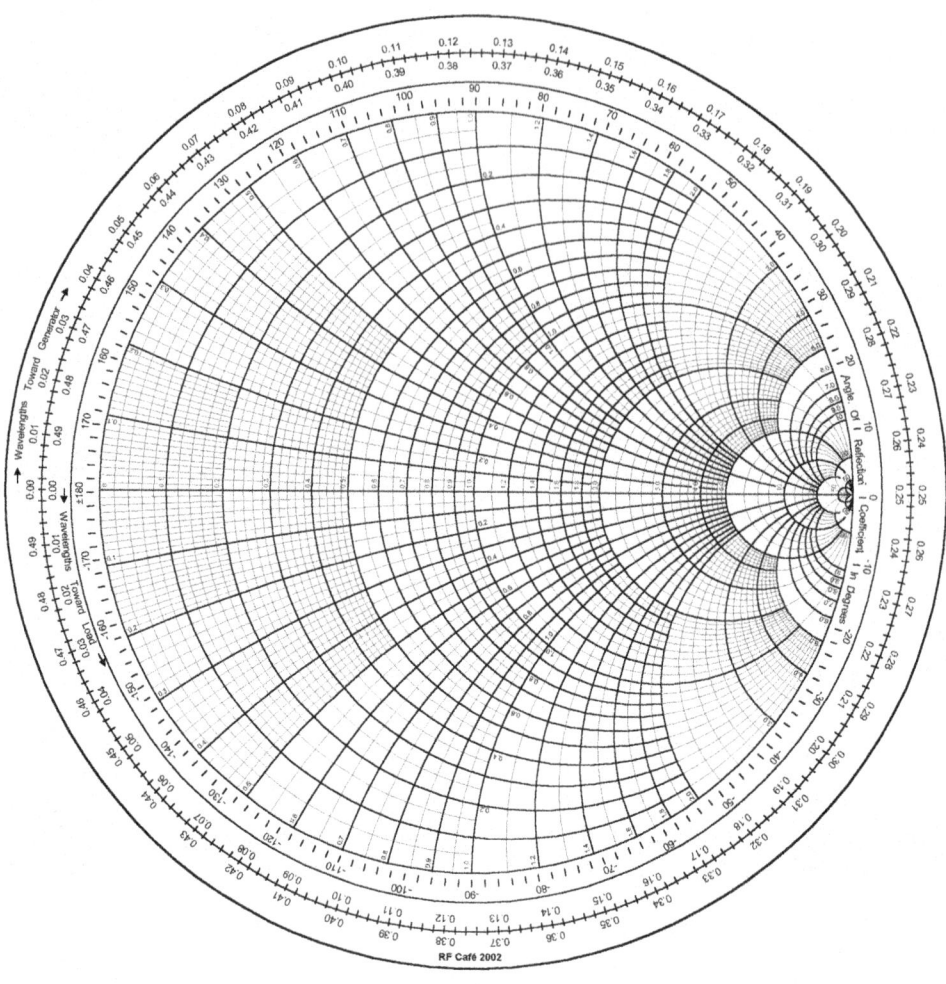

Figure 33.2 Smith chart.[23]

Key features of the Smith chart are:

- All values on the Smith chart are normalized to the characteristic impedance of the system (Z_0) and the ½ wavelength of the signal frequency. This makes the graph independent of the actual impedance or frequency.
- The horizontal center line of the graph represents pure resistance.
- The circumference of the graph represents pure reactance. Positive values (above the resistance line) are inductive; negative values are capacitive.
- The intersection of the horizontal line (between 0° and ± 180°) and the circle labeled 1.0 (the unit circle) represents the theoretically perfect complex impedance of the system—all resistance and no reactance.

- The circumference of the graph is ½ wavelength.

- When a point is plotted on the chart, the distance from the center represents the magnitude of the reflection coefficient. The center is minimum reflection because it represents Z_0, and the circumference is complete reflection, as from a perfect open or short.

- When a point is plotted on the chart, a line drawn from the center through that point and continuing to the edge will show the phase angle of the reflection coefficient.

- There are three scales around the circumference of the graph. The innermost one is the phase angle of the reflection coefficient. The other two are graduated in fractions of a wavelength toward the load (inner) and the generator (outer).

- Some printed versions of the Smith chart include additional horizontal scales above or below the graph. Most commonly, these are used to quickly plot or determine values such as standing wave ratio, or return loss in decibels, as well as other values. Automated systems can make these transformations internally.

In common metrology applications, it does not make much difference whether voltage or power is measured to determine SWR. That is because we have the capability to make the measurements at the reference planes of the devices being measured. In other applications, though, it makes a difference. If the measurement is made at the source end of a transmission line, the measurement will be correct only if a power ratio measurement is used. A voltage or current-based measurement is only correct at the load or at exact half-wavelength intervals from it along the transmission line. At any other point, it is useful only as a relative indicator of load match—to adjust the load for a minimum—because the actual displayed value has very little meaning.[24] A common situation where this type of measurement may be made is when testing a transmitter antenna.

RF Voltage Parameters

RF voltage measurement is used in two major ways. Many instruments use voltage measurements internally, even if they display a measurement result in other terms. If an instrument uses a diode-type sensor, it is likely making voltage-based measurements. The other principal method is the use of a thermal voltage converter for high accuracy AC/DC transfer measurement at radio frequencies. See Table 33.6 for more details on RF voltage.

As discussed in the section on RF power, many voltage measurements are made using diode detectors. The detectors are commonly part of measurement systems for other parameters such as power, attenuation, or reflection coefficient. An important manual use was measuring SWR by means of voltage measurements along a slotted line, but now this is rarely done in the majority of calibration labs.

The most important manual method of measuring RF voltage uses *thermal voltage converters* (TVC), which transfer direct current or voltage values to alternating current or voltage. There are two types in wide commercial use. The

Table 33.6 RF voltage parameters.

Parameter name	RF voltage
SI units	Volt (W / A)
Typical primary standards	RF voltage comparator
Typical transfer standards	Thermal voltage converter
Typical calibration standards	Thermal voltage converter
Typical IM&TE workload	Signal generators, RF voltmeters
Typical measurement method	AC/DC substitution at radio frequencies
Measurement considerations	Excessive power, frequency limits, AC-DC difference
Major uncertainty sources	*Coaxial connection repeatability.* Always use a torque wrench if the connector design allows it. *Reflection coefficient.* Affected by connector cleanliness, pin recession or protrusion dimensions, and overall connector wear. *Ambient temperature.* Standards and workload must be protected from drafts and temperature changes during the calibration process.

most common is a TVC that uses a vacuum thermoelement and is usable up to 100 MHz. The other type, called a *micro potentiometer*, uses a ceramic disk resistor and is usable up to 1 GHz. These devices have measurement uncertainties of about 1 %. The operating principle is that an AC RMS level is equal to a DC level if it produces the same heating effect. A newer device, with uncertainties in the ppm range, is the multi-junction thin film thermal voltage converter system developed by Sandia National Laboratory and NIST.

Modulation Parameters

Modulation is an essential property for communication because it is the modulation of a carrier signal that carries the information. The simplest form of modulation is turning the carrier signal on and off, but modern communication systems require much more complex methods. See Table 33.7 for more details on modulation.

Modulation is the area where low-frequency AC and RF-microwave come together. A modulated signal is a single-frequency signal, usually in the RF-microwave region, which is altered in a known manner to convey information. The modulation source can be audio, video, digital data, or the simple on-off action of a Morse code key. The modulation always adds energy to the carrier and increases either its power or its bandwidth. The modulation can alter the carrier by changing amplitude (AM), frequency (FM), phase (PM), or any combination. For example, one of the signals transmitted by an aviation very high frequency omnidirectional range (VOR) navigation base station is an amplitude-modulated signal at the assigned frequency. The modulating signal is an audio frequency that is itself frequency-modulated with a second audio frequency.

Table 33.7 Modulation parameters.

Parameter name	Modulation
SI units	Ratio
Typical primary standards	Derived from mathematical description of the modulation type and applied to RF and microwave primary standards
Typical transfer standards	Voltage, frequency, power, mathematics
Typical calibration standards	RF, microwave and audio signal generators, signal analyzers, spectrum analyzers, modulation theory
Typical IM&TE workload	Test sets for RF and microwave communication systems, radar systems, radio navigation systems
Typical measurement method	Comparison
Measurement considerations	—
Major uncertainty sources	*Coaxial connection repeatability.* Always use a torque wrench if the connector design allows it. *Reflection coefficient.* Affected by connector cleanliness, pin recession or protrusion dimensions, and overall connector wear. Linearity of audio detection system.

Radar systems and microwave communications systems use pulse modulation, which is a form of amplitude modulation. Pulse modulation can be as simple as on–off, but most systems now use methods to vary the pulse position, width, amplitude, or frequency as part of the method of carrying information.

Modulation is a ratio: 0 % is the absence of a modulating signal, and 100 % is the maximum amount that can be applied without causing distortion of the signal. It is sometimes referred to as the modulation index, which is a number between 0 and 1. Modulation, along with frequency and power, is one of the regulated parameters of a radio, television, or other transmitter.

Modulation is frequently measured by analyzers designed for that purpose, such as AM/FM modulation meters or video modulation analyzers. Amplitude modulation can be measured using an oscilloscope. Frequency modulation can be measured with a spectrum analyzer, using the Bessel null technique. Some modulation sources, such as the VOR signal mentioned earlier, require specialized test equipment.

Noise Figure, Excess Noise Ratio Parameters

Noise is an important parameter to measure in communication systems. All electronic devices generate some amount of noise from the random motion of electrons, which is related to heat. The total amount of self-generated noise in a system sets a limit to the weakest signal that can be detected. See Table 33.8 for details on noise.

The thermal noise (Johnson noise) in a resistor is

$$N = kTB$$

where N is the noise power in watts, k is Boltzmann's constant, T is the absolute temperature in Kelvin, and B is the bandwidth of the measurement system.

Table 33.8 Noise figure, excess noise ratio parameters.

Parameter Name	Noise Temperature
SI units	Kelvin
Typical primary standards	Primary thermal noise standards, total-power radiometer
Typical transfer standards	Thermal noise standard
Typical calibration standards	Thermal noise standard
Typical IM&TE workload	Thermal noise standards, amplifiers, noise figure meters
Typical measurement method	Comparison
Measurement considerations	Temperature, bandwidth, connector quality
Major uncertainty sources	*Reflection coefficient.* Affected by connector quality, condition and cleanliness, pin recession or protrusion dimensions, and overall connector wear. *Coaxial connection repeatability.* Always use a torque wrench if the connector design allows it. *Source stability.*

The thermal noise source calibration system used at standards laboratories has two resistive noise sources. One is immersed in liquid nitrogen (77.5 K or –195.7 °C) and the other is at laboratory ambient temperature. (Some systems may operate the second source at 373 K or 100 °C.)[25] For convenience in calculations, the laboratory ambient is usually taken to be a conventional value of 290 K (16.8 °C) and labeled T_0.

The value reported by NIST is the available noise temperature, defined as the available noise power spectral density at the measurement plane, divided by Boltzmann's constant. The temperature units are reported in kelvins. For noise temperatures over T_0, the excess noise ratio (ENR) is also reported.

MEASUREMENT METHODS

There are three important methods used for measuring RF and microwave parameters: spectrum analysis, scalar network analysis, and vector network analysis. Although they are not measurement parameters, the understanding of these methods is important because of their common use and wide application.

Spectrum Analysis

Like an oscilloscope, a spectrum analyzer measures signal amplitude on the vertical scale. The horizontal scale, however, is calibrated in terms of frequency instead of time. A spectrum analyzer displays the individual frequency components of a signal source. The essential parts of a spectrum analyzer are a swept-frequency local oscillator, a superheterodyne receiver, and a display unit.[26] The local oscillator sweep is coupled to the horizontal axis of the display, and the receiver output is coupled to the vertical axis. Spectrum analysis is not exclusive to RF; there are models that start in the sub-audio range as well as those measuring up to 110 GHz or more. The measurement from a spectrum analyzer is voltage-based, but the display is normally referenced to a power level and scaled in decibels.

Scalar Network Analysis

A scalar network analyzer, like a spectrum analyzer, has a display that shows amplitude versus frequency. There are more inputs, though, and the principal use is to make ratio measurements. The essential parts of a scalar network analyzer are a swept-frequency signal source, a multi-input broadband receiver, a signal splitter or directional bridge, diode detectors, and a display unit.

The sweep ramp of the signal source also controls the horizontal display of the network analyzer. The RF output of the source is connected in different ways depending on the application:

- To measure a two-port device such as an attenuator, the swept RF output goes through a power splitter. One side of the splitter is connected to the reference input of the analyzer. The other side of the splitter is connected to the measurement input of the analyzer through the device to be tested. Before use, the system is set up by measuring with a direct connection to set the zero level and with a check standard to verify performance.

- To measure a one-port device such as a termination, the swept RF output goes through a bridge or directional coupler. The system is set up by using short and open standards to set the zero-return loss level and verified by using mismatch standards. Then the device under test is attached to the measurement port and measured.

Because of the diode detectors, the scalar network analyzer is a broadband device. This is either an advantage or not a problem for many measurements, but does have some limitations. The sensitivity is limited to about −60 dB and cannot be improved by averaging.[27]

A scalar network analyzer requires a normalization procedure whenever it is started and whenever the test setup or parameters are changed. The normalization (often called *self-calibration*) is done using high-quality transmission and reflection standards. The scalar network analyzer computer makes and stores a set of response measurements, which are used as a reference for measurements of the devices being tested. This minimizes the effect of any power versus frequency variations. Because the measurements are scalar quantities, the error model cannot account for any other source of systematic error.[28]

To help understand the difference, look at measurements of attenuation made by scalar and vector-swept frequency measurement systems (see Figures 33.3 and 33.4).

As shown in the figures, the scalar measurement has an uncertainty circle (the dashed line) around it. This defines the uncertainty introduced because the reflection coefficient of the system is unknown. At any given frequency, the reflection coefficient could be anywhere from completely added to the magnitude to completely subtracted from it, but the exact value is unknown. A vector measurement system includes phase in the measurement and therefore produces a more exact result. There is still some residual uncertainty, but it is one or two orders of magnitude less than a scalar measurement.

Scalar and Vector: What Is the Difference?

A scalar is a real number. In measurement, a *scalar* is a quantity that only has a magnitude. Most of the numbers encountered are scalars. For example, if an object is moved, the measurement of the distance is a scalar.

A *vector* is a quantity that has both magnitude and a direction. If plotted on a graph, a vector starts at a defined point (its origin), has a length equal to the magnitude, and is at a specific angle to the reference line. Continuing the previous example, a vector would describe not only how far the object was moved, but also in what direction.

A scalar network analyzer measures magnitude only. The reflection coefficient (or return loss or VSWR) adds uncertainty. Its phase with respect to the incident signal changes with frequency and is unknown. So, at a given frequency, the actual magnitude is uncertain but somewhere in the range:

(Measured value ± Reflection coefficient)

This uncertainty is the reason the reflection coefficient of RF and microwave components, connections, transmission lines, and systems should be as low as possible.

A vector network analyzer measures magnitude and phase at each frequency point. The measured value has a reduced uncertainty because both the magnitude and phase of the reflection coefficient are known.

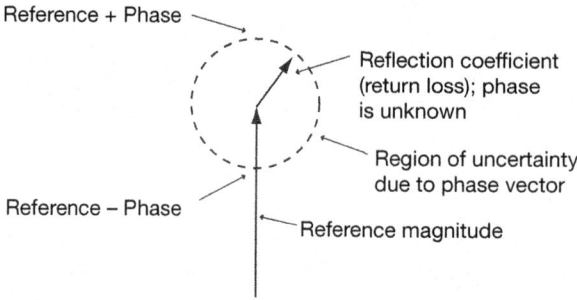

Figure 33.3 A scalar network analyzer measures magnitude only. The reflection coefficient phase varies with frequency and causes an area of uncertainty.

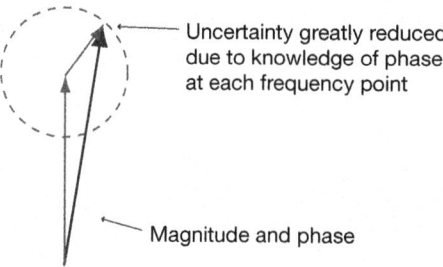

Figure 33.4 A vector network analyzer measures magnitude and phase at each frequency, so magnitude uncertainty is reduced.

Vector Network Analysis

A *vector network analyzer* (VNA) makes the same kind of measurements as a scalar analyzer, but in a manner that preserves the phase relationships of the signals. The essential parts of a vector network analyzer are similar to those of a scalar analyzer, but there are important differences. The major differences are use of narrow-band tuned receivers and the addition of an S-parameter test set or a transmission-reflection test set.

Conceptually, the S-parameter test set contains two dual directional couplers, four diode sensors, and some high-speed switching devices. (Actual implementation is more complex, of course!) The signal from the source is sent in one direction, and measurements are made of the four coupled outputs. The signal direction is reversed, and the measurements repeated. The values are mathematically combined to determine the four S-parameters of the device under test.[29] Some units use a *transmission-reflection* (T/R) test set. The transmission-reflection test set is very similar to an S-parameter test set, except that the signal only goes one direction. This makes it simpler because there is only one dual directional coupler, half the detectors, and no need for the switching arrangement. However, the T/R test set is limited to one-port measurements.[30]

A vector network analyzer requires a normalization procedure whenever it is started and whenever the test setup or parameters are changed. The normalization (often called *self-calibration*) is done using very high-quality transmission and reflection standards, along with a mathematical model of each standard. The VNA computer makes a set of measurements, compares them to the known model of the standard, and stores the difference as a correction. This minimizes the effect of any systematic error sources within the analyzer. Because the measurements are vector quantities, the error model accounts for all the major sources of systematic error.[31]

Some important advantages of vector network analysis are:[32]

- Using a tuned-receiver design provides better sensitivity and more dynamic range than simple diode detection.

- Phase information is preserved during the detection process. This results in lower measurement uncertainty. It also allows more complete error correction to compensate for the test setup.

- Dynamic range can be improved by averaging, since the averaging is done with the vector data. Another advantage is the ability to display the data in forms that may be more useful than an amplitude-versus-frequency display. For example, data can also be displayed on a polar graph or a Smith chart.

S-Parameters

S-parameter measurements are developed from the fundamental theorems of electronic networks. Recall that if a component or system is modeled as a single block with no knowledge of what is inside it, its behavior can be characterized by measuring the response of its outputs when various inputs are applied. In

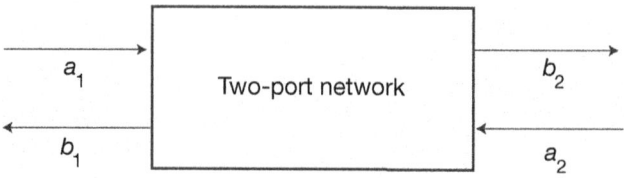

Figure 33.5 A general two-port network. This can be used to represent a device such as a coaxial attenuator.

a simple case such as a coaxial attenuator, the device has two ports, which are called the input and output, or 1 and 2 for convenience. Each port has two nodes, usually labeled a and b. A *node* is a point for measuring voltage or current passing into or out of the network. Figure 33.5 is a diagram of a basic two-port network.

The operation of a vector network analyzer is based on measurements of a two-port network and application of the mathematics of vector measurements and scattering or S-parameters. A basic understanding of those subjects is essential to understanding the VNA and using it effectively. The references by Adam and Keysight (formerly Agilent) have more detailed information.[33]

A vector quantity is one that has both magnitude and phase information. (If the phase is not measured, then the quantity is a scalar.) By definition, the VNA makes vector measurements. Also, all the raw measurements are voltages. S-parameters are a set of measurements of energy flow into and out of a device when the device is terminated in its characteristic impedance (Z_0). S-parameters are easy to measure at RF and microwave frequencies; directly represent values that are used by metrologists, technicians, and engineers; and are used to model components in design software. S-parameters are typically illustrated using a network flow graph diagram, which is a visual method of analyzing voltage traveling waves. A flow graph separates each port (input or output) into two nodes, labeled a_n and b_n. Node a always represents waves entering the network and node b always represents waves leaving the network. The number n is the port number that the node represents. All nodes are connected with directional lines that illustrate the possible signal flow paths. A network with N ports has $2N$ nodes and N^2 S-parameters. For example, a three-port network (such as a directional coupler) has three ports (input, output, coupled output) and nine S-parameters.

In Figure 33.6, a_1, b_1, a_2, and b_2 are the nodes. Node a_1 represents the incident wave entering port 1, and node b_1 represents the reflected wave leaving port 1.

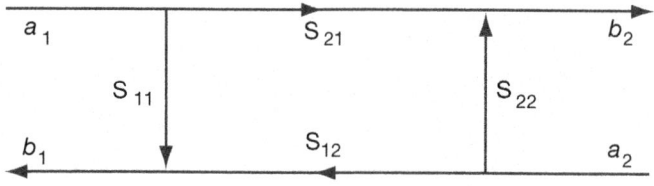

Figure 33.6 S-parameter flowgraph of a two-port network.

Nodes a_2 and b_2 represent the same things for port 2. The four S-parameters are s_{21}, s_{11}, s_{12}, and s_{22}. The convention for the subscript numbers is that the first digit is the port number of the output node, and the second digit is the port number of the input node.

An advantage of S-parameters is that they directly represent the transmission and reflection values that are commonly used in metrology. Any one S-parameter can be measured from the values of one a node and one b node, provided the other a node value is zero. If the device is terminated in its Z_0 impedance, this is assumed to be true. The relationships are shown as follows:

Input reflection coefficient $\qquad s_{11} = \dfrac{b_1}{a_1} \qquad$ when $a_2 = 0$

Forward transmission ratio $\qquad s_{21} = \dfrac{b_2}{a_1} \qquad$ when $a_2 = 0$

Reverse transmission ratio $\qquad s_{12} = \dfrac{b_1}{a_2} \qquad$ when $a_1 = 0$

Output reflection coefficient $\qquad s_{22} = \dfrac{b_2}{a_2} \qquad$ when $a_1 = 0$

All the S-parameters are voltage ratios. The input and output reflection coefficients are the same as G, described earlier. The forward and reverse transmission ratios can be converted to attenuation (or gain) as described earlier:

$$dB = 20 \log \dfrac{V_{out}}{V_{in}}$$

Four-port VNAs were introduced in 2001. Design and test engineers use this in developing equipment that uses differential inputs and outputs. A differential input is one that has both sides of the input or output floating—not connected to circuit common, or ground. This has a number of technical advantages that are not relevant here. High-speed data communication systems are working at RF frequencies, where RF behavior is important, so there is increasing use of RF and microwave test tools. An example of where the four-port VNA might be used is in design and testing of a high-speed serial data bus to connect a peripheral device such as a video camera to a computer. The four-port VNA can operate as the two-port described earlier, using only four S-parameters. In the four-port mode, it requires 16 S-parameters (eight each for differential mode and common mode) to fully describe the device under test.[34]

Chapter 34
Mass and Weighing

Most people are familiar with weighing instruments—they are everywhere! You might see a hanging scale at a roadside stand or farmers market or at the grocery store. Scales are found in the local doctor's office and hospital. You will find scales in museums and kitchens. Weighing instruments have been used in trade for thousands of years. The accuracy and resolution of the weighing instruments used in commerce have been specified by weights and measures regulators and the scales seem to inspire confidence in measurement results, so that the end user generally does not think about potential errors in weighing. Someone shopping for produce at the local market rarely thinks about measurement uncertainty of the weighing results.

People often laugh when they see the Leslie Thrasher[1] illustration called "Tipping the Scales" that was published on the cover of the October 3, 1936, *Saturday Evening Post*. You might have seen the picture framed and hung in a local weights and measures office. In the illustration, a consumer is pushing up on one side of a scale with a chicken hanging off the pan and the butcher is pushing down on the other side and both are looking at the scale indication. Though fraud or bad weighing such as illustrated in that picture certainly occurs (else we would have no need for legal metrology), consumers and even patients generally have confidence in weighing results and trust that *someone* ensures that, as an old saying goes: "a pound is a pound the world around." We should say a kilogram is a kilogram, but that does not rhyme! The fact that metrology is invisible to most consumers is a double-edged sword: While consumers have confidence in the measurement, they also have little understanding of the overall impact of weighing on their lives or about the infrastructure that allows them to trust most commercial measurements.

Because weighing has been around for thousands of years and is everywhere, most people think weighing is relatively "easy." However, once we get to higher precision levels of weighing, we find that there are different types of weighing instruments, different types of mass standards, and different levels of uncertainty. All these levels are affected by location, environment, the items being weighed, and whether the user follows procedures and good weighing and handling methods. This chapter presents information about the International System of Units (SI) unit of mass, and provides a brief look at some of the physics of mass and weighing, but also discusses the specifications and performance limits for weighing instruments and mass standards, and the procedures that (when followed carefully) help ensure that we can all continue to have confidence in weighing results at all levels.

SI UNIT OF MASS

In one sense it is easy to see why mass and weighing seem simple. The unit of mass is the kilogram. The kilogram maintained as just a hunk of metal was the last remaining artifact in the International System of Units (SI)—mostly because it, and others like it, were measurable with readily available precision weighing instruments and stable enough to meet international scientific and legal requirements for such a long period of time. When we think about the history of mass measurements, we find that the kilogram was originally defined in the late 1700s. For example, one such definition from that era was that of the Kilogram of the Archives, which was manufactured as a prototype in 1799: Its mass was equal to the mass of 1 dm^3 of water at its maximum density at approximately 4 °C. When we consider the many factors associated with using water as a standard, we realize that we would have to consider evaporation, a container, the temperature that affects the density and causes changes in volume and buoyancy, and maintenance of temperature stability, as well as the technology available in the 1700s used to manufacture this defined kilogram; thus, we can easily see the challenges that were faced in identifying a good mass standard. The kilogram, even before the latest redefinition, was more than just a hunk of metal.

In fact, in 1875, the meter and kilogram were identified in the Treaty of the Meter (Metre Convention) partly because mass and length measurements had been around for centuries, so there was a need for measurement agreement among countries; among other trading partners; among the burgeoning global scientific community, the global marketplace, and worldwide manufacturing industries. The signing of that treaty led to the production of the International Prototype of the kilogram in 1879 and its adoption in 1889. Many copies were made with the same materials and processes available at the time and were distributed to national laboratories and a few other parties (e.g., original signatories to the Metre Convention). The International Prototype Kilogram (IPK) was then maintained, with its copies, as the foundation of the SI unit of mass at the International Bureau of Weights and Measures (BIPM).

The IPK is a cylinder made of an alloy consisting of platinum (90 %) and iridium (10 %) (Pt-Ir) and is stored in an air-filled triple bell jar (not vacuum) in a secure safe at the BIPM (see Figure 34.1). Interestingly, the BIPM and the other prototype kilograms were securely maintained through two world wars without damage! The IPK really is a testament to mass and weighing being perceived as easy. The kilogram (and its copies) kept at the BIPM and IPK was the worldwide standard for mass until 2019. In fact, all National Metrology Institutes traced their measurements back to the IPK at the BIPM, and that was good enough for nearly 150 years. It would still be good enough for most of the commercial marketplace, but it was not good enough to ensure progress in the definitions of the SI, for the science using derived units, nor for current global measurement requirements.

But, to get to the current generation of mass with traceability to the SI, a lot of critical work was needed for many years. That work is ongoing to ensure that the realization and disseminations remain good enough for current and future requirements and scientific developments. In 2019, the kilogram was given a new definition and is now defined in terms of the Planck constant as approved by the General Conference on Weights and Measures (CGPM) on 16 November 2018.

Figure 34.1 The International Prototype Kilogram (IPK).

Defined

"The definition of the kilogram, SI base unit of mass, is as follows [2.1]:[2]

The kilogram, symbol kg, is the SI unit of mass. It is defined by taking the fixed numerical value of the Planck constant h to be $6.626\ 070\ 15 \times 10^{-34}$ when expressed in the unit J s, which is equal to kg m^2 s^{-1}, where the metre and the second are defined in terms of c and $\Delta\nu_{Cs}$.

Thus, the Planck constant *h* is exactly h = 6.626 070 15 × 10⁻³⁴ J s. This numerical value of *h* defines the unit joule second in the SI and, in combination with the SI second and metre, defines the kilogram. The second and metre are themselves defined by exact values of the hyperfine transition frequency Δv_{Cs} of the cesium 133 atom and the speed of light in vacuum *c*. The numerical value of *h* given in the definition of the kilogram has ensured the continuity of the unit of mass with the previous definition, as explained in Section 5."[3]

Realized

Well, we are not likely to see produce and meat at the local shop sold on the direct basis of the Planck constant! (Not for some time, anyway.) So, how do we get from a definition of the kilogram based on the Planck constant to mass standards? The definition must be *realized*: that is, made real, or put into practice. There is a recipe of sorts, or what is called the "*Mise en pratique* for the definition of the kilogram in the SI,"[4] that provides instructions for transferring the definition from the Planck constant to the embodiment of an artifact, i.e., a primary physical mass standard. This process is called *realizing the definition*.

There are two common current realization methods described in the *mise en pratique*:[5]

- Realization by comparing electrical power to mechanical power (using the Kibble balance, what was previously called the watt balance; see Figure 34.2); and

- Realization by the X-ray-crystal-density method (using a silicon sphere as a primary mass standard; see Figure 34.3).

The reader is encouraged to review the latest version of the "*Mise en pratique* for the definition of the kilogram" as posted on the BIPM website (https://www.bipm.org). Numerous technical references are listed in that document to provide additional background regarding the physics and technical requirements for realization and dissemination of the kilogram.

Disseminated

Once the kilogram has been realized, the unit is then transferred to other mass standards through a series of calibrations—comparisons among known standards, producing measurement results with stated uncertainties—with evidence for ensuring metrological traceability at each step.

The *mise en pratique*[6] describes dissemination this way: "The definition of the kilogram ensures that the unit of mass is constant in time and that the definition can be realized by any laboratory, or collaboration of laboratories, with the means to do so. Any National Metrology Institute (NMI), Designated Institute (DI), the Bureau International des Poids et Mesures (BIPM), or collaboration among them, that realizes the kilogram definition can *disseminate the SI kilogram from its primary mass standards to any other laboratory or, more generally, to any user of secondary mass standards*" (see Figure 34.4).

Figure 34.2 NIST Kibble balance.

Figure 34.4 shows a traceability pyramid like the figure shown in the *mise en pratique*, to illustrate the process from definition to primary mass standards through realization and then to secondary mass standards through dissemination.

Figure 34.3 Silicon sphere.

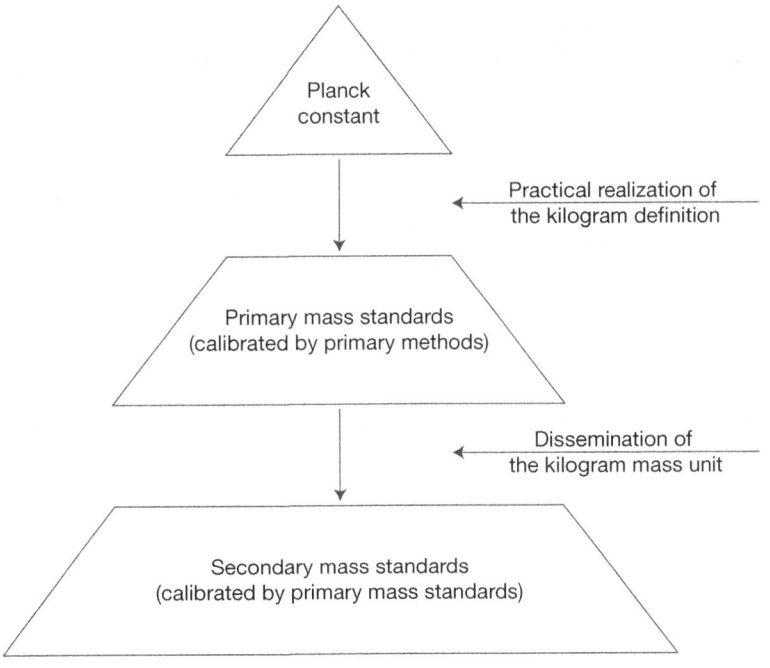

Figure 34.4 Kilogram metrological traceability pyramid.

MASS AND WEIGHING

The SI unit of mass is the kilogram and the *mise en pratique* guides us from definition to realization, and on to the dissemination to mass standards and the physics associated with that level of work. However, let us take a step back and look at some of the classical physics associated with the definition of mass and with the process that most of us are readily familiar with: that of weighing.

The mass of an object is often simply defined as the quantity of matter that comprises the object. Mass is commonly measured by weighing. This is step one of the process. Mass (as later used in this document) is defined in vacuum; if it is not in vacuum, then it must be "corrected mathematically" from the weighing result to account for forces acting on matter in air. Weighing something is affected by several forces: most critically a downward force due to gravity, but also (to a smaller extent) a lifting force due to air buoyancy.

Force Due to Gravity

Mass is one of the base quantities in classical mechanics. The concept of mass constitutes a universal characteristic of bodies. As a physical quantity, the mass could be determined in relationship with other physical quantities. Such relationships are Newton's second law of motion and the law of universal gravitation. Isaac Newton defined the concept of mass in his second law of motion, which can be stated: The rate of change of the velocity of a particle, or its acceleration, is directly proportional with the resultant of all external forces exerted on the particle and is in the same direction as the resultant force; their ratio is constant.

Experimentally, it was found that if the same force is applied to different bodies, they show different accelerations. Therefore, it was concluded that the accelerations of the bodies are generated not only by the external forces, but also by a physical property, characteristic for each body: the mass.

The mathematical expression of Newton's second law of motion is

$$\sum F = m \cdot a$$

where

$\sum F$ = summation of all the external forces applied to the particle
a = the acceleration of the particle; and
m = the mass of the particle.

From this equation we can write $m = F/a$.

The mass is numerically determined by the ratio between force and acceleration. Newton's law of gravitation, published in 1686, may be stated: Every particle of matter in the universe attracts every other particle with a force, which is directly proportional to the product of the masses of the particles and inversely proportional to the square of the distance between them. Or, in a mathematical form:

$$F = G \frac{m_1 m_2}{r^2}$$

where

F = The attraction force between the two bodies m_1 and m_2
m_1, m_2 = The masses of the two bodies
r = The distance between the centers of the two bodies
G = Gravitational constant

The 2018 Committee on Data for Science and Technology (CODATA) recommended value of the gravitational constant in SI units (with standard uncertainty in parentheses) is:[7]

$$G = 6.67430(15) \times 10^{-11} \text{ m}^3 \text{ kg}^{-1} \text{ s}^{-2}$$

An immediate application of the universal attraction to be observed on Earth is the weight of the bodies. The force with which an object is attracted to the Earth is called its *weight*. What appears to be weight based on a balance indication is a function of both gravity and air buoyancy (unless in vacuum). Weight is different from mass, which is a measure of the inertia an object displays. Although they are different physical quantities, mass and weight are closely related.

The weight of an object is the force that causes it to be accelerated when it is dropped. All objects in free fall near the Earth's surface have a downward acceleration of approximately $g = 9.8$ m/s², the acceleration due to gravity. If the object's mass is m, then the downward force on it, which is the weight w, can be found from the second law of motion where $F = W$ and $a = g$. So, for weight in a vacuum, we have: $W = m \cdot g$, or weight equals mass times acceleration of gravity.

The weight of any object is equal to its mass multiplied by the acceleration of gravity. Because g is a constant at any specific location near the Earth's surface, the weight w of an object is always directly proportional to its mass m: a large mass is heavier than a small one. The gravitational acceleration varies with latitude (because the Earth is not a perfect sphere), elevation, and local variations in subsurface density. So, the weight, when measured with a sufficiently sensitive weighing instrument, will vary from one location to another; however, the mass of an item remains constant.

The mass of an object is a more fundamental property than its weight because its mass, when at rest, is constant and is the same everywhere in the universe, whereas the gravitational force on it depends on its position relative to the Earth or to some other astronomical body (at the Earth's poles, $g = 9.832\ 17$ m/s², which is the maximum value, and at the equator $g = 9.780\ 39$ m/s², which is the minimum value).

While most of us are weighing things on Earth, the force on a mass on the moon will be significantly less. But the mass (i.e., matter of the item) does not change! Mass is the property of the item: how much matter it contains. So, although the forces on the items are certainly different on the moon, if you were to adjust a weighing instrument on the moon with a calibrated 1 kg mass standard and then weigh another 1 kg item, the instrument will read 1 kg. This principle of comparison weighing (also called substitution weighing) holds true on Earth as well.

This property of a different force of gravity on masses at various locations on Earth is one of the reasons it is recommended, for precision applications, that weighing instruments should always be calibrated with calibrated mass standards at the location of use. In that way, the major effects of gravity differences "cancel out." When weighing instruments are not calibrated at their location of use, the effects of gravity variations can introduce weighing errors, of course depending on the load and resolution of the weighing instrument. The principle of comparing calibrated mass standards with unknown standards at the same location on the same weighing instrument, as will be described later in this chapter, is why the effects of gravity *generally* cancel for most calibrations.

However, in precision mass calibrations like those used when disseminating from Pt-Ir mass standards to stainless steel mass standards, there are minor effects due to the center of gravity in each of the mass standards being slightly different. These differences must be measured and corrected for. In this example, the center of gravity difference between a Pt-Ir kilogram and a stainless-steel kilogram might be about 3 µg/kg. This amount is not significant for most routine weighing applications, but the effects of gravity must be considered at the highest levels of realizing and disseminating mass due to the relatively small uncertainties (about 20×10^{-9}) at that level.

Force Due to Buoyancy

A smaller force when weighing in air is that of air buoyancy. Air buoyancy causes a lifting effect, opposite to the effect of gravity, due to the mass of air displaced by the volume of the item being weighed. In scales used in large manufacturing or commerce, this effect is generally small enough to be insignificant. But in some analytical applications and certainly in mass calibrations, it must be considered. In simplest terms, the lifting force is the result of the mass of air displaced by an object, which depends on the density of the air, ρ_a, and the volume, V, of the object. The air buoyancy effect is represented by air density times volume, or as an equation $\rho_a V$. Another way to look at it is the mass, M, minus the air buoyancy force: $M - \rho_a V$. We will touch on the measurement and calculation of air density later.

A researcher who thinks that a milliliter of water will weigh 1 gram and uses that value in their research will introduce errors in their weighings. In fact, when weighing water or other solutions and considering buoyancy, several factors are in play: 1) the purity of the water and the saturation of air affects its density; 2) the density of water changes based on its temperature, thus changing its volume; and 3) weighing the water displaces an equal volume of air that has a mass and causes the lifting effect. The researcher must consider the purity, air saturation, and density of the water at the time of the measurement as well as the effects of air density and the time and location of weighing that influence the air buoyancy. This combination of factors must also be considered in the gravimetric calibration of glassware.

Here is one example to illustrate what happens when weighing. Let us say we have a set of density blocks, and all have a volume of 1 cm³. Figure 34.5 shows

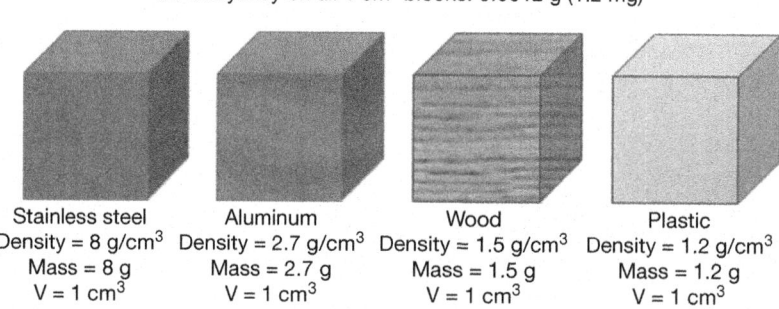

Figure 34.5 Density block and air buoyancy illustration.

Table 34.1 Example air buoyancy on 1 cm³ density blocks.

Material	Mass (g)	Density (g/cm³)	Volume (cm³)	Air Buoyancy Effect (g)
Stainless steel	8.0	8.0	1	−0.0012
Aluminum	2.7	2.7	1	−0.0012
Wood	1.5	1.5	1	−0.0012
Polycarbonate plastic	1.2	1.2	1	−0.0012
Air-saturated distilled water at 20 °C	0.998204	0.998204	1	−0.0012

example blocks and their attributes: the blocks are made from stainless steel, aluminum, wood, and plastic. Each of these items has different densities and each will weigh differently, as shown in Figure 34.5 and in Table 34.1. But remember their volumes are all 1 cm³, and for this example, we will consider a normal air density of 0.0012 g/cm³. The effect of gravity will affect the weighing of all blocks equally. Given the equation for the air buoyancy effect, we can calculate that the lifting effect due to air buoyancy is identical on all these blocks because they are all the same volume (i.e., 1 cm³) and the same mass of air is displaced (0.0012 g or 1.2 mg).

Now let us say we have several items that are perfect 1 kg mass standards (as measured in a vacuum). The volumes are all different due to their material densities. The lifting effect of air buoyancy on each kilogram mass when weighed in air is shown in Table 34.2. In a laboratory with a balance that has a resolution of even 0.001 g (1 mg), these differences could be significant if one wanted accurate mass measurement results. Looking at the tabulated values, we can see the air buoyancy effect when weighing an aluminum weight is almost three times that of stainless steel and the air buoyancy effect when weighing water is about eight times that of stainless steel. We will discuss this more when we discuss international conventions associated with weighing in air shortly, but since the late 1970s balance indications have been adjusted to provide *conventional mass* values *at reference conditions* which are based on comparison with a weight with

Table 34.2 Example air buoyancy on different materials.

Material	Mass (g)	Density (g/cm³)	Volume (cm³)	Air Buoyancy Effect (g)	Air Buoyancy Effect (mg)
Stainless steel	1000	8.0	125.0	−0.150	−150
Aluminum	1000	2.7	370.4	−0.444	−444
Wood	1000	1.5	400.0	−0.48	−480
Polycarbonate plastic	1000	1.2	833.3	−1.00	−1000
Air-saturated distilled water at 20 °C	1000	0.998204	1001.8	−1.202	−1202

426 *Part VI: Measurement Parameters*

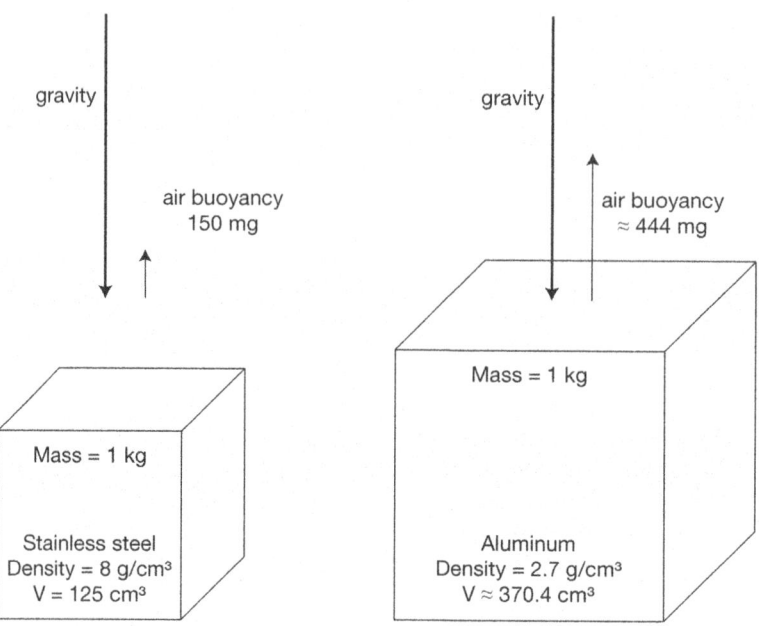

Figure 34.6 Density difference by block illustration.

a stainless-steel reference density of 8 000 kg m^{-3} and reference air density of 1.2 kg m^{-3}. But, unless you know these differences exist, the balance readings will not reflect mass; they will reflect conventional mass and will have buoyancy errors in these amounts. Further, if the weighing conditions are not at an air density of 1.2 kg m^{-3}, there will be additional buoyancy errors that must be mathematically corrected.

A visual comparison of the lift on two kilograms made of stainless steel and aluminum is shown in Figure 34.6. The illustration provides insight as to the effects of gravity and buoyancy for the two weights.

The photograph in Figure 34.7 provides insight as to the relative sizes of stainless steel (on the left) and aluminum (on the right) kilograms. Seeing the magnitude of the air buoyancy effects, we might ask whether these amounts are significant.

We can answer that by looking up the maximum permissible error for an OIML Class M_1 weight (to be discussed later), which is 50 mg. Because stainless steel is the reference basis for conventional mass, we are only interested in the difference between the air buoyancy effects of the aluminum weight compared to stainless steel, and the difference is just under 295 mg (almost six times larger than the tolerance for our M_1 class weight). It is easy to see why aluminum is not considered acceptable as a precision mass standard at this level.

Determining Volume and Density

The references provided in this chapter contain formulas that can be used for calculating air density based on knowledge of the laboratory temperature, air

Figure 34.7 Stainless steel vs. aluminum kilograms.

pressure, and relative humidity. The accuracy of the air density can be further improved with knowledge of the carbon dioxide concentration in the air when determining the mass of air, but that is not normally needed except at the highest levels of mass dissemination. The normal air density used in this example is based on values provided as conventions in the conventional mass definition (presented shortly).

However, we also need to know the volume of an item being weighed so that we can calculate the mass of air that is displaced during weighing, causing the buoyancy effect. If we had a perfectly machined block, we could measure it and use a formula for calculating the volume (e.g., length times width times height). We could calculate the volume with a perfect sphere or a cylinder, too. In these cases, we could measure the dimensions with varying accuracy using a ruler for coarse measurements to a coordinate measuring machine for finer measurements and refine our measurement results and the calculations of volume, and reduce the uncertainty along the way.

Of course, most things we weigh, including mass standards, are not perfectly dimensioned, so we use additional methods to determine the density of mass standards and then calculate volume using the equation: $V = M/\rho_m$, where M is the mass, and ρ_m is the density of the mass.

Six methods for determining density are described in OIML R 111[8] for mass standards. One method uses dimensional calculations, another uses calculations based on alloy properties, and the remaining four all involve multiple weighings

Table 34.3 Reference conditions for conventional mass.

Conventional conditions	Reference value
Temperature, t_{ref}	20 °C
Air density, ρ_a	1.2 kg m^{-3} or 0.0012 g/cm^3
Reference density, ρ_{ref}	8 000 kg m^{-3} or 8.0 g/cm^3

in/out of liquids with known densities. Beyond the measurement methods described, it is most common for calibration laboratories to use reference tables for the material in question or to use manufacturer's stated densities (especially for mass standards). Once we know the density, we can calculate the volume using the equation: $V = M/\rho_m$, where M is the mass, and ρ_m is the density of the mass.

Weighing in Air—Reference Conditions and Conventional Mass

The definition of conventional mass—or more formally, the conventional value of the result of weighing in air—can be found in OIML D 28.[9] Because most weighing is done in air and the previously described forces affect our weighing, international references have been selected that are based on the following defined conventions: For a weight taken at a reference temperature (t_{ref}) of 20 °C, the conventional mass is the mass of a reference weight of a density (ρ_{ref}) of 8 000 kg m^{-3} which it balances in air of a reference density (ρ_a) of 1.2 kg m^{-3}.[10] These conventions are shown in Table 34.3.

As noted, weighing items with densities significantly different from the reference basis for stainless steel (8 000 kg m^{-3}) or significantly different from the reference conditions can introduce errors in weighing and measuring results. Example standard conditions that will result in an air density close to 1.2 kg m^{-3} could be: 20 °C, 101325 Pa, and 42 % relative humidity. Deviation from these "weight in air" conventional mass reference conditions for an item being weighed also has to be mathematically corrected. The documentary standards for mass and mass calibration procedures referenced in this chapter provide equations for converting between conventional mass and true mass (mass in vacuum) values if the user needs to determine accurate mass values.

GOOD WEIGHING PROCEDURES

Whether weighing ingredients in a kitchen, weighing trucks to ensure that weight limits are maintained to protect the road infrastructures, weighing ultra-micro samples of reactive or controlled materials in a pharmaceutical research application or forensic laboratory, weighing samples in an analytical environment, or performing a calibration in a mass laboratory, most people use a weighing instrument because they want accurate results. Following good weighing procedures and understanding the countless influence factors that affect mass and weighing results is critical if you want accurate and consistent results. This section provides steps you can take to ensure good weighing results and presents many influence factors that will affect your weighing processes in one way or another.

Here are several steps that users of weighing instruments and mass standards should take to ensure the validity of their weighing results. Most pharmaceutical developers, medical devices producers, and clinical industries are regulated and require good measurement practices to be in place. They follow good Installation Qualification (IQ), Operational Qualification (OQ), and Performance Qualification (PQ) standards, as has been discussed in previous chapters (e.g., Chapter 18). These principles and concepts are useful beyond the pharmaceutical and clinical settings; they are important for anyone wanting to ensure good weighing results. It is also wise to make sure you retain adequate documents and records of each of the following steps to demonstrate that you are following good measurement concepts. Records are especially important if you are audited or undergoing an assessment, but also useful if you want to minimize risk or use data to support and measure improvement efforts.

1. Choose the right weighing instrument and mass standards based on application requirements. Consider the classification system of weighing instruments and mass standards as noted in later sections of this chapter.

2. Install and maintain equipment and standards properly, considering operating manuals and influence factors noted in this section.

3. Calibrate and maintain instruments and standards based on the environment and use considerations. ISO/IEC 17025:2017 requires appropriate calibration and maintenance for all equipment and standards that can influence measurement validity. Weighing instruments and mass standards are no exception. A suitable calibration program will include both the weighing instrument and associated mass standards.

4. Read and follow the operating manual when using each weighing instrument.

5. Understand the operating characteristics of the instrument and influence factors on the instrument and mass standards. Be especially aware of the nominal capacity to avoid overloading a weighing instrument and always be sure it is level and clean prior to use.

6. Evaluate how using the weighing instrument for a measurement process and/or calibration of mass standards affects the uncertainty of the verification or calibration results.

The following paragraphs provide some additional insight as to how measurement results might be affected by the location or environment of the instruments, effects due to care and handling, influence factors of and on the weighing instruments themselves, and the effects of the items being weighed. The care and design of the weighing procedures using weighing instruments and mass standards must take these factors into consideration to achieve accurate and

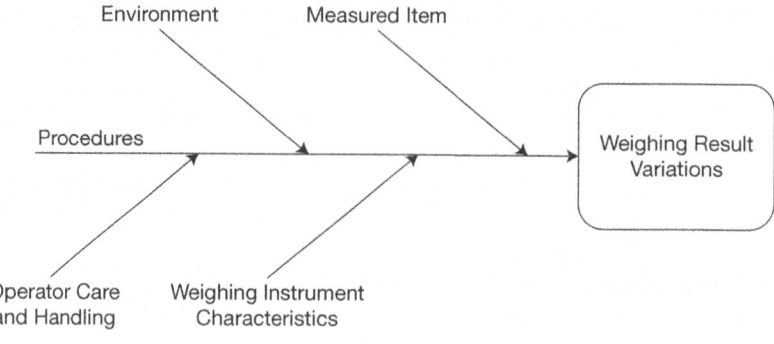

Figure 34.8 Cause-and-effect diagram.

traceable measurement results. Figure 34.8 shows a cause/effect diagram (also called Ishikawa diagram or fishbone diagram) which provides a graphical way to consider many of the influence factors described here and how they might influence weighing results. It does not list every item described in this section, but does provide a concept one should keep in mind while weighing: Every little thing matters and contributes to the measurement results (or errors) and associated uncertainties!

Operator Effects: Following Procedures and Correct Care and Handling

Proper following of procedures and correct care and handling of items being weighed is important for custody transfer in a legal setting and important to obtaining accurate measurement results. Taking the steps described earlier is an important part of obtaining accurate measurement results. Avoiding errors and minimizing variability is also important. Personnel performing measurements with weighing instruments or mass standards should take care to consider the following:

- Ensure adequate equilibration of the items to minimize convection currents (Table 10, ASTM E617-18; Table B.2, OIML R 111-1 2004).

- Use appropriate safety devices to lift and handle materials as needed.

- Use good *mise en place*[11] techniques to keep everything organized and readily available, and to keep similar items from being mixed up during a measurement.

- Minimize contamination transfer by using gloves or lifters to handle items to be weighed; at the same time, take care not to contaminate the gloves or lifters by touching balance tables, instruments, pens, computers, etc.

- Ensure that items being weighed, especially mass standards, are not dirty or contaminated, because cleaning (other than a light brush) will change the mass values and require recalibration.

- Avoid leaning on balance tables where weighing instruments are located, to minimize vibration and to prevent the balance from becoming unlevel.

Environmental Effects

A pristine environmentally controlled laboratory, with measurements conducted in a cleanroom using robotically controlled mass comparators, is not the typical weighing environment. While this type of environment might be common for most National Metrology Institutes and some accredited standards laboratories, in the usual weighing environment there are many environmental influence factors that must be considered in both the design and operation of the laboratory to ensure good measurement results.

Vibration can affect precision weighing and must be considered for high-precision mass calibrations. Consider the location of the facility and proximity to vibration sources during design of a laboratory or production process. Locations near a major highway, frequency and prevalence of earthquakes, closeness to a blasting zone, nearness to a major production environment with heavy equipment in use, or even the proximity of an elevator, hoist, electric forklift, or pallet jack can negatively influence the stability and repeatability of weighing instruments. In very high precision measurements, earthquakes that are hundreds and thousands of miles away have affected measurement results. Some laboratory designs include physically isolated rooms and isolated pads for balances. Furniture such as massive anti-vibration balance tables can minimize or isolate effects due to vibration.

Animal weighing presents some interesting vibration concerns as well as other potential measurement errors. From a stockyard where thousands of animals might be weighed in a day, to a research environment, animal weighing presents unique challenges related to both animal movement that causes vibration as well as a need to understand the stabilization features built into many scales or balances. Animal weighing can also introduce other errors in measurement results or instability due to things like shock-loading and off-center loading.

Temperature affects weighing results and mass calibration results in many ways. While working in a laboratory, performing air buoyancy corrections might correct the effects of temperature on the air density, but it does not correct the effects from many other thermal influence factors. Balance and scale manufacturers provide recommended temperature operating ranges for weighing instruments. Mass calibration procedures provide even narrower allowable ranges of temperature variation during a calibration. The temperature of an item being weighed, and even what appears to be a stable mass standard, will affect the volume and density to varying degrees, thus affecting the buoyancy correction.

Further, if the temperature of the item being weighed is even slightly different from the temperature inside the weighing chamber, there will be convection currents causing additional forces up (lower mass indication) or down (higher mass indications). The effect of convection currents is measurable and significant in precision mass calibrations or gravimetric glassware calibrations, where equilibration of the standards for 24 hours prior to a calibration is common. The influence of temperature and resulting convection currents could be even more significant in a production environment or analytical laboratory where thermally hot materials are being weighed. Temperature also affects the quality of weighing instrument verification when mass standards are not equilibrated prior to use, especially when brought to an installation area after being outside under hot or freezing conditions (depending on the season).

Air currents and air exchange rates affect the ability to control the air temperature within recommended limits and affect weighing results. A large vehicle scale located outside is often affected by wind (and precipitation and dirt). But high-precision mass calibration results can even be affected by what seems to be an innocent opening and closing of a laboratory door. Wind, opening/closing doors, and even ceiling vents affect the temperature stability and contribute air currents at varying levels. Weighing instruments are extremely sensitive to air flow and many laboratory balances and mass comparators come with integrated draft shields (or they can be purchased) for this reason. Some balances are sensitive enough that even a draft shield is not enough to minimize the effects of air flow in the environment. Good laboratory design must consider temperature stability and air flow for all weighing instruments, but especially for high-precision applications.

Related to temperature stability is heat sources from surrounding equipment and personnel. A computer located next to a balance can cause instability due to heat. The controllers and indicators on many balances are located separately from the weighing chamber to minimize the heat effects. Heat from personnel can also influence measurement results: From body heat generated while sitting in front of a balance to insertion of hands inside the weighing chamber when placing items on a weighing platform, manual operations will influence the temperature effects of weighing. Heat of the item being weighed (for example, heated solutions or animals) will also affect the results.

Cleanliness in weighing is important, as everything that has matter will weigh something and can affect measurement results. Problems in cleanliness can arise with particulates from skin cells, hair, pollutants in the air, dust produced from paper in printers, materials like powders that are being weighed, and contaminants on the items being weighed. The adsorption of humidity and pollutants adds mass to mass standards even when stored in a pristine laboratory environment. Minimizing the *extra material* that is weighed will be important to achieving accurate weighing results. Weighing in an environment with chemically reactive or corrosive materials, as well as excess humidity, can damage the weighing instruments as well as the mass standards.

Good identification and storage of mass reference standards are essential to keep from mixing the standards with submitted calibration items, to ensure that they remain clean and protected, and to make sure they have up-to-date calibration values. Labels should never be used on mass standards!

Weighing Instrument Effects

Balances, and resulting measurement uncertainties, are affected by all the physics associated with gravity and buoyancy discussed earlier. Taking appropriate care and using suitable procedures are essential to minimize errors that are possible with weighing instruments, as described in this section. Measurement results and uncertainties are also affected by the items on the following list and we can minimize errors and effects on our weighing results by following good measurement practices.

- Centering/off-center loading errors: place items gently on the weighing pan so that the center of gravity is over the center of the weighing pan.

- Force/shock due to sudden placement of items on the weighing pan can damage the instrument; gently place items on the weighing pan.

- Exceeding the load capacity can damage the instrument; be aware of instrument capacities and estimate items to be weighed by pre-weighing on an instrument of greater capacity.

- Excessive vibration can cause problems with repeatability and problems with stabilization; use a well-designed laboratory, anti-vibration measures, or isolated balance tables, and do not lean or write on the tables.

- Magnetism of items being weighed and even magnetic items in and around the table can cause errors in weighing; use appropriate shielding and replace magnetized mass standards.

- Electrostatic charges can cause instability in the balance readings and errors in measurement results; maintain relative humidity in the laboratory according to manufacturer or procedural recommendations.

- Avoid measurement errors at the extreme limits of balance indications; always follow recommended guidelines regarding minimum and maximum loads for weighing, as some balances are not as accurate near zero or near capacity.

It is important for users of weighing instruments to be aware of the physics, these influence factors, and the unique operating characteristics of each balance. A regular calibration and maintenance program for weighing instruments is required by ISO/IEC 17025:2017, but frequency is usually determined by factors such as operator care and handling, environmental conditions, and the cleanliness of the items being weighed.

Measured Item Effects

Mass standards are relatively stable if stored and handled properly, though a regular calibration interval and ongoing evaluation are essential. Mass standards are very stable compared to something like a liquid that can evaporate (mass loss) or a powder or other material that absorbs moisture from the air as it is weighed (mass gain). However, the use of a weighing vessel or weighing paper can degrade the repeatability of a measurement process over that of a simple repeated indication of a stainless-steel mass standard, so the measurement process must be considered when evaluating repeatability. Additional factors that should be considered include the presence of drift or instability due to electrostatic charges on the items being weighed. For example, the plastic on candy wrappers when check-weighing a production lot, or the plastic of lottery balls when they are being weighed for accuracy and fraud prevention will introduce static and potential drift or instability errors. Electrostatic charges can also be present on the weighing instruments themselves, especially in the case where the relative humidity is outside the ideal range (typically 40 % to 60 %).

Weighing magnetic materials can be challenging due to the electromagnetic force compensation technology used in many electronic weighing instruments. Shielding of magnetic materials is recommended for weighing. Magnetism will cause an additional force that can affect the accuracy of weighing in either direction (attracting or repelling). In fact, when possible, one way to test for magnetic influence effects is to turn the item being weighed by 180 degrees in short segments, by turning the item upside down or by placing the weight on top of a lightweight nonmagnetic spacer which is treated as tare. Some older high-precision mass standards may also be magnetized and cause weighing errors or lack of repeatability in a measurement process; they should be replaced, as noted earlier. Cast-iron weights are often magnetized—sometimes to a level where they seem out of tolerance in one direction and in tolerance when turned 180 degrees or upside down. Consistently quantifying and correcting for the effects of magnetism is not possible.

For high-precision mass calibrations, the surface finish quality and surface moisture or contaminants can affect the stability of the calibrated mass values. Specifications for various classes of mass standards, as described in this chapter, include surface finish and magnetism requirements to limit these influence factors as much as possible, so that the desired level of accuracy and stability in the measurement results can be achieved. In addition, the acceptable materials and material density ranges are limited for mass standards to ensure that errors in the calibrated mass values and buoyancy effects are minimized.

WEIGHING INSTRUMENTS—TYPES, DOCUMENTARY STANDARDS, CALIBRATION PROCEDURES

Specific terminology used in documentary standards includes a definition for weighing instruments.

Weighing instrument:[12] "Measuring instrument that serves to determine the mass of a body by using the action of gravity on this body.

Note: In this Recommendation [OIML R 76], 'mass' (or 'weight value') is preferably used in the sense of 'conventional mass' or 'conventional value of the result of weighing in air' according to R 111 and D 28, whereas 'weight' is preferably used for an embodiment (i.e., material measure) of mass that is regulated in regard to its physical and metrological characteristics."

In the *International Vocabulary in Metrology* (VIM 3), we find a definition important to the discussion of documentary standards associated with weighing instruments and mass standards: maximum permissible error.

Maximum permissible measurement error (tolerance, limit of error) is the extreme value of measurement error, with respect to a known reference quantity value, permitted by specifications or regulations for a given measurement, measuring instrument, or measuring system.

NOTE 1: Usually, the term "maximum permissible errors" or "limits of error" is used where there are two extreme values.

NOTE 2: The term "tolerance" should not be used to designate "maximum permissible error."

While the *VIM* 3 definition specifically states that the term *tolerance* should not be used to designate the maximum permissible error (*m.p.e.*), it is often used so in the reference publications. In context, the plus/minus values of the tolerance can be considered equivalent to the maximum permissible error listed within the references provided here.

Types of Weighing Instruments

Describing types of weighing instruments is somewhat arbitrary and depends on the source. What one laboratory might call a scale another might call a balance. The same might be true of mass comparators and balances. The confusion with lack of clear definitions is exacerbated when comparing weighing instrument specifications during purchasing. What follows are some generic concepts regarding the types of weighing instruments.

Equal-arm balances. As the name implies, these are two-pan balances with a beam that has equal arms on both sides of a central pivot point. These types of balances are historically the most common weighing instrument. Use of these balances usually requires comparison with a known standard on one side and an unknown item on the other. For precision comparisons, a transposition weighing might be performed to average the effects of minor inequalities in beam lengths on each side of a central pivot point. Alternatively, "ballast" weight can be applied to one side and substitution weighing can be performed using just one of the pans/arms. Equal-arm balances are increasingly rare in routine use, though some remain in a few calibration or analytical laboratories.

Scales and balances. ASTM E898[13] notes the following regarding non-automatic weighing instruments: "The measuring principle is usually based on the force compensation principle. This principle is realized either by elastic deformation, where the gravitational force of the object being weighed is measured by a strain gauge that converts the deformation into electrical resistance [lower accuracy weighing instruments], or by electromagnetic force compensation, where the gravitational force is compensated for by an electromagnetic counterforce that holds the load cell in equilibrium [higher-accuracy weighing instruments]."

Single-pan mechanical balances and electronic balances are designed to be calibrated with known mass standards or internal mass standards with specifications given for accuracy, linearity, and repeatability based on capacity and resolution. For many applications in analytical and testing laboratories, production environments, and legal metrology, these instruments are used as a direct reading instrument. But recall that the balance indication is the conventional mass value. Deviations from standard conditions specified for conventional mass will yield errors in measurement results. Most balances in service today provide a conventional mass indication as defined earlier based on the specified reference conditions and situations. In the case of weighing tomatoes or other produce (i.e., items that are not as dense as stainless steel), the balance resolution is coarse enough so that the impact of these buoyancy differences is not significant. In a laboratory or production setting, these instruments are often used with check standards to monitor accuracy between calibration or service intervals. When used in a mass calibration laboratory, scales and balances are used as *mass comparators* where the instrument is used simply to *compare* calibrated mass standards to lower-level mass standards using substitution methods.

Mass comparators. Mass comparators are weighing instruments designed to compare minor differences between a calibrated mass standard and an unknown item (perhaps another mass standard). These instruments may not be linear from zero to the nominal capacity and may have intermediate weighing ranges so that it is not possible to get an accurate weight by placing something on the weighing pan. Sometimes balances, as previously described, are used as mass comparators, though it is rare for a mass comparator be used as a direct measurement device even if it does have full-scale readability, if for no other reason than that the cost associated with mass comparators is generally higher and they are often not available without a justified requirement to work with this type of precision instrument.

Automatic weighing instruments and system. These systems are less common in laboratories. In automated systems, the entire weighing process might use robotics or computer controls without human intervention once the system is put in place. These systems range from in-motion weighing systems used in a mining operation all the way to the automation and robotics used in transferring the mass unit from a Kibble balance to a Pt-Ir mass standard during the realization of the unit of mass.

Selection of Weighing Instruments

Selecting the weighing instrument for a weighing application depends on several factors such as:

- Required level of capacity, resolution, and repeatability
- Availability of equipment
- Availability of trained operators
- Cost

Sometimes a lower-level instrument or procedure may be sufficient for a specific application rather than a laborious state-of-the-art weighing.

Documentary Standards and Classifications for Weighing Instruments

The ASTM E898 standard provides the following warning when considering legal metrology: "This standard does not purport to be suitable as the sole testing process for weighing systems designated for commercial service under weights and measures regulation. The legal requirements for such instruments vary from region to region and depend on specific applications. To determine applicable legal requirements, contact the weights and measures authority in the region where the device is located." As noted earlier, users should consider their accuracy and uncertainty requirements as well as any legal obligations they might have for weighing instruments.

Documentary standards in common use for evaluating weighing instruments include those on the following list (it is a sample list and not all-inclusive).

These documents may contain specifications and instrument requirements, but also contain test methodologies and evaluation criteria used in verification and calibration. Each documentary standard includes calibration or verification, repeatability, and user requirements. Some documentary standards also include uncertainty evaluation and conformity assessment methods. What might be most useful is that these documents describe different classifications based on the application, use, and required measurement uncertainty; of course, there is a cost associated with the better classifications in terms of purchasing costs as well as the consideration of good weighing procedures and care that must be taken, as described earlier:

- *OIML R 76, Non-automatic weighing instruments, Part 1: Metrological and technical requirements—Tests,* 2006
- EURAMET Calibration Guide No. 18, Version 4.0 (11/2015), *Guidelines on the Calibration of Non-Automatic Weighing Instruments*
- ASTM E898-20, *Standard Practice for Calibration of Non-Automatic Weighing Instruments*
- NIST *Handbook 44, Specifications, Tolerances, and Other Technical Requirements for Weighing and Measuring Devices as adopted by the National Conference on Weights and Measures* (current edition)

Weighing instruments are often categorized according to classes based on use-accuracy, capacity, and the number of divisions. These classifications are common in legal metrology. In that application, the maximum permissible errors (*m.p.e.*), or tolerances, are typically specified based on the number of divisions. In the case of legal metrology, the *m.p.e.* might be one, two, or three divisions, depending on capacity, to be considered to meet the in-service requirements. Outside of legal metrology, it is the users' responsibility to establish their *m.p.e.* based on the needs of their process. Each documentary standard also specifies the classification of mass standard that should be used for verification. See Table 34.4

Table 34.4 Weighing instrument classifications common in legal metrology.

Classes	Type of Weighing Instruments	Recommended Mass Standards for Verification[14]
I	Special accuracy; precision laboratory weighing and mass calibrations	OIML F_1 ASTM 1
II	High accuracy; laboratory weighing, precious metals and gem weighing, pharmaceutical research, forensics	OIML F_1, F_2 ASTM 1, 2
III	Medium accuracy; commercial weighing equipment not otherwise specified	M_1, M_{1-2}, M_2, M_{2-3}, M_3 ASTM 3, 4, 5, 6 NIST Class F
IIIL*	Medium accuracy; legal metrology	
IV	Ordinary accuracy; roadway weight enforcement	

OIML R 76 contains classes I to IV. NIST *Handbook 44* contains the additional class IIIL for unique legal metrology applications.

for a quick comparison of types of weighing and mass standards typically used for verification. User requirements are provided in documents such as NIST *Handbook 44*, which break down the uses based on the types of commodity costs/prices that are used with them (i.e., more expensive items require better controls for both the seller and consumer).

Calibration Procedures—Scales, Balances, Mass Comparators

The specifications and tolerance documents listed earlier may also include calibration procedures and methods for evaluating calibration uncertainties or for legal metrology verification. Methods for service personnel evaluation might consider adjustments as well as calibrations, but calibrations or verifications include some or all the following items:

- Instrument identification requirements
- Test load considerations
- User requirements (e.g., location of indications, level conditions, fit for purpose)
- Selection of suitable mass standards
- Consideration of possible buoyancy effects and equilibration
- Exercising the instrument in addition to ensuring a powered-on period prior to use or evaluation
- Increasing and decreasing load tests (e.g., evaluating linearity, drift, and hysteresis)
- Repeatability test
- Test for errors of indication
- Sensitivity tests
- Eccentricity test (e.g., off-center loading errors)
- Calculation of calibration uncertainties and conformity assessment where applicable
- Reporting verification and/or calibration results

MASS CALIBRATION STANDARDS (WEIGHTS)

Documentary Standards Specifications and Tolerances

There are several classifications of mass standards available, some of which were listed in Table 34.4. Both OIML and ASTM have international documentary standards for mass artifact standards (commonly called weights). Table 34.5 shows a comparison of 20 classes of weights and their typical uses. The table is broken into three echelons that loosely correspond to the specifications, as well as calibration requirements for the classes of mass standards in that level. The

Table 34.5 Comparison of mass standards classifications.

Echelon (Level)	Typical Use[15]	OIML R 111 Classes	ASTM E617 Classes	HB 105-1 (1990) Class F
I	Laboratory reference standards and working standards for calibration of lower-level mass standards	E_1, E_2	000, 00, 0, 1	Not applicable
II	Laboratory working standards for calibration of lower-level mass standards; evaluation of weighing instruments, e.g., Class I and II	F_1, F_2	2, 3	Not applicable
III	Evaluation of weighing instruments, Class III and lower	M_1, M_{1-2}, M_2, M_{2-3}, M_3	4, 5, 6, 7	Class F*

*No new Class F mass standards were to be approved after January 2020. *Handbook 105-1* (2019) points to the OIML and ASTM classifications only. Decision rules from *Handbook 105-1* (2019) also apply to all in-service Class F mass standards (see Calibration Procedures in the next section).

documentary standards contain specifications for materials, density, surface finish, magnetism, and calibration, all of which affect the requisite uncertainty and maximum permissible errors (tolerances).

- OIML Recommendation R 111 (Parts 1 and 2), "Weights of classes E_1, E_2, F_1, F_2, M_1, M_{1-2}, M_2, M_{2-3} and M_3" (2004). This document includes nine classes of weights. Designs and specifications requirements for weights of intermediate combination classes are similar, but the *m.p.e.* values vary.

- ASTM E617, *Standard Specification for Laboratory Weights and Precision Mass Standards*, 2018 (or current edition). This standard contains requirements for 11 classes of weights numerical from 000 through class 7.

Within the United States, NIST *Handbook 105-1* identifies weights used in legal metrology; it is widely used for Class III, IIIL, and IV weighing instruments both within and outside the legal metrology community. NIST's Office of Weights and Measures has been working to harmonize specifications and tolerances with internationally accepted classes of weights for many years. Thus, the 2019 version of NIST *Handbook 105-1* specifies that no new Class F mass standards should be put into service after January 1, 2020, and it explicitly points to OIML R 111 and ASTM E617 for the selection of suitable standards, including those for use in legal metrology.

- NIST *Handbook 105-1*, 1990, *Specifications and Tolerances for Field Standard Weights (Class F)*. These Class F mass standards are for use in legal metrology specified as in circulation prior to January 1, 2020. After January 1, 2020, the 2019 version is to be applied for new mass standards.

- NIST *Handbook 105-1, 2019, Specifications and Tolerances for Field Standard Weights*. The applications of standards in this handbook are referenced for use in legal metrology and general applications, as the handbook points to the OIML and ASTM mass standard classifications in addition to Class F standards in circulation prior to January 1, 2020. This publication is available free of charge from https://doi.org/10.6028/NIST.HB.105-1-2019.

As if 20 weight classifications were not enough, out-of-date publications of the U.S. National Bureau of Standards (now NIST) included the documentary standards in the following list that included various additional weight classifications (some of which are still in use or requested today)! These documentary standards were all superseded long ago, and users should not continue to reference them or these older weight classes in procedures and publications. An article[16] published in *ASTM Standardization News* in 1993 recommended that these out-of-date standards no longer be used. But use persists, often due to the failure to conduct regular reviews and update obsolete procedures.

- NBS Circular 3, Design and Test of Standards of Mass, 1918. This document included specifications and tolerances for commercial classes A, B, C, and T, as well as for scientific classes M and S. It also included extensive guidance on mass calibration procedures, buoyancy corrections, and calculation methods for procedures in addition to the material, design, and other specifications.

- NBS Circular 547, 1954. The foreword in 1954 stated: "National Bureau of Standards Circular 3, Design and Test of Standards of Mass, has been a basic reference on mass standards and weighing since its publication in 1918. The expanding needs of science, industry, and commerce call for the replacement of Circular 3 by a larger and more comprehensive document." A 1978 reprint of portions of Circular 547 can also be found reprinted as NBSIR 78-1476, Precision Laboratory Standards of Mass and Laboratory Weights.

However, given all these classifications, OIML R 111 is the most common documentary standard for mass standards used around the world. It defines the maximum permissible error (*m.p.e.*) as follows: "Maximum absolute value of the difference allowed by national regulation, between the measured conventional mass and the nominal value of a weight, as determined by corresponding reference weights." This definition applies the *VIM* 3 definition explicitly to mass standards. Keep in mind that you may see the term *tolerances* interchanged with *m.p.e.* in some documentary standards. Weights should be selected at the levels needed, as the cost and maintenance requirements rise significantly from the lowest levels to the highest levels due to factors discussed in good weighing procedures, the expected stability due to design and material differences, as well as the calibration uncertainties that must be maintained for standards to remain within the expected *m.p.e.* values.

Attempts have been made to harmonize ASTM E617 as much as possible with OIML R 111, with several exceptions. E617 includes weights with nominal values or multiples of 3 and includes tolerances and calculations for avoirdupois units of mass (e.g., pound, ounce), which are still in common use.

Calibration Procedures—Mass Standards (Weights)

Selection of mass calibration procedures depends on several factors associated with physics: the environment, handling, the item being weighed, and how all these factors affect the calibration uncertainties, which then also affect the decision rules and conformity assessments for each classification and associated tolerances. Table 34.6 provides recommendations as a starting point in the selection of procedures; however, the factors noted previously must be considered with respect to the level of accuracy and uncertainty required by the classification, tolerances, and final user requirements. The documentary standards already noted for mass standards specify requirements for procedures, environments, equilibration, measurement assurance, uncertainty evaluation, reporting, and stating conformity to specifications and maximum permissible errors.

Nearly all procedures used in mass calibrations are based on substitution methods, where the unknown mass standard is substituted on the balance before or after a calibrated mass standard observation is made. In lower-level mass calibrations, a mass standard might be used to verify calibration of the balance indications immediately prior to use; thus the balance becomes the substitution standard in the verified range of use, with all other calibration requirements being

Table 34.6 Mass calibration procedure recommendations.

Echelon	Suitable Calibration Procedures
I	Weighing designs consisting of redundant comparisons with known standards (replicate weighings) with built-in buoyancy corrections and statistical process controls for evaluating the standards and measurement process of Accuracy Classes OIML E_1 and ASTM 000, 00, 0. Example procedures or equivalents: OIML R 111, Annex B[17]; *NISTIR 5672*[18], SOP 28, Advanced Weighing Designs (for calibrating standards used for mass calibrations) OIML E_2 and ASTM 1; *NISTIR 5672*, SOP 5, Using a 3-1 Weighing Design (for calibrating standards used for balance calibrations) *NBS Technical Notes 844*[19], *952*[20]
II	Documented comparison calibrations with known standards that incorporate methods to minimize balance drift, provide buoyancy corrections, and incorporate suitable measurement assurance practices for OIML, Classes F_1, F_2 and ASTM, Classes 2, 3. Example procedures, equivalent or better: OIML R 111, Annex B[21]; *NISTIR 5672*, SOP 5, or *NISTIR 6969*[22], SOP 4, Weighing by Double Substitution.
III	Documented comparison calibrations with known standards that incorporate methods for suitable measurement assurance practices for OIML Classes M_1, M_{1-2}, M_2, M_{2-3}, M_3 and ASTM Classes 4, 5, 6, 7 and NIST Class F (1990). Buoyancy corrections to be evaluated for significance and incorporated as corrections or as uncertainty components when deemed significant. Example procedures, equivalent or better: OIML R 111, Annex B; *NISTIR 6969*, SOP 8, "Medium Accuracy Calibrations of Mass Standards by Modified Substitution."

Additional internationally accepted calibration procedures are referenced in OIML R 111.

met. However, mass calibration procedures increase in complexity from a basic one-to-one comparison of standard and unknown, to increasing redundancy in the comparisons, additional corrections for buoyancy and the inclusion of check standards or other measurement assurance methodologies. At the highest level of comparisons (e.g., dissemination from the realization methods described earlier), additional corrections and influence factors must be considered. For example, the transfer of mass from typical Pt-Ir reference standards to stainless steel working standards might require special buoyancy corrections and may also require a correction of something close to 3 µg/kg due to differences in the center of gravity. Corrections like this are common at the NMI level, but unlikely elsewhere.

Calibration procedures for mass generally include all the following components:

- Prerequisites—selecting proper reference and working standards with current metrological traceability and sufficiently small uncertainties, ensuring that the environment is within operational limits and environmental measurement standards are calibrated, ensuring that staff are trained on good handling techniques in the laboratory, treatment of the unknown standard for equilibration, and perhaps cleaning and stabilization.

- A step-by-step substitution procedure for comparing reference or working standards to the unknown standard or standards.

- Calculation methods that include, as a minimum, values of the reference standards and buoyancy corrections.

- Statistical control methods, which include repeatability assessments or more complex measurement assurance methods, with the integration of check standards for monitoring the stability of the standards and process over time and assessing reproducibility.

- Uncertainty components and calculations that must or should be considered.

- Conformity assessments.

- Reporting measurement results, uncertainties, and additional unique factors critical for mass calibrations.

- OIML R 111 also includes methods for cleaning, measuring surface roughness, measuring density, and determining magnetism and/or magnetic susceptibility of mass standards.

Mass standard specifications include two decision rules that are required to be met for conformity assessment of weights within applicable classes (providing specifications are also met or noted as appropriate):

1. First, the expanded uncertainty of the calibration must be less than one-third of the maximum permissible error (*m.p.e.*) or tolerance. This requirement is in place to ensure that there is an *m.p.e.* to uncertainty ratio of 3:1 so that the uncertainty is not taking up too much of the *m.p.e.* Presented as an equation: Uncertainty < 1/3 *m.p.e.*

2. Second, the sum of the absolute value of the conventional mass correction plus the expanded uncertainty must be less than the *m.p.e.* to confidently state that mass standards are in or out of tolerance. Another way to say this is exactly as stated in OIML R 111, section 5.3.1: "For each weight, the conventional mass, m_c (determined with an expanded uncertainty, U, according to 5.2) shall not differ from the nominal value of the weight, m_0, by more than the maximum permissible error, d_m, minus the expanded uncertainty." As an equation: $m_0 - (d_m - U) \leq m_c \leq m_0 + (d_m - U)$.

SUMMARY

From defining, realizing, and disseminating the kilogram, all the way down to commerce, we have covered much in the hierarchy of mass and weighing. We briefly covered some of the physics concepts and good weighing practices needed for achieving good mass and weighing results. Documentary standards and classifications of weighing instruments and mass standards guide much of what metrologists should implement and consider. Extensive treatment of this topic is provided in the references noted in this chapter. A final word of advice regarding mass and weighing measurements: Even with the best initial calibration, proper care, handling, and use, regularly scheduled periodic calibrations are essential as part of a good calibration program. Further, the use of measurement assurance methods such as repeatability evaluations and check standards will further enable confidence in producing good mass and weighing results.

Chapter 35

Force

BASIC INTRODUCTION

The big bang theory proposes that, way before an apple fell on Sir Isaac Newton's head, objects of a certain mass accelerated to form the universe and its many galaxies, pioneering the definition of force. The basic definition of *force* is mass times acceleration. Here is the mathematical expression of Newton's second law of motion:

$$\sum F = m \times a$$

where

$\sum F$ = summation of all external forces applied to the particle
m = the mass of the particle
a = the acceleration of the particle

A more formal definition of *force* is found in the CIPM/BIPM (International Bureau of Weights and Measures) Unit of Force: "The unit of force [in the MKS (metre, kilogram, second) system] is the force which gives to a mass of 1 kilogram an acceleration of 1 metre per second, per second." It is no surprise that the SI unit of force is expressed in Newtons (N).[1]

The BIPM SI brochure states that "the weight of a body is the product of its mass and the acceleration due to gravity; in particular, the standard weight of a body is the product of its mass and the standard acceleration due to gravity; The value adopted in the International Service of Weights and Measures for the standard acceleration due to gravity is 980.665 cm/s^2 (9.806 65 m/s^2); a value already stated in the laws of some countries."[2]

A major benefit to performing force measurements is that force is essentially the same anywhere in the world. However, weights used for force do require correction for local gravity, air density, and material density. Any weights adjusted for force cannot be used elsewhere without further correction to the new gravity and air buoyancy. Force-measuring instruments calibrated using force weights can be used anywhere in the world without any further corrections.

There are several published standards for force measurements and many other standards that reference these standards. The most recognized standards for force measurement are ISO 376 for most of Europe, and ASTM E74 for North America.

The ASTM International *E74-18: Standard Practices for Calibration and Verification for Force-Measuring Instruments* gives guidance on how to correct materials like those used to manufacture weights for force. ASTM International states, "The force exerted by a weight in air is calculated as follows:

$$\text{Force} = Mg\ /\ 9.80665\ S\ (1 - d/D)$$

where

    M = mass of the weight,
    g = local acceleration due to gravity, m/s2,
    d = air density (approximately 0.0012 Mg/m3),
    D = density of the weight in the same units as d, and
    9.80665 = the factor converting SI units of force into the customary units of force. For SI units, this factor is not used."[3]

The recommendation for force weights is to use steel with a uniform density, because part of the equation for force relies on knowing the material density and the associated uncertainties of that measurement. Knowing a theoretical density is different from having the density tested. Furthermore, the less uniform the density, the higher the probability of distorting the load path when using weights adjusted for force.

THE SI UNITS FOR FORCE

Force is a derived quantity value. It is derived from the SI units of length, mass, and time. The base units are defined by the BIPM as:

> The *second*, symbol s, is the SI unit of time. It is defined by taking the fixed numerical value of the cesium frequency ΔvCs, the unperturbed ground-state hyperfine transition frequency of the cesium 133 atom, to be 9,192,631,770 when expressed in the unit Hz, which is equal to s^{-1}.

> The *meter*, symbol *m*, is the SI unit of length. It is defined by taking the fixed numerical value of the speed of light in vacuum *c* to be 299,792,458 when expressed in the unit m s^{-1}, where the second is defined in terms of ΔvCs.

> The *kilogram*, symbol kg, is the SI unit of mass. It is defined by taking the fixed numerical value of the Planck constant h to be 6.626 070 15 × 10^{-34} when expressed in the unit J s, which is equal to kg m^2 s^{-1}, where the meter and the second are defined in terms of c and ΔvCs.[4]

WHY FORCE MEASUREMENT IS IMPORTANT

Force measurement is vital because when appropriately designed and tested, bridges, buildings, and other structures are less likely to be subject to possible failure when various forces are exerted upon them. When building a bridge, it is essential to get the concrete strength measurement correct; it is essential that the rebar steel be tested, and that the cables be appropriately checked and verified for pre-stress or post-tension. When these measurements are not done correctly, failures can happen, resulting in structure failures, injury, and quite possibly loss of life.

Force measurement is performed so frequently that we take it for granted. Almost every material item is tested using some form of traceable force measurement. For example, the packaging on most materials, including soaps, shampoos, cosmetics, and processed foods, is checked for tensile, peel, compressive, and tear strength to ensure that the product can be opened yet remain protected and aesthetically appealing. Foodstuffs are tested for crispiness, chewability, cutting resistance, penetrability, and more, all of which measure the application of force. Clothing, fabrics, shirts, trousers, undergarments, tablecloths, napkins, etc., are checked for tensile strength, tear strength, and seam slippage. The seams, buttons, and snaps are checked for tensile strength and pull-off forces. Towels are tested for strength in both weft and warp directions, as well as seam slippage. Yarn is tested for tensile strength, as is fishing line and composite filaments. The list goes on and on.

Building materials such as concrete, glass, rebar, beams, wood, and structural composite material are commonly tested for compressive strength, tensile strength, flexural strength, shear strength, rupture strength, and impact strength. Examples continue with the materials in vehicles, such as lamps, interior trim, oil filters, gaskets, etc. These are tested for tensile strength, flexural strength, ductility, shear strength, tear strength, rupture strength, and impact strength. Computers, including monitors and peripherals, are tested for tensile, flexure, and shear strength, and impact resistance. Keyboards are tested for click and operational forces.

Sample testing on manufactured lots might include anything from the materials used to build your house to the cardboard that holds a toilet paper roll.

Chances are, if it is being manufactured, some portion of the lot is being tested. When done properly, this testing helps to ensure that the products we use every day are safe. In addition to our safety, proper force measurement can result in increased productivity, reduction in waste, and improvement of a company's reputation. When things fail that should not, company reputation is often tarnished, if not lost.

TYPES OF FORCE INSTRUMENTS

Any mechanical instrument that measures force can be thought of as a transducer. In the broad sense of the term, a *transducer* is a device that turns one type of energy into another type.

Around the mid-1920s, the proving ring (Figure 35.1) was invented. A proving ring is a steel ring that can perform as a near-perfect elastic member when manufactured with the proper steel alloy. As such, a proving ring is a transducer, and when forces are exerted upon it, the material is deflected. A reed and micrometer with a dial attached to it are used to read the deflection. After backing off the dial, when force is applied the dial is raised until there is slight contact between the dial and the reed tip. Morehouse Instrument Company, Inc. worked with the National Bureau of Standards, known today as the National Institute of Standards and Technology (NIST), to create and refine the proving ring. It was developed to improve material hardness testing, specifically Brinell harness machines. The first calibration of a Brinell machine with a proving ring was performed around 1925. In 1925, the proving ring was the most accurate instrument for the calibration of force. In later years, the load cell (Figure 35.2) would change this.

Figure 35.1 Proving rings.
Source: Reproduced by permission from Morehouse Instrument Company, Inc.

A load cell is a transducer that converts mechanical energy into electrical signals. When a compressive or tensile force is exerted on a load cell, the mechanical energy is converted into equivalent electrical signals. As force is exerted on a load cell, the material deflects. The deflection is typically measured by several strain gauges typically placed on the material inside the load cell. When placed appropriately, the strain gauges will measure the change in resistance as force is applied. The ideal load cell only measures force in defined directions and ignores force components in all other directions. Approaching the ideal involves optimizing many design choices, including the mechanical structure, the gage pattern, gage placement, and the number of gages.

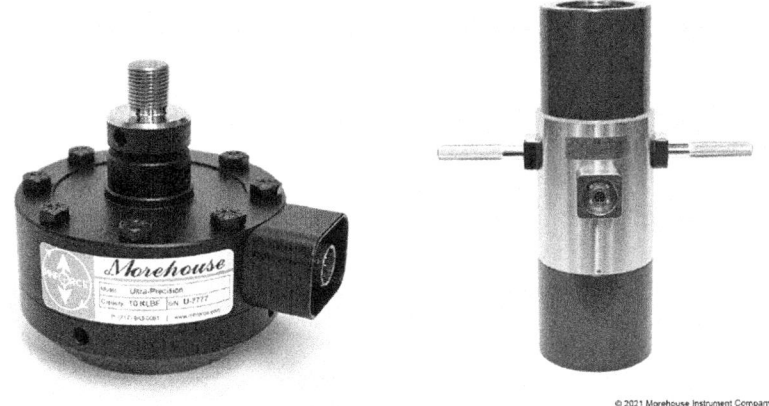

Figure 35.2 Load cells.
Source: Reproduced by permission from Morehouse Instrument Company, Inc.

448 Part VI: Measurement Parameters

Figure 35.3   Force gauge.
Source: Reproduced by permission from Morehouse Instrument Company, Inc.

When a meter or indicator is hooked up to a load cell, it often displays a force measurement value. A load cell may be calibrated using primary deadweight primary standards known to be within 0.002 % of applied force. The machine's deadweights are adjusted for local gravity, air density, and material density to apply the force accurately. The weights are used to calibrate the load cell, which may be used to calibrate and verify a testing machine.

A force gauge (Figure 35.3) is a measuring instrument used to measure forces in push or pull testing. Most force gauges have either an analog or a digital display and usually have an accuracy anywhere from 0.1 % of full scale to 2 % of full scale. They might be used for calibration of certain testing machines, weighing devices, assembly presses, control instruments, cable tension, soil testing, or other equipment measuring force. They can also be used as a prime weighing device or permanent load-sensing component in testing or production equipment. Other instruments, such as dynamometers, crane scales, tension links, Brinell calibrators, Rockwell calibrators, aircraft scales, and many more, are used to measure compression and tension forces for various applications. It is essential to understand the common types of load cells used in force measurement and choose the load cell suitable for your application.

The common types of load cells typically used in force measurement are bending beam, shear web, miniature, and column. Many other load cells are used in commercial applications, such as scales used at supermarket checkouts, weight-sensing devices, weighing, and other scales.

S-BEAM (S-TYPE)

The S-beam is a bending beam load cell that is typically used in weighing applications less than 50 lbf (222 N). These load cells work by placing a weight

Figure 35.4 S-beam load cell.
Source: Reproduced by permission from Morehouse Instrument Company, Inc.

or generating a force on the load cell's metal spring element, which causes elastic deformation. The strain gauges in the load cell measure the fractional change in length of the deformation. Typically, there are four strain gauges mounted in the load cell (Figure 35.4).

Advantages:

- In general, linearity will be enhanced by minimizing the ratio of deflection at the rated load to the length of the sensing beam, thus minimizing the change in the shape of the element.
- Load cells are typically used in crane scales, hanging scales, and some weighing scales.
- They are also frequently used in tension applications.

Disadvantages

- The load cell is susceptible to off-axis loading.
- Compression output will be different if the load cell is loaded through the threads versus flat against each base.
- Because of the susceptibility to off-axis loading, S-beams are typically not the right choice for force applications requiring calibration to the following standards: ASTM E74, ASTM E4, ISO 376, and ISO 7500.

SHEAR WEB

The shear web is a shear beam load cell that is ideal as a calibration reference standard up to 100,000 lbf (444,822 N). Shear web load cells are typically the most

Figure 35.5 Shear web load cells.
Source: Reproduced by permission from Morehouse Instrument Company, Inc.

accurate when installed on a tapered base with an integral threaded rod installed (Figure 35.5).

Advantages:

- Typically have very low creep and are not as sensitive to off-axis loading as other types of load cells.
- Recommended choice for force applications from 100 lbf (444 N) through 100,000 lbf (444,822 N).

Disadvantages:

- After 100,000 lbf (444,822 N), the cell's weight makes it exceedingly difficult to use as a reference standard in the field. A 100,000 lbf (444,822 N) shear web load cell weighs approximately 57 lbs (26 kg), and a 200,000 lbf (889,644) shear web load cell weighs over 120 lbs (54 kg). Note: If the weight of larger shear web type cells does not present a logistic or ergonomic issue, the higher capacities make great reference standards.

BUTTON LOAD CELL

The button load cell (Figure 35.6) is a miniature load cell that is typically used when space is limited. It is a compact strain gauge-based sensor with a spherical radius that is often used in weighing applications.

Advantages:

- The button load cell is suitable for applications where there is minimal room to perform a test.

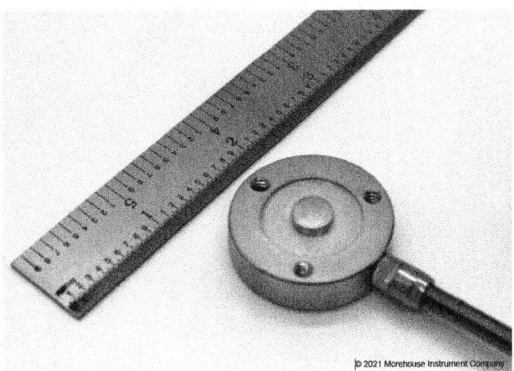

Figure 35.6 Button load cell.
Source: Reproduced by permission from Morehouse Instrument Company, Inc.

Disadvantages:

- Because there is a high sensitivity to off-axis or side loading, the load cell will produce high errors from any misalignment. For example, a 0.1 % misalignment can produce a significant cosine error. Some have errors anywhere from 1 % to 10 % of rated output.
- Do not repeat well in rotation.

SINGLE-COLUMN OR HIGH-STRESS LOAD CELLS

The single column is a column load cell that is good for general testing (Figure 35.7). The spring element is intended for axial loading and typically has a minimum of four strain gauges, with two in the longitudinal direction. Two are oriented transversely to sense the Poisson strain.

Advantages:

- Physical size and weight: It is common to have a 1,000,000 lbf (4,448,221 N) column cell weigh less than 100 lbs (45 kg).

Disadvantages:

- Reputation for inherent nonlinearity. This deviation from linear behavior is commonly ascribed to the change in the column's cross-sectional area (due to Poisson's ratio), which occurs with deformation under load.
- Sensitivity to off-center loading can be high.
- It has larger creep characteristics than other load cells and often does not return to zero as well as other load cells. ASTM Method A typically yields a larger lower limit factor (LLF).

452 *Part VI: Measurement Parameters*

Figure 35.7 Single-column load cell.
Source: Reproduced by permission from Morehouse Instrument Company, Inc.

- Different thread engagement can change the output.
- The design of this load cell requires a top adapter to be purchased with it. Varying the hardness of the top adapter will almost always change the output.

MULTI-COLUMN LOAD CELLS

The multi-column (Figure 35.8) is a column load cell that is often used from 100,000 lbf (444,822 N) through 1,000,000-plus lbf (4,448,221 N). The load is carried by four or more small columns in this design, each with its complement of strain gauges. The corresponding gauges from all the columns are connected in a series in the appropriate bridge arms. A 600,000 lbf (2,668,932 N) multi-column load cell can weigh as low as 27 lbs (12 kg) and has an accuracy of better than 0.02 % of full scale.

Advantages:

- It can be more compact than single-column cells.
- There is improved discrimination against the effects of off-axis load components.

© 2021 Morehouse Instrument Company

Figure 35.8 Lightweight 600k (26 lbf) multi-column load cell.
Source: Reproduced by permission from Morehouse Instrument Company, Inc.

- There is typically less creep and better zero returns than single-column cells.
- In many cases, a properly designed shear web spring element can offer greater output, better linearity, lower hysteresis, and faster response.

Disadvantages:

- The design of this load cell requires a top adapter to be purchased with it. Varying the hardness of the top adapter will change the output.

Picking the appropriate load cell for the application is as important as picking the right indicator to meet the requirements. The correct load cell with an unstable meter with low resolution may not yield satisfactory results.

The book *Force Calibration for Technicians and Quality Managers* describes force calibration in compression and tension modes:

"What Is a Compression Calibration?

When discussing compression calibration, we should think about something being compressed or something being squeezed. Compression calibration can be described as pushing or squeezing something.

Figure 35.9 shows two examples of a compression setup in a calibrating machine. The machine on the left is compressing both load cells by creating an upward force. The picture on the right is a compression setup in the deadweight machine where a downward force compresses the load cell.

454 Part VI: Measurement Parameters

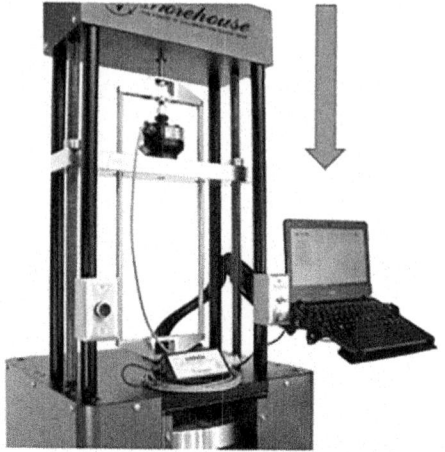

© 2021 Morehouse Instrument Company

Figure 35.9 Compression calibration examples.
Source: Reproduced by permission from Morehouse Instrument Company, Inc.

The key to this type of calibration is making sure everything is aligned and that the line of force is as straight as possible. I like to say free from eccentric or side forces. The key to proper alignment is using the right adapters in the calibrating machine, from alignment plugs to top adapters.

What Is a Tension Calibration?

When discussing tension calibration, we should think of something being stretched. Some may describe tension calibration as a pull.

Figure 35.10 shows multiple examples of tension setups in calibrating machines. The machine on the left is the benchtop calibrating machine.

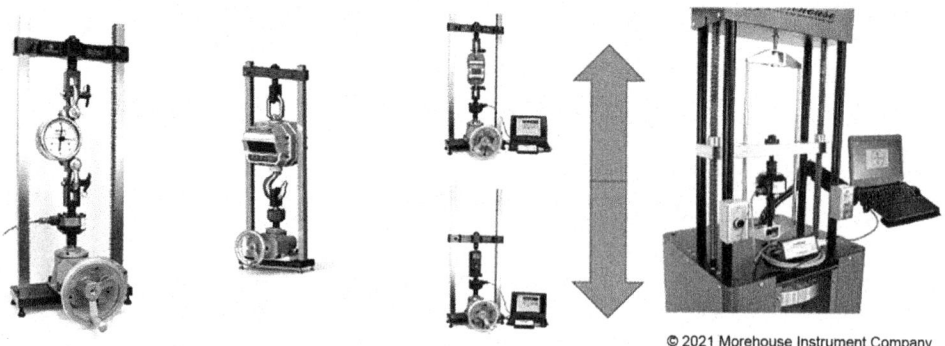

© 2021 Morehouse Instrument Company

Figure 35.10 Tension calibration examples.
Source: Reproduced by permission from Morehouse Instrument Company, Inc.

A dynamometer is fixed to a stationary beam, and force is generated by pulling on the load cell and the dynamometer. More examples are shown with different instruments, from crane scales to hand-held force gauges. The picture on the right shows a load cell fixtured for tension calibration in a deadweight machine. The load cell is fixtured to the frame, and the weights are applied and hung, which stretches the material. The key to getting great results in tension calibration is also adapters."[5]

"The ISO 376:2011 Annex gives excellent guidance on adapters that help keep the line of force pure. It states, "Loading fittings should be designed in such a way that the line of force application is not distorted. As a rule, tensile force transducers should be fitted with two ball nuts, two ball cups, and, if necessary, with two intermediate rings, while compressive force transducers should be fitted with one or two compression pads."[6]

Standardizing the design of adapters is essential to help technicians and end users replicate and reproduce calibration results.

In addition to compression and tension, forces are usually thought of as being either static or dynamic. They can be measured by incrementally increasing the force (ascending) or by decreasing the force (descending).

A force is static if the force remains constant. A static force does not change the size, position, or direction of the object acted upon. The most common example of a static force is suspending or placing weight on an object. A good example is a car parked in a parking garage. When it comes to measuring force, deadweight primary standards are what most force measurements trace back to, and they are static (Figure 35.11). The unit under test (UUT) is placed in a

Figure 35.11 Deadweight calibrating machine.
Source: Reproduced by permission from Morehouse Instrument Company, Inc.

Table 35.1 Basic force conversions.

Conversion	Multiply by
Newton (N) to lbf	0.224 808 943 870 96
Newton (N) to kgf	0.101 971 621 297 79
Newton (N) to ozf	3. 596 943 089 6
lbf to Newton (N)	4.448 221 615 3
lbf to kgf	0.453 592 368 443 86
lbf to ozf	16.0
kgf to lbf	2.204 622 629 412 2
kgf to Newton (N)	9.806 65
kgf to ozf	35.273 962 070 595
ozf to Newton (N)	0.278 013 851
ozf to lbf	0.062 5

machine, and weights are lowered onto the UUT. The UUT then supports the weights under the static load.

Dynamic forces are different from static forces, as they cause the object they are acting on to change its size, position, or direction. A good example is the weight of a moving car on the road. There are several methods of measuring dynamic forces, perhaps the most common being the use of a dynamic material testing machine to ascertain materials' physical and mechanical properties. Table 35.1 lists some basic force conversions.

Nonlinearity, nonrepeatability, hysteresis, and static error band are common load cell terms typically found on a load cell specification sheet. There are more terms to describe the performance and characteristics of a load cell. These four are the most common terms that are included on calibration certificates.[7]

"When broken out individually, these terms can help you select the suitable load cell for an application. Some of these terms may not be as important today as they were years ago because better meters are available that overcome inadequate specifications. One example is non-linearity. An indicator capable of multiple span points can significantly reduce the impact of a load cell's non-linear behavior."[8]

Nonlinearity can be described as the quality of a function that expresses a relationship that is not one of direct proportion. For force measurements, nonlinearity is defined as the algebraic difference between the output at a specific load and the corresponding point on the straight line drawn between the outputs at minimum load and maximum load. It usually is expressed in units of percent of full scale (Figure 35.12), and it is usually calculated between 40 to 60 % of the full scale.[9]

"Non-linearity is one of the specifications that would be particularly important if the indicating device or meter used with the load cell only has a two-point span, such as capturing values at zero and capacity or close to capacity. The specification gives the end-user an idea of the anticipated error or deviation from the best fit straight line. However, suppose the end-user has an indicator capable of multiple span points and uses coefficients from an ISO 376 or ASTM E74 type calibration. In that case, the non-linear behavior can be corrected, and the error significantly reduced.

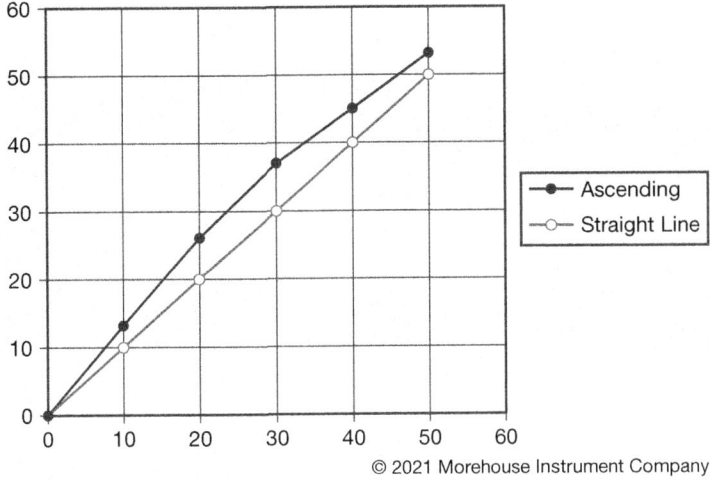

Figure 35.12 Non-linearity expressed graphically.
Source: Reproduced by permission from Morehouse Instrument Company, Inc.

"One way to calculate non-linearity is to use the slope formula or manually perform the calibration by using the load cell output at full scale minus zero and dividing it by force applied at full scale and 0. For example, a load cell reads 0 at 0 and 2.00010 mV/V at 1000 lbf. The formula would be (2.00010 − 0) / (1000 − 0) = 0.002. This formula gives you the slope of the line, assuming a straight-line relationship."[10] A less conservative approach would be to use higher-order quadratic equations.

"Plot the non-linearity baseline as shown . . . using the formula of force applied * slope + Intercept or y = mx + b.

If we look at the 50 lbf point, this becomes

$$50 * 0.0020001 + 0 = 0.100005.$$

Thus at 50 lbf, the non-linearity baseline is

$$0.100005.$$

To find the non-linearity percentage, take the mV/V value at 50 lbf minus the calculated value and divide by the full-scale output multiplied by 100 to convert to a percentage. Thus, the numbers become

$$((0.10008 − 0.100005)/2.00010) * 100) = 0.004\ \%."[11]$$

HYSTERESIS

Hysteresis is described as "the phenomenon in which the value of a physical property lags changes in the effect causing it." An example is when magnetic induction lags the magnetizing force. For force measurements, *hysteresis* is often defined as the algebraic difference between output at a given load descending from the maximum load and output at the same load ascending from the minimum load.[12]

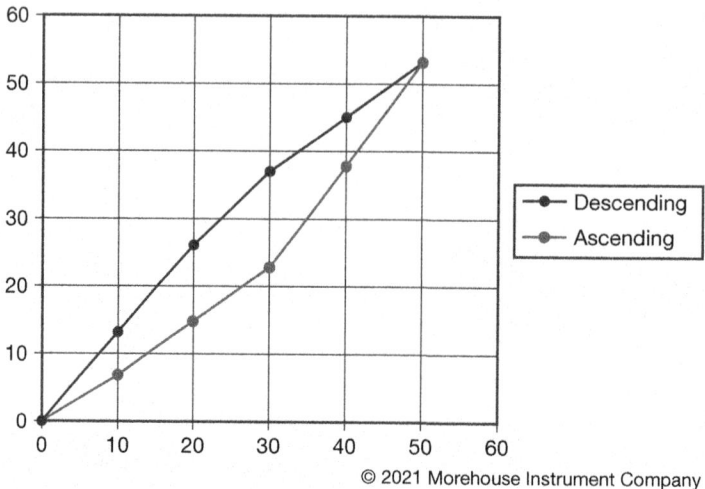

Figure 35.13 Hysteresis example.
Source: Reproduced by permission from Morehouse Instrument Company, Inc.

Hysteresis is normally expressed in units of % full-scale and is usually calculated between 40 to 60 % of full scale. The graph in Figure 35.13 shows a typical hysteresis curve where the descending measurements have a slightly higher output than the ascending curve. If the end user uses this load cell to make descending measurements, they may want to consider the effect of hysteresis. Errors from hysteresis can be high enough that if a load cell is used to make descending measurements, then it must be calibrated with a descending range. The difference in output on an ascending curve versus a descending curve can be significant.[13]

For example, an exceptionally good 100K precision shear web load cell had an output of –2.03040 on the ascending curve and –2.03126 on the descending curve. Using only the ascending curve would result in an additional error of 0.042 %.[14]

NONREPEATABILITY

Nonrepeatability: "The maximum difference between output readings for repeated loadings under identical loading and environmental conditions. Normally this is expressed in units as a % of rated output (RO)."[15] Only a portion of the expected performance of the load cell is captured in nonrepeatability. It does not account for the reproducibility or performance under different loading conditions (randomizing the loading conditions).

"The calculation on non-repeatability is straightforward. First, compare each observed force point's output and run a difference between those points. The formula would look something like this: non-repeatability = ABS(Run1 – Run2)/ AVERAGE (Run1, Run2, Run3) *100. Do this for each combination or runs, and then take the maximum of the three calculations."[16]

STATIC ERROR BAND

Static error band: "The band of maximum deviations of the ascending and descending calibration points from a best-fit line through zero output. It includes the effects of non-linearity, hysteresis, and non-return to minimum load and is usually expressed in units of % of full scale. Static error band might be the most exciting term. If the load cell is always used to make ascending and descending measurements, this term best describes the load cell's actual error from the straight line drawn between the ascending and descending curves. Earlier, I noted that the end-user might want to consider the effects of hysteresis unless they use the load cell described above because the static error band would be the better specification to use. The end-user could ignore non-linearity and hysteresis and focus on static error band as well as non-repeatability.

"However, we find that some calibration laboratories primarily operate using ascending measurements, and on occasion, may have a request for descending data. When that is the case, the user may want to evaluate non-linearity and hysteresis separately. When developing an uncertainty budget, use different budgets for each type of measurement, i.e., ascending and descending.

"What needs to be avoided is a situation where a load cell is calibrated following a standard such as ASTM E74 or ISO 376, and additional uncertainty contributors for non-linearity and hysteresis are added. ASTM E74 has a procedure and calculations that, when followed, uses a method of least squares to fit a polynomial function to the data points. The standard uses a specific term called the Lower Limit Factor (LLF), which is a statistical estimate of the error in forces computed from a force-measuring instrument's calibration equation when the instrument is calibrated following the ASTM E74 practice."[17]

METHODS/STANDARDS

The calibration method, such as compression, tension, ascending, descending, and the number of test points, is critical in using a force-measuring instrument. If the force-measuring instrument is to be used for compression (push) and tension (pull), then it must be calibrated in both modes. After the basics are discussed, the question becomes one of needing calibration to a documented metrology standard such as ASTM E74 or ISO 376. Most load cell responses are not symmetrical, and compression and tension calibration differences can be quite significant. Additionally, a force-measuring device should only be used within the range it was calibrated.

For example, consider a 10,000 lbf load cell that is calibrated at 10 % force increments. The device has not been calibrated below 1,000 lbf and may not be accurate from 0.1 lbf through close to 1,000 lbf. The easiest solution is to discuss the requirements with your calibration provider, because expecting a 10,000 lbf load cell to measure 20 lbf of force may not be realistic. However, using two load cells to measure from 20 lbf through 10,000 lbf is achievable.

A common error source is the assumption that the force-measuring instrument can be used to make descending or decremental measurements when only an ascending or incremental calibration was performed. Ascending and descending calibration is typically required for low cycle fatigue machines, nuclear requirements, and universities conducting research and development.

Another common error arises when a force-measuring device does not match the calibration results because the end user is using mass weights for the verification and not weights adjusted for force. Force is force anywhere in the world, and a force weight requires adjustment for material density, local gravity, and air buoyancy. Therefore, when mass weights are used to perform force measurement, the errors can be quite high, and the end user may not recognize this error.

If a load cell is calibrated following the ASTM E74 standard and a combined curve is used, the end user could use the load cell anywhere in the verified range of forces. The downside to this method is that the combined curve will produce a lower limit factor (LLF) larger than using separate curves. However, the larger LLF will include any point within the verified range of forces for ascending and descending forces. Suppose the end user cannot always load the reference standard to capacity and wants a smaller LLF. In that case, they will need to have the load cell tested with several hysteresis loops for every capacity they wish to calibrate.

COMMON ERROR SOURCES IN FORCE CALIBRATION

The end user of the force-measuring instrument must ensure that the laboratory performing the calibration replicates how the instrument will be used. Calibration laboratories are capable of replicating use if they use the customer's adapters and independent setups for compression and tension. However, because this takes more time and raises the cost, it is done infrequently.

Fixturing and adapters used with a force-measuring instrument may significantly contribute to the force-measuring instrument's overall uncertainty. These errors can be higher than 0.05 % of the output using top blocks of different hardness. Common error sources for force calibration include:

- Not replicating via calibration how the equipment is being used
- Not using independent setups for compression and tension when calibrating to ASTM E74 or ISO 376
- Alignment, which can be overcome with proper adapters
- Using a different hardness of adapter than what was used for calibration
- Using a different size adapter than what was used for calibration
- Loading against the threads instead of the shoulder
- Loading through the bottom threads in compression
- Not accounting for temperature effects on noncompensated force-measuring instruments
- Not accounting for temperature effect coefficients on zero and rated output
- Making cable length errors on a four-wire system
- Using electronic instruments (indicators) that were not used during calibration
- Using an excitation voltage that is different from the voltage used at the time of calibration

- Variations in bolting a force transducer to a base for calibration while the application is different
- Improper electronic cabling regarding shielding, proper grounding, use or nonuse of sensing lines, or cable length
- Failing to exercise the force-measuring instrument to the capacity it was calibrated at before use
- Ignoring the difference between the output of a high-quality force transducer when compared to the current machine and realized value from the deadweight calibration
- Having too much "pre-load" or tare weight on force transducer (this is transducer dependent and most start to show significant errors when the "pre-load" is over 5 % of the rated capacity)

The primary focus is not using independent setups for compression and tension when that is how the force-measuring devices are being used.

CALIBRATION USING DIFFERENT SETUPS FOR COMPRESSION AND TENSION

What replicates field use is often best practice. Different setups may have to be made with different adapters for tension and compression to best replicate field use. Figure 35.14 shows a machine capable of different setups for compression and tension. These are different from a compression/tension machine, which may not replicate field use, where compression and tension are done using the same setup.

The committee that drafted the ISO 376 standard has written specific guidance on adapters. Because they understand the importance of replicating field use, their recommendations include separate setups for compression and tension (Figures 35.15 and 35.16).

Many testing machines calibrated to ISO 7500 or ASTM E4 are calibrated in compression and tension, and the technician will use different setups for each mode (Figure 35.17). Many will use adapters as recommended in ISO 376, which states, "Loading fittings should be designed in such a way that the line of force application is not distorted. As a rule, tensile force transducers should be fitted with two ball nuts, two ball cups and, if necessary, with two intermediate rings, while compressive force transducers should be fitted with one or two compression pads."[18]

Replicating field use gives the end users confidence in their measurements. In many universal calibrating machine setups, when using the recommended adapters, the user is forced to change setups for compression and tension. Different setups help reduce common measurement errors, such as when using adapters other than what was used during calibration and when loading through the bottom threads in compression.

Other machines performing tension and compression in the same setup may not wait to apply the force for at least 30 seconds, which is specified in the time/loading profile in ISO 7500-1[19] and ISO 376.[20] Most National Metrology Institute

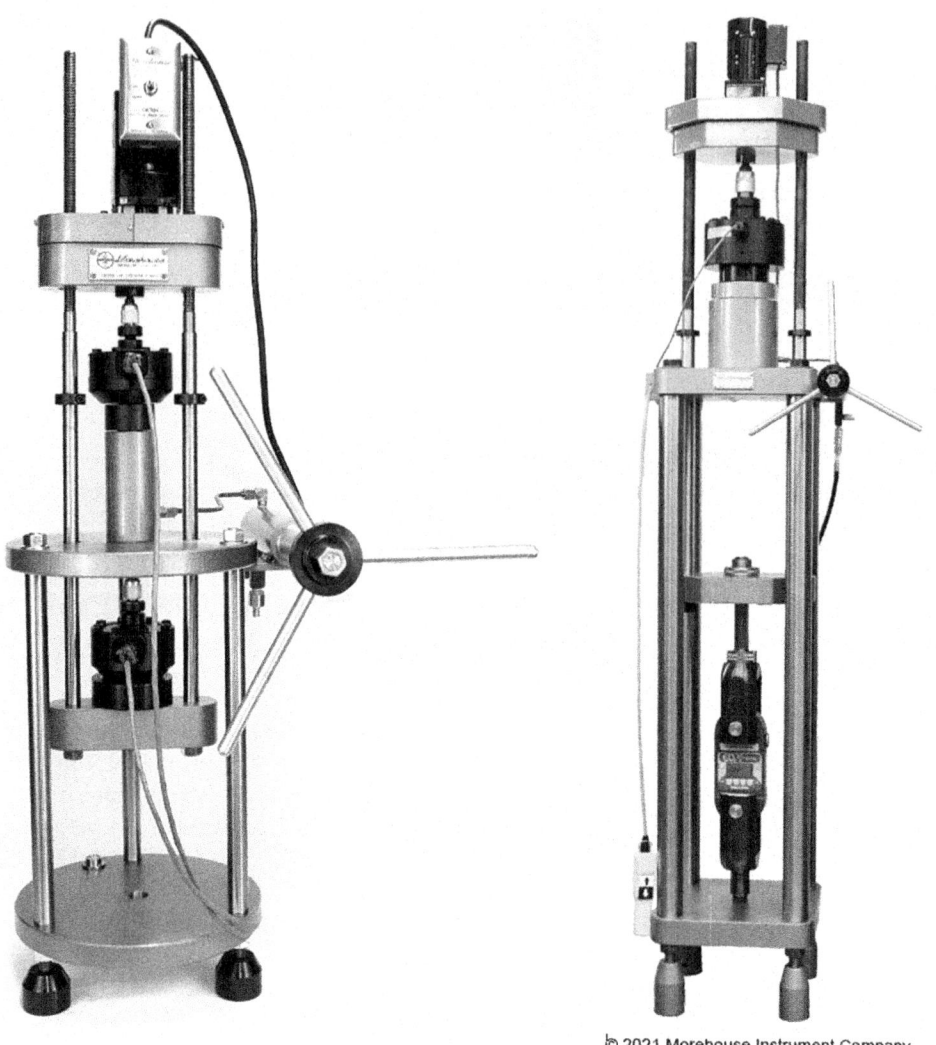

© 2021 Morehouse Instrument Company

Figure 35.14 Universal calibrating machine with a unit under test in compression (left) and in tension (right).

Source: Reproduced by permission from Morehouse Instrument Company, Inc.

(NMI) force standard machines are designed with separate areas for compression and tension setups.

It is important to note that dynamic calibration is not supported per the standard, and a dynamic machine should not be used for calibration following ISO 376. Per ISO 376: "This International Standard concerns only *static* force measurement. If the force-proving instrument is used under dynamic conditions, additional contributions should be taken into account. For example, the frequency responses of the force transducer and indicator, and the interaction with the mechanical structure, can strongly influence the measurement results. This requires a detailed analysis of dynamic measurement, which is not part of this International Standard."[21]

To achieve the best replication of field use for calibration, following ASTM E74[22] or ISO 376,[23] the minimum should be performed:

- The calibration laboratory should not perform compression and tension calibration in the same setup unless it replicates end-use conditions. This can be a customary practice for calibration suppliers because it is much quicker.
- The calibration laboratory should use the customer's top blocks and use separate compression setups.

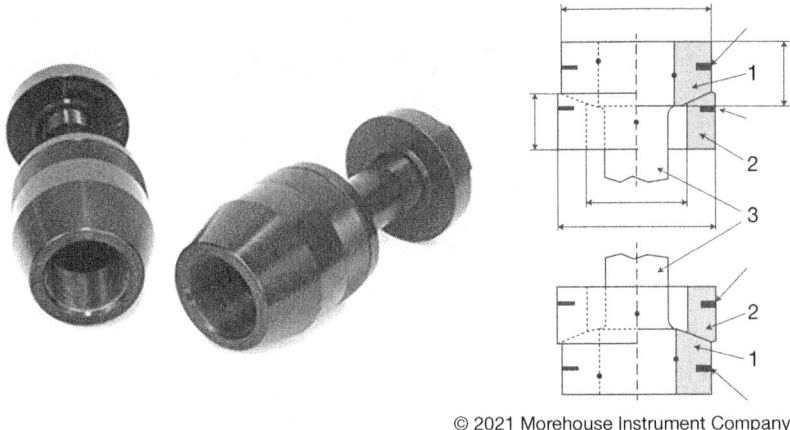

Figure 35.15 Tension adapters designed using recommendations from ISO 376.[24]
Source: Reproduced by permission from Morehouse Instrument Company, Inc.

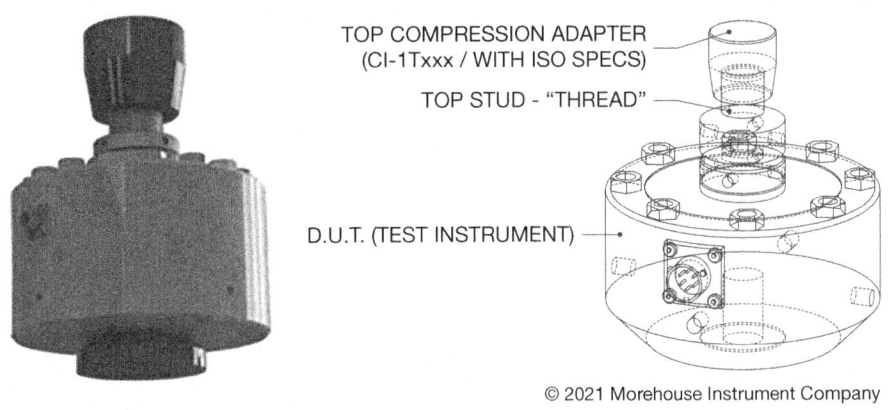

Figure 35.16 Compression adapters designed using recommendations from ISO 376.[25]
Source: Reproduced by permission from Morehouse Instrument Company, Inc.

Figure 35.17 Universal testing machine.
Source: Reproduced by permission from Tinius Olsen.

- In compression, the calibration laboratory should require a baseplate to load against.
- For tension calibration, if the end user is calibrating per ISO 7500, then the calibration laboratory should use adapters recommended per the ISO Annex, which would be different than what was shown earlier in this chapter.
- During contract review, the calibration laboratory should verify how the end user is using the device.

ADAPTERS AFFECT MECHANICAL INTERACTIONS

Several force measurement errors can result from using adapters different from what the force-measuring instrument was calibrated with. This is because, during force calibration, mechanical measurements are being made. Most adapters used at a laboratory level are manufactured to keep the line of force free from eccentric error and apply the same stresses from the adapter interface to the force-measuring instrument as were used at the time of calibration (Figure 35.18).

Figure 35.18 Ultra-precision shear web load cell showing eccentric forces.
Source: Reproduced by permission from Morehouse Instrument Company, Inc.

Using the wrong adapters to calibrate load cells, truck scales, aircraft scales, tension links, dynamometers, and other force-measuring devices can produce significant measurement errors and pose serious safety concerns. Different adapters can change the stress distribution on the force-measuring instrument and produce errors that range from minimal to an output difference more significant than the allowable tolerance.

If the calibration laboratory does not use the appropriate adapters, or is not using similar adapters, there could be substantial errors. It is common to find errors as high as 2 % of the full-scale output resulting from varying the loading condition and adapters. Not all force-measuring instruments are created equal, and replicating use is essential to providing proper force measurements for all equipment. Other important considerations are safety, and adapters that are not machined correctly, which may not allow for a distortion-free load path.

Why is it critical to reduce misalignment error? Pictured in Figure 35.19 is a test showing the spherical adapter without an alignment plug. The error observed is 0.752 % on S-beam load cells with less than 1/8-inch misalignment. Note: Not all S-beam load cells have similar performance characteristics. The example is why testing your equipment is critical as far as understanding potential measurement errors.

When the load cell was aligned and calibrated correctly, the Expanded Uncertainty was calculated at about 10 lbf (44 N); when the load cell was

Output in mV/V
Aligned in machine
−1.96732 mV/V

Output in mV/V
Slightly misaligned in machine
−1.98211 mV/V

© 2021 Morehouse Instrument Company

Figure 35.19 S-beam load cell with slight misalignment producing a 0.752 % error.
Source: Reproduced by permission from Morehouse Instrument Company, Inc.

misaligned, the Expanded Uncertainty was approximately 90 lbf (400 N), which is significant in a 10,000 lbf (44,482 N) S-beam load cell. If the technician misaligns the load cell in a testing machine, they might end up adjusting a machine that is actually "in tolerance," and a recall may result from this simple error. Alignment plugs and base plates with alignment holes (Figure 35.20) can drastically reduce misalignment errors.

When using alignment plugs that thread into the bottom of your load cells, make sure they are threaded flush to the load cell's bottom. Once they are flush, thread the adapter an extra turn into the cell. Make sure that none of the threads are exposed below the load cell base. If one or more threads is exposed, the load will be generated through the load cell's internal threads and not its base. The thread loading can result in an additional calibration error of about 0.012 % on shear web load cells and can often damage the alignment plug. On other types of load cells, the errors may be even larger.

The number-one complaint with button and washer load cells is how to get them to repeat between rotations. These load cells are notoriously sensitive in rotation, and any misalignment will produce significant errors. The sensitivity to off-axis or sideloading conditions is relatively high. It is high enough that misalignment of 0.1 % is going to produce a relatively large error. The error can sometimes be as large as 10 % of the rated output. Typically, this error is between 1 % and 2 % in a well-aligned deadweight machine.

The button and washer load cell adapters shown in Figure 35.21 can improve alignment and yield better calibration results, usually by a factor of five compared

Figure 35.20 Proper way to thread an alignment plug: Thread is past flush and into cell.
Source: Reproduced by permission from Morehouse Instrument Company, Inc.

to a technician trying to center the load cell. There are different recommended adapters for compression versus tension applications (Figure 35.22).

The ISO 376 standard says, "Loading fittings should be designed in such a way that the line of force application is not distorted. As a rule, tensile force transducers should be fitted with two ball nuts, two ball cups and, if necessary, with two intermediate rings, while compressive force transducers should be fitted with one or two compression pads."[26]

Figure 35.21 Button and washer load cell adapters.
Source: Reproduced by permission from Morehouse Instrument Company, Inc.

Figure 35.22 Tension members with two ball nuts and two ball cups.
Source: Reproduced by permission from Morehouse Instrument Company, Inc.

TENSION CLEVIS ADAPTERS FOR TENSION LINKS, CRANE SCALES, AND DYNAMOMETERS

If a calibration lab decides to use a pin that differs from the manufacturer's recommendations or what the end-user is currently using, there will be a larger than expected bias. However, most manufacturers agree on the following:

- Using correctly sized pins is critical.
- Do not use pins that are worn or bent.
- If the links are damaged, overly used, or worn, then decrease the time between calibrations.
- The same size and style of shackle and pin used during operation should be used for calibration.

The pictures in Figure 35.23 show a test with two different-size load pins. A static deadweight machine with an accuracy of better than 0.002 % of applied force is loaded to 50,000 lbf. When loaded with a smaller pin of 1.85 inches, the device reads 49,140 lbf. However, when loaded with a 2-inch pin, the device reads 50,000 lbf.

Knowing these issues exist, Morehouse Instrument Company, Inc. has patented pin and clevis kits (Figure 35.24) to replicate the appropriate pin size. These assemblies cross-reference the various manufacturers' recommended pin sizes and allow the calibration laboratory to calibrate hundreds of tension links, crane scales, dynamometers, and rod-end load cells, all using the same clevis. Most manufacturers will call out the recommended pin size on their drawings. Using these kits can simplify the logistics of having the proper adapter; it improves cycle time and standardizes the calibration process.[27]

Difference of 860 LBF or 1.72% error **at 50,000 LBF** from not using the proper size load pins.

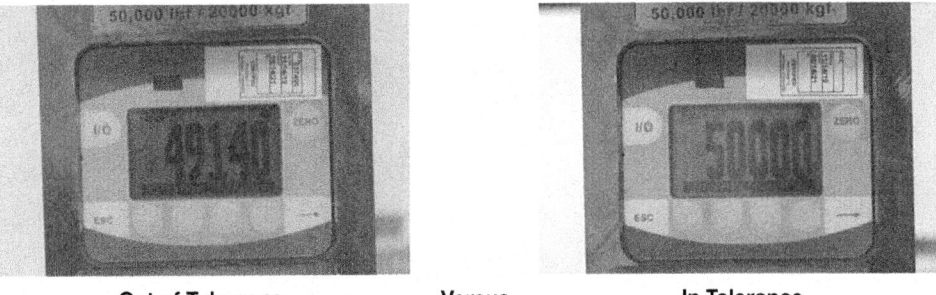

Out of Tolerance   Versus   In Tolerance

© 2021 Morehouse Instrument Company

Figure 35.23 Tension link difference in output with pin size.
Source: Reproduced by permission from Morehouse Instrument Company, Inc.

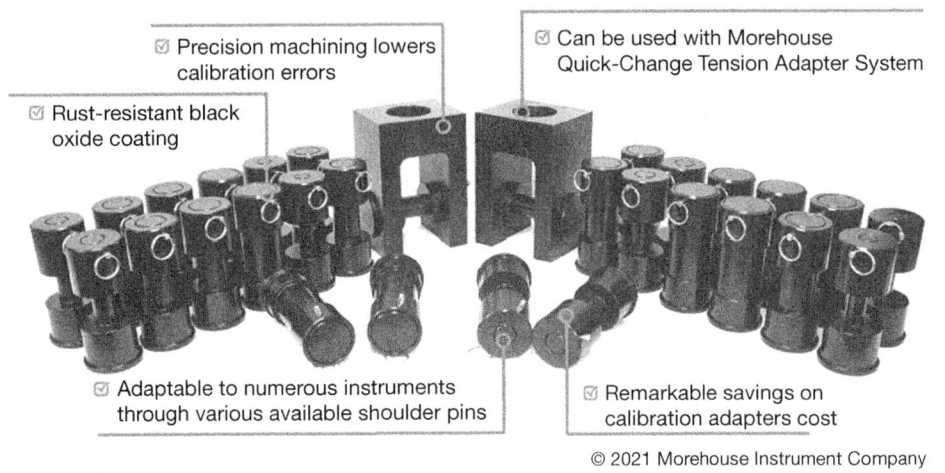

© 2021 Morehouse Instrument Company

Figure 35.24 Clevis kits (U.S. Patent 11,078,052).
Source: Reproduced by permission from Morehouse Instrument Company, Inc.

FOUR-WIRE AND SIX-WIRE SYSTEMS

4-Wire Systems

In understanding the errors associated with a 4-wire cable, we must first understand why this error exists. In general, cable resistance is a function of temperature, and the temperature change on a cable affects the thermal span characteristics of the load cell/cable system. On a 4-wire cable, this will affect thermal span performance, meaning that, as the temperature changes, the resistance of the cable changes and can cause a voltage drop over the

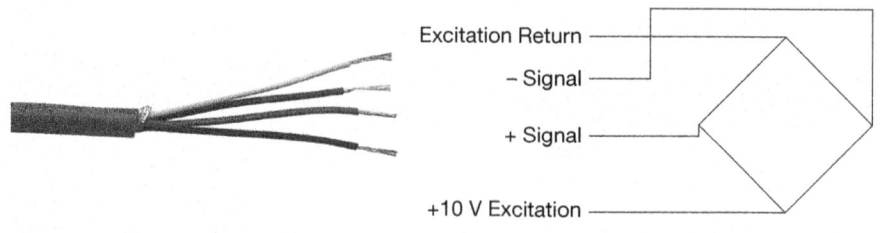

Figure 35.25 4-wire cable and diagram.
Source: Reproduced by permission from Morehouse Instrument Company, Inc.

cable length. A 4-wire setup simply cannot compensate for variations in lead resistance (Figure 35.25).

Substituting a cable of a different gauge or a different length will produce additional errors. A known example of this involves changing a 28-gauge or 22-gauge cable. On a 28-gauge cable, there will be a loss of sensitivity of approximately 0.37 % per 10 feet of 28-gauge cable. On a 22-gauge cable, there will be a loss of sensitivity of around 0.09 % per 10 feet of 22-gauge cable.

Considerations for 4-wire systems:

1. If you damage or replace your cable, the system may have to be calibrated immediately following replacement or repair.

2. Operating at different temperatures will change the resistance, which will cause a voltage drop, resulting in a change of measured output.

3. Cable substitution will result in an additional error and should be avoided.

4. Cables used for 4-wire systems should have a serial number (S/N) or a way to ensure that the same cable stays with the system with which it was calibrated. ISO 376 Clause 6.1 has a specific requirement for labeling cables, and it is good measurement practice technique.[28]

6-Wire Systems

A 6-wire cable that is run to the end of a load cell cable or connector and used with an indicator with sense lead capability will eliminate the errors associated with a 4-wire system. With a 6-wire system (Figure 35.26), the sense lines are separate from the excitation lines, eliminating effects due to variations in lead resistance. It also allows for long cable runs in outdoor environments with extreme temperatures.

Wiring a 6-wire cable for sense is easy. Simply run two lines from the load cell's positive excitation pin and two wires from the load cell's negative excitation pin. The remaining two wires are run in a positive and negative sense. The six wires then feed into the indicator with positive excitation and positive sense running to the indicator. Negative excitation and sense are run to the appropriate indicator connections and positive and negative signals.[29]

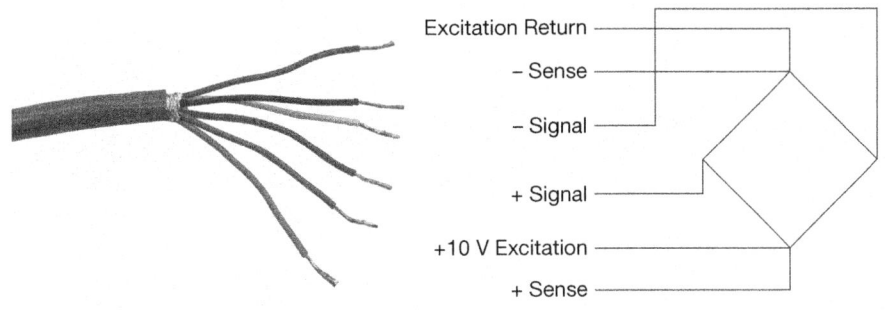

Figure 35.26 6-wire cable and diagram.
Source: Reproduced by permission from Morehouse Instrument Company, Inc.

EQUIPMENT DESIGN

The right equipment for force will be made to minimize off-center loading, bending, and torsion. To do this, force machines need to be:

1. **Plumb**—they should be exactly vertical or true.

2. **Level**—there should be a device for establishing a horizontal line or plane by means of a bubble in a liquid that shows the adjustment to the horizontal by the movement to the center of a slightly bowed glass tube.

3. **Square**—for force machines, this is about having four right angles.

4. **Rigid**—not flexible; if the loading surface starts to bend, all sorts of alignment errors can happen, which will affect the results.

5. **Free of Torsion**—free of being twisted when forces are applied. *Torsion* is the action of twisting or the state of being twisted.

UNCERTAINTY CONTRIBUTORS

ISO/IEC 17025:2017, Clause 7.6.1, states: "Laboratories shall identify the contributions to measurement uncertainty. When evaluating measurement uncertainty, all contributions that are of significance, including those arising from sampling, shall be taken into account using appropriate methods of analysis." ISO/IEC 17025:2017, Clause 7.6.2, states: "A laboratory performing calibrations, including of its own equipment, shall evaluate the measurement uncertainty for all calibrations."[30]

In general, when you calculate measurement uncertainties following JCGM 100-2008, *Guide to the Expression of Uncertainty in Measurement (GUM)*[31] and *ILAC* (International Laboratory Accreditation Cooperation) *P-14 Policy for Measurement Uncertainty in Calibration*,[32] as required by the ISO/IEC 17025 and ISO/IEC 17025 accrediting bodies, you will need to consider the following:

- Repeatability (Type A)
- Reproducibility

- Resolution of UUT
- Reference standard uncertainty
- Reference standard stability
- Resolution of reference standard (if not already considered in its calibration)
- Environmental factors

TYPE A UNCERTAINTY CONTRIBUTIONS

The *GUM* states that all data that is analyzed statistically is treated as a Type A contribution with a normal statistical distribution.[33] Typical examples are:

1) Repeatability
2) Reproducibility
3) Reference Standard(s) Stability/Drift*
4) Others (this would include ASTM E74 LLF, ISO 376 Uncertainty, nonlinearity, or static error band for commercial calibrations)

*Note 1: For our example, stability shall be treated as type B because we are taking values over a range using previous measurement data.

*Note 2: Stability data may be treated as Type A if an evaluation is made using statistical methods.

Repeatability contribution is required by the *GUM* and multiple ISO/IEC 17025 accreditation bodies around the world.

TYPE B UNCERTAINTY CONTRIBUTIONS

Per the *GUM*, Type B evaluation of standard uncertainty may include:

- Previous measurement data
- Experience with or general knowledge of the behavior and properties of relevant materials and instruments
- Manufacturer's specifications
- Data provided in calibration and other certificate(s)
- Uncertainties assigned to reference data taken from handbooks[34]

OTHER ERROR SOURCES

When evaluating other error sources, it is important that the end user of the force-measuring instrument replicate how the instrument was calibrated or that the laboratory performing the calibration replicate how the instrument is going to be used. Fixturing and adapters used with the force-measuring instrument may make a significant contribution to the overall uncertainty of the force-measuring instrument.

NOTE 1: For the parameter of force, some laboratories have top-quality force calibration machines such as deadweight machines. These machines are classified as primary standards and, if correctly designed, can reduce some of the above error sources to insignificance.

NOTE 2: Several laboratories using primary standards have found the repeatability of a top-quality force-measuring instrument in a deadweight machine to be less than 0.0002 %. Resolution of a top-quality force-measuring instrument can be better than 0.0001 % if high-quality indicators reading six or more decimal places are used. Some error sources, which may be insignificant using deadweight primary standards, may become significant at the next measurement tier.

GUIDANCE FOR TYPE A UNCERTAINTY CONTRIBUTIONS WHEN ASTM E74 IS USED AS THE STANDARD

The numbered sections here correlate with the numbered contributors in Figure 35.27.

1, 2 Repeatability and reproducibility between technicians

> Note: Repeatability and reproducibility are from an R & R study and should not be confused with repeatability conducted with the best existing force-measuring instrument. The end user must determine if these errors are significant and should be included in the final uncertainty budget.

3 Repeatability conducted with the best existing force-measuring instrument

4 ASTM LLF must be reduced to 1 standard deviation ($k = 1$) as the ASTM LLF is reported on calibration certificates with $k = 2.4$.

Note: ASTM LLF is called out because many reports do not list the standard deviation. In actuality, the standard deviation per section 8 of the ASTM E74 standard is what is required.

	Uncertainty Contributor	Magnitude	Type	Distribution	Divisor	df	Std. Uncert	Variance (Std. Uncert^2)	% Contribution
1	Repeatability Between Techs	0.032435888	A	Normal	1.000	1	32.44E-3	1.05E-3	0.24%
2	Reproducibility Between Techs	0.006481823	A	Normal	1.000	10	6.48E-3	42.01E-6	0.01%
3	Repeatability	577.3503E-3	A	Normal	1.000	3	577.35E-3	333.33E-3	75.66%
4	ASTM LLF at 1 Standard Deviation	104.1667E-3	A	Normal	1.000	32	104.17E-3	10.85E-3	2.46%
5	Resolution of UUT	25.0000E-3	B	Resolution	3.464	200	7.22E-3	52.08E-6	0.01%
6	Environmental Factors	75.0000E-3	B	Rectangular	1.732	200	43.30E-3	1.88E-3	0.43%
7	Reference Standard Stability	500.0000E-3	B	Rectangular	1.732	200	288.68E-3	83.33E-3	18.91%
8	Ref Standard Resolution	25.0000E-3	B	Resolution	3.464	200	7.22E-3	52.06E-6	0.01%
9	Other Error Sources	150.0000E-3	B	Rectangular	1.732	200	86.60E-3	7.50E-3	1.70%
10	Reference Standard Uncertainty	100.0000E-3	B	Expanded (95.45% k=2)	2.000	200	50.00E-3	2.50E-3	0.57%
				Combined Uncertainty (u_c) =			663.77E-3	440.59E-3	100.00%
				Effective Degrees of Freedom			5		
				Coverage Factor (k) =			2.57		
				Expanded Uncertainty (U) K =			1.71	0.03413%	

Figure 35.27 Example uncertainty budget contributions.

Source: Reproduced by permission from Morehouse Instrument Company, Inc.

GUIDANCE FOR TYPE B UNCERTAINTY CONTRIBUTORS WHEN ASTM E74 IS USED AS THE STANDARD

5. Resolution of the best existing force-measuring instrument
6. Environmental factors
7. Reference standard stability
8. Reference standard resolution (if applicable)
9. Other error sources (for example, side load sensitivity)
10. Reference standard uncertainty

All uncertainty contributions should be combined. If appropriate, the Welch-Satterthwaite equation, as described in JCGM 100:2008, should be used to determine the effective degrees of freedom for the appropriate coverage factor for a 95 % confidence interval.[35]

Force-measuring instruments calibrated according to the ASTM E74 standard are typically continuous reading force-measuring instruments, and uncertainty analysis should be conducted on several test points used throughout the loading range.[36] The % Contribution column in Figure 35.27 is useful in determining significant contributors to uncertainty. By using the significance of uncertainty contributors to the overall uncertainty, one can focus improvement efforts where they will make the most impact (i.e., the largest percentage contributors).

Other error sources for force include:

1. Alignment
2. Using a different hardness of adapter than was used for calibration
3. Using different-size adapters than what were used for calibration
4. Loading against the threads instead of the shoulder
5. Loading through the bottom threads in compression
6. Temperature effects on noncompensated force-measuring instruments
7. Temperature-effect coefficients on zero and rated output
8. Cable length errors on a 4-wire system
9. Using electronic instruments (indicators) that were not used during calibration
10. Using an excitation voltage that is different from the voltage used at the time of calibration
11. Variations in bolting a force transducer to a base for calibration while application is different
12. Not replicating via calibration how the equipment is being used
13. Electronic cabling mistakes regarding shielding, proper grounding, use or nonuse of sensing lines, cable length

14. Failure to exercise the force-measuring instrument to the capacity it was calibrated at prior to use

15. Difference between the output of a high-quality force transducer when compared against the current machine and the realized value from the deadweight calibration

16. Having too much "pre-load" or tare weight on force transducer (this is transducer dependent and most start to show significant errors when the "pre-load" is over 5 % of the rated capacity)

CONCLUSION

Our world is constantly changing, and our mass continues to move with the Earth at a speed of 2.1 million kilometers per hour away from where it all started.

Measuring forces may seem simple. One is either pushing or pulling on something. Seems easy enough, right? However, mechanical interactions between different materials and designs can lead to noticeable sources of error when the calibration does not replicate actual use.

Often, companies that set up commercial calibration laboratories or perform in-house calibrations do not have all the knowledge or experience presented throughout this chapter. What happens when these concepts are not followed, or errors are not detected? What are the effects on the measurement equipment used to test medical equipment, airplanes, cars, and bridges? Suppose these concepts are not followed by those using measurement equipment; how comfortable might you feel when sitting in a traffic jam on a bridge, experiencing mechanical problems on your next flight, or living or working in a high-rise building?

This chapter does not include or discuss all the error sources for force measurement. We are constantly learning and improving because the truths we speak are today's truths. Knowledge is power, and the more you know, the more tools you have to help ensure that proper measurements are being made. With the advent of new technologies and advancements in new equipment, continuous learning is required for those making measurements.

Calculations and software may help us make better designs and measurements as we progress into the digital age. However, it is doubtful that software will replace the need for calibration and testing. With any mechanical measurement, it remains the responsibility of the lab performing the calibration to ask the essential questions, which helps them replicate use and provide meaningful measurements that can reduce measurement risk. When someone correctly implements the force measurement information in this chapter, they help make the world safer for all of us.

Chapter 36
Dimensional and Mechanical Parameters

Between 1889 and 1960, the definition of the meter was based on the international prototype of platinum-iridium, the so-called *archive meter*. In 1960, the 11th General Conference on Weights and Measures (CGPM) replaced the old definition with a new one based on the wavelength of the krypton-86 atomic radiation. In 1983, at the 17th CGPM, in order to satisfy the industry's growing demand, an increase in the precision of materialization of the meter was considered a necessity; therefore, a new definition of the meter was adopted. "[T]he meter is the length of the path traveled by light in vacuum during a time interval of 1/299 792 458 of a second."[1]

Length, surface, and angle do not differ only from the quality point of view; they differ quantitatively as well. They could be smaller or larger, and we can ascertain how small or how large they are. In other words, we can *measure* them. Therefore, we can say that these geometrical quantities are part of the greater category of physical quantities.

From the metrological point of view, geometrical quantities could be classified, with respect to their dimension, as follows:

- Length (dimension L)
- Plane surface (dimension L^2)
- Plane angle (dimension L^0)
- Geometric dimensioning and tolerances (dimension L or L^0)
- Convergence of optical systems (dimension L^{-1})

This classification does not include volume (dimension L^3), which constitutes a separate branch in metrology.

Conversion Factors

1 inch = 0.0254 m = 25.4 mm

1 foot = 0.3048 m = 304.8 mm

1 yard = 0.9144 m = 914.4 mm

LENGTH MEASURES

The length measures category may be subdivided into categories.

Length measures:
- Graduated measures (graduated lines, tapes, and so on)
- End measures (slide calipers, gage blocks)
- Combined measures (rules, tapes)
- Radiation measures (Kr 86 radiation)

Angular measures:
- Graduated angular measures (protractor)
- End-angular measures (angular gage blocks, calipers, polygons)

Plane measures:
- Plan meters
- Area measuring machines

Geometric dimensioning and tolerances measures:
- Straightness measures (knife-edge rules)
- Flatness measures (surface plates, autocollimators)
- Roughness measures (specimen blocks, profilometers)

Length-measuring instruments of general designation for measuring:
- Length
- Angle
- Surface
- Geometric dimensioning
- Convergence of optical systems

Length-measuring instruments of special designation for measuring:
- Gears
- Threads
- Conicity
- Coating thickness
- Wires
- Trees (dendrometry or hypsometry)
- Human body (anthropometry)

Choosing the design principle as a criterion, the length-measuring instruments may be classified as:

- Mechanical
- Optical
- Combined optical and mechanical
- Pneumatic
- Hydraulic
- Interferential
- Electronic
- Magnetic
- Photoelectric
- Ultrasound
- Laser
- X-ray
- Nuclear radiation

ANGLE MEASUREMENT—DEFINITIONS

Angle (geometric). The figure formed by two straight lines (sides) emanating from one point (vertex), or by two or more planes.

Slope (m). The inclination of a straight line versus the horizontal plane. The slope could be evaluated through trigonometric tangent of the angle. Note: The slope is usually expressed in mm/m (inch/foot) for machinist levels, or in percentage for road construction. For example, 15 % slope corresponds to a level difference of 15 feet for 100 feet road horizontal projection.

Conicity. The angular size expressed by the ratio between the difference of two diameters, reduced to unit, and the axial distance between those diameters.

Angle Units of Measure

Radian (rad). The ratio between the length of the arc subtended by the central angle of a circle and the radius of that circle. The radian is the supplementary unit of measure for a plane angle in the International System of Units (SI). Considering that the plane angle is expressed as a ratio between two lengths, the supplementary unit, *radian*, is a dimensionless derived unit. The right angle has p/2 radians.

Sexagesimal degree. The sexagesimal system divides the circle into 360 degrees (°); each degree is divided into 60 minutes of arc ('), and each minute is divided into 60 seconds of arc ("). The right angle has 90°.

Centesimal system. The centesimal system divides the circle into 400 grades (g); each grade is divided into 100 minutes ('), and each minute is divided into 100 seconds ("). A right angle has 100g.

Millième or mil system. This system uses as an angular unit a thousandth of a radian. The length of a circle is 2B radians, or 6.283185 radians, or 6283 milliradians.

The practical mil system. This system uses as an angular unit the radian divided, for practical purposes, into 6400 equal portions.

Rimailho millième. This system uses as an angular unit the radian divided exactly into 6000 portions.

The conversion factors among these angular units are given in Table 36.1.

Angle Measurement—Instrumentation

Line graduated measures. Graduated discs of circular shape, semicircular, or a quarter of a circle with divisions of 1° or 1g. A vernier scale is often used for finer divisions. This type of measure is used for educational purposes, in drafting, cartography, or inside optical instruments such as rotary tables, dividing heads, theodolites, goniometers, and so on.

Table 36.1 Conversion factors for plane angle units.

Units	(rad)	(L)	(°)	(')	(")	(g)	(m)	(mR)
(rad)	1	2/p	180/p	10800/p	648000/p	200/p	3200/p	3000/p
(L)	p/2	1	90	5400	324000	100	1600	1500
(°)	p/180	1/90	1	60	3600	10/9	160/9	50/3
(')	p/10800	1/5400	1/60	1	60	1/54	8/27	5/18
(")	p/648000	1/324000	1/3600	1/60	1	1/3240	2/405	1/216
(g)	p/200	1/100	9/10	54	3240	1	16	15
(m)	p/3200	1/1600	9/160	27/8	405/2	1/16	1	15/16
(mR)	p/3000	1/1500	3/50	18/5	216	1/15	16/15	1

Where:

rad = radian

L = right angle (quadrant)

(°) = sexagesimal degree

(') = sexagesimal minute

(") = sexagesimal second

(g) = centesimal grade

(m) = the practical mil unit

(m_R) = the Rimailho millième

End measures. Angle gages usually employ integer values of angles. They are used in mechanical workshops mostly for the verification of cutting-tool angles. The angle gages are made of tool or hardened steel, or have the active surfaces plated with metallic carbide. The angle gage accuracy is verified either with fixed taper gages or other angle measurement instruments.

Angle squares. Angle squares are materializing unique plane angle values, usually 90° (although there are angle squares of 30°, 45°, 60°, or 120°). Angle squares are used along with machine tools for verification of external or internal angles, or for angular tracing (marking). They are made of tool steel entirely or of regular steel that has had the active measuring surfaces hardened. The measuring surfaces' flatness error is less than 0.01 mm (0.0004 in).

Angle squares are classified as a function of the longer side length, in four classes of accuracy, in respect to the perpendicularity error. The permissible errors are given by the following formulas:

$$E_1 = \pm \left[0.002 + \frac{h}{100} \right]$$

$$E_2 = \pm \left[0.002 + \frac{h}{100} \right]$$

where h (expressed in meters), is the length of the longer side of the square.

The angle accuracy for a right-angle square is verified by the two-square method and the three-square method. Both methods compare the squares to be verified between them versus a standard right-angle square. Also, an angle tester instrument may be employed for the same purpose.

Taper gages. Taper gages are fixed angular measures usually fabricated for standard dimensions. The most utilized systems in taper gage are the metric (1:20 taper) and the Morse (0.625 = 5/8 inch per foot). Taper gages are used either for the verification of internal conicity, in which case they are called *taper plugs*; or for the verification of external conicity, in which case they are called *sleeve gages*. In both cases, taper gages are made for the verification of a specific reference diameter.

Angle gage blocks. These blocks are the most common angular measuring device. In fact, angle gage blocks are end measures for a plane angle, having the shape of a rectangular prism. They are usually made of tempered steel.

The main components of an angle gage block are:

- Measuring surfaces, producing the active angle
- Nonmeasuring surfaces

The angle gage block may employ one active angle or more. The size of the active angle of an angle gage block may vary between 10″ (seconds of arc) and 100° (degrees of arc) combined in sets having different configuration depending on manufacturer and destination. For angles bigger than 100°, the angle gage blocks are combined suitably using special accessories.

Angle gage blocks are classified by one of the principal U.S. manufacturers in three classes of accuracy:

- Laboratory master, with an accuracy grade of ± ¼ second of arc
- Inspection master, with an accuracy grade of ± ½ second of arc
- Tool room, with an accuracy grade of ± 1 second of arc

In the metric system there are also three classes of accuracy:

- Class 0, with an admissible angular deviation of ± 3 seconds of arc
- Class 1, with an admissible angular deviation of ± 10 seconds of arc
- Class 2, with an admissible angular deviation of ± 30 seconds of arc

The number of angle gage blocks in a set varies from six to 16 pieces. The calibration of angle gage blocks is accomplished directly by using standard goniometers or by comparison with other standard angle measures with the assistance of an autocollimator.

Angle gage blocks have multiple applications, such as:

- Direct measuring of angles or angular variation
- Calibration of angular instrumentation (mechanical and optical protractor, etc.)
- Inspecting rotary tables and dividing heads

Polygons. Polygons are angle end measuring devices made of metal or optical glass (quartz) having three to 72 active surfaces; therefore, their angles range in magnitude between 120° and 5°. The polygons are standard measures of angles used for verification and calibration of angle measurement instrumentation. They are manufactured in two classes: reference class, with an accuracy grade of ± 0.25 second of arc, and calibration class, with an accuracy grade of ± 0.5 second of arc. The maximum allowable error of calibration varies between 0.05 second and 3 seconds of arc.

Complex Angle Measurement Instrumentation

Mechanical protractor. This is one of the most common angle measuring devices in a workshop. The main components of a mechanical protractor are a semicircular sector, divided in 180° with a resolution of 1°, and a mobile ruler provided with an indicator.

Mechanical protractor with vernier. Has a design similar to that of the basic mechanical protractor, although the resolution is improved by the vernier to 5 minutes of arc. The vernier is shaped as a circular sector divided in 12 even intervals on either side of the "0" division. The value of division of the vernier is given by:

$$\frac{1°}{12'} = \frac{60°}{12'} = 5$$

Optical protractor. This is very similar to the mechanical protractor with vernier using an optical system for reading. The smallest value of division is 5 minutes of arc.

Sine bars and plates. Both sine bars and sine plates are based on the trigonometric function of sine:

$$\sin\alpha = \frac{h}{L}$$

where h is the gage block height and L is the fixed length between the two axes of the rolls.

Sine plates are similar to sine bars but have a bigger reference surface. The angle generated by the sine bars and plates is from 0 to 60°. The distance between the two rolls varies from 100 mm to 200 mm (5 in to 10 in). The reference surface is up to 25 × 220 mm (1 × 10.75 in) for sine bars and up to 500 × 1200 mm (12 × 24 in) for sine plates.

The angle accuracy depends on the accuracy of the distance between the axes of the two rolls, the diameter of the rolls, and the combined height of the gage blocks. The achieved accuracy is within seconds of arc.

Tangent bar. The tangent bar has also two rolls, but of different diameters. Between the two rolls the distance L is determined by gage blocks.

The desired angle is given by the formula

$$tg\frac{\alpha}{2} = \frac{D-d}{D+d+2L}$$

The obtained angle accuracy is within seconds of arc and depends on the accuracy of the two rolls, gage blocks, and the parallelism between the top and bottom surfaces of the bar.

LEVELS

Levels with vials and bubble. This type of level, invented in 1666 by Thévenot, is mostly used for determining small angles of deviation from horizontal and vertical direction. This type of level has a vial with a curvature given by the radius R. The curvature is obtained by lapping, for precision levels, or by bending the glass tubing, for lower precision levels. The inside liquid utilized is ethanol (ethyl alcohol) with a bubble of the respective liquid. The vial is graduated outside within bubble motion range.

The graduations are equidistant.

$$\sin\alpha = \frac{d}{radian}$$

If we consider the level inclination a, the bubble moves with interval d. The corresponding inclination angle will be:

$$\sin\alpha = \frac{1}{1000}\frac{d}{radian} mm/m (inch/foot)$$

Customarily, the angle is expressed by mm/m or inch/foot.

The value of division could vary from 0.01 mm/m to 20 mm/m or higher, in the metric system, or from 0.0005 in/ft to 1 in/ft or higher, in the U.S. customary system.

There are few distinct designs of vial levels differentiated by shape and scope:

- **Cylindrical level.** The vial is mounted inside a cylindrical tube. It is used for verifying the horizontality of V-shaped guides (grooves).
- **Precision block level.** The bottom is the active surface, with a good flatness, also V-shaped for checking cylindrical surfaces.
- **Block level with micrometer.** Inside the block, the vial has an articulation on one end, the other end being adjusted with a micrometer.
- **Frame level.** Frame levels are equipped with cross-vials. The V shape accommodates cylindrical surfaces.
- **Level with microscope.** This is a combination between a level and an optical protractor.
- **Level with coincidence.** This type of level covers a range between ± 10 mm/m, in the metric system, or ± 1/8 in/ft in the U.S. customary system. The value of division achieved is 0.01 mm/m (2 seconds of arc).
- **Levels with flexible tubing.** This type of level employs the principle of communicating vessels. The main components are two graduated vessels with a bottom flange as a reference surface. Flexible tubing connects the two vessels. These levels are mainly used in construction or big workshops for mounting components at the same level.

Electronic levels. In an electronic level, an inductive transducer determines the angular position of a pendulum versus the case level. The value of division varies between 0.05 mm/m and 0.01 mm/m, with a corresponding range of ± 0.75 mm/m and ± 0.15 mm/m. The electronic circuits are battery powered. In the U.S. customary system of measurement, the value of division varies from 5×10^{-6} in/in to 1×10^{-4} in/in with a corresponding range of 1×10^{-4} in/in to 2×10^{-3} in/in. The high precision makes electronic levels an indispensable tool in the aircraft industry and in thermal, hydro, or nuclear power plants.

Autocollimators. Autocollimators are optical instruments for measuring small angles, such as the deviation from rectilinear, flatness; for alignment of different components; and for calibrating angular polygons and angle gage blocks. The value of division for an autocollimator varies between 0.1 second of arc and 10 seconds of arc for a corresponding range of 5 minutes of arc to 30 minutes of arc. Autocollimators are calibrated with a standard plane angle generator, using the tangent bar principle.

Optical dividing heads. Optical dividing heads are manufactured for a range of 360° and a value of division varying from 1 second of arc to 60 seconds of arc. The calibration of optical dividing heads is performed with standard polygons and autocollimators.

Optical rotary tables. Optical rotary tables are similar to optical dividing heads, although the tables have T-grooves for easier mounting of parts and may have a value of division of 0.5 second of arc.

Electronic encoders. Electronic encoders are not measuring devices, although they are used in conjunction with appropriate devices, and can be used for measuring and recording circular divisions. Electronic encoders use circular capacity transducers and may have an accuracy of 1 second of arc.

Theodolites. These instruments are capable of measuring angles in a horizontal or vertical plane. The value of division varies from 1 second of arc to 1 minute of arc. By using the measured angle and trigonometry, locations and elevations can be determined to a very high accuracy level.

Goniometers. Goniometers are instruments capable of measuring angles in a horizontal plane. They are also used in topography along with a leveling staff.

Both theodolites and goniometers are operated mounted on a strong tripod.

Chapter 37

Other Parameters: Electro-Optical and Radiation

As explained in Chapter 31, this part can only supplement the information found in greater detail in other documents, books, reference manuals, and so on. With this chapter's focus being general in nature, the hope is to explain some of the areas that most are unfamiliar to or have limited usage in the metrology community, or to give knowledge to help the reader understand areas which are focused in nature, and sometimes not even considered when referring to calibration or inspection, measurement, and test equipment (IM&TE). These areas should be considered, though. For example, chemistry has been around as long as any measurement. See Chapter 38, "Chemical and Biological Measurements and Uncertainties," for more explanation of chemical (and biological) measurements.

We will begin by discussing optics. The mere mention of optics brings a myriad of ideas to mind, including deep space exploration, the Hubble telescope, and the local surveying crew with their pole and transit.

OPTICS

The first subject discussed is the history behind optics, and how their importance continues into our future. Let us begin with some basics. Figures 37.1–37.5 show optical principles.

Applications of Optics

In many places, real estate boundaries might be defined by "metes and bounds," a system in which the property boundaries were identified by the area's natural landmarks, such as rivers; or by human-made structures, such as roads; or by

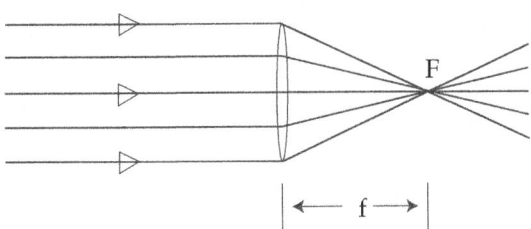

Figure 37.1 Converging (positive) lens. Convex lenses are those that are wider in the center than they are on the top and bottom. They are used to converge light.

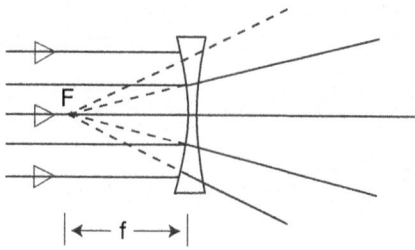

Figure 37.2 Diverging (negative) lens. Concave lenses are those that are wider at the top and bottom and narrower in the center. They are used to diverge light.

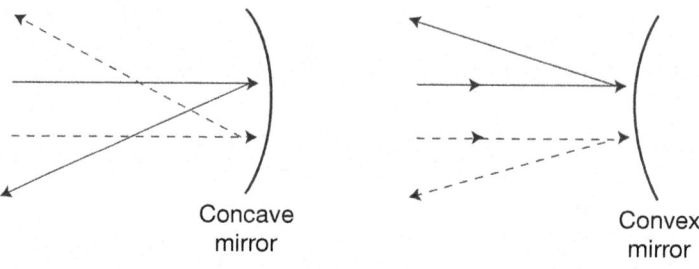

Figure 37.3 Concave mirrors reflect light inward, and convex mirrors reflect light outward.

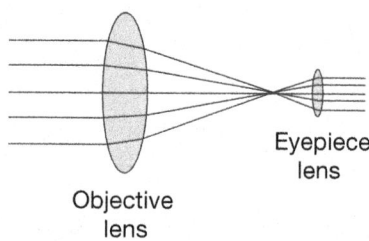

Figure 37.4 Refracting telescopes use lenses to focus light.

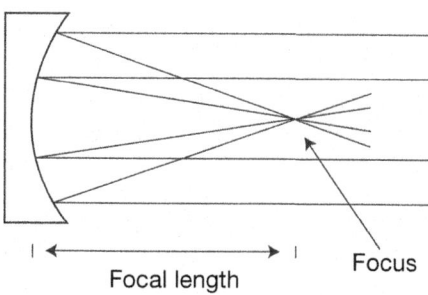

Figure 37.5 Reflecting telescopes use mirrors to focus light.

stakes or other markers. According to www.Britannica.com, the "land boundaries are run out by courses and distances, and monuments, natural or artificial, are fixed at the corners, or angles. A course is the direction of a line, usually with respect to a meridian but sometimes with respect to the magnetic north. Distance is the length of a course measured in some well-known unit, such as feet or chains."[1] www.Britannica.com continues by providing an example of a metes-and-bounds description of a triangular-shaped tract of land, as follows: "Beginning at a point from which the north quarter corner of Section 4, T. 1 N (township 1 north), R. 70 W (range 70 west) of the 6th PM (the sixth principal meridian, a north–south reference line) in Boulder County, Colorado, bears N 45° W 1,320 feet, at which point of beginning an iron stake has been placed; thence south 600 feet to a point also marked by an iron stake; thence N 45° W 700 feet to a large oak tree; thence northeasterly to the point of beginning."[2] One can see how angles and distances are critical to all parties having the same definition of land property boundaries.

The use of optics in surveying these boundaries is one of its applications. In the United States, the progression in the use of how metes and bounds were determined is described in the following (excerpted from http://www.surveyhistory.org/changing_chains.htm).

A grand scale. Metes and bounds were adequate in the original 13 colonies, Kentucky, Tennessee, and parts of other eastern states. But the vast western territories required a simple scheme of grand scale. So, the Congress of a young United States approved the subdivision of public lands into a grid of 36-square-mile townships, and square-mile sections. The corners of each were to be measured from the intersection of north/south meridian and east/west baselines.

To that end, teams of surveyors were dispatched by the U.S. Coast and Geodetic Survey. For the better part of 100 years, they took the measure of the land, using instruments that had been invented in the sixteenth and eighteenth centuries and techniques that evolved as they went along. The results were less than perfect. Today's land surveyors spend much of their time finding and correcting the errors of their forerunners in the interest of clearing title. Of course, today's surveyor has the advantage of soaring technology. Surveying instruments have changed more in the past 10 years than in the previous 200. And those changes will impact the way that property is measured, described, and held.

Little change. The instruments used to survey the USA changed little from the early 1800s well into the twentieth century. An 1813 surveying text notes that, in New England, most work was done with a magnetic compass and a surveyor's chain. The compass, invented in 1511, was in wide use until 1894. The chain was invented in 1620 by Edmund Gunter, an Englishman. It was made of 100 iron or steel links and was 66 feet long. Eighty chains made up one mile. Ten square chains made one acre. Gunter's chain was in universal use until the steel tape measure replaced it in the last decades of the eighteenth century. The transit was first made in 1831 by Philadelphian William J. Young. It was an adaptation of the theodolite invented in 1720 by John Sisson of England. Sisson had combined a telescope (invented circa 1608), a vernier—a device for subdividing measurements by 10ths (1631), and a spirit level (1704) into a single instrument. Young's improvement was to permit the telescope to revolve, or transit, upon its axis—a useful feature when prolonging straight lines or taking repeated readings to confirm accuracy. Improved versions of Young's transit were used for land surveying in the 1950s and are still broadly used in the construction trade.

Inaccuracies. Early surveys were often grossly inaccurate. The iron chains stretched with use. An error of one link (about 8 inches) in 3 to 5 chains was considered normal. The magnetic compass was a major source of error. It was subject to daily, annual, and lunar variations in the Earth's magnetic field, solar magnetic storms, local attractions, and static electricity in the compass glass. Optical glass varied in quality. There were no standards for equipment and many manufacturers. And, of course, no way to recalibrate equipment damaged 100 miles from nowhere.

Survey procedures were often less than precise. If a tree blocked a line of sight, a surveyor might sight to the trunk, walk around it, and approximate the continuing line. One modern text referring to nineteenth-century surveys cautions that "No line more than one-half mile can be regarded as straight." Of course, precision seemed unimportant when the land seemed endless and, at $1.25 an acre, cheap, especially to a surveyor who was paid by the mile. Other factors contributing to inaccuracy included a lack of supervision, a shortage of trained surveyors, an abundance of bears, wolves, wind, rain, snow, burning sun, and rugged terrain.

Technology soars. The technological boom of the past 15 years has greatly increased both the accuracy and precision of land surveys. In the 1860s an error of only 1 foot in 1500 feet was considered highly accurate [*even though it was worse than the dimensional accuracy of the pyramids in Egypt constructed 8000 years ago*—Editor]. By today's standards, an error of 1 foot in 10,000 feet is reasonably accurate, and measurements accurate to 1 foot in 1,000,000 feet are possible. A nineteenth-century compass measurement that came within 60 seconds was acceptably accurate. Modern electronic instruments are accurate to within one second (a second is 1/3600 of a degree).

Until recently, modern surveying was accomplished using manual devices known as *theodolites* along with plumb bobs and both measuring tapes and rods. The theodolite is an optical instrument that works similarly to a set of binoculars in that focusing the lenses will provide an approximate verification of distance. The instrument also typically has an angle function to allow a fairly good reading of the angle, or elevation. More exact distance measurements are obtained by use of the calibrated measuring tape and the rod. The tape can be very long, even into the hundreds of feet or meters. The theodolite is mounted on a tripod and the plumb bob is used to level the instrument.

The most modern versions of the theodolites can include lasers and/or GPS—allowing much more precise measurements to be made.

The most stunning breakthrough of modern technology has been in the measurement of distance. Electronic distance meters (EDMs) have replaced the steel tapes. EDMs operate on the basis of the time it takes a signal to travel from an emitter to a receiver, or to reflect back to the emitter. Short-range EDMs use infrared signals. EDMs designed for distances from 2 to 20 miles use microwaves. "They are accurate to within 3 millimeters on a clear day and adjust for atmospheric haze distortion and curvature of the earth," says Duke Dutch of Hadco Instruments, a major southern California distributor of survey equipment. Surveyor's transits now incorporate digital electronics that read down to one second.

Instead of jotting their calculations and notes in a field book, today's surveyors plug an electronic field book into their electronic transit. The electronic field book is a magnetic tape recorder with a digital display and keyboard. It automatically records each observation made by the transit.

Before the electronics boom, surveyors used a bulky plane table to manually plot maps in the field. Today's surveyors can connect their electronic field book to a computer. Special coordinate geometry software speeds the process of checking observations and, through an interface with a plotter, making maps. During the past several years, the EDM, electronic transit, and electronic field book have been combined into a single unit called a "total station."

> **A coming change?** The almost microscopic accuracy of electronic surveying equipment, when combined with advances in astronomy and satellite technology, may soon change the way property is described. It may even stimulate basic changes in title law, according to Paul Cuomo, section chief with the Orange County, California, Surveyor's Office, and treasurer of the California Land Surveyors' Association.
>
> Cuomo foresees a day when boundaries will be precise and permanent—with or without monuments. "All it would take is to tie all future surveys to the State Plane Coordinates System," he says. That system is a nationwide grid of survey stations established over the past 40 years. Each station's location was precisely plotted astronomically. Each is in sight of others, so triangulations are convenient. "Any survey of record tied to this system could be recreated on the ground, exactly, forever. All that would be required would be a record of the northing and easting," says Cuomo.
>
> In surveyor's parlance, that means a notation of the exact longitude and latitude of at least one corner of the property. That notation would be included in the property description in the deed.
>
> Formerly, such a notation required time-consuming observations and calculations that were prohibitively expensive. The advent of rapid operating, optically superior, and highly precise electronic survey devices changes the picture.
>
> "It will take major changes in the law," says Cuomo. "And we will have to get everyone to agree: the courts, the lawyers, the title companies. But it will happen someday."
>
> And, when it does, the line that someone's ancestors drew from the rock by the river to the distant tree will hold. Forever.[3]

Since the preceding material was written (1986), an even bigger change has come to surveying. Although optical transits may still be used, many survey systems now use GPS receivers, technology that was in its infancy in the mid-1980s. The GPS system allows determination of three-dimensional position (latitude, longitude, and elevation) completely independent of other references. As of 2014, even a GPS-enabled smartphone is accurate within a 4.9-m (16-ft.) radius under open sky.[4] Combined with optical triangulation and electronic distance measuring, this discipline has been taken to a new level.

COLORIMETRY

Colorimetry is the science of the measurement of color; it replaces subjective responses to and descriptions of colors with an objective numerical system. Colorimetry is the measurement of the wavelength and the intensity of electromagnetic radiation in the visible region of the spectrum. It is used extensively for identification and determination of concentrations of substances that absorb light.[5]

Color is often a characteristic of an object by which its quality is judged. This could be fruit at the grocery store, clothes in the department store, or a new car. The appearance of an object is the result of a complex interaction of the light field incident upon the object, the scattering and absorption properties of the object, and human perception. The measurable attributes of appearance are divided into color (hue, saturation, and lightness) and geometry (gloss, haze). The appearance of an object cannot be measured; only the specific reflectance characteristics of the surface are measured.[6] Further definitions and explanations of surface color, specular gloss, special effects coatings, and transmission density can be found at the NIST website (https://www.nist.gov/programs-projects/color-and-appearance).[7]

IONIZING RADIATION

Ionizing radiation sources can be found in a wide range of occupational settings, including healthcare facilities, research institutions, nuclear reactors and their support facilities, nuclear weapon production facilities, and other various manufacturing settings, to name just a few. These radiation sources can pose a considerable health risk to affected workers if not properly controlled. This chapter provides a starting point for technical and regulatory information regarding the recognition, evaluation, and control of occupational health hazards associated with ionizing radiation.

Ionizing radiation is radiation that has sufficient energy to remove electrons from atoms. In this chapter, it will be referred to simply as *radiation*. One source of radiation is the nuclei of unstable atoms. For these radioactive atoms (also referred to as radionuclides or radioisotopes) to become more stable, the nuclei eject or emit subatomic particles and high-energy photons (gamma rays). This process is known as *radioactive decay*. Unstable isotopes of radium, radon, uranium, and thorium (among others) exist naturally. Others are continually being made naturally or by human activities such as the splitting of atoms in a nuclear reactor. Either way, they release ionizing radiation. The major types of radiation emitted as a result of spontaneous decay are alpha and beta particles, and gamma rays. X-rays, another major type of radiation, arise from processes outside of the nucleus.

Alpha Particles

Alpha particles are energetic, positively charged particles (helium nuclei) that rapidly lose energy when passing through matter. They are commonly emitted in the radioactive decay of the heaviest radioactive elements such as uranium and radium, as well as of some human-made elements. Alpha particles lose energy rapidly in matter and do not penetrate far; however, they can cause damage over their short path through tissue. These particles are usually completely absorbed by the outer dead layer of the human skin, so alpha-emitting radioisotopes are not a hazard outside the body. However, they can be very harmful if they are ingested or inhaled. Alpha particles can be stopped completely by a sheet of paper.

X-rays

X-rays are high-energy photons produced by the interaction of charged particles with matter. X-rays and gamma rays have essentially the same properties, but differ in origin; that is, X-rays are emitted from processes outside the atomic nucleus, whereas gamma rays originate inside the nucleus. X-rays are generally lower in energy and therefore less penetrating than gamma rays. Thousands of X-ray machines are used daily in medicine and industry for examinations, inspections, and process controls. X-rays are also used for cancer therapy to destroy malignant

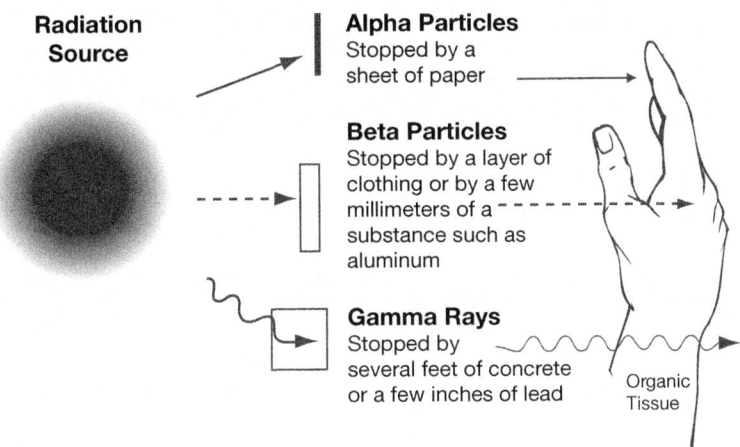

Figure 37.6   Penetrating powers of alpha and beta particles, and gamma rays.

cells. Because of their many uses, X-rays are the single largest source of radiation exposure made by humans. A few millimeters of lead can stop medical X-rays.

Sources of Radiation

Natural radiation. Humans are primarily exposed to natural radiation from the sun, cosmic rays, and naturally occurring radioactive elements found in the Earth's crust. Granite, for example, usually contains small grains of minerals that include radioactive elements. Radon, which emanates from the ground, is another important source of natural radiation. Cosmic rays from space include energetic protons, electrons, gamma rays, and X-rays. The primary radioactive elements found in the Earth's crust are uranium, thorium, and potassium, and their radioactive derivatives. These elements emit alpha and beta particles, or gamma rays.

Human-made radiation. Radiation is used on an ever-increasing scale in medicine, dentistry, and industry. Main users of human-made radiation include medical facilities such as hospitals and pharmaceutical facilities; research and teaching institutions; nuclear reactors and their supporting facilities such as uranium mills and fuel preparation plants; and federal facilities involved in nuclear weapons production as part of their normal operation.

Many of these facilities generate some radioactive waste, and some release a controlled amount of radiation into the environment. Radioactive materials are also used in common consumer products such as digital and luminous-dial wristwatches, ceramic glazes, artificial teeth, and smoke detectors.

Health Effects of Radiation Exposure

Depending on the level of exposure, radiation can pose a health risk. It can adversely affect individuals directly exposed as well as their descendants. Radiation can affect cells of the body, increasing the risk of cancer or harmful genetic mutations that can be passed on to future generations. If the dosage is large enough to cause massive tissue damage, it may lead to death within a few weeks of exposure.

OPTICAL RADIATION

Optical radiation is all around us every day. Generally, we are not harmed by it. Optical radiation is typically measured by wavelength rather than by frequency. For example, the red laser pointer that is often used in meetings has a wavelength of approximately 632.8 nm (nanometers).

Optical radiation is light. Visible light occupies a very small portion of the light spectrum, yet we also are often aware of light effects outside the visible spectrum. For example, we apply sunscreen lotion to cut down on the harmful ultraviolet (UV) rays from the sun. The UV rays are a lower wavelength than visible, typically in the 200 nm to 400 nm range.

Light measurements are performed using devices known as optical detectors and power meters. The detectors have specified wavelength ranges and power ranges. Measuring the output of a home lightbulb would only require a low-power detector coupled to a power meter, operating in the visible range. Measuring the output of a high-powered laser would require a high-power detector and power meter.

Power meters typically use electronic calibration methods, as they are electronic. The detectors require calibration using a certified light source and a system known as a monochromator. The light source can be one of two types: either the pulse type, which is a specific wavelength or very small band; or a continuous wave that has a very wide wavelength range. The continuous-wave light sources for the monochromator system require selectable mirrors inside the monochromator for wavelength selection. The wavelength must be selected in order to plot the calibration response of the detector. Detector response is typically plotted as power at given wavelengths. There are several detector types, based on the material used for what is known as the *active surface*, which is the portion that is actually excited by the light. Each material has a different response range.

A device that is often coupled with a detector for actual use is the integrating sphere. The name is descriptive, although some other shapes are also used. The integrating sphere is used to ensure that all the electrons from the light source are applied to the active surface.

Generally, there are three different negative effects that can occur with optical radiation, either in use or during calibration. These effects are diffusion, reflection, and refraction. *Diffusion* is simply the effect of electrons escaping from the focused beam that is being sent to the receiving device. A simple way to demonstrate the effect is to shine a flashlight on an object where the beam path has a dark background. If you can see the actual beam of light, then the beam is experiencing diffusion. If the beam cannot be seen, and only the impact spot of the light can be seen, then there is no diffusion. *Reflection* is exactly what the term suggests: The beam or some portion thereof is reflected back toward the source. In the microwave realm, this is known as voltage standing wave ratio (VSWR). *Refraction* is the dissemination of some of the electrons at an angle that is not desired. A simple way to demonstrate refraction is by using a half-full glass of water. Put a pencil in the glass and note that part of the pencil is exactly as expected, and part appears to be bent in a different direction.

Sometimes directed and amplified light (often meaning lasers) requires specific safety precautions. This description will address lasers, but any directed and amplified light should be treated with the utmost regard for safety. Lasers are classified according to their potential to cause biological damage. The determining parameters are:

- Laser output energy or power
- Radiation wavelengths
- Exposure duration
- Cross-sectional area of the laser beam at the point of interest[8]

In addition to these general parameters, lasers are classified in accordance with the accessible emission limit (AEL), which is the maximum accessible level of laser radiation permitted within a particular laser class. Safety thresholds for lasers are expressed in terms of maximum permissible exposure (MPE). The higher the classification numbers, the greater potential risk the laser or laser system presents. Two bodies are involved in laser hazard classification in the United States. The Center for Devices and Radiological Health (CDRH) a part of the Food & Drug Administration and The American National Standards Institute Z136.1 Safe Use of Laser Standard. The IEC 60825-series of standards provide international guidance, classification, and requirements for safety of laser products. The classifications are as follows.

Class 1

This class is eye-safe under all operating conditions. A Class 1 laser is safe for use under all reasonably anticipated conditions of use; in other words, it is not expected that the MPE can be exceeded.

Class 1 Product

This is a laser product or device which may include lasers of a higher class whose beams are confined within a suitable enclosure so that access to laser radiation is physically prevented. Such products do not require a laser warning label on the exterior (an example is a laser printer).

Class 1M

This class is safe for viewing directly with the naked eye but may be hazardous to view with the aid of optical instruments. In general, the use of magnifying glasses increases the hazard from a widely diverging beam (e.g., LEDs and bare laser diodes), and binoculars or telescopes increase the hazard from a wide, collimated beam (such as those used in open-beam telecommunications systems).

Class 1M lasers produce large-diameter beams, or beams that are divergent. The MPE for a Class 1M laser cannot normally be exceeded unless focusing or imaging optics are used to narrow down the beam. If the beam is refocused, the hazard of Class 1M lasers may be increased and the product class may be changed.

Class 2

A Class 2 laser emits light in the visible region (400 nm to 700 nm). It is presumed that the natural aversion response to the very bright light will be sufficient to prevent damaging exposure, although prolonged viewing may be dangerous. Eye protection is recommended.

Class 2M

These are also visible lasers, between 400 nm to 700 nm. This class is safe for accidental viewing with the naked eye, as long as the natural aversion response is not overcome (as with Class 2), but may be hazardous (even for accidental viewing) when viewed with the aid of optical instruments, as with class 1M.

Classes 1M and 2M broadly replace the old class 3A under IEC and EN classification. Prior to the 2001 amendment, there were also lasers which were Class 3B but were eye-safe when viewed without optical instruments. These lasers are Class 1M or 2M under the current classification system.

Class 3R (Replacement for Class 3A)

A Class 3R laser is a *continuous wave* (CW) laser, which may produce up to five times the emission limit of Class 1 or Class 2 lasers. Although the MPE can be exceeded, the risk of injury is low. The laser can produce no more than 5 mW in the visible region.

Visible Class 3R is similar to Class IIIA in the U.S. regulations.

Class 3B

A Class 3B laser produces light of intensity such that the MPE for eye exposure may be exceeded and direct viewing of the beam is potentially serious. Diffuse radiation (i.e., that which is scattered from a diffusing surface) should not be hazardous. Continuous wave (CW) emission from such lasers at wavelengths above 315 nm must not exceed 0.5 watts. For a pulse laser system to be Class 3B, its output cannot exceed 125 mJ in less than 0.25 seconds.

Class 4

This is the highest class of laser radiation. These are the most hazardous and the most unsafe to view at any time. Class 4 lasers may cause devastating and permanent eye damage, may have sufficient energy to ignite materials, and may cause significant skin damage. Exposure of the eye or skin to both the direct laser beam and to scattered beams, even those produced by reflection from diffusing surfaces, must be avoided at all times. In addition, they may pose a fire risk and may generate hazardous fumes. Class 4 output levels for CW lasers start at 500 mW, and for pulse systems they *can* produce over 125 mJ in less than 0.25 seconds.[9]

pH

Definition

The pH scale is a method for representing how acidic or basic a substance is. It is widely used in biology, chemistry, geology, and other physical sciences. pH translates the values of the concentration of the hydrogen ion (pH = potential of hydrogen) into numbers on a scale of between 0 and 14. Water has a pH of 7, which is neutral. The lower the number on the scale, the more acidic the substance is. The higher the number on the scale, the more basic, or alkaline, the substance is. Substances that are extremely acidic or basic tend to be corrosive or burn-causing. The scale is logarithmic, meaning that it is based on tens. Thus, a substance with a pH of 4 is 10 times more acidic than a substance with a pH of 5.[10] Because of uncertainty about the physical significance of the hydrogen ion concentration, the definition of the pH is an operational one; that is, it is based on a method of measurement. NIST has defined pH values in terms of the electromotive force existing between certain standard electrodes in specified solutions.[11]

The water molecule. All substances are made up of millions of tiny atoms. These atoms form small groups called *molecules*. In water, for example, each molecule is made up of two hydrogen atoms and one oxygen atom. The formula for a molecule of water is H_2O. "H" means hydrogen, "2" means two hydrogen atoms, and "O" means oxygen.

Acids and bases in water. When an acid is poured into water, it gives up H (hydrogen) to the water. When a base is poured into water, it gives up OH (hydroxide) to the water.[12]

A Bit of History

It all began with food. Food is truly of the essence, and good food is highly valued. In earlier years people used to taste the food to establish the quality of a product. Working for the food industry could be a hard job. Little did they know at the time that the pH of food/drinks can often yield information about its state, such as whether fruit is fresh or not, or whether wine will taste sweet or bitter. Lemon juice is acidic. Bleach is alkaline.

Potentiometry was developed, which enabled people to directly measure the pH instead of tasting. Not only was this the birth of proper quality control in the food industry, but it also, in conjunction with further developments, allowed other industries to grow.

Modern pH measurement

The method: Potentiometry. In pH, potentiometry measures (via meter) the voltage (potential) caused by hydrogen ion activity in solutions. This method gives more accurate, reliable, and faster results than the tasting ability of any human being. It also saved some people from early death and gave the rest of the chemical industry a chance to prosper. Finally, scientists all over the world could measure things that were previously unmeasurable.

Tools for measuring pH. As mentioned, potentiometry is a measurement of voltage. The tools used for this are: a pH meter (to accurately measure and transform the voltage caused by our hydronium ion into a pH value); a pH electrode (to sense all the hydrogen ions and to produce a potential); and a reference electrode (to give a constant potential no matter the concentration of the hydrogen ion).

The pH meter

A pH meter measures the potential between the pH electrode (which is sensitive to the hydrogen ions) and the reference electrode (which does not care what's in the solution).

The pH electrode

A glass pH probe (the most common type) contains two electrodes: a sensor electrode and a reference electrode. When placed into a solution to measure pH, the probe measures the difference in voltage potentials between the two electrodes. The pH electrode's potential changes with the hydrogen ion concentration in the solution. The clever bit is that the pH electrode only senses the hydrogen ions. This means that any voltage produced is from hydrogen ions only. This way we can relate the potential directly to the hydrogen concentration.

The reference electrode

The reference electrode supplies a "constant" value against which we measure the potential of the pH electrode. That is the curious thing about potentials: They have to be in pairs to produce a voltage.

The classical setup

The classical setup for measuring pH consisted of a pH meter, a pH electrode, and a separate reference electrode.

The modern setup

Although one could easily measure pH using the "classical" setup, it was soon realized that the two electrodes could be built into the same probe (although the electrodes are still completely separated inside the same probe). This is now called the "combined" pH electrode, which is, of course, much more practical.

Calibration of pH Instruments

Calibration of pH meters depends on their construction. In many cases, the meter and probe are separate items, and the meter can use interchangeable probes. The meter can be calibrated electrically because it is basically a millivolt meter. The probe itself—and the entire instrument if the probe and meter are a single unit—must be standardized using pH standards before use.

A pH measuring system is a classic example of a measuring system that must be standardized before every use, even if it has a separate calibrated meter. To calibrate a pH measuring system, the operator needs at least two (preferably three) pH buffer solutions and some distilled water, all at the same temperature. The buffer solutions are liquids that are known to have a specific pH because of their chemical composition. These may be purchased commercially in sealed single-use packages or as larger containers of buffer solutions. If using a larger container of buffer solution, dispense the buffer to be used into a small, clean, and dry container. After use, DO NOT return the buffer to the original container, but dispose of it according to manufacturer instructions and safety protocols!

Calibrating with pH buffer solutions. Follow the instructions of the pH meter to determine how many buffers to use for calibration. One of the buffers will have a pH of ~7.0, and the other(s) a pH toward the 0 and 14 ends of the pH scale.

The process is:

1. Confirm that the stock buffer solutions are still within their period of validity.

2. Dispense the buffer solutions into smaller, clean, and dry inert containers that can fit the pH probe.

3. Measure the temperature of each of the dispensed pH buffers to confirm that they are all at the same temperature (ideally ~25 °C). Some models of pH meters may have automatic or manual temperature compensations. Follow the manufacturer's instructions for compensating for temperature.

4. Thoroughly rinse the probe in distilled or deionized water and blot dry with Kimwipes™ or other equivalent tissue before proceeding to the next solution. By blotting the probe dry, the pH buffer solution will not be diluted by the distilled or deionized water.

5. Place the probe in the 7.0 buffer, completely immersing the electrode tip into the buffer solution. Allow time for the reading to settle and then record the reading.

6. Remove the probe from the 7.0 buffer and thoroughly rinse the probe in distilled or deionized water and blot dry with Kimwipes™ or other equivalent tissue before proceeding to the next solution.

7. Place the probe in the other buffer, completely immersing the electrode tip into the buffer solution. Allow the reading to settle and then record the reading.

8. Remove the probe from the other buffer, thoroughly rinse the probe in distilled or deionized water, and return the probe to its storage container.

Dispensed buffer solutions should be discarded after use, as carbon dioxide from the air dissolves into the solution and changes the pH over time.

Two notes about pH probes should be taken to heart. First, they have a limited lifetime and thus should be considered a consumable item. Second, the design of some probes requires that they be stored wet; follow the manufacturer's directions for the solution the probe should be stored in and for how long. Other designs may be stored dry. Be sure to read the instructions.

Chapter 38

Chemical and Biological Measurements and Uncertainties

INTRODUCTION

This handbook is intended as a practical reference for metrology (and calibration) professionals as well as for other types of professionals who deal with measurements. The purpose of this chapter is to identify when and how metrological principles and calibration are used and how they affect the work of those testing and measurement laboratories that carry out chemical and some biological measurements. This chapter refers to tests based on chemical properties as *chemical tests* and refers to tests based on biological properties as *biological tests*; this nomenclature is also applied to those testing and measurement laboratories conducting these types of measurements.

It is helpful to consider that tests can be based on physical properties, such as melting point, which uses temperature, or based on chemical properties, such as an ELISA (enzyme-linked immunosorbent assay), which uses the interaction of molecules in the immune response. As an example, ELISA can detect and measure antibodies in blood to determine if one has antibodies related to certain infectious conditions. Metrological principles have been applied to physical properties and have been well-characterized. For instance, the melting-point experiment uses equipment that is well-defined, carefully controlled, and calibrated. However, for tests based on chemical properties this has not always been the case. This is changing, and metrological concepts such as measurement uncertainty and traceability are being applied. In particular, ISO/IEC 17025:2017[1] and the *United States Pharmacopeia (USP)*, General Chapter <1220>, Analytical Procedures Life Cycle,[2] incorporate measurement uncertainty and decision rules. The International Union of Pure and Applied Chemistry (IUPAC) provides standardized methods for measurement and atomic weights.[3] In addition, Eurachem, which is a network of organizations in Europe, has the objective of establishing a system for the international traceability of chemical measurements and the promotion of good quality practices.[4] Eurachem publishes many guidance documents, such as *Quantifying Uncertainty in Analytical Measurement, 3rd Edition* (2012)[5] and *Metrological Traceability in Chemical Measurement, 2nd Edition* (2019).[6]

Chemical tests often use measuring and test equipment (MT&E). In the chemical labs, these types of equipment are often referred to as *analytical instruments* or *instruments*. The physical properties of the instruments, such as wavelength or voltage, are supported by metrologically traceable MT&E.

In addition, these analytical instruments are carefully calibrated using chemical reference materials. The use and availability of certified reference materials (CRM) is increasing, and their associated measurement uncertainty is becoming more frequently evaluated and reported with better confidence. However, the chemical properties of many of these tests may not be metrologically well-supported or well-defined. This presents challenges. Careful attention is given to how the tests are performed following guidelines such as the Organisation for Economic Co-operation and Development's (OECD) *Good Laboratory Practice*,[7] to help ensure repeatability and reproducibility. However, the reference materials and their values are often not well-defined or well-characterized. The analytical laboratories rely on judgment, experience, consensus among analysts and laboratories, and taking special care when selecting and using the references. In addition, different types of tests have different purposes. Some tests confirm identity of a substance. Some tests change the analyte or its quantity value as it is being tested. Some tests are operationally defined and must be conducted following a specific process and strict instructions with set parameters. These challenges make it a complex task to identify the requirements for traceability, to demonstrate adequate traceability, and to evaluate the uncertainty. Because best practices, such as handling chemicals properly to avoid contamination, are generally understood and the quality systems described elsewhere in this book also apply to chemical measurement laboratories, this chapter focuses on how testing laboratories are approaching metrological (chemical) traceability and measurement uncertainty. It is worthwhile to note that the approach to metrological traceability and evaluating measurement uncertainty for chemical tests is similar to that for physical measurements.

Eurachem is "a network of organizations in Europe having the objective of establishing a system for the international traceability of chemical measurements and the promotion of good quality practices."[8] Eurachem provides guides for metrology in analytical chemistry, including traceability and measurement uncertainty. These guides are invaluable for the analytical chemistry community. Many of the guides are referenced in this chapter and can be consulted for detailed discussions and guidance. In celebration of the twentieth anniversary of the publication of the JCGM 100:2008 *Guide to the Expression of Uncertainty in Measurement* (*GUM*), a special issue of *Metrologia* was published by BIPM. It includes the paper, "Implementing measurement uncertainty for analytical chemistry: the Eurachem Guide for measurement uncertainty" by Stephen L R Ellison,[9] which provides a history of the development of uncertainty concepts in chemistry. It reviews the early use of spreadsheet methods, the incorporation of method validation, guidance for uncertainties near zero, and Monte Carlo methods. We invite you to study this paper for a useful and interesting history of measurement uncertainty in analytical measurements.

ORGANIZATIONS

There are international organizations providing oversight and guidance for measurements and uncertainty. (The bibliography of this handbook has a list of some sources cited in this chapter. Also, Chapter 29 of this handbook is dedicated solely to uncertainty of measurement.) Some of these international organizations include:

- BIPM: International Bureau of Weights and Measures (https://ww.bipm.org)
    - o JCTLM: Joint Committee on Traceability in Laboratory Medicine (https://www.jctlm.org)
    - o JCGM: Joint Committee for Guides in Metrology (https://www.bipmorg/en/committees/jc)
- ISO: International Standards Organization (https://iso.org)
- ILAC: International Laboratory Accreditation Cooperation (www.ilac.org)
- CITAC: Cooperation on International Traceability in Analytical Chemistry (http://www.citac.cc)
- IUPAC: International Union of Pure and Applied Chemistry (www.iupac.org)
- EURACHEM: Analytical Chemistry in Europe (https://eurachem.org)
- AOAC: Association of Official Analytical Collaboration (https://www.aoac.org)
- NORDTEST (https://www.nordtest.info)
- United States Pharmacopeia (USP) (https://usp.org)
- European Pharmacopoeia/Pharmeuropa (EDQM) (https://www.edqm.eu/en)
- ASTM International (https://www.astm.org)

TRACEABILITY FUNDAMENTALS

Chemical tests rely on the SI (International System of Units) base unit, the mole. Other measurement units may be required; however, the approach in this chapter is to express results in the SI base unit of the mole. Biological tests are not directly traceable to SI units; however, studies are being conducted to develop traceability based on criteria such as characteristics, methods, and techniques.

Method development for a chemical test provides an analytical procedure to estimate the value of the measurand as well as an equation to calculate the measurement result. This equation identifies the relations with other measured quantities. The capability of the chemical test is defined. The conditions under which the test must be conducted are defined.

Method validation is then performed to demonstrate that the analytical procedure, the equation, and the set of conditions are adequate and fit for the intended purpose. This process is defined in detail in *USP* General Chapter <1220>, "Analytical Procedure Lifecycle."[10]

When the analytical procedure is used to produce the measured quantity, that measured quantity and the specified conditions can be related to appropriate standards. This is achieved by calibration for the critical quantities in the measurement. Control of less critical values may be less strict. A *critical quantity* is one that has a large effect on the result, as discussed later in this chapter.

Before describing the process to establish traceability, we will discuss some aspects that chemical tests utilize.

SI Base Unit, the Mole

The mole is defined by the General Conference on Weights and Measures (CGPM):

> The mole, symbol *mol*, is the SI unit of amount of substance. One mole contains exactly 6.022 140 76 × 10 23 elementary entities. This number is the fixed numerical value of the Avogadro constant, N_A, when expressed in mol–1, and is called the Avogadro number. The amount of substance, symbol n, of a system is a measure of the number of specified elementary entities. An elementary entity may be an atom, a molecule, an ion, an electron, or any other particle or specified group of particles.[11]

Note that the mole is the only SI base unit that requires further qualification when it is used; the entity must be stated. References such as the Eurachem/CITAC *Guide for Metrological Traceability in Chemical Measurement*[12] provide lists of typical and often-used derived units for chemical measurements. In chemistry, quantities such as mass fraction may not be "dimensionless" because they commonly refer to the fraction of one substance as part of a mixture of other substances. This means that traceability may be required for each measurement result or substance.

In chemistry, the periodic table of elements is a visual representation of each chemical element. Responsibility for approving or rejecting new elements lies with two sister organizations: the International Union of Pure and Applied Chemistry (IUPAC) and the International Union of Pure and Applied Physics (IUPAP). The periodic table of the elements, current as of November 2021, is shown in Figure 38.1.

ESTABLISHING TRACEABILITY FOR CHEMICAL TESTS

In many ways, the process to establish traceability for chemical tests is the same as that for physical tests. The discussion in this chapter will present the general process and emphasize the unique aspects for a chemical test. A detailed guide for establishing traceability is provided in the Eurachem/CITAC *Guide for Metrological Traceability in Chemical Measurement*.[13] Chemical tests often use physical quantities such as time, mass, and volume. These are well understood and adequately (often more than adequately) controlled. Historically, extreme control and care were used for these physical quantities, especially with classical analyses that relied on gravimetric and titrimetric procedures, endowing chemical tests with a legacy of excellent understanding and control of their physical quantities. With the introduction of new methods, such as instrumental and biological techniques, there is no such established discipline, so care must be taken to ensure traceability for the chemical aspects *of a method*.

Traceability requires the identification and assessment of influence quantities. An *influence quantity*, as defined in *VIM* 2.52, is:

502 Part VI: Measurement Parameters

IUPAC Periodic Table of the Elements

1	2											13	14	15	16	17	18
1 H hydrogen 1.008 [1.0078, 1.0082]																	2 He helium 4.0026
3 Li lithium 6.94 [6.938, 6.997]	4 Be beryllium 9.0122											5 B boron 10.81 [10.806, 10.821]	6 C carbon 12.011 [12.009, 12.012]	7 N nitrogen 14.007 [14.006, 14.008]	8 O oxygen 15.999 [15.999, 16.000]	9 F fluorine 18.998	10 Ne neon 20.180
11 Na sodium 22.990	12 Mg magnesium 24.305 [24.304, 24.307]	3	4	5	6	7	8	9	10	11	12	13 Al aluminium 26.982	14 Si silicon 28.085 [28.084, 28.086]	15 P phosphorus 30.974	16 S sulfur 32.06 [32.059, 32.076]	17 Cl chlorine 35.45 [35.446, 35.457]	18 Ar argon 39.95 [39.792, 39.963]
19 K potassium 39.098	20 Ca calcium 40.078(4)	21 Sc scandium 44.956	22 Ti titanium 47.867	23 V vanadium 50.942	24 Cr chromium 51.996	25 Mn manganese 54.938	26 Fe iron 55.845(2)	27 Co cobalt 58.933	28 Ni nickel 58.693	29 Cu copper 63.546(3)	30 Zn zinc 65.38(2)	31 Ga gallium 69.723	32 Ge germanium 72.630(8)	33 As arsenic 74.922	34 Se selenium 78.971(8)	35 Br bromine 79.904 [79.901, 79.907]	36 Kr krypton 83.798(2)
37 Rb rubidium 85.468	38 Sr strontium 87.62	39 Y yttrium 88.906	40 Zr zirconium 91.224(2)	41 Nb niobium 92.906	42 Mo molybdenum 95.95	43 Tc technetium	44 Ru ruthenium 101.07(2)	45 Rh rhodium 102.91	46 Pd palladium 106.42	47 Ag silver 107.87	48 Cd cadmium 112.41	49 In indium 114.82	50 Sn tin 118.71	51 Sb antimony 121.76	52 Te tellurium 127.60(3)	53 I iodine 126.90	54 Xe xenon 131.29
55 Cs caesium 132.91	56 Ba barium 137.33	57-71 lanthanoids	72 Hf hafnium 178.49(2)	73 Ta tantalum 180.95	74 W tungsten 183.84	75 Re rhenium 186.21	76 Os osmium 190.23(3)	77 Ir iridium 192.22	78 Pt platinum 195.08	79 Au gold 196.97	80 Hg mercury 200.59	81 Tl thallium 204.38 [204.38, 204.39]	82 Pb lead 207.2	83 Bi bismuth 208.98	84 Po polonium	85 At astatine	86 Rn radon
87 Fr francium	88 Ra radium	89-103 actinoids	104 Rf rutherfordium	105 Db dubnium	106 Sg seaborgium	107 Bh bohrium	108 Hs hassium	109 Mt meitnerium	110 Ds darmstadtium	111 Rg roentgenium	112 Cn copernicium	113 Nh nihonium	114 Fl flerovium	115 Mc moscovium	116 Lv livermorium	117 Ts tennessine	118 Og oganesson

57 La lanthanum 138.91	58 Ce cerium 140.12	59 Pr praseodymium 140.91	60 Nd neodymium 144.24	61 Pm promethium	62 Sm samarium 150.36(2)	63 Eu europium 151.96	64 Gd gadolinium 157.25(3)	65 Tb terbium 158.93	66 Dy dysprosium 162.50	67 Ho holmium 164.93	68 Er erbium 167.26	69 Tm thulium 168.93	70 Yb ytterbium 173.05	71 Lu lutetium 174.97
89 Ac actinium	90 Th thorium 232.04	91 Pa protactinium 231.04	92 U uranium 238.03	93 Np neptunium	94 Pu plutonium	95 Am americium	96 Cm curium	97 Bk berkelium	98 Cf californium	99 Es einsteinium	100 Fm fermium	101 Md mendelevium	102 No nobelium	103 Lr lawrencium

Key: atomic number / Symbol / name / conventional atomic weight / standard atomic weight

INTERNATIONAL UNION OF PURE AND APPLIED CHEMISTRY

For notes and updates to this table, see www.iupac.org. This version is dated 1 December 2018.
Copyright © 2018 IUPAC, the International Union of Pure and Applied Chemistry.

United Nations Educational, Scientific and Cultural Organization

2019 IYPT · International Year of the Periodic Table of Chemical Elements

Figure 38.1 IUPAC Periodic Table of the Elements, 1 Dec 2018.[14]

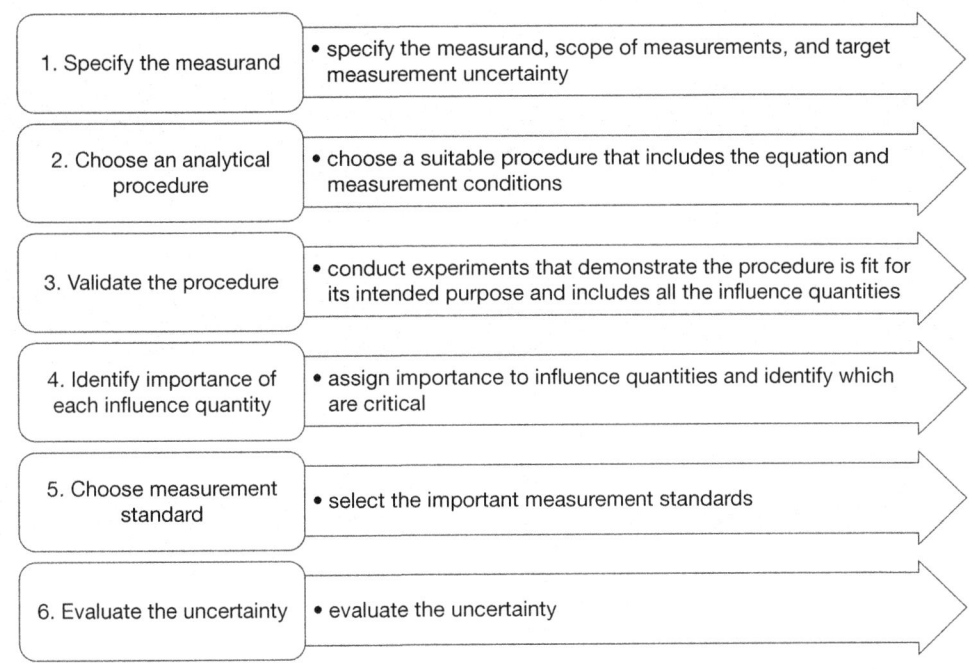

Figure 38.2 Basic process for establishing traceability in a chemical test.[15]

"Quantity that, in a direct measurement, does not affect the quantity that is actually measured, but affects the relation between the indication and the measurement result."[16]

The general process for establishing traceability for a chemical test is illustrated in Figure 38.2 and further described after Figure 38.2. This general process is similar to that undertaken for physical tests. The steps that require unique or special attention for a chemical test are discussed.

Step 1. Specify the measurand

The measurand for a chemical test is more than just the analyte. It must be clear and complete, so that the measured result is suitable for its intended purpose. It may have to include a description of the matrix in which the analyte exists. In some cases, such as operationally defined methods, the conditions used in the test must be included (e.g., the temperature used must be included if the amount of moisture lost depends upon the temperature). The measurand may specify that the measurement result refers to the laboratory sample or to the whole bulk. The target measurement uncertainty (TMU) must be defined to ensure that the method is fit for its intended purpose. The *Eurachem/CITAC Guide: Setting and Using Target Uncertainty in Chemical Measurement*[17] provides guidance on how to set the TMU.

Step 2. Choose an analytical procedure

An analytical procedure can be simple, such as measuring the pH of a clean liquid, or it can be complex, requiring many steps to comminute the sample, dissolve or extract the analyte, and finally measure the analyte with an instrument. The traceability and evaluation of measurement uncertainty must take the impact of all the steps into account. If the laboratory sample is collected in the field and the final reported result refers to the field, then the impact of sampling must be considered as well. The analytical procedure could be developed in-house, it could be an existing procedure that is transferred to the laboratory, it could be an existing procedure that is modified, or it could be a standard test method. The laboratory decides the suitability of a procedure for its intended use.

Step 3. Validate the procedure

Method validation is described in detail in many references such as the *Eurachem Guide: The Fitness for Purpose of Analytical Methods* (2014).[18] The output of a method validation is a demonstration that the method produces measurement results that are fit for purpose and verifies the equation is correct and complete. It also includes all influence quantities that are significant.

Step 4. Identify importance of each influence quantity

Influence quantities have to be assessed as to whether they are critical; that is, affecting the measurement result. Some may require calibration and others that are less critical may require less control.

Step 5. Choose the measurement standard

Calibration for a test method can be applied separately to individual parts of a procedure, but it can also be applied to an entire procedure within a method. The balances, timers, and volumetric apparatuses are calibrated independently. Other aspects of the test method are calibrated each time the method is used (e.g., gas chromatography instruments and inductively coupled plasma instruments). Such instruments can be calibrated with a pure chemical, a pure chemical with components of the sample matrix added, or the calibration standard which is a certified reference material (CRM),[19] as defined by ISO 17034:2016 *General requirements for the competence of reference material producers*:

> Reference material characterized by a metrologically valid procedure for one or more specified properties, accompanied by a reference material certificate that provides the value of the specific property, its associated uncertainty, and a statement of metrological traceability.

In ideal cases, a CRM exists that can be taken through the entire measurement process. As an example, consider the determination of silver in ore. The balance used to weigh the ore is calibrated on its own. The ore is then digested in acid. A CRM made from a silver ore is weighed and digested along with the samples. Then the CRM and samples are measured on an inductively coupled plasma-optical emission instrument (ICP-OES) which is calibrated using pure silver reference material. The parts, such as the balance and the ICP-OES, are calibrated independently. The CRM ensures that the acid digestion is traceable.

Certified reference material (CRM) basics:[20]

- Sources of CRMs with established valid metrological traceability:
  - CRMs produced by National Metrology Institutes (NMIs) using a service that is included in the BIPM Key Comparison Data Base (KCDB)
  - CRMs produced by an accredited reference material producer (RMP) under its scope of accreditation and where the Accreditation Body is covered by a recognized ILAC arrangement
  - CRMs with assigned certified values that are covered by entries in the Joint Committee for Traceability in Laboratory Medicine (JCTLM) database
- Not all CRMs are available from an NMI or an accredited RMP. Where metrological traceability to the SI is not technically possible, it is up to the user to either:
  - Use certified values of certified reference materials from a competent producer, or
  - Document the results of a suitable comparison to reference measurement procedures, specified methods, or consensus standards that are clearly described and accepted as providing measurement results fit for their intended use.

This discussion illustrates two principles:

- When a measurement result is calculated from a reference value, it is traceable to that reference value.
- Traceability to common references allows meaningful comparisons between results.

Figures 38.3 and 38.4 illustrate these principles. In Figure 38.3, the results are arrived at independently. Each result is traceable to separate reference materials used in the test, but there is no way to compare RM x_1 and Result y_1 with RM x_2 and Result y_2. In Figure 38.4, each test includes the use of the same CRM, with value x_0, so the results, y_1 and y_2, can be compared using the value x_0. The test results can now be compared through the CRM and its value, x_0.

Step 6. Evaluate the measurement uncertainty

Calibration reference materials affect the measurement uncertainty. The influence factors may contribute to the uncertainty, becoming uncertainty components. The uncertainty evaluation must take into account the entire method and all influence quantity components.

506　*Part VI: Measurement Parameters*

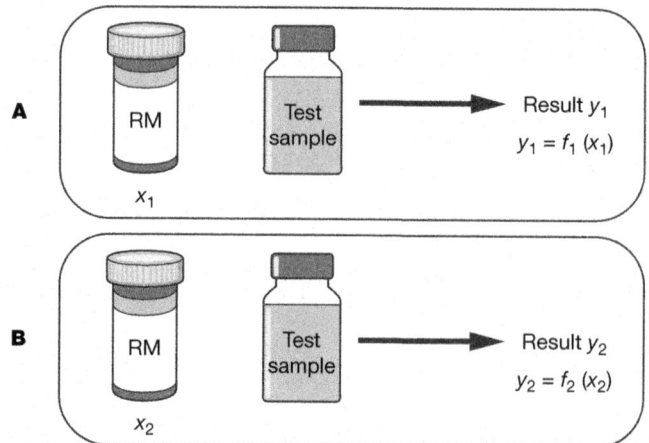

Figure 38.3 Two results obtained in two different tests using different reference materials.[21]

Source: Figure reproduced by permission of Eurachem from S L R Ellison and A Williams (Eds.), *Eurachem/CITAC Guide: Metrological Traceability in Analytical measurement* (2nd ed. 2019). ISBN: 978-0-948926-34-1. Available from http://www.eurachem.org

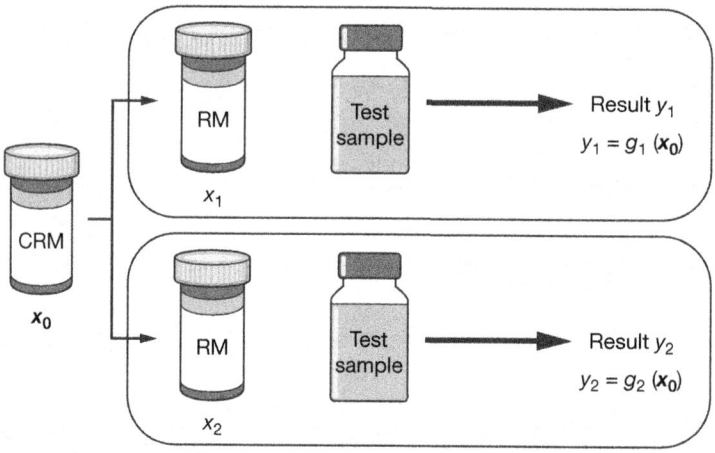

Figure 38.4 Two results obtained in two different tests using the same reference materials.[22]

Source: Figure reproduced by permission of Eurachem from S L R Ellison and A Williams (Eds.), *Eurachem/CITAC Guide: Metrological Traceability in Analytical measurement* (2nd ed. 2019). ISBN: 978-0-948926-34-1. Available from http://www.eurachem.org

EVALUATING MEASUREMENT UNCERTAINTY FOR CHEMICAL TESTS

Many in the analytical chemistry laboratories were initially introduced to measurement uncertainty (MU) by the 1999 revision of ISO/IEC 17025. The *Eurachem/CITAC Quantifying Uncertainty in Analytical Measurement* (*QUAM*) has gone through three revisions, and each revision illustrates the increased understanding of MU. At the time this handbook was written, the *QUAM* is in its third edition.[23] The use of MU in decision rules, method validation, and quality control is motivating more comprehensive and complete evaluations of MU. Fortunately, the overall process for evaluating MU is the same for chemical tests as for physical. A brief overview of the evaluation of MU for chemical tests is provided, with an emphasis on the unique aspects for chemistry tests.

When Is Testing of Measurement Uncertainty Necessary?

Measurement uncertainty analysis is *always* necessary for a measurement to be metrologically traceable, regardless of whether MU can be calculated or not. What determines whether measurement uncertainty can and should be calculated? Internationally accepted standards provide clarity around this question. Refer to ISO/IEC 17025:2017, 7.6.1: "Laboratories shall identify the contributions to measurement uncertainty. When evaluating measurement uncertainty, all contributions that are of significance, including those arising from sampling, shall be taken into account using appropriate methods of analysis."[24] This statement is broad enough to let the laboratory determine the significance of uncertainty contributors and specific enough to say that uncertainty evaluation is necessary (whether it is reported to the customer or not). Chemical laboratories may make this process an independent documented procedure or may make it part of the method validation process. Regardless of whether measurement uncertainty can be calculated, it is imperative that a laboratory have a process in which they:

- Identify contributors to measurement uncertainty
- Use appropriate methods of analysis to calculate measurement uncertainty
- Evaluate its measurement uncertainty

What Currently Happens if Testing Is Performed Following a Standard Test Method?

Let us begin by defining what a standard test method is. A *standard test method* is considered a definitive procedure that, when followed, produces a reproducible test result. These standard test methods allow results generated by different laboratories to be comparable with one another. When a lab follows a standard test method using competent people, adequate quality control, and appropriate (and calibrated) equipment to test the same item, it is then anticipated that the results should be repeatable and reproducible (and, ideally, valid). Examples of standard methods are test methods published by reputable international,

national, or regional standards and/or testing organizations. Some examples of these organizations are (though not limited to) ASTM, ASME, AOAC, EPA (in the United States), ISO, USP, SAE, NSF, UL, and TIA. More organizations can be found on www.standardsportal.org (hosted by ANSI). By using and following accepted standard test methods, consumer confidence and trade are supported.

When Does Measurement Uncertainty Have to Be Calculated if Testing Is Performed Following a Standard Test Method?

According to ISO/IEC 17025:2017, Clause 7.6.3, "A laboratory performing testing shall evaluate measurement uncertainty. Where the test method precludes rigorous evaluation of measurement uncertainty, an estimation shall be made based on an understanding of the theoretical principles or practical experience of the performance of the method." Clause 7.6.3 includes three notes worthy of reference:

> NOTE 1. In those cases where a well-recognized test method specifies limits to the values of the major sources of measurement uncertainty and specifies the form of presentation of the calculated results, the laboratory is considered to have satisfied 7.6.3 by following the test method and reporting instructions.
> NOTE 2. For a particular method where the measurement uncertainty of the results has been established and verified, there is no need to evaluate measurement uncertainty for each result if the laboratory can demonstrate that the identified critical influencing factors are under control.
> NOTE 3. For further information, see ISO/IEC Guide 98-3, ISO 21748, and the ISO 5725 series.[25]

ILAC G17:01/2021, Guidelines for Measurement Uncertainty in Testing, provides guidance and related references for the evaluation and reporting of measurement uncertainty in testing in order to fulfill expectations of the relevant clauses of ISO/IEC 17025:2017. The document also aims to assist laboratories in understanding the common approach taken by accreditation bodies when performing assessments against these requirements included in ISO/IEC 17025:2017.[26] ILAC G17:01/2021, Section 3: Guidance on Evaluation of Measurement Uncertainty in Testing, provides references to multiple recognized and accepted international guides on estimating and evaluating measurement uncertainty. ILAC G17 continues: "For quantitative measurements where the final results are expressed in a qualitative way (e.g., pass/fail), evaluation of measurement uncertainty is still applicable."[27]

What Happens if a Standard Test Method Does Not Contain Uncertainty Guidance?

If there are no uncertainty guidelines, the following must be determined: "In some areas of testing in which uncertainty cannot be expressed as an expanded uncertainty for the test result (e.g., *qualitative* testing or examinations), other means for evaluation of measurement uncertainty, such as a probability for false positive or false negative test results, may be more relevant."[28] ILAC G17 recommends the following references for uncertainty of qualitative tests:

1. When these measurement results are *qualitative,* are the results expressed based on something that cannot be counted, such as color, odor, texture, etc.? Then the lab must at least identify the possible contributors to measurement uncertainty. In these cases, the following has to be determined: "In some areas of testing in which uncertainty cannot be expressed as an expanded uncertainty for the test result (e.g., *qualitative* testing or examinations), other means for evaluation of measurement uncertainty, such as a probability for false positive or false negative test results, may be more relevant."[29] ILAC G17 recommends the following references for uncertainty of qualitative tests:

    a. "Quality assurance of qualitative analysis in the framework of the European project 'MEQUALAN'," *Accredited Quality Assurance* 8 (2003), 68–77.

    b. "IFCC-IUPAC Recommendations 2017 Vocabulary on nominal property, examination, and related concepts for clinical laboratory sciences," *Pure & Applied Chemistry* 90 (2018), 913–935.

2. ILAC G17 recommends the following references for measurement uncertainty of microbiological tests:[30]

    a. ISO 29201:2012 *Water Quality—The Variability of Test Results and the Uncertainty of Measurement of Microbiological Enumeration Methods*

    b. ISO 19036:2019 *Microbiology of the Food Chain—Estimation of Measurement Uncertainty for Quantitative Determinations*

3. Currently, no measurement uncertainty value can be calculated on qualitative results. What is sought is the ability to calculate a probability of occurrence/observations.

    a. This is covered in "Appendix H: Probability of Detection (POD) as a Statistical Model for the Validation of Qualitative Methods (2012)" in AOAC's electronic *Official Method of Analysis.*[31]

4. Follow the *GUM* or EURACHEM/CITAC's *QUAM:2012.*[32]

5. In certain cases, it may be sufficient to report only the reproducibility (sometimes referred to as precision).

Approach to Evaluating Measurement Uncertainty

The approach to calculating measurement uncertainty for chemical measurements is, in theory, still the same as physical measurements. In other words, begin by following the *GUM* (*Guide to Uncertainty in Measurement*)[33] approach, which is also further clarified in NIST SOP 29 (*SOP for Assignment of Uncertainty*):[34]

1. Specify the measurand.

2. Identify the scope of analysis.

3. Examine the measurement process and identify all possible sources of uncertainty.
4. Quantify the uncertainty components/contributors.
5. Combine the contributors.
6. Calculate the expanded uncertainty using the appropriate coverage (k factor) multiplier.
7. Review the entries and analyses.
8. Report measurement uncertainty in an uncertainty report with the appropriate information (add notes and comments for future reference).

According to EURACHEM/CITAC *QUAM:2012*, section 1.3, some approaches for estimating measurement uncertainty in chemical testing include:[35]

- Evaluation of the effect of the identified sources of uncertainty on the analytical result for a single method implemented as a defined measurement procedure in a single laboratory
- Information from method development and validation
- Results from defined internal quality control procedures in a single laboratory
- Results from collaborative trials used to validate methods of analysis in a number of competent laboratories
- Results from proficiency test schemes used to assess the analytical competency of laboratories

CONTRIBUTORS TO UNCERTAINTY IN CHEMICAL TESTING[36]

Measurement uncertainty in chemical testing is a combination of bias (method bias and laboratory bias) and within-laboratory reproducibility (repeatability and reproducibility over time). Use of control chart results can be combined with bias and reproducibility to calculate measurement uncertainty.

Bias is defined as the systematic effects (of the method used and of the individual laboratory). Typical sources of systematic effects may also include:

- Incomplete definition of the measurand
- Sampling
- Matrix effects and interferences
- Environmental conditions
- Uncertainties of masses and volumetric equipment
- Uncertainties of reference values
- Approximations and assumptions incorporated in the measurement method and procedure
- Random variation

Random effects are captured with repeatability and reproducibility data. Typical sources of repeatability and reproducibility data may include:

- Proficiency testing programs
- Method validation replicate data
- Reference sample
- Spike recovery data
- Sample duplicate

Combining Uncertainty

For chemistry tests, some experiments may include the impact of many uncertainty sources. Using good design of experiments, a well-designed experiment may include tests performed over multiple days by multiple analysts using different equipment. An ANOVA analysis of the results may provide an estimate of measurement uncertainty that includes the significant uncertainty components. For some chemical tests, such experiments are the only way to quantify an uncertainty component (e.g., the impact of different lots of antibodies in an ELISA test).

Decision Rules

Decision rules have not been used extensively in analytical chemistry in the past, but have been introduced by standards and regulations and are now being used more frequently. A *decision rule* is a rule that describes how measurement uncertainty is accounted for when stating conformity with a specified requirement.[37] A few such documents are:

- ILAC-G8:09/2019, *Guidelines on Decision Rules and Statements of Conformity*
- ISO/IEC 17025:2017
- *USP <1220> Analytical Procedure Lifecycle*
- The Eurachem/CITAC guide, *Use of Uncertainty Information in Compliance Assessment*

CONCLUSION

Analytical chemistry laboratories are adopting and using metrological approaches to measurements, including traceability and measurement uncertainty. Chemical tests are based on chemical properties which sometimes require processes for establishing traceability and evaluating measurement uncertainty that differ from those used in physical test methods.

Guides, references, standards, and regulations either require use of measurement uncertainty or provide guidance on how to estimate uncertainty and how to apply metrological concepts to these unique aspects of chemical tests.

One such organization, Eurachem, provides the most comprehensive, current, and detailed guidance for chemistry measurements. Chemical test laboratories have well-established quality systems, such as Good Laboratory Practices. However, the reference materials used may not be well defined. Many initiatives have been created and are underway to provide certified reference materials with well-defined certified reference values and accompanying standard uncertainties.

In this chapter, the fundamentals of traceability for a chemical test were presented. The mole, the SI unit for the amount of substance, is central to metrological traceability for chemistry tests. A six-step process for establishing traceability was described. This process includes the use of target measurement uncertainty, method validation, identification of influence quantities, and evaluation of measurement uncertainty. The need for and importance of certified reference materials for establishing traceability was discussed.

The process for evaluating measurement uncertainty for a chemical test was presented. The steps are the same as those for a physical test. The laboratory should have a process to identify contributors to measurement uncertainty, estimate the contributions, and combine these appropriately. Standard test methods were defined and the options for evaluating measurement uncertainty were presented.

The approach to calculating measurement uncertainty for chemical measurements was presented. In many cases a well-designed experiment will include all significant uncertainty components.

Finally, decision rules for chemistry tests were discussed, emphasizing that realistic measurement uncertainty evaluations are needed in order to make useful decisions. The metrological requirements for chemical tests are similar in many ways to those of physical tests. Chemical tests have some unique aspects that require specific approaches. The metrological approach ensures that traceability and measurement uncertainty are achieved so that the chemical test results are fit for their intended purpose and the decisions made are valid.

PART VII

Managing a Metrology Department or Calibration Laboratory

Chapter 39 Getting Started
Chapter 40 Best Practices
Chapter 41 Process Workflow
Chapter 42 Budgeting and Resource Management
Chapter 43 Vendors and Suppliers
Chapter 44 Housekeeping and Safety

Chapter 39
Getting Started

Part VII is an attempt to show how calibration and metrology functions are managed. There are no fixed rules for managing a successful calibration laboratory, metrology department, or combination of the two. There are, however, tried-and-true policies and procedures that have been successfully used to help managers and supervisors get the most out of their resources.

Different strategies and ideas incorporated into various work environments will be described, from large commercial calibration facilities to internal calibration labs to one-person calibration companies. These strategies and ideas have been used in calibration labs, metrology sections, groups and departments, machine shops, production lines, and the vast array of calibration functions around the world. How one incorporates them into one's own work environment is decided by the requirements of the metrology department or of the overall organization.

Within Chapters 39–44, you will find ideas on improving customer service, using metrics, preventive maintenance programs, how to use surveys to improve the organization's policies and procedures, and more. We also discuss different ideas on how to manage the workflow: from arrival of inspection, measurement, and test equipment (IM&TE) to its return, budgeting and resource management, vendors and suppliers, and housekeeping and safety.

One topic that must be discussed is ethics. Ethics is fundamental to metrology and its practices. Lives may depend on our measurements. We must make sure we make impartial, accurate, and traceable measurements. One of the best sets of ethical guidelines is the ASQ Code of Ethics.[1]

Code of Ethics

Introduction

The purpose of the American Society for Quality (ASQ) Code of Ethics is to establish global standards of conduct and behavior for its members, certification holders, and anyone else who may represent or be perceived to represent ASQ. . . .

Fundamental Principles

ASQ requires its representatives to be honest and transparent. Avoid conflicts of interest and plagiarism. Do not harm others. Treat them with respect, dignity, and fairness. Be professional and socially responsible. Advance the role and perception of the Quality professional.

Expectations of a Quality Professional:

A. Act with Integrity and Honesty

1. Strive to uphold and advance the integrity, honor, and dignity of the Quality profession.
2. Be truthful and transparent in all professional interactions and activities.
3. Execute professional responsibilities and make decisions in an objective, factual, and fully informed manner.
4. Accurately represent and do not mislead others regarding professional qualifications, including education, titles, affiliations, and certifications.
5. Offer services, provide advice, and undertake assignments only in your areas of competence, expertise, and training.

B. Demonstrate Responsibility, Respect, and Fairness

1. Hold paramount the safety, health, and welfare of individuals, the public, and the environment.
2. Avoid conduct that unjustly harms or threatens the reputation of the Society, its members, or the Quality profession.
3. Do not intentionally cause harm to others through words or deeds. Treat others fairly, courteously, with dignity, and without prejudice or discrimination.
4. Act and conduct business in a professional and socially responsible manner.
5. Allow diversity in the opinions and personal lives of others.

C. Safeguard Proprietary Information and Avoid Conflicts of Interest

1. Ensure the protection and integrity of confidential information.
2. Do not use confidential information for personal gain.
3. Fully disclose and avoid any real or perceived conflicts of interest that could reasonably impair objectivity or independence in the service of clients, customers, employers, or the Society.
4. Give credit where it is due.
5. Do not plagiarize. Do not use the intellectual property of others without permission. Document the permission as it is obtained.

Various standards highlight the requirements for acting impartially and maintaining confidentiality. Impartiality and confidentiality are such crucial factors that they have been added to ISO/IEC 17025:2017 in their own clauses; 4.1 and 4.2.[2] Good businesses evaluate and manage impartiality and confidentiality to reduce their own risk. A company, department, or laboratory must keep secure the proprietary information that is entrusted to it and conduct its business in a professional manner, which assumes that its conduct and ethics are above reproach. Not only is this good business, but also it should also be a carryover from the way calibrations are performed, data is collected, and traceability is ensured. Professional conduct and ethics form the foundation for any business to maintain a reputation for honesty, integrity, and truth in its dealings with its customers.

Naturally, one will not find all the answers to one's management questions in these chapters. The ideas and suggestions given here are to offer direction. Good managers are not born. Often, high-performing employees are rewarded with promotions into supervisory or management roles with little to no training on how to be successful in these roles. It is a failure of the organization not to provide the necessary resources and training skill sets for supervisory and management positions with the same support and resources as it does for budgeting, process, and technical skills and knowledge. Managers are molded, taught, and shaped by their experiences and training. An old axiom says, "It is better (and faster) to learn from others' mistakes than to make them yourself." These chapters are in the same vein.

Chapter 40

Best Practices

> *Originality is nothing but judicious imitation. The most original writers borrowed one from another. The instruction we find in books is like fire. We fetch it from our neighbor's, kindle it at home, communicate it to others, and it becomes the property of all.*
>
> —Voltaire

Such is the case with best practices. These are a compilation of success stories, policies, procedures, work instructions, military axioms, learning experiences, and, in some cases, simply good old common sense. If they can help anyone improve their process, production, profits, or performance, then the goals of this chapter have been met.

It is doubtful that any of the suggestions, policies, or practices that will be discussed here are original to the author who suggested them. The originality came in how they were applied to the situation that created their need. The phrase "Improvise, adapt, and overcome" has been attributed to the U.S. Marine Corps. It could also apply to best practices. Use them when needed or mix and match them to an organization's situation.

Paul Arden wrote in his book *It's Not How Good You Are, It's How Good You Want to Be*:

> Do not covet your ideas. Give away everything you know, and more will come back to you . . . [R]emember from school other students preventing you from seeing their answers by placing their arm around their exercise book or exam paper? It is the same at work, people are secretive with ideas. "Don't tell them that, they'll take the credit for it." The problem with hoarding is you end up living off your reserves. Eventually you'll become stale. If you give away everything you have, you are left with nothing. This forces you to look, to be aware, to replenish. Somehow the more you give away the more comes back to you. Ideas are open knowledge. Don't claim ownership. They're not your ideas anyway, they're someone else's. They are out there floating by on the ether. You just have to put yourself in a frame of mind to pick them up.[1]

Arden's bit of knowledge could be used either in the Preface or the Conclusion to this handbook. In writing this handbook, the authors are sharing their ideas, programs, procedures, and solutions in the hope that the metrology and calibration community can continue to grow and improve in the years to come. By sharing knowledge, experiences, failures, and successes, we hope to reduce the need to continually reinvent the wheel when good wheels are already out there being used by fellow practitioners.

We hope, as this book continues to be revised, that more best practices are added, and that the sharing of ideas and programs will continue. That is not to say that confidentiality should be broken by telling trade secrets or patented ideas. However (as an example), it is simple math that anyone can calibrate more items in a given period of time by doing like items instead of having to change setups and standards several times in the same time period.

CUSTOMER SERVICE (LAB LIAISONS)

When we think about customers, we must consider our perspective and ask the following questions: Who are your customers? What is the difference between good customer service and bad customer service? What are they paying you to do for them? What should they expect in return for their money? How do you know if you are providing the level of customer service they expect?

A *customer* is a person who buys products or services. A *service* is an act of work done for others (as an occupation or a business). One could say that *customer service* is an act of providing services to customers before, during, and after a purchase. Nothing in these definitions mentions quality or timeliness. There is also no mention of error-free satisfaction, "most bang for the buck," or best price. But if the customer never returns, something in the service might have been lacking.

Many questions hang on two simple words: "customer service." If the answers were simple, there would be no need for this section. But the answers are not simple. It is complicated, sometimes overanalyzed, and can make the difference between a company staying in business or closing its doors forever.

Most of us have been on both sides of this issue; we have been the customer and we have dealt with customers. We all know how good it feels to walk away from a satisfactory encounter at a store, shop, or establishment after we have made a purchase. We also know how upset we get when the experience was not up to our expectations. We may not always remember the good exchanges, but we have a tough time forgetting the bad ones. Let us explore ways to take care of our customers and keep them coming back for more quality service.

Customer service is not only a good idea from a buyer's point of view; it is also a requirement in international quality standards. Both ISO 9001:2015, Clause 8.2, "Requirements for products and services," and ISO/IEC 17025:2017, Clause 7.1, "Review of requests, tenders and contracts" have guidance about communicating with customers and requirements of what should be understood and documented about customers' request for service.

Customers value the maintenance of effective communication, advice, and guidance in technical matters, and opinions and interpretations based on results. Communication with the customer, especially in large assignments, should be maintained throughout the work. The laboratory should inform the customer of any delays or major deviations in the performance of the tests and/or calibrations.

ANSI/ISO/ASQ Q10012:2003 states, in paragraph 5.2, "Customer focus," "The management of the metrological function shall ensure that:

- Customer measurement requirements are determined and converted into metrological requirements,

- The measurement management system meets the customers' metrological requirements, and

- Compliance to customer-specified requirements can be demonstrated."[2]

If all these requirements are fulfilled, are you guaranteed a happy customer? Is that all you must do to keep them satisfied? During a presentation at a NIST seminar in November 2002, the author made the following statement: "The customer is always right. I do not believe that to be true. The customer is not always right—but . . . they are always the customer!"

This is not to say that calibration facilities must cater to clients who do not know what they are talking about or do not understand uncertainty or the time it takes to complete a complex calibration, but the customer still is the source of income for many companies, and those customers' idiosyncrasies, lack of knowledge, and/or limited understanding of metrology must be factored into any equation. Honest, intelligent communication with calibration customers about one's capabilities, scope, and products can only enhance the calibration facility's ability to deliver satisfactory service. The calibration facility is no better or worse than anybody else in the business community. It must find a way to interact with its customers in a professional manner and provide the type of service for which customers are willing to pay.

It is good practice to maintain an up-to-date customer database with contact points, telephone numbers, and email addresses. If lab technicians must contact the customer prior to picking up the equipment, can they quickly and easily do so? Do they have the correct contact point? Do they need to know the owner or user of the IM&TE? Depending on the function, calibration lab, or metrology department, the user could be either or both the user and the owner.

For example, the USAF Precision Measurement Equipment Laboratory (PMEL) had Owning Work Center (OWC) monitors assigned as the liaisons between the different squadrons (or work centers) and their supporting PMELs. If the OWC monitor was not the user of the IM&TE, they knew who the user's supervisor was to allow notification of IM&TE being out of tolerance or when approval was required for limited calibrations. The relationship between the PMEL scheduler and the OWC monitor was crucial from both perspectives. The OWC monitors received monthly updates to their master IM&TE listings; they were the ones who delivered and picked up the IM&TE on a regular basis and usually knew who to contact for limited calibrations or who to notify when their test equipment was out of tolerance. The PMEL scheduler had a contact point for doing business and was the interface between the OWC and PMEL's management and calibration technicians.

In most cases, PMEL schedulers represented the calibration function to those whom they supported. The attitude and customer service displayed by the scheduler represented all the individuals working behind the scenes. Such is the case with most laboratories and calibration functions. The person who communicates with the customer represents the organization's supervision, technicians, and its quality assurance personnel. When less than a positive and professional attitude is presented to the customer, that shortcoming still represents the entire company, group, department, or laboratory to that customer. Therefore, the hiring of competent and professional staff who will be dealing with clients, customers, or the public in general is vitally important.

To maintain a good working relationship with customers in a company setting, lab, division, or group, customer liaisons have filled the gap in some organizations. The liaison is the contact person for questions about the IM&TE and may even deliver the test equipment to the calibration function. Good liaisons need to receive training in what they are responsible for, such as how to properly maintain IM&TE and what the various listings and terminology mean in the metrology world. One cannot expect to communicate effectively with the customer if one perceives that customer to be speaking in a different language. Traceability, uncertainty, 4:1 ratios, decision rules, NIST, SI units, IM&TE, calibration, and reproducibility terms . . . these may seem like Greek to the untrained person. An orientation session can help improve vocabulary while giving the liaison the opportunity to answer the customers' questions and get a better feel for their needs and expectations.

Here are some suggestions on how to keep customers coming back for more:

- Regularly send them a list of their items supported, with the next due date, a few months ahead of their due dates. (They were asked what kind of calibration interval they wanted to have, right?) This will prompt them to contact you and schedule their calibrations.

- Maintain a website with a Frequently Asked Questions (FAQs) section.

    o Keep this site up to date with customers' actual questions. Sometimes it is easy to forget that new customers are always coming through the door and that communicating at a basic level can be a wise decision. This can help cut down on the time spent repeatedly answering the same questions and empower your customers through their own self-education.

- Have orientation sessions or tours of your facilities. Open houses can be a way to draw in new customers who are curious about what you do or how you do it. Open houses can also build loyalty with new and established customers by giving them the faces behind the names they deal with.

- Support and participate in local events, local charities, and other local businesses to help make your name or business more available to those who would not normally associate with calibration or metrology. Is this really a part of customer service? If the customer does not come to you, you must go to them. Once you have someone as your customer, then you need to shift gears to keep them as your customer.

- Benchmark other companies' practices, identifying and adapting to best industry practices.

- Survey customers for valuable feedback information. Reach out to customers who respond, whether they responded with positive or negative feedback.

- Make available the results of assessments by accreditation bodies.

Timely information, quality service, accurate data, and a friendly smile (or voice over the phone) can go a long way in keeping customers happy, satisfied, and coming back for more.

USING METRICS FOR DEPARTMENT/LABORATORY MANAGEMENT

Humans tend to be visual beings. We see most things as a picture, in color, with shape and form. Place a bunch of numbers in front of us, and we will start thinking. When a supervisor asks for data, which is easier to analyze: a column of 17 numbers or a graph with 17 bars of different heights, sorted by whatever common factor was used to collect the data? For most of us, the graph gives us immediate recognition of what is important, what is not, and where to focus our attention.

When we present data or give a presentation, we are either trying to sell it, persuade our audience, or entertain them. Metrics are a form of communication. They are visual, immediately tell a story, and help get your point across in a medium that is easy to understand and comprehend.

Metrics can help an organization forecast its workload, show production trends, and repair problems. How often IM&TE is out of tolerance can help determine future calibration intervals or replacement. Production totals can graphically show who should be promoted, rewarded, or given additional responsibilities.

Here are some examples of metrics. Keep in mind that they have been applied to a specific work environment, and when used outside of that situation, may not give the same results. When forecasting for future workload, make a list of all the items that are due for calibration over the next 12 months. Sort the list by the next calibration due dates. Once that is accomplished, total the number of calibrations due for each of the next 12 months. Then place those numbers into a graph. The graph might look something like Figure 40.1, which shows a projected forecast for the next year. The straight line is the average of all the months added together and divided by 12. As seen in the graph, October, January, May, and June could be smoothed out with increases in the other months to balance the workload. Each item in the inventory that shows up in this graph could have its calibration date moved up to accommodate the smoothing of the workload timing. None of the due dates could be moved to a later time, as their items would then be overdue for calibration. Naturally, there would be a mix of calibration intervals within this inventory and, because different groups of IM&TE would be coming due at different rates in the coming years, this one-time smoothing of the peaks and valleys would have to be repeated on a regular basis.

Analyzing historical data can also give an idea of how many calibrations are typically scheduled and how many are typically unscheduled. Whether the calibration lab is a commercial lab or an internal lab, this type of tracking can help

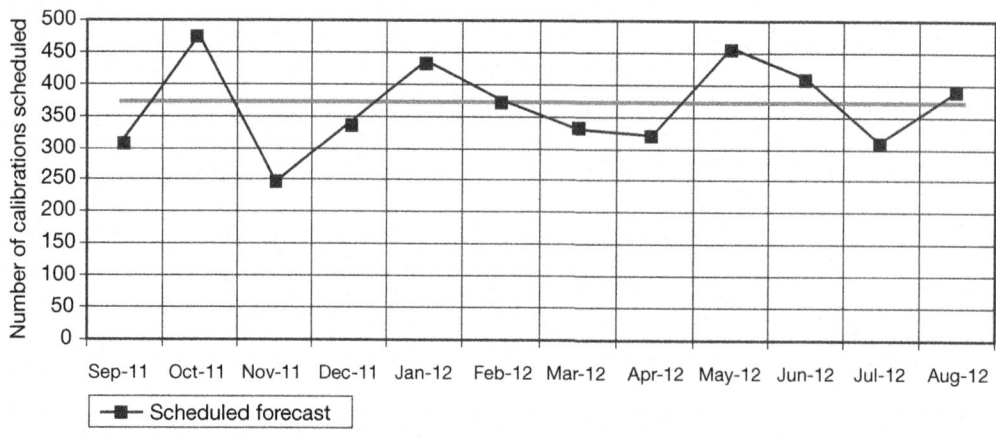

Figure 40.1 Forecast of scheduled calibrations due.

forecast growth and resource planning. To help in getting an accurate forecast of the coming workload, experience shows that a precalculated increase in the numbers would more accurately reflect the actual workload coming in the door. This number is found by comparing actual workload against forecasted workload over the previous year's collected data. With new IM&TE continuously being added to the inventory, items requiring repair also needing calibration, and other cases where calibrations were performed out of cycle, the actual number of IM&TE requiring calibration each month was significantly higher than the forecast for the next 12 months. By analyzing the historical data and projecting the difference into the yearly forecast, it is easier to forecast the future calibration workload.

Another metric that has a significant impact on the workload is the compilation of out-of-tolerance items versus calibrations performed. In addition, repeat out-of-tolerances are particularly worth tracking. It is found that actual numbers versus perceived data provide a far more accurate idea of which IM&TE are reliable compared to others that needed their calibration interval reduced. Usually, managers and supervisors are focused on the few bad pieces of equipment that required a longer time to repair, adjust, or calibrate after being down for maintenance when, in fact, more items passed calibration without any adjustment, but were easily forgotten in the daily grind to meet metrics. By tallying the yearly pass rate for each type of equipment and setting limits as to when it would increase, decrease, or allow the calibration interval to remain the same, an organization can continue being proactive instead of being continually plagued with IM&TE requiring adjustment and repair before they come due for calibration. The pass rate can be figured for each item by dividing the number of items found out of tolerance during their scheduled calibration by the total number of like items receiving calibration, and then multiplying by 100. If the pass rate exceeds 95 %, then that type of equipment is eligible for an increase in its calibration interval (95 % and 98 % receive different increase rates). If the type falls below 95 %, then there is discussion concerning whether the calibration interval should remain the same or if the time between calibrations should be decreased. Special circumstances should be considered. Many types of equipment may have their calibration intervals remain the same even though they had less than a 95 % pass rate.

Most numbers, of course, can work for or against an organization. As an example, let us look at production numbers. Mario and Maria both work in the same area, calibrating the same types of equipment. Mario produces 45 units a month, while Maria produces 25 units. One might think either that Mario is an outstanding producer or that Maria needs additional training or assistance. But by analyzing the production numbers, it is found that Mario is putting in 10 additional hours of work a week (two hours a day) because he is single and wants higher production totals. Maria is actually calibrating more of the items that have longer calibration procedures and is also waiting for Mario to use the standards they both have to share. With enough history and data, expected amounts of time per type of calibration can be determined (and accounting for instances when repair or adjustments are needed). These expected amounts of time can be figured into work scheduled for technicians. Well-researched and well-defined expected times to complete a calibration can also serve as a metric for new technicians. If a new technician takes 45 minutes to complete a calibration that takes seasoned technicians only 20 minutes, then the new technician has an idea of when they have reached a proficiency level. These times can also be used to quantify time savings when improvements are made in processes or equipment.

Further efficiencies can sometimes be realized when large volumes of the same types of calibrations can be batched together. Some of the efficiencies may come from the test or calibration setup only needing to be done once instead of multiple times during the day or week. It is also possible to equalize the differences in calibrations by assigning different types of calibrations a weight or factor in determining the value of that calibration based on complexity. As an example, one might give a weight factor of one to a multimeter calibration, whereas a spectrum analyzer might receive a weight factor of seven. The weight factor would be based on both the time required for calibration and/or adjustment and the sophistication of technology and training required to use the specific IM&TE. A more balanced schedule and variety of equipment can then be assigned across competent and authorized technicians.

Different types of calibration functions will report different types of metrics. The question might be what to report and what to keep secret. One may or may not wish to embarrass liaisons who do not get their IM&TE to the lab in a timely manner. Labs cannot post or publish confidential information on their customers, but they can post generic information about the number of items they support or calibrate without breaking confidentiality. Production totals, pass rates, overdue rates, trends in growth, productivity, or forecasts can all help an organization see where they are or where they wish to go.

Another metric that may be overlooked for other reasons is turnaround time. The customer wants its IM&TE back as quickly as possible. Calibration technicians (or supervisors or managers) may be graded on their ability to satisfy customers, and this is a valid indicator as to how responsive they are to customer needs. Like any metric, though, this one also has its good and bad points. One must consider the time IM&TE waits for parts, the adjustment to be made, and recalibration if required. This adds to the overall time in the hands of the technician, when in fact there was nothing anyone could do while the item was waiting for something or someone else to do their job. These factors should be added to the equation for an accurate and competitive look at a laboratory's actual turnaround time.

PREVENTIVE MAINTENANCE PROGRAMS

"You can pay me now or pay me later." Have you heard those words before? Was it from a TV commercial, or did a relative utter those words while showing you how to fix your car in the family garage? Or did you hear it from your supervisor the first time you started working with IM&TE? Whatever the source, it is true.

The optimum word here is *preventive*. The person accomplishing work is trying to prevent something negative from happening by doing something positive first. Invest resources, time, and money to prevent spending substantial amounts of time and money in the future on more expensive repairs. Preventive maintenance (PM) can really be effective.

Just because the equipment is a solid-state device does not mean it will last forever. The same is true for mechanical, dimensional, or any other category of IM&TE. Would a technician put away gage blocks without cleaning them and coating them with a protective film? At the very least, that is being preventive. We change the oil and oil filter in our cars on a regular basis. Why? Because it will save time and money on the investment in transportation.

Most of us do not work in a clean environment and neither does our equipment. Over time, air filters become contaminated with dust and dirt, then heat starts to build up inside the equipment, and it is not long before there is an equipment breakdown. Organizations should perform preventive maintenance on high-use items at least once a year—and in circumstances called for by the manufacturer, even more often. Returning to the automobile example, in many cases the owner's manual has different maintenance schedules for different types of operation. In some work environments, PMs are scheduled during routine calibration, or if the item does not require calibration, scheduled on a regular basis. Even if one is only cleaning filters, vacuuming the inside of an item, or lubricating bearings, it is another opportunity to check the unit for smooth operation and proper function.

Here are some areas that might require a technician's attention: Any IM&TE that has a fan usually has an air filter that should be checked. If the environment where the unit is used has above-average contamination, it would be appropriate to check it more often than one used in a cleaner area. If an item has ball bearings, an armature, or any type of moving parts that might require lubrication or replacement of parts, a regular check on their condition could save a breakdown when the item is most needed. Murphy's Law, "If anything can go wrong it will," is alive and well in the metrology world. Not only will it go wrong, but it will go wrong at the most inopportune time. Usually, an item is not used until it is most needed, which means it may not have been used for an extended period of time. When taken off the shelf and fired up, it has a better chance of breaking than if it has been used on a regular basis; the first time it is turned on, the sparks fly and electronic components short out. The common factors causing this are dried-up grease or lubricants in the unit, a clogged fan filter, or mysterious dust bunnies that have built a nest inside. There are many reasons why IM&TE breaks down, but a reliable PM program can help prevent this from happening if performed on a regular basis.

Another benefit of having a PM program is the availability of common parts being on hand. If filters are changed, bearings lubricated, and special parts cleaned on a regularly scheduled basis, technicians are more likely to have the required parts, components, or supplies in stock. It is one thing to have something break down when most needed and quite another to have to add days to the down time because the required parts or supplies have to be researched and then ordered.

An important part of any PM program is having the correct service manuals on hand. They often give specific instructions on how to perform the PMs, along with a recommended list of required parts. The manufacturer usually knows more about what can go wrong, what areas need special attention, and which parts or supplies should be on hand. This eliminates having to figure out on your own what is needed for every preventive maintenance situation.

Writing the instructions for performing PM inspections into procedures if service manuals are not readily available is advisable. This precludes having to memorize them, makes them available for training new personnel, and allows for updates and improvements as equipment and procedures are changed.

If performing a preventive maintenance inspection on a calibrated item, you should perform an *as-found* calibration prior to the PM. Cleaning parts, replacing components, or changing the settings on internal adjustments could lead to the wrong conclusions when calibrating the unit without prior clarification. However, minor cleaning may first be needed to protect your own measurement standards; for example, cleaning grease and dirt from the anvil and spindle faces of a micrometer before letting it near your gage blocks or by wiping a balance pan before placing precision weights on it. Most quality systems and standards require records of when repairs, adjustments, or calibration are performed on IM&TE. This is also true for preventive maintenance. Even if nothing is replaced or adjusted, a record that the PM was performed as scheduled should be documented. When it was accomplished, who did it, and the final results should also be in the record. ISO/IEC 17025:2017, Clause 6.4.13, states that records shall be retained for equipment which can influence laboratory activities and shall include the following, where applicable: "g) the maintenance plan and maintenance carried out to date, where relevant to the performance of the equipment; h) details of any damage, malfunction, modification to, or repair of, the equipment."[3]

Some IM&TE require that periodic checks be performed with check standards, or that self-tests be run on a regular basis. ISO/IEC 17025:2017, Clause 6.4.10, states: "when intermediate checks are necessary to maintain confidence in the performance of the equipment, these checks shall be carried out according to a procedure."[4] These should also be documented for easy retrieval to see what has been done on a particular piece of IM&TE. Schedules for accomplishing use of check standards would normally require updating of the computer system that generates the schedule. This would automatically update the database for recordkeeping purposes, but if a customer is performing the self-tests, or auto-calibrations, one needs to have a system in place for identifying who is performing the task, when it was accomplished, and the final results. As with any PM program, self-diagnostics and the use of check standards are activities that can catch minor problems before they become bigger issues and save both time and money overall.

SURVEYS AND CUSTOMER SATISFACTION

How does an organization know if its customers are satisfied? If they only use the organization's service once and never return, that could be one way of knowing they are not satisfied. There could be other reasons, too. There must be a less ambiguous way to know. There is: Ask them!

Everyone has seen the customer surveys next to the cash register or by the door exits at most stores. Have you ever filled one out, sent it in, or even thought of letting the store know you were not satisfied? Or better yet, let the store know that you received exceptional service from the summer helpers who only work a 20-hour week?

Unless we have our own business and understand the importance of knowing how our customers feel, we usually do not care. We receive surveys in the mail, as pop-ups on the internet, and during dinner in the evenings when the phone rings. Why are we being asked for our opinions? Does someone really care? Yes, they do. Surveys give vital information about our needs, how those needs can be satisfied, and what we most want out of our relationship with the seller. We are willing to spend our money. The seller is willing to take it. The seller needs to know what we are willing to buy and for how much, while we want to know what we can get for the least amount of money. The quickest and easiest way of exchanging that information is by using a survey.

When most of us hear the word *survey*, we think of someone asking us questions about things we are not interested in, or it concerns something we have already purchased (a service such as an oil change, medical care, or landscaping or a product such as a car, appliance, or furniture). If we are satisfied, we may just throw the survey in the trash. If we are not, there is more of a chance that we will complete and return it. What if we received the survey when we received the product? Can we realistically complete it at that time without knowing the quality of the product? How do we know if the item will even work properly until we return to work or home? What would the difference be if you received the survey one week after receiving the product? Each situation has different circumstances and should be analyzed using different types of surveys.

If an organization wants to hear from the disgruntled customer who visits their physical establishment, it might have customer satisfaction cards at the front door or reception desk. If it wants to know the quality of the IM&TE it is sending back to the customer, it would be better to send a survey back with the item after performing calibrations and/or repairs. A customer satisfaction form linked in an email signature and available online could provide an easier, more reliable form of passing on expectations to the customer. Whichever way they are used, surveys can be an invaluable way of communicating with those who like the service given and those who do not. The former will be returning on a regular basis; the latter will not. An organization needs to focus on the latter to ensure that they become the former. One unhappy customer can cost a company many future customers simply through word of mouth. And it could be something as simple as the customer-facing person having a bad day or the telephone operator accidentally hanging up on the customer.

Once you have a completed survey in hand, what can you do with it? A couple of ideas come to mind. One could perform root cause analysis (using the quality tools such as Five Whys) on the problems identified. But what do you do if there are multiple problems? Possibly sort them by importance to the customers, not to the company. Sort them by the number of clients concerned about a particular topic, how often they reoccur within the surveys, and the type of service provided. Some quality systems require preventive and/or corrective action plans to be in place. It would be appropriate to identify problem(s) in your corrective action plan, give a timeline for solving the problem, and follow up to see if the solution(s) implemented succeeded in satisfying the client. Documenting the problems, solution implementation, and final results can only make process improvements easier while showing your customers that the investment in completing the surveys was well worth the time and effort.

Sometimes, a survey will show that some customers have unrealistic expectations. Handling those expectations may require a follow-up conversation to determine if they really want that or if they simply do not know what is realistically possible. If the laboratory has a web page, it can be an opportunity to educate customers about expectations the customer should have.

Organizations with a captive customer base still need to check the pulse of their customers on a regular basis. Are they providing customers the quality service they require? How do they know? Are customer satisfaction forms available? Is there a formal complaint system in place for customer use? One might think that with a captive audience there is little or no need for surveys or complaint forms. Nothing could be further from the truth. The quality of the service or product an organization provides its customers should be foremost on its mind. The customer's ability to do their job could be determined by the quality of the IM&TE they use. The safety of flight, purity of drug manufacturing, or traceability of product could hinge on their willingness to produce a quality product, whether it has a captive audience or not.

Both ISO 9001 and ISO/IEC 17025 standards also have requirements for evaluation of their customer services. ISO 9001:2015, Clause 9.1.2, states: "The organization shall monitor customers' perceptions of the degree to which their needs and expectations have been fulfilled."[5] ISO/IEC 17025:2017, Clause 8.6.2, states: "The laboratory shall seek feedback, both positive and negative, from its customers. The feedback shall be analysed and used to improve the management system, laboratory activities and customer service."[6]

The adage "the squeaky wheel gets the grease" has never been more applicable than in relation to customer complaints. Being proactive in asking for their comments, suggestions, or ideas can only raise customers' opinions of an organization's department, group, or calibration lab. Everybody likes to feel wanted and that their input has value. When genuinely asked for opinions or comments, it is much easier to relate small or irrelevant problems instead of waiting for them to grow into large problems that take time when you can least afford it. In retail sales, it has been known for decades that whereas a happy customer may only tell two or three other people, an unhappy customer will tell an average of ten other people. It would not be surprising if this is found to be true of metrology customers as well, so it is beneficial to make all your customers happy ones.

Chapter 41

Process Workflow

One size does *not* fit all when it comes to how inspection, measurement, and test equipment (IM&TE) should flow through a facility for calibration. Depending on how the operation is set up, staffed, managed, or controlled, there are many ways to get the IM&TE in and back out with efficient, economical processes. By giving the reader a few examples of how this process workflow operates in different environments and organizations, it is hoped that improvements can be made with minimum impact on their current operations or cost in terms of money and/or resources.

One of the previous authors and previous editor of the *Metrology Handbook*, Jay Bucher, used to manage a metrology department for a biotechnology company. Each of the calibration technicians had responsibility for specific facilities, areas, and departments. They were taught the most efficient way to schedule their workload and, over a period of time, make improvements to the system that allowed them flexibility, innovation, and efficiency in the scheduling and management of their workload. The following is a detailed example of how they scheduled their work and accomplished their responsibilities.

After producing a list of calibrations that will be due in the next 30 days, they sort the list by type of equipment and the location of that equipment. By calibrating like items as much as possible, they reduce the time it normally takes to change calibration procedures, standards, and setup. Except for a couple of types of items that are calibrated in the technician's work area, everything else is calibrated on site, meaning in the actual environment where it is used. This translates to taking everything needed to calibrate a particular piece of IM&TE to the location where the item is used. With limited resources to maintain duplicate sets of standards, the efficient and coordinated use of available standards is critical to the success of the department. By coordinating with the other calibration technicians, time and money are saved in the sharing and use of department standards.

One might believe that calibrating like items is common sense or taught throughout the metrology community as a standard practice. Many years of experience in the field of metrology have shown that nothing can or should be taken for granted. The calibration of like items refers to setting up your work to produce (calibrate) the maximum number of items with the minimum amount of time and effort. By calibrating all the pH meters due for calibration during a specific time, such as in the next two weeks, or in a specific facility or area, one might greatly reduce the time it takes to set up for the calibrations (standards, procedures, forms, calibration labels, cables, buffers, and so on). By reducing the time for setup, teardown, changing of standards, accessories, forms, and procedures, your production increases while your customer receives a quicker turnaround time.

By allowing the calibration technicians to set their own schedules, the opportunity for boredom is exponentially reduced. They each decide how best to fit the most calibrations into their workday while keeping repetitive work to a minimum. Doing like items is encouraged, but doing dozens of the same calibration repeatedly can lower anyone's morale. The needs of the company and its customers come first, but the flexibility of doing their own scheduling has paid dividends in higher productivity, higher morale, outstanding customer service, and (most importantly) calibration technicians who have learned to think on their feet, become more self-sufficient, and require minimal supervision.

Once the scheduling of work is done, the technician prepares for the calibration of the IM&TE. This includes retrieving the appropriate calibration procedure; reading it to ensure that the proper standards, accessories, forms, and labels are available; collecting everything needed; and setting up for the actual calibration. When using a paperless system, a laptop computer or tablet is utilized for collection of the calibration data. Once they ensure that their laptop or tablet is charged up and the correct electronic forms are available, they proceed to where the IM&TE is located. If they are doing like items, they might collect all the day's workload into one area or move the cart on which all their required items are transported from room to room, calibrating each item where it is actually used by the customer.

Once the calibration is complete, the old calibration label is removed and a new one is placed on the unit. One of the established policies includes leaving the IM&TE in better condition than it was found. This has helped increase the usable life span of the IM&TE, raised the status of the metrology department in the eyes of the customer, and helped the calibration technicians become proactive in identifying problems before they affect the use of the test equipment. By cleaning or replacing filters, replacing batteries and bulbs, and inspecting for leaks, fluids, and missing parts or components, each technician has saved valuable time in either eliminating or reducing the number of future repairs or unscheduled maintenance. Also included in the service is inspecting the equipment from a safety aspect before leaving the lab. This could include checking for frayed power cords, inoperative safety locks, proper functioning of all lamps or indicating devices, and proper wiring of the power cord and fuse system (including being able to remove the fuse using the fuse cap when an inherently safe fuse holder is not in use).

The technician completes the calibration record and transfers it for review by the supervisor and returns the unit to service. Depending on the type of system that is in place, peer review of the calibration record may be needed before the certificate and activity can be approved or signed off. The technician must also update the software system to show that the unit has been calibrated, with comments, standards used, and any repair costs that might have been incurred. Remember, the phrase "The job isn't finished until the paperwork is done" is as applicable in metrology and calibration as in any other industry. Without the system updating for the next calibration due date, no one would know that the calibration had ever occurred. This is also an ideal time to record problems, discrepancies, or adjustments that were made, and track any trends, both good and bad, of that specific piece of equipment. If a quality assurance program of some type is used in an organization, it would probably be incorporated after the technician completes the calibration and before the IM&TE is returned to the user or customer. Some organizations perform a quality inspection of IM&TE as the items are received. This could be of additional benefit, as it might identify problems with the equipment that would prevent it from being calibrated (missing parts, cables, manuals, etc.) so that the customer could be contacted in a timely manner to resolve the problem. Also, if the customer delivers the items to the organization and problems are found, the technician could coordinate resolutions or take the units back without having to make additional trips. Incoming inspections are also a suitable time for capturing warranty dates, lease return dates, and any type of information that could accurately update your database and might be used in the future. In cases where multiple accessories are included with the incoming IM&TE, it is often helpful to take a digital picture of the items and archive it with the unit's incoming inspection history, if possible. This practice is visual insurance to protect both the customer and the calibration practitioner.

Within any organization that performs calibrations, certain accessories will be needed for completing calibrations, including adapters, cables, buffers, standards, work benches, etc. Two schools of thought come to mind for the process of distributing these assets. One idea is to have a central location for these items. As the items are needed, the technician retrieves and uses them, and returns them at the end of the calibration or the end of the workday. Another idea is to give each technician the minimum number of accessories needed to do the job and have the more expensive standards or accessories available at a central location. Both systems have their benefits and drawbacks. It is more expensive to purchase duplicate cables, loads, and standards, but it can save valuable time in retrieving and returning them. Also, if an accessory is used only on rare occasions, it is a waste of resources to purchase one for each technician to have available when needed. If a specialized setup is in place for ease in calibrating a large variety of equipment (for example, an oscilloscope package or microwave system), it would be counterproductive to remove a cable or load to complete another calibration while that system sat idle awaiting the return of the removed accessory. This has been seen in more than one organization. Time is money for most calibration functions, and the availability of resources is a double-edged sword. It costs money to purchase standards and accessories. It also costs money (in wasted time) for technicians to be idle while waiting to perform calibrations. Efficient scheduling of standards and the calibration of like items can help reduce both wasted time and duplication of standards.

A commercial calibration function might use the following processes in its workflow. The lab receives a customer inquiry. Then:

1. Determine if the request is a qualifying job.

    a. Routine job: qualifies.

    b. Nonroutine job: Obtain verbal and written (email, phone, mail) details, loading/unloading; discuss additional cost; provide an estimated uncertainty to see if will satisfy the potential client;

obtain management approval for overtime, extra labor force, and resources needed.

2. Tentatively schedule the job.

   a. Requester must provide a purchase order within an expected time (for example, two days) to confirm its place on the schedule. The purchase order will also contain all details discussed and agreed upon between the requester and the lab.

3. Schedule the job.

4. Job arrives at the calibration laboratory.

5. Items with associated paperwork ("paperwork" may be in electronic form) are received, checked in, and assigned a unique, nonrepetitive number in the tracking system. This number will follow the job at every step in the process and will be the main identifier.

6. Items are unpacked and inspected. Any further work stops if there is any discrepancy between what was received and what was expected or if there is damage. The customer is notified and any communications and decisions by the customer are documented.

7. Items are prepared for calibration, including the paperwork, initial cleaning, thermal soaking, etc.

8. Items are calibrated using the appropriate method, standards, environment, and operator.

9. Necessary calculations are performed.

10. Calibration certificate is generated containing as-found data, measurement uncertainty, and all other ISO/IEC 17025:2017 calibration certificate requirements.

11. Customer is contacted with final cost and pickup or shipping instructions.

12. Invoice is sent to the customer.

13. Payment is received and tracked against the job performed.

Measurement or calibration activities are characterized as being routine or nonroutine.

Routine activities generally:

- Have been done before.

- Follow regular published procedures with competent staff using calibrated equipment.

- Do not require deviating from normal workflow patterns or established laboratory policies.

- Have available all necessary support equipment, documentation, and technical competency to successfully perform them.

- Produce calibration and/or repair data consistent with the activity, and result in known equipment status, which is acceptable for a given application.

An activity that deviates from this list is considered *nonroutine*. Furthermore, any activity can be treated as nonroutine until it is determined to be otherwise. Nonroutine activities involving workflow, equipment status, and laboratory policy or practice issues should be addressed by an authorized person. Nonroutine activities requiring internal or external technical support should be addressed by an authorized person. This includes the use of alternative test methodologies, equipment, and specifications. Nonroutine activities may result in a corrective action (CA) being initiated.

Figure 41.1 is an example of a business process interaction diagram similar to that used by a calibration laboratory that is registered to ISO 9001:2015. It shows the core business processes and their main interactions, and other processes needed to support them and the quality management system.

The core processes for this laboratory are:

- **Receiving.** Unpacking shipped items, entering receipt data for them and customer-delivered items into the database.

- **Evaluation.** Determining the type of service required. If there is nothing that says otherwise, the assumption is that routine calibration is required and the item is staged on the ready-to-work shelves. Some items may be sent directly to external suppliers.

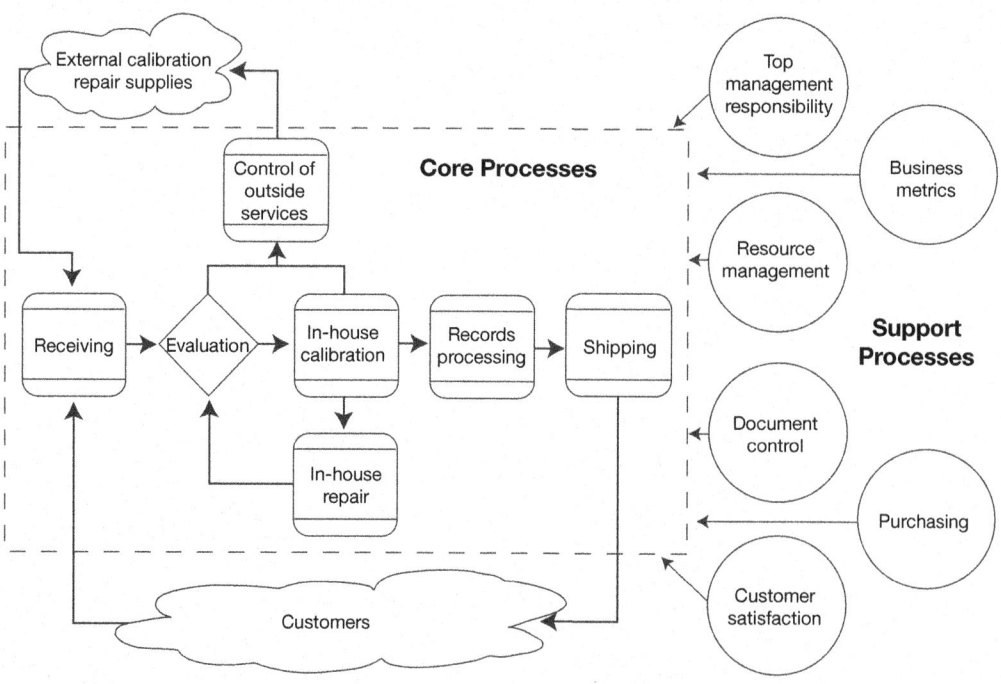

Figure 41.1 A sample business process interaction diagram.

- **Calibration.** Calibrate the equipment and place it on the completed-work shelf. If more than minor repair is needed, transfer it to the repair process.

- **Records processing.** Verify that all database entries have been made, any other records are updated as appropriate, and then apply the appropriate calibration label.

- **Shipping.** Notify local customers that equipment is ready for pickup. Package, prepare shipping documents, and dispatch completed work to other customers.

- **Control of outside services.** As needed, manage interactions with other calibration and repair providers.

- **In-house repair.** As needed, repair equipment and then return it to the beginning of the calibration process. (All repaired items are calibrated before they are released.)

This is a top-level view. This level of detail does not show exceptions to the normal flows, such as out-of-tolerance conditions or in-place calibrations. Those details are shown in quality procedures for each of the core processes. A laboratory information management system (LIMS) is used to manage and record the workflow, including special instructions for certain equipment, ensuring that the current calibration procedure is used and that all measurement standards are within their due dates, and automatically generating calibration certificates and printing labels. The database automatically records data such as calibration time per unit, days for outside supplier service, and even the temperature and humidity at the time of calibration.

Chapter 42

Budgeting and Resource Management

From one author's point of view, the greatest angst among managers comes from either training or budgeting and resource management. Training was discussed in Chapter 15. Like training, maintaining an accurate, up-to-date financial budget is critical to staying out of the red on your balance sheet. Additionally, by being proactive in meeting your needs, one can anticipate extra expenses before they actually occur.

Resource management is not only a part of any calibration function, but also a requirement listed in ISO/IEC 17025:2017, *General requirements for the competence of testing and calibration laboratories*; in ISO 9001:2015, *Quality management systems—Requirements*; and in ISO 10012-2003, *Measurement management systems—Requirements for measurement processes and measuring equipment*.

ISO/IEC 17025:2017 Clause 6 covers "Resource requirements," including 6.1 General, 6.2 Personnel, 6.3 Facilities and environmental conditions, 6.4 Equipment, 6.5 Metrological traceability, and 6.6 Externally provided products and services.[1]

ISO 9001:2015 Clause 7.1 covers "Resources," including 7.1.1 General, 7.1.2 People, 7.1.3 Infrastructure, 7.1.4 Environment for the operation of processes, 7.1.5 Monitoring and measuring resources, and 7.1.6 Organizational knowledge.[2]

ISO 10012-2003 Clause 6 covers "Resource management," including 6.1 Human resources, 6.2 Information resources, 6.3 Material resources, and 6.4 Outside suppliers.[3]

To operate any business, money is needed to pay for the people, places, location, things, and knowledge to perform the functions supporting the business.

BUDGET

What comes to mind when this word is spoken, written, or mentioned in any way, shape, or form? Money comes to mind. How to spend it also appears. By definition, *budget* is "an estimate of income and expenditure for a set period of time."[4] Financial budgets are a way to show where the organization expects to spend its resources for a given period of time. Calibration departments must pay their staff, purchase parts, maintain a facility, and keep their standards in a calibrated state. All these items require money. Whether you receive the funds from outside sources or internally from a company, it is expected that a valid estimate will be provided on a regular basis of the expenses expected to be incurred. Companies typically make budget forecasts within the realm of various assumptions and management directives. Assumptions may presume some inflationary figure or expected growth rate, personnel workforce increases or reductions, increased efficiencies, and so on. Directives from senior management are often used as guidance for keeping expenditures "flat": in other words, previous expenditures = projected forecast, or reducing them by some amount.

There are many ways to accomplish budget forecasting that is both accurate and reliable. Historical data are usually a good starting point. By maintaining records of past expenses, it is possible to accurately forecast future needs. Knowing when standards will come due for calibration and what it cost previously to have those calibrations accomplished will usually allow a close estimate. Some organizations automatically include small inflation increases, while others do not.

If an organization includes the cost of replacing parts in the inspection, measurement, and test equipment (IM&TE) it supports as part of its budget, then available computer software is helpful to calculate this information. Compiling the cost of spare parts over a specific time period can provide accurate data from a historical perspective. This can be done on a quarterly, semiannual, or yearly basis, depending on how the budget is set up.

Depending on the situation, the following areas might have to be included in a budget: salaries and wages, bonuses, general calibration supplies (cables, loads, filters, lubricants, fuses, etc.), dues and subscriptions (for professional organizations), telephone calls, hotel and lodging (for conferences, overnight on-site calibrations, trainings), airfare, transportation costs (taxis or rental cars), business meals, training (seminars, conferences, schooling, materials), repair expenses on IM&TE and/or standards, educational reimbursement, having standards calibrated off-site or at a higher-echelon laboratory, or the cost of maintaining a company van or truck. The list could be even longer depending on responsibilities, size, and location.

As previously mentioned, resource management can be an integral part of one's responsibilities. This could include being responsible for all personnel assigned to the calibration system in the organization. These responsibilities could be defined in an organization chart, individual job descriptions, quality management system manual, or work instruction and/or procedures.

Spreadsheets are often useful for developing, tracking, and updating budgets. They can also be cumbersome to program, inefficient to use, and have many chances for errors. Newer software, including some management systems software, can compile data from multiple software systems and provide easy-to-read and real-time visual dashboards of key performance indicators (KPI). Knowing what to measure, track, and report can be learned through trial and error, from upper management directives, and from business and finance training.

Chapter 43

Vendors and Suppliers

Does your organization have procedures in place for selecting a qualified supplier? Or do you ask your second cousin on your mother's side if Uncle Chris can get what you need? Most standards and requirements recognize this problem and address it in their manuals.

ISO/IEC 17025:2017, Clause 6.6, notes examples of products as "measurement standards and equipment, auxiliary equipment, consumable materials and reference materials" and examples of service as "calibration services, sampling services, testing services, facility and equipment maintenance services, proficiency testing services, and assessment and auditing services."[1]

ISO/IEC 17025:2017, Clauses 6.6.1–6.6.3, require three main actions by laboratories for any external supply or service that will affect calibration laboratory activities:

1. **Make sure that the products and services used by the calibration laboratory come from a suitable supplier.**

    "6.6.1 The laboratory shall ensure that only suitable externally provided products and services that affect laboratory activities are used . . ."[2]

An example of a suitable supplier is one that adheres to the requirements of ISO/IEC 17025:2017.

2. **Make sure that you have a procedure for ordering products and services, and make sure you retain records of the products and services supplied.**

    "6.6.2 The laboratory shall have a procedure and retain records for:

    a) defining, reviewing, and approving the laboratory's requirements for externally provided products and services,

    b) defining the criteria for evaluation, selection, monitoring of performance and re-evaluation of the external providers,

    c) ensuring that externally provided products and services conform to the laboratory's established requirements, or when applicable, to the relevant requirements of this document, before they are used or directly provided to the customer,

    d) taking any actions arising from evaluations, monitoring of performance and re-evaluations of the external providers."[3]

Notice that this procedure also covers how your product and service providers are selected, evaluated, monitored for performance, and then reevaluated over time. There is also a requirement on addressing situations when an action is for product and service providers. This could be for something as simple as a question about an order or as complex as remedial action for poor performance. This means you will need a process for performing these activities, including when, where, and how they are performed. Another component of the procedure is making sure that the products and services meet your requirements, and (where applicable) the requirements of ISO/IEC 17025:2017, before they are released for use in the laboratory.

3. **Make sure you clearly communicate your requirements to the supply and service provider.**

    "6.6.3 The laboratory shall communicate its requirements to external providers for:

    a. the products and services to be provided,

    b. the acceptance criteria,

    c. competence, including any required qualification of personnel,

    d. activities that the laboratory, or its customer, intends to perform at the external provider's premises."[4]

Whether your supplier is a service provider, a supplier vendor, a test lab, or a calibration lab, clear communication as to which products or service you are requesting must be included in your purchase order or other documented communication. Likewise, any acceptance criteria, especially when requesting a statement of conformity and desired decision rule (how measurement uncertainty is accounted for in pass/fail or in-tolerance/out-of-tolerance decisions) should also be clearly stated in your request. It is important to be specific about preferred methods as well as needed parameters, ranges, accuracies, and any desired test points.

When sending or arranging for inspection, measurement, and test equipment (IM&TE) to be calibrated, it is not enough to simply send a request to the calibration vendor to "Please calibrate this." It is important to set out the details of your equipment, an understanding of how it is used, and the overall requirements for an acceptable calibration. Does the calibration have to be ISO/IEC 17025-accredited? If not, what information and data should be documented? To whom should the equipment be returned and how? How will payment be made? What is the expected turnaround time? What is the requested calibration interval? Per ISO/IEC 17025:2017, Clause 7.8.4.3, the calibration vendor cannot recommend a calibration interval, except where it has been agreed with the customer.[5] Making use of a template that includes basic details and criteria and allows for customization can save time, money, and frustration for both the customer and the vendor.[6]

Any requirements for competence, including laboratory accreditation, as well as any activities you or your customer plan to perform at the service provider's location, should also be included in your communication to the service provider.

This communication is especially important when establishing metrological traceability for any equipment that requires calibration. ISO/IEC 17025:2017, Clause 6.5.1, states:

> The laboratory shall establish and maintain metrological traceability of its measurement results by means of a documented unbroken chain of calibrations, each contributing to the measurement uncertainty, linking them to an appropriate reference.[6]

The easiest way to ensure that all the links in the calibration chain are unbroken, and linked to an appropriate reference, is to require that the calibration report contain measurement data (before and after any adjustments), measurement uncertainty data, and the symbol of or reference to the accreditation scope of the ISO/IEC 17025-accredited calibration service provider in your request for service. Because the Accreditation Body reviewed the traceability links from the service provider to an appropriate reference for all calibrations listed on their scope of accreditation, inclusion of the accredited symbol or reference to accreditation scope, including scope number, by the service provider on your calibration report ensures metrological traceability for your calibration.

ANSI/ISO/ASQ Q10012:2003 states, in paragraph 6.4, "Outside suppliers," "The management of the metrological function shall define and document the requirements for products and services to be provided by outside suppliers for the measurement management system. Outside suppliers shall be evaluated and selected based on their ability to meet the documented requirements. Criteria for selection, monitoring and evaluation shall be defined and documented, and the results of the evaluation shall be recorded. Records shall be maintained of the products or services provided by outside suppliers."[7]

NCSL International Recommended Practice-6: *Recommended Practice for Calibration Quality Systems for the Healthcare Industries*, RP-6, lists requirements in several places. Paragraph 5.15, "Supplier Control," states: "Healthcare organizations that use outside supplier maintenance and calibration services suppliers should ensure that the supplier's calibration quality system complies with all the organization's requirements. To this end, organizations are advised to establish an agreement with outside supplier service organizations to:

- Use approved procedures outlining methodology used in maintaining and calibrating the client's M&TE

- Provide a certified report of calibration, complete with all necessary supporting documentation

- Perform the maintenance and calibration activities meeting the organization's specific requirements

- Provide a copy of a quality manual or alternate documentation of their quality system

- Provide objective evidence of quality records, environmental control, equipment history files, and any other relevant quality materials."[8]

NCSL International *RP-6*, paragraph 5.15.1, "Supplier Organization Responsibilities," states: "The organization should provide the supplier with a request or a specification document that can be used to ensure that calibration activities and desired results match the organization's requirements. This document may include, but is not limited to:

- Type of calibration service
- Applicable documents, technical reference, or standard
- Equipment specification
- Service performance requirements
- Physical, environmental, and data-format requirements
- Safety and biohazard requirements
- Confidentiality and security requirements
- Certificate of calibration specifications requirements
- Notification of out-of-tolerance"[9]

Finally, *RP-6*, paragraph 5.15.3, "Supplier Evaluation and Monitoring," states: "Each organization is responsible for conducting on a periodic basis a documented evaluation of all calibration service for adequacy of their quality systems and facilities, process capability and conformance to applicable regulatory and organization's requirements."[10]

Some organizations or regulatory agencies require that parts be replaced only by the original equipment manufacturer (OEM), whereas others only stipulate that the part must be a direct and equal replacement for the part being replaced. Depending on the quality system to which an organization conforms, the cost could be a factor in what to procure, or it might not affect the choice at all. If you must use the OEM, consider comparing prices at their various locations. Another possibility might be purchasing a service contract with an equipment vendor. For a fixed amount of money and for a predetermined amount of time (usually a year), all repairs, including parts and labor, can be arranged. If the IM&TE is getting old but is still serviceable, a service contract can save money overall. Money is saved by not having to purchase new equipment, the set amount of money needed is known and can be budgeted, and all costs are borne by the vendor. Plus, the supplier usually provides some type of warranty on the replaced parts and labor. Also, some vendors provide discounts if the total amount for all the service contracts exceeds a predetermined limit. This is something to consider when looking at the option of using service contracts.

Market consolidation and contraction has resulted in some manufacturers becoming the sole source for the type of products they make and the associated parts and service. In many cases the manufacturers in these situations continue to provide fast, efficient, quality, and reasonably priced service. However, in some instances, the sole-source equipment manufacturer may be a hindrance to providing quality service to your customer. There are situations in which any rational supplier qualification scheme would exclude the manufacturer as a supplier for one or more reasons, but you are forced to deal with the manufacturer anyway because there is no alternative. The laboratory must have a process for dealing with cases like this. A primary recommendation is to fully document every interaction with the supplier. The documentation can be useful in dealing with their complaint process, although a realistic view is that the typical calibration laboratory has little or no leverage in dealing with a huge sole-source corporation. If necessary, you can also use the documentation as a record in dealing with your customers and/or your auditors and assessors.

Chapter 44

Housekeeping and Safety

On a list of most important items, where is safety found? At the top, partway down, or close to the bottom? Does money make a difference in an organization's philosophy about safety? Should it?

It is easy to see where this is going, but do you really care? You should! No matter what type of calibration is performed, or in what environment it is performed, safety should be the driving force behind every task. Calibrations can be redone and money replaced. A person's eye, finger, limb, or life cannot.

Good housekeeping and safety should go together. It is hard to have one without the other. Different industries have their special hazards, and most are explained during orientation or safety briefings. If you work around high voltage or current, you should not wear wire-rimmed glasses, rings, watches, jewelry, or anything that could conduct electricity. Some organizations use the rule "no metal above the waist." When working with machinery, loose clothing or long hair can get a person in trouble very quickly, as can inattentive work habits. The authors have put together a list of items that specific industries should be cautious of when working with or around inspection, measurement, and test equipment (IM&TE).

ALL INDUSTRIES

- Everything should have a place to be when it is not in use. There are many ways to do this; some organizations have had good results from adopting the 5S method.[1] Building on the 5S methodology is 6S, which is 5S + Safety (as the sixth "S").

- Know what personal protective equipment (PPE) is required and when to use it. Use your PPE when it is required.

- Provide safe and appropriate storage for chemicals. All containers should be labeled with identification of their contents. Do not store oxidizers with combustibles or flammables.

- Know where the exit, fire alarm, and fire extinguisher are. Know how to use the fire extinguisher.

5S METHODOLOGY

Based on Japanese words that begin with S, the 5S philosophy focuses on effective workplace organization and standard work procedures. 5S simplifies your work environment, and reduces waste and non-value-added activity while improving quality efficiency and safety (sort—*seiri*, set in order—*seiton*, shine—*seiso*, standardize—*seiketsu*, and sustain—*shitsuke*). The sixth S is the addition of "Safety" to the methodology.

BIOTECHNOLOGY AND PHARMACEUTICAL INDUSTRY

- Always wear safety glasses/goggles/side shields or other eye and face protection that are certified for safety around chemicals and reagents.

- Always wear gloves, because it is not possible to know what chemicals or reagents have come into contact with the IM&TE being calibrated and/or repaired.

- Most companies require a smock or lab coat to be worn when in a lab or working with IM&TE.

- Special considerations arise when working with radioactive material or isotopes; these conditions require gloves, eye protection, and lab coats as a minimum.

- Know where the nearest emergency eyewash and emergency shower are and how to use them. Ensure that they are always in proper working order.

ELECTRONICS OR HIGH-VOLTAGE/CURRENT INDUSTRIES

- Practice the "one-hand rule." Work with only one hand and keep the other hand at your side or in your pocket, away from all conductive material.

- Remove rings, watches, jewelry, chains, metal-rimmed eyewear—anything that could conduct current or voltage. Do not wear metal above the waist.

- Do not work inside energized equipment unless there is no alternative. If you are working with high voltage or high current, have a qualified safety observer nearby.

- Everybody in the laboratory should be qualified in emergency first aid and cardiopulmonary resuscitation (CPR).

- Calibration procedures should be designed such that performance verification of the unit under test can be performed with all covers installed, using only the standard external inputs and outputs.

- The high-voltage calibration area should have limited access, so people not working on that equipment cannot accidentally wander in. This will help reduce accidental shocks. A good practice is to have—and use—a visible indicator, such as a flashing light, physical barriers, or prominent signage so that other people know when the high voltage is energized.

- If the laboratory has a repair area, it should have limited access, so that people not working on that equipment cannot accidentally wander in. This may help reduce accidental burns. The soldering station should have an air filtering exhaust hood or exhaust system to eliminate the fumes.

AIRLINE INDUSTRY

- Be sure the IM&TE meets all requirements before it is released for use. Because it will be used to perform maintenance on an airframe, engine, or aircraft component, the performance of those items can directly affect flight safety.

- If the workload includes aircraft system test sets, the laboratory will require 400 Hz single-phase and/or three-phase electrical power. The receptacles for this power should be clearly marked to indicate that they are not the normal utility power frequency.

PHYSICAL-DIMENSIONAL CALIBRATION

- Treat torque testers with respect. Although the overall probability is small, there is a possibility that a torque wrench or a part of the tester might break and fly off in a random direction.

- When performing pressure calibrations, first be sure all the hydraulic or pneumatic piping, hoses, and couplings are in excellent condition and properly fastened.

- Many items are heavy. Lift them properly to protect your back. Handle them with care to protect the equipment. Be especially careful when placing things on or removing them from a granite or steel surface plate so you do not damage the top of the surface plate.

- Do not stare into a laser beam, as the laser beam can injure the retina and cause blindness. It may be the last thing you see.

- Anything used for calibration of oxygen or breathing air equipment should be in a separate room dedicated for that purpose. That area and the equipment in it must be completely free of contamination from oil and other petroleum products.

- If you have a mercury manometer (a primary standard used for absolute pressure calibration), it must be in a room separate from everything else. That room must have a separate ventilation system, a mercury vapor detector and alarm, and a mercury spill cleanup kit.

- Treat thermal standards with care, as they can cause burn or freeze injuries, depending on the temperature.

COMPUTER INDUSTRY

- Practice the "one-hand rule." If you work with high-voltage circuits, you should keep one hand in your pocket or behind your back to prevent yourself from bridging the circuit with both hands.

- Use the buddy system. Never engage in work with hazardous material, hazardous electrical potentials, or work alone when there is even a remote possibility of injury due to falls, burns, contact with machinery, and so on. Always work with somebody nearby who can give aid in case of an accident.

- Never cheat safety interlocks unless specifically authorized and guided by the original equipment manufacturer (OEM) where doing so is needed to service its equipment.

- Any hazardous work should require engineering safeguards to prevent unauthorized contact or interference. Some examples of engineering safeguards include posting signs, roping off areas, and posting personnel to prevent access.

- Laboratory personnel should be aware of emergency numbers, locations of nearby emergency medical facilities, location of emergency exits and emergency equipment, evacuation procedures, and so on.

- Monthly assessments are encouraged to find possible hazardous conditions such as frayed power cords, blocked emergency exits/equipment access, improperly stored hazardous chemicals, and so on. A standardized checklist is recommended to document compliance and to note discrepancies. All safety discrepancies are escalated to the highest priority level for corrective action.

Appendix A

Further Resources and Publications

This appendix lists resources (some not referenced in this handbook) that are valuable and publicly available (some for a fee).

AOAC International. *Official Methods of Analysis, 21st Edition.* (2019). Comprised of more than 3,000 validated chemical and microbiological methods and consensus standards. Many *Official Methods* have been adopted as harmonized international reference methods by the International Organization for Standardization (ISO), International Dairy Federation (IDF), International Union of Pure and Applied Chemistry (IUPAC), and the Codex Alimentarius Commission. http://www.aoac.org

BIPM. International Metrology Resource Registry. A collaboration of BIPM Member Institutes. http://imrr.bipm.org

———. *Metrologia.* The leading international journal in pure and applied metrology, published by IOP Publishing on behalf of BIPM. https://iopscience.iop.org/journal/0026-1394

———. *SI Brochure—Appendix 2*. Practical realization of the definition of some important units, including ampere, candela, kelvin, kilogram, metre, mole, and second. https://www.bipm.org/en/publications/mises-en-pratique

———. Technical services. https://www.bipm.org/en/technical-areas

CIPM (International Committee for Weights and Measures). CIPM MRA documents. https://www.bipm.org/en/cipm-mra/cipm-mra-documents

CITAC: Cooperation on International Traceability in Analytical Chemistry. http://www.citac.cc

EURACHEM: https://www.eurachem.org

EURACHEM/CITAC Guide: Assessment of Performance and Uncertainty in Qualitative Chemical Analysis (1st ed., 2021). Edited by R. Bettencourt da Silva and S. L. R. Ellison. https://www.eurachem.org

EURACHEM/CITAC Guide: Guide to Quality in Analytical Chemistry—An Aid to Accreditation (3rd ed., 2016). Edited by Vicki Barwick. https://www.eurachem.org

EURACHEM/CITAC Guide: Quality Assurance for Research and Development and Non-Routine Analysis (1998). https://www.citac.cc/wp-content/uploads/2021/02/rdguide.pdf

EURACHEM/CITAC Guide: Setting and Using Target Uncertainty in Chemical Measurement (1st ed., 2015). Edited by Ricardo Bettencourt da Silva and Alex Williams. https://www.eurachem.org

EURACHEM/CITAC Guide: Traceability in Chemical Measurement (2003). A guide to achieving comparable results in chemical measurement. https://www.eurachem.org

EURACHEM/CITAC Guide: Use of Uncertainty Information in Compliance Assessment (2nd ed., 2021). Edited by Alex Williams (UK), and Bertil Magnusson (SE). ISBN 978-0-948926-38-9. https://www.eurachem.org

EURACHEM/CITAC. Setting Target Measurement Uncertainty (July 16, 2018) [leaflet]. Edited by Ricardo Bettencourt da Silva and Alex Williams. https://www.eurachem.org

EURACHEM/EUROLAB/CITAC/Nordtest/UK RSC. *Analytical Methods Committee Guide: Measurement Uncertainty Arising from Sampling* (2d ed., 2019). Edited by Michael H. Ramsey (University of Sussex, UK), Stephen L. R. Ellison (LGC, UK), and Peter Rostron (University of Sussex, UK). https://www.eurachem.org

Fluke Corporation. Cubyt. https://www.Cubyt.io
"Clear, connected measurement science for all, by all. Cubyt is a central community to access and share metrology content. Cubyt is a repository for instrument specifications and calibration methods. Cubyt provides instant access to CAL-SHEETS with test points and tolerances. Cubyt facilitates collaboration and continuous consensus."

ILAC: International Laboratory Accreditation Cooperation. https://ilac.org/publications-and-resources/ilac-guidance-series. The following items are available through this ILAC website:

ILAC G8:09/2019 Guidelines on Decision Rules and Statements of Conformity.

ILAC G18:12/2021 Guideline for Describing Scopes of Accreditation

ILAC G26:11/2018 Guidance for the Implementation of a Medical Accreditation Scheme

ILAC G27:07/2019 Guidance on Measurements Performed as Part of an Inspection Process

ILAC G28:07/2018 Guideline for the Formulation of Scopes of Accreditation for Inspection Bodies

ILAC P8:03/2019 ILAC Mutual Recognition Arrangement (Arrangement): Supplementary Requirements for the Use of Accreditation Symbols and for Claims of Accreditation Status by Accredited Conformity Assessment Bodies

ILAC P9:06/2014 ILAC Policy for Participation in Proficiency Testing Activities

ILAC P15:05/2020 Application of ISO/IEC 17020:2012 for the Accreditation of Inspection Bodies

ISO 8601-2:2019—Date and time—Representations for information interchange—Part 2: Extensions. (February, 2019). Geneva, Switzerland: ISO.

IUPAC: International Union of Pure and Applied Chemistry. https://iupac.org

IUPAC/CITAC Guide: Classification, modeling, and quantification of human errors in a chemical analytical laboratory (IUPAC Technical Report). By Ilya Kuselman and Francesca Pennecchi. *Pure & Applied Chemistry* 88 (5), 477–515 (2016). https://iupac.org/etoc-alert-pure-and-applied-chemistry-may-2016

IUPAC/CITAC Guide: Evaluation of risks of false decisions in conformity assessment of a multicomponent material or object due to measurement uncertainty (IUPAC Technical Report). By Ilya Kuselman, Francesca R. Pennecchi, Ricardo J. N. B. da Silva, and David Brynn Hibbert. *Pure & Applied Chemistry* 93(1), 113–154 (2021). https://www.degruyter.com/document/doi/10.1515/pac-2019-0906/html

IUPAC/CITAC Guide: Investigating out of specification test results of chemical composition based on metrological concepts (IUPAC Technical Report). By Ilya Kuselman, Francesca Pennecchi, Cathy Burns, Aleš Fajgelj, and Paolo de Zorzi. *Pure & Applied Chemistry* 84(9), 1939–1971 (9 July 2012). http://dx.doi.org/10.1351/divAC-REP-11-10-04

IUPAC/CITAC Guide: Selection and use of proficiency testing schemes for a limited number of participants—chemical analytical laboratories (IUPAC Technical Report). By Ilya Kuselman and Ales Fajgelj. *Pure & Applied Chemistry* 82(5), 1099–1135 (2010). http://publications.iupac.org/pac/82/5/1099/index.html

JCGM: Joint Committee for Guides in Metrology https://www.bipm.org/en/committees/jc/jcgm/publications

JCGM 101:2008 Supplement 1—Propagation of distributions using a Monte Carlo method. https://www.bipm.org/en/publications/guides

JCGM 102:2011 Supplement 2—Extension to any number of output quantities. https://www.bipm.org/en/publications/guides

JCGM 104:2009. An introduction to the "GUM" and related documents. https://www.bipm.org/en/publications/guides

JCGM GUM-6:2020. Guide to the expression of uncertainty in measurement—Part 6: Developing and using measurement models. https://www.bipm.org/en/publications/guides

JCGM Annotated VIM3. *VIM definitions with informative annotations. Developed exclusively by JCGM-WG2.* https://www.bipm.org/en/publications/guides

JCTLM: Joint Committee on Traceability in Laboratory Medicine. Home page. https://www.jctlm.org

Keysight Technologies. https://www.keysight.com/us/en/lib/resources/training-materials/fundamentals-of-rf-and-microwave-power-measurements-an1449—application-note-272209.html

NIST: National Institute for Standards and Technology. Unit Conversion | NIST. Gaithersburg, MD: NIST's Physical Measurement Laboratory - Weights and Measures - Metric Program. https://www.nist.gov/pml/weights-and-measures/metric-si/unit-conversion

Nordic Cooperation. NORDTEST. http://www.nordtest.info

NPL: National Physical Laboratory (United Kingdom). Good Practice Guides. https://www.npl.co.uk/resources/gpgs/all-gpgs

Quality assurance of qualitative analysis in the framework of the European project 'MEQUALAN'. By A. Rios, D. Barceló, L. Buydens, and S. Cárdenas. *Accreditation and Quality Assurance* 8(2): 68–77 (2003).

U.S. Department of Defense. *MIL-STD-1309D, Definitions of Terms for Testing, Measurement and Diagnostics* (1992). Washington, D.C.: Government Printing Office.

USP: U.S. Pharmacopeia. https://www.usp.org and https://www.uspnf.com/

USP General Chapter <41> Balances

USP General Chapter <1251> Weighing on an Analytical Balance.

World Metrology Day. https://www.worldmetrologyday.org

YouTube.com Videos

Doing a general search in YouTube for "metrology" can yield virtually endless sources of information.

BIPM. https://www.youtube.com/c/TheBIPM

BIPM. "Redefinition of the SI" (series of videos). https://youtube.com/playlist?list=PL-vj-3_a7wTAoKqKQxT1UWMKTrWpHS5EW

JCGM/WG2 Webinar. Joint Committee for Guides in Metrology (JCGM) Working Group on the International Vocabulary of Metrology (VIM) - WK2: An Overview—New Edition of the International Vocabulary of Metrology (VIM4). 6 May 2021. https://youtu.be/vBkefN0q9lY

Note: For the latest version of any referenced standards, regulations, and guidance, visit these websites.

ASME (American Society of Mechanical Engineers) standards @ https://www.asme.org

ASNT (American Society for Non-Destructive Testing) standards @ https://www.asnt.org

ASTM International standards @ https://www.astm.org

BIPM, CIPM, JCGM publications @ https://www.bipm.org

Eurachem publications @ https://www.eurachem.org

FDA regulations @ www.ecfr.gov

FDA guidance @ https://www.fda.gov/regulator-information/search-fda-guidance-documents

GIDEP database of calibration procedures and calibration data sheets @ https://www.gidep.org (only open to US and Canadian organizations)

IEEE documents @ https://www.ieee.org//

ILAC documents @ https://www.ilac.org

ISA documents @ https://www.isa.org

ISO standards @ www.iso.org

NCSL International publications @ https://www.ncsli.org

NIST documents @ https://www.nist.gov

NPL documents @ https://www.npl.gov

OIML documents @ https://www.oiml.org

Appendix B
Acronyms and Abbreviations

Acronym	Meaning	More information
A2LA	American Association for Laboratory Accreditation	https://www.a2la.org
AC	alternating current	
ACCET	Accrediting Council for Continuing Education and Training	https://accet.org
ACIL	American Council of Independent Laboratories	https://www.acil.org
AF	audio frequency	
AFC	automatic frequency control	
AFRAC	African Accreditation Cooperation	http://www.intra-afrac.com
AFRIMETS	Intra-Africa Metrology System	http://www.afrimets.org
AGC	automatic gain control	
AIAG	Automotive Industry Action Group	https://www.aiag.org
AIST	National Institutes of Advanced Industrial Science and Technology (Japan)	www.aist.go.jp
ALC	automatic level control	
AM	amplitude modulation	
A.M.	*ante meridian* (Latin for "before midday")—between midnight to before noon	
ANAB	ANSI National Accreditation Board	https://www.anab.org
ANOVA	analysis of variance	
ANSI	American National Standards Institute, Inc.	https://www.ansi.org
AOAC	Association of Analytical Communities, International	https://www.aoac.org
APAC	Asia Pacific Accreditation Cooperation	https://www.apac-accreditation.org
API	application programming interface (software)	

Continued

Continued

Acronym	Meaning	More information
APLMF	Asia-Pacific Legal Metrology Forum	https://www.aplmf.org
APMP	Asia Pacific Metrology Programme	https://www.apmpweb.org
ARAC	Arab Accreditation Cooperation	http://arab-accreditation.org
ARFTG	Automatic RF Techniques Group	https://www.arftg.org
ASCII	American Standard Code for Information Interchange (data processing)	
ASEAN	Association of Southeast Asian Nations	https://www.asean.org
ASIC	application-specific integrated circuit	
ASNT	American Society for Nondestructive Testing	https://www.asnt.org
ASQ	American Society for Quality	https://www.asq.org
ASQ CCT	ASQ Certified Calibration Technician	https://asq.org/cert/calibration-technician
ASQ MQD	ASQ Measurement Quality Division	https://my.asq.org/communities/home/156
ASTM	ASTM International (formerly American Society for Testing and Materials)	https://www.astm.org
ATE	automated (or automatic) test equipment	
ATM	asynchronous transfer mode (data communications)	
ATM	automatic teller machine (banking)	
AV	alternating voltage	
AVC	automatic volume control	
AWG	American wire gage	
BAAS	British Association for the Advancement of Science	
BAB	Bangladesh Accreditation Council	https://www.bac.gov.bd
BAS	Bulgarian Accreditation Service	https://www.nab-bas.bg/en
BASIC	Beginner's All-Purpose Symbolic Instruction Code (software)	
BCD	binary-coded decimal	
BELCERT	Belgian Accreditation System	https://economie.fgov.be/en/entreprises/life_enterprise/quality_policy/Accreditation
BELTEST	Belgian Organisation for Accreditation and Conformity Assessment	https://economie.fgov.be/en/entreprises/life_enterprise/quality_policy/Accreditation
BIM	Bulgarian Institute of Metrology	https://en.bim.government.bg
BIOS	basic input/output system (computers)	

Acronym	Meaning	More information
BIPM	International Bureau of Weights and Measures	https://www.bipm.org
BMC	See CMC	
BoA	Bureau of Accreditation (National Accreditation Body of Vietnam)	https://www.boa.gov.vn
BPSLAS	Bureau of Product Standards Laboratory Accreditation Scheme (Philippines)	https://www.bps.dti.gov.ph
BS	British Standard	
BSI	British Standards Institution	https://www.bsigroup.com
BSN	National Standardization Agency (Baden Standardisasi Nasional) (Indonesia)	https://www.bsn.go.id
CAD	computer-aided design	
CAI	Czech Accreditation Institute	https://www.cai.cz
CALA	Canadian Association for Laboratory Accreditation Inc.	https://www.cala.ca
CAM	computer-aided manufacturing	
CARICOM	Caribbean Community	
CASCO	ISO Committee on Conformity Assessment	https://www.iso.org/casco.html
CASE	Coordinating Agency for Supplier Evaluation	https://www.caseinc.org
CC	Consultative Committees (BIPM)	
CCEM	Consultative Committee for Electricity and Magnetism	
CCQM	Consultative Committee for Amount of Substance: Metrology in Chemistry and Biology	
CCT	Certified Calibration Technician	https://asq.org/cert/calibration-technician
CCT	Consultative Committee for Thermometry	
CCTF	Consultative Committee for Time and Frequency	https://www.bipm.org/en/committees/cc/cctf
CCU	Consultative Committee for Units	
CD	compact disk (disc)	
CDC	Centers for Disease Control and Prevention (United States)	https://www.cdc.gov
CDMA	code division multiple access (cellular telephones)	
CEM	Spanish Metrology Center (Centro Espanol de Metrologia)	https://www.cem.es

Continued

Continued

Acronym	Meaning	More information
CEN	European Committee for Standardization	https://www.cencenelec.eu
CENAM	National Center for Metrology (Mexico)	https://www.cenam.mx
CENELEC	European Committee for Electrotechnical Standardization	https://www.cenelec.eu
CGCRE/ INMETRO	Brazil General Coordination for Accreditation	http://www.inmetro.gov.br/
cGLP	current good laboratory practice	
cGMP	current good manufacturing practice	
CGPM	General Conference on Weights and Measures	https://www.bipm.org/en/committees/cg/cgpm
CIPM	International Committee for Weights and Measures	https://www.bipm.org/en/committees/ci/cipm
CIPM MRA	CIPM Mutual Recognition Arrangement	https://www.bipm.org/en/cipm-mra
CMC	calibration and measurement capabilities	
CMM	Component Maintenance Manual (aviation)	
CMM	coordinate measuring machine	
CNAS	China National Accreditation Service for Conformity Assessment	https://eng.cnas.org.cn
CNCA	China Certification and Accreditation Administration	http://www.cnca.gov.cn/cnca
CODATA	Committee on Data for Science and Technology	https://www.codata.org
COFRAC	National Accreditation Authority of France (Comite Francais d'Accreditation)	https://www.cofrac.fr/en
COHSASA	Council for Health Service Accreditation of Southern Africa	https://www.cohsasa.co.za
COOMET	Euro-Asian Cooperation of National Metrological Institutes	http://www.coomet.org
CPR	calibration problem report	
CPR	cardiopulmonary resuscitation	
CRT	cathode-ray tube	
CTA	Consumer Technology Association	https://www.ce.org
DA/DPA	Albania General Directorate of Accreditation (Drejtoria e Përgjithshme e Akreditimit Kërko)	http://dpa.gov.al/en/home-2
DAC	data acquisition and control	
DAC	digital to analog converter	

Acronym	Meaning	More information
DAkkS	Germany National Accreditation Body (Deutsch Akkreditierungsstelle GmbH)	https://www.dakks.de/en
DANAK	Danish Accreditation	https://danak.org
DARPA	Defense Advanced Research Projects Agency (USA)	https://www.darpa.mil
dB	decibel	
dBm	decibels relative to 1 milliwatt in a specified impedance	
DC	direct current	
DEW	directed-energy weapon	
DEW	distant early warning	
DIN	German Institute for Standardization	https://www.din.de
DMAIC	define, measure, analyze, improve, control	
DMM	digital multimeter	
DOD	Department of Defense (USA)	https://www.defense.gov
DOE	Department of Energy (USA)	https://www.energy.gov
DTMF	dual tone, multiple frequency	
DUC	device under calibration	
DUT	device under test	
DV	direct voltage	
DVA	Dutch Accreditation Council (Raad Voor Accreditatie)	https://www.rva.nl
DVD	digital video disk or digital versatile disk	
DVM	differential voltmeter	
DVM	digital voltmeter	
EA	European Cooperation for Accreditation	https://www.european-accreditation.org
EAK	National Accreditation Body of Estonia (Estonian Accreditation Centre)	https://www.eak.ee
ECA	Electric Cooperation Association	https://www.ec-central.org
ECA	National Accreditation Body of Costa Rica (Ente Costarricense de Acreditacion)	https://www.eca.or.cr
EDI	electronic data interchange	
EGAC	Egyptian Accreditation Council	http://www.egac.gov.eg
EIA	Electronic Industries Alliance	https://www.ecianow.org/eia-technical-standards
EMA	Mexico National Accreditation Body (Entidad Mexicana de Acreditacion a.c)	https://www.ema.org.mx/portal_v3

Continued

Continued

Acronym	Meaning	More information
EMF	electromotive force, voltage	
EMI	electromagnetic interference	
EMU	electromagnetic units (obsolete)	
EN	European Standard	https://www.en-standard.eu
EOTC	European Organisation for Conformity Assessment	
ESD	electrostatic discharge	
ESU	electrostatic units (obsolete)	
ESYD	Greece National Accreditation Body (Hellenic Accreditation System)	https://esyd.gr/main/en
ETSI	European Telecommunications Standards Institute	https://www.etsi.org
EUC	equipment under calibration	
Eurachem	Focus for Analytical Chemistry in Europe	https://www.eurachem.org
EURAMET	European Association of National Metrology Institutes	https://www.euramet.org
EUROLAB	European Federation of National Associations of Measurement, Testing and Analytical Laboratories	https://www.eurolab.org
EUT	equipment under test	
FAA	Federal Aviation Administration (USA)	https://www.faa.gov
FAQ	frequently asked questions	
FAX	facsimile (transmission of images by telephone)	
FDA	Food and Drug Administration (USA)	https://www.fda.gov
FFT	fast Fourier transform (mathematics)	
FINAS	Finnish Accreditation Service	https://www.finas.fi/sites/en/Pages/default.aspx
FM	frequency modulation	
FO	fiber optics	
FS	Federal Standard (USA)	
FTP	file transfer protocol (Internet)	
GEP	Good Experimentation Practices	
GIDEP	Government–Industry Data Exchange Program (USA)	https://www.gidep.org
GLONASS	global navigation satellite system (Russia)	
GLP	Good Laboratory Practices	
GMP	Good Manufacturing Practices	

Acronym	Meaning	More information
GMT	Greenwich mean time—obsolete, see UTC	
GPETE	general purpose electronic test equipment	
GPIB	general purpose interface bus (Tektronix term for IEEE-488)	
GPS	global positioning system (satellites) (short form of NAVSTAR GPS)	https://www.space.com/19794-navstar.html
GSM	Global System for Mobile (telecommunications)	
GUM	*Guide to the Expression of Uncertainty in Measurement* (JCGM 100 Series)	https://www.bipm.org/en/publications/guides
HAA	Croatian Accreditation (National Accreditation Body)	https://www.akreditacija.hr/EN
HIPOT	high potential (a type of electrical test)	
HKAS	Hong Kong Accreditation Service	https://www.itc.gov.hk/en/quality/hkas/about.htm
HPGL	Hewlett-Packard graphics language (plotters, printers)	
HPIB	Hewlett-Packard interface bus (Hewlett-Packard term for IEEE-488)	
HPML	Hewlett-Packard multimeter language	
HTML	hypertext markup language	https://www.w3.org
HTTP	hypertext transfer protocol (what makes the www work!)	https://www.w3.org
HV	high voltage	
HVAC	heating, ventilation, and air conditioning	
IAAC	Inter-American Accreditation Cooperation	http://www.iaac.org.mx
IACET	International Accreditors for Continuing Education and Training	https://iacet.org
IAF	International Accreditation Forum Inc.	https://www.iaf.nu
IANZ	International Accreditation New Zealand	https://www.ianz.govt.nz
IAQG	International Aerospace Quality Group	https://www.iaqg.org
IARM	Institute for Accreditation for the Republic of Macedonia	https://www.iarm.gov.mk
IAS	International Accreditation Service, Inc. (USA)	https://www.iasonline.org
IATF	International Automotive Task Force	
IC	integrated circuit	
ICC	International Code Council	https://www.iccsafe.org

Continued

Continued

Acronym	Meaning	More information
ICSCA	Industry Cooperation on Standards and Conformity Assessment (Australia)	
ICT	information and communications technologies	
IEC	International Electrotechnical Commission	https://www.iec.ch
IEEE	Institute of Electrical and Electronics Engineers	https://www.ieee.org
IFCC	International Federation of Clinical Chemistry and Laboratory Medicine	https://www.ifcc.org
ILAC	International Laboratory Accreditation Cooperation	https://www.ilac.org
ILC	inter-laboratory comparison	
IM&TE	inspection, measurement, and test equipment	
IMEKO	International Measurement Confederation	https://www.imeko.org
INDOCAL	Dominican Institute for Quality (Instituto Domincano Para la Calidad)	https://indocal.gob.do
INN	National Accreditation Body of Chile (Instituto Nacional de Normaizacion)	https://www.inn.cl
IPAC	National Accreditation Body of Portugal (Instituto Português de Acreditação)	https://www.ipac.pt
IR	infrared	
ISA	International Society of Automation	https://www.isa.org
ISDN	Integrated Services Digital Network	
ISO	International Organization for Standardization	https://www.iso.org
ISO	Insurance Services Office (insurance risk ratings)	https://www.iso.com
ISRAC	Israel Laboratory Accreditation Authority	https://www.israc.gov.il
IST	International Steam Table	
IT	information technology	
IT	In-tolerance	
ITS-90	International Temperature Scale of 1990	https://www.its-90.com
ITU	International Telecommunications Union	https://www.itu.int
IU	International Units	
IUC	instrument under calibration	
IUPAC	International Union of Pure and Applied Chemistry	https://www.iupac.org

Acronym	Meaning	More information
IUPAP	International Union of Pure and Applied Physics	https://www.iupap.org
IUT	instrument under test	
JAB	Japan Accreditation Board for Conformity Assessment	https://www.jab.or.jp/en
JAN	Joint Army-Navy (USA)	
JAS-ANZ	Joint Accreditation System of Australia and New Zealand	https://www.jas-anz.com.au
JCGM	Joint Committee for Guides in Metrology	https://www.bipm.org/en/committees/jc/jcgm
JIT	just-in-time	
JNLA	Japan National Laboratory Accreditation	https://www.nite.go.jp/en/iajapan/jnla/index.html
JUSE	Union of Japanese Scientists and Engineers	https://www.juse.or.jp
KAB	Korea Accreditation Board	https://www.eng.kab.or.kr
KCDB	Key Comparison Database	https://www.bipm.org/kcdb
KOLAS	Korea Laboratory Accreditation Scheme	https://kolas.go.kr
LAN	local area network	
LASER	light amplification by stimulated emission of radiation	
LATAK	Latvian National Accreditation Bureau	https://www.latak.lv
LCD	Liquid crystal display	
LCR	inductance-capacitance-resistance (meter) from the circuit symbols	
LED	light-emitting diode	
LIMS	laboratory information management systems	
LinkedIn	LinkedIn professional network	https://www.linkedin.com
LORAN	LOng-RAnge aid to Navigation	
LSD	least significant digit	
LSD	National Accreditation Body of Iceland (Loggildingarstofa)	https://neytendastofa.is/English
MAP	Measurement Assurance Program	
MASER	microwave amplification by stimulated emission of radiation	
MIL-HDBK	Military Handbook (USA)	
MIL-PRF	Military Performance-Based Standard (USA)	

Continued

Continued

Acronym	Meaning	More information
MIL-STD	Military Standard (USA)	
MOD	Ministry of Defence (UK)	https://www.mod.uk
MOU	memorandum of understanding	
MRA	mutual recognition arrangement	
MSA	measurement systems analysis	
MSC	Measurement Science Conference	https://www.msc-conf.com
NA	Norwegian Accreditation Body	https://www.akkreditert.no
NAB	Irish National Accreditation Board	https://www.inab.ie
NAB	National Accreditation Body of Italy (2009 combining of SIT, SINAL, and SINCERT) L'ente Italiano di Accreditamento	https://www.accredia.it/en
NABCB	National Accreditation Board for Certification Bodies (Quality Council of India)	http://www.qcin.org
NAC	National Accreditation Council (Thai Industrial Standards Institute)	https://www.tisi.go.th
NACLA	National Cooperation for Laboratory Accreditation	https://www.nacla.org
NAFTA	North American Free Trade Agreement	
NAPT	National Association for Proficiency Testing	https://www.proficiency.org
NASA	National Aeronautics and Space Administration (USA)	https://www.nasa.gov
NAT	National Accreditation Board of Hungary (Nemzeti Akkreditalo Testulet)	http://www.nat.hu
NATA	National Association of Testing Authorities, Australia	https://www.nata.com.au
NATO	North Atlantic Treaty Organization	https://www.nato.int
NAVSTAR GPS	navigation system with time and ranging global positioning system	https://www.space.com/19794-navstar.html
NBS	National Bureau of Standards; see NIST	https://www.nist.gov
NCA	National Centre of Accreditation (Republic of Kazakhstan)	http://www.nca.kz
NCO	non-commissioned officer (military rank)	
NCSL	National Conference of Standards Laboratories (see NCSLI)	https://www.ncsli.org
NCSLI	NCSL International	https://www.ncsli.org
NCWM	National Conference on Weights and Measures (USA)	https://www.ncwm.net

Acronym	Meaning	More information
NDE	nondestructive evaluation	https://www.ndt-ed.org
NDT	nondestructive testing	https://www.ndt-ed.org
NEPAD	New Partnership for Africa's Development	https://www.nepad.org
NIST	National Institute of Standards and Technology (USA)	https://www.nist.gov
NISTIR	National Institute of Standards and Technology Internal/Interagency Report	https://www.nist.gov
NITE	National Institute of Technology and Evaluation (Japan)	https://www.nite.go.jp
NLA	National Laboratory Association (South Africa)	https://www.nla.org.za
NLAB	National Laboratories Accreditation Bureau (Egypt)	https://www.egac.gov.eg
NMI	National Metrology Institute	
NMISA	National Metrology Institute of South Africa	https://www.nmisa.org
NORAMET	North American Cooperation in Metrology	https://www.sim-metrologia.org
NORDTEST	Nordic cooperation (for accreditation)	http://www.nordtest.info
NPL	National Physical Laboratory (UK)	https://www.npl.co.uk
NRC	National Research Council (Canada)	https://nrc.canada.ca/en
NRC	United States Nuclear Regulatory Commission (US NRC)	https://www.nrc.gov
NTSC	National Television System Committee (TV format in USA, Canada, Japan)	
NVLAP	National Voluntary Laboratory Accreditation Program	https://www.nist.gov/nvlap
OAA	National Accreditation Body of Argentina (Organismo Argentino de Acreditacion)	https://www.oaa.com.ar
OEM	original equipment manufacturer	
OIML	International Organization of Legal Metrology	https://www.oiml.org
OOT	out of tolerance	
PAL	phase alternation by line (TV format in Europe, China)	
PARD	periodic and random deviation	
PAVM	phase angle voltmeter	
PC	personal computer	

Continued

Continued

Acronym	Meaning	More information
PC	printed circuit (wiring board)	
PCA	Polish Centre for Accreditation (Polskie Centrum Akredytacji)	https://www.pca.gov.pl
PCS	personal communications services (telecommunications, USA)	
PDCA	plan–do–check–act (the Deming or Shewhart cycle)	
PDF	Portable Document Format (documents; trademark of Adobe Corp.)	
PDSA	plan–do–study–act (the Deming or Shewhart cycle, another version)	
PM	phase modulation	
PM	preventive maintenance	
P.M.	*post meridian* (Latin for "after midday")—from noon to before midnight	
PME	precision measuring equipment	
PME	professional military education (U.S. Armed Forces)	
PMEL	precision measurement equipment laboratory	
PMET	precision measuring equipment and tooling	
PNAC	Pakistan National Accreditation Council	https://www.pnac.org.pk
PO	purchase order	
POTS	plain old telephone service (a basic subscriber line with no extra features)	
PPB	parts per billion (preferred usage is "parts in 10 E 09")	
PPE	personal protective equipment	
PPM	parts per million (preferred usage is "parts in 10 E 06")	
PPT	parts per trillion (preferred usage is "parts in 10 E 12")	
PT	proficiency test	
PTB	Physikalisch-Technische Bundesanstalt (German NMI)	https://www.ptb.de
PXI	compact PCI eXtensions for Instrumentation	http://www.pxisa.org
QA	quality assurance	
QC	quality control	

Acronym	Meaning	More information
QHE	quantum Hall effect	
QMP-LS	Quality Management Program—Laboratory Service (Ontario Medical Association, Canada)	http://www.qmpls.org
QMS	quality management system	
QS	quality system (as in QS-9000)	
R&R	repeatability and reproducibility	
RADAR	radio detection and ranging	
RAM	random-access memory	
REMCO	ISO Committee on reference materials	https://www.iso.org/committee/55002.html
RENAR	Romanian Accreditation Association	https://www.renar.ro
RF	radio frequency	
RFID	radio frequency identification (tags)	
RMS	root-mean-square	
ROM	read-only memory (computers)	
RP	Recommended Practice (as in NCSL RP-1)	
RS	Recommended Standard (as in EIA RS-232)	
RSS	root-sum-square	
RTD	resistor temperature device	
RTF	rich text format (documents)	
RvA	Dutch Accreditation Council (Raad voor Accreditatie) (Netherlands)	https://www.rva.nl/en
SA	Slovenian National Accreditation Body	https://www.slo-akreditacija.si
SAAA	South African Accreditation Authority	https://www.saaa.gov.za
SABS	South African Bureau of Standards	https://www.sabs.co.za
SAC	Singapore Accreditation Council	https://www.sac-accreditation.gov/sg
SADCA	Southern African Development Community Accreditation	https://www.sadca.org
SADCMET	Southern African Development Community Cooperation in Measurement Traceability	www.sadcmet.org
SAE	Society of Automotive Engineers	https://www.sae.org
SANAS	South African National Accreditation System	https://www.sanas.co.za
SAS	Swiss Accreditation Service, State Secretariat for Economic Affairs (SECO)	https://www.sas.admin.ch/sas/en/home.html

Continued

Continued

Acronym	Meaning	More information
SAW	surface acoustic wave	
SCC	Standards Council of Canada	https://www.scc.ca
SCPI	standard commands for programmable instrumentation	https://www.ivifoundation.org/scpi
SI	International System of Units	
SIM	Intra-American Metrology System (Sistema Interamer de Metrol)	https://www.sim-metrologia.org
SNA	scalar network analyzer	
SNAS	Slovak National Accreditation Service	https://www.smi.sk
SOC	US Department of Labor Standard Occupational Classification	https://www.bls.gov/soc
SONAR	SOund Navigation And Ranging	
SOP	standard operating procedure	
SPC	statistical process control	
SPETE	special purpose electronic test equipment	
SQC	statistical quality control	
SQL	structured query language	
SQUID	superconducting quantum interference device	
Sr. NCO	senior Noncommissioned Officer (U.S. military rank)	
SRG	spinning rotor gage	
SRM	standard reference material (NIST)	https://www.nist.gov/srm
SWEDAC	Swedish Board for Accreditation and Conformity Assessment	https://www.swedac.se
SWR	standing wave ratio	
TAF	Taiwan Accreditation Foundation	https://www.taftw.org.tw/en
TAG	Technical Advisory Group (ISO)	
TAG	Truck Advisory Group (AIAG)	
TAI	International Atomic Time	https://www.bipm.org/documents/20126/59466374/6_establishment_TAR20.pdf/5b18b648-0d5a-ee02-643d-a60ed6c148fc
TAR	test accuracy ratio	
TCP/IP	transfer control protocol/Internet protocol	
TDMA	time division multiple access (cellular telephones)	
TDR	time-domain reflectometer	

Acronym	Meaning	More information
TELCO	telecommunications company	
THD	total harmonic distortion	
TI	test item	
TIA	Telecommunications Industry Association	https://www.tiaonline.org
TMDE	test, measurement, and diagnostic equipment	
TQL	total quality leadership	
TQLS	total quality lip service (humor)	
TQM	total quality management	
TRMS	true RMS	
TUR	test uncertainty ratio	
TURKAK	Turkish Accreditation Agency	https://www.turkak.org.tr
TVC	thermal voltage converter	
TWSTT	Two-Way Satellite Time Transfer	
UILI	International Union of Independent Laboratories	https://www.uili.org
UKAS	United Kingdom Accreditation Service	https://www.ukas.com
UMTS	universal mobile telecommunications system	
UNIDO	United Nations Industrial Development Organization	https://www.unido.org
USB	universal serial bus	https://www.usb.org
USNO	US Naval Observatory	https://www.usno.navy.mil/USNO
UT1	universal time scale 1	https://www.bipm.org/en/committees/cc/cctf
UTC	universal coordinated time (formerly GMT): time of day at 0° longitude (ITU Radiocommunication Assembly, 2002)	http://www.itu.int/dms_pubrec/itu-r/rec/tf/R-REC-TF.460-6-200202-I!!PDF-E.pdf
UUC	unit under calibration	
UUT	unit under test	
VIM	ISO *International vocabulary of basic and general terms in metrology* (JCGM 200)	https://www.bipm.org/en/publications/guides
VNA	vector network analyzer	
VOM	volt-ohm meter	
VTVM	vacuum tube voltmeter	
VXI	VME extensions for instrumentation	https://www.vxibus.org

Continued

Continued

Acronym	Meaning	More information
WADA	World Anti-Doping Agency	https://www.wada-ama.org
WAN	wide area network	
W-CDMA	wideband code division multiple access (cellular telephones)	
WELMEC	European Cooperation in Legal Metrology	https://www.welmec.org
WHO	World Health Organization	https://www.who.int
WLAN	wireless local area network	
WPAN	wireless personal area network (range approximately 10 m)	
WTO	World Trade Organization	https://www.wto.org
WWV	NIST time and frequency radio transmitter in Ft. Collins, Colorado	https://www.nist.gov/time-distribution/radio-station-wwv
WWVB	NIST 60 kHz digital time code radio transmitter in Ft. Collins, Colorado	https://www.nist.gov/pml/time-and-frequency-division/time-distribution/radio-station-wwvb
WWVH	NIST time and frequency radio transmitter in Kauai, Hawaii	https://tf.nist.gov/stations/wwvh.htm
WWW	World Wide Web Consortium	https://www.w3.org
Z	symbol for impedance	
ZABS	Zambia Bureau of Standards	https://www.zabs.org.zm
ZINQAP	Zimbabwe National Quality Assurance Programme	https://www.zinqap.org.zw
ZULU	(U.S. military) indicates reference to UTC time	
ZULU	phonetic for the last letter of the alphabet	

Appendix C
Glossary of Terms

INTRODUCTION

This glossary is a quick reference to the meaning of common terms. Terms are referenced from a number of documents and standards. International standards such as the *VIM*, the *GUM*, ISO and ISO/IEC standards, and BIPM references take precedence. Secondary standards and publications, such as NCSL International Recommended Practices and Lab Management, ANSI/NCSL Z540 standards, and NIST (U.S. National Metrology Institute) and U.S. Department of Defense standards, serve as supplemental references.

In technical, scientific, and engineering work (such as metrology) it is important to correctly use words that have a technical meaning. Definitions of these words are in relevant national, international, and industry standards, journals, and other publications, as well as publications of relevant technical and professional organizations. Those documents give the intended meaning of the word, so everyone in the business knows what it is. *In technical work, only the technical definitions should be used.*

Many of these definitions are adapted from the references. Where they are unique to a cited reference, the reference and clause or section are included. In some cases, several may have been merged to better clarify the meaning or adapt the wording to common metrology usage. The technical definitions may be different from the definitions published in common grammar dictionaries. However, the purpose of common dictionaries is to record the ways that people actually use words, not to standardize the way the words should be used. *If a word is defined in a technical standard, its definition from a common grammar dictionary should never be used in work where the technical standard can apply.*

The following is a list of publications referenced in this appendix. When referred to in the glossary, the respective reference identification and clause are included.

American National Standards Institute/American Society for Quality/International Organization for Standardization. 2003. *ANSI/ISO/ASQ Q10012:2003—Measurement Management Systems—Requirements for Measurement Processes and Measuring Equipment.* Milwaukee, WI: ASQ Quality Press. [ANSI/ISO/ASQ Q10012:2003]

American National Standards Institute/American Society for Quality/International Organization for Standardization. 2015. *ASQ/ANSI/ISO*

2015/ISO 9000:2015 Quality Management Systems—Fundamentals and Vocabulary. Milwaukee, WI: ASQ Quality Press. [ISO 9000:2015]

American National Standards Institute/NCSL International. 1994. *ANSI/NCSL Z540.1-1994 (R2002), American National Standards for Calibration—Calibration laboratories and measuring and test equipment—General requirements*. Boulder, CO: NCSL. [Z540.1]

American Society for Quality/American National Standards Institute/International Organization for Standardization. 2018. *ASQ/ANSI/ISO 19011:2018—Guidelines for auditing management systems*. Milwaukee, WI: ASQ Quality Press. [ASQ/ANSI/ISO 19011:2018]

Bureau International des Poids et Mesures. 2021. *The International System of Units (SI)*. 2019-05-20. Sèvres, France: BIPM. https://www.bipm.org/en/measurement-units. [BIPM]

Bureau International des Poids et Mesures. 2021, March 30. *CIPM MRA-G-13 Calibration and Measurement Capabilities in the Context of the CIPM MRA—Guidelines for Their Review, Acceptance and Maintenance. Version 1.1*. Sèvres, France: BIPM. [CIPM MRA-G-13]

Bureau International des Poids et Mesures, International Committee of Weight and Measures (CIPM). CIPM MRA Key Comparison Database (KCDB). Sèvres, France: BIPM. https://www.bipm.org/en/cipm-mra/kcdb [KCDB]

International Laboratory Accreditation Cooperation. n.d. *ILAC MRA and Signatories International Laboratory Accreditation Cooperation*. https://ilac.org/ilac-mra-and-signatories. [ILAC MRA]

International Organization for Standardization. 2015. *ISO Guide 30:2015 Reference materials—Selected terms and definitions*. Geneva, Switzerland: ISO. [*ISO Guide 30:2015*]

International Organization for Standardization/International Electrotechnical Commission. 2010. *ISO/IEC 17043:2010 Conformity assessment—General requirements for proficiency testing*. Geneva, Switzerland. [ISO/IEC 17043:2010]

International Organization for Standardization/International Electrotechnical Commission. 2017. *ISO/IEC 17025:2017—General requirements for the competence of testing and calibration laboratories*. Geneva, Switzerland. ISO. [*ISO/IEC 17025:2017*]

International Organization for Standardization/International Electrotechnical Commission. 2020. *ISO/IEC 17000:2020 Conformity assessment—Vocabulary and general principles*. Geneva, Switzerland. ISO. [*ISO/IEC 17000:2020*]

Joint Committee for Guides in Metrology. 2008. *JCGM 100:2008, Evaluation of measurement data—Guide to the expression of uncertainty in measurement*

(*GUM*); BIPM, IEC, IFCC, ILAC, ISO, IUPAC, IUPAP, and OIML. Geneva, Switzerland: BIPM. [*GUM*]

Joint Committee for Guides in Metrology. 2012. *JCGM 106:2012, Evaluation of measurement data—The role of measurement uncertainty in conformity decision*. BIPM, IEC, IFCC, ILAC, ISO, IUPAC, IUPAP, and OIML. Sèvres, France: BIPM. [JCGM 106:2012]

Joint Committee for Guides in Metrology. 2012. *JCGM 200:2012, International vocabulary of metrology—Basic and general concepts and associated terms (VIM)*; BIPM, IEC, IFCC, ILAC, ISO, IUPAC, IUPAP, and OIML. Geneva, Switzerland: BIPM. [*VIM*]

National Institutes of Standards and Technology/SEMATECH. 2013, April. *NIST/SEMATECH e-Handbook of Statistical Methods*. https://doi.org/10.18434/M32189 [NIST/SEMATECH]

NCSL International. 1999. *LM-3 Glossary of Metrology-Related Terms*, 2nd ed. Boulder, CO: NCSL International. [*LM-3*]

Yale University. 1998. *Sample Means*. New Haven, CT: Yale Department of Statistics and Data Science. http://www.stat.yale.edu/Courses/1997-98/101/sampmn.htm#clt [Yale]

Some terms may be listed in this glossary to expand on the definitions in these references but should be considered an *addition to* the references in the preceding list, *not a replacement* of them. (It is assumed that a calibration or metrology activity includes copies of these as part of its basic reference material.) Most acronyms and abbreviations are listed in Appendix B.

GLOSSARY

accreditation [of a laboratory]—third-party *attestation* related to a *conformity assessment body*, conveying formal demonstration of its competence, *impartiality*, and consistent operation in performing specific conformity assessment activities (ISO/IEC 17000:2020, 7.7).

Accreditation Body—authoritative body that performs *accreditation* (ISO/IEC 17000:2020, 4.7). An organization that conducts laboratory accreditation evaluations in conformance to ISO/IEC 17011:2017.

accreditation certificate—document issued by an *Accreditation Body* to a laboratory that has met the conditions and criteria for accreditation. The certificate, with the documented measurement parameters and their best uncertainties, serves as proof of accredited status for the time period listed. An accreditation certificate without the documented parameters is incomplete.

accreditation criteria—set of requirements used by an accrediting body that a laboratory must meet to be accredited.

adjustment—changing (by electronic, electrical, or physical means) a variable in an item to cause a change in its output characteristics.

analyte—the specific component measured in a chemical analysis.

assessment—examination, typically performed on site, of a testing or calibration laboratory to evaluate its conformance to conditions and criteria for accreditation.

attestation—issue of a statement, based on a *decision*, that fulfillment of *specified requirements* has been demonstrated (ISO/IEC 17000:2020, 7.3).

attribute—an identifiable feature capable of being measured.

audit—systematic, independent, and documented *process* for obtaining *objective evidence* and evaluating it objectively to determine the extent to which the *audit criteria* are fulfilled (ISO 19011:2018, 3.1 and ISO 9000:2015, 3.13.1)

audit criteria—set of *policies, procedures,* or requirements used as a reference against which *objective evidence* is compared (ISO 19011:2018, 3.7 and ISO 9000:2015, 3.13.7)

audit plan—description of the activities and arrangements for an *audit* (ISO 19011:2018, 3.6 and ISO 9000:2015, 3.13.6)

audit scope—extent and boundaries of an *audit* (ISO 19011:2018, 3.5 and ISO 9000:2015, 3.13.5)

best measurement capability—*see* calibration measurement capability (CMC)

bias—a systematic error inherent in a method or caused by some artifact or idiosyncrasy of the measurement system. *Estimate of a systematic measurement error* (*VIM*, 2.18).

   instrument bias—average of replicate indications minus a reference quality value (*VIM*, 4.20).

calibration—(1) (*See VIM*, 2.39, and NCSL pages 4–5 for primary and secondary definitions.) An operation that, under specified conditions, in a first step, establishes a relation between the quantity values with *measurement uncertainties* provided by *measurement standards* and corresponding *indications* with associated measurement uncertainties and, in a second step, uses this information to establish a relation for obtaining a measurement result from an indication.

Calibration is a term that has many different—but similar—definitions. Calibration is performed with the item being calibrated in its normal operating configuration; that is, as the normal operator would use it. Calibration without associated measurement uncertainties is not a valid calibration.

The result of a calibration is a determination of the performance quality of the instrument with respect to the desired specifications. This may be in the form of a pass/fail decision, determining or assigning one or more values, or the determination of one or more corrections.

The calibration process consists of comparing an inspection, measurement, and test equipment (IM&TE) unit with specified tolerances, but of unverified accuracy, to a measurement system or device of specified capability and known uncertainty in order to detect, report, or minimize by adjustment any deviations from the tolerance limits or any other variation in the accuracy of the instrument being compared.

Calibration is performed according to a specified documented calibration procedure, under a set of specified and controlled measurement conditions, and with a specified and controlled measurement system.

Notes:

- A requirement for calibration does *not* imply that the item being calibrated can or should be adjusted.

- The calibration process *may* include, if necessary, calculation of correction factors or adjustment of the instrument being compared to reduce the magnitude of the inaccuracy.

- In some cases, minor repair such as replacement of batteries, fuses, or lamps, or minor adjustment such as zero and span, may be included as part of the calibration activity, although these repairs are not calibration.

- Calibration does *not* include any maintenance or repair actions except as just noted. *See also:* calibration procedure, performance test. *Contrast with:* calibration (2A, 2B), repair.

calibration—(2A) Many manufacturers incorrectly use the term *calibration* to name the process of alignment or adjustment of an item that is either newly manufactured or is known to be out of tolerance or is otherwise in an indeterminate state. Many calibration procedures in manufacturers' manuals are actually factory alignment procedures that only have to be performed if a unit under test (UUT) is in an indeterminate state because it is being manufactured, is known to be out of tolerance, or after it is repaired. When used this way, *calibration* means the same as "alignment" or *adjustment*, which are repair activities and excluded from the metrological definition of calibration.

(2B) In many cases, IM&TE instruction manuals may use *calibration* to describe tasks normally performed by the operator of a measurement system. Examples include performing a self-test as part of normal operation or performing a self-calibration (normalizing) of a measurement system before use. When calibration is used to refer to tasks like this, the intent is that they are part of the normal work done by a trained user of the system. These and similar tasks are excluded from the metrological definition of calibration.

Contrast with: calibration (1). *See also:* self-calibration, standardization.

calibration activity or provider—a laboratory or facility (including personnel) that performs calibrations in an established location or at customer location(s). It may be external or internal, including subsidiary operations of a larger entity. It may be called a calibration laboratory, shop, or department; a metrology laboratory or department; or an industry-specific name; or any combination or variation of these.

calibration certificate—(1) A document which states that a specific item was calibrated by an organization. The certificate identifies the item calibrated, the organization presenting the certificate, and the effective date. A calibration certificate should provide other information to allow the user to judge the adequacy and quality of the calibration.

(2) In a laboratory database program, "certificate" often refers to the permanent record of the result of a calibration. A laboratory database certificate is a record that cannot be changed; if it is amended later, a new certificate is created. *See also:* calibration report.

calibration measurement capability (CMC)—a Calibration and Measurement Capability available to customers under normal conditions (CIPM MRA-G-13, p. 1)

- As published in the BIPM key comparison database (KCDB) of the CIPM MRA; or

- As described in the laboratory's scope of accreditation granted by a signatory to the ILAC Arrangement.

calibration method—defined technical procedure for performing a calibration or verification.

calibration procedure—a controlled document that provides a validated method for evaluating and verifying the essential performance characteristics, specifications, or tolerances for a model of measurement or testing equipment. A calibration procedure documents one method of verifying the actual performance of the item being calibrated against its performance specifications. It provides a list of recommended calibration standards to use for the calibration, a means to record quantitative performance data both before and after adjustments (where adjustments are possible), and information sufficient to determine if the unit being calibrated is operating within the necessary performance specifications. A calibration procedure always starts with the assumption that the unit under test is in good working order and only has to have its performance verified. Note: A calibration procedure need not include any maintenance or repair actions.

calibration program—a process of the quality management system that includes management of the use and control of calibrated IM&TE, and the process of calibrating IM&TE used to determine conformance to requirements or used in supporting activities and to maintain confidence in the status of calibration. A calibration program may also be called a "measurement management system" (ISO 10012:2003, 1).

calibration report—a document that provides details of the calibration of an item. In addition to the basic items of a calibration certificate, a calibration report includes details of the methods and standards used, the parameters checked, and the actual measurement results and associated measurement uncertainty. *See also:* calibration certificate.

calibration seal—a device, placard, or label that, when removed or tampered with, by virtue of its design and material, clearly indicates tampering. The purpose of a calibration seal is to ensure the integrity of the calibration. A calibration seal is usually imprinted with a legend similar to "Calibration Void if Broken or Removed" or "Calibration Seal—Do Not Break or Remove." A calibration seal provides a means of deterring the user from tampering with any adjustment point that can affect the calibration of an instrument and detecting any attempt to access controls that can affect the calibration of an instrument.

Note: A calibration seal may also be referred to as a "tamper-evident seal."

certification—third-party *attestation* related to an *object of conformity assessment*, with the exception of accreditation (ISO/IEC 17000:2020, 7.6).

certified reference material (CRM)—*reference material*, accompanied by documentation issued by an authoritative body and providing one or more specified property values with associated uncertainties and traceabilities, using valid procedures (*VIM*, 5.14). *See related*: *reference material*.

CRM is a reference material characterized by a metrologically valid procedure for one or more specified properties, accompanied by a reference material certificate that provides the value of the specified property, its associated uncertainty, and a statement of metrological traceability (International Organization for Standardization. 2016. *ISO 17034:2016 General requirements for competence of reference material producers*. Geneva, Switzerland: ISO 17034:2016, 3.2).

NIST also includes the following about certified reference materials (https://www.nist.gov/srm/srm-definitions):

1. The concept of value includes qualitative attributes such as identity or sequence. Uncertainties for such attributes may be expressed as probabilities.

2. Metrologically valid procedures for the production and certification of reference materials are given in, among others, ISO 17034 and ISO Guide 35.

3. ISO Guide 31 gives guidance on the contents of certificates.

characteristic—a distinguishing feature (ISO 9000:2015, 3.10.1). A physical, chemical, visual, functional, or any other identifiable property of a product of material.

coefficient of variation—the standard deviation divided by the value of the parameter measured.

combined standard uncertainty—standard uncertainty of the result of a measurement when that result is obtained from the values of a number of other quantities, equal to the positive square root of a sum of terms, the terms being the variances or covariances of these other quantities weighted according to how the measurement result varies with changes in those quantities (*GUM*, 2.3.4). *See also*: expanded uncertainty.

competence—for a laboratory, the demonstrated ability to perform the tests or calibrations within the accreditation scope and to meet other criteria established by the Accreditation Body. For a person, the demonstrated ability to apply knowledge and skills to achieve intended results (ISO 9000:2015, 3.10.4). Note: The word "qualification" is sometimes used in the personnel sense.

confidence interval—the estimated range of values which is likely to include the unknown population parameter, the estimated range being calculated from a given set of sample data (Yale). The confidence interval is the p value, or calculated probability, that is calculated from sample statistics. Confidence intervals can be calculated for points, lines, slopes, standard deviations, etc. For an infinite (or very large compared to the sample) population, the central limit theorem is used to calculate the confidence interval.

conformity assessment—demonstration that *specified requirements* are fulfilled. Conformity assessment includes, but is not limited to, testing, inspection, validation, verification, certification, and accreditation (ISO/IEC 17000:2020, 4.1).

conformity assessment body (CAB)—body that performs conformity assessment activities, excluding accreditation (ISO/IEC 17000:2020, 4.6).

continual improvement—recurring activity to enhance *performance* (ISO 9000:2015, 3.3.2).

correction—compensation for an estimated systematic effect (*VIM*, 2.53). The correction value is equal to the negative of the bias. An example is the value calculated to compensate for the calibration difference of a reference thermometer or for the calibrated offset voltage of a thermocouple reference junction. *Correction* is also defined as the measured error with the changed sign. *See also:* bias, random measurement error, systematic measurement error.

corrective action—action to eliminate the cause of a *nonconformity* and to prevent recurrence. *Compare with:* preventive action.

coverage factor—number larger than one by which a *combined standard uncertainty* is multiplied to obtain an *expanded uncertainty* (*VIM*, 2.38 [expanded measurement uncertainty]). The coverage factor is identified by the symbol k, typically in the range 2 to 3. It is usually given the value 2, which corresponds to an approximate probability of 95 % for degrees of freedom greater than 10.

coverage interval—interval containing the set of true quantity values of a *measurand* with a stated probability, based on the information available (*VIM*, 2.36). Note: A coverage interval should not be termed "confidence interval," to avoid confusion with the statistical concept (see *GUM* 6.2.2).

customer—person or *organization* that could or does receive a *product* or a *service* that is intended for or required by this person or organization (ISO 9000:2015, 3.2.4).

data—facts about an *object* (ISO 9000:2015, 3.8.1).

decision—conclusion, based on the results of *review*, that fulfillment of *specified requirements* has or has not been demonstrated (ISO/IEC 17000:2020, 7.2).

decision rule—rule that describes how measurement uncertainty is accounted for when stating conformity with a specified requirement (ISO/IEC 17025:2017, 3.7, p.2).

deficiency—nonfulfillment of conditions and/or criteria for accreditation; sometimes referred to as a "nonconformity."

departure value—a term used by a few calibration laboratories to refer to *bias*, "error," or *systematic measurement error*. The exact meaning can usually be determined from examination of the calibration certificate.

detection limit—the smallest concentration of amount of some component of interest that can be measured by a single measurement with a stated level of confidence.

document—information and the medium on which it is contained (ISO 9000:2015, 3.8.5).

effectiveness—extent to which planned activities are realized and planned results are achieved (ISO 9000:2015, 3.7.11).

end of period (EOP)—refers to the measurement reliability of an item at the end of its calibration interval (LM-3, p 10).

environment—the aggregate of all external and internal conditions (such as temperature, humidity, radiation, magnetic and electrical fields, shock, vibration) either natural, human-made, or self-induced, that influence the form, performance, reliability, or survival of an item (LM-3, p 11).

equipment (measuring and test)—all equipment used to measure, gauge, test, inspect, or otherwise determine compliance with prescribed technical requirements. It may generate, modify, or measure electrical, electronic, physical, or mechanical quantities or parameters to defined specifications or aid in the measurement process. Includes test fixtures, test devices, and measurement standards. (LM-3, p 10) Also referred to as TMDE (test, measurement, and diagnostic equipment). *See:* IM&TE. In chemistry fields, measuring and test equipment is often referred to as "instruments" where the supporting (nonreporting) components of a system are referred to as "equipment."

equivalence—(1) Acceptance of the competence of other National Metrology Institutes (NMIs), accreditation bodies, and/or accredited organizations in other countries as being equal to the NMI, Accreditation Body, and/or accredited organizations within the host country. (2) A formal, documented determination that a specific instrument or type of instrument is suitable for use in place of the one originally listed, for a particular application.

expanded uncertainty—quantity defining an interval about the result of a measurement that may be expected to encompass a large fraction of the distribution of values that could reasonably be attributed to the measurand (*GUM* 2.3.5).

external provider/external supplier—*provider* that is not part of the *organization* (ISO 9000:2015, 3.2.6).

functional test—a test that determines whether the unit under test is functioning properly. The operational environment (such as stimuli and loads) can be actual or simulated.

gages and measuring equipment, classification of (LM-3, p. 16):

length standards: standards of length and angle from which all measurements of gages are derived. Consists of precision gage blocks, end mastering rods, line graduated standards, master disks, calibrated wires and rolls, precision squares, graduated circles, and similar items.

master gages: gages that are made to their basic dimensions as accurately as possible and are used in reference, such as for checking or setting inspection or manufacturer's gages.

inspection gages: The representative of the purchasing agency uses inspection gages to inspect product for acceptance. These gages are made in accordance with established design requirements.

manufacturer's gages: manufacturer's gages are used by the manufacturer or contractor for inspection of parts during "production." To ensure that the product will be within the limits of the inspection gages, manufacturer's or working gages should have dimensional limits, resulting from gage tolerances and wear allowances, slightly farther from the specified limits of the parts inspected.

gage R&R—gage repeatability and reproducibility study, which (typically) employs numerous instruments, personnel, and measurements over a period of time to capture quantitative observations. The data captured are analyzed statistically to obtain best measurement capability, which is expressed as an uncertainty with a coverage factor of $k = 2$ to approximate 95%. The number of instruments, personnel, measurements, and length of time are established to be statistically valid consistent with the size and level of activity of the organization.

GUM—an acronym commonly used to identify the ISO *Guide to the Expression of Uncertainty in Measurement* (JCGM 200:2012, ISO/IEC Guide 99:2007). In the United States, the related documents are NIST Technical Note 1297–1994, *Guidelines for Evaluating and Expressing the Uncertainty of NIST Measurement Results*; and ANSI/NCSL Z540-2-1997, *U.S. Guide to the Expression of Uncertainty in Measurement*.

Hipot (test)—"Hipot" is an abbreviation for high potential (voltage) test. A Hipot test is the most common type of electrical safety test. Hipot is a deliberate application of extreme high voltage, direct or alternating, to test the insulation system of an electrical product well beyond its normal limits. An accepted guideline for the applied value is to double the highest operating voltage plus one kilovolt. Current through the insulation is measured while the voltage is applied. If the current exceeds a specified value, a failure is indicated. Hipot testing is normally done during research and development, factory production and inspection, and sometimes after repair. A synonym is "dielectric strength testing."

A high-potential tester normally has meters to display the applied voltage and the leakage current at the same time.

Caution! Hipot testing involves lethal voltages.

Caution! Hipot testing is a potentially destructive test. If the insulation system being tested fails, the leakage creates a path of permanently lowered resistance. This may damage the equipment and may make it unsafe to use.

Routine use of Hipot testing must be carefully evaluated. Note: *Hypot* is a registered trademark of Associated Research Corp. and should not be used as a generic term.

IEC (International Electrochemical Commission)—the world's leading organization that prepares and publishes international standards for all electrical, electronic, and related technologies, collectively known as "electrotechnology."

ILAC (International Laboratory Accreditation Cooperation)—"ILAC is the international organisation for accreditation bodies operating in accordance with ISO/IEC 17011 and involved in the accreditation of conformity assessment bodies including calibration laboratories (using ISO/IEC 17025), testing laboratories (using ISO/IEC 17025), medical testing laboratories (using ISO 15189), inspection bodies (using ISO/IEC 17020), proficiency testing providers (using ISO/IEC 17043) and reference material producers (using ISO 17034)."

ILAC-MRA (ILAC Mutual Recognition Arrangement)—"provides significant technical underpinning to the calibration, testing, medical testing and inspection results, provision of proficiency testing programs and production of the reference materials of the accredited conformity assessment bodies that in turn delivers confidence in the acceptance of services and results. In addition, the ILAC MRA enhances the acceptance of products across national borders. By removing the need for additional calibration, testing, medical testing and/or inspection of imports and exports, technical barriers to trade are reduced. In this way the ILAC MRA promotes international trade and the free-trade goal of 'accredited once, accepted everywhere' can be realized" (ILAC MRA).

impartiality—objectivity with regard to the outcome of a conformity assessment activity (ISO/IEC 17000:2020). Presence of objectivity (ISO/IEC 17025:2017, 1).

improvement—activity to enhance *performance* (ISO 9000:2015, 3.3.1).

IM&TE—an acronym for "inspection, measurement, and test equipment." This term includes all items that fall under a calibration or measurement management program. IM&TE items are typically used in applications where the measurement results are used to determine conformance to technical or quality requirements before, during, or after a process. Some organizations do not include instruments used solely to check for the presence or absence of a condition (such as voltage, pressure, and so on) where a tolerance is not specified, and the indication is not critical to safety. Note: Organizations may refer to IM&TE items as MTE (measuring and testing equipment), TMDE (test, measuring, and diagnostic equipment), GPETE (general-purpose electronic test equipment), PME (precision measuring equipment), PMET (precision measuring equipment and tooling), or SPETE (special purpose electronic test equipment).

indication—quantity value provided by a measuring instrument or a measuring system (*VIM*, 4.1).

instrument—a device intended to make a measurement, alone or in conjunction with supplementary equipment (Yale). In chemistry fields, measuring and test equipment is often referred to as "instruments" where the supporting (nonreporting) components of a system are referred to as "equipment" (*ANSI/NCSL Z540.1*).

insulation resistance (test)—an insulation resistance test provides a qualitative measure of the performance of an insulation system. Resistance is measured in megohms. The applied voltage can be as low as 10 volts DC, but use of 500 or 1000 volts is more common. Insulation resistance can be a predictor of potential failure, especially when measured regularly and plotted over time on a trend chart. The instrument used for this test may be called an *insulation resistance tester* or a *megohmmeter*. An insulation tester displays the insulation resistance in megohms and may display the applied voltage. Note: *Megger* is a registered trademark of AVO International and should not be used as a generic term.

interlaboratory comparison—organization, performance, and evaluation of measurements or tests on the same or similar items or materials by two or more laboratories in accordance with predetermined conditions (ISO/IEC 17043:2010, 3.4).

internal audit—a systematic and documented process for obtaining audit evidence and evaluating it objectively to verify that a laboratory's operations comply with the requirements of its quality system. An internal audit is done by or on behalf of the laboratory itself, so it is a first-party audit (ISO 9000:2015, 3.13.1).

international measurement standard—measurement standard recognized by signatories to an international agreement and intended to serve worldwide (*VIM*, 5.2).

International Organization for Standardization (ISO)—an international nongovernmental organization chartered by the United Nations in 1947, with headquarters in Geneva, Switzerland. The mission of ISO is "to promote the development of standardization and related activities in the world with a view to facilitating the international exchange of goods and services, and to developing cooperation in the spheres of intellectual, scientific, technological and economic activity." The scope of ISO's work covers all fields of business, industry, and commerce except electrical and electronic engineering. The members of ISO are the designated national standards bodies of each country. (ANSI represents the United States.) ISO is not really an acronym. The founding delegates chose the word "ISO," derived from the Greek "isos," meaning equal (International Organization for Standardization. *ISO—About Us*. Geneva, Switzerland: ISO. https://www.iso.org/about-us.html). *See also:* ISO.

International System of Units (SI)—the recommended practical system of units of measurement, with the international abbreviation **SI**. (The acronym SI is from the French *Système International d'Unités*.) From 20 May 2019 on, all SI units are defined in terms of constants that describe the natural world. The SI was defined in terms of seven base units and derived units defined as products of powers of the base units. The seven base units were chosen for historical reasons, and were, by convention, regarded as dimensionally independent: the meter, the kilogram, the second, the ampere, the kelvin, the mole, and the candela. This role for the base units continues in the present SI even though the SI itself is now defined in terms of the defining natural physical constants (BIPM).

The SI system is popularly known as the *metric system*.

intra-laboratory comparison—organization, performance, and evaluation of measurements or tests on the same or similar items within the same laboratory in accordance with predetermined conditions (ISO/IEC 17025:2017, 3.4).

intrinsic measurement standard—measurement standard based on an inherent and reproducible property of a phenomenon or substance (*VIM*, 5.10). Also called "intrinsic standard."

ISO—"iso" is a Greek word root meaning "equal." The International Organization for Standardization chose the word as the short form of its name so the name would be a constant in all languages. In this context, ISO is not an acronym. (If the acronym based on the full name were used, it would be different in each language.) The name also symbolizes the mission of the organization: to equalize standards worldwide.

key comparison database (KCDB)—outcomes of the CIPM Mutual Recognition Arrangement (CIPM MRA). The CIPM MRA is the framework through which National Metrology Institutes (NMIs) demonstrate the international equivalence of their measurement standards and mutual acceptance of the calibration and measurement certificates they issue.

The outcomes of this KCDB MRA include:

1. the internationally recognized calibration and measurement capabilities (CMCs) of the participating institutes

2. the results of scientific comparisons underpinning these CMCs (KCDB).

laboratory—body that performs one or more of the following activities: testing, calibration, sampling (associated with subsequent testing or calibration) (ISO/IEC 17025:2017, 3.6).

level of confidence—defines an interval about the measurement result that encompasses a large fraction p of the probability distribution characterized by that result and its combined standard uncertainty, where p is the coverage probability or level of confidence of the interval. Effectively, the coverage level expressed as a percentage.

maintenance of a measurement standard (conservation of a measurement standard)—set of operations necessary to preserve the metrological properties of a *measurement standard* within stated limits (*VIM*, 5.11).

management—coordinated activities to direct and control an *organization* (ISO 9000:2015, 3.3.3).

management review—the planned, formal, periodic, and scheduled examination of the status and adequacy of the quality management system in relation to its quality policy and objectives by the organization's top management.

management system—set of interrelated or interacting elements of an *organization* to establish *policies* and objectives, and *processes* to achieve those objectives (ISO 9000:2015, 3.5.3).

measurand—quantity intended to be measured (*VIM*, 2.3).

measurement—process of experimentally obtaining one or more quantity values that can reasonably be attributed to a quantity (*VIM*, 2.1).

measurement accuracy (of a measurement)—the closeness of agreement between a measured quantity value and a true quantity value of a *measurand* (*VIM*, 2.13). Because the true value is always unknown, accuracy of a measurement is always an estimate. An accuracy statement by itself has no meaning other than as an indicator of quality. It has quantitative value only when accompanied by information about the uncertainty of the measuring system.

measurement bias—estimate of a *systematic measurement error* (*VIM*, 2.18). The value and direction of the bias is determined by calibration and/or gage repeatability and reproducibility (R&R) studies. Adding a correction, which is always the negative of the bias, compensates for the bias. *See also:* correction, systematic measurement error.

measurement error—measured quantity value minus a reference quantity value (*VIM*, 2.16). The error can never be known exactly; it is always an estimate. Error may be systematic and/or random. Systematic error (also known as *bias*) may be corrected. Note: Measurement error should not be confused with production error or mistake.

Also called: error, error of measurement. *See also:* bias, correction (of error), random error, systematic error.

measurement precision—closeness of agreement between *indications* or measured quantity values obtained by replicate *measurements* on the same or similar objects under specified conditions (*VIM*, 2.15); also called "precision." Precision is a property of a measuring system or instrument. Precision is a measure of the repeatability of a measuring system: how much agreement there is within a group of repeated measurements of the same quantity under the same conditions.

measurement standard—(*See VIM*, 5.1 through 5.12 and NCSL LM-3 pages 36–38.) "Realization of the definition of a given **quantity**, with stated **quantity value** and associated **measurement uncertainty**, used as a reference" (*VIM*, 5.1).

The highest-level standards, found in national and international metrology laboratories, are the realizations or representations of SI units. . . .

The value and uncertainty of the standard define a limit to the measurements that can be made: a laboratory can never have better precision or accuracy or uncertainty than its standards. Measurement standards are generally used in calibration laboratories. Items with similar uses in a production shop are generally regarded as working-level instruments by the calibration program.

A measurement standard is an IM&TE item, artifact, standard reference material, or measurement transfer standard that is designated as being used only to perform calibrations of other IM&TE items. As calibration standards are used to calibrate other IM&TE items, they are more tightly controlled and characterized than the workload items for which they are used. Calibration standards generally have lower uncertainty and better resolution than general-purpose items. Designation as a calibration standard is based on the use of the specific instrument, however, not on any other consideration. For example, in a group of identical instruments, one might be designated as a calibration standard while the others are all general-purpose IM&TE items.

measurement system—the set of equipment, conditions, people, methods, and other quantifiable factors that combine to determine the success of a measurement process. The measurement system includes at least the test and measuring instruments and devices, associated materials and accessories, the personnel, the procedures used, and the physical environment.

measurement uncertainty—non-negative parameter characterizing the dispersion of the quantity value being attributed to a *measurand*, based on the information used (*VIM*, 2.26).

Uncertainty is an estimate of the range of values that the true value of the measurement is within, with a specified level of confidence. After an item that has a specified tolerance has been calibrated using an instrument with a known accuracy, the result is a value with a calculated uncertainty. *See also:* Type A evaluation of measurement uncertainty, Type B evaluation of measurement uncertainty.

metrological traceability—property of a measurement result whereby the result can be related to a reference through a documented unbroken chain of *calibrations*, each contributing to the *measurement uncertainty* (*VIM*, 2.41). Also referred to as "traceability" or "traceable."

Traceability is the ability to relate individual measurement results, through a contiguous sequence of measurement accuracy verifications, to national or internationally accepted measurement systems (LM-3, p. 42). The stated references are normally the base or supplemental SI units as maintained by a National Metrology Institute, fundamental or physical natural constants that are reproducible and have defined values, ratio type comparisons, certified standard reference materials, or industry or other accepted consensus reference standards.

Traceability provides the ability to demonstrate the accuracy of a measurement result in terms of the stated reference. Measurement assurance methods applied to a calibration system include demonstration of traceability. A calibration system operating under a program controls system only implies traceability. Evidence of traceability includes the calibration report (with values and uncertainty) of calibration standards, but the report alone is not sufficient. The laboratory must also apply and use the data.

The evidence of metrological traceability on an ISO/IEC 17025-accredited lab's calibration certificate or test report is the Accreditation Body's logo with scope number (or narrative reference to accreditation scope number by the accreditation body). This is because an ISO/IEC 17025-accredited lab has been assessed by its accreditation body and the metrological traceability of the measurement results has been assessed by the accreditation body to ensure metrological traceability of measurement results.

A calibration laboratory, a measurement system, a calibrated IM&TE, a calibration report, or any other thing is not and cannot be traceable to a national standard. Only the *result* of a specific measurement can be said to be metrologically traceable, provided all the conditions just listed are met. The term "NIST-traceable" is an incorrect statement of metrological traceability. Additionally, reference to a NIST (test) number is specifically *not* evidence of traceability. A NIST number is merely a catalog number of the specific service provided by NIST to a customer so it can be identified on a purchase order. A "NIST number" is not evidence of what happened, if anything, nor is it evidence of traceability of measurements.

metrology—science of *measurement* and its application (*VIM*, 2.2).

mobile or field operations—operations that are independent of an established calibration laboratory facility. Mobile operations may include working from an office space, home, vehicle, or the use of a virtual office.

monitoring—determining the status of a *system,* a *process,* a *product,* a *service,* or an activity (ISO 9000:2015, 3.11.3).

national measurement standard—measurement standard recognized by national authority to serve in a state or economy as the basis for assigning quantity values to other measurement standards for the kind of quantity concerned. (*VIM*, 5.3) Also called *national measurement standard*.

natural (physical) constant—a fundamental value that is accepted by the scientific community as valid and defined by the SI. Each of the seven base units has no assigned uncertainty until it is realized by a laboratory. Natural constants are used in the basic theoretical descriptions of the universe.

Other examples of natural physical constants important in metrology are the triple point of water (273.16 K), the quantum charge ratio (h/e), the gravitational constant (G), the ratio of a circle's circumference to its diameter (pi), and the base of natural logarithms (e). Defined by CODATA Task Group on Fundamental Physical Constants. CODATA Recommended Values and literature detailing the latest least squares adjustment are available on the NIST website on Fundamental Physical Constants.

nonconformity—nonfulfillment of a requirement (ISO 9000:2015, 3.6.9).

nondestructive testing (NDT)—the field of science and technology dealing with the testing of materials without damaging the material or impairing its future usefulness. The purposes of NDT include discovering hidden defects, quantifying quality attributes, or characterizing the properties of the material, part, structure, or system. NDT uses methods such as X-ray and radioisotopes, dye penetrant, magnetic particles, eddy current, ultrasound, and more. NDT specifically applies to physical materials, not biological specimens.

object—anything perceivable or conceivable (ISO 9000:2015, 3.6.1). Also called "entity" or "item."

object of conformity assessment—entity to which *specified requirements* apply (ISO/IEC 17000:2020, 4.2).

objective evidence—*data* supporting the existence or verity of something (ISO 9000:2015, 3.8.3).

offset—the difference between a nominal value (for an artifact) or a target value (for a process) and the actual measured value.

For example, if the thermocouple alloy leads of a reference junction probe are formed into a measurement junction and placed in an ice point cell, and the reference junction itself is also in the ice point, then the theoretical thermoelectric emf measured at the copper wires should be zero. Any value other than zero is an offset created by inhomogeneity of the thermocouple wires combined with other uncertainties. *Compare with:* bias.

on-site operations—operations that are based in or directly supported by an established calibration laboratory facility, but are actually performed (the calibration actions) at customer locations. This includes climate-controlled mobile laboratories. Also referred to as "field operations."

organization—person or group of people that has its own functions with responsibilities, authorities, and relationships to achieve its objectives (ISO 9000:2015, 3.2.1).

performance—measurable result (ISO 9000:2015, 3.7.8).

performance test—(also known as a performance verification) is the activity of verifying the performance of an item of measuring and test equipment to provide assurance that the instrument is capable of making correct measurements when it is properly used. A performance test is done with the item in its normal operating configuration. A performance test is the same as a *calibration* (1). *See also:* calibration (1).

policy—within an organization, intentions and direction of an *organization* as formally expressed by its top management (ISO 9000:2015, 3.5.8).

A policy defines and sets out the basic objectives, goals, vision, or general management position on a specific topic. A policy describes what management intends to do regarding a given portion of business activity. Policy statements relevant to the quality management system are generally stated in the quality manual. Policies can also be in the organization's policy/procedure manual. *See also:* procedure.

preventive action—action to eliminate the cause of a potential *nonconformity* or other potential undesirable situation. Preventive action practices are part of continuous improvement (ISO 9000:2015, 3.12.1). *Contrast with:* corrective action.

primary measurement standard—measurement standard established using a primary reference measurement procedure, or created as an artifact, chosen by convention (*VIM*, 5.4). Also called "primary standard."

Accepted as having the highest metrological qualities and whose value is accepted without reference to other standards of the same quantity. Examples: Triple point of water cell and caesium beam frequency standard.

Probability of False Accept (PFA)—a statement of conformity evaluation where the chance of accepting a failing result as a pass is estimated. Also referred to as "consumer's risk"—the risk of problems with a measurement (or product) that does not meet quality and will go undetected and thus be reported (or enter the market) (JCGM 106:2012, p. 25).

Probability of False Reject (PFR)—a statement of conformity evaluation where the chance of rejecting a passing result as a fail is estimated. Also referred to as "producer's risk"—the risk that a good measurement (or product) will be rejected or marked as a bad measurement (or product by the consumer) (JCGM 106:2012, p. 25).

procedure—specified way to carry out an activity or *process* (ISO 9000:2015, 3.4.5). A procedure describes a specific process for implementing all or a portion of a policy. There may be more than one procedure for a given policy. A procedure has more detail than a policy but less detail than a work instruction. The level of detail supplied should correlate with the level of education and training of the people with the usual qualifications to do the work and the amount of judgment normally allowed to them by management. Some policies may be implemented by detailed procedures, whereas others may only have a few general guidelines.

process—set of interrelated or interacting activities that use inputs to deliver an intended result (ISO 9000:2015, 3.4.1).

product—output of an *organization* that can be produced without any transaction taking place between the organization and the *customer* (ISO 9000:2015, 3.7.6).

proficiency testing—evaluation of participant performance against pre-established criteria by means of interlaboratory comparisons (ISO/IEC 17043:2010, 3.7).

provider/supplier—*organization* that provides a *product* or a *service* (ISO 9000:2015, 3.2.5).

quality—degree to which a set of inherent characteristics of an *object* fulfills requirements (ISO 9000:2015, 3.6.2).

quality management system (QMS)—part of a *management system* with regard to *quality* (ISO 9000:2015, 3.5.4).

quality manual—*specification* for the *quality management system* of an *organization* (ISO 9000:2015, 3.8.8).

Document that describes the quality management policy of an organization with respect to a specified conformance standard. The quality manual briefly defines the general policies as they apply to the specified conformance standard and affirms the commitment of the organization's top management to the policy. In addition to its regular use by the organization, auditors use the quality manual when they audit the quality management system. The quality manual is generally provided to customers on request; therefore, it does not usually contain any detailed policies and never contains any procedures, work instructions, or proprietary information.

random measurement error—component of a *measurement error* that in replicate *measurements* varies in an unpredictable manner (*VIM*, 2.19).

Random error causes scatter in the results of a sequence of readings and, therefore, is a measure of dispersion. Random error is usually evaluated by Type A methods, but Type B methods are also used in some situations. Note: Contrary to widespread belief, the *GUM* specifically does not replace *random error* with either Type A or Type B methods of evaluation (JCGM 100:3008 *GUM* 3.2.2, note 2). *See also:* random error, random measurement error. *Compare with* systematic measurement error.

record—*document* stating results achieved or providing evidence of activities performed (ISO 9000:2015, 3.8.10).

reference material (RM)—material, sufficiently homogeneous and stable with respect to one or more specified properties, which has been established to be fit for its intended use in measurement or in examination of nominal properties (*VIM*, 5.13).

According to *ISO Guide 30:2015*, the following should be noted regarding reference materials:

1. RM is a generic term.

2. Properties can be quantitative or qualitative, e.g., identity of substance or species.

3. Uses may include the calibration of a measurement system, assessment of a measurement procedure, assigning values to other materials, and quality control.

4. International Vocabulary of Metrology—Basic and General Concepts and Associated Terms has an analogous definition (*VIM*; *GUM*, 5.13), but restricts the term "measurement" to apply to quantitative values and not to qualitative properties. However, Note 3 of *GUM* 5.13, specifically includes the concept of qualitative attributes, called "nominal properties."

When using reference materials, a single RM cannot be used for both calibration and validation of results in the same measurement procedure. *See related:* certified reference material

reference measurement standard (standard)—measurement standard designated for the calibration of other measurement standards for quantities of a given kind in a given organization or at a given location (*VIM*, 5.6). Also called "reference standard."

repair—the process of returning an unserviceable or nonconforming item to serviceable condition. The instrument is opened, or has covers removed, or is removed from its case and may be disassembled to some degree. Repair includes adjustment or alignment of the item as well as component-level repair. (Some minor adjustment such as zero and span may be included as part of the calibration process, but adjustments and repair are different from calibration.) The need for repair may be indicated by the results of a calibration. For calibratable items, recording of calibration data before repair may or may not be possible. For calibratable items, repair is always followed by calibration of the item. Passing of the calibration test indicates success of the repair. *Contrast with:* calibration (1), repair (minor).

repair (minor)—*Minor repair* is the process of quickly and economically returning an unserviceable item to serviceable condition by doing simple work using parts that are in stock in the calibration lab. Examples include replacement of batteries, fuses, or lamps; minor cleaning of switch contacts; repairing a broken wire; or replacing one or two in-stock components. The need for repair may be indicated by the results of a calibration. For calibratable items, recording of calibration data before repair may or may not be possible. For calibratable items, repair is always followed by calibration of the item. Passing of the calibration test indicates success of the repair. *Minor repairs* are repairs that take no longer than a short time, as defined by laboratory management, where no parts have to be ordered from external suppliers, and where substantial disassembly of the instrument is not required. *Contrast with:* calibration (1), repair.

repeatability—*measurement precision* under a set of *repeatability conditions of measurement* (*VIM*, 2.21).

repeatability condition (of measurement)—condition of *measurement* out of a set of conditions that include the same measurement procedure, same operators, same *measurement system*, and replicate measurements on the same or similar objects over a short period of time (*VIM*, 2.20).

reported value—quantity value representing a measurement result (*VIM*, 2.10). One or more numerical results of a calibration process, with the associated measurement uncertainty, as recorded on a calibration report or certificate. The specific type and format vary according to the type of measurement being made. In general, most reported values will be in one of these formats:

Measurement result and uncertainty. The reported value is usually the mean of a number of repeat measurements. The uncertainty is usually expanded uncertainty as defined in the *GUM*.

Deviation from the nominal (or reference) value and uncertainty. The reported value is the difference between the nominal value and the mean of a number of repeat measurements. The uncertainty of the deviation is usually *expanded uncertainty* as defined in the *GUM*.

Estimated systematic error and uncertainty. The value may be reported this way when it is known that the instrument is part of a measuring system, and the systematic error will be used to calculate a correction that will apply to the measurement system results.

requirement—need or expectation that is stated, generally implied or obligatory (ISO 9000:2015, 3.6.4).

review—consideration of the suitability, adequacy, and effectiveness of selection and determination activities, and the results of these activities, with regard to fulfillment of *specified requirements* by an *object of conformity assessment* (ISO/IEC 17000:2020, 7.1).

risk—effect of uncertainty (ISO 9000:2015, 3.7.9).

round-robin—an interlaboratory test (measurement, analysis, or experiment) performed independently several times (NIST/SEMATECH). *See:* interlaboratory comparison.

sampling—selection and/or collection of material or data regarding an *object of conformity assessment* (ISO/IEC 17000:2020, 6.1).

scope of accreditation—for an accredited calibration or testing laboratory, the scope is a documented list of calibration or testing fields, parameters, specific measurements, or calibrations and their *calibration measurement capability (CMC)*. The scope document is an attachment to the certificate of accreditation, and the certificate is incomplete without it. Only the calibration or testing areas that the laboratory is accredited for are listed in the scope document, and only the listed areas may be offered as accredited calibrations or tests. The accreditation body usually defines the format and other details.

secondary measurement standard—measurement standard established through calibration with respect to a primary measurement standard for a quantity of the same kind. Also called "secondary standard."

self-calibration—a process performed by a user or automatically by an instrument (based on predetermined criteria or parameters) for the purpose of making an IM&TE instrument or system ready for use. The process may be required at intervals such as every power-on sequence; if the ambient temperature changes by a specified amount; or once per shift, day, or week of continuous operation. Once initiated, the process may be performed totally by the instrument or may require user intervention and/or use of external calibrated reference standards. The usual purpose is accuracy enhancement by characterization of errors inherent in the measurement system before the item to be measured is connected.

Self-calibration is *not* equivalent to periodic calibration (performance verification) because it is not performed using a calibration procedure and does not meet the metrological traceability requirements for calibration. Also, if an instrument requires self-calibration before use, then that will also be accomplished at the start of a calibration procedure. Self-calibration may also be called *standardization*. *Contrast with:* calibration (1).

service—output of an *organization* with at least one activity necessarily performed between the organization and the *customer* (ISO 9000:2015, 3.7.7).

specification—a quantitative description of the specified characteristics of an instrument, device system, product, or process. A *document* stating requirements (ISO 9000:2015, 3.8.7). In metrology, a specification is a documented statement of the expected performance capabilities of a large group of substantially identical measuring instruments, given in terms of the relevant parameters and including the accuracy or uncertainty. Customers use specifications to determine the suitability of a product for their own applications. A product that performs outside the specification limits when tested (calibrated) is rejected for later adjustment, repair, or scrapping.

specified requirement—need or expectation that is stated (ISO/IEC 17000:2020, 5.1).

standard (document)—a standard (industry, national, government, or international standard; a "norme") is a document that describes the processes and methods that must be performed in order to achieve a specific technical or management objective, or the methods for evaluation of any of these. An example is ANSI/ NCSL Z540-1-1994, a national standard that describes the requirements for the quality management system of a calibration organization and the requirements for calibration and management of the measurement standards used by the organization.

standard operating procedure (SOP)—a term used by some organizations to identify policies, procedures, or work instructions. A procedure adopted for repetitive use when performing a specific measurement or one developed by the user (LM-3, p. 39). *See also:* policy.

standard reference data—*reference material* issued by a recognized authority (*VIM*, 5.17). Example 1—values of the fundamental physical constants as regularly recommended by CODATA of ICSU. Example 2—relative atomic mass values, also called atomic weight values, of the elements, as evaluated every two years by IUPAC-CIAAW, approved by the IUPAC General Assembly, and published in *Pure and Applied Chemistry*.

standard uncertainty—uncertainty of the result of a measurement expressed as a standard deviation (*GUM*, 2.3.1).

standardization—in analytical chemistry, the assignment of a compositional value to one standard on the basis of another standard.

system—set of interrelated or interacting elements (ISO 9000:2015, 3.5.1).

systematic measurement error—component of *measurement error* that in replicate *measurements* remains constant or varies in a predictable manner (*VIM*, 2.17).

> NOTE 2: Systematic measurement error, and its causes, can be known or unknown. A *correction* can be applied to compensate for a known systematic measurement error (*VIM*, 2.17).

> NOTE 3: Systematic measurement error equals measurement error minus random measurement error (*VIM*, 2.17).

See also: bias, correction. *Compare with:* random measurement error.

test accuracy ratio—(1) In a calibration procedure, the test accuracy ratio (TAR) is the ratio of the accuracy tolerance of the unit under calibration to the accuracy tolerance of the calibration standard used (JCGM 106:2012, p. 18).

$$TAR = \frac{UUT\ tolerance}{STD\ tolerance}$$

The TAR must be calculated using identical parameters and units for the unit under test (UUT) and the calibration standard. If the accuracy tolerances are expressed as decibels, percentage, or another ratio, they must be converted to absolute values of the basic measurement units. (2) In the normal use of IM&TE items, the TAR is the ratio of the tolerance of the parameter being measured to the accuracy tolerance of the IM&TE. Note: TAR may also be referred to as the "accuracy ratio." TAR is *different* from the test uncertainty ratio (TUR).

test uncertainty ratio (TUR)—ratio of the span of the tolerance of a measurement quantity subject to calibration, to twice the 95% expanded uncertainty of the measurement process used for calibration. Note: This applies to two-sided tolerances (JCGM 106:2012, p. 18).

$$TUR = \frac{S_t}{2U_{95\%}}$$

where:

S_t = the span of the tolerance limits for the UUT. The span, or range, is the width of the acceptance tolerance for the measurement quantity.
$U_{95\%}$ = the 95% expanded uncertainty for the calibration process.

Note: The *expanded uncertainty* is already at k = 2 (or at approximately 2 standard deviations, where the level of confidence is approximately 95%). The denominator is *twice* the *expanded uncertainty* (or approximately four times the standard uncertainty).

Note: The expanded uncertainty is of the measurement process, not just that of the reference standards (M&TE) used for calibration (as was used in the previous definition of TUR).

Units of measurement: The TUR must be calculated using identical parameters and units for the UUT and the expanded uncertainty. If the units are expressed as decibels, percentage, or another ratio, they must be converted to absolute values of the basic measurement units.

testing—determination of one or more characteristics of an *object of conformity assessment,* according to a *procedure* (ISO/IEC 17000:2020, 6.2). An element of inspection or testing generally denotes the determination by technical means of the properties or elements of supplies or components thereof, including functional operation, and involves the application of established scientific principles and procedures (LM-3, p 40).

tolerance—the total permissible deviation of a measurement from a designated value. A tolerance is a design feature that defines limits within which a quality characteristic is supposed to be on individual parts; it represents the maximum allowable deviation from a specified value. Tolerances are applied during design and manufacturing. A tolerance is a property of the item being measured (LM-3, p41). *Compare with:* specification, uncertainty.

transfer measurement device—device used as an intermediary to compare *measurement standards*. Note: Sometimes measurement standards are used as transfer devices (*VIM*, 5.9). Also called "transfer device."

Typical applications of transfer standards are to transfer a measurement parameter from one organization to another, from a primary standard to a secondary standard, or from a secondary standard to a working standard to create or maintain measurement traceability.

Examples of typical transfer standards are DC volt sources (standard cells or Zener sources), and single-value standard resistors, capacitors, or inductors.

travelling measurement standard—measurement standard, sometimes of special construction, intended for transport between different locations (*VIM*, 5.8). Also called "travelling standard."

Type A evaluation of measurement uncertainty—evaluation of a component of *measurement uncertainty* by a statistical analysis of measured quantity values obtained under defined measurement conditions (*VIM*, 2.28). *See also:* Type B evaluation of measurement uncertainty.

Method of evaluation of uncertainty by the statistical analysis of series of observations (*GUM*, 2.3.2).

Type B evaluation of measurement uncertainty—evaluation of a component of *measurement uncertainty* determined by means other than a *Type A evaluation of measurement uncertainty* (*VIM*, 2.29). *See also:* Type A evaluation of measurement uncertainty.

Method of evaluation of uncertainty by means other than the statistical analysis of series of observations (*GUM*, 2.3.3).

uncertainty budget—statement of a *measurement uncertainty*, of the components of that measurement uncertainty, and of their calculation and combination (*VIM*, 2.33).

UUC, UUT—the *unit under calibration* or the *unit under test*: the instrument being calibrated. These are standard generic labels for the IM&TE item that is being calibrated, which are used in the text of the calibration procedure for convenience. Also may be called "device under test" (DUT) or "equipment under test" (EUT).

validation—*verification*, where the specified requirements are adequate for an intended use (*VIM*, 2.45).

verification—provision of objective evidence that a given item fulfills specified requirements (*VIM*, 2.44).

VIM—an acronym commonly used to identify the BIPM JCGM 200:2012, *International Vocabulary of Metrology—Basic and General Concepts and Associated Terms*. (The acronym comes from the French title, *Vocabulaire international de métrologie—Concepts fondamentaux et généraux et termes associés*.)

work instruction—in a quality management system, a work instruction defines the detailed steps necessary to carry out a procedure. Work instructions are used only where they are needed to ensure the quality of the product or service. The level of education and training of the people with the usual qualifications to do the work must be considered when writing a work instruction. In a metrology laboratory, a calibration procedure is a type of work instruction.

working measurement standard—measurement standard that is used routinely to calibrate or verify measuring instruments or measuring systems (*VIM*, 5.7). Also called "working standard."

Appendix D

Common Conversions

Note: Table is alphabetical within "From/Symbol" column.
The "Multiply By" numbers are expressed in SI recommended form.

To Convert Measurement	From Symbol	From Name	To Symbol	To Name	Multiply By
angle, plane	'	minute (arc)	°	degree of angle	0.016 666 667
angle, plane	'	minute (arc)	rad	radian	0.000 290 888
angle, plane	"	second	°	degree of angle	0.000 277 778
angle, plane	"	second	rad	radian	4.848 14 E-06
angle, plane	°	degree of angle	rad	radian	0.017 453 293
energy	10^6 kcal	ton (from energy equivalent of one ton TNT)	J	joule	4 184 000 000
electrical conductance	ab ℧	abmho (EMU conductance, obsolete)	S	siemens	1 000 000 000
electric current	abA, aA	abampere (EMU current, obsolete)	A	ampere	10
charge, electric, electrostatic, quantity of electricity	abC	abcoulomb (EMU charge, obsolete)	C	coulomb	10
electric capacitance	abF	abfarad (EMU capacitance, obsolete)	F	farad	1 000 000 000
electric inductance	abH	abhenry (EMU inductance, obsolete)	H	henry	1.0 E-09

Continued

To Convert Measurement	From Symbol	From Name	To Symbol	To Name	Multiply By
electric potential difference, electromotive force	abV	abvolt (EMU voltage, obsolete)	V	volt	100 000 000
electric resistance	abΩ	abohm (EMU resistance, obsolete)	Ω	ohm	1 000 000 000
area, plane	acre	acre (U.S. survey)	ha	hectare	0.404 687 261
area, plane	acre	acre (U.S. survey)	m²	square meter	4046.872 61
volume, capacity	acre-ft	acre-foot (U.S. survey)	m³	cubic meter	1233.481 837 547 52
charge, electric, electrostatic, quantity of electricity	A·h	ampere hour	C	coulomb	3600
pressure, stress	at	atmosphere (technical)	kPa	kilopascal	98.0665
pressure, stress	at	atmosphere (technical)	Pa	pascal	98 066.5
magnetomotive force	At	ampere turn	A	ampere	1
mass	AT	ton (assay)	g	gram	29.166 67
mass	AT	ton (assay)	kg	kilogram	0.029 166 67
magnetic field strength	At / in	ampere-turn per inch	A / m	ampere per meter	39.370 078 74
magnetic field strength	At / m	ampere-turn per meter	A / m	ampere per meter	1
pressure, stress	atm	atmosphere (std)	bar	bar	1.013 25
pressure, stress	atm	atmosphere (std)	kPa	kilopascal	101.325
pressure, stress	atm	atmosphere (std)	Pa	pascal	101 325
area, plane	b	barn	m²	square meter	1E-28
pressure, stress	bar	bar	kPa	kilopascal	100
pressure, stress	bar	bar	Pa	pascal	100 000

Continued

Continued

To Convert Measurement	From Symbol	From Name	To Symbol	To Name	Multiply By
volume, capacity	bbl	barrel (oil, 42 U.S. gallons)	L	liter	158.987 294 9
volume, capacity	bbl	barrel (oil, 42 U.S. gallons)	m^3	cubic meter	0.158 987 295
electric current	Bi	biot (see also abampere)	A	ampere	10
activity	Bq	becquerel	s^{-1}	per second (disintegration)	1
energy	Btu	Btu (39 degree Fahrenheit)	J	joule	1059.67
energy	Btu	Btu (59 degree Fahrenheit)	J	joule	1054.8
energy	Btu	Btu (60 degree Fahrenheit)	J	joule	1054.68
energy	Btu	Btu (mean)	J	joule	1055.87
irradiance, heat flux density, heat flow rate / area	Btu / (h·ft^2)	Btu per hour square foot	W / m^2	watt per square meter	3.154 590 745
energy, molar	Btu / lb·mol	Btu per pound-mole	J / kmol	joule per kilomole	2326
molar entropy, molar heat capacity	Btu / (lb·mol·°F)	Btu per pound-mole degree Fahrenheit	J / (kmol·K)	joule per kilomole kelvin	4186.8
energy	Btu_{IT}	Btu (International Table)	J	joule	1055.055 853
heat capacity, entropy	Btu_{IT} / °F	Btu (International Table) per degree Fahrenheit	J / K	joule per kelvin	1899.100 535
energy per area	Btu_{IT} / ft^2	Btu (International Table) per square foot	J / m^2	joule per square meter	11 356.526 68
heat flow per area	Btu_{IT} / (ft^2·h)	Btu (International Table) per square foot hour	W / (m^2 K)	watt per square meter kelvin	3.154 590 745
heat flow per area	Btu_{IT} / (ft^2·s)	Btu (International Table) per square foot second	W / (m^2 K)	watt per square meter kelvin	11 356.526 68
conductivity, thermal	Btu_{IT} / ft^3	Btu (International Table) per cubic foot	J / m^3	joule per cubic meter	37 258.945 81

594 Appendix D

To Convert Measurement	From Symbol	From Name	To Symbol	To Name	Multiply By
heat flow rate	Btu_{IT} / h	Btu (International Table) per hour	W	watt	0.293 071 07
coefficient of heat transfer	$Btu_{IT} / (h \cdot ft^2 \cdot °F)$	Btu (International Table) per hour square foot degree Fahrenheit	$W / (m^2 \cdot K)$	watt per square meter kelvin	5.678 263 341
energy per mass, specific energy	Btu_{IT} / lb	Btu (International Table) per pound-mass	J / kg	joule per kilogram	4183.998 95
specific heat capacity, specific entropy	$Btu_{IT} / (lbm \cdot °R)$	Btu (International Table) per pound mass degree Rankine	$J / (kg \cdot K)$	joule per kilogram kelvin	4186.8
coefficient of heat transfer	$Btu_{IT} / (s \cdot ft^2 \cdot °F)$	Btu (International Table) per second square foot degree Fahrenheit	$W / (m^2 \cdot K)$	watt per square meter kelvin	20 441.748 03
conductivity, thermal	$Btu_{IT} \cdot ft / (h \cdot ft^2 \cdot °F)$	Btu (International Table) foot per hour square foot degree Fahrenheit	$W / (m \cdot K)$	watt per meter kelvin	1.730 734 666 371 39
conductivity, thermal	$Btu_{IT} \cdot in / (h \cdot ft^2 \cdot °F)$	Btu (International Table) inch per hour square foot degree Fahrenheit	$W / (m \cdot K)$	watt per meter kelvin	0.144 227 889
conductivity, thermal	$Btu_{IT} \cdot in / (s \cdot ft^2 \cdot °F)$	Btu (International Table) inch per second square foot degree Fahrenheit	$W / (m \cdot K)$	watt per meter kelvin	519.220 3999
energy	Btu_{th}	Btu (thermochemical)	J	joule	1054.35
heat capacity, entropy	$Btu_{th} / °F$	Btu (thermochemical) per degree Fahrenheit	J / K	joule per kelvin	1897.83
heat capacity, entropy	$Btu_{th} / °R$	Btu (thermochemical) per degree Rankine	J / K	joule per kelvin	1897.83
energy per area	Btu_{th} / ft^2	Btu (thermochemical) per square foot	J / m^2	joule per square meter	11 348.928 95

Continued

Continued

To Convert	From		To		Multiply By
Measurement	Symbol	Name	Symbol	Name	
heat flux density	Btu_{th} / $(ft^2 \cdot h)$	Btu (thermochemical) per square foot hour	$W / (m^2)$	watt per square meter	3.152 480 263
heat flux density	Btu_{th} / $(ft^2 \cdot min)$	Btu (thermochemical) per square foot minute	$W / (m^2)$	watt per square meter	189.148 815 8
heat flux density	Btu_{th} / $(ft^2 \cdot s)$	Btu (thermochemical) per square foot second	$W / (m^2)$	watt per square meter	1 634 245.7685
conductivity, thermal	Btu_{th} / ft^3	Btu (thermochemical) per cubic foot	J / m^3	joule per cubic meter	37 258.945 81
heat flow rate	Btu_{th} / h	Btu (thermochemical) per hour	W	watt	0.292875
coefficient of heat transfer	Btu_{th} / $(h \cdot ft^2 \cdot {}^\circ F)$	Btu (thermochemical) per hour square foot degree Fahrenheit	$W / (m^2 K)$	watt per square meter kelvin	5.674 464 474
energy per mass, specific energy	Btu_{th} / lb	Btu (thermochemical) per pound-mass	J / kg	joule per kilogram	2326
specific heat capacity, specific entropy	Btu_{th} / $(lbm \cdot {}^\circ R)$	Btu (thermochemical) per pound mass degree Rankine	$J / (kg \cdot K)$	joule per kilogram kelvin	4183.998 95
heat flow rate	Btu_{th} / min	Btu per minute (thermochemical)	W	watt	17.5725
heat flow rate	Btu_{th} / s	Btu (thermochemical) per second	W	watt	1054.35
coefficient of heat transfer	Btu_{th} / $(s \cdot ft^2 \cdot {}^\circ F)$	Btu (thermochemical) per second square foot degree Fahrenheit	$W / (m^2 K)$	watt per square meter kelvin	20 428.072 11
conductivity, thermal	$Btu_{th} \cdot ft$ / $(h \cdot ft^2 \cdot {}^\circ F)$	Btu (thermochemical) foot per hour square foot degree Fahrenheit	$W / (m \cdot K)$	watt per meter kelvin	1.729 576 771 653 54

To Convert Measurement	From Symbol	From Name	To Symbol	To Name	Multiply By
conductivity, thermal	Btu$_{th}$·in / (h·ft^2·°F)	Btu (thermochemical) inch per hour square foot degree Fahrenheit	W / (m·K)	watt per meter kelvin	0.144 131 398
conductivity, thermal	Btu$_{th}$·in / (s·ft^2·°F)	Btu (thermochemical) inch per second square foot degree Fahrenheit	W / (m·K)	watt per meter kelvin	518.873 0315
mass	bu	bushel (barley)	kg	kilogram	21.8
mass	bu	bushel (corn, shelled)	kg	kilogram	25.4
mass	bu	bushel (oats)	kg	kilogram	14.5
mass	bu	bushel (potatoes)	kg	kilogram	27.2
mass	bu	bushel (soybeans)	kg	kilogram	27.2
mass	bu	bushel (wheat)	kg	kilogram	27.2
volume, capacity	bu	bushel (U.S.)	m^3	cubic meter	0.0352 390 7
volume, capacity	bu	bushel (U.S.)	L	liter	35.239 070 17
temperature	°C	Celsius (interval) degree	°F	Fahrenheit (interval) degree	1.8
temperature	°C	Celsius (tempeature) degree	°F	Fahrenheit (temperature) degree	formula: $t_{°F} = (t_{°C} * 1.8) + 32$
temperature	°C	Celsius (interval) degree	K	kelvin	1
temperature	°C	Celsius (temperature) degree	K	kelvin	formula: $t_K = t_{°C} + 273.15$
temperature	°C	centigrade (interval) degree	°C	Celsius (interval) degree	1
temperature	°C	centigrade (temperature) degree	°C	Celsius (temperature) degree	1

Continued

APPENDIX D 597

Continued

To Convert	From		To		Multiply By
Measurement	Symbol	Name	Symbol	Name	
velocity / speed	c	speed of light in a vacuum	m / s	meter per second	299 792 458
energy per area	cal / cm^2	langley	kJ / m^2	kilojoule per square meter	41.84
energy	cal$_{15\,°C}$	calorie (15 °C)	J	joule	4.1858
energy	cal$_{20C}$	calorie (20 °C)	J	joule	4.1819
energy	cal$_{IT}$	calorie (International Table)	J	joule	4.1868
energy per mass	cal$_{IT}$ / g	calorie (International Table) per gram	J / kg	joule per kilogram	4186.8
specific heat capacity, specific entropy	cal$_{IT}$ / (g·°C)	calorie (International Table) per gram degree Celsius	J / (kg·K)	joule per kilogram kelvin	4186.8
specific heat capacity, specific entropy	cal$_{IT}$ / (g·K)	calorie (International Table) per gram degree kelvin	J / (kg·K)	joule per kilogram kelvin	4186.8
energy	Cal$_{IT}$, kcal	Calorie (nutrition, International Table) (kilocalorie)	J	joule	4186.8
energy	cal$_{th}$	calorie (thermochemical)	J	joule	4.184
heat energy per area	cal$_{th}$ / cm^2	calorie (thermochemical) per square centimeter	J / m^2	joule per square meter	41840
heat flow rate per area	cal$_{th}$ / (cm^2·min)	calorie (thermochemical) per square centimeter minute	W / m^2	watt per square meter	697.333 333 3
heat flow rate per area	cal$_{th}$ / (cm^2·s)	calorie (thermochemical) per square centimeter second	W / m^2	watt per square meter	41 840
specific heat capacity, specific entropy	cal$_{th}$ / (g·K)	calorie (thermochemical) per gram degree kelvin	J / (kg·K)	joule per kilogram kelvin	4184

To Convert Measurement	From Symbol	From Name	To Symbol	To Name	Multiply By
heat flow rate	cal_{th} / min	calorie (thermochemical) per minute	W	watt	0.069 733 333
energy, molar	cal_{th} / mol	calorie (thermochemical) per mole	J / mol	joule per mole	4.184
molar entropy, molar heat capacity	cal_{th} / (mol·°C)	calorie (thermochemical) per mole degree Celsius	J / (mol·K)	joule per mole kelvin	4.184
heat flow rate	cal_{th} / s	calorie (thermochemical) per second	W	watt	4.184
energy	Cal_{th}, kcal	Calorie (nutrition, thermochemical) (kilocalorie)	J	joule	4184
luminance	cd / in^2	candela per square inch	cd / m^2	candela per square meter	1550.0031
volume / time, (flowrate)	cfm	cubic foot per minute	L / s	liter per second	0.471 947 443
volume / time, (flowrate)	cfm	cubic foot per minute	m^3 / s	cubic meter per second	0.000 471 947
length	ch	chain (U.S. survey)	m	meter	20.116 840 23
activity	Ci	curie	Bq	becquerel	37 000 000 000
resistance, thermal	clo	clo	K-m^2 / W	kelvin square meter per watt	0.155
pressure, stress	cm H_2O	centimeter water (conventional)	Pa	pascal	98.0665
pressure, stress	cm H_2O (4 °C)	centimeter water (4 °C)	Pa	pascal	98.063 754 14
pressure, stress	cm Hg	centimeter mercury (conventional)	Pa	pascal	1333.224
pressure, stress	cm Hg	centimeter mercury (conventional)	kPa	kilopascal	1.333 224
pressure, stress	cm Hg (°C)	centimeter mercury (0 °C)	Pa	pascal	1333.221 913
pressure, stress	cm Hg (°C)	centimeter mercury (0 °C)	kPa	kilopascal	1.333 221 913
area, plane	cmil	circular mil	m^2	square meter	5.067 07 E-10

Continued

Continued

To Convert Measurement	From Symbol	From Name	To Symbol	To Name	Multiply By
area, plane	cmil	circular mil	mm^2	square millimeter	0.000 506 707
volume, capacity	cord	cord	m^3	cubic meter	3.624 556 364
viscosity, dynamic	cP	centipoise	Pa·s	pascal second	0.001
viscosity, kinematic	cSt	centistokes	s / m^2	per second square meter	0.000001
mass	ct	carat (metric)	g	gram	0.2
mass	ct	carat (metric)	kg	kilogram	0.0002
mass	ct	carat (metric)	mg	milligram	200
volume, capacity	cup	cup (U.S. liquid)	L	liter	0.236 588 237
volume, capacity	cup	cup (U.S. liquid)	m^3	cubic meter	0.000 236 588
volume, capacity	cup	cup (U.S. liquid)	mL	milliliter	236.588 236 5
time	d	day (24 h)	h	hour	24
time	d	day (24 h)	s	second	86 400
time	d	day (ISO)	s	second	86 400
time	d	day (sidereal)	s	second	86 164.090 57
electric dipole moment	D	debye	C·m	coulomb meter	3.335 64 E-30
mass / length	D	denier (den)	kg / m	kilogram per meter	1.111 11 E-07
mass / length	D	denier (den)	g / m	gram per meter	0.000 111 111
mass / length	D	denier (den)	mg / m	milligram per meter	0.111 111 111
mass	Da	dalton	kg	kilogram	1.660 54 E-27
permeability	darcy	darcy	m^2	square meter	9.869 23 E-13
velocity, angular	deg / s	degree per second	rad / s	radian per second	0.017 453 293

To Convert Measurement	From Symbol	From Name	To Symbol	To Name	Multiply By
acceleration, angular	deg / s²	degree per second squared	rad / s²	radian per second squared	0.017 453 293
volume, capacity	dry pt	pint (U.S. dry)	L	liter	0.550 610 471
volume, capacity	dry qt	quart (U.S. dry)	m³	cubic meter	0.001 101 221
volume, capacity	dry qt	quart (U.S. dry)	L	liter	1.101 220 943
mass	dwt	pennyweight	g	gram	1.555 173 84
mass	dwt	pennyweight	kg	kilogram	0.001 555 174
force	dyn	dyne	N	newton	0.000 01
pressure, stress	dyn / cm²	dyne per square centimeter	Pa	pascal	0.1
moment of force, torque, bending moment	dyn·cm	dyne centimeter	N·m	newton meter	0.000 000 1
energy	erg	erg	J	joule	0.000 000 1
power density, power / area	erg / cm²	erg per square centimeter	W / m²	watt per square meter	0.001
power	erg / s	erg per second	W	watt	0.000 000 1
temperature	°F	Fahrenheit (interval) degree	°C	Celsius (interval) degree	0.555 555 556
temperature	°F	Fahrenheit (interval) degree	K	kelvin	0.555 555 556
temperature	°F	Fahrenheit (temperature) degree	°C	Celsius (temperature) degree	formula: $t_{°C} = (t_{°F} - 32) / 1.8$
temperature	°F	Fahrenheit (temperature) degree	K	kelvin	formula: $t_K = (t_{°F} + 459.67) / 1.8$
insulance, thermal	°F·ft²·h / Btu$_{th}$	degree Fahrenheit square foot hour per Btu (thermochemical)	K m² / W	kelvin square meter per watt	0.176 228 084

Continued

APPENDIX D **601**

Continued

To Convert Measurement	From Symbol	From Name	To Symbol	To Name	Multiply By
resistivity, thermal	°F·ft²·h / (Btu$_{th}$·in)	degree Fahrenheit square foot hour per Btu (thermochemical) inch	K m / W	kelvin meter per watt	6.933 471 799
resistance, thermal	°F·h / Btu$_{th}$	degree Fahrenheit hour per Btu (thermochemical)	K / W	kelvin per watt	1.896 903 305
resistance, thermal	°F·s / Btu$_{IT}$	degree Fahrenheit second per Btu (International Table)	K / W	kelvin per watt	0.000 526 565
resistance, thermal	°F·s / Btu$_{th}$	degree Fahrenheit second per Btu (thermochemical)	K / W	kelvin per watt	0.000 526 918
charge, electric, electrostatic, quantity of electricity	faraday	faraday (based on) carbon 12	C	coulomb	96 485.3415
length	fathom	fathom (6 ft)	m	meter	1.8288
illuminance	fc	footcandle	lx	lux	10.763 910 42
length	fermi	fermi	fm	femtometer	1
length	fermi	fermi	m	meter	1E-15
luminance	fL	footlambert	cd / m²	candela per square meter	3.426 259 1
volume, capacity	fl oz	ounce (UK liquid)	m³	cubic meter	2.841 31 E-05
volume, capacity	fl oz	ounce (UK liquid)	mL	milliliter	28.413 0625
length	ft	foot	ft	foot, U.S. survey	1.000 002
length	ft	foot	m	meter	0.3048
length	ft	foot, U.S. survey	m	meter	0.304 800 61
pressure, stress	ft H$_2$O	foot water (conventional)	kPa	kilopascal	2.98 906 692
pressure, stress	ft H$_2$O	foot water (conventional)	Pa	pascal	2989.06692

To Convert Measurement	From Symbol	From Name	To Symbol	To Name	Multiply By
pressure, stress	ft H$_2$O (39.2 °F)	foot water (39.2 °F)	kPa	kilopascal	2.988 983 226
pressure, stress	ft H$_2$O (39.2 °F)	foot water (39.2 °F)	Pa	pascal	2988.983 226
pressure, stress	ft H$_g$	foot mercury (conventional)	Pa	pascal	40.636 667 52
velocity / speed	ft / h	foot per hour	m / s	meter per second	8.466 67 E-05
velocity / speed	ft / min	foot per minute	m / s	meter per second	0.005 08
velocity / speed	ft / s	foot per second	km / h	kilometer per hour	1.097 28
velocity / speed	ft / s	foot per second	m / s	meter per second	0.3048
acceleration, linear	ft / s^2	foot per second squared	m / s^2	meter per second squared	0.3048
energy	ft·lbf	foot pound-force	J	joule	1.355 817 948
energy per area	ft·lbf / ft^2	foot pound-force per square foot	J / m^2	joule per square meter	14.593 902 94
power density, power / area	ft·lbf / ft^2·s	foot pound-force per square foot second	W / m^2	watt per square meter	14.593 902 94
energy density	ft·lbf / ft^3	pound force - foot per cubic foot	J / m^3	joule per cubic meter	47.880 258 98
power	ft·lbf / h	foot pound-force per hour	W	watt	0.000 376 616
power	ft·lbf / min	foot pound-force per minute	W	watt	0.022 596 966
power	ft·lbf / s	foot pound-force per second	W	watt	1.355 817 948
energy	ft·pdl	foot poundal	J	joule	0.042 140 11
area, plane	ft^2	square foot	cm^2	square centimeter	929.0304
diffusivity, thermal	ft^2 / h	square foot per hour	m^2 / s	square meter per second	2.580 64 E-05
diffusivity, thermal	ft^2 / s	square foot per second	m^2 / s	square meter per second	0.092 903 04
volume, capacity	ft^3	cubic foot	m^3	cubic meter	0.028 316 847

Continued

Continued

To Convert Measurement	From Symbol	From Name	To Symbol	To Name	Multiply By
furlong	fur	Furlong (660 ft = 10 ch = 10 rd)	m	meter	201.168
magnetic flux density, induction	g	gamma	nT	nanotesla	1
magnetic flux density, induction	g	gamma	T	tesla	0.000 000 001
magnetic flux density, induction	G	gauss	T	tesla	0.0001
density, mass / volume	g / cm^3	gram per cubic centimeter	kg / m^3	kilogram per cubic meter	1000
volume, capacity	gal	gallon (Imperial)	m^3	cubic meter	0.004 546 09
volume, capacity	gal	gallon (Imperial)	L	liter	4.546 09
volume, capacity	gal	gallon (U.S. liquid)	m^3	cubic meter	0.003 785 412
volume, capacity	gal	gallon (U.S. liquid)	L	liter	3.785 411 784
acceleration, linear	Gal	gal (galileo)	cm / s^2	centimeter per second squared	1
acceleration, linear	Gal	gal (galileo)	m / s^2	meter per second squared	0.01
volume / time, (flow rate)	gal / d	gallon (U.S. liquid) per day	m^3 / s	cubic meter per second	4.38 126 E-08
volume / energy	gal / (hp·h)	gallon per horsepower hour	L / J	liter per joule	1.410 09 E-06
volume / energy	gal / (hp·h)	gallon per horsepower hour	m^3 / J	cubic meter per joule	1.410 09 E-09
pressure, stress	g$_f$ / cm^2	gram force per square centimeter	Pa	pascal	98.0665
volume, capacity	gi	gill (Imperial)	m^3	cubic meter	0.000 142 065
volume, capacity	gi	gill (Imperial)	L	liter	0.142 065 313
volume, capacity	gi	gill (U.S.)	m^3	cubic meter	0.000 118 294
volume, capacity	gi	gill (U.S.)	L	liter	0.118 294 118
magnetomotive force	Gi	gilbert	A	ampere	0.795 774 715

To Convert	From		To		Multiply By
Measurement	Symbol	Name	Symbol	Name	
acceleration, linear	g$_n$	gravity, standard acceleration due to	m / s^2	meter per second squared	9.806 65
angle, plane	gon	gon	°	degree of angle	0.9
angle, plane	gon	gon	rad	radian	0.015 707 963
volume / time, (flow rate)	gpm	gallon (U.S. liquid) per minute	m^3 / s	cubic meter per second	6.30 902 E-05
volume / time, (flow rate)	gpm	gallon (U.S. liquid) per minute	L / s	liter per second	0.063 090 196
mass	gr	grain	kg	kilogram	6.479 89 E-05
mass	gr	grain	mg	milligram	64.798 91
density, mass / volume	gr / gal	grain per gallon (U.S. liquid)	kg / m^3	kilogram per cubic meter	0.017 118 061
density, mass / volume	gr / gal	grain per gallon (U.S. liquid)	mg / L	milligram per liter	17.118 061 05
angle, plane	grad	grad	°	degree of angle	0.9
angle, plane	grad	grad	rad	radian	0.015 707 963
angle, plane	grade	grade	°	degree of angle	0.9
angle, plane	grade	grade	rad	radian	0.015 707 963
time	h	hour	s	second	3600
time	h	hour (sidereal)	s	second	3590.170 44
area, plane	ha	hectare	hm^2	square hectometer	1
area, plane	ha	hectare	m^2	square meter	10000
power	hp	horsepower (550 foot pound - force per second)	W	watt	745.699 8716
power	hp	horsepower (boiler, approx 33470 Btu per hour)	W	watt	9809.5

Continued

Continued

To Convert Measurement	From Symbol	Name	To Symbol	Name	Multiply By
power	hp	horsepower (electric)	W	watt	746
power	hp	horsepower (metric)	W	watt	735.4988
power	hp	horsepower (UK)	W	watt	745.7
power	hp	horsepower (water)	W	watt	746.043
length	in	inch	cm	centimeter	2.54
length	in	inch	m	meter	0.0254
length	in	inch	mm	millimeter	25.4
length	in	inch, U.S. survey	m	meter	0.025 400 051
speed, velocity	in / s	inch per second	m / s	meter per second	0.0254
acceleration, linear	in / s^2	inch per second squared	m / s^2	meter per second squared	0.0254
pressure, stress	in H$_2$O (39.2 °F)	inch water (39.2 °F)	Pa	pascal	249.0819 355
pressure, stress	in H$_2$O (60 °F)	inch water (60 °F)	Pa	pascal	248.843 2087
pressure, stress	in Hg	inch mercury (conventional)	kPa	kilopascal	3.386 388 96
pressure, stress	in Hg (0 °C)	inch mercury, 0 °C	Pa	pascal	3386.383 659
pressure, stress	in Hg (0 °C)	inch mercury, 0 °C	kPa	kilopascal	3.386 383 659
pressure, stress	in Hg (60 °F)	inch mercury, 60 °F	Pa	pascal	3376.846 044
pressure, stress	in Hg (60 °F)	inch mercury, 60 °F	kPa	kilopascal	3.376 846 044
area, plane	in^2	square inch	m^2	square meter	0.000 645 16
area, plane	in^2	square inch	cm^2	square centimeter	6.4516
volume, capacity	in^3	cubic inch	m^3	cubic meter	1.638 71 E-05

To Convert Measurement	From Symbol	From Name	To Symbol	To Name	Multiply By
volume / time, (flowrate)	in³ / min	cubic inch per minute	m³ / s	cubic meter per second	2.731 18 E-07
moment of section	in⁴	inch to the fourth power	m⁴	meter to the fourth power	4.162 31 E-07
surface tension	J / m²	joule per square meter	N / m	newton per meter	1
time	Jiffy	Jiffy	ns	nanosecond	16 666 666.67
time	Jiffy	Jiffy	s	second	0.0167
density, luminous flux	Jy	jansky	W / (m²Hz)	watt per square meter hertz	1 E-26
per length	K	kayser	m⁻¹	per meter	100
temperature	K	kelvin	°C	Celsius (temperature) degree	formula: $t_{°C} =$ K − 273.15
energy	kcal$_{IT}$	kilocalorie (International Table)	J	joule	4186.8
energy	kcal$_{mean}$	kilocalorie (mean)	J	joule	4190.02
energy	kcal$_{th}$	kilocalorie (thermochemical)	J	joule	4184
heat flow rate	kcal$_{th}$ / min	kilocalorie (thermochemical) per minute	W	watt	69.733 333 33
heat flow rate	kcal$_{th}$ / s	kilocalorie (thermochemical) per second	W	watt	4184
force	kgf	kilogram force	N	newton	9.806 65
pressure, stress	kgf / cm²	kilogram-force per square centimeter	kPa	kilopascal	98.066 5
pressure, stress	kgf / cm²	kilogram-force per square centimeter	Pa	pascal	98 066.5
pressure, stress	kgf / m²	kilogram-force per square meter	Pa	pascal	9.806 65
pressure, stress	kgf / mm²	kilogram-force per square millimeter	Pa	pascal	9 806 650

Continued

Continued

To Convert Measurement	From Symbol	From Name	To Symbol	To Name	Multiply By
pressure, stress	kgf / mm²	kilogram-force per square millimeter	MPa	megapascal	9.806 65
mass	kgf s² / m	kilogram-force second squared per meter	kg	kilogram	9.806 65
moment of force, torque, bending moment	kgf·m	kilogram force meter (torque)	N·m	newton meter	9.806 65
force	kip	kip (1000 lbf)	N	newton	4448.221 615
force	kip	kip (1000 lbf)	kN	kilonewton	4.448 221 615
volume, capacity	kL	kiloliter (or, m³)	L	liter	1000
velocity / speed	kn	knot (nautical mile per hour)	m / s	meter per second	0.514 444 444
velocity / speed	kph, km / h	kilometer per hour	m / s	meter per second	0.277 777 778
energy	kW·h	kilowatt-hour	J	joule	3 600 000
energy	kW·h	kilowatt-hour	MJ	megajoule	3.6
luminance	L	lambert	cd / m²	candela per square meter	3183.098 862
volume, capacity	L	liter	m³	cubic meter	0.001
mass	lb	pound (avoirdupois)	kg	kilogram	0.453 592 37
mass	lb	pound (troy or apothecary)	kg	kilogram	0.373 241 722
mass / area	lb / ft²	pound per square foot	kg / m²	kilogram per square meter	4.882 427 636
viscosity, dynamic	lb / (ft·h)	pound per foot hour	Pa·s	pascal second	0.000 413 379
viscosity, dynamic	lb / (ft·s)	pound per foot second	Pa·s	pascal second	1.488 163 944
mass / time	lb / h	pound per hour	kg / s	kilogram per second	0.000 125 998
mass / energy	lb / (hp·h)	pound per horsepower hour	kg / J	kilogram per joule	1.689 66 E-07
mass / length	lb / in	pound per inch	kg / m	kilogram per meter	17.857 967 32

To Convert Measurement	From Symbol	From Name	To Symbol	To Name	Multiply By
mass / area	lb / in²	pound per square inch	kg / m²	kilogram per square meter	703.069 579 6
mass per mole	lb / (lb·mol)	pound per pound mole	kg / mol	kilogram per mole	0.001
mass / time	lb / min	pound per minute	kg / s	kilogram per second	0.007 559 873
mass / time	lb / s	pound per second	kg / s	kilogram per second	0.453 592 37
mass / length	lb / yard	pound per yard	kg / m	kilogram per meter	0.496 054 648
mass	lb·mol	pound-mole	mol	mole	0.453 592 37
force	lbf	pound force	N	newton	4.448 221 615
force / length	lbf / ft	pound force per foot	N / m	newton per meter	14.593 902 94
force / length	lbf / in	pound force per inch	N / m	newton per meter	175.126 835 2
thrust / mass	lbf / lb	pound force per pound	N / kg	newton per kilogram	9.806 65
other	lbf·ft / in	pound-force foot per inch	N·m / m	newton meter per meter	53.378 659 38
moment of force, torque, bending moment	lbf·ft	pound force foot (torque)	N·m	newton meter	1.355 817 948
moment of force, torque, bending moment	lbf·in	pound force inch (torque)	N·m	newton meter	0.112 984 829
viscosity, dynamic	lbf·s / in²	pound-force second per square inch	Pa·s	pascal second	6 894.757 293
mass / length	lbm / ft	pound mass per foot	kg / m	kilogram per meter	1.488 163 944
density, mass / volume	lbm / ft³	pound mass per cubic foot	kg / m³	kilogram per cubic meter	16.018 463 37
density, mass / volume	lbm / gal	pound mass per gallon	kg / m³	kilogram per cubic meter	119.826 427 3
density, mass / volume	lbm / gal	pound mass per gallon	kg / L	kilogram per liter	0.119 826 427
density, mass / volume	lbm / gal (UK)	pound mass per gallon (Imperial)	kg / m³	kilogram per cubic meter	99.776 372 66

Continued

Continued

To Convert	From		To		Multiply By
Measurement	Symbol	Name	Symbol	Name	
density, mass / volume	lbm / gal (UK)	pound mass per gallon (Imperial)	kg / L	kilogram per liter	0.099 776 373
density, mass / volume	lbm / in^3	pound mass per cubic inch	kg / m^3	kilogram per cubic meter	27 679.904 71
density, mass / volume	lbm / yd^3	pound mass per cubic yard	kg / m^3	kilogram per cubic meter	0.593 276 421
moment of inertia	lbm ft^2	pound mass foot squared	kg·m^2	kilogram meter squared	0.042 140 11
moment of inertia	lbm in^2	pound mass inch squared	kg·m^2	kilogram meter squared	0.000 292 64
illuminance	lm / ft^2	lumen per square foot	lm / m^2	lumen per square meter	10.763 910 42
mass	long cwt	hundredweight, long	kg	kilogram	50.802 345 44
length	l.y.	light year	m	meter	9.460 53 E+15
energy per area	ly$_{15}$	langley	J / m^2	joule per square meter	41 855
energy per area	ly$_{IT}$	langley International Table	J / m^2	joule per square meter	41 868
energy per area	ly$_{th}$	langley thermochemical	J / m^2	joule per square meter	41 840
length	m	micron, μ	m	meter	0.000 001
length	m	micron, μ	mm	micrometer	1
viscosity, dynamic	lbf·s / ft^2	pound-force second per foot squared	Pa·s	pascal second	47.880 258 98
pressure, stress	mbar	millibar	hPa	hectopascal	1
pressure, stress	mbar	millibar	kPa	kilopascal	0.1
pressure, stress	mbar	millibar	Pa	pascal	100
electrical conductance	mho, ℧	mho	S	siemens	1
length	mi	mile (international)	ft	foot	5280
length	mi	mile (international)	km	kilometer	1.609 344

To Convert Measurement	From Symbol	From Name	To Symbol	To Name	Multiply By
length	mi	mile (international)	m	meter	1609.344
length	mi	mile (U.S. survey)	km	kilometer	1.609 347 219
length	mi	mile (U.S. survey)	m	meter	1609.347 219
velocity / speed	mi / min	mile per minute	m / s	meter per second	26.8224
velocity / speed	mi / min	mile (U.S. survey) per minute	m / s	meter per second	26.822 453 64
velocity / speed	mi / s	mile per second	m / s	meter per second	1609.344
area, plane	mi²	square mile (International)	km²	square kilometer	2.589 988 110 336
area, plane	mi²	square mile (International)	m²	square meter	2 589 988.110 336
area, plane	mi²	square mile (U.S. survey)	km²	square kilometer	2.589 998 470 319 521
area, plane	mi²	square mile (U.S. survey)	m²	square meter	2 589 998.470 319 521
volume, capacity	mi³	cubic mile (international)	m³	cubic meter	4 168 181 825
angle, plane	mil	mil (angle)	rad	radian	0.000 981 748
length	mil	mil (= 0.001 in)	m	meter	0.000 025 4
length	mil	mil (= 0.001 in)	mm	millimeter	0.0254
length	min	microinch	m	meter	2.54 E-08
length	min	microinch	mm	micrometer	0.0254
time	min	minute	s	second	60
time	min	minute (sidereal)	s	second	59.836 174 01
angle, plane	min (arc)	minute (arc)	rad	radian	0.000 290 888
volume, capacity	mL	microliter (= 1 mm³)	L	liter	0.000 001
volume, capacity	mL	milliliter	L	liter	0.001

Continued

Continued

To Convert Measurement	From Symbol	From Name	To Symbol	To Name	Multiply By
length	mm	millimicron	m	meter	0.000 000 001
pressure, stress	mm H$_2$O	millimeter water (conventional)	Pa	pascal	9.806 65
pressure, stress	mm Hg	millimeter mercury (conventional)	Pa	pascal	133.322 4
pressure, stress	mm Hg (0 °C)	millimeter mercury (0 °C)	Pa	pascal	133.322 191 3
length / volume	mpg	mile per gallon (U.S.)	km / L	kilometer per liter	0.425 143 707
fuel efficiency	mpg	mile per gallon (U.S.)	L / (100 km)	liter per hundred kilometer	formula: L / 100 km = 235.214583 / mpg
length / volume	mpg	mile per gallon (U.S.)	m / m^3	meter per cubic meter	425 143.707 4
velocity / speed	mph	mile per hour	m / s	meter per second	0.447 04
velocity / speed	mph	mile (international) per hour	km / h	kilometer per hour	1.609 344
velocity / speed	mph	mile (U.S. survey) per hour	m / s	meter per second	0.447 040 894
magnetic flux	Mx	maxwell	Wb	weber	0.000 000 01
length	nmi	mile, nautical	m	meter	1852
magnetic field strength	Oe	oersted	A / m	ampere per meter	79.577 471 55
magnetomotive force	Oe - cm	oersted centimeter	A	ampere	0.795 774 715
mass	oz	ounce (avoirdupois)	g	gram	28.349 523 13
mass	oz	ounce (avoirdupois)	kg	kilogram	0.028 349 523
mass	oz	ounce (troy or apothecary)	g	gram	31.103 476 8
mass	oz	ounce (troy or apothecary)	kg	kilogram	0.031 103 477
volume, capacity	oz	ounce (U.S. liquid)	m^3	cubic meter	2.957 35 E-05

To Convert Measurement	From Symbol	From Name	To Symbol	To Name	Multiply By
volume, capacity	oz	ounce (U.S. liquid)	mL	milliliter	29.573 529 56
mass / area	oz / ft²	ounce (avoirdupois) per square foot	kg / m²	kilogram per square meter	0.305 151 727
density, mass / volume	oz / gal	ounce (avoirdupois) per gallon (U.S.)	kg / m³	kilogram per cubic meter	7.489 151 707
mass per volume	oz / gal	ounce (avoirdupois) per gallon (U.S.)	kg / m³	kilogram per cubic meter	7.489 151 707
mass per volume	oz / gal	ounce (avoirdupois) per gallon (U.S.)	g / L	gram per liter	7.489 151 707
mass per volume	oz / gal (UK)	ounce (avoirdupois) per gallon (Imperial)	kg / m³	kilogram per cubic meter	6.236 023 291
mass per volume	oz / gal (UK)	ounce (avoirdupois) per gallon (Imperial)	g / L	gram per liter	6.236 023 291
density, mass / volume	oz / in³	ounce (avoirdupois) per cubic inch	kg / m³	kilogram per cubic meter	1729.994 044
force	ozf	ounce-force	N	newton	0.278 013 851
moment of force, torque, bending moment	ozf·in	ounce force inch (torque)	N·m	newton meter	0.007 061 552
moment of force, torque, bending moment	ozf·in	ounce force inch (torque)	mN·m	millinewton meter	7.061 551 814
length	p	point (computer, 1/72 in)	m	meter	0.000 352 778
length	p	point (computer, 1/72 in)	mm	millimeter	0.352 777 778
length	p	point (printer's)	m	meter	0.000 351 46
length	p	point (printer's)	mm	millimeter	0.351 46
viscosity, dynamic	p	poise	dyn·s / cm²	dyne - second per centimeter squared	1

Continued

Continued

To Convert Measurement	From Symbol	From Name	To Symbol	To Name	Multiply By
viscosity, dynamic	p	poise	Pa·s	pascal second	0.1
length	pc	parsec	m	meter	3.085 68 E+16
pressure, stress	pdl / ft²	poundal per square foot	Pa	pascal	1.488 163 944
permeability	perm	perm (0 °C)	kg / (N·s)	kilogram per newton second	5.721 35 E-11
permeability	perm	perm (23 °C)	kg / (N·s)	kilogram per newton second	5.745 25 E-11
permeability	perm in	perm inch (0 °C)	kg / (Pa·s·m)	kilogram per pascal second meter	1.453 22 E-12
permeability	perm in	perm inch (23 °C)	kg / (Pa·s·m)	kilogram per pascal second meter	1.459 29 E-12
illuminance	ph	phot	lm / m²	lumen per square meter	10 000
illuminance	ph	phot	lx	lux	10 000
length	pi	pica (computer, ⅙ in)	m	meter	0.004 233 333
length	pi	pica (computer, ⅙ in)	mm	millimeter	4.233 333 333
length	pi	pica (printer's)	m	meter	0.004 217 518
length	pi	pica (printer's)	mm	millimeter	4.217 517 6
volume, capacity	pk	peck (U.S.)	m³	cubic meter	0.008 809 768
volume, capacity	pk	peck (U.S.)	L	liter	8.809 767 542
volume, capacity	pottle	pottle (U.S. liquid)	m³	cubic meter	0.001 892 706
volume, capacity	pottle	pottle (U.S. liquid)	L	liter	1.892 705 892
pressure, stress	psf	pound-force per square foot	Pa	pascal	47.880 258 98
pressure, stress	psi	pound-force per square inch	kPa	kilopascal	6.894 757 293
pressure, stress	psi	pound-force per square inch	Pa	pascal	6 894.757 293

To Convert Measurement	From Symbol	From Name	To Symbol	To Name	Multiply By
volume, capacity	pt	pint (Imperial)	m³	cubic meter	0.000 568 261
volume, capacity	pt	pint (Imperial)	L	liter	0.568 261 25
volume, capacity	pt	pint (U.S. liquid)	m³	cubic meter	0.000 473 176
volume, capacity	pt	pint (U.S. liquid)	L	liter	0.473 176 473
volume, capacity	qt	quart (U.S. liquid)	m³	cubic meter	0.000 946 353
volume, capacity	qt	quart (U.S. liquid)	L	liter	0.946 352 946
energy	quad	quad	J	joule	1.055 06 E+18
temperature	°R	Rankine (interval) degree	K	kelvin	0.555 555 556
temperature	°R	Rankine (temperature) degree	K	kelvin	0.555 555 556
exposure (X and gamma rays)	R	roentgen	C / kg	coulomb per kilogram	0.000 258
acceleration, angular	r	revolution	rad / s²	radian per second squared	6.283 185 307
angle, plane	r	revolution	rad	radian	6.283 185 307
velocity, angular	r / s	revolution per second	rad / s	radian per second	6.283 185 307
absorbed dose	rad	rad	Gy	gray	0.01
absorbed dose	rad	rad	J / kg	joule per kilogram	0.01
absorbed dose rate	rad / s	rad per second	Gy / s	gray per second	0.01
length	rd	rod	ft	foot, U.S. survey	16.5
length	rd	rod	m	meter	5.029 210 058
dose equivalent	rem	rem	Sv	sievert	0.01
viscosity, dynamic	rhe	rhe	1 / (Pa·s)	per pascal second	10

Continued

Continued

To Convert Measurement	From Symbol	From Name	To Symbol	To Name	Multiply By
velocity, angular	rpm	revolution per minute	rad / s	radian per second	0.104 719 755
time	s	second (sidereal)	s	second	0.997 269 567
luminance	sb	stilb	cd / cm²	candela per square centimeter	1
luminance	sb	stilb	cd / m²	candela per square meter	10 000
angle, plane	sec	second	rad	radian	4.848 14 E-06
time	shake	shake	ns	nanosecond	10
time	shake	shake	s	second	0.000 000 01
mass	slug	slug	kg	kilogram	14.593 902 94
mass / length	slug / ft	slug per foot	kg / m	kilogram per meter	47.880 258 98
viscosity, dynamic	slug / (ft·s)	slug per foot second	Pa·s	pascal second	47.880 258 98
mass / area	slug / ft²	slug per square foot	kg / m²	kilogram per square meter	157.087 463 8
density, mass / volume	slug / ft³	slug per cubic foot	kg / m³	kilogram per cubic meter	515.378 818 4
volume, capacity	st	stere	m³	cubic meter	1
viscosity, kinematic	St	stokes	cm² / s	centimeter squared per second	1
viscosity, kinematic	St	stokes	m² / s	meter squared per second	0.0001
electrical conductance	st ℧	statmho (ESU conductance, obsolete)	S	siemen	1.112 65 E-12
electric current	statA	ESU current (statampere)	A	ampere	3.335 64 E-10
electric current	statA	statampere (ESU current, obsolete)	A	ampere	3.335 64 E-10
charge, electric, electrostatic, quantity of electricity	statC	statcoulomb (ESU charge, obsolete)	C	coulomb	3.335 64 E-10

To Convert Measurement	From Symbol	From Name	To Symbol	To Name	Multiply By
electric capacitance	statF	ESU capacitance (statfarad)	F	farad	1.112 65 E-12
electric capacitance	statF	statfarad (ESU capacitance, obsolete)	F	farad	1.112 65 E-12
inductance, electrical	statH	ESU inductance (stathenry)	H	henry	8.987 55 E+11
inductance, electrical	statH	stathenry (ESU inductance, obsolete)	H	henry	8.987 55 E+11
electric potential difference, electromotive force	statV	ESU electric potential (statvolt)	V	volt	299.792 458
electric potential difference, electromotive force	statV	statvolt (ESU voltage, obsolete)	V	volt	299.792 458
electric resistance	statΩ	ESU resistance (statohm)	Ω	ohm	8.987 55 E+11
electric resistance	statΩ	statohm (ESU resistance, obsolete)	Ω	ohm	8.987 55 E+11
mass	t	ton (metric)	kg	kilogram	1000
mass	t	ton (metric)	Mg	megagram	1
mass	t	ton (metric), tonne	kg	kilogram	1000
mass	t	tonne (metric)	kg	kilogram	1000
mass / length	tex	tex	kg / m	kilogram per meter	0.000 001
energy	therm	therm (EEC)	J	joule	105 506 000
energy	therm	therm (U.S.)	J	joule	105 480 400
heat flow rate	ton	ton of refrigeration (U.S.)	W	watt	3 516.852 842
mass	ton	ton (short)	kg	kilogram	907.184 74
mass	ton	ton, long (2240 pound)	kg	kilogram	1 016.046 909
mass	ton	ton, long (2240 pound)	lb	pound (avoirdupois)	2240

Continued

Continued

To Convert Measurement	From Symbol	From Name	To Symbol	To Name	Multiply By
volume, capacity	ton	ton (register)	m³	cubic meter	2.831 684 659
mass / time	ton / h	ton (short) per hour	kg / s	kilogram per second	0.251 995 761
density, mass / volume	ton / yd³	ton (long) per cubic yard	kg / m³	kilogram per cubic meter	1 328.939 184
density, mass / volume	ton / yd³	ton (short) per cubic yard	kg / m³	kilogram per cubic meter	1 186.552 843
force	tonf	ton-force (short) U.S.	N	newton	8 896.443 231
force	tonf	ton-force (short) U.S.	kN	kilonewton	8.896 443 231
pressure, stress	Torr	torr (mm Hg 0 °C)	Pa	pascal	133.322 368 4
volume, capacity	tsp	teaspoon (U.S., dry or liquid)	mL	milliliter	4.928 921 594
length	ua	astronomical unit	m	meter	1.495 98E+11
magnetic flux	unit pole	unit pole	Wb	weber	1.256 64 E-07
power density, power / area	W / cm²	watt per square centimeter	W / m²	watt per square meter	10 000
power density, power / area	W / in²	watt per square inch	W / m²	watt per square meter	1550.0031
resistance length	Ω·cm	ohm centimeter	Ω·m	ohm meter	0.01
energy, electrical	Ω·h	watt hour	J	joule	3600
resistance length	Ω·mil / ft	ohm circular mil per foot	Ω·m	ohm meter	1.662 43 E-09
resistance length	Ω·mil / ft	ohm circular mil per foot	Ω·mm² / m	ohm square millimeter per meter	0.001 662 426
energy	W s	watt second	J	joule	1
magnetic flux density, induction	Wb / m²	weber per square meter	T	tesla	1
length	yd	yard	ft	foot	3
length	yd	yard	m	meter	0.9144
area, plane	yd²	square yard	m²	square meter	0.836 127 36

Continued

To Convert	From		To		Multiply By
Measurement	Symbol	Name	Symbol	Name	
volume, capacity	yd^3	cubic yard	m^3	cubic meter	0.764 554 858
volume / time, (flowrate)	yd^3 / min	cubic yard per minute	m^3 / s	cubic meter per second	0.012 742 581
time	yr	year (365-day)	s	second	31 536 000
time	yr	year (sidereal)	s	second	31 558 150
time	yr	year (tropical)	s	second	31 556 930

Endnotes

Chapter 1: History and Philosophy of Metrology/Calibration

1. Nemeroff, Ed. 1996. "The Story of the Egyptian Cubit." *Cal Lab: The International Journal of Metrology* 3, no. 5 (September-October): 16–17.
2. Bureau International des Poids et Mesures. *Metre Convention—BIPM*. Sèvres, France: BIPM. https://www.bipm.org/en/metre-convention
3. National Bureau of Standards (NBS). 1975. *The International Bureau of Weights and Measures 1875–1975* (NBS Special Publication 420). Washington, D.C.: U.S. Government Printing Office.
4. Juran, Joseph M. 1997. "Early SQC: A Historical Supplement." *Quality Progress* 30 (September): 73–81. https://www.itl.nist.gov/div898/handbook/pmc/section7/pmc7.htm
5. U.S. Food & Drug Administration. 2018. *Milestones in U.S. Food and Drug Law | FDA*. January 31. https://www.fda.gov/about-fda/fda-history/milestones-us-food-and-drug-law

Chapter 2: The Basics of a Quality System

1. International Organization for Standardization/International Electrotechnical Commission. 2017. *ISO/IEC 17025:2017—General requirements for the competence of testing and calibration laboratories*. Geneva, Switzerland: ISO, p. 20.
2. American National Standards Institute/American Society for Quality/International Organization for Standardization. 2015. *ASQ/ANSI/ISO 2015/ISO 9001:2015 Quality Management Systems—Requirements*. Milwaukee, WI: ASQ Quality Press, p. 2.
3. American National Standards Institute/NCSL International. 1994. *ANSI/NCSL Z540.1-1994 (R2002), American National Standards for Calibration—Calibration laboratories and measuring and test equipment—General requirements*. Boulder, CO: NCSL International.
4. American National Standards Institute/NCSL International. 2006. *ANSI/NCSL Z540.3:2006 (R2013), American National Standards for Calibration—Requirements for the Calibration of Measuring and Test Equipment*. Boulder, CO: NCSL International.
5. American National Standards Institute/American Society for Quality/International Organization for Standardization. 2003. *ANSI/ISO/ASQ Q10012:2003—Measurement Management Systems—Requirements for Measurement Processes and Measuring Equipment*. Milwaukee, WI: ASQ Quality Press, p. 2.
6. NCSL International. 2015. *Recommended Practice-6: Recommended Practice for Calibration Quality Systems for the Healthcare Industries, RP-6*. 4th ed. Boulder, CO: NCSL International, p. 3.

7. Bucher, Jay L. 2000. *When Your Company Needs a Metrology Program, But Can't Afford to Build a Calibration Laboratory . . . What Can You Do?* NCSL International Workshop & Symposium, 2000. Boulder, CO: NCSL International.
8. Pinchard, Corinne. 2001. *Training a Calibration Technician . . . in a Metrology Department?* NCSL International Workshop & Symposium, 2001. Boulder, CO: NCSL International.
9. Tague, Nancy R. 2005. *The Quality Toolbox* (2d ed.). Milwaukee, WI: ASQ Quality Press.
10. Ibid.

Chapter 3: Quality Standards and Their Evolution

1. American National Standards Institute/American Society for Quality/International Organization for Standardization. 2015. *ASQ/ANSI/ISO 2015/ISO 9000:2015 Quality Management Systems—Fundamentals and Vocabulary*. Milwaukee, WI: ASQ Quality Press.
2. Health Canada. n.d. *Health Canada—Canada.Ca*. Canada.Ca. / Gouvernement du Canada. https://www.canada.ca/en/health-canada.html
3. European Medicines Agency. 2020. *What We Do | European Medicines Agency*. Amsterdam, The Netherlands: EMA. https://www.ema.europa.eu/en/about-us/what-we-do#what-we-don't-do-section
4. International Organization for Standardization. *ISO—About Us*. Geneva, Switzerland: ISO. https://www.iso.org/about-us.html
5. Ibid.
6. Ibid.
7. International Organization for Standardization. *ISO—Certification*. Geneva, Switzerland: ISO. https://www.iso.org/certification.html
8. International Accreditation Forum. *Home—IAF*. Quebec, Ontario, Canada: IAF. https://iaf.nu/en/home
9. International Laboratory Accreditation Cooperation. *About ILAC International Laboratory Accreditation Cooperation*. Silverwater, Australia. https://ilac.org/about-ilac
10. International Organization for Standardization. *ISO's Committee on Conformity Assessment (CASCO)*. Geneva, Switzerland: ISO. https://www.iso.org/casco.html
11. International Organization for Standardization. *ISO—Organizations in Cooperation with ISO*. Geneva, Switzerland: ISO. https://www.iso.org/organizations-in-cooperation-with-iso.html
12. Bureau International des Poids et Mesures. 2019. *SI Brochure: The International System of Units (SI)*. 9th ed. Vol. v1.08. Sèvres, France: BIPM. https://www.bipm.org/en/measurement-units
13. International Laboratory Accreditation Cooperation. n.d. "About the International Laboratory Accreditation Cooperation." Silverwater, Australia: ILAC. https://ilac.org/about-ilac
14. Juran, Joseph M. 1997. "Early SQC: A Historical Supplement." *Quality Progress* 30 (September): 73–81. https://www.itl.nist.gov/div898/handbook/pmc/section7/pmc7.htm
15. American Society for Quality. 2021. *ASQ: Who We Are—History*. Milwaukee, WI: ASQ. https://asq.org/about-asq/how-we-do-it/history
16. Kilian, Cecelia S. 1992. *The World of W. Edwards Deming*. Knoxville, TN: SPC Press, pp. 357–376.
17. Marash, Stanley A. 2003. What's Good for Defense . . . How Security Needs of the Cold War Inspired the Quality Revolution. *Quality Digest*, Issue 23 (June): 18.

18. British Standards Institution. 1974. *BS-5179: Guide to the Operation and Evaluation of Quality Assurance Systems.* (Withdrawn without replacement on Oct. 1, 1997). London, UK: BSI.
19. Marquardt, Donald W. 1997. "Background and Development of ISO 9000 Standards." In Robert W. Peach, ed., *ISO 9000 Handbook.* 3rd ed. New York: McGraw-Hill, p. 21.
20. Marash, Stanley A. 2003. What's Good for Defense . . . How Security Needs of the Cold War Inspired the Quality Revolution. *Quality Digest,* Issue 23 (June): 18.
21. Pellegrino, Charles R., and Joshua Stoff. 1985. *Chariots for Apollo.* New York: Atheneum, pp. 91–93.
22. The W. Edwards Deming Institute. 2015, 19 November. "If Japan Can, Why Can't We?" Ketchum, ID. https://deming.org. https://www.youtube.com/watch?v=vcG_Pmt_Ny4
23. National Institutes of Standards and Technology. *NIST—Malcolm Baldrige National Quality Award.* Gaithersburg, MD: NIST. https://www.nist.gov/baldrige/how-baldrige-works/about-baldrige/baldrige-faqs

Chapter 4: Quality Documentation

1. International Organization for Standardization/International Electrotechnical Commission. 2017. *ISO/IEC 17025:2017—General requirements for the competence of testing and calibration laboratories.* Geneva, Switzerland: ISO, p. 20.
2. American National Standards Institute/American Society for Quality/International Organization for Standardization. 2015. *ASQ/ANSI/ISO 2015/ISO 9001:2015 Quality Management Systems—Requirements.* Milwaukee, WI: ASQ Quality Press, p. 8.
3. American National Standards Institute/American Society for Quality/International Organization for Standardization. 2015. *ASQ/ANSI/ISO 2015/ISO 9001:2015 Quality Management Systems—Requirements.* Milwaukee, WI: ASQ Quality Press, p. 9.

Chapter 5: Calibration Procedures

1. American National Standards Institute/American Society for Quality/International Organization for Standardization. 2003. *ANSI/ISO/ASQ Q10012:2003—Measurement Management Systems—Requirements for Measurement Processes and Measuring Equipment.* Milwaukee, WI: ASQ Quality Press, p. 4.
2. American National Standards Institute/American Society for Quality Control. 1996. *ANSI/ASQC M1-1996 American National Standard for Calibration Systems.* Milwaukee, WI: ASQC Quality Press, Clause 4.9.
3. American National Standards Institute/American Society for Quality/International Organization for Standardization. 2015. *ASQ/ANSI/ISO 2015/ISO 9001:2015 Quality Management Systems—Requirements.* Milwaukee, WI: ASQ Quality Press, p. 7.
4. International Organization for Standardization/International Electrotechnical Commission. 2017. *ISO/IEC 17025:2017—General requirements for the competence of testing and calibration laboratories.* Geneva, Switzerland: ISO, p. 4.
5. NCSL International. 2007. *Recommended Practice-3: Calibration Procedure Requirements, RP-3.* Boulder, CO: NCSL International, pp. 1–2.
6. Joint Committee for Guides in Metrology. 2012. *JCGM 200:2012, International vocabulary of metrology—Basic and general concepts and associated terms (VIM 3);* BIPM, IEC, IFCC, ILAC, ISO, IUPAC, IUPAP, and OIML. Sèvres, France: BIPM, p. 28. https://www.bipm.org/documents/20126/2071204/JCGM_200_2012.pdf/f0e1ad45-d337-bbeb-53a6-15fe649d0ff1

7. International Organization for Standardization/International Electrotechnical Commission. 2017. *ISO/IEC 17025:2017—General requirements for the competence of testing and calibration laboratories.* Geneva, Switzerland: ISO, p. 10.
8. Joint Committee for Guides in Metrology. 2012. *JCGM 200:2012, International vocabulary of metrology—Basic and general concepts and associated terms (VIM 3);* BIPM, IEC, IFCC, ILAC, ISO, IUPAC, IUPAP, and OIML. Sèvres, France: BIPM, p. 31. https://www.bipm.org/documents/20126/2071204/JCGM_200_2012.pdf/f0e1ad45-d337-bbeb-53a6-15fe649d0ff1
9. Ibid.
10. International Organization for Standardization/International Electrotechnical Commission. 2017. *ISO/IEC 17025:2017—General requirements for the competence of testing and calibration laboratories.* Geneva, Switzerland: ISO, pp. 10–11.
11. International Organization for Standardization/International Electrotechnical Commission. 2017. *ISO/IEC 17025:2017—General requirements for the competence of testing and calibration laboratories.* Geneva, Switzerland: ISO, p. 14.
12. International Organization for Standardization/International Electrotechnical Commission. 2017. *ISO/IEC 17025:2017—General requirements for the competence of testing and calibration laboratories.* Geneva, Switzerland: ISO, p. 11.
13. Ibid.
14. International Organization for Standardization/International Electrotechnical Commission. 2017. *ISO/IEC 17025:2017—General requirements for the competence of testing and calibration laboratories.* Geneva, Switzerland: ISO, p. 12.

Chapter 6: Calibration Records

1. International Organization for Standardization/International Electrotechnical Commission. 2017. *ISO/IEC 17025:2017—General requirements for the competence of testing and calibration laboratories.* Geneva, Switzerland: ISO, p. 13.
2. American National Standards Institute/American Society for Quality/International Organization for Standardization. 2003. *ANSI/ISO/ASQ Q10012:2003—Measurement Management Systems—Requirements for Measurement Processes and Measuring Equipment.* Milwaukee, WI: ASQ Quality Press, p. 5.
3. NCSL International. 2015. *Recommended Practice-6: Recommended Practice for Calibration Quality Systems for the Healthcare Industries, RP-6.* 4th ed. Boulder, CO: NCSL International, p. 14.
4. American National Standards Institute/American Society for Quality Control. 1996. *ANSI/ASQC M1-1996 American National Standard for Calibration Systems.* Milwaukee, WI: ASQC Quality Press, Clause 4.7.
5. International Organization for Standardization/International Electrotechnical Commission. 2017. *ISO/IEC 17025:2017—General requirements for the competence of testing and calibration laboratories.* Geneva, Switzerland: ISO, p. 21.

Chapter 7: Calibration Certificates

1. International Organization for Standardization/International Electrotechnical Commission. 2017. *ISO/IEC 17025:2017—General requirements for the competence of testing and calibration laboratories.* Geneva, Switzerland: ISO, p. 14.
2. Ibid.
3. Ibid.

4. International Organization for Standardization/International Electrotechnical Commission. 2017. *ISO/IEC 17025:2017—General requirements for the competence of testing and calibration laboratories*. Geneva, Switzerland: ISO, p. 15.
5. Ibid.
6. International Organization for Standardization/International Electrotechnical Commission. 2017. *ISO/IEC 17025:2017—General requirements for the competence of testing and calibration laboratories*. Geneva, Switzerland: ISO, p. 16.
7. International Organization for Standardization/International Electrotechnical Commission. 2017. *ISO/IEC 17025:2017—General requirements for the competence of testing and calibration laboratories*. Geneva, Switzerland: ISO, p. 17.
8. Ibid.
9. International Organization for Standardization/International Electrotechnical Commission. 2017. *ISO/IEC 17025:2017—General requirements for the competence of testing and calibration laboratories*. Geneva, Switzerland: ISO, p. 15.
10. International Organization for Standardization/International Electrotechnical Commission. 2017. *ISO/IEC 17025:2017—General requirements for the competence of testing and calibration laboratories*. Geneva, Switzerland: ISO, p. 17.
11. International Laboratory Accreditation Cooperation. 2020. *ILAC P14:09/2020 ILAC Policy for Measurement Uncertainty in Calibration*. Silverwater, Australia: ILAC, pp. 7–8.
12. Joint Committee for Guides in Metrology. 2008. *JCGM 100:2008, Evaluation of measurement data—Guide to the expression of uncertainty in measurement (GUM)*; BIPM, IEC, IFCC, ILAC, ISO, IUPAC, IUPAP, and OIML. Sèvres, France: BIPM, p. 26.
13. International Laboratory Accreditation Cooperation. 2020. *ILAC P14:09/2020 ILAC Policy for Measurement Uncertainty in Calibration*. Silverwater, Australia: ILAC, p. 7.
14. International Organization for Standardization/International Electrotechnical Commission. 2017. *ISO/IEC 17025:2017—General requirements for the competence of testing and calibration laboratories*. Geneva, Switzerland: ISO, p. 8.
15. International Laboratory Accreditation Cooperation. 2020. *ILAC P14:09/2020 ILAC Policy for Measurement Uncertainty in Calibration*. Silverwater, Australia: ILAC, p. 7.
16. Ibid.
17. Joint Committee for Guides in Metrology. 2008. *JCGM 100:2008, Evaluation of measurement data—Guide to the expression of uncertainty in measurement (GUM)*; BIPM, IEC, IFCC, ILAC, ISO, IUPAC, IUPAP, and OIML. Sèvres, France: BIPM, pp. 70–78.

Chapter 8: Management Systems and Quality Manuals

1. International Organization for Standardization/International Electrotechnical Commission. 2017. *ISO/IEC 17025:2017—General requirements for the competence of testing and calibration laboratories*. Geneva, Switzerland: ISO, p. 1.
2. American National Standards Institute/American Society for Quality/International Organization for Standardization. 2015. *ASQ/ANSI/ISO 2015/ISO 9001:2015 Quality Management Systems—Requirements*. Milwaukee, WI: ASQ Quality Press, p. 1.
3. International Organization for Standardization/International Electrotechnical Commission. 2017. *ISO/IEC 17025:2017—General requirements for the competence of testing and calibration laboratories*. Geneva, Switzerland: ISO, pp. 19–20.
4. Ibid.

624 *Endnotes*

5. Ibid.
6. American National Standards Institute/American Society for Quality/International Organization for Standardization. 2015. *ASQ/ANSI/ISO 2015/ISO 9001:2015 Quality Management Systems—Requirements*. Milwaukee, WI: ASQ Quality Press, p. 4.
7. International Organization for Standardization/International Electrotechnical Commission. 2017. *ISO/IEC 17025:2017—General requirements for the competence of testing and calibration laboratories*. Geneva, Switzerland: ISO, p. 20.
8. Ibid.

Chapter 9: Metrological Traceability

1. Joint Committee for Guides in Metrology. 2012. *JCGM 200:2012, International vocabulary of metrology—Basic and general concepts and associated terms (VIM 3)*. BIPM, IEC, IFCC, ILAC, ISO, IUPAC, IUPAP, and OIML. Sèvres, France: BIPM, pp. 29–30. https://www.bipm.org/documents/20126/2071204/JCGM_200_2012.pdf/f0e1ad45-d337-bbeb-53a6-15fe649d0ff1
2. Ibid.
3. Ibid.
4. Ibid.
5. Possolo, Antonio, Sally S. Bruce, and Robert L. Watters, Jr. 2021, May. *NIST Technical Note 2156: Metrological Traceability Frequently Asked Questions and NIST Policy*, p. 21. Gaithersburg, MD: NIST. https://doi.org/10.6028/NIST.TN.2156
6. Bureau International des Poids et Mesures, International Committee of Weight and Measures (CIPM). 2021. *CIPM MRA-P-11 Overview and Implementation of the CIPM MRA, Version 1.1, 30/03/2021*. Sèvres, France: BIPM. https://www.bipm.org/en/cipm-mra/cipm-mra-documents
7. Bureau International des Poids et Mesures, International Committee of Weight and Measures (CIPM). *CIPM MRA Key Comparison Database (KCDB)*. Sèvres, France: BIPM. https://www.bipm.org/en/cipm-mra/kcdb
8. International Laboratory Accreditation Cooperation. *ILAC MRA and Signatories International Laboratory Accreditation Cooperation*. https://ilac.org/ilac-mra-and-signatories
9. International Laboratory Accreditation Cooperation. December 2020. *Facts & Figures—International Laboratory Accreditation Cooperation (ILAC)*. https://ilac.org/about-ilac/facts-and-figures
10. Bureau International des Poids et Mesures, International Committee of Weight and Measures (CIPM). CIPM MRA Key Comparison Database (KCDB). Sèvres, France: BIPM. https://www.bipm.org/en/cipm-mra/kcdb
11. International Laboratory Accreditation Cooperation. *ILAC MRA and Signatories International Laboratory Accreditation Cooperation*. https://ilac.org/ilac-mra-and-signatories
12. Bureau International des Poids et Mesures. 2018. *Joint BIPM, OIML, ILAC, and ISO Declaration on Metrological Traceability*. BIPM, OIML, ILAC, ISO. https://www.bipm.org/documents/20126/42177518/BIPM-OIML-ILAC-ISO_joint_declaration_2018.pdf/7f1a4834-da36-b012-2a81-fc51a79b0726
13. International Laboratory Accreditation Cooperation. 2020. *ILAC P10:07/2020 ILAC Policy on Metrological Traceability of Measurement Results*. Silverwater, Australia: ILAC.
14. Harris, Georgia L. 2019. *NISTIR 6969, Selected Laboratory and Measurement Practices, and Procedures to Support Basic Mass Calibrations*. Gaithersburg, MD: NIST. https://doi.org/10.6028/NIST.IR.6969-2019

Chapter 10: Calibration Program

1. Ehrlich, C. D., and S. D. Rasberry. 1998. Metrological Timelines in Traceability. *Journal of Research (NIST JRES) 103*, no. 5: 93–105. https://www.ncbi.nlm.nih.gov/pmc/articles/PMC4891962
2. International Organization for Standardization/International Electrotechnical Commission. 2017. *ISO/IEC 17025:2017—General requirements for the competence of testing and calibration laboratories*. Geneva, Switzerland: ISO, p. 7.
3. Joint Committee for Guides in Metrology. 2012. *JCGM 200:2012, International vocabulary of metrology—Basic and general concepts and associated terms (VIM 3)*; BIPM, IEC, IFCC, ILAC, ISO, IUPAC, IUPAP, and OIML. Sèvres, France: BIPM, p. 30. https://www.bipm.org/documents/20126/2071204/JCGM_200_2012.pdf/f0e1ad45-d337-bbeb-53a6-15fe649d0ff1
4. International Laboratory Accreditation Cooperation. 2020. *ILAC P10:07/2020 ILAC Policy on Metrological Traceability of Measurement Results*. Silverwater, Australia: ILAC.
5. International Organization for Standardization/International Electrotechnical Commission. 2017. *ISO/IEC 17025:2017—General requirements for the competence of testing and calibration laboratories*. Geneva, Switzerland: ISO, p. 17.
6. Ibid.
7. Oxford Languages. n.d. "Oxford Languages and Google." *Oxford Languages | The Home of Language Data*. https://languages.oup.com/google-dictionary-en
8. Ibid.
9. International Organization for Standardization/International Electrotechnical Commission. 2017. *ISO/IEC 17025:2017—General requirements for the competence of testing and calibration laboratories*. Geneva, Switzerland: ISO, p. 7.
10. Harris, Georgia L. 2021. *Informal and Unpublished Pre-work for Training on Calibration Programs from Laboratories Participating in 2021*. Gaithersburg, MD: NIST.
11. International Laboratory Accreditation Cooperation and OIML: International Organization of Legal Metrology. 2007. *ILAC G24:2007/OIML D10:2007 (E) Guidelines for the Determination of Calibration Intervals in Measuring Instruments*. [under revision at the time of this writing.] Silverwater, Australia: ILAC; Paris, France: OIML.
12. Harris, Georgia L. 2019. *NISTIR 6969, Selected Laboratory and Measurement Practices, and Procedures to Support Basic Mass Calibrations*. Gaithersburg, MD: NIST. https://doi.org/10.6028/NIST.IR.6969-2019
13. NCSL International. 2010. *Recommended Practice-1: Establishment and Adjustment of Calibration Intervals, RP-1*. 4th ed. Boulder, CO: NCSL International, pp. 113–115.
14. NCSL International. 2018. *Laboratory Management-19: Implementation of a Delayed Dating Approach to Calibration, LM-19*. Boulder, CO: NCSL International, p. ii.
15. NCSL International. 2018. *Laboratory Management-19: Implementation of a Delayed Dating Approach to Calibration, LM-19*. Boulder, CO: NCSL International, p. 1.
16. NCSL International. 2018. *Laboratory Management-19: Implementation of a Delayed Dating Approach to Calibration, LM-19*. Boulder, CO: NCSL International, pp. 2–4.
17. Ibid.
18. Ibid.
19. Ibid.
20. NCSL International. 2018. *Laboratory Management-19: Implementation of a Delayed Dating Approach to Calibration, LM-19*. Boulder, CO: NCSL International, pp. 4, 17.
21. NCSL International. 2018. *Laboratory Management-19: Implementation of a Delayed Dating Approach to Calibration, LM-19*. Boulder, CO: NCSL International, pp. e 5–6.
22. NCSL International. 2018. *Laboratory Management-19: Implementation of a Delayed Dating Approach to Calibration, LM-19*. Boulder, CO: NCSL International, p. 5.

626 *Endnotes*

23. NCSL International. 2018. *Laboratory Management-19: Implementation of a Delayed Dating Approach to Calibration, LM-19*. Boulder, CO: NCSL International, p. 7.
24. NCSL International. 2018. *Laboratory Management-19: Implementation of a Delayed Dating Approach to Calibration, LM-19*. Boulder, CO: NCSL International, p. 8.
25. NCSL International. 2018. *Laboratory Management-19: Implementation of a Delayed Dating Approach to Calibration, LM-19*. Boulder, CO: NCSL International, p. 11.

Chapter 11: The International System of Units (SI) and Measurement Standards

1. Bureau International des Poids et Mesures. 2019. *SI Brochure: The International System of Units (SI)*. 9th ed. Vol. v1.08. Sèvres, France: BIPM. https://www.bipm.org/en/publications/si-brochure
2. Bureau International des Poids et Mesures. 2019. *SI Brochure: The International System of Units (SI)*. 9th ed. Vol. v1.08. Sèvres, France: BIPM, p. 122. https://www.bipm.org/en/publications/si-brochure
3. U.S. Government. 2019. *Title 15—COMMERCE AND TRADE CHAPTER 7—NATIONAL INSTITUTE OF STANDARDS AND TECHNOLOGY Sec. 272—Establishment, Functions, and Activities*. Washington, D.C.: U.S. Government Publishing Office. https://www.govinfo.gov/app/details/USCODE-2019-title15/USCODE-2019-title15-chap7-sec272
4. U.S. Government. 2006. *United States Code, 2006 Edition, Supplement 5, Title 15—COMMERCE AND TRADE—Declaration of Policy (15 U.S.C. 205b)*. Washington, D.C.: U.S. Government Publishing Office. https://www.govinfo.gov/app/details/USCODE-2011-title15/USCODE-2011-title15-chap6-subchapII-sec205b
5. Newell, David B., and Eite Tiesinga, eds. 2019, August. *NIST Special Publication 330 | The International System of Units (SI) 2019 Edition*. Gaithersburg, MD: NIST—Physical Measurement Laboratory. https://www.nist.gov/pml/special-publication-330
6. Joint Committee for Guides in Metrology. 2012. *JCGM 200:2012, International vocabulary of metrology—Basic and general concepts and associated terms (VIM 3)*; BIPM, IEC, IFCC, ILAC, ISO, IUPAC, IUPAP, and OIML. Sèvres, France: BIPM. https://www.bipm.org/documents/20126/2071204/JCGM_200_2012.pdf/f0e1ad45-d337-bbeb-53a6-15fe649d0ff1
7. Bureau International des Poids et Mesures. 2019. *SI Brochure: The International System of Units (SI)*. 9th ed. Vol. v1.08. Sèvres, France: BIPM. https://www.bipm.org/en/publications/si-brochure
8. Ibid.
9. Ibid.
10. Joint Committee for Guides in Metrology. 2012. *JCGM 200:2012, International vocabulary of metrology—Basic and general concepts and associated terms (VIM 3)*; BIPM, IEC, IFCC, ILAC, ISO, IUPAC, IUPAP, and OIML. Sèvres, France: BIPM, p. 6. https://www.bipm.org/documents/20126/2071204/JCGM_200_2012.pdf/f0e1ad45-d337-bbeb-53a6-15fe649d0ff1
11. Bureau International des Poids et Mesures. 2019. *SI Brochure: The International System of Units (SI)*. 9th ed. Vol. v1.08. Sèvres, France: BIPM, pp. 137–138. https://www.bipm.org/en/publications/si-brochure
12. Bureau International des Poids et Mesures. 2019. *SI Brochure: The International System of Units (SI)*. 9th ed. Vol. v1.08. Sèvres, France: BIPM, p. 143. https://www.bipm.org/en/publications/si-brochure
13. Bureau International des Poids et Mesures. 2019. *SI Brochure: The International System of Units (SI)*. 9th ed. Vol. v1.08. Sèvres, France: BIPM, p. 147. https://www.bipm.org/en/publications/si-brochure

14. Gentry, Elizabeth J. (Benham), and Georgia L. Harris. 2016. "Accuracy Matters: Understanding and Communicating Measurement Uncertainties." *Quality Progress* 49, Issue 5 (May).
15. Gentry, Elizabeth J. (Benham), and Georgia L. Harris. 2016. "Write It Right: Understanding Nuances of Metrics in Technical Writing." *Quality Progress* 49, no. 7 (July).
16. Joint Committee for Guides in Metrology. 2012. *JCGM 200:2012, International vocabulary of metrology—Basic and general concepts and associated terms (VIM 3)*; BIPM, IEC, IFCC, ILAC, ISO, IUPAC, IUPAP, and OIML. Sèvres, France: BIPM, p. 46. https://www.bipm.org/documents/20126/2071204/JCGM_200_2012.pdf/f0e1ad45-d337-bbeb-53a6-15fe649d0ff1
17. Ibid.
18. Joint Committee for Guides in Metrology. 2012. *JCGM 200:2012, International vocabulary of metrology—Basic and general concepts and associated terms (VIM 3)*; BIPM, IEC, IFCC, ILAC, ISO, IUPAC, IUPAP, and OIML. Sèvres, France: BIPM, p. 29. https://www.bipm.org/documents/20126/2071204/JCGM_200_2012.pdf/f0e1ad45-d337-bbeb-53a6-15fe649d0ff1
19. Joint Committee for Guides in Metrology. 2012. *JCGM 200:2012, International vocabulary of metrology—Basic and general concepts and associated terms (VIM 3)*; BIPM, IEC, IFCC, ILAC, ISO, IUPAC, IUPAP, and OIML. Sèvres, France: BIPM, p. 28. https://www.bipm.org/documents/20126/2071204/JCGM_200_2012.pdf/f0e1ad45-d337-bbeb-53a6-15fe649d0ff1
20. Joint Committee for Guides in Metrology. 2012. *JCGM 200:2012, International vocabulary of metrology—Basic and general concepts and associated terms (VIM 3)*; BIPM, IEC, IFCC, ILAC, ISO, IUPAC, IUPAP, and OIML. Sèvres, France: BIPM. https://www.bipm.org/documents/20126/2071204/JCGM_200_2012.pdf/f0e1ad45-d337-bbeb-53a6-15fe649d0ff1
21. Joint Committee for Guides in Metrology. 2012. *JCGM 200:2012, International vocabulary of metrology—Basic and general concepts and associated terms (VIM 3)*; BIPM, IEC, IFCC, ILAC, ISO, IUPAC, IUPAP, and OIML. Sèvres, France: BIPM. https://www.bipm.org/documents/20126/2071204/JCGM_200_2012.pdf/f0e1ad45-d337-bbeb-53a6-15fe649d0ff1
22. Ibid.
23. Bureau International des Poids et Mesures. 2019. *SI Brochure: The International System of Units (SI)*. 9th ed. Vol. v1.08. Sèvres, France: BIPM. https://www.bipm.org/en/publications/si-brochure
24. Newell, David B., and Eite Tiesinga, eds. 2019. *NIST Special Publication 330 | The International System of Units (SI) 2019 Edition*. Gaithersburg, MD: NIST—Physical Measurement Laboratory. https://www.nist.gov/pml/special-publication-330
25. Thompson, Amber, and Barry N. Taylor. 2008, March. *NIST Special Publication 811 | Guide for the Use of the International System of Units (SI)—2008 Edition*. Gaithersburg, MD: NIST. https://www.nist.gov/pml/special-publication-811
26. U.S. Metric Association. 2015, September 18. *Unit Mixups—US Metric Association*. Windsor, CO: U.S. Metric Association. https://usma.org/unit-mixups#
27. Ibid.
28. Newell, David B., and Eite Tiesinga, eds. 2019. *NIST Special Publication 330 | The International System of Units (SI) 2019 Edition*. NIST—Physical Measurement Laboratory. https://www.nist.gov/pml/special-publication-330
29. Thompson, Amber, and Barry N. Taylor. 2008, March. *NIST Special Publication 811 | Guide for the Use of the International System of Units (SI)—2008 Edition*. Gaithersburg, MD: NIST. https://www.nist.gov/pml/special-publication-811
30. National Institutes of Standards and Technology. 1997. *Metric Style Guide for the News Media*. Gaithersburg, MD: U.S. Department of Commerce, Technology

Administration, NIST. https://www.nist.gov/system/files/documents/2017/05/09/LC-1137-Metric-Style-Guide-News-Media-1997.pdf

31. Benham, Elizabeth. 2021. "Writing with the SI." *Weights and Measures Connection* 11, Issue 1 (January). Article Ref: J-031. https://www.nist.gov/system/files/documents/2021/05/25/J-032%20Writing%20with%20the%20SI.pdf
32. Thompson, Amber, and Barry N. Taylor. 2008, March. *NIST Special Publication 811 | Guide for the Use of the International System of Units (SI)—2008 Edition.* Gaithersburg, MD: NIST. https://www.nist.gov/pml/special-publication-811
33. Lloyd, Robin. 1999. "Metric Mishap Caused Loss of NASA Orbiter."—*CNN*, September 30. http://edition.cnn.com/TECH/space/9909/30/mars.metric/
34. *Mars Climate Orbiter—Mishap Investigation Board, Phase I Report.* 1999, November. http://sunnyday.mit.edu/accidents/MCO_report.pdf
35. National Institutes of Standards and Technology. "Unit Conversion." Gaithersburg, MD: NIST. https://www.nist.gov/pml/weights-and-measures/metric-si/unit-conversion
36. Frysinger, James R. 2000. "SI Crosses All Language Barriers." https://metricmethods.com/Multilingual_SI_unit_names.gif quantum Integrative Medicine
37. Bureau International des Poids et Mesures. 2019. *SI Brochure: The International System of Units (SI).* 9th ed. Vol. v1.08. Sèvres, France: BIPM. https://www.bipm.org/en/publications/si-brochure
38. Barton, John W., et al., eds. 2021. *NIST Handbook 44: Specifications, Tolerances, and Other Technical Requirements for Weighing and Measuring Devices, 2022 Edition.* Gaithersburg, MD: U.S. Department of Commerce—NIST. https://nvlpubs.nist.gov/nistpubs/hb/2022/NIST.HB.44-2022.pdf
39. If you have a question about metric system use, style, or related publications, send it to thesi@nist.gov and visit http://www.nist.gov/metric

Chapter 12: Audit Requirements

1. Bucher, Jay L. 2000. *When Your Company Needs a Metrology Program, But Can't Afford to Build a Calibration Laboratory . . . What Can You Do?* NCSLI Workshop & Symposium, 2000. Boulder, CO: NCSL International, p. 2.
2. International Organization for Standardization/International Electrotechnical Commission. 2020. *ISO/IEC 17000:2020 Conformity assessment—Vocabulary and general principles:* Clause 6, Terms relating to selection and determination. Geneva, Switzerland: ISO, Clause 6.4.
3. American Society for Quality/American National Standards Institute/International Organization for Standardization. 2018. *ASQ/ANSI/ISO 19011:2018—Guidelines for auditing management systems.* Milwaukee, WI: ASQ Quality Press, p. 1.
4. International Organization for Standardization/International Electrotechnical Commission. 2017. *ISO/IEC 17025:2017—General requirements for the competence of testing and calibration laboratories.* Geneva, Switzerland: ISO, p. 23.
5. Ibid.
6. American National Standards Institute/American Society for Quality/International Organization for Standardization. 2015. *ASQ/ANSI/ISO 2015/ISO 9001:2015 Quality Management Systems—Requirements.* Milwaukee, WI: ASQ Quality Press, p. 17.
7. American National Standards Institute/American Society for Quality/International Organization for Standardization. 2003. *ANSI/ISO/ASQ Q10012:2003—Measurement Management Systems—Requirements for Measurement Processes and Measuring Equipment.* Milwaukee, WI: ASQ Quality Press, p. 12.
8. NCSL International. 2015. *Recommended Practice-6: Recommended Practice for Calibration Quality Systems for the Healthcare Industries, RP-6.* 4th ed. Boulder, CO: NCSL International, p. 21.

9. American Society for Quality/American National Standards Institute/International Organization for Standardization. 2018. *ASQ/ANSI/ISO 19011:2018—Guidelines for auditing management systems*. Milwaukee, WI: ASQ Quality Press, p. vi.
10. International Organization for Standardization/International Electrotechnical Commission. 2015. *ISO/IEC 17021:2015 Conformity assessment—Requirements for bodies providing audit and certification of management systems—Part 1: Requirements*. Geneva, Switzerland: ISO, p. 1.
11. American Society for Quality/American National Standards Institute/International Organization for Standardization. 2018. *ASQ/ANSI/ISO 19011:2018—Guidelines for auditing management systems*. Milwaukee, WI: ASQ Quality Press, p. vi.

Chapter 14: Labels and Equipment Status

1. International Organization for Standardization/International Electrotechnical Commission. 2017. *ISO/IEC 17025:2017—General requirements for the competence of testing and calibration laboratories*. Geneva, Switzerland: ISO, p. 7.
2. American National Standards Institute/American Society for Quality/International Organization for Standardization. 2003. *ANSI/ISO/ASQ Q10012:2003—Measurement Management Systems—Requirements for Measurement Processes and Measuring Equipment*. Milwaukee, WI: ASQ Quality Press, p. 6.
3. American National Standards Institute/American Society for Quality/International Organization for Standardization. 2015. *ASQ/ANSI/ISO 2015/ISO 9001:2015 Quality Management Systems—Requirements*. Milwaukee, WI: ASQ Quality Press, p. 7.
4. NCSL International. 2015. *Recommended Practice-6: Recommended Practice for Calibration Quality Systems for the Healthcare Industries, RP-6*. 4th ed. Boulder, CO: NCSL International, p. 16.
5. NCSL International. 2015. *Recommended Practice-6: Recommended Practice for Calibration Quality Systems for the Healthcare Industries, RP-6*. 4th ed. Boulder, CO: NCSL International, p. 17.
6. Ibid.
7. International Organization for Standardization/International Electrotechnical Commission. 2017. *ISO/IEC 17025:2017—General requirements for the competence of testing and calibration laboratories*. Geneva, Switzerland: ISO, p. 7.
8. American National Standards Institute/American Society for Quality/International Organization for Standardization. 2003. *ANSI/ISO/ASQ Q10012:2003—Measurement Management Systems—Requirements for Measurement Processes and Measuring Equipment*. Milwaukee, WI: ASQ Quality Press, p. 7.

Chapter 15: Training and Competency

1. International Organization for Standardization/International Electrotechnical Commission. 2017. *ISO/IEC 17025:2017—General requirements for the competence of testing and calibration laboratories*. Geneva, Switzerland: ISO, p. 5.
2. Ibid.
3. U.S. Food and Drug Administration, HHS. 2011. 21 CFR pt. 820.25 Subpart B—Quality System Regulation, Section 820.25—Personnel. Rockville, MD: US FDA, p. 165. https://www.govinfo.gov/content/pkg/CFR-2021-title21-vol8/pdf/CFR-2021-title21-vol8-sec820-25.pdf
4. NCSL International. 2010. *Laboratory Management-14: Metrology Human Resources Handbook, LM-14*. Boulder, CO: NCSL International.
5. NCSL International. 2012. *Laboratory Management-13: Calibration Laboratory Personnel Qualifications, LM-13*. Boulder, CO: NCSL International.

6. NCSL International. 2007. *Recommended Practice-17: Documenting Metrology Education, Training and On the Job Training, RP-17*. Boulder, CO: NCSL International.
7. Larson, Miriam B., and Barbara B. Lockee. 2020. *Streamlined ID: A Practical Guide to Instructional Design*. 2nd ed. New York, NY: Routledge, p. 94.
8. Ibid.
9. Ibid.
10. Doyle, Alison. 2013, August 8. "Hard Skills vs. Soft Skills: What's the Difference?" *The Balance Careers*. https://www.thebalancecareers.com/hard-skills-vs-soft-skills-2063780
11. Larson, Miriam B., and Barbara B. Lockee. 2020. *Streamlined ID: A Practical Guide to Instructional Design*. 2d ed. New York, NY: Routledge, p. 94.
12. Doyle, Alison. 2013, August 8. "Hard Skills vs. Soft Skills: What's the Difference?" *The Balance Careers*. https://www.thebalancecareers.com/hard-skills-vs-soft-skills-2063780
13. U.S. Government. 2017, November 28. Standard Occupational Classification (SOC) System—Revision for 2018. *Federal Register* https://www.federalregister.gov/documents/2017/11/28/2017-25622/standard-occupational-classification-soc-system-revision-for-2018
14. Ibid.
15. U.S. Department of Labor, Office of Administrative Law Judges. 1991. *OALJ Law Library, DOT, Professional, Technical, and Managerial Occupations 001.061-010 to 024.364-010 | Metrologist (profess. & kin.) 012.067-010*. Washington, DC: DOL-OALJ. https://www.dol.gov/agencies/oalj/PUBLIC/DOT/REFERENCES/DOT01A
16. U.S. Department of Labor, Office of Administrative Law Judges. 1991. *OALJ Law Library, DOT, Professional, Technical, and Managerial Occupations 001.061-010 to 024.364-010 | Calibration Laboratory Technician (aircraft mfg.; electron. comp.) 019.281-010*. Washington, DC: DOL-OALJ. https://www.dol.gov/agencies/oalj/PUBLIC/DOT/REFERENCES/DOT01A
17. U.S. Government. 2017. *2018 Standard Occupational Classification System*. Washington, DC: U.S. Bureau of Labor Statistics, Office of Employment and Unemployment Statistics, Occupational Employment Statistics. https://www.bls.gov/soc/2018/major_groups.htm#17-0000
18. Grachanen, Christopher. 2007. "The Metrology Job Description Initiative: NCSLI and ASQ Partnering for the Future." *NCSL International Measure* 2, no. 2 (June), 26–31. https://doi.org/10.1080/19315775.2007.11721368
19. U.S. Military MOS Database. [a website that provides a searchable database for all of the U.S. military services occupational specialty job codes and descriptions.] http://www.mosdb.com
20. American Society for Quality. *Calibration Technician Certification CCT*. Milwaukee, WI: ASQ. https://asq.org/cert/calibration-technician
21. A2LA WorkPlace Training. *AWPT | ISO and Metrology Training and Consulting*. Frederick, MD: A2LA WPT. https://www.a2lawpt.org

Chapter 16: Environmental Controls

1. International Organization for Standardization/International Electrotechnical Commission. 2017. *ISO/IEC 17025:2017—General requirements for the competence of testing and calibration laboratories*. Geneva, Switzerland: ISO, p. 4.
2. American National Standards Institute/NCSL International. 2006. *ANSI/NCSL Z540.3:2006 (R2013), American National Standards for Calibration—Requirements for*

the Calibration of Measuring and Test Equipment. Boulder, CO: NCSL International, paragraph 5.3.6.
3. American National Standards Institute/American Society for Quality/International Organization for Standardization. 2003. *ANSI/ISO/ASQ Q10012:2003—Measurement Management Systems—Requirements for Measurement Processes and Measuring Equipment*. Milwaukee, WI: ASQ Quality Press, p. 5.
4. American National Standards Institute/American Society for Quality Control. 1996. *ANSI/ASQC M1-1996 American National Standard for Calibration Systems*. Milwaukee, WI: ASQC Quality Press, Clause 4.4.
5. American National Standards Institute/American Society for Quality/International Organization for Standardization. 2015. *ASQ/ANSI/ISO 2015/ISO 9001:2015 Quality Management Systems—Requirements*. Milwaukee, WI: ASQ Quality Press, p. 6.
6. NCSL International. 2015. *Recommended Practice-6: Recommended Practice for Calibration Quality Systems for the Healthcare Industries, RP-6*. 4th ed. Boulder, CO: NCSL International, p. 12.
7. Liang Mok, Wei. 2003. "Power of a Human." In Glenn Elert, ed., *The Physics Factbook*. [Hypertextbook.] 2003. https://hypertextbook.com/facts/2003/WeiLiangMok.shtml
8. OIML: International Organization of Legal Metrology. 1989, July. *OIML G13 (Ex P 7): Planning of Metrology and Testing Laboratories*. Paris, France: International Bureau of Legal Metrology. https://www.oiml.org/en/files/pdf_g/g013-e89.pdf
9. Doiron, Ted. 2007. "20 °C—A Short History of the Standard Reference Temperature for Industrial Dimensional Measurements." *Journal of Research of the National Institute of Standards and Technology* 112 (1): 1–23. https://nvlpubs.nist.gov/nistpubs/jres/112/1/V112.N01.A01.pdf
10. Ibid.
11. International Organization for Standardization. 2016. ISO 1:2016(en), *Geometrical Product Specifications (GPS)—Standard reference temperature for the specification of geometrical and dimensional properties*. Geneva, Switzerland.
12. *ISA-TR52.00.01-2006 Recommended Environments for Standards Laboratories*. 2006. International Society of Automation, p. 10. https://www.isa.org/products/isa-tr52-00-01-2006-recommended-environments-for-s
13. *ISA-TR52.00.01-2006 Recommended Environments for Standards Laboratories*. 2006. International Society of Automation, pp. 10–15. https://www.isa.org/products/isa-tr52-00-01-2006-recommended-environments-for-s
14. Ibid.
15. UKAS. 2019. *LAB 36—Laboratory Accommodation and Environment in Measurement of Length, Angle, and Form*. 4th ed. Staines-upon-Thames, England: United Kingdom Accreditation Service, p. 4. https://www.ukas.com/wp-content/uploads/filebase/publications/publications-relating-to-laboratory-accreditation/LAB-36-Laboratory-Environment-in-measurement-of-length-angle-and-form-Edition-4-October-2019.pdf
16. Ibid.
17. National Geodetic Survey. 2018, January 25. Surface Gravity Prediction. Washington, D.C. https://geodesy.noaa.gov/cgi-bin/grav_pdx.prl
18. International Organization for Standardization/International Electrotechnical Commission. 2017. *ISO/IEC 17025:2017—General requirements for the competence of testing and calibration laboratories*. Geneva, Switzerland: ISO, p. 13.
19. International Organization for Standardization/International Electrotechnical Commission. 2017. *ISO/IEC 17025:2017—General requirements for the competence of testing and calibration laboratories*. Geneva, Switzerland: ISO, p. 14.

Chapter 17: Industry-Specific Requirements

1. Shewhart, Walter A. 1939. *Statistical Method from the Viewpoint of Quality Control* (with the editorial assistance of W. E. Deming). Edited by W. Edwards Deming. Washington, D.C.: The Graduate School, Department of Agriculture, Washington.
2. International Organization for Standardization/International Electrotechnical Commission. 2017. *ISO/IEC 17025:2017—General requirements for the competence of testing and calibration laboratories*. Geneva, Switzerland: ISO, p. v.
3. International Organization for Standardization/International Electrotechnical Commission. 2017. *ISO/IEC 17025:2017—General requirements for the competence of testing and calibration laboratories*. Geneva, Switzerland: ISO, p. 3.
4. Joint Committee for Guides in Metrology. 2012. *JCGM 200:2012, International vocabulary of metrology—Basic and general concepts and associated terms (VIM 3)*; BIPM, IEC, IFCC, ILAC, ISO, IUPAC, IUPAP, and OIML. Sèvres, France: BIPM, p. 29. https://www.bipm.org/documents/20126/2071204/JCGM_200_2012.pdf/f0e1ad45-d337-bbeb-53a6-15fe649d0ff1
5. American National Standards Institute/American Society for Quality/International Organization for Standardization. 2015. *ASQ/ANSI/ISO 2015/ISO 9001:2015 Quality Management Systems—Requirements*. Milwaukee, WI: ASQ Quality Press, p. 21.
6. American National Standards Institute/American Society for Quality/International Organization for Standardization. 2015. *ASQ/ANSI/ISO 2015/ISO 9001:2015 Quality Management Systems—Requirements*. Milwaukee, WI: ASQ Quality Press, p. 7.
7. Stein, Philip. 2000. "Don't Whine—Calibrate." *Quality Progress* 33 (11, November): 85 [Measure for Measure article]. https://asq.org/quality-progress/articles/dont-whinecalibrate?id=feebc3ed046a4687a7c257d22115f502
8. American National Standards Institute/American Society for Quality/International Organization for Standardization. 2015. *ASQ/ANSI/ISO 2015/ISO 9001:2015 Quality Management Systems—Requirements*. Milwaukee, WI: ASQ Quality Press, p. 7.
9. American National Standards Institute/American Society for Quality/International Organization for Standardization. 2003. *ANSI/ISO/ASQ Q10012:2003—Measurement Management Systems—Requirements for Measurement Processes and Measuring Equipment*. Milwaukee, WI: ASQ Quality Press, p. 1.
10. Lowery, Andrew, Judy Strojny, and Joseph Puleo. 1996. *Medical Device Quality Systems Manual: A Small Entity Compliance Guide*. Rockville, MD: US Department of Health and Human Services, Public Health Service, Food and Drug Administration.
11. U.S. Food and Drug Administration. 2021, October 1. 21 CFR ch. 1 Subchapter H, pt. 820, Subpart G, Section 820.72, (b) Calibration. https://www.accessdata.fda.gov/scripts/cdrh/cfdocs/cfcfr/cfrsearch.cfm?fr=820.72
12. Lowery, Andrew, Judy Strojny, and Joseph Puleo. 1996. *Medical Device Quality Systems Manual: A Small Entity Compliance Guide*. Rockville, MD: U.S. Department of Health and Human Services, Public Health Service, Food and Drug Administration.
13. Ibid.
14. Ibid.
15. Ibid.
16. U.S. Food and Drug Administration. 2016, April. *Data Integrity and Compliance With Drug CGMP Questions and Answers Guidance for Industry*. Rockville, MD: U.S. Food and Drug Administration. https://www.fda.gov/regulatory-information/search-fda-guidance-documents/data-integrity-and-compliance-drug-cgmp-questions-and-answers-guidance-industry
17. U.S. Food and Drug Administration. 2021. *Warning Letters*. Rockville, MD: U.S. Food and Drug Administration. https://www.fda.gov/inspections-compliance-enforcement-and-criminal-investigations/compliance-actions-and-activities/warning-letters

18. U.S. Food and Drug Administration. 2021. *ORA FOIA Electronic Reading Room*. Rockville, MD: U.S. Food and Drug Administration. https://www.fda.gov/about-fda/office-regulatory-affairs/ora-foia-electronic-reading-room
19. The AS prefix is used in North and South America. The EN prefix is used in Europe and Africa, and the JIS prefix is used in the Asia/Pacific region.
20. International Aerospace Quality Group. *About IAQG*. Bruxelles, Belgium. https://iaqg.org/about-us
21. Ibid.
22. Hills, Graham. 2003. "The Effect of ISO/TS 16949:2002." *InsideStandards* (November).
23. Bird, Malcolm. 2002. "A Few Small Miracles Give Birth to an ISO Quality Management Systems Standard for the Automotive Industry." *ISO Bulletin* (August).
24. Harral, William M. 2003. "What Is ISO/TS 16949:2002?" *Automotive Excellence Newsletter* (Summer/Fall 2003), pp. 5–6.
25. Kymal, Chad, and Dave Watkins. 2003. "How Can You Move from QS-9000 to ISO/TS 16949:2002? More Key Challenges of QS-9000 and ISO/TS 16949:2002 Transition." *Automotive Excellence Newsletter* (Summer/Fall), pp. 9–13.
26. Bird, Malcolm. 2002. "A Few Small Miracles Give Birth to an ISO Quality Management Systems Standard for the Automotive Industry." *ISO Bulletin* (August).
27. International Organization for Standardization. 2009. ISO/TS 16949:2009 *Quality management systems—Particular requirements for the application of ISO 9001:2008 for automotive production and relevant service part organizations*, Clause 7.6.3. Geneva, Switzerland.
28. In 1925 Congress passed the Contract Air Mail Act. This and the Air Commerce Act of 1926 formed the basis of future laws and also helped establish airlines as viable companies. This law also got at least six branches of the federal government involved in regulating or providing services for air transport. The Air Mail Act of 1934, in addition to its principal intent, cut the number of agencies regulating the industry and required aircraft manufacturers to divest their ownership of air transport companies. In 1938, a new law created the Civil Aviation Administration (CAA) and Civil Aeronautics Board (CAB), which became the only agencies with regulatory power over aviation. In 1958 the CAA became the Federal Aviation Agency, which was reorganized again by Congress in 1967 as the Federal Aviation Administration (FAA).
29. Millbrooke, Anne Marie. 1999. *Aviation History*. Englewood, CO: Jeppesen Sanderson, pp. 9–53; Wells, Alexander T. 1998. *Air Transportation: A Management Perspective*. 4th ed. Belmont, CA: Brooks/Cole, pp. 58–69.
30. U.S. Government Code of Federal Regulations. 2021, January 1. *Code of Federal Regulations Title 14, Aeronautics and Space, Chapter I: Federal Aviation Administration, Parts 1–199*. Washington, D.C.: National Archives.
31. U.S. Government Code of Federal Regulations. 2021, January 1. *Code of Federal Regulations Title 14, Aeronautics and Space, Chapter I: Federal Aviation Administration, Department of Transportation, Subchapter G Air Carriers and Operators for Compensation or Hire: Certification and Operation, Part 121 Operating Requirements: Domestic, Flag, and Supplemental Operations* (per amended date 2021-10-06). Washington, D.C.: National Archives.
32. U.S. Government Code of Federal Regulations. 2021, January 1. *Code of Federal Regulations Title 14, Aeronautics and Space, Chapter I: Federal Aviation Administration, Department of Transportation, Subchapter H Schools and Other Certified Agencies, Part 145 Repair Stations* (per amended date 2021-10-06). Washington, D.C.: National Archives.
33. U.S. Government Code of Federal Regulations. 2021, January 1. *Code of Federal Regulations Title 14, Aeronautics and Space, Chapter I: Federal Aviation Administration,*

Department of Transportation, Subchapter G Air Carriers and Operators for Compensation or Hire: Certification and Operation, Part 121 Operating Requirements: Domestic, Flag, and Supplemental Operations (per amended date 2021-10-06). Washington, D.C.: National Archives.
34. U.S. Government Code of Federal Regulations. 2021, January 1. *Code of Federal Regulations Title 14, Aeronautics and Space, Chapter I: Federal Aviation Administration, Department of Transportation, Subchapter H Schools and Other Certified Agencies.* Washington, D.C.: National Archives.
35. Ibid.

Chapter 18: Computers, Software, and Software Validation

1. International Organization for Standardization/International Electrotechnical Commission. 2017. *ISO/IEC 17025:2017—General requirements for the competence of testing and calibration laboratories.* Geneva, Switzerland: ISO, p. 6.
2. International Organization for Standardization/International Electrotechnical Commission. 2017. *ISO/IEC 17025:2017—General requirements for the competence of testing and calibration laboratories.* Geneva, Switzerland: ISO, p. 19.
3. American National Standards Institute/American Society for Quality/International Organization for Standardization. 2003. *ANSI/ISO/ASQ Q10012:2003—Measurement Management Systems—Requirements for Measurement Processes and Measuring Equipment.* Milwaukee, WI: ASQ Quality Press, p. 4.
4. NCSL International. 2015. *Recommended Practice-6: Recommended Practice for Calibration Quality Systems for the Healthcare Industries, RP-6.* 4th ed. Boulder, CO: NCSL International, p. 10.
5. U.S. Food and Drug Administration. 2003. *21 CFR, Part 11, Electronic Records; Electronic Signatures—Scope and Application | FDA.* Rockville, MD: U.S. Food and Drug Administration. https://www.fda.gov/regulatory-information/search-fda-guidance-documents/part-11-electronic-records-electronic-signatures-scope-and-application
6. NCSL International. 1996. *Recommended Practice-13: Computer Systems in Metrology, RP-13.* Boulder, CO: NCSL International, p. 4.
7. National Institute of Standards and Technology. 2019. *Calibration Procedures for Weights and Measures Laboratories—GLP 15 Software Quality Assurance.* Edited by G. Harris. Gaithersburg, MD: Author, p. 3.
8. National Institute of Standards and Technology. 2019. *Calibration Procedures for Weights and Measures Laboratories—GLP 15 Software Quality Assurance.* Edited by G. Harris. Gaithersburg, MD: Author, p. 1.
9. Ibid., p. 1.
10. Ibid., p. 3.
11. National Institute of Standards and Technology. 2019. *Calibration Procedures for Weights and Measures Laboratories—GLP 15 Software Quality Assurance.* Edited by G. Harris. Gaithersburg, MD: Author.
12. Wichmann, Brian. 2000. *Software Support for Metrology Best Practice Guide No.1—Measurement System Validation: Validation of Measurement Software.* April 2000. Teddington, Middlesex, England: National Physical Laboratory. http://www.npl.co.uk
13. International Society of Pharmaceutical Engineering. 2008, February. *Good Automated Manufacturing Practice: GAMP 5: A Risk-based Approach to Compliant GxP Computerized Systems.* North Bethesda, MD: ISPE.

14. U.S. Food and Drug Administration. 2002, January. *Guidance Document—General Principles of Software Validation: Guidance for Industry and FDA Staff | FDA*. Rockville, MD: U.S. Food and Drug Administration. https://www.fda.gov/regulatory-information/search-fda-guidance-documents/general-principles-software-validation
15. Joint Committee for Guides in Metrology. 2012. *JCGM 200:2012, International vocabulary of metrology—Basic and general concepts and associated terms (VIM 3)*; BIPM, IEC, IFCC, ILAC, ISO, IUPAC, IUPAP, and OIML. Sèvres, France: BIPM, p. 31. https://www.bipm.org/documents/20126/2071204/JCGM_200_2012.pdf/f0e1ad45-d337-bbeb-53a6-15fe649d0ff1
16. Ibid.
17. Ibid.
18. U.S. Food and Drug Administration. 2002, January. *Guidance Document—General Principles of Software Validation: Guidance for Industry and FDA Staff | FDA*. Rockville, MD: U.S. and Drug Administration. https://www.fda.gov/regulatory-information/search-fda-guidance-documents/general-principles-software-validation
19. Ibid.
20. Ibid.
21. U.S. Food and Drug Administration. 2003. *21 CFR, Part 11, Electronic Records; Electronic Signatures—Scope and Application | FDA*. Rockville, MD: U.S. Food and Drug Administration. https://www.fda.gov/regulatory-information/search-fda-guidance-documents/part-11-electronic-records-electronic-signatures-scope-and-application
22. Nowocin, W. 2021. "Selecting a Calibration Management Software System in a Regulated Environment." *Cal Lab Magazine: The International Journal of Metrology 28*, no. 1 (Jan-Feb-Mar): 29. https://www.callabmag.com/wp-content/uploads/2021/03/jan21_web.pdf
23. Nowocin, W. 2021. "Implementing a Calibration Management Software System in a Regulated Environment." *Cal Lab Magazine: The International Journal of Metrology 28*, no. 2 (Apr-May-Jun): 27. https://www.callabmag.com/wp-content/uploads/2021/06/apr21_web.pdf
24. Ibid.
25. Ibid.
26. Nowocin, W. 2021. "Implementing a Calibration Management Software System in a Regulated Environment." *Cal Lab Magazine: The International Journal of Metrology 28*, no. 2 (Apr-May-Jun): 28. https://www.callabmag.com/wp-content/uploads/2021/06/apr21_web.pdf
27. National Institute of Standards and Technology. *System Development Life Cycle (SDLC)*. https://csrc.nist.gov/glossary/term/system_development_life_cycle
28. U.S. Food and Drug Administration. 2012. *21 CFR Part 820.70(i), Production and process controls: Automated processes*. Rockville, MD: U.S. Food and Drug Administration, 21 CFR § 820.70(i). https://www.ecfr.gov/current/title-21/chapter-I/subchapter-H/part-820/subpart-G/section-820.70#p-820.70(i)
29. U.S. Food and Drug Administration. 2021. "Computer Software Assurance for Production and Quality System Software" (Draft Guidance Topic for FY2022). https://www.fda.gov/medical-devices/guidance-documents-medical-devices-and-radiation-emitting-products/cdrh-proposed-guidances-fiscal-year-2022-fy2022
30. Nowocin, W. 2021. "Maintaining a Calibration Management Software System in a Regulated Environment." *Cal Lab Magazine: The International Journal of Metrology 28*, no. 3 (July): 35. https://www.callabmag.com/maintaining-a-calibration-management-software-system-in-a-regulated-environment
31. Nowocin, W. 2021. "Maintaining a Calibration Management Software System in a Regulated Environment." *Cal Lab Magazine: The International Journal of Metrology 28*,

no. 3 (July): 37. https://www.callabmag.com/maintaining-a-calibration-management-software-system-in-a-regulated-environment
32. U.S. Food and Drug Administration. 2003. *21 CFR, Part 11, Electronic Records; Electronic Signatures—Scope and Application | FDA*. Rockville, MD: U.S. Food and Drug Administration. https://www.fda.gov/regulatory-information/search-fda-guidance-documents/part-11-electronic-records-electronic-signatures-scope-and-application
33. International Organization for Standardization/International Electrotechnical Commission. *ISO/IEC 27001: Information Security Management System (ISMS)*. Geneva, Switzerland: ISO. https://www.iso.org/isoiec-27001-information-security.html
34. International Society of Pharmaceutical Engineering. 2008, February. *Good Automated Manufacturing Practice: GAMP 5: A Risk-based Approach to Compliant GxP Computerized Systems*. North Bethesda, MD: ISPE, pp. 71–105.
35. U.S. Food and Drug Administration. 2021. *Code of Federal Regulations Title 21, Volume 4, Chapter I, Subchapter C, Part 211.68 Automatic, mechanical, and electronic equipment, (b)* (April 1, 2011).
36. Yen, James, Dennis Leber, and Letici Pibida. 2020, September. *NIST Technical Note 2106: Comparing Instruments*, p. 27. Gaithersburg, MD: NIST. https://doi.org/10.6028/NIST.TN.2106
37. U.S. Food and Drug Administration. 2021. *Warning Letters | FDA*. Rockville, MD: U.S. Food and Drug Administration. https://www.fda.gov/inspections-compliance-enforcement-and-criminal-investigations/compliance-actions-and-activities/warning-letters
38. Hermans, F., and E. Murphy-Hill. 2015. *Enron's Spreadsheets and Related Emails: A Dataset and Analysis*. Paper presented at the IEEE/ACM 37th IEEE International Conference on Software Engineering, Florence, Italy.
39. Panko, Raymond R. 2014. *Audits of Operational Spreadsheets*. Raymond Panko Spreadsheet and Human Error. http://panko.com/ssr/Audits.html
40. International Society of Pharmaceutical Engineering. 2008, February. *Good Automated Manufacturing Practice: GAMP 5: A Risk-based Approach to Compliant GxP Computerized Systems*. North Bethesda, MD: ISPE, pp. 291–293.
41. U.S. Food and Drug Administration, Office of Regulatory Affairs. 2019. *ORA Laboratory Manual Vol. III Section 4: Basic Statistics and Data Presentation*. https://www.fda.gov/media/73535/download
42. Cantellops, D., E. Bonnin, and A. Reid. 2010, October 19. *Laboratory Information Bulletin Number 4317: Spreadsheet Design and Validation for the Multi-User Application for the Chemistry Laboratory*. Rockville, MD: FDA, Office of Regulatory Affairs. http://www.spreadsheetvalidation.com/pdf/LIB_Design_Multi-User2A.pdf
43. National Institute of Standards and Technology. 2019. *Calibration Procedures for Weights and Measures Laboratories—GLP 15 Software Quality Assurance*. Edited by G. Harris. Gaithersburg, MD: Author, pp. 6–8.
44. National Institute of Standards and Technology. 2019. *Calibration Procedures for Weights and Measures Laboratories—GLP 15 Software Quality Assurance*. Edited by G. Harris. Gaithersburg, MD: Author, pp. 9–11.

Chapter 19: A General Understanding of Metrology

1. Joint Committee for Guides in Metrology. 2012. *JCGM 200:2012, International vocabulary of metrology—Basic and general concepts and associated terms (VIM 3)*; BIPM, IEC, IFCC, ILAC, ISO, IUPAC, IUPAP, and OIML. Sèvres, France: BIPM, p. 6. https://www.bipm.org/documents/20126/2071204/JCGM_200_2012.pdf/f0e1ad45-d337-bbeb-53a6-15fe649d0ff1

2. Ibid.
3. National Institute of Standards and Technology. 2018. *Fundamental Physical Constants from NIST*. 2018. Gaithersburg, MD: Physical Measurement Laboratory | NIST. https://physics.nist.gov/cuu/Constants/index.html
4. Joint Committee for Guides in Metrology. 2012. *JCGM 200:2012, International vocabulary of metrology—Basic and general concepts and associated terms (VIM 3)*; BIPM, IEC, IFCC, ILAC, ISO, IUPAC, IUPAP, and OIML. Sèvres, France: BIPM, p. 34. https://www.bipm.org/documents/20126/2071204/JCGM_200_2012.pdf/f0e1ad45-d337-bbeb-53a6-15fe649d0ff1

Chapter 20: Measurement Methods, Systems, Capabilities, and Data

1. Joint Committee for Guides in Metrology. 2012. *JCGM 200:2012, International vocabulary of metrology—Basic and general concepts and associated terms (VIM 3)*; BIPM, IEC, IFCC, ILAC, ISO, IUPAC, IUPAP, and OIML. Sèvres, France: BIPM. https://www.bipm.org/documents/20126/2071204/JCGM_200_2012.pdf/f0e1ad45-d337-bbeb-53a6-15fe649d0ff1
2. Ibid.
3. Castrup, H., et al. 1994. *NASA Reference Publication 1342, Metrology—Calibration and Measurement Processes Guidelines*. Pasadena, CA: Jet Propulsion Laboratory California Institute of Technology.
4. Kimothi, Shri Krishna. 2002. *The Uncertainty of Measurements: Physical and Chemical Metrology Impact and Analysis*. Milwaukee, WI: ASQ Quality Press.
5. Automotive Industry Action Group. 2010. *Measurement Systems Analysis (MSA) Reference Manual*. 4th ed. Southfield, MI: AIAG.
6. Castrup, H., et al. 1994. *NASA Reference Publication 1342, Metrology—Calibration and Measurement Processes Guidelines*. Pasadena, CA: Jet Propulsion Laboratory California Institute of Technology.
7. Fluke. 1994. *Calibration: Philosophy in Practice*. 2nd ed. Everett, WA: Fluke Corporation.
8. American National Standards Institute/American Society for Quality/International Organization for Standardization. 2015. *ASQ/ANSI/ISO 2015/ISO 9001:2015 Quality Management Systems—Requirements*. Milwaukee, WI: ASQ Quality Press, p. 7.
9. American National Standards Institute/American Society for Quality/International Organization for Standardization. 2015. *ASQ/ANSI/ISO 2015/ISO 9001:2015 Quality Management Systems—Requirements*. Milwaukee, WI: ASQ Quality Press, p. 17.
10. Joint Committee for Guides in Metrology. 2012. *JCGM 200:2012, International vocabulary of metrology—Basic and general concepts and associated terms (VIM 3)*; BIPM, IEC, IFCC, ILAC, ISO, IUPAC, IUPAP, and OIML. Sèvres, France: BIPM, p. 23. https://www.bipm.org/documents/20126/2071204/JCGM_200_2012.pdf/f0e1ad45-d337-bbeb-53a6-15fe649d0ff1
11. Joint Committee for Guides in Metrology. 2012. *JCGM 200:2012, International vocabulary of metrology—Basic and general concepts and associated terms (VIM 3)*; BIPM, IEC, IFCC, ILAC, ISO, IUPAC, IUPAP, and OIML. Sèvres, France: BIPM, p. 22. https://www.bipm.org/documents/20126/2071204/JCGM_200_2012.pdf/f0e1ad45-d337-bbeb-53a6-15fe649d0ff1
12. National Institutes of Standards and Technology/SEMATECH. 2013, April. *NIST/SEMATECH e-Handbook of Statistical Methods*. https://doi.org/10.18434/M32189
13. Joint Committee for Guides in Metrology. 2012. *JCGM 200:2012, International vocabulary of metrology—Basic and general concepts and associated terms (VIM 3)*; BIPM, IEC, IFCC, ILAC, ISO, IUPAC, IUPAP, and OIML. Sèvres, France: BIPM,

p. 22. https://www.bipm.org/documents/20126/2071204/JCGM_200_2012.pdf/f0e1ad45-d337-bbeb-53a6-15fe649d0ff1
14. Joint Committee for Guides in Metrology. 2012. *JCGM 200:2012, International vocabulary of metrology—Basic and general concepts and associated terms (VIM 3)*; BIPM, IEC, IFCC, ILAC, ISO, IUPAC, IUPAP, and OIML. Sèvres, France: BIPM, p. 24. https://www.bipm.org/documents/20126/2071204/JCGM_200_2012.pdf/f0e1ad45-d337-bbeb-53a6-15fe649d0ff1
15. Joint Committee for Guides in Metrology. 2012. *JCGM 200:2012, International vocabulary of metrology—Basic and general concepts and associated terms (VIM 3)*; BIPM, IEC, IFCC, ILAC, ISO, IUPAC, IUPAP, and OIML. Sèvres, France: BIPM, p. 23. https://www.bipm.org/documents/20126/2071204/JCGM_200_2012.pdf/f0e1ad45-d337-bbeb-53a6-15fe649d0ff1
16. Joint Committee for Guides in Metrology. 2012. *JCGM 200:2012, International vocabulary of metrology—Basic and general concepts and associated terms (VIM 3)*; BIPM, IEC, IFCC, ILAC, ISO, IUPAC, IUPAP, and OIML. Sèvres, France: BIPM, p. 25. https://www.bipm.org/documents/20126/2071204/JCGM_200_2012.pdf/f0e1ad45-d337-bbeb-53a6-15fe649d0ff1
17. Joint Committee for Guides in Metrology. 2012. *JCGM 200:2012, International vocabulary of metrology—Basic and general concepts and associated terms (VIM 3)*; BIPM, IEC, IFCC, ILAC, ISO, IUPAC, IUPAP, and OIML. Sèvres, France: BIPM, p. 24. https://www.bipm.org/documents/20126/2071204/JCGM_200_2012.pdf/f0e1ad45-d337-bbeb-53a6-15fe649d0ff1
18. Joint Committee for Guides in Metrology. 2012. *JCGM 200:2012, International vocabulary of metrology—Basic and general concepts and associated terms (VIM 3)*; BIPM, IEC, IFCC, ILAC, ISO, IUPAC, IUPAP, and OIML. Sèvres, France: BIPM, p. 43. https://www.bipm.org/documents/20126/2071204/JCGM_200_2012.pdf/f0e1ad45-d337-bbeb-53a6-15fe649d0ff1
19. Ibid.
20. Juran, Joseph M., Frank M. Gryna, and Richard S. Bingham. 1974. *Quality Control Handbook*. 4th ed. Wilton, CT: McGraw-Hill.
21. Ibid.
22. Ritter, Diane, and Michael Brassard. 2018. *GOAL/QPC Memory Jogger 2*. 2nd ed. (2018 Revision). Methuen, MA: GOAL/QPC.
23. Ibid.
24. Juran, Joseph M., Frank M. Gryna, and Richard S. Bingham. 1974. *Quality Control Handbook*. 4th ed. Wilton, CT: McGraw-Hill.
25. Griffith, Gary K. 1986. *The Quality Technician's Handbook*. 3rd ed. Englewood Cliffs, NJ: Prentice Hall.
26. Ibid.
27. Joint Committee for Guides in Metrology. 2012. *JCGM 200:2012, International vocabulary of metrology—Basic and general concepts and associated terms (VIM 3)*; BIPM, IEC, IFCC, ILAC, ISO, IUPAC, IUPAP, and OIML. Sèvres, France: BIPM, p. 42. https://www.bipm.org/documents/20126/2071204/JCGM_200_2012.pdf/f0e1ad45-d337-bbeb-53a6-15fe649d0ff1

Chapter 21: Specifications

1. Joint Committee for Guides in Metrology. 2012. *JCGM 200:2012, International vocabulary of metrology—Basic and general concepts and associated terms (VIM 3)*; BIPM, IEC, IFCC, ILAC, ISO, IUPAC, IUPAP, and OIML. Sèvres, France: BIPM,

p. 17. https://www.bipm.org/documents/20126/2071204/JCGM_200_2012.pdf/f0e1ad45-d337-bbeb-53a6-15fe649d0ff1
2. Joint Committee for Guides in Metrology. 2012. *JCGM 200:2012, International vocabulary of metrology—Basic and general concepts and associated terms (VIM 3)*; BIPM, IEC, IFCC, ILAC, ISO, IUPAC, IUPAP, and OIML. Sèvres, France: BIPM, p. 44. https://www.bipm.org/documents/20126/2071204/JCGM_200_2012.pdf/f0e1ad45-d337-bbeb-53a6-15fe649d0ff1
3. International Organization for Standardization/International Electrotechnical Commission. 2020. *ISO/IEC 17000:2020 Conformity assessment—Vocabulary and general principles*. Geneva, Switzerland: ISO, Clause 5.1.
4. American National Standards Institute/American Society for Quality/International Organization for Standardization. 2015. *ASQ/ANSI/ISO 2015/ISO 9000:2015 Quality Management Systems—Fundamentals and Vocabulary*. Milwaukee, WI: ASQ Quality Press, Clause 3.8.7.
5. NCSL International. 2016. *Recommended Practice-5: Recommended Practice for Measuring and Test Equipment (MTE) Specifications, RP-5*. 3rd ed. Boulder, CO: NCSL International, p. 11.
6. Barton, John W., et al., eds. 2021. *NIST Handbook 44: Specifications, Tolerances, and Other Technical Requirements for Weighing and Measuring Devices. 2022 Edition*. Gaithersburg, MD: U.S. Department of Commerce—NIST, p. D-26. https://doi.org/10.6028/NIST.HB.44-2022
7. NCSL International. 1999. *Laboratory Management-3: Glossary of Metrology Related Terms, LM-3* 2d ed. Boulder, CO: NCSL International.
8. Joint Committee for Guides in Metrology. 2012. *JCGM 200:2012, International vocabulary of metrology—Basic and general concepts and associated terms (VIM 3)*; BIPM, IEC, IFCC, ILAC, ISO, IUPAC, IUPAP, and OIML. Sèvres, France: BIPM, p. 44. https://www.bipm.org/documents/20126/2071204/JCGM_200_2012.pdf/f0e1ad45-d337-bbeb-53a6-15fe649d0ff1
9. NCSL International. 2016. *Recommended Practice-5: Recommended Practice for Measuring and Test Equipment (MTE) Specifications, RP-5*. 3rd ed. Boulder, CO: NCSL International, p. 12.
10. Barton, John W., et al., eds. 2021. *NIST Handbook 44: Specifications, Tolerances, and Other Technical Requirements for Weighing and Measuring Devices. 2022 Edition*. Gaithersburg, MD: U.S. Department of Commerce—NIST, p. D-28. https://doi.org/10.6028/NIST.HB.44-2022
11. Shewhart, Walter A. 1939. *Statistical Method from the Viewpoint of Quality Control* (with the editorial assistance of W. E. Deming). Edited by W. Edwards Deming. Washington, D.C.: The Graduate School, Department of Agriculture, Washington, p. 51.
12. Kimothi, Shri Krishna. 2002. *The Uncertainty of Measurements: Physical and Chemical Metrology Impact and Analysis*. 2002. Milwaukee, WI: ASQ Quality Press, p. 15.
13. Griffith, Gary K. 1986. *The Quality Technician's Handbook*. 3rd ed. Englewood Cliffs, NJ: Prentice Hall, p. 82.
14. Kimothi, Shri Krishna. 2002. *The Uncertainty of Measurements: Physical and Chemical Metrology Impact and Analysis*. 2002. Milwaukee, WI: ASQ Quality Press, p. 15.
15. Griffith, Gary K. 1986. *The Quality Technician's Handbook*. 3rd ed. Englewood Cliffs, NJ: Prentice Hall, p. 82.
16. The American Society of Mechanical Engineers. n.d. *ASME B89.1.9-2002: Gage Blocks*. R2012 ed. New York, NY: American National Standards Institute (ANSI).
17. Fluke. 1994. *Calibration: Philosophy in Practice*. 2nd ed. Everett, WA: Fluke Corporation, ch. 31.

18. The American Society of Mechanical Engineers. n.d. *ASME B89.1.9-2002: Gage Blocks.* R2012 ed. New York, NY: American National Standards Institute (ANSI).
19. Ibid.
20. Fluke. 1994. *Calibration: Philosophy in Practice.* 2nd ed. Everett, WA: Fluke Corporation, ch. 31-9.
21. Joint Committee for Guides in Metrology. 2012. *JCGM 200:2012, International vocabulary of metrology—Basic and general concepts and associated terms (VIM 3);* BIPM, IEC, IFCC, ILAC, ISO, IUPAC, IUPAP, and OIML. Sèvres, France: BIPM, p. 43. https://www.bipm.org/documents/20126/2071204/JCGM_200_2012.pdf/f0e1ad45-d337-bbeb-53a6-15fe649d0ff1
22. Ibid.
23. "What Is Phase Noise: Phase Jitter." n.d. Electronics Notes: Reference Site for Electronics, Radio & Wireless. https://www.electronics-notes.com/articles/basic_concepts/electronic-rf-noise/phase-noise-jitter-what-is.php
24. Blair, Thomas. 2003. Letter to author (Payne), September 28.
25. Ibid.

Chapter 22: Substituting Calibration Standards

1. Niagara Falls Info. 2017, February 3. "Niagara Falls History of Power Development—The Ontario Power Company." https://www.niagarafallsinfo.com/niagara-falls-history/niagara-falls-power-development/the-history-of-power-development-in-niagara/the-ontario-power-company
2. *Boeing: Select Products in Boeing History.* n.d. The Boeing Company. https://www.boeing.com/history/products/index.page
3. Fluke. 1994. *Calibration: Philosophy in Practice.* 2d ed. Everett, WA: Fluke Corporation, ch. 31.
4. Ibid.

Chapter 23: Proficiency Testing, Interlaboratory Comparisons, and Measurement Assurance Programs

1. International Organization for Standardization/International Electrotechnical Commission. 2010. *ISO/IEC 17043:2010 Conformity assessment—General requirements for proficiency testing.* Geneva, Switzerland: ISO.
2. Ibid.
3. International Organization for Standardization/International Electrotechnical Commission. 2017. *ISO/IEC 17025:2017—General requirements for the competence of testing and calibration laboratories.* Geneva, Switzerland: ISO, p. 13.
4. National Institute of Standards and Technology. *1.3.5.9. F-Test for Equality of Two Variances.* Gaithersburg, MD: Information Technology Laboratory | NIST. https://www.itl.nist.gov/div898/handbook/eda/section3/eda359.htm
5. ASTM International. 2021. *ASTM E2587-16(2021)e1, Standard Practice for Use of Control Charts in Statistical Process Control.* West Conshohocken, PA: ASTM International, pp. 7–8.
6. Ibid.

Chapter 24: Number Formatting

1. Committee on Data for Science and Technology. 2018. *Fundamental Physical Constants—CODATA, The Committee on Data for Science and Technology.* Paris, France: CODATA. https://codata.org/initiatives/data-science-and-stewardship/fundamental-physical-constants

2. International Organization for Standards. 1973. *ISO 17:1973 Guide to the use of preferred numbers and of series of preferred numbers*. Geneva, Switzerland: ISO.
3. Thompson, Amber, and Barry N. Taylor. 2008, March. *NIST Special Publication 811 | Guide for the Use of the International System of Units (SI)—2008 Edition*. Gaithersburg, MD: NIST. https://www.nist.gov/pml/special-publication-811
4. Wallard, Andrew. 2004. "News from the BIPM—2003: Resolution 10 of the 22nd CGPM (2003)—Symbol for the Decimal Marker." *Metrologia* 41(1 January): 99–108. https://doi.org/10.1088/0026-1394/41/1/m01
5. Ibid.
6. Ibid.
7. Newell, David B., and Eite Tiesinga, eds. 2019. *NIST Special Publication 330 | The International System of Units (SI) 2019 Edition*. Gaithersburg, MD: NIST, Physical Measurement Laboratory. https://www.nist.gov/pml/special-publication-330
8. Thompson, Amber, and Barry N. Taylor. 2008, March. *NIST Special Publication 811 | Guide for the Use of the International System of Units (SI)—2008 Edition*. Gaithersburg, MD: NIST. https://www.nist.gov/pml/special-publication-811
9. U.S. Department of Transportation, Bureau of Transportation Statistics. 2021, March 12. *History of Daylight Savings Time | Bureau of Transportation Statistics*. Washington, D.C.: Bureau of Transportation Statistics, U.S. Department of Transportation. https://www.bts.gov/geospatial/daylight-savings-time (Includes material from: Phillips, Charles. "A Day to Remember: November 18, 1883." *American History*, Dec. 2004, pp. 16–18.)
10. U.S. Department of Transportation, Bureau of Transportation Statistics. 2021, March 12. *History of Daylight Savings Time | Bureau of Transportation Statistics*. Washington, D.C.: Bureau of Transportation Statistics. U.S. Department of Transportation. https://www.bts.gov/geospatial/daylight-savings-time (Includes material from Gordon, John Steele. "Standard Time: We All Live By What Happened on November 18, 1883." *American Heritage*, Jul./Aug. 2001, pp. 22–23.)
11. Ibid.
12. Bonikowsky, Laura Neilson. 2013, October 18. *Invention of Standard Time | The Canadian Encyclopedia*. Home | The Canadian Encyclopedia. https://www.thecanadianencyclopedia.ca/en/article/invention-of-standard-time-feature
13. U.S. Library of Congress. n.d. *Today in History—November 18: Time! | Library of Congress*. The Library of Congress. Washington, D.C.: Library of Congress. https://www.loc.gov/item/today-in-history/november-18
14. Bureau International des Poids et Mesures. 2019. *SI Brochure: The International System of Units (SI)*. 9th ed. Vol. v1.08, *Appendix 1. Decisions of the 10th CIPM—Decision on the first six base units*. Sèvres, France: BIPM, p. 163. https://www.bipm.org/documents/20126/41483022/si-brochure-9-App1-EN.pdf/3f415ca7-130b-7ebb-757a-bf8c0e87d9ce
15. Bureau International des Poids et Mesures. 2019. *SI Brochure: The International System of Units (SI)*. 9th ed. Vol. v1.08, *Appendix 1. Decisions of the 10th CGPM—Decision to adopt the name "International System of Units."* Sèvres, France: BIPM, p. 164. https://www.bipm.org/documents/20126/41483022/si-brochure-9-App1-EN.pdf/3f415ca7-130b-7ebb-757a-bf8c0e87d9ce
16. UTC (Universal Time Coordinated). greenwichmeantime.com
17. International Organization for Standardization—Technical Committee 154. 2019, August 26. "Introduction to the New ISO 8601-1 and ISO 8601-2—ISO/TC 154." https://www.isotc154.org/posts/2019-08-27-introduction-to-the-new-8601
18. International Organization for Standardization. 2019, February. *ISO 8601-1:2019—Date and time—Representations for information interchange—Part 1: Basic rules*. Geneva, Switzerland: ISO.
19. Ibid.
20. Ibid.

Chapter 25: Unit Conversions

1. Bureau International des Poids et Mesures. 2019. *SI Brochure: The International System of Units (SI)*. 9th ed. Vol. v1.08. Sèvres, France: BIPM, p. 130. https://www.bipm.org/en/publications/si-brochure
2. Bureau International des Poids et Mesures. 2019. *SI Brochure: The International System of Units (SI)*. 9th ed. Vol. v1.08. Sèvres, France: BIPM, p. 145. https://www.bipm.org/en/publications/si-brochure
3. Bureau International des Poids et Mesures. 2019. *SI Brochure: The International System of Units (SI)*. 9th ed. Vol. v1.08. Sèvres, France: BIPM, p. 139. https://www.bipm.org/en/publications/si-brochure
4. Bureau International des Poids et Mesures. 2019. *SI Brochure: The International System of Units (SI)*. 9th ed. Vol. v1.08. Sèvres, France: BIPM, pp. 137–138. https://www.bipm.org/en/publications/si-brochure
5. Bureau International des Poids et Mesures. 2019. *SI Brochure: The International System of Units (SI)*. 9th ed. Vol. v1.08. Sèvres, France: BIPM, p. 139. https://www.bipm.org/en/publications/si-brochure
6. Bureau International des Poids et Mesures. 2019. *SI Brochure: The International System of Units (SI)*. 9th ed. Vol. v1.08. Sèvres, France: BIPM, p. 143. https://www.bipm.org/en/publications/si-brochure
7. Butcher, Kenneth S., et al., eds. *NIST SP 1038—The International System of Units (SI)—Conversion Factors for General Use*. 2006 ed. Gaithersburg, MD: NIST—Weights and Measures Division Technology Services, 2006, pp. 4–5. https://nvlpubs.nist.gov/nistpubs/Legacy/SP/nistspecialpublication1038.pdf
8. National Institute for Standards and Technology. 2019, May. *Fundamental Physical Constants from NIST*. Gaithersburg, MD: Physical Measurement Laboratory, NIST. https://physics.nist.gov/cuu/Constants/index.html

Chapter 27: Statistics

1. Turan, Meltem Sönmez, Elaine Barker, John Kelsey, Kerry McKay, Mary L. Baish, and Mike Boyle. 2018, January. *NIST SP 800-90B Recommendation for the Entropy Sources Used for Random Bit Generation*. Gaithersburg, MD: U.S. Department of Commerce: NIST, p. 63. https://doi.org/10.6028/NIST.SP.800-90B
2. Yale University. 1998. *Sample Means*. New Haven, CT: Yale Department of Statistics and Data Science. http://www.stat.yale.edu/Courses/1997-98/101/sampmn.htm#clt
3. Ibid.
4. Yale University. 1998. *Confidence Intervals*. New Haven, CT: Yale Department of Statistics and Data Science. http://www.stat.yale.edu/Courses/1997-98/101/confint.htm
5. Yale University. 1998. *Sample Means*. New Haven, CT: Yale Department of Statistics and Data Science. http://www.stat.yale.edu/Courses/1997-98/101/sampmn.htm#clt
6. National Institute for Standards and Technology. *1.3.5.11. Measures of Skewness and Kurtosis*. Information Technology Laboratory | NIST. Gaithersburg, MD: NIST. https://www.itl.nist.gov/div898/handbook/eda/section3/eda35b.htm
7. National Institute for Standards and Technology. *Skewness*. Statistical Engineering Division – Dataplot | NIST. Gaithersburg, MD: NIST. https://www.itl.nist.gov/div898/software/dataplot/refman2/auxillar/skewness.htm
8. Ibid.
9. National Institute for Standards and Technology. 2021. *DLMF: Chapter 15 Hypergeometric Function*. DLMF: NIST Digital Library of Mathematical Functions. Gaithersburg, MD: NIST. https://dlmf.nist.gov/15

10. National Institute for Standards and Technology. *1.3.6.6.18. Binomial Distribution.* Information Technology Laboratory | NIST. Gaithersburg, MD: NIST. https://www.itl.nist.gov/div898//handbook/eda/section3/eda366i.htm
11. Ibid.

Chapter 29: Measurement Uncertainty

1. Joint Committee for Guides in Metrology. 2012. *JCGM 200:2012, International vocabulary of metrology—Basic and general concepts and associated terms (VIM 3)*; BIPM, IEC, IFCC, ILAC, ISO, IUPAC, IUPAP, and OIML. Sèvres, France: BIPM, p. 25. https://www.bipm.org/documents/20126/2071204/JCGM_200_2012.pdf/f0e1ad45-d337-bbeb-53a6-15fe649d0ff1
2. Joint Committee for Guides in Metrology. 2012. *JCGM 200:2012, International vocabulary of metrology—Basic and general concepts and associated terms (VIM 3)*; BIPM, IEC, IFCC, ILAC, ISO, IUPAC, IUPAP, and OIML. Sèvres, France: BIPM, p. 20. https://www.bipm.org/documents/20126/2071204/JCGM_200_2012.pdf/f0e1ad45-d337-bbeb-53a6-15fe649d0ff1
3. Joint Committee for Guides in Metrology. 2012. *JCGM 200:2012, International vocabulary of metrology—Basic and general concepts and associated terms (VIM 3)*; BIPM, IEC, IFCC, ILAC, ISO, IUPAC, IUPAP, and OIML. Sèvres, France: BIPM, p. 22. https://www.bipm.org/documents/20126/2071204/JCGM_200_2012.pdf/f0e1ad45-d337-bbeb-53a6-15fe649d0ff1
4. Joint Committee for Guides in Metrology. 2012. *JCGM 200:2012, International vocabulary of metrology—Basic and general concepts and associated terms (VIM 3)*; BIPM, IEC, IFCC, ILAC, ISO, IUPAC, IUPAP, and OIML. Sèvres, France: BIPM, p. 19. https://www.bipm.org/documents/20126/2071204/JCGM_200_2012.pdf/f0e1ad45-d337-bbeb-53a6-15fe649d0ff1
5. Joint Committee for Guides in Metrology. 2012. *JCGM 200:2012, International vocabulary of metrology—Basic and general concepts and associated terms (VIM 3)*; BIPM, IEC, IFCC, ILAC, ISO, IUPAC, IUPAP, and OIML. Sèvres, France: BIPM, p. 53. https://www.bipm.org/documents/20126/2071204/JCGM_200_2012.pdf/f0e1ad45-d337-bbeb-53a6-15fe649d0ff1
6. Joint Committee for Guides in Metrology. 2012. *JCGM 200:2012, International vocabulary of metrology—Basic and general concepts and associated terms (VIM 3)*; BIPM, IEC, IFCC, ILAC, ISO, IUPAC, IUPAP, and OIML. Sèvres, France: BIPM, p. 12. https://www.bipm.org/documents/20126/2071204/JCGM_200_2012.pdf/f0e1ad45-d337-bbeb-53a6-15fe649d0ff1
7. Joint Committee for Guides in Metrology. 2012. *JCGM 200:2012, International vocabulary of metrology—Basic and general concepts and associated terms (VIM 3)*; BIPM, IEC, IFCC, ILAC, ISO, IUPAC, IUPAP, and OIML. Sèvres, France: BIPM, p. 19. https://www.bipm.org/documents/20126/2071204/JCGM_200_2012.pdf/f0e1ad45-d337-bbeb-53a6-15fe649d0ff1
8. Joint Committee for Guides in Metrology. 2012. *JCGM 200:2012, International vocabulary of metrology—Basic and general concepts and associated terms (VIM 3)*; BIPM, IEC, IFCC, ILAC, ISO, IUPAC, IUPAP, and OIML. Sèvres, France: BIPM, p. 17. https://www.bipm.org/documents/20126/2071204/JCGM_200_2012.pdf/f0e1ad45-d337-bbeb-53a6-15fe649d0ff1
9. Joint Committee for Guides in Metrology. 2012. *JCGM 200:2012, International vocabulary of metrology—Basic and general concepts and associated terms (VIM 3)*; BIPM, IEC, IFCC, ILAC, ISO, IUPAC, IUPAP, and OIML. Sèvres, France: BIPM, p. 2. https://www.bipm.org/documents/20126/2071204/JCGM_200_2012.pdf/f0e1ad45-d337-bbeb-53a6-15fe649d0ff1

10. Joint Committee for Guides in Metrology. 2012. *JCGM 200:2012, International vocabulary of metrology—Basic and general concepts and associated terms (VIM 3);* BIPM, IEC, IFCC, ILAC, ISO, IUPAC, IUPAP, and OIML. Sèvres, France: BIPM, p. 3. https://www.bipm.org/documents/20126/2071204/JCGM_200_2012.pdf/f0e1ad45-d337-bbeb-53a6-15fe649d0ff1
11. Joint Committee for Guides in Metrology. 2012. *JCGM 200:2012, International vocabulary of metrology—Basic and general concepts and associated terms (VIM 3);* BIPM, IEC, IFCC, ILAC, ISO, IUPAC, IUPAP, and OIML. Sèvres, France: BIPM, p. 29. https://www.bipm.org/documents/20126/2071204/JCGM_200_2012.pdf/f0e1ad45-d337-bbeb-53a6-15fe649d0ff1
12. Joint Committee for Guides in Metrology. 2012. *JCGM 200:2012, International vocabulary of metrology—Basic and general concepts and associated terms (VIM 3);* BIPM, IEC, IFCC, ILAC, ISO, IUPAC, IUPAP, and OIML. Sèvres, France: BIPM, p. 28. https://www.bipm.org/documents/20126/2071204/JCGM_200_2012.pdf/f0e1ad45-d337-bbeb-53a6-15fe649d0ff1
13. Joint Committee for Guides in Metrology. 2012. *JCGM 200:2012, International vocabulary of metrology—Basic and general concepts and associated terms (VIM 3);* BIPM, IEC, IFCC, ILAC, ISO, IUPAC, IUPAP, and OIML. Sèvres, France: BIPM, p. 46. https://www.bipm.org/documents/20126/2071204/JCGM_200_2012.pdf/f0e1ad45-d337-bbeb-53a6-15fe649d0ff1
14. Joint Committee for Guides in Metrology. 2012. *JCGM 200:2012, International vocabulary of metrology—Basic and general concepts and associated terms (VIM 3);* BIPM, IEC, IFCC, ILAC, ISO, IUPAC, IUPAP, and OIML. Sèvres, France: BIPM, p. 37. https://www.bipm.org/documents/20126/2071204/JCGM_200_2012.pdf/f0e1ad45-d337-bbeb-53a6-15fe649d0ff1
15. Joint Committee for Guides in Metrology. 2012. *JCGM 200:2012, International vocabulary of metrology—Basic and general concepts and associated terms (VIM 3);* BIPM, IEC, IFCC, ILAC, ISO, IUPAC, IUPAP, and OIML. Sèvres, France: BIPM, p. 24. https://www.bipm.org/documents/20126/2071204/JCGM_200_2012.pdf/f0e1ad45-d337-bbeb-53a6-15fe649d0ff1
16. Joint Committee for Guides in Metrology. 2012. *JCGM 200:2012, International vocabulary of metrology—Basic and general concepts and associated terms (VIM 3);* BIPM, IEC, IFCC, ILAC, ISO, IUPAC, IUPAP, and OIML. Sèvres, France: BIPM, p. 23. https://www.bipm.org/documents/20126/2071204/JCGM_200_2012.pdf/f0e1ad45-d337-bbeb-53a6-15fe649d0ff1
17. Joint Committee for Guides in Metrology. 2012. *JCGM 200:2012, International vocabulary of metrology—Basic and general concepts and associated terms (VIM 3);* BIPM, IEC, IFCC, ILAC, ISO, IUPAC, IUPAP, and OIML. Sèvres, France: BIPM, p. 25. https://www.bipm.org/documents/20126/2071204/JCGM_200_2012.pdf/f0e1ad45-d337-bbeb-53a6-15fe649d0ff1
18. Joint Committee for Guides in Metrology. 2012. *JCGM 200:2012, International vocabulary of metrology—Basic and general concepts and associated terms (VIM 3);* BIPM, IEC, IFCC, ILAC, ISO, IUPAC, IUPAP, and OIML. Sèvres, France: BIPM, p. 24. https://www.bipm.org/documents/20126/2071204/JCGM_200_2012.pdf/f0e1ad45-d337-bbeb-53a6-15fe649d0ff1
19. Ibid.
20. Ibid.
21. International Organization for Standardization/International Electrotechnical Commission. 2017. *ISO/IEC 17025:2017—General requirements for the competence of testing and calibration laboratories.* Geneva, Switzerland: ISO, p. 7.
22. Ibid.
23. Joint Committee for Guides in Metrology. 2012. *JCGM 200:2012, International vocabulary of metrology—Basic and general concepts and associated terms (VIM 3);*

BIPM, IEC, IFCC, ILAC, ISO, IUPAC, IUPAP, and OIML. Sèvres, France: BIPM, p. 25. https://www.bipm.org/documents/20126/2071204/JCGM_200_2012.pdf/ f0e1ad45-d337-bbeb-53a6-15fe649d0ff1
24. Ibid.
25. Dilip Shah, E = mc^3 Solutions.
26. Joint Committee for Guides in Metrology. 2008. *JCGM 100:2008, Evaluation of measurement data—Guide to the expression of uncertainty in measurement (GUM)*; BIPM, IEC, IFCC, ILAC, ISO, IUPAC, IUPAP, and OIML. Sèvres, France: BIPM, p. 27.
27. Ibid.
28. National Institute for Standards and Technology. 2019. *SOP 29 SOP for the Assignment of Uncertainty*. Gaithersburg, MD: NIST. https://doi.org/10.6028/NIST.IR.6969-2019
29. Joint Committee for Guides in Metrology. 2008. *JCGM 100:2008, Evaluation of measurement data—Guide to the expression of uncertainty in measurement (GUM)*; BIPM, IEC, IFCC, ILAC, ISO, IUPAC, IUPAP, and OIML. Sèvres, France: BIPM, p. 6.
30. Joint Committee for Guides in Metrology. 2008. *JCGM 100:2008, Evaluation of measurement data—Guide to the expression of uncertainty in measurement (GUM)*; BIPM, IEC, IFCC, ILAC, ISO, IUPAC, IUPAP, and OIML. Sèvres, France: BIPM, pp. 10–11.
31. Dilip Shah, E = mc^3 Solutions.
32. Joint Committee for Guides in Metrology. 2008. *JCGM 100:2008, Evaluation of measurement data—Guide to the expression of uncertainty in measurement (GUM)*; BIPM, IEC, IFCC, ILAC, ISO, IUPAC, IUPAP, and OIML. Sèvres, France: BIPM, pp. 11–14.
33. Joint Committee for Guides in Metrology. 2008. *JCGM 100:2008, Evaluation of measurement data—Guide to the expression of uncertainty in measurement (GUM)*; BIPM, IEC, IFCC, ILAC, ISO, IUPAC, IUPAP, and OIML. Sèvres, France: BIPM, pp. 15–18.
34. Joint Committee for Guides in Metrology. 2008. *JCGM 100:2008, Evaluation of measurement data—Guide to the expression of uncertainty in measurement (GUM)*; BIPM, IEC, IFCC, ILAC, ISO, IUPAC, IUPAP, and OIML. Sèvres, France: BIPM, p. 64.
35. American Association for Laboratory Accreditation. 2020. *A2LA G129—Measurement Uncertainty Budget Template, version 1.1*. Frederick, MD: A2LA. https://mktg.a2la.org/l/273522/2020-02-26/3yzty65
36. Joint Committee for Guides in Metrology. 2008. *JCGM 100:2008, Evaluation of measurement data—Guide to the expression of uncertainty in measurement (GUM)*; BIPM, IEC, IFCC, ILAC, ISO, IUPAC, IUPAP, and OIML. Sèvres, France: BIPM, pp. 14, 82.
37. Joint Committee for Guides in Metrology. 2008. *JCGM 100:2008, Evaluation of measurement data—Guide to the expression of uncertainty in measurement (GUM)*; BIPM, IEC, IFCC, ILAC, ISO, IUPAC, IUPAP, and OIML. Sèvres, France: BIPM, pp. 18–22.
38. Boromir in *Lord of the Rings* (2001).
39. Joint Committee for Guides in Metrology. 2008. *JCGM 100:2008, Evaluation of measurement data—Guide to the expression of uncertainty in measurement (GUM)*; BIPM, IEC, IFCC, ILAC, ISO, IUPAC, IUPAP, and OIML. Sèvres, France: BIPM, pp. 21–23.
40. Joint Committee for Guides in Metrology. 2008. *JCGM 100:2008, Evaluation of measurement data—Guide to the expression of uncertainty in measurement (GUM)*; BIPM, IEC, IFCC, ILAC, ISO, IUPAC, IUPAP, and OIML. Sèvres, France: BIPM, pp. 23–24.
41. Joint Committee for Guides in Metrology. 2008. *JCGM 100:2008, Evaluation of measurement data—Guide to the expression of uncertainty in measurement (GUM)-Annex G, Section 1*; BIPM, IEC, IFCC, ILAC, ISO, IUPAC, IUPAP, and OIML. Sèvres, France: BIPM, p. 70.
42. Joint Committee for Guides in Metrology. 2008. *JCGM 100:2008, Evaluation of measurement data—Guide to the expression of uncertainty in measurement (GUM)*; BIPM, IEC, IFCC, ILAC, ISO, IUPAC, IUPAP, and OIML. Sèvres, France: BIPM, pp. 24–27.
43. International Laboratory Accreditation Cooperation. 2020. *ILAC P14:09/2020 ILAC Policy for Measurement Uncertainty in Calibration*. Silverwater, Australia: ILAC.

44. International Laboratory Accreditation Cooperation. 2020. *ILAC P14:09/2020 ILAC Policy for Measurement Uncertainty in Calibration.* Silverwater, Australia: ILAC, p. 8.
45. Zumbrun, Henry, and Dilip Shah. 2020. "The New Dimension to Resolution: Can It Be Resolved?" NCSLI Workshop & Symposium, 2000 (Best paper). Boulder, CO: NCSL International.
46. Joint Committee for Guides in Metrology. 2008. *JCGM 100:2008, Evaluation of measurement data—Guide to the expression of uncertainty in measurement (GUM)*; BIPM, IEC, IFCC, ILAC, ISO, IUPAC, IUPAP, and OIML. Sèvres, France: BIPM, p. 19.
47. Joint Committee for Guides in Metrology. 2008. *JCGM 100:2008, Evaluation of measurement data—Guide to the expression of uncertainty in measurement (GUM)—Annex G, Section 4*; BIPM, IEC, IFCC, ILAC, ISO, IUPAC, IUPAP, and OIML. Sèvres, France: BIPM, p. 73.
48. Brownlee, K. A. 1960. *Statistical Theory and Methodology in Science and Engineering.* New York: John Wiley & Sons.
49. Joint Committee for Guides in Metrology. 2008. *JCGM 100:2008, Evaluation of measurement data—Guide to the expression of uncertainty in measurement (GUM)—Annex G, Section 3*; BIPM, IEC, IFCC, ILAC, ISO, IUPAC, IUPAP, and OIML. Sèvres, France: BIPM, p. 72.
50. National Institutes of Standards and Technology/SEMATECH. *1.3.6.7.2. Critical Values of the Student's t Distribution.* Gaithersburg, MD: NIST, Information Technology Laboratory. https://www.itl.nist.gov/div898/handbook/eda/section3/eda3672.htm
51. Joint Committee for Guides in Metrology. 2008. *JCGM 100:2008, Evaluation of measurement data—Guide to the expression of uncertainty in measurement (GUM)—Annex G, Table G.2*; BIPM, IEC, IFCC, ILAC, ISO, IUPAC, IUPAP, and OIML. Sèvres, France: BIPM, p. 78.

Chapter 30: Measurement Risk and Decision Rules

1. Dilip Shah, E = mc^3 Solutions. 2021.
2. SAE International. 2016, September 20. *AS9100D—Quality Management Systems—Requirements for Aviation, Space, and Defense Organizations.* Warrendale, PA: SAE International, p. 8.
3. Crowder, Stephen, Collin Delker, Eric Forrest, and Nevin Martin. 2020. *Introduction to Statistics in Metrology.* Springer Nature Switzerland AG, pp. 81–82.
4. International Organization for Standardization/International Electrotechnical Commission. 2017. *ISO/IEC 17025:2017—General requirements for the competence of testing and calibration laboratories.* Geneva, Switzerland: ISO, p. 2.
5. International Laboratory Accreditation Cooperation. 2019. *ILAC G8:09/2019 Guidelines on Decision Rules and Statements of Conformity.* Silverwater, Australia: ILAC, p. 16.
6. Joint Committee for Guides in Metrology. 2012. *JCGM 106:2012, Evaluation of measurement data—The role of measurement uncertainty in conformity decision.* BIPM, IEC, IFCC, ILAC, ISO, IUPAC, IUPAP, and OIML. Sèvres, France: BIPM, p. 19.
7. UKAS. 2020. *LAB 48—Decision Rules and Statements of Conformity.* June 2020. UKAS. Com. 3rd ed. Staines-upon-Thames, England: United Kingdom Accreditation Service, p. 6. https://www.ukas.com/wp-content/uploads/filebase/publications/publications-relating-to-laboratory-accreditation/LAB-48-Decision-Rules-Edition-3-June-2020.pdf
8. NCSL International. 2017. *Handbook for the Application of ANSI/NCSL Z540.3-2006: Requirements for the Calibration of Measuring and Test Equipment.* Boulder, CO: NCSL International, pp. 143–148.
9. Joint Committee for Guides in Metrology. 2012. *JCGM 200:2012, International vocabulary of metrology—Basic and general concepts and associated terms (VIM 3)*;

BIPM, IEC, IFCC, ILAC, ISO, IUPAC, IUPAP, and OIML. Sèvres, France: BIPM, p. 26. https://www.bipm.org/documents/20126/2071204/JCGM_200_2012.pdf/f0e1ad45-d337-bbeb-53a6-15fe649d0ff1
10. Dilip Shah, E = mc^3 Solutions. 2021.
11. Ibid.
12. Morehouse Instruments and Dilip Shah, E = mc^3 Solutions. 2021.
13. International Laboratory Accreditation Cooperation. 2019. *ILAC G8:09/2019 Guidelines on Decision Rules and Statements of Conformity.* Silverwater, Australia: ILAC, p. 5.
14. International Laboratory Accreditation Cooperation. 2019. *ILAC G8:09/2019 Guidelines on Decision Rules and Statements of Conformity.* Silverwater, Australia: ILAC, p. 8.
15. UKAS. 2020. *LAB 48—Decision Rules and Statements of Conformity.* June 2020. *UKAS.Com.* 3rd ed. Staines-upon-Thames, England, United Kingdom: United Kingdom Accreditation Service, p. 33. https://www.ukas.com/wp-content/uploads/filebase/publications/publications-relating-to-laboratory-accreditation/LAB-48-Decision-Rules-Edition-3-June-2020.pdf
16. Joint Committee for Guides in Metrology. 2012. *JCGM 106:2012, Evaluation of measurement data—The role of measurement uncertainty in conformity decision.* BIPM, IEC, IFCC, ILAC, ISO, IUPAC, IUPAP, and OIML. Sèvres, France: BIPM, p. 20.
17. International Organization for Standardization/International Electrotechnical Commission. 2017. *ISO/IEC 17025:2017—General requirements for the competence of testing and calibration laboratories.* Geneva, Switzerland: ISO, p. 17.
18. International Laboratory Accreditation Cooperation. 2019. *ILAC G8:09/2019 Guidelines on Decision Rules and Statements of Conformity.* Silverwater, Australia: ILAC, p. 9.
19. Ibid.
20. Morehouse Instruments and Dilip Shah, E = mc^3 Solutions. 2021.
21. Ibid.
22. International Organization for Standardization/International Electrotechnical Commission. 2017. *ISO/IEC 17025:2017—General requirements for the competence of testing and calibration laboratories.* Geneva, Switzerland: ISO, pp. 9–10.
23. American National Standards Institute/NCSL International. 2006. *ANSI/NCSL Z540.3:2006 (R2013), American National Standards for Calibration—Requirements for the Calibration of Measuring and Test Equipment.* Boulder, CO: NCSL International, paragraph 3.11.
24. Zumbrun, Henry, and Dilip Shah. 2021, October. "The New Dimension to Resolution: Can It Be Resolved?" York, PA: Morehouse Instrument Company, p. 8. https://mhforce.com/wp-content/uploads/2021/10/The-new-dimension-to-resolution-Can-it-be-resolved.pdf
25. NCSL International. 2017. *Handbook for the Application of ANSI/NCSL Z540.3-2006: Requirements for the Calibration of Measuring and Test Equipment.* Boulder, CO: NCSL International, p. 11.
26. Ibid.
27. Zumbrun, Henry, and Dilip Shah. 2021, October. "The New Dimension to Resolution: Can It Be Resolved?" York, PA: Morehouse Instrument Company, p. 8. https://mhforce.com/wp-content/uploads/2021/10/The-new-dimension-to-resolution-Can-it-be-resolved.pdf
28. NCSL International. 2017. *Handbook for the Application of ANSI/NCSL Z540.3-2006: Requirements for the Calibration of Measuring and Test Equipment.* Boulder, CO: NCSL International, p. 48.
29. Ibid.
30. International Laboratory Accreditation Cooperation. 2020. *ILAC P14:09/2020 ILAC Policy for Measurement Uncertainty in Calibration.* Silverwater, Australia: ILAC, p. 8.

648 *Endnotes*

31. Zumbrun, Henry, and Dilip Shah. 2021, October. "The New Dimension to Resolution: Can It Be Resolved?" York, PA: Morehouse Instrument Company, p. 8. https://mhforce.com/wp-content/uploads/2021/10/The-new-dimension-to-resolution-Can-it-be-resolved.pdf
32. Zumbrun, Henry, and Dilip Shah. 2021, October. "The New Dimension to Resolution: Can It Be Resolved?" York, PA: Morehouse Instrument Company, p. 9. https://mhforce.com/wp-content/uploads/2021/10/The-new-dimension-to-resolution-Can-it-be-resolved.pdf
33. Zumbrun, Henry, and Dilip Shah. 2021, October. "The New Dimension to Resolution: Can It Be Resolved?" York, PA: Morehouse Instrument Company, p. 2. https://mhforce.com/wp-content/uploads/2021/10/The-new-dimension-to-resolution-Can-it-be-resolved.pdf
34. Zumbrun, Henry. 2021, December. *Why a 4:1 TUR is not Enough: The Importance of Analyzing the Probability of False Accept Risk.* York, PA: Morehouse Instrument Company, pp. 12–15. https://mhforce.com/wp-content/uploads/2021/04/Why-a-4-to-1-TUR-is-not-enough.pdf
35. Zumbrun, Henry. 2021, December. *Why a 4:1 TUR is not Enough: The Importance of Analyzing the Probability of False Accept Risk.* York, PA: Morehouse Instrument Company, p. 1. https://mhforce.com/wp-content/uploads/2021/04/Why-a-4-to-1-TUR-is-not-enough.pdf
36. American National Standards Institute/NCSL International. 2006. *ANSI/NCSL Z540.3:2006 (R2013), American National Standards for Calibration—Requirements for the Calibration of Measuring and Test Equipment.* Boulder, CO: NCSL International.
37. Crowder, Stephen, Collin Delker, Eric Forrest, and Nevin Martin. 2020. *Introduction to Statistics in Metrology.* Springer Nature Switzerland AG, p. 83.
38. NCSL International. 2017. *Handbook for the Application of ANSI/NCSL Z540.3-2006: Requirements for the Calibration of Measuring and Test Equipment.* Boulder, CO: NCSL International, pp. 143–148.

Chapter 31: Introduction to Measurement Parameters

1. NCSL International. 2014. *Recommended Practice-9: Measurement Capability Description, RP-9.* 3rd ed. Boulder, CO: NCSL International, p. 19.

Chapter 32: DC and Low Frequency

1. Bureau International des Poids et Mesures. 2019. *SI Brochure: The International System of Units (SI).* 9th ed. Vol. v1.08. Sèvres, France: BIPM, p. 132. https://www.bipm.org/en/publications/si-brochure
2. Ibid.
3. Ibid.
4. Bureau International des Poids et Mesures. 2019. *SI Brochure: The International System of Units (SI).* 9th ed. Vol. v1.08. Sèvres, France: BIPM, Appendix 2. https://www.bipm.org/en/publications/si-brochure
5. Keithley, A Tektronix Company. 2014. *Low Level Measurements Handbook—Precision DC Current, Voltage, and Resistance Measurements.* 7th ed. pp. 1–7 to 1–8. https://download.tek.com/document/LowLevelHandbook_7Ed.pdf
6. Bureau International des Poids et Mesures. 2019. *SI Brochure: The International System of Units (SI).* 9th ed. Vol. v1.08. Sèvres, France: BIPM, p. 138. https://www.bipm.org/en/publications/si-brochure

7. Calhoun, Richard. 1994. *Calibration & Standards: DC to 40 GHz*. Louisville, KY: SS&S, p. 18.
8. Fluke. 1994. *Calibration: Philosophy in Practice*. 2nd ed. Everett, WA: Fluke Corporation, pp. 7–9.
9. Bureau International des Poids et Mesures. 2019. *SI Brochure: The International System of Units (SI)*. 9th ed. Vol. v1.08. Sèvres, France: BIPM, pp. 137–138. https://www.bipm.org/en/publications/si-brochure
10. Bureau International des Poids et Mesures. 2019. *SI Brochure: The International System of Units (SI)*. 9th ed. Vol. v1.08, Appendix 2. Sèvres, France: BIPM, p. 2. https://www.bipm.org/en/publications/si-brochure
11. Ibid.
12. Keithley, A Tektronix Company. 2014. *Low Level Measurements Handbook—Precision DC Current, Voltage, and Resistance Measurements*. 7th ed. Table 3-1, pp. 3–4. https://download.tek.com/document/LowLevelHandbook_7Ed.pdf
13. Fluke. 1994. *Calibration: Philosophy in Practice*. 2nd ed. Everett, WA: Fluke Corporation, p. 33-3.
14. Fluke. 1994. *Calibration: Philosophy in Practice*. 2nd ed. Everett, WA: Fluke Corporation, p. 8-8.
15. Bureau International des Poids et Mesures. 2019. *SI Brochure: The International System of Units (SI)*. 9th ed. Vol. v1.08. Sèvres, France: BIPM, p. 178. https://www.bipm.org/en/publications/si-brochure
16. Bureau International des Poids et Mesures. 2019. *SI Brochure: The International System of Units (SI)*. 9th ed. Vol. v1.08. Sèvres, France: BIPM, Appendix 2, p. 3. https://www.bipm.org/en/publications/si-brochure
17. Ibid.
18. Fluke. 1994. *Calibration: Philosophy in Practice*. 2d ed. Everett, WA: Fluke Corporation, p. 33-3.
19. Fluke. 1994. *Calibration: Philosophy in Practice*. 2d ed. Everett, WA: Fluke Corporation, p. 8-8.
20. Keithley, A Tektronix Company. 2014. *Low Level Measurements Handbook—Precision DC Current, Voltage, and Resistance Measurements*. 7th ed., p. 3-23. https://download.tek.com/document/LowLevelHandbook_7Ed.pdf
21. Fluke. 1994. *Calibration: Philosophy in Practice*. 2d ed. Everett, WA: Fluke Corporation, pp. 10-3 to 10-11; Calhoun, Richard. 1994. *Calibration & Standards: DC to 40 GHz*. Louisville, KY: SS&S, pp. 159–160.
22. Fluke. 1994. *Calibration: Philosophy in Practice*. 2d ed. Everett, WA: Fluke Corporation, p. 10-10.
23. Fluke. 1994. *Calibration: Philosophy in Practice*. 2d ed. Everett, WA: Fluke Corporation, p. 10-6; Calhoun, Richard. 1994. *Calibration & Standards: DC to 40 GHz*. Louisville, KY: SS&S, p. 160.
24. Fluke. 1994. *Calibration: Philosophy in Practice*. 2nd ed. Everett, WA: Fluke Corporation, pp. 10-6 to 10-7.d
25. Fluke. 1994. *Calibration: Philosophy in Practice*. 2nd ed. Everett, WA: Fluke Corporation, pp. 10-9 to 10-10, 11-11.
26. Calhoun, Richard. 1994. *Calibration & Standards: DC to 40 GHz*. Louisville, KY: SS&S, p. 174.
27. Fluke. 1994. *Calibration: Philosophy in Practice*. 2nd ed. Everett, WA: Fluke Corporation, pp. 12-7 to 12-8; Calhoun, Richard. 1994. *Calibration & Standards: DC to 40 GHz*. Louisville, KY: SS&S, pp. 99–100.
28. Fluke. 1994. *Calibration: Philosophy in Practice*. 2nd ed. Everett, WA: Fluke Corporation, pp. 12-3 to 12-4; Calhoun, Richard. 1994. *Calibration & Standards: DC to 40 GHz*. Louisville, KY: SS&S, pp. 115–116.

29. Bureau International des Poids et Mesures. 2019. *SI Brochure: The International System of Units (SI)*. 9th ed. Vol. v1.08. Sèvres, France: BIPM, p. 127. https://www.bipm.org/en/publications/si-brochure
30. Bureau International des Poids et Mesures. 2019. *SI Brochure: The International System of Units (SI)*. 9th ed. Vol. v1.08. Sèvres, France: BIPM, p. 130. https://www.bipm.org/en/publications/si-brochure
31. Lombardi, Michael A. 2014. "Time Measurement." In *Measurement, Instrumentation, and Sensors Handbook*, edited by John G. Webster and Halit Eren, ch. 41. Boca Raton, FL: CRC Press/Taylor and Francis Group, LLC. https://tf.nist.gov/general/pdf/2488.pdf
32. Lombardi, Michael A. 2014. "Time Measurement." In *Measurement, Instrumentation, and Sensors Handbook*, edited by John G. Webster and Halit Eren, ch. 42. Boca Raton, FL: CRC Press/Taylor and Francis Group, LLC. https://tf.nist.gov/general/pdf/2488.pdf
33. Lombardi, Michael A. 2016. "Evaluating the Frequency and Time Uncertainty of GPS Disciplined Oscillators and Clocks." *NCSL International Measure 11*, no. 3–4 (October): 30–44. https://doi.org/10.1080/19315775.2017.1316696
34. Matsakis, D., J. Levine, and M. Lombardi. 2018. "Metrological and Legal Traceability of Time Signals." 2018 Precise Time and Time Interval (PTTI) Meeting, Reston, VA, [online]. https://tsapps.nist.gov/publication/get_pdf.cfm?pub_id=924939
35. Systema Interamercano de Metrologia (SIM). n.d. SIM Time and Frequency Metrology Working Group. https://tf.nist.gov/sim
36. "SOP 24-1 Calibration of Field Standard Stopwatches." In Georgia L. Harris, ed., *NISTIR 8250: Calibration Procedures for Weights and Measures Laboratories*. Gaithersburg, MD: NIST, p. 1. https://doi.org/10.6028/NIST.IR.8250
37. "SOP 24-1 Calibration of Field Standard Stopwatches." In Georgia L. Harris, ed., *NISTIR 8250: Calibration Procedures for Weights and Measures Laboratories*. Gaithersburg, MD: NIST, p. 3. https://doi.org/10.6028/NIST.IR.8250
38. Gust, Jeff C., Robert M. Graham, and Michael A. Lombardi. 2009, January. *NIST Special Publication 960-12, Stopwatch and Timer Calibrations*. Gaithersburg, MD: US Government Printing Office. https://tf.nist.gov/general/pdf/1930.pdf
39. Bureau International des Poids et Mesures. n.d. BIPM Technical Services: Time Metrology. https://www.bipm.org/en/time-metrology
40. Bureau International des Poids et Mesures. 2019. *SI Brochure: The International System of Units (SI)*. 9th ed. Vol. v1.08. Sèvres, France: BIPM, p. 151. https://www.bipm.org/en/publications/si-brochure
41. Turgel, Raymond S., and N. Michael Oldham. 1978. "High-Precision Audio-Frequency Phase Calibration Standard." *IEEE Transactions on Instrumentation and Measurement* IM-27, Issue 4 (December): 460–464.
42. Clarke, Kenneth K., and Donald T. Hess. 1990. "Phase Measurement, Traceability, and Verification Theory and Practice." *IEEE Transactions on Instrumentation and Measurement*, IM-39, Issue 1 (February): 53.
43. Ibid.
44. Stenbakken, Gerard N. 1984. "A Wideband Sampling Wattmeter." *IEEE Transactions on Power Apparatus and Systems*, PAS-103 (October): 2919–2926.

Chapter 33: Radio Frequency and Microwave

1. Bin Zikria, Yousaf, Rashid Ali, Muhammad Khalil Afzal, and Sung Won Kim. 2021, February 7. *Next-Generation Internet of Things (IoT): Opportunities, Challenges, and Solutions*. PubMed Central (PMC). U.S. National Library of Medicine, National

Center for Biotechnology Information. https://www.ncbi.nlm.nih.gov/pmc/articles/PMC7915959
2. Federal Standard 1037C, *Telecommunications: Glossary of Telecommunication Terms*. 1996. Washington, D.C.: General Services Administration. https://www.its.bldrdoc.gov/fs-1037/fs-1037c.htm
3. National Telecommunications and Information Administration. 1996. *United States Frequency Allocation Chart*. July 2003. Washington, D.C.: NTIA, Office of Spectrum Management. www.ntia.doc.gov/osmhome/allochrt.pdf
4. Adapted from Shrader, Robert L. 1980. *Electronic Communication*. 4th ed. New York: McGraw-Hill.
5. Saad, Theodore S., Robert C. Hansen, and Gershon J. Wheeler. 1971. *Microwave Engineer's Handbook*, Vol. 1. Dedham, MA: Artech House.
6. Adam, Stephen F. 1969. *Microwave Theory and Applications*. Englewood Cliffs, NJ: Prentice-Hall, pp. 58–59.
7. EIA is the Electronics Industries Alliance. IEC is the International Electrotechnical Commission. JAN refers to the U.S. military Joint Army-Navy specifications system. A useful reference for other current waveguide designations can be found at Agilent Technologies. On their website (www.agilent.com), search for *"Agilent Waveguide Overview."*
8. Adam, Stephen F. 1969. *Microwave Theory and Applications*. Englewood Cliffs, NJ: Prentice-Hall.
9. Adam, Stephen F. 1969. *Microwave Theory and Applications*. Englewood Cliffs, NJ: Prentice-Hall, pp. 211–212.
10. Laverghetta, Thomas S. 1981. *Handbook of Microwave Testing*. Dedham, MA: Artech House, p. 34.
11. Keysight. n.d. *Fundamentals of RF and Microwave Power Measurements (Part 2), Power Sensors and Instrumentation* (formerly Agilent AN 1449-2). Literature number 5988-9214EN, p. 5. Colorado Springs, CO: Keysight.
12. Larsen, Neil T. 1976. "A New Self-Balancing DC-Substitution RF Power Meter." *IEEE Transactions on Instrumentation and Measurement*, no. 4 (December): 343–347.
13. Carr, Joseph J. 1996. *Elements of Electronic Instrumentation and Measurement*. 3rd ed. Upper Saddle River, NJ: Prentice-Hall.
14. Keysight. n.d. *Fundamentals of RF and Microwave Power Measurements (Part 2), Power Sensors and Instrumentation* (formerly Agilent AN 1449-2). Literature number 5988-9214EN, pp. 19–50. Colorado Springs, CO: Keysight.
15. Agilent Technologies. 2002. "Power Measurement Basics." In *Agilent 2002, Back to Basics Seminar*. Publication number 5988-6641EN. Palo Alto, CA: Agilent, p. 4-23. (Keysight now owns Agilent.)
16. Agilent Technologies. 2002. "Power Measurement Basics." In *Agilent 2002, Back to Basics Seminar*. Publication number 5988-6641EN. Palo Alto, CA: Agilent, pp. 4-24 to 4-30. (Keysight now owns Agilent.)
17. Keysight. n.d. *Fundamentals of RF and Microwave Power Measurements (Part 3), Power Measurement Uncertainty per International Guides* (formerly Agilent AN 1449-3). Literature number 5988-9215EN, pp. 5–19. Colorado Springs, CO: Keysight.
18. Keysight. n.d. *Fundamentals of RF and Microwave Power Measurements (Part 2), Power Sensors and Instrumentation* (formerly Agilent AN 1449-2). Literature number 5988-9214EN, p. 8. Colorado Springs, CO: Keysight.
19. American National Standards Institute/Institute of Electrical and Electronics Engineers. 1973. *ANSI/IEEE 474-1973, Specifications and Test Methods for Fixed and Variable Attenuators, DC to 40 GHz*. New York, NY: IEEE.
20. Keysight. n.d. *Application Note 1287-1: Understanding the Fundamental Principles of Vector Network Analysis*. Publication number 5965-7707E, p. 10. Colorado Springs,

CO: Keysight; Adam, Stephen F. 1969. *Microwave Theory and Applications.* Englewood Cliffs, NJ: Prentice-Hall, pp. 58–59.
21. Carr, Joseph J. 1996. *Elements of Electronic Instrumentation and Measurement.* 3rd ed. Upper Saddle River, NJ: Prentice-Hall.
22. Keysight. n.d. *Fundamentals of RF and Microwave Power Measurements (Part 1), Introduction to Power, History, Definitions, International Standards, and Traceability* (formerly Agilent AN 1449-1). Literature number 5988-9213EN, p. 31. Colorado Springs, CO: Keysight.
23. The Smith Chart. 2010, September 19. Wikimedia Commons, by author Wdwd. https://commons.wikimedia.org/wiki/File:Smith_chart_gen.svg
24. Carr, Joseph J. 1996. *Elements of Electronic Instrumentation and Measurement.* 3rd ed. Upper Saddle River, NJ: Prentice-Hall, p. 495.
25. Laverghetta, Thomas S. 1981. *Handbook of Microwave Testing.* Dedham, MA: Artech House, pp. 34, 142.
26. Carr, Joseph J. 1996. *Elements of Electronic Instrumentation and Measurement.* 3rd ed. Upper Saddle River, NJ: Prentice-Hall, p. 535.
27. Keysight. 2019, October 17. *Application Note 1287-2: Exploring the Architectures of Network Analyzers.* Publication number 5965-7708E. Keysight Technologies, p. 6.
28. Agilent Technologies. 2002a. "Power Measurement Basics." In *Agilent 2002 Back to Basics Seminar.* Publication number 5988-6641EN. Palo Alto, CA: Agilent, pp. 1–54. (Keysight now owns Agilent.)
29. Keysight. 2020, July 23. *Application Note AN 154: S-Parameter Design.* Publication number 5952-1087. Colorado Springs, CO: Keysight. For a comprehensive review of S-parameters and their measurement, see Adam, Stephen F. 1969. *Microwave Theory and Applications.* Englewood Cliffs, NJ: Prentice-Hall, Chapter 6.1. Another useful reference is Keysight AN 1287-1/2/3.
30. Keysight. 2014/2019. *Application Note 1287-2: Exploring the Architectures of Network Analyzers.* Publication number 5965-7708E. Keysight Technologies.
31. Agilent Technologies. 2002. "Power Measurement Basics." In *Agilent 2002, Back to Basics Seminar.* Publication number 5988-6641EN. Palo Alto, CA: Agilent, pp. 1-48 to 1-77. (Keysight now owns Agilent.)
32. Keysight. 2014/2019. *Application Note 1287-2: Exploring the Architectures of Network Analyzers.* Publication number 5965-7708E. Keysight Technologies.
33. Adam, Stephen F. 1969. *Microwave Theory and Applications.* Englewood Cliffs, NJ: Prentice-Hall, pp. 86–106, 351–457; Keysight. n.d. *Application Note 1287-1: Understanding the Fundamental Principles of Vector Network Analysis.* Publication number 5965-7707E. Colorado Springs, CO: Keysight, pp. 1–14; Keysight. 2020, July 23. *Application Note AN 154: S-Parameter Design.* Publication number 5952-1087. Colorado Springs, CO: Keysight, 1–44.
34. Keysight. 2002, April 15. *Application Note 1382-7: VNA-Based System Tests the Physical Layer.* Publication number 5988-5075EN. Colorado Springs, CO: Keysight, pp. 1–16.

Chapter 34: Mass and Weighing

1. Thrasher, Leslie. 1936, October 3. "Tipping the Scales." *Saturday Evening Post.* Indianapolis, IN: Saturday Evening Post Society/Curtis Publishing Co. This illustration is often incorrectly attributed to Norman Rockwell.
2. Bureau International des Poids et Mesures. 2021, July 7. "*Mise en pratique* for the Definition of the Kilogram in the SI." In Bureau International des Poids et Mesures. *SI Brochure: The International System of Units (SI).* 9th ed. Vol. v1.08, Appendix 2. Sèvres, France: BIPM. https://www.bipm.org/documents/20126/41489673/SI-App2-kilogram.pdf/5881b6b5-668d-5d2b-f12a-0ef8ca437176

3. Ibid.
4. Ibid.
5. Ibid.
6. Ibid.
7. National Institute of Standards and Technology. 2019, May 20. *CODATA Value: Newtonian Constant of Gravitation*. Gaithersburg, MD: Physical Measurement Laboratory | NIST. https://physics.nist.gov/cgi-bin/cuu/Value?bg
8. OIML: International Organization of Legal Metrology. 2004. *International Recommendation; OIML R 111-1: Weight classes of E_1, E_2, F_1, F_2, M_1, M_{1-2}, M_2, M_{2-3}, and M_3. Part 1: Metrological and technical requirements. Annex B.* pp. 28–60. 2004. Paris, France: Bureau International de Métrologie Légale.
9. OIML: International Organization of Legal Metrology. 2004. *International Document; OIML D 28: Conventional value of the result of weighing in air.* Paris, France: Bureau International de Métrologie Légale. (Note: D 28 was previously published as OIML R 33).
10. Ibid.
11. *Mise en place* means to "put in place" and refers to keeping everything in its place and organized. It is most often used in a cooking environment, but is an applicable concept for laboratories.
12. OIML: International Organization of Legal Metrology. 2006. *International Recommendation; OIML R 76-1: Non-automatic weighing instruments.* Paris, France: Bureau International de Métrologie Légale.
13. ASTM International. 2020. *ASTM E898-20: Standard Practice for Calibration of Non-Automatic Weighing Instruments.* West Conshohocken, PA: ASTM International.
14. See weight classifications in the next section.
15. Harris, Georgia L. 1993, April. "Ensuring Accuracy and Traceability of Weighing Instruments." *ASTM Standardization News.* West Conshohocken, PA: ASTM International.
16. OIML: International Organization of Legal Metrology. 2004. *International Recommendation; OIML R 111-1: Weight classes of E_1, E_2, F_1, F_2, M_1, M_{1-2}, M_2, M_{2-3}, and M_3. Part 1: Metrological and technical requirements,* p. 4. Paris, France: Bureau International de Métrologie Légale.
17. OIML: International Organization of Legal Metrology. 2004. *International Recommendation; OIML R 111-1: Weight classes of E_1, E_2, F_1, F_2, M_1, M_{1-2}, M_2, M_{2-3}, and M_3. Part 1: Metrological and technical requirements. Annex B.* pp. 28–60. Paris, France: Bureau International de Métrologie Légale.
18. Fraley, Kenneth L, and Georgia L. Harris. 2019. *NISTIR 5672: Advanced Mass Calibrations and Measurements Assurance Program for the State Calibration Laboratories.* Gaithersburg, MD: NIST.
19. Cameron, Joseph M., and Geraldine E. Hailes. 1974. *NBS Technical Note 844: Designs for the Calibration of Small Groups of Standards in the Presence of Drift.* Gaithersburg, MD: National Bureau of Standards (later renamed as National Institute for Standards and Technology [NIST]).
20. Cameron, J. M., M. C. Croarkin, and R. C. Raybold. 1977. *NBS Technical Note 952: Designs for the Calibration of Standards of Mass.* Gaithersburg, MD: National Bureau of Standards (later renamed as National Institute for Standards and Technology [NIST]).
21. OIML: International Organization of Legal Metrology. 2004. *International Recommendation; OIML R 111-1: Weight classes of E_1, E_2, F_1, F_2, M_1, M_{1-2}, M_2, M_{2-3}, and M_3. Part 1: Metrological and technical requirements. Annex B.* pp. 28–60. Paris, France: Bureau International de Métrologie Légale.
22. Harris, Georgia L. 2019. *NISTIR 6969: Selected Laboratory and Measurement Practices, and Procedures to Support Basic Mass Calibrations.* Gaithersburg, MD: NIST. https://doi.org/10.6028/NIST.IR.6969-2019

Chapter 35: Force

1. Bureau International des Poids et Mesures. 2019. *SI Brochure: The International System of Units (SI).* 9th ed. Vol. v1.08, p. 160. Sèvres, France: BIPM. https://www.bipm.org/en/publications/si-brochure
2. Bureau International des Poids et Mesures. 2019. *SI Brochure: The International System of Units (SI).* 9th ed. Vol. v1.08, p. 159. Sèvres, France: BIPM. https://www.bipm.org/en/publications/si-brochure
3. ASTM International. 2018. *ASTM E74-18e1, Standard Practices for Calibration and Verification for Force-Measuring Instruments.* West Conshohocken, PA: ASTM International.
4. Bureau International des Poids et Mesures. 2019. *SI Brochure: The International System of Units (SI).* 9th ed. Vol. v1.08, p. 199. Sèvres, France: BIPM. https://www.bipm.org/en/publications/si-brochure
5. Zumbrun, Henry. 2021. *Force Calibration for Technicians and Quality Managers: Top Conditions, Methods, and Systems That Impact Force Calibration Results.* 1st ed. Kindle Direct Publishing, pp. 10–12.
6. Zumbrun, Henry. 2021. *Force Calibration for Technicians and Quality Managers: Top Conditions, Methods, and Systems That Impact Force Calibration Results.* 1st ed. Kindle Direct Publishing, pp. 11–12.
7. Zumbrun, Henry. 2021. *Force Calibration for Technicians and Quality Managers: Top Conditions, Methods, and Systems That Impact Force Calibration Results.* 1st ed. Kindle Direct Publishing, p. 16.
8. Zumbrun, Henry. 2021. *Force Calibration for Technicians and Quality Managers: Top Conditions, Methods, and Systems That Impact Force Calibration Results.* 1st ed. Kindle Direct Publishing, pp. 16–17.
9. Zumbrun, Henry. 2021. *Force Calibration for Technicians and Quality Managers: Top Conditions, Methods, and Systems That Impact Force Calibration Results.* 1st ed. Kindle Direct Publishing, p. 17.
10. Zumbrun, Henry. 2021. *Force Calibration for Technicians and Quality Managers: Top Conditions, Methods, and Systems That Impact Force Calibration Results.* 1st ed. Kindle Direct Publishing, p. 18.
11. Ibid.
12. Zumbrun, Henry. 2021. *Force Calibration for Technicians and Quality Managers: Top Conditions, Methods, and Systems That Impact Force Calibration Results.* 1st ed. Kindle Direct Publishing, p. 19.
13. Zumbrun, Henry. 2021. *Force Calibration for Technicians and Quality Managers: Top Conditions, Methods, and Systems That Impact Force Calibration Results.* 1st ed. Kindle Direct Publishing, pp. 19–20.
14. Zumbrun, Henry. 2021. *Force Calibration for Technicians and Quality Managers: Top Conditions, Methods, and Systems That Impact Force Calibration Results.* 1st ed. Kindle Direct Publishing, p. 20.
15. Ibid.
16. Ibid.
17. Zumbrun, Henry. 2021. *Force Calibration for Technicians and Quality Managers: Top Conditions, Methods, and Systems That Impact Force Calibration Results.* 1st ed. Kindle Direct Publishing, pp. 21–22.
18. International Organization for Standardization. 2011. *ISO 376:2011 Metallic materials—Calibration of force-proving instruments used for verification of uniaxial testing machines.* Geneva, Switzerland: ISO.
19. International Organization for Standardization. 2018. *ISO 7500-1:2018 Metallic materials—Calibration and verification of static uniaxial testing machines—PART 1:*

Tension/compression testing machines—Calibration and verification of the force-measuring system. Geneva, Switzerland. ISO.
20. International Organization for Standardization. 2011. *ISO 376:2011 Metallic materials—Calibration of force-proving instruments used for verification of uniaxial testing machines*. Geneva, Switzerland: ISO.
21. Ibid.
22. ASTM International. 2018. *ASTM E74-18e1: Standard Practices for Calibration and Verification for Force-Measuring Instruments*. West Conshohocken, PA: ASTM International.
23. International Organization for Standardization. 2011. *ISO 376:2011 Metallic materials—Calibration of force-proving instruments used for verification of uniaxial testing machines*. Geneva, Switzerland: ISO.
24. Zumbrun, Henry. 2021. *Force Calibration for Technicians and Quality Managers: Top Conditions, Methods, and Systems That Impact Force Calibration Results*. 1st ed. Kindle Direct Publishing, p. 73; International Organization for Standardization. 2011. *ISO 376:2011 Metallic materials—Calibration of force-proving instruments used for verification of uniaxial testing machines; Annex A.4.a*. Geneva, Switzerland: ISO.
25. Zumbrun, Henry. 2021. *Force Calibration for Technicians and Quality Managers: Top Conditions, Methods, and Systems That Impact Force Calibration Results*. 1st ed. Kindle Direct Publishing, p. 73; International Organization for Standardization. 2011. *ISO 376:2011 Metallic materials—Calibration of force-proving instruments used for verification of uniaxial testing machines; Clause C.2.11*. Geneva, Switzerland: ISO.
26. International Organization for Standardization. 2011. *ISO 376:2011 Metallic materials—Calibration of force-proving instruments used for verification of uniaxial testing machines*. Geneva, Switzerland: ISO, p. 12.
27. Zumbrun, Henry. 2021. *Force Calibration for Technicians and Quality Managers: Top Conditions, Methods, and Systems That Impact Force Calibration Results*. 1st ed. Kindle Direct Publishing, p. 95.
28. Zumbrun, Henry. 2021. *Force Calibration for Technicians and Quality Managers: Top Conditions, Methods, and Systems That Impact Force Calibration Results*. 1st ed. Kindle Direct Publishing, p. 116; International Organization for Standardization. 2011. *ISO 376:2011 Metallic materials—Calibration of force-proving instruments used for verification of uniaxial testing machines; Clause 6.1*. Geneva, Switzerland: ISO.
29. Zumbrun, Henry. 2021. *Force Calibration for Technicians and Quality Managers: Top Conditions, Methods, and Systems That Impact Force Calibration Results*. 1st ed. Kindle Direct Publishing, pp. 116–117.
30. International Organization for Standardization/International Electrotechnical Commission. 2017. *ISO/IEC 17025:2017—General requirements for the competence of testing and calibration laboratories*. Geneva, Switzerland: ISO, p. 13.
31. Joint Committee for Guides in Metrology. 2008. *JCGM 100:2008, Evaluation of measurement data—Guide to the expression of uncertainty in measurement (GUM)*. BIPM, IEC, IFCC, ILAC, ISO, IUPAC, IUPAP, and OIML. Sèvres, France: BIPM, p. 78.
32. International Laboratory Accreditation Cooperation. 2020. *ILAC P14:09/2020 ILAC Policy for Measurement Uncertainty in Calibration*. Silverwater, Australia: ILAC, p. 8.
33. Joint Committee for Guides in Metrology. 2008. *JCGM 100:2008, Evaluation of measurement data—Guide to the expression of uncertainty in measurement (GUM)*. BIPM, IEC, IFCC, ILAC, ISO, IUPAC, IUPAP, and OIML. Sèvres, France: BIPM, p. 11.

34. Joint Committee for Guides in Metrology. 2008. *JCGM 100:2008, Evaluation of measurement data—Guide to the expression of uncertainty in measurement (GUM)*. BIPM, IEC, IFCC, ILAC, ISO, IUPAC, IUPAP, and OIML. Sèvres, France: BIPM, p. 73.
35. ASTM International. 2018. *ASTM E74-18e1: Standard Practices for Calibration and Verification for Force-Measuring Instruments*. West Conshohocken, PA: ASTM International, p. 17.
36. Ibid.

Chapter 36: Dimensional and Mechanical Parameters

1. Bureau International des Poids et Mesures. 1983. *17th CGPM, 1983*, Resolution 1.

Chapter 37: Other Parameters: Electro-Optical and Radiation

1. Britannica Group, Inc. 1998, July 20. "metes and bounds." *Encyclopedia Britannica*. https://www.britannica.com/topic/metes-and-bounds
2. Ibid.
3. Virtual Museum of Surveying. n.d. "Changing Chains." http://www.surveyhistory.org/changing_chains.htm
4. van Diggelen, Frank, and Per Enge. 2015. "The World's First GPS MOOC and Worldwide Laboratory Using Smartphones." *Proceedings of the 28th International Technical Meeting of the Satellite Division of The Institute of Navigation* (ION GNSS+ 2015), Tampa, Florida, September 2015, pp. 361–369. https://www.ion.org/publications/abstract.cfm?articleID=13079
5. Britannica Group, Inc. 2014, 17 February. "Colorimetry." *Encyclopedia Britannica*. https://www.britannica.com/science/colorimetry
6. National Institute for Standards and Technology. 2021, March 8. *Color and Appearance*. Gaithersburg, MD: NIST. https://www.nist.gov/programs-projects/color-and-appearance
7. Ibid.
8. Lawrence Berkeley National Lab. 2018, July 2. "Laser Classification Explanation." Berkeley, CA: Berkeley Lab EH&S—Environment, Health & Safety. https://ehs.lbl.gov/resource/documents/radiation-protection/laser-safety/laser-classification-explanation
9. Ibid.
10. Britannica Group, Inc. 2020, 3 June. "pH." *Encyclopedia Britannica*. https://www.britannica.com/science/pH
11. Ibid.
12. U.S. Environmental Protection Agency. 2017. *SOP: EQ-01-08 - SOP for Calibration and Maintenance of PH Meters, Rev 05-03-2017*. Ft. Meade, MD: U.S. EPA, Office of Pesticide Programs. https://www.epa.gov/sites/default/files/2018-01/documents/eq-01-08.pdf

Chapter 38: Chemical and Biological Measurements and Uncertainties

1. International Organization for Standardization/International Electrotechnical Commission. 2017. *ISO/IEC 17025:2017—General requirements for the competence of testing and calibration laboratories*. Geneva, Switzerland: ISO.
2. United States Pharmacopeia and National Formulary. 2022. *<1220> Analytical Procedures Life Cycle*. USP-NF 2022 Issue 1. Rockville, MD: United

States Pharmacopeial Convention; Official May 1, 2022. Current DocID: GUID-35D7E47E-65E5-49B7-B4CC-4D96FA230821_2_en-US
3. International Union of Pure and Applied Chemistry. 2021. *What We Do—IUPAC | International Union of Pure and Applied Chemistry*. Research Triangle Park, NC: IUPAC Secretariat. https://iupac.org/what-we-do
4. Eurachem. 2021. *Welcome to Eurachem*. Teddington, Middlesex, UK: Eurachem Secretary. https://www.eurachem.org
5. Ellison, S. L. R., and A. Williams, eds. 2012. *Eurachem/CITAC Guide: Quantifying Uncertainty in Analytical Measurement (QUAM)*. 3rd ed. Teddington, Middlesex, UK: Eurachem Secretary. https://www.eurachem.org
6. Ellison, S. L. R., and A. Williams, eds. 2019. *Eurachem/CITAC Guide: Metrological Traceability in Analytical measurement*. 2nd ed. Teddington, Middlesex, UK: Eurachem Secretary. https://www.eurachem.org
7. Environment Directorate Organisation for Economic Co-operation and Development. 1998. *OECD Series on Principles of Good Laboratory Practice and Compliance Monitoring, Number 1. ENV/MC/CHEM(98)17*. Paris, France: OECD. https://one.oecd.org/document/ENV/MC/CHEM(98)17/en/pdf
8. Eurachem. 2021. *Welcome to Eurachem*. Teddington, Middlesex, UK: Eurachem. https://www.eurachem.org
9. Ellison, Stephen L. R. 2014, July. "Implementing Measurement Uncertainty for Analytical Chemistry: The *Eurachem Guide* for Measurement Uncertainty." *Metrologia* 51, no. 4, pp. S199–205. https://doi.org/10.1088/0026-1394/51/4/s199
10. United States Pharmacopeia and National Formulary. 2022. *<1220> Analytical Procedure Life Cycle*. USP-NF 2022 Issue 1. Rockville, MD: United States Pharmacopeial Convention; Official May 1, 2022. Current DocID: GUID-35D7E47E-65E5-49B7-B4CC-4D96FA230821_2_en-US
11. Bureau International des Poids et Mesures. 2019. *SI Brochure: The International System of Units (SI)*. 9th ed. Vol. v1.08, p. 134. Sèvres, France: BIPM. https://www.bipm.org/en/publications/si-brochure
12. Ellison, S. L. R., and A. Williams, eds. 2019. *Eurachem/CITAC Guide: Metrological Traceability in Analytical Measurement*. 2nd Ed. Teddington, Middlesex, UK: Eurachem Secretary. https://www.eurachem.org
13. Ellison, S. L. R., and A. Williams, eds. 2019. *Eurachem/CITAC Guide: Metrological Traceability in Analytical Measurement*. 2nd Ed. Teddington, Middlesex, UK: Eurachem Secretary.
14. International Union of Pure and Applied Chemistry. 2018. *IUPAC Periodic Table of the Elements* (dated 1 Dec 2018). Research Triangle Park, NC: IUPAC Secretariat. https://iupac.org/what-we-do/periodic-table-of-elements
15. Ellison, S. L. R., and A. Williams, eds. 2019. *Eurachem/CITAC Guide: Metrological Traceability in Analytical Measurement*. 2nd Ed. Teddington, Middlesex, UK: Eurachem Secretary, p. 15.
16. Joint Committee for Guides in Metrology. 2012. *JCGM 200:2012, International vocabulary of metrology—Basic and general concepts and associated terms (VIM 3)*; BIPM, IEC, IFCC, ILAC, ISO, IUPAC, IUPAP, and OIML. Sèvres, France: BIPM, p. 33. https://www.bipm.org/documents/20126/2071204/JCGM_200_2012.pdf/f0e1ad45-d337-bbeb-53a6-15fe649d0ff1
17. Bettencourt da Silva, R., and A. Williams, eds. 2015. *Eurachem/CITAC Guide: Setting and Using Target Uncertainty in Chemical Measurement*. 1st ed. Teddington, Middlesex, UK: Eurachem Secretary.
18. Magnusson, B., and U. Örnemark, eds. 2014. *Eurachem Guide: The Fitness for Purpose of Analytical Methods—A Laboratory Guide to Method Validation and Related Topics*. 2nd ed. Teddington, Middlesex, UK: Eurachem Secretary.

19. International Organization for Standardization. 2016. *ISO 17034:2016 General requirements for competence of reference material producers*. Geneva, Switzerland: ISO.
20. Ibid.
21. Ellison, S. L. R., and A. Williams, eds. 2019. *Eurachem/CITAC Guide: Metrological Traceability in Analytical Measurement*. 2nd Ed. Teddington, Middlesex, UK: Eurachem Secretary. https://www.eurachem.org
22. Ibid.
23. Ellison, S. L. R., and A. Williams, eds. 2012. *Eurachem/CITAC Guide: Quantifying Uncertainty in Analytical Measurement (QUAM)*. 3rd ed. Teddington, Middlesex, UK: Eurachem Secretary. https://www.eurachem.org
24. International Organization for Standardization/International Electrotechnical Commission. 2017. *ISO/IEC 17025:2017—General requirements for the competence of testing and calibration laboratories*. Geneva, Switzerland: ISO, p. 13.
25. Ibid.
26. International Laboratory Accreditation Cooperation. 2019. *ILAC G17:01/2021 Guidelines for Measurement Uncertainty in Testing*. Silverwater, Australia: ILAC, p. 6.
27. Ibid.
28. Ibid.
29. Ibid.
30. International Laboratory Accreditation Cooperation. 2019. *ILAC G17:01/2021 Guidelines for Measurement Uncertainty in Testing*. Silverwater, Australia: ILAC, p. 10.
31. AOAC International. 2012. Appendix H: Probability of Detection (POD) as a Statistical Model for the Validation of Qualitative Methods, Official Method 2008.01. *Official Methods of Analysis of AOAC INTERNATIONAL*. 19th ed. Gaithersburg, MD: AOAC International.
32. Ellison, S. L. R., and A. Williams, eds. 2012. *Eurachem/CITAC Guide: Quantifying Uncertainty in Analytical Measurement (QUAM)*. 3rd ed. Teddington, Middlesex, UK: Eurachem Secretary. https://www.eurachem.org
33. Joint Committee for Guides in Metrology. 2008. *JCGM 100:2008, Evaluation of measurement data—Guide to the expression of uncertainty in measurement (GUM)*. BIPM, IEC, IFCC, ILAC, ISO, IUPAC, IUPAP, and OIML. Sèvres, France: BIPM, p. 27.
34. National Institute for Standards and Technology. 2019. *SOP 29 SOP for the Assignment of Uncertainty*. Gaithersburg, MD: NIST, pp. 1–2. https://doi.org/10.6028/NIST.IR.6969-2019
35. Ellison, S. L. R., and A. Williams, eds. 2012. *Eurachem/CITAC Guide: Quantifying Uncertainty in Analytical Measurement (QUAM)*. 3rd ed. Teddington, Middlesex, UK: Eurachem Secretary. https://www.eurachem.org
36. Ibid.
37. International Organization for Standardization/International Electrotechnical Commission. 2017. *ISO/IEC 17025:2017—General requirements for the competence of testing and calibration laboratories*. Geneva, Switzerland: ISO, p. 2.

Chapter 39: Getting Started

1. American Society for Quality. 2021. Code of ethics. https://asq.org/about-asq/conferences-events-policies/code-of-ethics
2. International Organization for Standardization/International Electrotechnical Commission. 2017. *ISO/IEC 17025:2017—General requirements for the competence of testing and calibration laboratories*. Geneva, Switzerland: ISO, p. 3.

Chapter 40: Best Practices

1. Arden, Paul. 2003. *It's Not How Good You Are, It's How Good You Want to Be: The World's Best-Selling Book.* New York, NY: Phaidon Press.
2. American National Standards Institute/American Society for Quality/International Organization for Standardization. 2003. *ANSI/ISO/ASQ Q10012:2003—Measurement Management Systems—Requirements for Measurement Processes and Measuring Equipment.* Milwaukee, WI: ASQ Quality Press, p. 3.
3. International Organization for Standardization/International Electrotechnical Commission. 2017. *ISO/IEC 17025:2017—General requirements for the competence of testing and calibration laboratories.* Geneva, Switzerland: ISO, p. 8.
4. International Organization for Standardization/International Electrotechnical Commission. 2017. *ISO/IEC 17025:2017—General requirements for the competence of testing and calibration laboratories.* Geneva, Switzerland: ISO, p. 7.
5. American National Standards Institute/American Society for Quality/International Organization for Standardization. 2015. *ASQ/ANSI/ISO 2015/ISO 9001:2015 Quality Management Systems—Requirements.* Milwaukee, WI: ASQ Quality Press, p. 17.
6. International Organization for Standardization/International Electrotechnical Commission. 2017. *ISO/IEC 17025:2017—General requirements for the competence of testing and calibration laboratories.* Geneva, Switzerland: ISO, p. 22.

Chapter 42: Budgeting and Resource Management

1. International Organization for Standardization/International Electrotechnical Commission. 2017. *ISO/IEC 17025:2017—General requirements for the competence of testing and calibration laboratories.* Geneva, Switzerland: ISO, pp. 5–9.
2. American National Standards Institute/American Society for Quality/International Organization for Standardization. 2015. *ASQ/ANSI/ISO 2015/ISO 9001:2015 Quality Management Systems—Requirements.* Milwaukee, WI: ASQ Quality Press, pp. 6–7.
3. American National Standards Institute/American Society for Quality/International Organization for Standardization. 2003. *ANSI/ISO/ASQ Q10012:2003—Measurement Management Systems—Requirements for Measurement Processes and Measuring Equipment.* Milwaukee, WI: ASQ Quality Press, pp. 4–6.
4. Oxford Languages English Dictionary. https://languages.oup.com/google-dictionary-en

Chapter 43: Vendors and Suppliers

1. International Organization for Standardization/International Electrotechnical Commission. 2017. *ISO/IEC 17025:2017—General requirements for the competence of testing and calibration laboratories.* Geneva, Switzerland: ISO, p. 8.
2. Ibid.
3. International Organization for Standardization/International Electrotechnical Commission. 2017. *ISO/IEC 17025:2017—General requirements for the competence of testing and calibration laboratories.* Geneva, Switzerland: ISO, p. 9.
4. Ibid.
5. International Organization for Standardization/International Electrotechnical Commission. 2017. *ISO/IEC 17025:2017—General requirements for the competence of testing and calibration laboratories.* Geneva, Switzerland: ISO, p. 8.
6. Wade, Heather A. 2011. "Effective Communication Between Customers and Their Labs." *Cal Lab Magazine: The International Journal of Metrology* 18, no. 4 (October):

36–40. https://www.callabmag.com/wp-content/uploads/2011/11/issue_oct2011.pdf#page=38
7. American National Standards Institute/American Society for Quality/International Organization for Standardization. 2003. *ANSI/ISO/ASQ Q10012:2003—Measurement Management Systems—Requirements for Measurement Processes and Measuring Equipment*. Milwaukee, WI: ASQ Quality Press, p. 6.
8. NCSL International. 2015. *Recommended Practice-6: Recommended Practice for Calibration Quality Systems for the Healthcare Industries, RP-6*. 4th ed. Boulder, CO: NCSL International, p. 19.
9. NCSL International. 2015. *Recommended Practice-6: Recommended Practice for Calibration Quality Systems for the Healthcare Industries, RP-6*. 4th ed. Boulder, CO: NCSL International, p. 20.
10. Ibid.

Chapter 44: Housekeeping and Safety

1. Hirano, Hiroyuki. 1995. *5 Pillars of the Visual Workplace: The Sourcebook for 5S Implementation*. Bruce Talbot, translator. Shelton, CT: Productivity Press.

Bibliography/References

ASQE does not endorse any one particular reference source.

PUBLICLY AVAILABLE

A2LA WorkPlace Training. *AWPT | ISO and Metrology Training and Consulting*. Frederick, MD: A2LA WPT. https://www.a2lawpt.org.

Adam, Stephen F. 1969. *Microwave Theory and Applications*. Englewood Cliffs, NJ: Prentice-Hall.

Agilent Technologies. 2002a. *Power Measurement Basics*. Agilent 2002 Back to Basics Seminar. Publication number 5988-6641EN. Palo Alto, CA: Agilent. [Note: Keysight now owns Agilent.]

American Association for Laboratory Accreditation. 2020. *A2LA G129—Measurement Uncertainty Budget Template, version 1.1*. Frederick, MD: A2LA. https://mktg.a2la.org/l/273522/2020-02-26/3yzty65

American National Standards Institute/American Society for Quality/International Organization for Standardization. 2003. *ANSI/ISO/ASQ Q10012:2003—Measurement Management Systems—Requirements for Measurement Processes and Measuring Equipment*. Milwaukee, WI: ASQ Quality Press.

American National Standards Institute/American Society for Quality/International Organization for Standardization. 2015. *ASQ/ANSI/ISO 2015/ISO 9000:2015 Quality Management Systems—Fundamentals and Vocabulary*. Milwaukee, WI: ASQ Quality Press.

American National Standards Institute/NCSL International. 1994. *ANSI/NCSL Z540.1-1994 (R2002), American National Standards for Calibration—Calibration laboratories and measuring and test equipment—General requirements*. NCSL International. Boulder, CO: NCSL.

American Society for Quality. n.d. *Calibration Technician Certification—Become CCT Certified | ASQ*. Milwaukee, WI: ASQ. https://asq.org/cert/calibration-technician

American Society for Quality. 2021. About ASQ. https://asq.org/about-asq/conferences-events-policies/code-of-ethics

American Society for Quality. 2021. ASQ: Who We Are—History | ASQ. Milwaukee, WI: ASQ. https://asq.org/about-asq/how-we-do-it/history

American Society for Quality/American National Standards Institute/International Organization for Standardization. 2018. *ASQ/ANSI/ISO 19011:2018—Guidelines for auditing management systems*. Milwaukee, WI: ASQ Quality Press.

Barton, John W., et al. (eds.). 2021. *Specifications, Tolerances, and Other Technical Requirements for Weighing and Measuring Devices. 2022 Edition, NIST Handbook 44.* Gaithersburg, MD: U.S. Dept of Commerce/NIST. https://doi.org/10.6028/NIST.HB.44-2022

Benham, Elizabeth. 2021, January. "Writing with the SI. Weights and Measures." *Connection* 11, no. 1 (NIST). Article Ref: J-031. https://www.nist.gov/system/files/documents/2021/05/25/J-032%20Writing%20with%20the%20SI.pdf

Bettencourt da Silva, R., and A. Williams. (eds.). 2015. *Eurachem/CITAC Guide: Setting and Using Target Uncertainty in Chemical Measurement, First Ed.* Teddington, Middlesex, UK: Eurachem Secretary. Available from https://www.eurachem.org

Bin Zikria, Yousaf, Rashid Ali, Muhammad Khalil Afzal, and Sung Won Kim. 2021, February 7. "Next-Generation Internet of Things (IoT): Opportunities, Challenges, and Solutions." *Sensors (Basel)* 21, no. 4, 1174. https://www.ncbi.nlm.nih.gov/pmc/articles/PMC7915959
Licensee MDPI, Basel, Switzerland. This article is an open-access article distributed under the terms and conditions of the Creative Commons Attribution (CC BY) license (http://creativecommons.org/licenses/by/4.0).

Bird, Malcolm. 2002, August. "A Few Small Miracles Give Birth to an ISO Quality Management Systems Standard for the Automotive Industry." *ISO Bulletin* (Geneva, Switzerland).

The Boeing Company. "Boeing: Select Products in Boeing History." https://www.boeing.com/history/products/index.page

Bonikowsky, Laura Neilson. 2013, October 13. "Invention of Standard Time." *The Canadian Encyclopedia.* https://www.thecanadianencyclopedia.ca/en/article/invention-of-standard-time-feature

Britannica Group, Inc. 1998, July 20. "Metes and bounds." *Encyclopedia Britannica.* https://www.britannica.com/topic/metes-and-bounds

Britannica Group, Inc. 2014, February 17. "Colorimetry." *Encyclopedia Britannica.* https://www.britannica.com/science/colorimetry

Britannica Group, Inc. 2020, June 3. "pH." *Encyclopedia Britannica.* https://www.britannica.com/science/pH

British Standards Institution. 1974. *BS-5179: Guide to the Operation and Evaluation of Quality Assurance Systems.* (Withdrawn without replacement on Oct. 1, 1997). London, UK: BSI. https://www.document-center.com/standards/show/BS-5179/history/

Bureau International des Poids et Mesures. n.d. BIPM Technical services: Time Metrology. https://www.bipm.org/en/time-metrology

Bureau International des Poids et Mesures. n.d. The Convention of the Metre. https://www.bipm.org/en/metre-convention

Bureau International des Poids et Mesures. n.d. *Metre Convention—BIPM.* Sèvres, France: BIPM. https://www.bipm.org/en/metre-convention

Bureau International des Poids et Mesures. 2018. *Joint BIPM, OIML, ILAC, and ISO Declaration on Metrological Traceability.* BIPM, OIML, ILAC, ISO. https://www.bipm.org/documents/20126/42177518/BIPM-OIML-ILAC-ISO_joint_declaration_2018.pdf/7f1a4834-da36-b012-2a81-fc51a79b0726

Bureau International des Poids et Mesures. 2019. *SI Brochure: The International System of Units (SI).* 9th ed. Vol. v1.08. Sèvres, France: BIPM. https://www.bipm.org/en/measurement-units

Bureau International des Poids et Mesures. 2019. *SI Brochure: The International System of Units (SI)*. 9th ed. Vol. v1.08, *Appendix 1. Decisions of the 10th CIPM—Decision on the first six base units*. Sèvres, France: BIPM. https://www.bipm.org/documents/20126/41483022/si-brochure-9-App1-EN.pdf/3f415ca7-130b-7ebb-757a-bf8c0e87d9ce

Bureau International des Poids et Mesures. 2019. *SI Brochure: The International System of Units (SI)*. 9th ed. Vol. v1.08, *Appendix 1. Decisions of the 10th CGPM—Decision to adopt the name "International System of Units."* Sèvres, France: BIPM. https://www.bipm.org/documents/20126/41483022/si-brochure-9-App1-EN.pdf/3f415ca7-130b-7ebb-757a-bf8c0e87d9ce

Bureau International des Poids et Mesures. 2021. *The International System of Units (SI)*. 2019-05-20. Sèvres, France: BIPM. https://www.bipm.org/en/measurement-units

Bureau International des Poids et Mesures. 2021, March 30. *CIPM MRA-G-13 Calibration and Measurement Capabilities in the Context of the CIPM MRA—Guidelines for Their Review, Acceptance and Maintenance. Version 1.1*. Sèvres, France: BIPM.

Bureau International des Poids et Mesures. 2021, July 7. "*Mise en pratique* for the Definition of the Kilogram in the SI." In Bureau International des Poids et Mesures. *SI Brochure: The International System of Units (SI)*. 9th ed. Vol. v1.08, Appendix 2. Sèvres, France: BIPM. https://www.bipm.org/documents/20126/41489673/SI-App2-kilogram.pdf/5881b6b5-668d-5d2b-f12a-0ef8ca437176

Bureau International des Poids et Mesures, International Committee of Weight and Measures (CIPM). n.d. *CIPM MRA Key Comparison Database (KCDB)*. Sèvres, France: BIPM. https://www.bipm.org/en/cipm-mra/kcdb

Bureau International des Poids et Mesures, International Committee of Weight and Measures (CIPM). 2021. *CIPM MRA-P-11 Overview and Implementation of the CIPM MRA, Version 1.1, 30/03/2021*. Sèvres, France: BIPM. https://www.bipm.org/en/cipm-mra/cipm-mra-documents

Butcher, Kenneth S., et al. (eds.). *NIST SP 1038—The International System of Units (SI)—Conversion Factors for General Use, 2006 ed*. Gaithersburg, MD: NIST—Weights and Measures Division Technology Services, 2006. https://nvlpubs.nist.gov/nistpubs/Legacy/SP/nistspecialpublication1038.pdf

Cameron, J. M., M. C. Croarkin, and R. C. Raybold. 1977. *NBS Technical Note 952: Designs for the Calibration of Standards of Mass*. Gaithersburg, MD: National Bureau of Standards (later renamed as National Institute for Standards and Technology [NIST]).

Cameron, Joseph M., and Geraldine E. Hailes. 1974. *NBS Technical Note 844: Designs for the Calibration of Small Groups of Standards in the Presence of Drift*. Gaithersburg, MD: National Bureau of Standards (later renamed as National Institute for Standards and Technology [NIST]).

Cantellops, D., E. Bonnin, and A. Reid. 2010, October 19. *Laboratory Information Bulletin Number 4317: Spreadsheet Design and Validation for the Multi-User Application for the Chemistry Laboratory*. Rockville, MD: FDA, Office of Regulatory Affairs. http://www.spreadsheetvalidation.com/pdf/LIB_Design_Multi-User2A.pdf

Castrup, H., et al. 1994. *NASA Reference Publication 1342, Metrology—Calibration and Measurement Processes Guidelines*. Pasadena, CA: Jet Propulsion Laboratory California Institute of Technology.

Clarke, Kenneth K., and Donald T. Hess. 1990. "Phase Measurement, Traceability, and Verification Theory and Practice." *IEEE Transactions on Instrumentation and Measurement*, IM-39, Issue 1 (February).

Committee on Data for Science and Technology. 2018. *Fundamental Physical Constants—CODATA, The Committee on Data for Science and Technology*. Paris, France: CODATA. https://codata.org/initiatives/data-science-and-stewardship/fundamental-physical-constants

Diggelen, Frank, and Per Enge. 2015. "The World's First GPS MOOC and Worldwide Laboratory Using Smartphones." *Proceedings of the 28th International Technical Meeting of the Satellite Division of The Institute of Navigation* (ION GNSS+ 2015), Tampa, Florida, September 2015, pp. 361–369. https://www.ion.org/publications/abstract.cfm?articleID=13079

Doiron, Ted. 2007. "20 °C—A Short History of the Standard Reference Temperature for Industrial Dimensional Measurements." *Journal of Research of the National Institute of Standards and Technology 112*, no. 1: 1–23. https://nvlpubs.nist.gov/nistpubs/jres/112/1/V112.N01.A01.pdf

Doyle, Alison. 2013, August 8. "Hard Skills vs. Soft Skills: What's the Difference?" *The Balance Careers*. https://www.thebalancecareers.com/hard-skills-vs-soft-skills-2063780

Ehrlich, C. D., and S. D. Raspberry. 1998. "Metrological Timelines in Traceability." *Journal of Research (NIST JRES) 103*, no. 5: 93–105. https://www.ncbi.nlm.nih.gov/pmc/articles/PMC4891962

Ellison, S. L. R., and A. Williams. (eds.). 2012. *Eurachem/CITAC Guide: Quantifying Uncertainty in Analytical Measurement (QUAM)*. 3rd ed. Teddington, Middlesex, UK: Eurachem Secretary. https://www.eurachem.org

Ellison, S. L. R., and A. Williams. (eds.). 2019. *Eurachem/CITAC Guide: Metrological Traceability in Analytical measurement*. 2nd ed. Teddington, Middlesex, UK: Eurachem Secretary. https://www.eurachem.org

Ellison, Stephen L. R. 2014, July. "Implementing Measurement Uncertainty for Analytical Chemistry: The *Eurachem Guide* for Measurement Uncertainty." *Metrologia 51*, no. 4, pp. S199–205. https://doi.org/10.1088/0026-1394/51/4/s199

Environment Directorate Organisation for Economic Co-operation and Development. 1998. *OECD Series on Principles of Good Laboratory Practice and Compliance Monitoring, Number 1*. ENV/MC/CHEM(98)17. Paris, France: OECD. https://one.oecd.org/document/ENV/MC/CHEM(98)17/en/pdf

Eurachem. 2021. *Welcome to Eurachem*. Teddington, Middlesex, UK: Eurachem. https://www.eurachem.org

European Medicines Agency. 2020. *What We Do | European Medicines Agency*. Amsterdam, The Netherlands: EMA. https://www.ema.europa.eu/en/about-us/what-we-do#what-we-don't-do-section

Evans, James R., and William M. Lindsay. 1993. *The Management and Control of Quality*. 2nd ed. St. Paul, MN: West.

Federal Standard 1037C, *Telecommunications: Glossary of Telecommunication Terms*. 1996. Washington, D.C.: General Services Administration. https://www.its.bldrdoc.gov/fs-1037/fs-1037c.htm

Fraley, Kenneth L, and Georgia L. Harris. 2019. *NISTIR 5672: Advanced Mass Calibrations and Measurements Assurance Program for the State Calibration Laboratories*. Gaithersburg, MD: NIST.

Frysinger, James R. 2000. "SI Crosses All Language Barriers." https://metricmethods.com/Multilingual_SI_unit_names.gif

Gentry, Elizabeth J. (Benham), and Georgia L. Harris. 2016. "Accuracy Matters: Understanding and Communicating Measurement Uncertainties." *Quality Progress 49*, no. 5 (May).

Gentry, Elizabeth J. (Benham), and Georgia L. Harris. 2016. "Write It Right: Understanding Nuances of Metrics in Technical Writing." July 2016. *Quality Progress 49*, no. 7 (July).

Grachanen, Christopher. 2007. "The Metrology Job Description Initiative: NCSLI and ASQ Partnering for the Future." *NCSL International Measure* 2, no. 2 (June), 26–31. https://doi.org/10.1080/19315775.2007.11721368

Gust, Jeff C., Robert M. Graham, and Michael A. Lombardi. 2009, January. *NIST Special Publication 960-12, Stopwatch and Timer Calibrations.* Gaithersburg, MD: U.S. Government Printing Office. https://tf.nist.gov/general/pdf/1930.pdf

Harral, William M. 2003. "What Is ISO/TS 16949:2002?" *Automotive Excellence Newsletter* (Summer/Fall 2003), pp. 5–6.

Harris, Georgia L. 1993, April. "Ensuring Accuracy and Traceability of Weighing Instruments." *ASTM Standardization News.* West Conshohocken, PA: ASTM International.

Harris, Georgia L. 2019. *NISTIR 6969, Selected Laboratory and Measurement Practices, and Procedures to Support Basic Mass Calibrations.* Gaithersburg, MD: NIST. https://doi.org/10.6028/NIST.IR.6969-2019

Health Canada. n.d. *Health Canada—Canada.Ca.* Canada.Ca./Gouvernement du Canada. https://www.canada.ca/en/health-canada.html

Hermans, F., & E. Murphy-Hill. 2015. *Enron's Spreadsheets and Related Emails: A Dataset and Analysis.* Paper presented at the IEEE/ACM 37th IEEE International Conference on Software Engineering, Florence, Italy.

Hills, Graham. 2003. "The Effect of ISO/TS 16949:2002." *InsideStandards* (November).

International Accreditation Forum. *Home—IAF.* Quebec, Ontario, Canada: IAF. https://iaf.nu/en/home

International Aerospace Quality Group. *About IAQG.* Bruxelles, Belgium. https://iaqg.org/about-us

International Laboratory Accreditation Cooperation. n.d. *About ILAC International Laboratory Accreditation Cooperation.* Silverwater, Australia. https://ilac.org/about-ilac

International Laboratory Accreditation Cooperation. n.d. *ILAC MRA and Signatories International Laboratory Accreditation Cooperation.* https://ilac.org/ilac-mra-and-signatories

International Laboratory Accreditation Cooperation. 2019. *ILAC G8:09/2019 Guidelines on Decision Rules and Statements of Conformity.* Silverwater, Australia: ILAC.

International Laboratory Accreditation Cooperation. 2019. *ILAC G17:01/2021 Guidelines for Measurement Uncertainty in Testing.* Silverwater, Australia: ILAC.

International Laboratory Accreditation Cooperation. 2020. *ILAC P10:07/2020 ILAC Policy on Metrological Traceability of Measurement Results.* Silverwater, Australia: ILAC.

International Laboratory Accreditation Cooperation. 2020. *ILAC P14:09/2020 ILAC Policy for Measurement Uncertainty in Calibration.* Silverwater, Australia: ILAC.

International Laboratory Accreditation Cooperation. 2020, December. *Facts & Figures—International Laboratory Accreditation Cooperation (ILAC).* https://ilac.org/about-ilac/facts-and-figures

International Laboratory Accreditation Cooperation and OIML: International Organization of Legal Metrology. 2007. *ILAC G24:2007/OIML D10:2007 (E) Guidelines for the Determination of Calibration Intervals in Measuring Instruments.* [under revision at the time of this writing.] Silverwater, Australia: ILAC; Paris, France: OIML.

International Organization for Standardization. n.d. *ISO—About Us.* Geneva, Switzerland: ISO. https://www.iso.org/about-us.html

International Organization for Standardization. n.d. *ISO—Certification.* Geneva, Switzerland: ISO. https://www.iso.org/certification.html

International Organization for Standardization. n.d. *ISO—Organizations in Cooperation with ISO.* Geneva, Switzerland: ISO. https://www.iso.org/organizations-in-cooperation-with-iso.html

International Organization for Standardization. n.d. *ISO's Committee on Conformity Assessment (CASCO).* Geneva, Switzerland: ISO. https://www.iso.org/casco.html

International Organization for Standardization. 2015. *ISO Guide 30:2015 Reference materials—Selected terms and definitions.* Geneva, Switzerland: ISO.

International Organization for Standardization. 2016. *ISO 17034:2016 General requirements for competence of reference material producers.* Geneva, Switzerland: ISO.

International Organization for Standardization—Technical Committee 154. 2019, August 26. "Introduction to the New ISO 8601-1 and ISO 8601-2—ISO/TC 154." https://www.isotc154.org/posts/2019-08-27-introduction-to-the-new-8601

International Organization for Standardization/International Electrotechnical Commission. 2010. *ISO/IEC 17043:2010 Conformity assessment—General requirements for proficiency testing.* Geneva, Switzerland: ISO.

International Organization for Standardization/International Electrotechnical Commission. 2017. *ISO/IEC 17025:2017—General requirements for the competence of testing and calibration laboratories.* Geneva, Switzerland: ISO.

International Union of Pure and Applied Chemistry. 2018. *IUPAC Periodic Table of the Elements* (dated 1 Dec. 2018). Research Triangle Park, NC: IUPAC Secretariat. https://iupac.org/what-we-do/periodic-table-of-elements

International Union of Pure and Applied Chemistry. 2021. *What We Do—IUPAC | International Union of Pure and Applied Chemistry.* Research Triangle Park, NC: IUPAC Secretariat. https://iupac.org/what-we-do

Jackson, Peter (director). 2001. *The Lord of the Rings: The Fellowship of the Ring.* New Line Cinema and WingNut Films, distributed by New Line Cinema.

Joint Committee for Guides in Metrology. 2008. *JCGM 100:2008, Evaluation of measurement data—Guide to the expression of uncertainty in measurement (GUM);* BIPM, IEC, IFCC, ILAC, ISO, IUPAC, IUPAP, and OIML. Sèvres, France: BIPM.

Joint Committee for Guides in Metrology. 2012. *JCGM 106:2012, Evaluation of measurement data—The role of measurement uncertainty in conformity decision.* BIPM, IEC, IFCC, ILAC, ISO, IUPAC, IUPAP, and OIML. Sèvres, France: BIPM.

Joint Committee for Guides in Metrology. 2012. *JCGM 200:2012, International vocabulary of metrology—Basic and general concepts and associated terms (VIM 3);* BIPM, IEC, IFCC, ILAC, ISO, IUPAC, IUPAP, and OIML. Sèvres, France: BIPM. https://www.bipm.org/documents/20126/2071204/JCGM_200_2012.pdf/f0e1ad45-d337-bbeb-53a6-15fe649d0ff1

Juran, Joseph M. 1997. "Early SQC: A Historical Supplement." *Quality Progress* 30 (September): 73–81. https://www.itl.nist.gov/div898/handbook/pmc/section7/pmc7.htm

Juran, Joseph M. 1999. *Juran's Quality Handbook*. Wilton, CT: McGraw-Hill.

Keithley, A Tektronix Company. 2014. *Low Level Measurements Handbook—Precision DC Current, Voltage, and Resistance Measurements*. 7th ed. https://download.tek.com/document/LowLevelHandbook_7Ed.pdf

Keysight. n.d. *Application Note 1287-1: Understanding the Fundamental Principles of Vector Network Analysis*. Publication number 5965-7707E, p. 10. Colorado Springs, CO: Keysight.

Keysight. n.d. *Fundamentals of RF and Microwave Power Measurements (Part 1) Introduction to Power, History, Definitions, International Standards, and Traceability* (formerly Agilent AN 1449-1), literature number 5988-9213EN. Colorado Springs, CO: Keysight.

Keysight. n.d. *Fundamentals of RF and Microwave Power Measurements (Part 2), Power Sensors and Instrumentation* (formerly Agilent AN 1449-2). Literature number 5988-9214EN5. Colorado Springs, CO: Keysight.

Keysight. n.d. *Fundamentals of RF and Microwave Power Measurements (Part 3), Power Measurement Uncertainty per International Guides* (formerly Agilent AN 1449-3). Literature number 5988-9215EN. Colorado Springs, CO: Keysight.

Keysight. 2002, April 15. *Application Note 1382-7: VNA-Based System Tests the Physical Layer*. Publication number 5988-5075EN. Colorado Springs, CO: Keysight.

Keysight. 2019, October 17. *Application Note 1287-2: Exploring the Architectures of Network Analyzers*. Publication number 5965-7708E. Colorado Springs, CO: Keysight Technologies.

Keysight. 2020, July 23. *Application Note AN 154: S-Parameter Design*. Publication number 5952-1087. Colorado Springs, CO: Keysight.

Kymal, Chad, and Dave Watkins. 2003. "How Can You Move from QS-9000 to ISO/TS 16949:2002? More Key Challenges of QS-9000 and ISO/TS 16949:2002 Transition." *Automotive Excellence Newsletter* (Summer/Fall), pp. 9–13.

Larsen, Neil T. 1976. "A New Self-Balancing DC-Substitution RF Power Meter." *IEEE Transactions on Instrumentation and Measurement*, no. 4 (December): 343–347.

Lawrence Berkeley National Lab. 2018, July 2. "Laser Classification Explanation." Berkeley, CA: Berkeley Lab EH&S—Environment, Health & Safety. https://ehs.lbl.gov/resource/documents/radiation-protection/laser-safety/laser-classification-explanation

Liang Mok, Wei. 2003. "Power of a Human." In Glenn Elert, ed., *The Physics Factbook*. [Hypertextbook.] https://hypertextbook.com/facts/2003/WeiLiangMok.shtml

Lloyd, Robin. 1999, September 30. "Metric Mishap Caused Loss of NASA Orbiter." *CNN*. http://www.cnn.com/TECH/space/9909/30/mars.metric.02

Lombardi, Michael A. 2014. "Time Measurement." In *Measurement, Instrumentation, and Sensors Handbook*, edited by John G. Webster and Halit Eren, ch. 41. Boca Raton, FL: CRC Press/Taylor and Francis Group, LLC. https://tf.nist.gov/general/pdf/2488.pdf

Lombardi, Michael A. 2016. "Evaluating the Frequency and Time Uncertainty of GPS Disciplined Oscillators and Clocks." *NCSL International Measure 11*, no. 3–4 (October): 30–44. https://doi.org/10.1080/19315775.2017.1316696

Lowery, Andrew, Judy Strojny, and Joseph Puleo. 1996. *Medical Device Quality Systems Manual: A Small Entity Compliance Guide*. Rockville, MD: U.S. Department of Health and Human Services, Public Health Service, Food and Drug Administration.

Magnusson, B., and U. Örnemark. (eds.). 2014. *Eurachem Guide: The Fitness for Purpose of Analytical Methods—A Laboratory Guide to Method Validation and Related Topics, 2nd Ed.* Teddington, Middlesex, UK: Eurachem Secretary. ISBN 978-91-87461-59-0. https://www.eurachem.org

Marash, Stanley A. 2003. "What's Good for Defense . . . How Security Needs of the Cold War Inspired the Quality Revolution." *Quality Digest*, Issue 23 (June).

Mars Climate Orbiter—Mishap Investigation Board, Phase I Report. 1999, November. http://sunnyday.mit.edu/accidents/MCO_report.pdf

Matsakis, D., J. Levine, and M. Lombardi. 2018. "Metrological and Legal Traceability of Time Signals." 2018 Precise Time and Time Interval (PTTI) Meeting, Reston, VA [online]. https://tsapps.nist.gov/publication/get_pdf.cfm?pub_id=924939

National Bureau of Standards (NBS). 1975. *The International Bureau of Weights and Measures 1875–1975* (NBS Special Publication 420). Washington, D.C.: U.S. Government Printing Office.

National Geodetic Survey. 2018, January 25. Surface Gravity Prediction. Washington, D.C. https://geodesy.noaa.gov/cgi-bin/grav_pdx.prl

National Institute of Standards and Technology. n.d. *NIST—Malcolm Baldrige National Quality Award.* Gaithersburg, MD: NIST. https://www.nist.gov/baldrige/how-baldrige-works/about-baldrige/baldrige-faqs

National Institute of Standards and Technology. n.d. *System Development Life Cycle (SDLC).* https://csrc.nist.gov/glossary/term/system_development_life_cycle

National Institute of Standards and Technology. n.d. "Unit Conversion." Gaithersburg, MD: NIST. https://www.nist.gov/pml/weights-and-measures/metric-si/unit-conversion

National Institute of Standards and Technology. 1997. *Metric Style Guide for the News Media.* Gaithersburg, MD: U.S. Department of Commerce, Technology Administration, NIST. https://www.nist.gov/system/files/documents/2017/05/09/LC-1137-Metric-Style-Guide-News-Media-1997.pdf

National Institute of Standards and Technology. 2018. *Fundamental Physical Constants from NIST.* 2018. Gaithersburg, MD: Physical Measurement Laboratory | NIST. https://physics.nist.gov/cuu/Constants/index.html

National Institute of Standards and Technology. 2019. *Calibration Procedures for Weights and Measures Laboratories—GLP 15 Software Quality Assurance.* Edited by G. Harris. Gaithersburg, MD: NIST.

National Institute of Standards and Technology. 2019, May 20. *CODATA Value: Newtonian Constant of Gravitation.* Gaithersburg, MD: Physical Measurement Laboratory | NIST. https://physics.nist.gov/cgi-bin/cuu/Value?bg

National Institute of Standards and Technology. 2019. *SOP 29 SOP for the Assignment of Uncertainty.* Gaithersburg, MD: NIST. https://doi.org/10.6028/NIST.IR.6969-2019

National Institute of Standards and Technology. 2021. "Baldrige FAQs." https://www.nist.gov/baldrige/how-baldrige-works/about-baldrige/baldrige-faqs

National Institute of Standards and Technology. 2021. *DLMF: Chapter 15 Hypergeometric Function.* DLMF: NIST Digital Library of Mathematical Functions. Gaithersburg, MD: NIST. https://dlmf.nist.gov/15

National Institute of Standards and Technology. 2021. *NIST Policy on Metrological Traceability.* Gaithersburg, MD: NIST. https://www.nist.gov/calibrations/traceability

National Institute of Standards and Technology. 2021, March 8. *Color and Appearance*. Gaithersburg, MD: NIST. https://www.nist.gov/programs-projects/color-and-appearance

National Institute of Standards and Technology/SEMATECH. 2013, April. *NIST/SEMATECH e-Handbook of Statistical Methods*. https://doi.org/10.18434/M32189

National Telecommunications and Information Administration. 1996. *United States Frequency Allocation Chart*. July 2003. Washington, D.C.: NTIA, Office of Spectrum Management. www.ntia.doc.gov/osmhome/allochrt.pdf

Nemeroff, Ed. 1996. "The Story of the Egyptian Cubit." *Cal Lab: The International Journal of Metrology* 3, no. 5 (September-October): 16–17.

Newell, David B., and Eite Tiesinga. (eds.). 2019, August. *NIST Special Publication 330 | The International System of Units (SI) 2019 Edition*. Gaithersburg, MD: NIST, National Institute of Standards and Technology—Physical Measurement Laboratory. https://www.nist.gov/pml/special-publication-330

Niagara Falls Info. 2017, February 3. "Niagara Falls History of Power Development—The Ontario Power Company." https://www.niagarafallsinfo.com/niagara-falls-history/niagara-falls-power-development/the-history-of-power-development-in-niagara/the-ontario-power-company

Nowocin, W. 2021. "Implementing a Calibration Management Software System in a Regulated Environment." *Cal Lab Magazine: The International Journal of Metrology* 28, no. 2 (Apr-May-Jun): 27. https://www.callabmag.com/wp-content/uploads/2021/06/apr21_web.pdf

Nowocin, W. 2021. "Maintaining a Calibration Management Software System in a Regulated Environment." *Cal Lab Magazine: The International Journal of Metrology* 28, no. 3 (July): 35. https://www.callabmag.com/maintaining-a-calibration-management-software-system-in-a-regulated-environment

Nowocin, W. 2021. "Selecting a Calibration Management Software System in a Regulated Environment." *Cal Lab Magazine: The International Journal of Metrology* 28, no. 1 (Jan-Feb-Mar): 29. https://www.callabmag.com/wp-content/uploads/2021/03/jan21_web.pdf

OIML: International Organization of Legal Metrology. 1989, July. *OIML G13 (Ex P 7): Planning of Metrology and Testing Laboratories*. Paris, France: International Bureau of Legal Metrology. https://www.oiml.org/en/files/pdf_g/g013-e89.pdf

OIML: International Organization of Legal Metrology. 2004. *International Document; OIML D 28: Conventional value of the result of weighing in air*. Paris, France: Bureau International de Métrologie Légale.

OIML: International Organization of Legal Metrology. 2004. *International Recommendation; OIML R 111-1: Weight classes of E_1, E_2, F_1, F_2, M_1, $M_{1\text{-}2}$, M_2, $M_{2\text{-}3}$, and M_3. Part 1: Metrological and technical requirements*. Paris, France: Bureau International de Métrologie Légale.

OIML: International Organization of Legal Metrology. 2006. *International Recommendation; OIML R 76-1: Non-automatic weighing instruments*. Paris, France: Bureau International de Métrologie Légale.

Oxford Languages English Dictionary. https://languages.oup.com/google-dictionary-en

Panko, Raymond R. 2014. *Audits of Operational Spreadsheets*. Raymond Panko Spreadsheet and Human Error. http://panko.com/ssr/Audits.htmlvan

Shewhart, Walter A. 1939. *Statistical Method from the Viewpoint of Quality Control* (with the editorial assistance of W. E. Deming). Edited by W. Edwards Deming. Washington, D.C.: The Graduate School, Department of Agriculture, Washington.

The Smith Chart. 2010, September 19. Wikimedia Commons, by author Wdwd. https://commons.wikimedia.org/wiki/File:Smith_chart_gen.svg

"SOP 24-1 Calibration of Field Standard Stopwatches." In Georgia L. Harris (ed.), *NISTIR 8250: Calibration Procedures for Weights and Measures Laboratories*. Gaithersburg, MD: NIST, p. 1. https://doi.org/10.6028/NIST.IR.8250

Stein, Philip. 2000. "Don't Whine—Calibrate." *Quality Progress* 33 (November): 85 [Measure for Measure article]. http://download.caltech.se/download/validering/diverse/17025/17025comparisons/Don't_whine_be_smart_11=00.doc

Stenbakken, Gerard N. 1984. "A Wideband Sampling Wattmeter." *IEEE Transactions on Power Apparatus and Systems*, PAS-103 (October): 2919–2926.

Swade, Doron. 2000. *The Difference Engine: Charles Babbage and the Quest to Build the First Computer*. London: Viking Penguin Books.

Systema Interamercano de Metrologia (SIM). n.d. SIM Time and Frequency Metrology Working Group. https://tf.nist.gov/sim

Thompson, Amber, and Barry N. Taylor. 2008, March. *NIST Special Publication 811 | Guide for the Use of the International System of Units (SI)—2008 Edition*. Gaithersburg, MD: NIST, National Institute of Standards and Technology. https://www.nist.gov/pml/special-publication-811

Thrasher, Leslie. 1936, October 3. "Tipping the Scales." *Saturday Evening Post*. Indianapolis, IN: Saturday Evening Post Society/Curtis Publishing Co.

Turan, Meltem Sönmez, Elaine Barker, John Kelsey, Kerry McKay, Mary L. Baish, and Mike Boyle. 2018, January. *NIST SP 800-90B Recommendation for the Entropy Sources Used for Random Bit Generation*. Gaithersburg, MD: US Department of Commerce: NIST. https://doi.org/10.6028/NIST.SP.800-90B

Turgel, Raymond S., and N. Michael Oldham. 1978. "High-Precision Audio-Frequency Phase Calibration Standard." *IEEE Transactions on Instrumentation and Measurement* IM-27, Issue 4 (December): 460–464.

UKAS [United Kingdom Accreditation Service]. 2019. *LAB 36—Laboratory Accommodation and Environment in Measurement of Length, Angle, and Form*. 4th ed. Staines-upon-Thames, England: United Kingdom Accreditation Service. https://www.ukas.com/wp-content/uploads/filebase/publications/publications-relating-to-laboratory-accreditation/LAB-36-Laboratory-Environment-in-measurement-of-length-angle-and-form-Edition-4-October-2019.pdf

UKAS. 2020. *LAB 48—Decision Rules and Statements of Conformity*. June 2020. *UKAS. Com.* 3rd ed. Staines-upon-Thames, England: United Kingdom Accreditation Service. https://www.ukas.com/wp-content/uploads/filebase/publications/publications-relating-to-laboratory-accreditation/LAB-48-Decision-Rules-Edition-3-June-2020.pdf

U.S. Department of Labor, Office of Administrative Law Judges. 1991. *OALJ Law Library, DOT, Professional, Technical, and Managerial Occupations 001.061-010 to 024.364-010 | Metrologist (profess. & kin.) 012.067-010*. Washington, DC: DOL-OALJ. https://www.dol.gov/agencies/oalj/PUBLIC/DOT/REFERENCES/DOT01A

U.S. Department of Transportation, Bureau of Transportation Statistics. 2021, March 12. *History of Daylight Savings Time | Bureau of Transportation Statistics*. Washington, D.C.:

Bureau of Transportation Statistics, U.S. Department of Transportation. https://www.bts.gov/geospatial/daylight-savings-time (Includes material from: Phillips, Charles. "A Day to Remember: November 18, 1883." *American History*, Dec. 2004, pp. 16–18.)

U.S. Department of Transportation, Bureau of Transportation Statistics. 2021, March 12. *History of Daylight Savings Time | Bureau of Transportation Statistics.* Washington, D.C.: Bureau of Transportation Statistics. U.S. Department of Transportation. https://www.bts.gov/geospatial/daylight-savings-time (Includes material from Gordon, John Steele. "Standard Time: We All Live By What Happened on November 18, 1883." *American Heritage*, Jul./Aug. 2001, pp. 22–23.)

U.S. Environmental Protection Agency. 2017. *SOP: EQ-01-08 - SOP for Calibration and Maintenance of PH Meters, Rev 05-03-2017.* Ft. Meade, MD: US EPA, Office of Pesticide Programs. https://www.epa.gov/sites/default/files/2018-01/documents/eq-01-08.pdf

U.S. Food and Drug Administration. 2002, January. *Guidance Document—General Principles of Software Validation: Guidance for Industry and FDA Staff | FDA.* Rockville, MD: U.S. Food and Drug Administration. https://www.fda.gov/regulatory-information/search-fda-guidance-documents/general-principles-software-validation

U.S. Food and Drug Administration. 2003. *21 CFR, Part 11, Electronic Records; Electronic Signatures—Scope and Application | FDA.* Rockville, MD: U.S. Food and Drug Administration. https://www.fda.gov/regulatory-information/search-fda-guidance-documents/part-11-electronic-records-electronic-signatures-scope-and-application

U.S. Food and Drug Administration. 2011. *21 CFR pt. 820.25 Subpart B—Quality System Regulation, Section 820.25—Personnel.* Rockville, MD: U.S. Food and Drug Administration. https://www.govinfo.gov/content/pkg/CFR-2021-title21-vol8/pdf/CFR-2021-title21-vol8-sec820-25.pdf

U.S. Food and Drug Administration. 2012. *21 CFR Part 820.70(i), Production and process controls: Automated processes.* Rockville, MD: U.S. Food and Drug Administration. https://www.ecfr.gov/current/title-21/chapter-I/subchapter-H/part-820/subpart-G/section-820.70#p-820.70(i)

U.S. Food and Drug Administration. 2016, April. *Data Integrity and Compliance With Drug CGMP Questions and Answers Guidance for Industry.* Rockville, MD: U.S. Food and Drug Administration. https://www.fda.gov/regulatory-information/search-fda-guidance-documents/data-integrity-and-compliance-drug-cgmp-questions-and-answers-guidance-industry

U.S. Food and Drug Administration. 2018. *Milestones in U.S. Food and Drug Law | FDA.* January 31. https://www.fda.gov/about-fda/fda-history/milestones-us-food-and-drug-law

U.S. Food and Drug Administration. 2021. "Computer Software Assurance for Production and Quality System Software" (Draft Guidance Topic for FY2022). https://www.fda.gov/medical-devices/guidance-documents-medical-devices-and-radiation-emitting-products/cdrh-proposed-guidances-fiscal-year-2022-fy2022

U.S. Food and Drug Administration. 2021. *ORA FOIA Electronic Reading Room.* Rockville, MD: U.S. Food and Drug Administration. https://www.fda.gov/about-fda/office-regulatory-affairs/ora-foia-electronic-reading-room

U.S. Food and Drug Administration. 2021. *Code of Federal Regulations Title 21, Volume 4, Chapter I, Subchapter C, Part 211.68 Automatic, mechanical, and electronic equipment, (b)* (April 1, 2011).

U.S. Food and Drug Administration. 2021. *Warning Letters.* Rockville, MD: U.S. Food and Drug Administration. https://www.fda.gov/inspections-compliance-enforcement-and-criminal-investigations/compliance-actions-and-activities/warning-letters

U.S. Food and Drug Administration. 2021, October 1. *21 CFR ch. 1 Subchapter H, pt. 820, Subpart G, Section 820.72, (b) Calibration.* https://www.accessdata.fda.gov/scripts/cdrh/cfdocs/cfcfr/cfrsearch.cfm?fr=820.72

U.S. Food and Drug Administration, Office of Regulatory Affairs. 2019. *ORA Laboratory Manual Vol. III Section 4: Basic Statistics and Data Presentation.* https://www.fda.gov/media/73535/download

U.S. Government. 2006. *United States Code, 2006 Edition, Supplement 5, Title 15—COMMERCE AND TRADE—Declaration of Policy (15 U.S.C. 205b).* Washington, D.C.: U.S. Government Publishing Office. https://www.govinfo.gov/app/details/USCODE-2011-title15/USCODE-2011-title15-chap6-subchapII-sec205b

U.S. Government. 2017. *2018 Standard Occupational Classification System.* Washington, DC: U.S. Bureau of Labor Statistics, Office of Employment and Unemployment Statistics, Occupational Employment Statistics. https://www.bls.gov/soc/2018/major_groups.htm#17-0000

U.S. Government. 2017, November 28. Standard Occupational Classification (SOC) System—Revision for 2018. *Federal Register.* https://www.federalregister.gov/documents/2017/11/28/2017-25622/standard-occupational-classification-soc-system-revision-for-2018

U.S. Government. 2019. *Title 15—COMMERCE AND TRADE CHAPTER 7—NATIONAL INSTITUTE OF STANDARDS AND TECHNOLOGY Sec. 272—Establishment, Functions, and Activities.* Washington, D.C.: U.S. Government Publishing Office. https://www.govinfo.gov/app/details/USCODE-2019-title15/USCODE-2019-title15-chap7-sec272

U.S. Government Code of Federal Regulations. 2021, January 1. *Code of Federal Regulations Title 14, Aeronautics and Space, Chapter I: Federal Aviation Administration, Department of Transportation, Subchapter G Air Carriers and Operators for Compensation or Hire: Certification and Operation, Part 121 Operating Requirements: Domestic, Flag, and Supplemental Operations* (per amended date 2021-10-06). Washington, D.C.: National Archives.

U.S. Government Code of Federal Regulations. 2021, January 1. *Code of Federal Regulations Title 14, Aeronautics and Space, Chapter I: Federal Aviation Administration, Department of Transportation, Subchapter H Schools and Other Certified Agencies, Part 145 Repair Stations* (per amended date 2021-10-06). Washington, D.C.: National Archives.

U.S. Library of Congress. n.d. *Today in History—November 18: Time! | Library of Congress.* The Library of Congress. Washington, D.C.: Library of Congress. https://www.loc.gov/item/today-in-history/november-18

U.S. Metric Association. 2015, September 18. *Unit Mix-ups—U.S. Metric Association.* Windsor, CO: U.S. Metric Association. https://usma.org/unit-mixups#

U.S. Military MOS Database. [a website that provides a searchable database for all of the U.S. military services occupational specialty job codes and descriptions.] http://www.mosdb.com

U.S. Navy. 1966. *Why Calibrate?* Washington, D.C.: Department of the Navy. Training film number MN-10105. https://archive.org/details/gov.ntis.AVA14173VNB1

U.S. Navy. 2021. USS Constitution. Washington, D.C.: Department of the Navy. https://www.navy.mil/uss-constitution

United States Pharmacopeia and National Formulary. 2022. *<1220> Analytical Procedures Life Cycle*. USP-NF 2022 Issue 1. Rockville, MD: United States Pharmacopeial Convention; Official May 1, 2022. Current DocID: GUID-35D7E47E-65E5-49B7 -B4CC-4D96FA230821_2_en-US

Virtual Museum of Surveying. n.d. "Changing Chains." http://www.surveyhistory.org/changing_chains.htm

The W. Edwards Deming Institute. 2015, 19 November. "If Japan Can, Why Can't We?" Ketchum, ID. https://deming.org. https://www.youtube.com/watch?v=vcG_Pmt_Ny4

Wade, Heather A. 2011. "Effective Communication Between Customers and Their Labs." *Cal Lab Magazine: The International Journal of Metrology 18*, no. 4 (October): 36–40. https://www.callabmag.com/wp-content/uploads/2011/11/issue_oct2011.pdf #page=38

Wallard, Andrew. 2004. "News from the BIPM—2003: Resolution 10 of the 22nd CGPM (2003)—Symbol for the Decimal Marker." *Metrologia* 41(1 January): 99–108. https://doi.org/10.1088/0026-1394/41/1/m01

"What Is Phase Noise: Phase Jitter." n.d. Electronics Notes: Reference Site for Electronics, Radio & Wireless. https://www.electronics-notes.com/articles/basic_concepts/electronic-rf-noise/phase-noise-jitter-what-is.php

Wichmann, Brian. 2000. *Software Support for Metrology Best Practice Guide No.1—Measurement System Validation: Validation of Measurement Software*. April 2000. Teddington, Middlesex, England: National Physical Laboratory. http://www.npl.co.uk

Yale University. 1998. *Confidence Intervals*. New Haven, CT: Yale Department of Statistics and Data Science. http://www.stat.yale.edu/Courses/1997-98/101/confint.htm

Yale University. 1998. *Sample Means*. New Haven, CT: Yale Department of Statistics and Data Science. http://www.stat.yale.edu/Courses/1997-98/101/sampmn.htm#clt

Yen, James, Dennis Leber, and Letici Pibida. 2020, September. *NIST Technical Note 2106: Comparing Instruments*, p. 27. Gaithersburg, MD: NIST. https://doi.org/10.6028/NIST.TN.2106

Zumbrun, Henry. 2021. *Force Calibration for Technicians and Quality Managers: Top Conditions, Methods, and Systems That Impact Force Calibration Results*. 1st ed. Kindle Direct Publishing.

Zumbrun, Henry. 2021, December. *Why a 4:1 TUR is not Enough: The Importance of Analyzing the Probability of False Accept Risk*. York, PA: Morehouse Instrument Company. https://mhforce.com/wp-content/uploads/2021/04/Why-a-4-to-1-TUR-is-not-enough.pdf

Zumbrun, Henry, and Dilip Shah. 2021, October. "The New Dimension to Resolution: Can It Be Resolved?" York, PA: Morehouse Instrument Company. https://mhforce.com/wp-content/uploads/2021/10/The-new-dimension-to-resolution-Can-it-be-resolved.pdf

CONFERENCE PAPERS

Blair, Thomas. 2003. Letter to author (Payne), September 28.

Bucher, Jay L. 2000. "When Your Company Needs a Metrology Program, But Can't Afford to Build a Calibration Laboratory . . . What Can You Do?" NCSLI Workshop & Symposium, 2000. Boulder, CO: NCSL International.

Pinchard, Corinne. 2001. "Training a Calibration Technician ... in a Metrology Department?" NCSLI Workshop & Symposium, 2001. Boulder, CO: NCSL International.

Zumbrun, Henry, and Dilip Shah. 2020. "The New Dimension to Resolution: Can It Be Resolved?" NCSLI Workshop & Symposium, 2000 (Best paper). Boulder, CO: NCSL International.

STANDARDS and OTHER PUBLISHED DOCUMENTS

AOAC International. 2012. Appendix H: Probability of Detection (POD) as a Statistical Model for the Validation of Qualitative Methods, Official Method 2008.01. *Official Methods of Analysis of AOAC INTERNATIONAL*. 19th ed. Gaithersburg, MD: AOAC International.

ASTM International. 2018. *ASTM E74-18e1 Standard Practices for Calibration and Verification for Force-Measuring Instruments*. West Conshohocken, PA: ASTM International.

ASTM International. 2020. *ASTM E898-20: Standard Practice for Calibration of Non-Automatic Weighing Instruments*. West Conshohocken, PA: ASTM International.

ASTM International. 2021. *ASTM E2587-16(2021)e1, Standard Practice for Use of Control Charts in Statistical Process Control*. West Conshohocken, PA: ASTM International.

International Organization for Standardization. 1973. *ISO 17:1973 Guide to the use of preferred numbers and of series of preferred numbers*. Geneva, Switzerland: ISO.

International Organization for Standardization. 2009. *ISO/TS 16949:2009 Quality management systems—Particular requirements for the application of ISO 9001:2008 for automotive production and relevant service part organizations*. 2009. Geneva, Switzerland: ISO.

International Organization for Standardization. 2011. *ISO 376:2011 Metallic materials—Calibration of force-proving instruments used for verification of uniaxial testing machines*. Geneva, Switzerland: ISO.

International Organization for Standardization. 2015. *ISO Guide 30:2015 Reference materials—Selected terms and definitions*. Geneva, Switzerland: ISO.

International Organization for Standardization. 2016. *ISO 1:2016(en), Geometrical Product Specifications (GPS)—Standard reference temperature for the specification of geometrical and dimensional properties*. Geneva, Switzerland.

International Organization for Standardization. 2016. *ISO 17034:2016 General requirements for competence of reference material producers*. Geneva, Switzerland: ISO.

International Organization for Standardization. 2018. *ISO 7500-1:2018 Metallic materials—Calibration and verification of static uniaxial testing machines—PART 1: Tension/compression testing machines—Calibration and verification of the force-measuring system*. Geneva, Switzerland. ISO.

International Organization for Standardization. 2019. *ISO 8601-1:2019—Date and time—Representations for information interchange—Part 1: Basic rules*. February 2019. Geneva, Switzerland: ISO.

International Organization for Standardization/International Electrotechnical Commission. 2005. *ISO/IEC 17025:2005—General requirements for the competence of testing and calibration laboratories*. Geneva, Switzerland: ISO.

International Organization for Standardization/International Electrotechnical Commission. 2010. *ISO/IEC 17043:2010 Conformity assessment—General requirements for proficiency testing*. Geneva, Switzerland: ISO.

International Organization for Standardization/International Electrotechnical Commission. 2015. *ISO/IEC 17021:2015 Conformity assessment—Requirements for bodies providing audit and certification of management systems—Part 1: Requirements.* Geneva, Switzerland: ISO.

International Organization for Standardization/International Electrotechnical Commission. 2017. *ISO/IEC 17025:2017—General requirements for the competence of testing and calibration laboratories.* Geneva, Switzerland: ISO.

International Organization for Standardization/International Electrotechnical Commission. 2020. *ISO/IEC 17000:2020 Conformity assessment—Vocabulary and general principles.* Geneva, Switzerland. ISO.

International Organization for Standardization/International Electrotechnical Commission. *ISO/IEC 27001: Information Security Management System (ISMS).* Geneva, Switzerland: ISO. https://www.iso.org/isoiec-27001-information-security.html

International Society of Automation. 2006. *ISA-TR52.00.01-2006 Recommended Environments for Standards Laboratories.* 2006. Durham, NC: The International Society of Automation. https://www.isa.org/products/isa-tr52-00-01-2006-recommended-environments-for-s

NCSL International. 1996. *Recommended Practice-13: Computer Systems in Metrology, RP-13.* [First Edition.] Boulder, CO: NCSL International.

NCSL International. 1999. *Laboratory Management-3: Glossary of Metrology Related Terms, LM-3.* 2d ed. Boulder, CO: NCSL International.

NCSL International. 2007. *Recommended Practice-3: Calibration Procedure Requirements, RP-3.* Boulder, CO: NCSL International.

NCSL International. 2007. *Recommended Practice-17: Documenting Metrology Education, Training and On the Job Training, RP-17.* [First Edition.] Boulder, CO: NCSL International.

NCSL International. 2010. *Laboratory Management-14: Metrology Human Resources Handbook, LM-14.* [First Edition.] Boulder, CO: NCSL International.

NCSL International. 2010. *Recommended Practice-1: Establishment and Adjustment of Calibration Intervals, RP-1.* 4th ed. Boulder, CO: NCSL International.

NCSL International. 2012. *Laboratory Management-13: Calibration Laboratory Personnel Qualifications, LM-13.* [First Edition.] Boulder, CO: NCSL International.

NCSL International. 2014. *Recommended Practice-9: Measurement Capability Description, RP-9.* 3rd ed. Boulder, CO: NCSL International.

NCSL International. 2015. *Recommended Practice-6: Recommended Practice for Calibration Quality Systems for the Healthcare Industries, RP-6.* 4th ed. Boulder, CO: NCSL International.

NCSL International. 2016. *Recommended Practice-5: Recommended Practice for Measuring and Test Equipment (MTE) Specifications.* 3rd ed. Boulder, CO: NCSL International.

NCSL International. 2017. *Handbook for the Application of ANSI/NCSL Z540.3-2006—Requirements for the Calibration of Measuring and Test Equipment.* Boulder, CO: NCSL International.

NCSL International. 2018. *Laboratory Management-19: Implementation of a Delayed Dating Approach to Calibration, LM-19.* [First Edition.] Boulder, CO: NCSL International.

OIML: International Organization of Legal Metrology. 1989, July. *OIML G13 (Ex P 7): Planning of Metrology and Testing Laboratories.* Paris, France: International Bureau of Legal Metrology. https://www.oiml.org/en/files/pdf_g/g013-e89.pdf

OIML: International Organization of Legal Metrology. 2004. *International Document; OIML D 28: Conventional value of the result of weighing in air.* Paris, France: Bureau International de Métrologie Légale.

OIML: International Organization of Legal Metrology. 2004. *International Recommendation; OIML R 111-1: Weight classes of E_1, E_2, F_1, F_2, M_1, $M_{1\text{-}2}$, M_2, $M_{2\text{-}3}$, and M_3. Part 1: Metrological and technical requirements.* Paris, France: Bureau International de Métrologie Légale.

OIML: International Organization of Legal Metrology. 2006. *International Recommendation; OIML R 76-1: Non-automatic weighing instruments.* Paris, France: Bureau International de Métrologie Légale.

SAE International. 2016. *AS9100D—Quality Management Systems—Requirements for Aviation, Space, and Defense Organizations.* Warrendale, PA: SAE International.

BOOKS

Adam, Stephen F. 1969. *Microwave Theory and Applications.* Englewood Cliffs, NJ: Prentice-Hall.

Arden, Paul. 2003. *It's Not How Good You Are, It's How Good You Want to Be: The World's Best-Selling Book.* New York, NY: Phaidon Press.

Automotive Industry Action Group. 2010. *Measurement Systems Analysis (MSA) Reference Manual.* 4th ed. Southfield, MI: AIAG.

Brownlee, K. A. 1960. *Statistical Theory and Methodology in Science and Engineering.* New York, NY: John Wiley & Sons.

Calhoun, Richard. 1994. *Calibration & Standards: DC to 40 GHz.* Louisville, KY: SS&S.

Carr, Joseph J. 1996. *Elements of Electronic Instrumentation and Measurement.* 3rd ed. Upper Saddle River, NJ: Prentice-Hall.

Crowder, Stephen, Collin Delker, Eric Forrest, and Nevin Martin. 2020. *Introduction to Statistics in Metrology.* Springer Nature Switzerland AG.

Fluke. 1994. *Calibration: Philosophy in Practice.* 2nd ed. Everett, WA: Fluke Corporation.

Griffith, Gary K. 1986. *The Quality Technician's Handbook.* 3rd ed. Englewood Cliffs, NJ: Prentice Hall.

Hirano, Hiroyuki. 1995. *5 Pillars of the Visual Workplace: The Sourcebook for 5S Implementation.* Bruce Talbot, translator. Shelton, CT: Productivity Press.

International Society of Pharmaceutical Engineering. 2008, February. *Good Automated Manufacturing Practice: GAMP 5: A Risk-Based Approach to Compliant GxP Computerized Systems.* North Bethesda, MD: ISPE.

Juran, Joseph M., Frank M. Gryna, and Richard S. Bingham. 1974. *Quality Control Handbook.* 4th ed. Wilton, CT: McGraw-Hill.

Kilian, Cecelia S. 1992. *The World of W. Edwards Deming.* Knoxville, TN: SPC Press.

Kimothi, Shri Krishna. 2002. *The Uncertainty of Measurements: Physical and Chemical Metrology Impact and Analysis.* Milwaukee, WI: ASQ Quality Press.

Larson, Miriam B., and Barbara B. Lockee. 2020. *Streamlined ID: A Practical Guide to Instructional Design.* 2nd ed. New York, NY: Routledge.

Laverghetta, Thomas S. 1981. *Handbook of Microwave Testing.* Dedham, MA: Artech House.

Marquardt, Donald W. 1997. "Background and Development of ISO 9000 Standards." In Robert W. Peach (ed.), *ISO 9000 Handbook*. 3rd ed. New York: McGraw-Hill.

Millbrooke, Anne Marie. 1999. *Aviation History*. Englewood, CO: Jeppesen Sanderson.

Pellegrino, Charles R., and Joshua Stoff. 1985. *Chariots for Apollo*. New York: Atheneum.

Ritter, Diane, and Michael Brassard. 2018. *GOAL/QPC Memory Jogger 2*. 2nd ed. (2018 Revision). Methuen, MA: GOAL/QPC.

Saad, Theodore S., Robert C. Hansen, and Gershon J. Wheeler. 1971. *Microwave Engineer's Handbook*, Vol. 1. Dedham, MA: Artech House.

Shewhart, Walter A. 1939. *Statistical Method from the Viewpoint of Quality Control*. Edited by W. Edwards Deming. Washington, D.C.: The Graduate School, Department of Agriculture, Washington.

Shrader, Robert L. 1980. *Electronic Communication*. 4th ed. New York: McGraw-Hill. [Now in its sixth edition, 1991.]

Tague, Nancy R. 2005. *The Quality Toolbox*. 2d ed. Milwaukee, WI: ASQ Quality Press.

Wells, Alexander T. 1998. *Air Transportation: A Management Perspective*. 4th ed. Belmont, CA: Brooks/Cole.

Index

A

Abbreviations (Appendix B), 550–565
Accessible emission limit (AEL), 493
Accreditation, 25, 568
Accreditation Body, 25, 26, 51, 52, 65, 66, 67, 328, 472, 505, 508, 521, 538, 568
Accuracy, 96
    ground control to, 99
Acids, in water, 495
Acme Widget Company, 78
Acoustics, electronics *vs.*, 373
Acronyms (Appendix B), 550–565
Active surface, 492
Adams, John Quincy, 2
Adapters
    and mechanical interactions, 464–468
    for tension links, crane scales, and dynamometers, 468–469
Addition
    significant digits rule, 245
    using scientific notation, 250–251
Adjustment, 36, 39, 45–46, 109, 140, 198, 220, 221, 438, 460, 522, 523 568
    of calibration intervals, 77, 79–84, 84
Aerospace manufacturing, industry-specific requirements, 147–148
Air, weighing in, 428
Air-handling system, 128
Airline Deregulation Act of 1978, 150
Airline industry, housekeeping and safety in, 543
Allied Quality Assurance Procedures (AQAP), 29
Alpha particles, 490
Alternating current (AC)
    direct current *vs.*, 372
    measurement parameters, 381–382
Alternating voltage (AV), 372
    measurement parameters, 380–381

American Society for Quality (ASQ)
    Code of Ethics, 514–515
American Society for Quality Control (ASQC), 27
    Certified Quality Engineer program, 30
Ampere (A), 9
    realization of, 374
Analysis of variance (ANOVA), in laboratory proficiency testing, 230, 231–232
    Number formatting, 255
    F-distribution (statistics), 302
    Sum of squares (Statistics), 289
    Uncertainty of measurement, 330, 511
Analyte, 499, 503, 504, 569
Analytical instruments, 498
Ancient measurement, 2–5
Andromeda galaxy, 247, 248
Angle
    formulas for, 312–314
    geometric, 478
    symbol for, 257
    units of measure, 478–479
Angle gage blocks, 480–481
Angle measurement
    definitions, 478–482
    instrumentation, 479–481
        complex, 481–482
Angle squares, 480
ANSI/ASQC M1-1987 standard, 28
ANSI/ASQC M1-1996 standard
    and calibration procedures, 35
    and calibration records, 44
    and environmental controls, 123
ANSI/ISO/ASQ Q10012-2003 standard, 138
    and calibration records, 44
    and customer service, 519
    and vendors and suppliers, 538
ANSI/NCSL Z540.1-1994 standard, 18, 28
ANSI/NCSL Z540.3-2006 standard, 18, 28
    and environmental controls, 123

Apparatus, 38
Apprenticeship, 120
Archive meter, 476
Area, 3
    confidence intervals, 288
    formulas of, 308, 314–317
    SI derived unit, 263
    under Normal frequency distribution curve, 184, 301
Assessment, 569
Assize of Bread, 11
Assumptions, measurement, 284
Attenuation
    defined, 403
    measurement parameters, 403–404
Attribute, 183, 185, 200, 385, 489, 569
Audio frequencies, 373
Audit
    of calibration system, 144–146
    criteria, 569
    defined, 101, 569
    external, 104–105
    first-party, 25, 104
    internal, 102
    plan, 569
    requirements, 101–105
    second-party, 25, 104
    third-party, 25
Autocollimators, 477, 481, 483
Autocorrelation, 306–307
Autocorrelation coefficient, 306
Autocorrelation series, 307
Automated calibration procedure software, 165–166
Automated comparison method, 166
Automatic weighing instruments and system, 436
Automotive manufacturing, industry-specific requirements, 148–149
Average (mean), 285–286
Avogadro constant, 86, 87, 173, 247, 275, 501
Avogadro's number, 248, 249
AVSQ 94 standard, 148

B

Babylonian units, 3–5
Bar graph, 282
Baron Kelvin of Largs, 370
Barretter, 398
Base, scientific notation, 249
Baseline specifications, 202, 203–205
Bases, in water, 495
Basic format, for time, 260
Bel, 278
Bell, Alexander Graham, 278

Best-fit line, 296–297
Best practices, in calibration operations, 517–527
Bias
    definition, 569
    measurement system, 186, 188
Big bang theory, 444
Bilateral specification, 201–202
Bimodal distributions, 286, 287
Binomial distribution, 303–304
Biological tests, 498
Biologics Control Act, 1902, 11
Biotechnology industry, housekeeping and safety in, 542
Black book, 12
Black dots, 96–97
Block level with micrometer, 483
Bolometer, 398
Brainstorming, 21–22
British Standards Institution, 29
Brownlee, K. A., 345
BS 5179 standards, 29
BS 5750 standards, 29
Bubble graphs, 282–283
Budgeting, in calibration operations, 534–535
Bureau International des Poids et Mesures (BIPM), 65
Bureau of Chemistry, 11, 12
Button load cell, 450–451

C

Calculators, recent advances in, 393
Calibrate before use (CBU) labels, 110
Calibration, 142
    audit of, 144–146
    circular, 71–72
    defined, 39, 93
    documenting hierarchies, 69–71
    equipment, 141–144
    general-purpose laboratories, 127
    history and philosophy of, 1–15
    methods, validation and verification of, 41–42
    of pH instruments, 496–497
    procedures. *See* Calibration procedures
    recall system for, 107
    recommended providers, 65–67
    scheduling system for, 106
Calibration certificates, 48–55
Calibration due date, 109
Calibration laboratories
    managing
        using metrics for, 521–524
Calibration management software system, 155–162
    implementation of software, 156–160

maintenance of software, 160–162
    selection process, 156
Calibration methods, 197–198
Calibration: Philosophy in Practice (Fluke), 185
Calibration procedures, 35–42
    contents, 37–38
    end-of-calibration activities, 40
    environmental and stabilization
        requirements, 38
    procedural steps, 39–40
    purpose, 37
    references (bibliography), 40–41
    safety and handling considerations, 38–41
    SOPs and/or competencies, 38
    title and revision control, 37
Calibration programs, 75–84
    components of, 76, 77
    delay dating, 82–84
Calibration records, 43–47, 142
    integrity and retention, 46–47
Calibration standards, 142
    determining substitute for obsolete,
        219–221
    equipment substitution and, 221
    substituting, 213–221
Calibration stickers, 108
Calibration techniques, 197–198
Candela (cd), 9
Capacitance, measurement parameters,
    383–384
Capacity, 3
Capitalization, 99–100
Cause-and-effect diagram, 20, 430
Centesimal system, 479
Central tendency, 285
Certification, 25
Certified Quality Engineer program (ASQC),
    30
Certified reference materials (CRM), 499,
    504–505
Check sheet, 19
Chemical tests, 498–499
    establishing traceability for, 501–506
    evaluating measurement uncertainty for,
        507–510
Chi-squared distribution, 302
CIMP (Mutual Recognition Arrangement),
    65–67
Circle
    area of, 316
    general formula, 311
    perimeter of, 318
Circular calibration, 71–72
Civil aviation, industry-specific requirements,
    149–150
Code of Federal Regulations (CFR), 139–140
Coherence, of SI units, 262–263
Colorimetry, 489–490

Column graph, 282
Combined standard uncertainty,
    determination of, 338–339
Commerce, and metrology, 6–8
Commercial off-the-shelf (COTS) software
    packages, 153
Committee on Data for Science and
    Technology (CODATA), 423
Common cause variation, 20
Competence/competency
    levels, 119
    requirements, 113–115
Complex instrumentation software, 165
Compression, 461–464
Compression calibration, 453–454
Computer industry, housekeeping and safety
    in, 544
Computers/software, 151–169
    automated calibration procedure software,
        165–166
    calibration management software system,
        155–162
    instrumentation software classifications,
        164–165
    overview, 151–155
    security management, 162–164
Cone
    truncated, volume of, 322
    volume of, 321–322
Confidence interval, 288, 290–291, 329–330,
    340–341, 358, 474, 573
Confidentiality, 132
Confidentiality, measurement data,
    196–197
Conicity, 478
Constants, under SI, 173
Consumer Bill of Rights, 13
Consumer risk, 353
Continuous wave (CW) laser, 494
Control charts, 20–21, 188–192
Control number, 33
Conventional mass, 428
Convention of the Metre, 7
Conversion factors, unit, 273
    for converting between equivalent
        measurement units, 276
    table (Appendix D), 591–618
Correlation, 293–295
Correlation coefficients, 340
Correlogram, 307
Cosines, law of, 314
Coverage factor, 329–330, 340–341, 573
Crosby, Philip, 30
Cubit, 4, 5
Customary units, U.S., 98, 100, 257, 261–262,
    273–274, 445
Customer satisfaction, in calibration
    operations, 526–527

Customer service, in calibration operations, 518–521
Cylindrical level, 483

D

Data
    proficiency testing, acceptance of, 230, 232–233
Data conversion, 157
Data Quality Act, 2000, 15
Date and time formats, 258–260
Davy, Humphrey, 9
Decibel measures, 277–281
Decibel scales in use, 279–281
Decimal marker, 246
Decimal place, 246
Decimal point, 246
Decision rules, 351–352
Degrees, 257, 390
Degrees of freedom (df, DoF), 55, 284–285, 337, 340–341, 345–347, 474, 573
Degree to radian, 312
Delaney proviso, 13
Delay dating, 82–84
Deming, W. Edwards, 27, 30
Deming Prize, 27
Density, determination of, 426–428
Designated Institutes (DI), 64–65
Dictionary of Occupational Titles (DOT), 115, 116
Difference, in laboratory proficiency testing, 232
Diffusion, of optical radiation, 492
Digit, 4–5
Digital display size, 204
Digital multimeter (DMM), 335
Digits, significant, 244–246
Dimensional measurement parameters, 476–484
Diode sensor meters, 400–401
Diode sensors, 400
Direct current (DC)
    alternating current *vs.*, 372
    measurement parameters, 374, 375
Direct voltage (DV), 372
    measurement parameters, 375–376
Display size, digital, 204
Dissemination, 419–421
Distance
    in general plane between two points, 310
    between two x–y points, 310
Distributions
    associated with measurement uncertainty, 334–336
    types of, 301–306
Distribution skewness, 292

Division
    significant digits rule, 245–246
    using scientific notation, 251–252
Document control, 32–33
Dodge, Harold, 27
Double-sided specification, 201–202
Doughnut chart, 282
Drift (stability), 188, 206–207, 241, 334, 376, 433, 438, 472
Drug Importation Act, 1848, 11
Drug Quality Safety and Security Act (DQSA), 2012, 15
Drug Safety Board, 15
Durham-Humphrey Amendment, 12

E

EAQF 94 standard, 148
Edgar the Peaceful, 5
Edward I, king of England, 5
Egypt, 8
    and ancient measurement, 2–5
Egyptian calendar, 2
Electrical power, measurement parameters, 391–392
Electronic distance meters (EDMs), 488
Electronic encoders, 484
Electronic levels, 483
Electronics industry, housekeeping and safety in, 542–543
Electronics *vs.* acoustics, 373
ELISA (enzyme-linked immunosorbent assay), 498
Elixir of Sulfanilamide, 12
Ellipse
    area of, 317
    major axis horizontal, 311
    major axis vertical, 312
    perimeter of, 318
Ellipsoid
    volume of, 320
Ellipsoid of revolution
    oblate, surface area of, 324
    prolate, surface area of, 324
End measures, 480
Engineering notation, 248–250
E_n number, in laboratory proficiency testing, 233
Environmental controls, 123–130
Equal-arm balances, 435
Equipment calibration, 141–144
Equipment GMP controls, 140
Equipment status, labels for, 108–111
Equipment substitution, and calibration procedures, 221
estimation, using scientific notation, 253

Eurachem, 499
European Association of Aerospace Companies (AECMA), 147
European Medicines Agency (EMA), 24
European Union (EU), 24
Excess noise, measurement parameters, 409–410
Exponent, scientific notation, 249
Extended format, for time, 260
External audits, 104–105
Extrapolation, 297

F

Fahrenheit, Daniel Gabriel, 9
False Accept (FA) condition, 350
False Reject (FR) condition, 350
Faraday, Michael, 10
F-distribution, 302
Federal Aviation Regulations (FARs), 150
Federal Communications Commission (FCC), 394
Federal Food, Drug, and Cosmetic (FDC) Act of 1938, 12
Federal Register, 8, 13
Federal Tea Tasters Repeal Act, 1996, 14
Federal Trade Commission, 12
Feigenbaum, Armand, 30
Fines Enhancement Laws of 1984 and 1987, 13–14
First-party audits, 104
Fishbone diagram, 20, 430
Fission, nuclear, 10
5S philosophy, 541–542
Floor term, of baseline specification, 205
Flowcharts, 20
Food, Drug, and Cosmetic Act, 15
Food, Drug, and Insecticide Administration, 12
Food Additives Amendment, 13
Food and Drug Administration (FDA), 12, 13, 14, 15, 24
    quality system regulations, 139–147
    software testing principles, 159
Food and Drug Administration Act of 1988, 14
Food and Drug Administration Modernization Act, 1997, 14
Food and Drugs Act, 1906, 11, 12
Food Safety and Modernization Act (FSMA), 2011, 15
Force, 444–475
    calibration, common error sources in, 460–461
    defined, 444
    due to buoyancy, 424–426
    due to gravity, 422–424
    equipment design, 471
    instruments, types of, 446–448
    measurement, importance of, 445–446
    measurement uncertainty, 471–475
    methods/standards, 459–460
    SI units for, 445
Force gauge, 448
Formal education, 120
Formal training, 119, 120–121
Format, measurement data, 196
Formulas, for calibration procedures, 38
Four-wire (4-wire) systems, 469–470
Frame level, 483
Franklin, Benjamin, 10
Frequency, measurement parameters, 385–386
Frequency standard, calibrating, 386–388
Fusion, nuclear, 10

G

Gage repeatability and reproducibility assessment, 192–195
GAMP 5 Guide: A Risk-Based Approach to Compliant GxP Computerized Systems, 153–154
Gaussian probability distribution, 301
General Conference on Weights and Measures (CGPM), 7, 10, 417, 476
General plane triangle relationships, right triangle, 313
General Principles of Software Validation: Guidance for Industry and FDA Staff, 154
Gentry, Elizabeth J. (Benham), 91, 94, 98
Geodetic Survey, 487
George IV, king of England, 6
Gin, 3
Global Positioning System (GPS), 386, 489
Global trade, industry-specific requirements for, 134–135
Goniometers, 484
Good Manufacturing Practice (GMP)
    calibration requirements, 141
    equipment controls, 140
Government law, 23
Graphs, 282–283
Gravity, force due to, 422–424
Greenwich Mean Time (GMT), 259
Griffith, Gary, 189
Guard band, 129, 351–352, 357–358, 367–368

Guide to the Expression of Uncertainty in Measurement (GUM), 55, 329, 499
Gunter, Edmund, 487

H

Harris, Georgia L., 91, 94, 98
Health Canada, 24
Henry (inductance), 89, 264, 269, 276, 384
Henry III, king of England, 5
Hertz (frequency), 89, 175, 254, 264, 373, 383–385, 395
Hertz, Heinrich R., 10
High-stress load cells, 451–452
High voltage/current industries, housekeeping and safety in, 542–543
Histograms, 20, 282
Housekeeping, in calibration operations, 541–544
Human-made radiation, 491
Hypergeometric distribution, 303
Hysteresis, 377, 456–459

I

If Japan Can . . . Why Can't We?, 30
ILAC (International Laboratory Accreditation Cooperation), 26
ILAC-G8, 351–352, 356–359, 361, 366
ILAC-G24, 79
ILAC Mutual Recognition Arrangement (ILAC-MRA), 26, 65–67
ILAC P-10, 75
ILAC P14:09/2020, 50–52, 345, 362, 471
Impartiality, 132
Inductance, measurement parameters, 384–385
Inductor, 384
Industrial Revolution, 9–11
Industry-specific requirements, 131–150
  civil aviation, 149–150
  global trade, 134–135
  ISO 10012, 138–139
  ISO 9000 family of standards, 147–149
  ISO 9001 standards, 133–134, 135–136
  ISO/IEC 17025, 131–133
  ISO/TS 16949:2002 standard, 148–149
  QS-9000 standard, 148
Influence quantity, 501, 504
Informal training, 119, 121–122
Infrared, 394
Infrasonic audio frequencies, 373
Insertion loss, measurement parameters, 403–404
Installation qualification (IQ), 159

Interlaboratory comparison defined, 222
Interlaboratory testing scheme, 228, 229
Internal audit, 102, 104
International Accreditation Forum (IAF), 26
International Aerospace Quality Group (IAQG), 147
International Automotive Task Force (IATF), 148
International Bureau of Weights and Measures (BIPM), 417
  SI Brochure, 85–86
International Committee of Weights and Measures (CIPM), 65
  Mutual Recognition Arrangement, 65–67
International Electrotechnical Commission (IEC), 26
International Laboratory Accreditation Cooperation (ILAC)
  ILAC-G24, 79
  Mutual Recognition Arrangement, 65–67
  policy, 63–64
International Organization for Legal Metrology (OIML)
  D 10, *Guidelines for the determination of calibration intervals of measuring instruments*, 79
International Organization for Standardization (ISO)
  role in quality standards development, 24–26
International organizations, 499–500
International Prototype Kilogram (IPK), 417, 418
International Recommended Practice (NCSL), 18, 36–37, 44
International standards, national versions of, 135
International System of Units (SI), 8, 9, 26, 85–91, 500
  base units in, 86, 87
  defined, 88
  defining constants of, 86–88
  diagram, 90
  disseminating, 88
  prefixes for multiples and submultiples, 90
  realizing, 88
  with special names and symbols, 89
  units and symbols, 91
  units not to be used within, 266–268
International Telecommunications Union (ITU), 26
  Radiocommunication Sector, 394
International Temperature Scale, 10
International Vocabulary of Metrology (VIM), 154–155, 173
  frequently used terms, 179
  measurement systems definitions, 182

International Vocabulary of Metrology—Basic and general concepts and associated terms (2012) (VIM 3), 63–64, 86
Interpolation, 297
    linear methods, 298–299
    methods for nonlinear data, 299–300
Intrinsic, 67,
    definition, 92
    error, 212
    standard, 67, 262, 578
Ionizing radiation, 490–492
ISA-TR52.00.01-2006 standard
    and environmental controls, 126
Ishikawa, Kaoru, 30
Ishikawa diagram, 20, 430
ISO 9000, 136
ISO 9001, 25
    features of, 135–136
    industry-specific requirements, 133–134, 135–136
    management system, 56–57
ISO 9000 family of standards
    industry-specific requirements, 147–149
ISO GUIDE 25:1990 standard, 28
ISO/IEC 17000:2020 standard
    and audit requirements, 101
ISO/IEC 17011 standard, 26
ISO/IEC 17025 standard, 112–113
    industry-specific requirements, 131–133
ISO/IEC 17025:1999 standard, 28
ISO/IEC 17025:2005 standard, 28
ISO/IEC 17025:2017 standard, 25, 28, 56
    and audit requirements, 101
    and calibration methods, 41, 42
    and calibration procedures, 35
    and calibration programs, 75
    and calibration records, 44, 46–47, 48–50, 51
    Clause 8.2.1, 18
    Clause 8.3, control of management system documents, 32
    and computer systems/software, 151–152
    customer service, 518
    document control, 32, 33
    and environmental controls, 123, 129
    and equipment status labeling, 108, 109
    industry-specific requirements, 131–133
    and quality manuals, 56–57, 58–62
    reporting statements of conformity, 357
    and traceability, 63
    and vendors and suppliers, 536–540
ISO 10012-Part 1-1992 standard, 28
ISO 10012-Part 2-1997 standard, 28
ISO preferred numbers, 254–255
ISO 1:2016 standard
    and environmental controls, 126
ISO 8601-1:2019 standard, 259

ISO 9001:1987 standard, 58
ISO 9001:2008 standard
    and calibration, 136–138
ISO 9001-2015 standard
    Clause 4.4.1, 18
    Clause 7.5, 32–33
    and quality manuals, 56, 59
ISO 9001:2015 standard, 25
    and audit requirements, 102–103
    and calibration procedures, 35
    document control, 32, 33
    and environmental controls, 123
    and quality manuals, 58, 59
ISO 10012:2003 standard, 28
    and audit requirements, 103
    and calibration procedures, 35
    Clause 4, 18
    and computer systems/software, 152
    and environmental controls, 123
    and equipment status labeling, 108, 109
ISO 19011:2018 standard, 101
    and audit requirements, 103
    and first-party audits, 104
ISO/TC 176, 148–149
ISO/TS 16949:2009 standard, 136
    industry-specific requirements, 147–149
ISPE
    GAMP 5 Guide: A Risk-Based Approach to Compliant GxP Computerized Systems, 153–154

J

Jefferson, Thomas, 6
jitter, in specifications, 207
Job descriptions, 115–118
John, King of England, 5, 11
Joint Aviation Authorities (JAA) regulations (JAR), 150
Juran, Joseph, 27, 30

K

Kefauver-Harris Drug Amendments, 13
Kelsey, Frances, 13
Kelvin (K), 9
Kennedy, John F., 13
Kent, King of, 5
Kilogram (kg), 9, 418–419, 445
Kimothi, S. K., 182
Knowledge, attitudes, skills, and interpersonal skills (KASI), 113–115
Kurtosis, 292–293

L

Labels, and equipment status, 108–111
Lab liaisons, in calibration operations, 518–521
Laboratory information management systems (LIMS), 151
Laboratory Management-19, Implementation of a Delayed Dating Approach to Calibration (LM-19), 82
Laboratory Management (LM-13): Calibration Laboratory Personnel Qualifications, 113
Language arts, 99–100
Lasers, 10, 493–494
Law of cosines, 314
Law of sines, 313–314
Leading zero, 98
Leading zeros, 245, 257
Least-squares regression line, 296–297
Length, 3
    conversion factors, 476
    defined, 2
    formulas for, 308–310
Length measures, categories of, 477–478
Levels, 482–484
    with coincidence, 483
    electronic, 483
    with flexible tubing, 483
    with microscope, 483
    with vials and bubble, 482–483
Limited calibration labels, 110
Lincoln, Abraham, 11
Linear interpolation methods, 298–299
Linearity, 186–187
Linearization, calibration method, 197
Linearizing transformations, 282
Linear regression, best-fit line through a data set, 296–297
Linear relationships, 295–297
Line charts, 283
Line graduated measures, 479
Line power term, of modifier specification, 208
Load cells, 447
    button, 450–451
    multi-column, 452–457
    S-beam, 448–449
    shear web, 449–450
    single-column/high-stress, 451–452
Load term, of modifier specification, 208
Logarithms, 277–278
    in Microsoft Excel, 281–282
Long-term stability, 207
Lord Kelvin, 9, 370
Loschmidt constant, 248
Louis XVI, king of France, 6
Low frequency alternating current (AC), 372–373

M

Magazines/books/articles, 121–122
Magnetic moment, 248
Malcolm Baldrige National Quality Award (MBNQA) program, 30
Mana, 4
Management systems, and quality manuals, 56–62
Manual comparison method, 166
Marash, Stanley, 29
Mass
    conventional, 428
    SI unit of, 417–421
    and weighing, 422–428
Mass calibration standards, 438–443
Mass comparators, 436, 438
Mass standards, 433–434
Mathematical operations, using scientific notation, 250–252
Maximum permissible exposure (MPE), 493
Maximum permissible measurement error, 352, 434
Mean, 285–286
Measurement
    ancient, 2–5
    area and volume, 3
    capacity, 3
    data, 195–197
    digit, 4–5
    equipment calibration, 141–144
    and Industrial Revolution, 9–11
    length. *See* Length
    pillars of, 349–350
    progress over millennia, 5–6
    results, 96
    royal cubit, 4
    standards. *See* Measurement standards
    terms and communication, 95–96
    uncertainties, 94–97
    units, 173
    weight. *See* Weight
Measurement assumptions, 284
Measurement assurance program (MAP), 234–241
Measurement capabilities, 185–195
Measurement comparison scheme, in laboratory proficiency testing, 223–228
Measurement confidence, 349
Measurement data, 195–197
Measurement error, 199
    measurement uncertainty *vs.*, 326

Measurement instrument
    comparing specifications to another instrument, 214–221
    comparing to measurement task, 212
Measurement methods, 179–181
Measurement parameters, 177–178, 373–392
    introduction to, 370–371
    under SI, 173–175
Measurement precision, 187–188
Measurement repeatability, 187
Measurement reproducibility, 187–188
measurement requirements, stages in defining, 183
Measurement risk, 349–350
    understanding, 350–351
Measurement standards
    chemical measurements, 503–504
    definitions and examples of, 92–93, 327
    hierarchies, 17, 69
    historical development of, 6–8
    International System of Units (SI), 87
    in measurement uncertainty, 330
    quantum, 262
    substitution, 213–221
    types of, 91–94
Measurement systems, 181–185
    properties of, 182
Measurement task, comparing an instrument to, 212
Measurement uncertainty, 326–348
    budget, documenting, 336–338
    considerations, 342–344
    contributors in chemical testing, 510–511
    defined, 326
    distributions associated with, 334–336
    effect of resolution on, 345
    evaluating for chemical tests, 507–510
    evaluation of, 329–344
    identify sources of, 330
    managing, 346–348
    measurement error *vs.*, 326
    report, 341–342
Measuring instrument, 174
Measuring instrument stability, 188, 207
Meat Inspection Act, 1906, 11
Mechanical measurement parameters, 476–484
Mechanical protractor, 481
Mechanical protractor with vernier, 481
Median, 286
Medical Device Amendments, 13
Medium instrumentation software, 164–165
Mensuration, 308
Meter (m), 9, 445
Metes and bounds, 487
Metre Convention, 1875, 7–8
Metric Act of 1866, 6
Metric Association, 7–8
Metric Conversion Act, 85

Metric system, 6
Metrological confirmation, 138–139
Metrological function, 138, 139
Metrological traceability. *See* Traceability
Metrological traceability chain, 64
Metrological traceability to a measurement unit, 64
Metrology
    and commerce, 6–8
    general understanding of, 172–178
    history and philosophy of, 1–15
    and standards, 6–8
Metrology—Calibration and Measurement Processes Guidelines, 181–182
Metrology department, managing using metrics for, 521–524
Microcomputers, recent advances in, 394
Micro potentiometer, 408
Microwave frequency spectrum, 393–395
Microwaves
    defined, 394
    measurement methods, 410–415
MIL-C-45662 standard, 28
MIL-HDBK-50 standard, 28
MIL-H-110 standard, 28
Military Occupational Specialty (MOS) codes, 117–118
Military specification, 27
Military standard, 27, 28
Millième/mil system, 479
MIL-Q-21549B standard, 29
MIL-Q-5923 standard, 28
MIL-Q-9858 standard, 28
MIL-STD-45662 standard, 28
Minutes, symbol for, 257
Mode, 286
Modifier specification, 202, 206–208
Modulation, measurement parameters, 408–409
Mole, 501
Mole (mol), 9
Mouton, Gabriel, 6
Multi-column load cells, 452–457
Multiplication
    significant digits rule, 245–246
    using scientific notation, 251
Mutual Recognition Arrangement
    (CIPM-MRA), 65–67
    (ILAC-MRA), 26, 65–67

N

Napoleon, 6
National Aeronautics and Space Administration (NASA), 99
    Reference Publication 1342, 185

National Institute of Standards and
 Technology (NIST), 85, 98
 SP 811, 98
 Special Publication 330, 2019 edition, 86, 91
National Metrology Institute (NMI)
 as source of traceability, 64–65
National Telecommunications and
 Information Administration (NTIA), 394
Natural physical constants, 326
Natural radiation, 491
NCSL International
 Laboratory Management-19, Implementation of a Delayed Dating Approach to Calibration (LM-19), 82
 Laboratory Management (LM-13): Calibration Laboratory Personnel Qualifications, 113
 Recommended Practice-1 (RP-1): Establishment and Adjustment of Calibration Intervals, 79–80
 Recommended Practice-3 (RP-3): Calibration Procedure Requirements, 36–37
 Recommended Practice-5 (RP-5): Measuring and Test Equipment Specifications, 200
 Recommended Practice-6 (RP-6): Recommended Practice for Calibration Quality Systems for the Healthcare Industries, 44, 103, 108, 113, 123–124, 151–152, 538–539
 Recommended Practice-7 (RP-7): Laboratory Design, 125
 Recommended Practice-13 (RP-13), Computer Systems in Metrology—1996, 151–152
 Recommended Practice-14 (RP-14): Guide to Selecting Standards Laboratory Environments, 125–126
 Recommended Practice-16 (RP-16): Verification of Laboratory Environments, 125
 Recommended Practice (RP-17), Documenting Metrology Education, Training and On-the-Job Training, 113
Negative skewness, 292
Neutron mass, 248
Newton, Isaac, 444
Newton's second law of motion, 422
Niagara Falls, hydroelectric power plant, 213
Nindan, 3
NIST GLP 15, 152–153, 168–169
NISTIR 6969-2019, GMP 11: Good Measurement Practice 11, 79
No calibration required (NCR) stickers, 108, 109
Noise figure, measurement parameters, 409–410
Nonlinear data, interpolation methods for, 299–300

Nonlinearity, 456–457
Nonrepeatability, 458
Normal distribution, 301, 334
North Atlantic Treaty Organization (NATO), 29
Notation methods, 246–254
Nulling, calibration method, 198
Number bases, 274
Number formatting, 244–260
 ISO preferred numbers, 254–255
 notation methods, 246–254
 other issues, 257–258
 significant digits, 244–246
 U.S. customary, 257–258
number rounding methods, 255–257

O

Oblate ellipsoid of revolution, surface area of, 324
Oblique triangle
 area of, 316
Ohm, George Simon, 10
Ohm's law of electrical resistance, 10
Omnibus Trade and Competitiveness Act of 1988, 85
One-way, n-point interpolation—nonlinear (Lagrangian), 300
One-way, three-point interpolation— quadratic, 300
One-way, two-point linear interpolation, 298
One-way specification, 201
One-way tabulations, 297
On-line training courses or information, 121
On-the-job training, 120
Operational qualification (OQ), 159
Optical dividing heads, 483
Optical protractor, 481
Optical radiation, 492–497
Optical rotary tables, 483
Optics, 485–489
Original equipment manufacturer (OEM), 35–36, 83
Orphan Drug Act, 1983, 13
Output term, of baseline specification, 203
Over-the-counter drug, 13
Owning Work Center (OWC), 519

P

Pandemic and All-Hazards Preparedness Reauthorization Act (PAHPRA), 2013, 15
Parallelogram
 area of, 315
Pareto chart, 19–20

Pass rate, 78
Peak sensor meters, 401
Percent difference, in laboratory proficiency testing, 232–233
Percent of range, 204
Performance qualification (PQ), 159
Perimeter, formulas for, 317–318
Perpendicular lines, slope relationship, 310
pH, 495–497
　instruments, calibration of, 496–497
Pharmaceutical industry, housekeeping and safety in, 542
Phase angle, 390–391
pH electrode, 496
pH meter, 496
Photoelectric effect, 10
Physical-dimensional calibration, 543
Pie chart, 282
Planck, Max, 10
Planck constant, 418, 419
Planck Radiation Law, 10
Plan–do–check–act (PDCA) cycle, 21
Plan–do–study–act (PDSA) cycle, 21
Plane area, formulas for, 314–317
Point-slope equation of line, 309
Poisson distribution, 304–305
Policy on Traceability (NIST), 64
Polygons, 481
Population standard deviation, 290
Population variance, 290
Positive skewness, 292
Potentiometry, 495–496
Power frequencies, 373
Power meters, 402
Power sensors, 402–403
Practical mil system, 479
Precision, measurement, 187–188
Precision block level, 483
Precision Measurement Equipment Laboratory (PMEL), 519
Prescription Drug Marketing Act, 14
Prescription Drug User Fee Act of 1992, 14
Preventive maintenance programs, in calibration operations, 524–525
Prism, rectangular
　surface area of, 323
　volume of, 319
Probability of false accept (PFA), 356
Process description chart, 20
Process improvement techniques, 21–22
Process workflow, in calibration operations, 528–533
Producer risk, 354
Professional certification, 120
Professional training events, 121
Proficiency testing, of laboratories, 222–234
　acceptability of data, 230, 232–233

Program
　calibration. *See* Calibration programs
　defined, 76
Prolate ellipsoid of revolution, surface area of, 324
Pronunciation of measurement units, 99
Protractors, 481
Public Health Service Act, 15
Pure Food and Drugs Act, 11
Pyramid
　truncated, volume of, 321
　volume of, 320–321
Pythagorean (right-angle triangle) theorem, 313

Q

Qualifier specifications, 202, 208–209
Quality assurance (QA) system, 146
Quality documentation, 32–34
Quality management system, 23
Quality management system standards, 23
Quality manuals, 56–62
Quality Progress (magazine), 91
Quality standards
　classes of, 23
　defined, 23
　evolution of, 26
　history of, 8–9, 27–31
　importance of, 23–24
Quality system, basics of, 18–22
Quality system regulations (FDA), 139–147
　automated production and QA systems, 141
　equipment GMP controls, 140
　manufacturing materials, 140–141
The Quality Technician's Handbook (Griffith), 189
Quality tools, 19–21
Quantum Hall, 92, 374, 378, 562

R

Radar chart, 282
Radian, 261, 390, 478
Radiation
　health effects of, 491–492
　optical, 492–497
　sources of, 491
Radioactive decay, 490
Radio frequency (RF) power
　measurement methods, 410–415
　measurement parameters, 397–398
　sensors and their meters, 398–401
　　calibration of, 401–402

Radio frequency (RF) spectrum, 393–395
Radio frequency (RF) voltage, measurement parameters, 408–409
Radio waves, 394
raising to powers, using scientific notation, 252–253
Random variation, 20
Ratios, 277–283
   decibel measures, 277–281
R-bar control chart, 190, 192
Readability, measurement data, 196
Realization
   definition, 88, 93
   of SI Units, 65, 67, 374, 376, 386, 417, 422
   methods, 419, 420
   metrological traceability, 64, 91, 327
   product, 137–139
Recall systems, for calibration, 107
Recommended Practice-1 (RP-1), Establishment and Adjustment of Calibration Intervals, 79–80
Recommended Practice-6: Recommended Practice for Calibration Quality Systems for the Healthcare Industries, 108, 123–124, 152, 538–539
Recommended Practice-13 (RP-13), Computer Systems in Metrology—1996, 152
Recommended Practice (RP-17), Documenting Metrology Education, Training and On-the-Job Training, 113
Recommended Practice for Calibration Quality Systems for the Healthcare Industries (NCSL International Recommended Practice–6), 18
Rectangle
   area of, 314
   perimeter of, 317
Rectangular distribution, 305, 334–335
Rectangular prism
   surface area of, 323
   volume of, 319
Reference electrode, 496
Reference standards, 38
Reference value, assigning, in laboratory proficiency testing, 223–228
Reflection, of optical radiation, 492
Reflection coefficient, 404–407
Refraction, of optical radiation, 492
Relative uncertainty, 306
Renard, Charles, 254
Renard series, 254
Renew always method, 80
Renew-as-needed method, 80
Renew-if-failed method, 80
Repeatability, 95, 331
Repeatability, measurement, 187
Reproducibility, 331
   measurement, 187–188

Residuals, 285
Resistance, measurement parameters, 378–380
Resolution, measurement data, 196
Resolution distribution, 336
Resource management, in calibration operations, 534–535
Resources and publications (Appendix A), 545–549
Reverse traceability, 72–73, 137
   and recalls, 107
Right-angle triangle (Pythagorean) theorem, 313
Right triangle
   area of, 315–316
   general plane relationships, 313
   perimeter of, 317–318
Rimailho millième, 479
Risk
   defined, 350
   types of, 353–360
Risk appetite/tolerance, 79
Roman Empire, 8
Roosevelt, Theodore, 11
Root mean square (RMS), 289
Root sum of squares (RSS), 289, 338–339
Royal cubit, 4
Run charts, 283

S

Safe Medical Devices Act, 1990, 14
Safety, in calibration operations, 541–544
Safety and handling considerations, calibration procedures, 38–41
Sample standard deviation, 291
Sample variance, 290–291
Sar, 3
S-beam, 448–449
Scalar network analysis, 411–412
Scalar *vs.* vector, 412
Scales and balances, 435–436, 438
Scale term, of baseline specification, 203–205
Scatter diagrams, 20
Scatter plots, 282
Scheduling systems, for calibration, 106
Scientific notation, 247–253
Second (s), 9, 385, 445
Second-party audits, 104
Seconds, symbol for, 257
Security management, 162–164
Seebeck, Thomas Johann, 9
Semiconductors, 10
Seminars, 120–121
Sensitivity coefficients, 345
Sexagesimal degree, 478

She (barleycorn), 3
Shear web, 449–450
Shewhart, Walter, 27
Shewhart, Walter A., 188
Short-term stability, 207
SI-derived units, 263–265
Significant digits, 244–246
Sila, 3
Simple acceptance decision rule, 354, 355–356
Simple instrumentation software, 164
Sine bars, 482
Sine plates, 482
Sines, law of, 313–314
Single-column load cells, 451–452
Single measurement bliss, 354
Single-sided specification, 201
SI prefix system, 253–254, 265–266, 267
Sirius, 2
Sisson, John, 487
SI units
    arranged by unit category, 268–272
    coherence of, 262–263
    conversions, 261–262
    for force, 445
    of mass, 417–421
    mole, 501
    multiples of SI base/derived units, 267
Six (6-wire) wire systems, 470–471
Skewness, 292
Sleeve gages, 480
Slope (m), 478
Slope-intercept equation of line, 309
Slope of line, 309
Smith, Phillip H. S, 405
Smith chart, 405–407
Society of Japanese Aerospace Companies (SJAC), 147
Society of Motor Manufacturers and Traders, 29
Soft-skill proficiency, 114
Softwares. *See* Computers/softwares
Solenoid, 384
Space exploration, 10–11
Spanning, calibration method, 198
S-parameters, 413–415
Specifications, 199–200, 352–353
Specifications, of measurement instruments, 199–212
    application of, 201
    characteristics of, 202–203
    comparing, 212, 214–221
    forms of writing, 205–206
    limits, types of, 201–212
    tables, 209–211
    unusual terminology, 212
Spectrum analysis, 410
Spelling, of measurement units, 99

Sphere
    surface area of, 323
    volume of, 319
Spider chart, 282
Split-sample testing scheme, in laboratory proficiency testing, 228, 230, 231
Spot frequency, calibration method, 198
Spreadsheets, 166–169
Stability
    measuring instrument, 188, 207
Standard deviation, 331–332
Standard error of the mean (SEM), 291
Standardize before use labels, 110
Standard normal (gaussian) distribution, 301
Standard notation, 246–247
Standard Occupational Classification System (SOC), 115
    Job Code 17-3028 Calibration Technologist and Technicians, 116–117
Standard operating procedures (SOPs), 38
Standard Railway Time (SRT), 259
Standards of measurement. *See* Measurement standards
Standard uncertainty, 247
Standing wave ratio (SWR), 404–407
Stanford Applied Engineering (SAE) International, 147
Static error band, 459
Statistical process control (SPC), 188–192
Statistical quality control (SQC), 27
Statistics, 284–307
    autocorrelation, 306–307
    bimodal distributions, 286, 287
    central limit theorem, 286–288
    central tendency, 285
    confidence interval, 288
    correlation, 293–295
    degrees of freedom, 284–285
    interpolation, 297
    kurtosis, 292–293
    linear relationships, 295–297
    mean, 285–286
    median, 286
    mode, 286
    population standard deviation, 290
    population variance, 290
    residuals, 285
    root mean square, 289
    root sum of squares, 289
    sample standard deviation, 291
    sample variance, 290–291
    skewness, 292
    standard error of the mean, 291
    sum of squares, 289
Steradian, 261

692 Index

Stevin, Simon, 6
Stopwatches, 388–389
Substances Generally Recognized as Safe (GRAS), 13
Subtraction
  significant digits rule, 245
  using scientific notation, 250–251
Suitability, measurement data, 196
Sum of squares (SS), 289
Suppliers, in calibration operations, 536–540
Surface area, formulas for, 323–324
Surface graphs, 283
Surveys, in calibration operations, 526–527
Sweet spots, 198
Symbols, for calibration procedures, 38
Systematic measurement error, 186
Systematic offset, 186
System Development Life Cycle (SDLC), 158–159

T

Tabular data, formats of, 297–298
Tangent bar, 482
Taper gages, 480
Taper plugs, 480
Target measurement uncertainty (TMU), 503
t-distribution, 302, 346
Tea Importation Act of 1897, 14
Technical writing, 98–100
Telephone service, advances in, 393
Temperature
  stability, 127
  uniformity, 128
Temperature term, of modifier specification, 207–208
Tension, 461–464
Tension calibration, 454–457
The Tenth (Stevin), 6
terminology, unusual, in specifications, 212
Test accuracy ratio (TAR), 28
Test scripts, 159
Test uncertainty ratio (TUR), 28, 146, 360–367
  calculation of, 360–367
Thalidomide, 13
Theodolites, 484
Thermistor sensor meters, 399
Thermistor sensors, 398–399
Thermocouples, 9
Thermocouple sensor meters, 400–401
Thermocouple sensors, 399–400
Thermometers, 9
Thomson, J. J., 10
Thomson, William, 370

Thrasher, Leslie, 416
3-D graphs, 283
Three-way interpolation, 299
Three-way tabulations, 298
Time
  defined, 2
Time interval, measurement parameters, 385–386
Time of day, 389–390
Time term, of modifier specification, 206–207
Timing devices, 388–389
"Tipping the Scales," 416
Tolerance
  application of, 201
  definition, 199–201
  of measurement instrument, 199
Tolerance limits, 352–353
Traceability, 63–74, 93, 133
  defined, 63–64
  elements of, 67–69
  establishing for chemical tests, 501–506
  evaluation of, 67
  fundamentals, 500–51
  ILAC policy on, 75
  importance of, 73–74
  international agreements and arrangements, 65–67
  National Metrology Institute (NMI) as source of, 64–65
  requirement of, 51
  reverse, 72–73
  tools for assessing, 73
Training, 112, 132
  on document control, 33–34
  documents, 112–113
  formal, 119, 120–121
  informal, 119, 121–122
  levels, 119
  matrix, 118
  plan, 118–119
  sources, 119
Transducer, 446
Transit, 487
Transmission-reflection test set, 413
Trapezoid
  area of, 315
Treaty of the Meter (Metre Convention), 7, 417
Triangular distributions, 305–306, 335–336
Truncated cone, volume of, 322
Truncated pyramid
  volume of, 321
Two-point slope-intercept relationship, in linear data sets, 295–296
Two-way, two-point interpolation, 299
Two-way specification, 201–202
Two-way tabulations, 298

Type I error, 354
Type II error, 353

U

Ultrasonic audio frequencies, 373
Ultrasonic frequencies, 373
Uncertainty
    for fundamental units, 273–274
    measurement. *See* Measurement uncertainty
    normal distribution, 305
    overlap, 228
    rectangular distribution, 305
    relative, 306
    standard, 247
    triangular distribution, 305–306
    type of, 330–334
The Uncertainty of Measurements (Kimothi), 182
Uniform distribution, 305
Unilateral specification, 201
Union of Japanese Scientists and Engineers (JUSE), 27
Unit conversions, 261–276
    errors, 98
    factors, 273
United States v. Dotterweich, 12
Unit of measurement, 173
Units, uncertainties for, 273–274
Unit under test (UUT), 386–387, 455–456
Universal Coordinated Time (UTC), 387, 388
Universal coordinated time (UTC), 262
U.S. customary units, 100
U.S. Department of Agriculture, 11
U.S. Department of Defense (DoD), 117
U.S. Department of Labor Office of Administrative Law Judges (OALJ) Law Library, 116
    Dictionary of Occupational Titles (DOT), 115, 116
U.S. food and drug laws, 11–15
U.S. Metric Association (USMA), 8
U.S. Mint, 6
U.S. Pharmacopeia, 11
U-shaped distributions, 306, 336
USS Constitution, 6–7

V

Validation
    of calibration methods, 41–42
    defined, 155
    software, 152
VDA 6.1 standard, 148
Vector, scalar *vs.*, 412

Vector network analysis, 413
Vector network analyzer (VNA), 413, 414–415
Vendors, in calibration operations, 536–540
Verification
    of calibration methods, 41–42
    defined, 154–155
Verification, validation, and testing (VV&T), 154
Victoria, queen of England, 7
Vitamins and Minerals Amendments (Proxmire Amendments), 13
Volt, 10
Volta, Alessandro, 10
Voltage standing wave ratio (VSWR), 405
Voltaire, 517
Volume, 3
    determination of, 426–428
    formulas for, 319–322
Voluntary quality standards, 23–24

W

Water molecule, 495
Watt, 397
Waveguides, 395–396
Weber, 384
Webinars, 122
Weibull distributions, 302
Weighing
    in air, 428
    environmental effects, 431–432
    good procedures, 428–434
    instrument effects, 432–433
    instruments. *See* Weighing instruments
    mass and, 422–428
    procedures and correct care and handling, 430–431
Weighing instruments, 434–435
    calibration procedures, 438
    documentary standards and classifications for, 436–438
    selection of, 436
    types of, 435–436
Weight, 4
    defined, 2
    mass calibration standards, 438–443
Weights and Measures Act, 6
Welch-Satterthwaite formula, 346
Western Electric Company, 27
Wetherill, Charles M., 11
Wheeler-Lea Act, 12
Whitney, Eli, 9
Wiley, Harvey W., 11
Workflows, 158
Workspace setup, for calibration procedures, 39
www.ClinicalTrials.gov, 14

X

X-bar control chart, 190, 191
X-rays, 490–491

Y

y-intercept of line, 308
Young, William J., 487

Z

Z-distribution, 301
Zener diode, 376–378
Zero and span relationships, 297
Zeroing, calibration method, 198
Z-score, in laboratory proficiency testing, 233

About the Contributors

Georgia L. Harris

Georgia Harris earned a bachelor's degree in biology (minor in chemistry) from the University of Minnesota, Moorhead, and a master's degree in technical management from the Johns Hopkins University, Whiting School of Engineering. She began her metrology career working for the Minnesota Metrology Laboratory, where she was also active in NCSL International as a Twin Cities section coordinator. After five years, she left Minnesota for Maryland and began working for the National Institute of Standards and Technology (NIST) Office of Weights and Measures (OWM). After 29 years of service, Georgia retired from NIST. As a part of her responsibilities at NIST OWM, she oversaw the Laboratory Metrology Program of recognition (accreditation) for weights-and-measures laboratories, effectively ensuring quality measurement results for nearly 400,000 annual calibrations, supporting the foundation for the U.S. legal metrology system. She actively designed and conducted training for more than 1700 participants in 65 unique seminars and webinars and led the accreditation efforts of the OWM training program according to the International Association for Continuing Education and Training (IACET). She conducted training on technical topics related to measurement, calibration, laboratory accreditation, and ISO/IEC 17205. Her primary focus in training has been metrology fundamentals, legal metrology, mass, and volume.

During her work at NIST, Georgia wrote or co-authored more than 50 technical publications and papers presented at conferences, including topics on metrology, weights and measures, and education and training. Georgia has published and presented papers and conducted metrology and adult education training throughout the United States; won best paper awards from Measurement Science Conference, NCSL International, and the American Society for Engineering Education; and has presented papers and conducted training at measurement conferences in Canada, Mexico, South Africa, and Colombia. She was a NIST liaison to the Measurement Science Conference for NIST Seminars for 12 years and a member of the NCSL International board of directors for nearly 20 years, including holding the position of president. She has won awards from MSC (1997 Andrew J. Woodington Award), NCSL International (2015 Education and Training Award, 2019 Wildhack Award), ASQ Measurement Quality Division (2011 Max J. Unis Award), two Bronze Medal awards from NIST, and was awarded a Fulbright Specialist grant in 2016.

Heather A. Wade

Heather A. Wade, CCT, CQA, ASQ Senior Member, is a recognized metrology subject matter expert, internationally sought speaker, and proud metrology nerd. A graduate of University of Michigan with a BS in biology, she has 30+ years of professional lab and testing experience. Heather worked as a microbiologist, filter test specialist, laboratory compliance officer, extraction chemist, and analytical chemist before moving full-time into metrology. She was calibration officer at NSF International, in Ann Arbor, Michigan, where, from beginning as a one-person department (including managing the calibration lab), she built and led a metrology team serving NSF's microbiology, chemistry, and physical engineering test labs as well as its field auditors and growing global labs. She dealt with a wide spectrum of measurements, including temperature, mass and weight, force, torque, hardness, electrical, time, dimensional, luminous intensity, pressure, flow, humidity, irradiance, and sound.

Heather has been active in ASQ-Measurement Quality Division (MQD) leadership since 2009 and has held each elected position, from secretary, to treasurer, and to Office of the Chair. In addition, she is MQD chair of the Joe D. Simmons Memorial Scholarship and is also an MQD distinguished speaker. She was a leader and subject-matter expert for the ongoing development of the ASQ-CCT exam from 2012 to 2020, and also served as CCT Exam Subcommittee chair from 2012 to 2015. She has served as ASQ World Conference on Quality and Improvement Technical Paper Reviewer multiple times.

She has been published in multiple publications, including *Cal Lab Magazine* and *The International Journal of Metrology*. Heather was featured in the *Cool Careers in Metrology* DVD and on the MetrologyCareers.com website. Heather is one of the co-authors of *The Metrology Handbook* (Second Edition), edited by Jay Bucher. She is also co-author, with Diana Baldi, of a self-published ebook titled *Demystifying ISO/IEC 17025 Requirements* (Baldi & Wade, Innovation Training & Consulting, 2020 [updated in 2022]). Heather is editor, co-author wrangler, and one of the co-authors of *The Metrology Handbook* (Third Edition).

She has been a frequent presenter at Measurement Science Conference, NCSL International symposia, and local and regional ASQ and NCSL International meetings; for webinars with instrument manufacturers; at ASQ sections and technical divisions; and for accreditation bodies. She presented about metrology at ASQ-WCQI 2022. She has also presented about metrology at the 2019 Cannabis Quality Conference, 2019 Cannabis Labs Virtual Conference, the 21st Asia-Pacific Quality Conference, the 2014 China Conference on Quality, and the 10th Shanghai International Symposium on Quality. She is an American Association for Laboratory Accreditation (A2LA) lead and technical assessor for ISO/IEC 17025 for calibration and testing labs. She is also an active member of A2LA's Measurement Advisory Council and Accreditation Council. With NCSL International, she is an active member of NCSL International's Testing Lab committee and Test Equipment Asset Management committee; where she is an unlisted contributor to NCSL International's *Lab Management Guide: LM-19 Implementation of a Delayed Dating Approach to Calibration*. Heather is a voting member of ASTM Committees D37 (Cannabis: Quality Management Systems, Laboratory, Terminology), E11 (Quality and Statistics: Metrology), E20 (Temperature Measurement: Fundamentals in Thermometry), E28 (Mechanical Testing: Calibration of Mechanical Testing Machines and Apparatus, Uniaxial Testing), E36 (Accreditation and Certification: Laboratory/Inspection Standards), and E41 (Laboratory Apparatus: Laboratory Instruments and Equipment). Heather is a member of the U.S. Pharmacopeia (USP) Expert Committee on Measurement and Data Quality for the 2020–2025 cycle. This committee has written the new USP General Chapter <1220> on Analytical Procedure Lifecycle that includes the metrological concepts of measurement uncertainty and metrological traceability. Other professional memberships include A2LA, ASTM, ASQ, and NCSL International.

As president of Heather Wade Group, LLC, Heather provides auditing, ISO/IEC 17025 accreditation assistance, root cause analysis and corrective action training, measurement uncertainty, gap analysis, and technical consulting. She brings all her professional experiences together to provide "Pain Relief for Measurement Headaches" for her consulting clients.

Henry A. Zumbrun II

Henry A. Zumbrun has more than 25 years of industry experience in metrology, specifically in force and torque measurements. He started working at Morehouse Instrument Company in 1995, polishing Proving Rings and has worked his way up the ladder as the company's current president. Today, Morehouse is one of the most respected names in the calibration and measurement world. Henry likes to think he has helped make an impact with force and torque measurements by helping several organizations be better and make better measurements.

Henry has a passion for metrology, training, outreach, and running a business based on doing what's right. He supports continuous learning and, along with several degrees, has earned a Six Sigma Black Belt, become a lean champion, and completed LMI leadership program Sandler Management Program, Sales Training. He considers Dilip A. Shah one of his best friends, mentor, and overall curmudgeon. He is a current Vistage member and is constantly learning from several mentors and his business coach, Mr. Chad Harvey, on how to make a meaningful impact beyond what a typical business would do.

Henry has taught various classes and sessions at NCSL International, MSC, A2LA, etc.; authored several published papers; written a book titled *Force Calibration for Technicians and Quality Managers: Top Conditions, Methods, and Systems That Impact Force Calibration Results*, and was the primary author on the "G126—Guidance on Uncertainty Budgets for Force Measuring Devices" for A2LA. He has been part of the ASTM E28 committee for well over a decade and made several contributions to the ASTM E74 and E2428 standards through the committee.

Jane Weitzel

Jane Weitzel has worked in analytical chemistry for more than 40 years, in the mining and the pharmaceutical industries. She has been a director of large quality control laboratories supporting pharmaceutical manufacturing producing transdermal patches, pills, capsules, biologics, and blood plasma products. Jane is currently a consultant to ISO/IEC 17025 on good manufacturing practices, quality control laboratory management, and for laboratory accreditation to ISO/IEC 17025. She provides education and training on risk analysis, evaluation of measurement uncertainty, and the use of decision rules. Jane is a lead assessor for laboratory accrediting bodies for ISO/IEC 17025, ISO 17034, and ISO/IEC 17043.

Jane has been a member of ASQ and active with the California and Manitoba sections. She has obtained ASQ-CQE and ASQ-CMQ/OE certifications.

Jane is active with the U.S. Pharmacopeia (USP) and is a member of the USP Council of Experts. She is chair of the USP Expert Committee on Measurement and Data Quality for the 2020–2025 cycle. This committee has written the new USP General Chapter <1220> on Analytical Procedure Lifecycle that includes the metrological concepts of measurement uncertainty and metrological traceability. Jane has been on the USP Expert Committees for Statistics and for Reference Materials and on the USP Expert Panel for Method Validation and Verification.

Jane has served on many committees that have written guides and standards for analytical testing laboratories. These include the Standard Council of Canada ISO/IEC 17025 amplification guide for the mineral analysis industry; the AOAC Analytical Laboratory Accreditation Criteria Committee (ALACC) Guidelines for Laboratories Performing Microbiological and Chemical Analyses of Food, Dietary Supplements, and Pharmaceuticals; and USP General Chapter <1210> Statistical Tools for Procedure Validation. Jane has authored and co-authored many papers.

Pamela Wright

Pamela Wright is currently employed as a learning experience designer II for IngenioRx, where she designs and develops training curricula. Prior to joining IngenioRx, she gained more than a combined 20 years of professional experience in the laboratory, quality, accreditation, and training fields. Pamela worked as an instructional design manager for A2LA WorkPlace Training (AWPT), updating, developing, and teaching standards training courses and quality topics including *ISO/IEC 17025:2017 General Requirements for the Competence of Testing and Calibration Laboratories, ISO 19011: 2018 Guidelines for Auditing Management Systems, ISO 9000:2015 Quality Management Systems—Fundamental and Vocabulary,* and *ISO 20387 Biotechnology—Biobanking—General Requirements for Biobanking.* She formerly worked for the American Association for Laboratory Accreditation (A2LA) as quality manager, responsible for maintaining the international Mutual Recognition Arrangement/Multilateral Recognition Arrangement (ILAC-MRA/IAF-MLA) signatory status requirements of ILAC, Asia Pacific Accreditation Cooperation Incorporated (APAC), and Inter American Accreditation Cooperation (IAAC); and as calibration accreditation manager, supervising the A2LA Calibration Accreditation Program. She previously also worked in several laboratories, either to manufacture biological products or to perform clinical testing for mutations of the p53 gene.

While at A2LA, Pamela served as chair of the NCSL International Calibration System Resources 171 committee, as a voting member of the NCSL International ASC Z540 Standards Writing Group 174 committee, as a member of the NCSL International Accreditation Resources 146 committee, as a member of ILAC AIC Working Group 2, as a moderator for the A2LA Measurement Advisory Committee, as a lead quality system assessor for A2LA, and as a peer evaluator for APAC and IAAC. She has been published in *Cal Lab Magazine* and the *16th International Congress of Metrology*, is a former committee member of the Association for Talent Development (ATD), the International Society for Performance Improvement (ISPI), and the IACET Accredited Provider.

Pamela holds a B.S. in biology from Towson University, and an M.S. in education with a specialization in instructional design for online learning systems from Capella University.

Paul Keep

Paul Keep began his career in metrology in 1979, enlisting in the U.S. Army as a calibration specialist. Upon completion of MOS35H training at Lowry Air Force Base, Paul served as part of the U.S. Army Test, Measurement, and Diagnostic Equipment (TMDE) Support Group based in Redstone Arsenal, Alabama. While serving at Redstone Arsenal, Paul taught fellow soldiers as part of the 35B Conversion Course at New Equipment Technical Training (NETT). He also served in South Korea and Italy, and completed his service at Fort Riley, Kansas.

Paul spent the next 28-plus years in various technical and management roles at the largest commercial service provider in the USA, a career that culminated in the role of director of corporate quality.

While working for the commercial service provider, Paul authored several papers related to accreditation processes and laboratory performance metrics that were presented at NCSL International symposia. The first published papers included a measurement uncertainty budget template that was designed to ease the process of managing uncertainty budgets.

Paul is currently providing international metrology consulting services through Keep Metrology Services and often works as an ISO/IEC 17025 calibration and testing technical assessor for the American Association for Laboratory Accreditation (A2LA). He also serves A2LA as chair of the Measurement Advisory Council and is an active member of the Accreditation Council.

Paul also works as an instructor for A2LA WorkPlace Training (AWPT), teaching a variety of courses that include "Introduction to Measurement Uncertainty," "Applied Measurement Uncertainty for Calibration Laboratories," and "Leading an Effective ISO/IEC 17025:2017 Audit Team."

Tony Hamilton

Tony Hamilton has more than 20 years of experience in the metrology field. He received his technical training in the U.S. Navy as a nuclear qualified electronics technician. During his six-year stint, he served four years on the *USS Dwight D. Eisenhower* (CVN 69), qualifying as a reactor operator, shutdown reactor operator, and reactor technician. After leaving the Navy in 1997, Tony worked as a nuclear instrument technician at the Surry Power Station in Virginia. In the summer of 2000, he accepted a position as a process instrument technician for the central maintenance group in the Lilly Corporate Center. With a strong background in physics and math, Tony was asked to assist in the Corporate Standards Laboratory (CSL) during its restructuring, which began in the summer of 2002. This position became permanent by the end of that year, and he has been with the lab since that time. In 2006, he was given oversight of the pressure and vacuum measurement systems in the lab, and in 2008 he was promoted to the position of engineer. Tony has calibration experience in water chemistry, electrical, mass, scale/balances, temperature, hygrometry, pressure, force, fluid flow, velocity, and torque. As a metrologist, he was also responsible for the validation, work instruction, maintenance, and training. Tony has more than 16 years of experience using the Guide to the Expression of Uncertainty (GUM) in the development of complex uncertainty budgets. As of June 2021, Tony accepted a senior quality engineer role as the quality assurance representative for the Corporate Standards Laboratory, which has been accredited by the ISO/IEC 17025 since 2008. In addition to his accomplishments at Lilly, Tony earned his bachelor of science in business management from Indiana Wesleyan University in 2005, and more recently in 2018 became a qualified assessor for the ISO/IEC 17025:2017 through the American Association of Laboratory Accreditation (A2LA).

Walter E. Nowocin

Walter E. Nowocin is currently the life sciences product manager for IndySoft Corporation in Charleston, South Carolina. Walter works with development, marketing, and sales to ensure that IndySoft is optimized to support calibration quality systems in regulated industries while being compliant with FDA, GMP, and ISO requirements. Walter has 38 years of calibration experience as a technician, technical instructor, production supervisor, and laboratory manager, and as a metrology and equipment services department manager.

Walter started his distinguished career with the U.S. Marine Corps as a fire control technician working on the A6E Intruder jet aircraft where he serviced the radar, inertial navigation, ballistic computer, and armament electronic systems. He was a collateral duty inspector and attained A6E Safety of Flight sign-off certification representing the commanding officer. After six years, he attended the Precision Measurement Equipment Specialist training at Lowry Air Force Base, Colorado, transitioning into calibration and became an honor graduate. After working as a calibration technician in a PMEL laboratory for three years, Walter was selected to return to Lowry Air Force Base to become a precision measurement equipment laboratory (PMEL) instructor, where he earned the Air Force Commendation Medal for meritorious service. After Lowry AFB, Walter worked in several Marine Corps calibration laboratories in the United States and in Japan, in ever-increasing roles as a production supervisor, lab manager, and then attaining master sergeant rank and designation as a PMEL chief. Walter earned numerous awards during his 22-year Marine Corps career: the President's Meritorious Service Medal, three Navy Commendation medals, and a Navy/Marine Corps Achievement Medal. Additionally, he won the Sergeant Major of the Marine Corp's Leadership Writing Award and the Commandant of the Marine Corps Achievement Award while attending the Advanced Staff Non Commissioned Officer Academy at Camp Lejeune, North Carolina, where he graduated with honors.

Walter joined Medtronic, the world's largest medical device manufacturer, and served 21 years as a Metrology and Equipment Services Department senior engineering manager. During his tenure, Walter became very active in the calibration community; he has served as the chair of the NCSL International Healthcare Metrology Committee since 2012 and serves on the NCSL International Twin Cities Steering Committee. Walter led the project team that updated the NCSL International Recommended Practice for Calibration Quality Systems in the Healthcare Industries in 2008 and 2015. Walter has presented numerous papers and tutorials at NCSL International, ASQ, and Measurement Science Conference (MSC), earning an NCSL International's Best Paper award and an MSC Best Developer award.

He earned a master's degree in engineering management from St. Cloud State University (SCSU), Minnesota, and has taught for 11 years in SCSU's Executive Master's in Engineering Management graduate degree program. Walter is a certified professional engineering manager and a fellow of the American Society of Engineering Management (ASEM). He is the associate editor for the ASEM *Engineering Body of Knowledge* (4th and 5th eds.) and is an ASEM board member.

Notes

Notes

Notes

Notes